VITAMINS AND HORMONES

VOLUME 76

VITAMIN E

VITAMINS AND HORMONES
ADVANCES IN RESEARCH AND APPLICATIONS

Editor-in-Chief

GERALD LITWACK

Former Professor and Chair
Department of Biochemistry and Molecular Pharmacology
Thomas Jefferson University Medical College
Philadelphia, Pennsylvania

Former Visiting Scholar
Department of Biological Chemistry
David Geffen School of Medicine at UCLA
Los Angeles, California

VOLUME 76

AMSTERDAM • BOSTON • HEIDELBERG • LONDON
NEW YORK • OXFORD • PARIS • SAN DIEGO
SAN FRANCISCO • SINGAPORE • SYDNEY • TOKYO

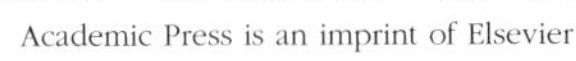
Academic Press is an imprint of Elsevier

Cover Photograph credit: **PDB ID: 1OIZ and 1OIP**
Meier, R., Tomizaki, T., Schulze-Briese, C., Baumann, U., Stocker, A.
The Molecular Basis of Vitamin E Retention: Structure of Human
Alpha-Tocopherol Transfer Protein *J. Mol. Biol. V331 pp. 725, 2003*

Academic Press is an imprint of Elsevier
525 B Street, Suite 1900, San Diego, California 92101-4495, USA
84 Theobald's Road, London WC1X 8RR, UK

This book is printed on acid-free paper. ♾

For information on all Elsevier Academic Press publications
visit our Web site at www.books.elsevier.com

ISBN-13: 978-0-12-373592-8

PRINTED IN THE UNITED STATES OF AMERICA
07 08 09 10 9 8 7 6 5 4 3 2 1

Former Editors

CONTENTS

1

VITAMIN E

2

Structure and Function of α-Tocopherol Transfer Protein: Implications for Vitamin E Metabolism and AVED

K. Christopher Min

3

The α-Tocopherol Transfer Protein

D. Manor and S. Morley

4

Molecular Associations of Vitamin E

Peter J. Quinn

6

VITAMIN E AND NF-κB ACTIVATION: A REVIEW

HOWARD P. GLAUERT

7

SYNTHESIS OF VITAMIN E

THOMAS NETSCHER

8

TOCOTRIENOLS: THE EMERGING FACE OF NATURAL VITAMIN E

CHANDAN K. SEN, SAVITA KHANNA, CAMERON RINK, AND SASHWATI ROY

9

VITAMIN E BIOTRANSFORMATION IN HUMANS

FRANCESCO GALLI, M. CRISTINA POLIDORI, WILHELM STAHL,
PATRIZIA MECOCCI, AND FRANK J. KELLY

10

α-TOCOPHEROL STEREOISOMERS

SØREN KROGH JENSEN AND CHARLOTTE LAURIDSEN

11

ADDITION PRODUCTS OF α-TOCOPHEROL WITH LIPID-DERIVED FREE RADICALS

RYO YAMAUCHI

12

Vitamin E and Apoptosis

Paul W. Sylvester

13

Vitamin E During Pre- and Postnatal Periods

Cathy Debier

14

α-Tocopherol: A Multifaceted Molecule in Plants

Sergi Munné-Bosch

15

VITAMIN E AND MAST CELLS

JEAN-MARC ZINGG

16

TOCOTRIENOLS IN CARDIOPROTECTION

SAMARJIT DAS, KALANITHI NESARETNAM, AND DIPAK K. DAS

17

VITAMIN E AND CANCER

KIMBERLY KLINE, KARLA A. LAWSON, WEIPING YU, AND BOB G. SANDERS

18

Vitamin E Analogues and Immune Response in Cancer Treatment

Marco Tomasetti and Jiri Neuzil

19

The Roles of α-Vitamin E and Its Analogues in Prostate Cancer

Jing Ni and Shuyuan Yeh

20

VITAMIN E: INFLAMMATION AND ATHEROSCLEROSIS

U. SINGH AND S. DEVARAJ

21

VITAMIN E IN CHRONIC LIVER DISEASES AND LIVER FIBROSIS

ANTONIO DI SARIO, CINZIA CANDELARESI, ALESSIA OMENETTI, AND ANTONIO BENEDETTI

Contributors

Numbers in parentheses indicate the pages on which the authors' contributions begin.

Antonio Benedetti (551) Department of Gastroenterology, Università Politecnica delle Marche, Polo Didattico III, Piano, Via Tronto 10, 60020 Torrette, Ancona, Italy

Richard S. Bruno (1) Department of Nutritional Sciences, University of Connecticut, Storrs, Connecticut 06269

Cinzia Candelaresi (551) Department of Gastroenterology, Università Politecnica delle Marche, Polo Didattico III, Piano, Via Tronto 10, 60020 Torrette, Ancona, Italy

David G. Cornwell (99) Department of Molecular and Cellular Biochemistry, The Ohio State University College of Medicine, Columbus, Ohio 43210

Dipak K. Das (419) Cardiovascular Research Center, University of Connecticut School of Medicine, Farmington, Connecticut 06030

Samarjit Das (419) Cardiovascular Research Center, University of Connecticut School of Medicine, Farmington, Connecticut 06030

Cathy Debier (357) Institut des Sciences de la Vie, Unité de Biochimie de la Nutrition, Université catholique de Louvain, Croix du Sud 2/8, B-1348 Louvain-la-Neuve, Belgium

S. Devaraj (519) Department of Pathology, Laboratory for Atherosclerosis and Metabolic Research, UC Davis Medical Center, Sacramento, California 95817

Francesco Galli (263) Department of Internal Medicine, Section of Applied Biochemistry and Nutritional Sciences, University of Perugia, Italy

Howard P. Glauert (135) Graduate Center for Nutritional Sciences, University of Kentucky, Lexington, Kentucky 40506

Søren Krogh Jensen (281) Department of Animal Health, Welfare and Nutrition, Faculty of Agricultural Sciences, University of Aarhus, DK-8830 Tjele, Denmark

Frank J. Kelly (263) Pharmaceutical Science Division, School of Biomedical and Health Sciences, King's College London, London, United Kingdom

Savita Khanna (203) Laboratory of Molecular Medicine, Department of Surgery, Davis Heart and Lung Research Institute, The Ohio State University Medical Center, Columbus, Ohio 43210

Kimberly Kline (435) Division of Nutrition, University of Texas at Austin, Austin, Texas 78712

Charlotte Lauridsen (281) Department of Animal Health, Welfare and Nutrition, Faculty of Agricultural Sciences, University of Aarhus, DK-8830 Tjele, Denmark

Karla A. Lawson (435) Cancer Prevention Fellowship Program, National Cancer Institute, National Institutes of Health, Bethesda, Maryland 20892

Jiyan Ma (99) Department of Molecular and Cellular Biochemistry, The Ohio State University College of Medicine, Columbus, Ohio 43210

D. Manor (45) Division of Nutritional Sciences, Cornell University, Ithaca, New York 14853; Department of Nutrition, Case School of Medicine, Case Western Reserve University, Cleveland, Ohio 44106

Patrizia Mecocci (263) Institute of Gerontology and Geriatrics, University of Perugia, Italy

K. Christopher Min (23) Department of Neurology, Columbia University, New York, New York 10032

S. Morley (45) Division of Nutritional Sciences, Cornell University, Ithaca, New York 14853

Sergi Munné-Bosch (375) Departament de Biologia Vegetal, Facultat de Biologia, Universitat de Barcelona, Avinguda Diagonal 645, E-08028 Barcelona, Spain

Debbie J. Mustacich (1) Linus Pauling Institute, Oregon State University, Corvallis, Oregon 97331

Kalanithi Nesaretnam (419) Malaysian Palm Oil Board, Kuala Lumpur, Malaysia

Thomas Netscher (155) Research and Development, DSM Nutritional Products, CH-4002 Basel, Switzerland

Jiri Neuzil (463) Apoptosis Research Group, School of Medical Science, Griffith University, Southport, Qld, Australia; Molecular Therapy Group, Institute of Molecular Genetics, Czech Academy of Sciences, Prague, Czech Republic

Jing Ni (493) Department of Urology and Department of Pathology, University of Rochester, Rochester, New York 14642

Alessia Omenetti (551) Department of Gastroenterology, Università Politecnica delle Marche, Polo Didattico III, Piano, Via Tronto 10, 60020 Torrette, Ancona, Italy

M. Cristina Polidori (263) Institute of Biochemistry and Molecular Biology I, Heinrich-Heine University, Düsseldorf, Germany

Peter J. Quinn (67) Department of Biochemistry, King's College London, London SE2 9NH, United Kingdom

Cameron Rink (203) Laboratory of Molecular Medicine, Department of Surgery, Davis Heart and Lung Research Institute, The Ohio State University Medical Center, Columbus, Ohio 43210

Sashwati Roy (203) Laboratory of Molecular Medicine, Department of Surgery, Davis Heart and Lung Research Institute, The Ohio State University Medical Center, Columbus, Ohio 43210

Bob G. Sanders (435) School of Biological Sciences, University of Texas at Austin, Austin, Texas 78712

Antonio Di Sario (551) Department of Gastroenterology, Università Politecnica delle Marche, Polo Didattico III, Piano, Via Tronto 10, 60020 Torrette, Ancona, Italy

Chandan K. Sen (203) Laboratory of Molecular Medicine, Department of Surgery, Davis Heart and Lung Research Institute, The Ohio State University Medical Center, Columbus, Ohio 43210

U. Singh (519) Department of Pathology, Laboratory for Atherosclerosis and Metabolic Research, UC Davis Medical Center, Sacramento, California 95817

Wilhelm Stahl (263) Institute of Biochemistry and Molecular Biology I, Heinrich-Heine University, Düsseldorf, Germany

Paul W. Sylvester (329) College of Pharmacy, University of Louisiana at Monroe, Monroe, Louisiana 71209

Marco Tomasetti (463) Department of Molecular Pathology and Innovative Therapies, Polytechnic University of Marche, Ancona, Italy

Maret G. Traber (1) Department of Nutrition and Exercise Sciences, Linus Pauling Institute, Oregon State University, Corvallis, Oregon 97331

Ryo Yamauchi (309) Department of Applied Life Science, Faculty of Applied Biological Sciences, Gifu University, 1-1 Yanagido, Gifu City, Gifu 501-1193, Japan

Shuyuan Yeh (493) Department of Urology and Department of Pathology, University of Rochester, Rochester, New York 14642

Weiping Yu (435) School of Biological Sciences, University of Texas at Austin, Austin, Texas 78712

Jean-Marc Zingg (393) Institute of Biochemistry and Molecular Medicine, University of Bern, 3012 Bern, Switzerland

Preface

Vitamin E is an important topic because many benefits and some risks have been attributed to it when used as a nutritional supplement. Moreover, there has been considerable progress in the basic science of this vitamin in recent years. This volume reviews recent aspects of the biochemistry and molecular biology of this vitamin, associated macromolecules, metabolism of the vitamin and its derivatives, and its many roles in health and disease. The various contributions comprising this volume are listed here in the order in which they are presented in the volume.

The first chapter is entitled "Vitamin E," an overview contributed by D. J. Mustacich, R. S. Bruno, and M. G. Traber. The next is "Structure and Function of α-Tocopherol Transfer Protein: Implications for Vitamin E Metabolism and AVED," by K. C. Min. This is followed by "The α-Tocopherol Transfer Protein," by D. Manor and S. Morley. P. J. Quinn offers "Molecular Associations of Vitamin E," and then D. G. Cornwell and J. Ma follow with "Studies in Vitamin E: Biochemistry and Molecular Biology of Tocopherol Quinones." Continuing with considerations of the molecular biology of this vitamin, H. P. Glauert presents "Vitamin E and NF-κB Activation: A Review." T. Netscher contributes a complete chapter on the "Synthesis of Vitamin E." C. K. Sen, S. Khanna, C. Rink, and S. Roy offer "Tocotrienols: The Emerging Face of Natural Vitamin E." Next is "Vitamin E Biotransformation in Humans" by F. Galli, M. C. Polidori, W. Stahl, P. Mecocci, and F. J. Kelly. Chemical considerations are added by S. K. Jensen and C. Lauridsen with "α-Tocopherol Stereoisomers" followed by "Addition Products of α-Tocopherol with Lipid-Derived Free Radicals," by R. Yamauchi. The timely topic entitled "Vitamin E and Apoptosis" is reviewed by P. W. Sylvester. C. Debier contributes "Vitamin E During

Pre- and Postnatal Periods." Vitamin E in plants is covered by S. Munné-Bosch with a chapter entitled "α-Tocopherol: A Multifaceted Molecule in Plants."

Switching to a section on the consideration of cells, organs, and diseases, J.-M. Zingg contributes "Vitamin E and Mast Cells." "Tocotrienols in Cardioprotection" is reviewed by S. Das, K. Nesaretam, and D. K. Das. Three cancer-related chapters follow. The first of these is "Vitamin E and Cancer" by K. Kline, K. A. Lawson, W. Yu, and B. G. Sanders. M. Tomasetti and J. Neuzil offer "Vitamin E Analogues and Immune Response in Cancer Treatment" and J. Ni and S. Yeh review "The Roles of α-Vitamin E and Its Analogues in Prostate Cancer." Two final chapters focus on vitamin E in diseases other than cancer "Vitamin E: Inflammation and Atherosclerosis," by U. Singh and S. Devaraj, and "Vitamin E in Chronic Liver Diseases and Liver Fibrosis," by A. Di Sario, C. Candelaresi, A. Omenetti, and A. Benedetti.

Several recent volumes, including this one, and some in the planning stages emphasize the vitamins, topics that have been subordinated to the hormones in recent years; in endocrinology, enormous strides have been made. In keeping with the title of this volume, it is intended to bring up-to-date discussions of the advances in the vitamin field as well.

The Editor-in-Chief is grateful for the cooperation of Renske van Dijk and Tari Broderick of Academic Press/Elsevier in bringing these volumes to completion.

Gerald Litwack
Toluca Lake, California
December, 2006

1

VITAMIN E

DEBBIE J. MUSTACICH,* RICHARD S. BRUNO,[†,‡]
AND MARET G. TRABER*,[‡]

*Linus Pauling Institute, Oregon State University, Corvallis, Oregon 97331
[†]Department of Nutritional Sciences, University of Connecticut
Storrs, Connecticut 06269
[‡]Department of Nutrition and Exercise Sciences
Oregon State University, Corvallis, Oregon 97331

Vitamins and Hormones, Volume 76

0083-6729/07 $35.00
DOI: 10.1016/S0083-6729(07)76001-6

The term vitamin E is used to describe eight lipophilic, naturally occurring compounds that include four tocopherols and four tocotrienols designated as α-, β-, γ-, and δ-. The most well-known function of vitamin E is that of a chain-breaking antioxidant that prevents the cyclic propagation of lipid peroxidation. Despite its antioxidant function, dietary vitamin E requirements in humans are limited only to α-tocopherol because the other forms of vitamin E are poorly recognized by the hepatic α-tocopherol transfer protein (TTP), and they are not converted to α-tocopherol by humans. In attempts to gain a better understanding of vitamin E's health benefits, the molecular regulatory mechanisms of vitamin E have received increased attention. Examples of these mechanisms include: (1) the role of the hepatic α-TTP in preferentially secreting α-tocopherol into the plasma, (2) phase I and phase II metabolism of vitamin E and the potential impact for drug–vitamin E interactions, and (3) the regulation of biliary excretion of vitamin E by ATP-binding cassette protein(s). It is expected that the continued studies of these regulatory pathways will provide new insights into vitamin E function from which additional human health benefits will evolve.

I. INTRODUCTION

Vitamin E was discovered by Evans and Bishop (1922) as a dietary factor necessary for reproduction in rats. At present, it is among the most commonly consumed dietary supplements in the United States due to the belief that vitamin E, as an antioxidant, may attenuate morbidity and mortality. Despite the frequent use among millions of Americans, the health benefits beyond its classic antioxidant function remain an enigma because nonantioxidant functions have yet to illustrate vitamin E's required role in human nutrition. Moreover, it is now becoming evident that vitamin E concentrations in humans are tightly regulated such that high doses of vitamin E supplements do not enhance plasma concentrations by more than three- to fourfold. Therefore, this chapter seeks to define the forms and isomers of vitamin E, the molecular basis for their differences in biological activity, and the mechanisms responsible for tissue delivery and for the apparently strict regulation of hepatic vitamin E concentrations.

II. VITAMIN E STRUCTURES AND FUNCTION

Plants synthesize eight different molecules with vitamin E antioxidant activity, including α-, β-, γ-, and δ-tocopherols and the corresponding four tocotrienols (Fig. 1). The α-, β-, γ-, and δ-forms differ with respect to the number and position of the methyl groups on their chromanol ring. The tocotrienols have an unsaturated tail containing three double bonds, while the four tocopherols have a phytyl tail (Fig. 1). *RRR*-α-Tocopherol is the naturally occurring form of α-tocopherol, containing chiral carbons in the *R*-conformation at positions 2, 4′, and 8′. However, chemical synthesis of α-tocopherol results in an equal mixture of eight different stereoisomers (*RRR*, *RSR*, *RRS*, *RSS*, *SRR*, *SSR*, *SRS*, *SSS*) with half having position 2 in the *S*-conformation. The synthetic vitamin E is called all-*rac*-α-tocopherol. Position 2 is critical for *in vivo* α-tocopherol activity and only the 2*R*-forms are recognized to meet human requirements (Food and Nutrition Board and Institute of Medicine, 2000). (Note: For the purposes of this chapter, the term α-tocopherol will refer to *RRR*-α-tocopherol.)

Vitamin E functions *in vivo* as a potent peroxyl radical scavenger (Burton *et al.*, 1983). Peroxyl radicals (ROO•) react 1000 times more favorably with α-tocopherol (Vit E-OH) than with polyunsaturated fatty acids (RH). The tocopherol's phenolic hydroxyl group reacts with an organic peroxyl radical to form the corresponding organic hydroperoxide (ROOH) and the tocopheroxyl radical (Vit E-O•) (Burton *et al.*, 1985).

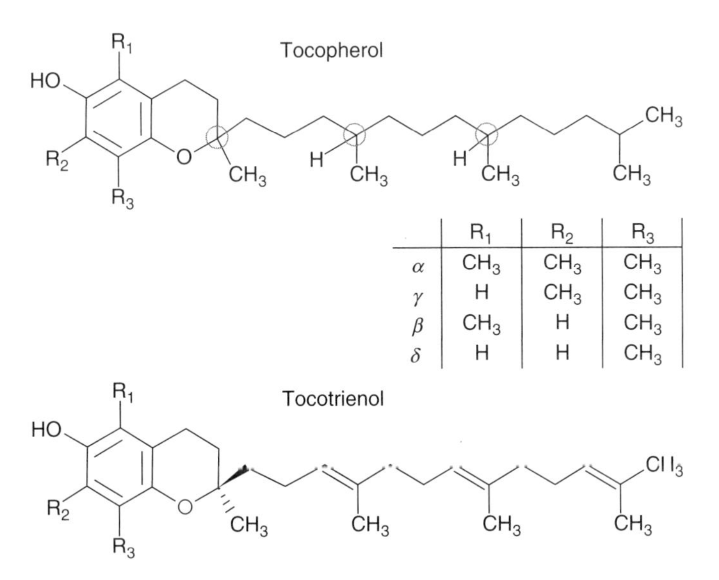

	R_1	R_2	R_3
α	CH_3	CH_3	CH_3
γ	H	CH_3	CH_3
β	CH_3	H	CH_3
δ	H	H	CH_3

FIGURE 1. Structures of tocopherols and tocotrienols. The circles mark the three chiral centers, 2, 4′, and 8′ in the tocopherols.

In the presence of vitamin E: ROO• + Vit E-OH → ROOH + Vit E-O•
In the absence of vitamin E: ROO• + RH → ROOH + R•

$$R\bullet + O_2 \rightarrow ROO\bullet$$

The tocopheroxyl radical (Vit E-O•) reacts with vitamin C (or other reductants serving as hydrogen donors, AH), thereby oxidizing the latter and returning vitamin E to its reduced state (Buettner, 1993).

$$\text{Vit E-O}\bullet + \text{AH} \rightarrow \text{Vit E-OH} + \text{A}\bullet$$

Biologically important hydrogen donors, which have been demonstrated *in vitro* to regenerate tocopherol from the tocopheroxyl radical, include ascorbate (vitamin C). Importantly, this regeneration of vitamin E by vitamin C has been demonstrated to occur in humans such that cigarette smokers have faster vitamin E turnover that can be normalized by vitamin C supplementation (Bruno *et al.*, 2005, 2006a).

III. ABSORPTION, TRANSPORT, AND DISTRIBUTION TO TISSUES

A. VITAMIN E ABSORPTION

Vitamin E is fat-soluble and therefore requires all of the processes needed for fat absorption. Specifically, intestinal absorption of vitamin E requires the secretion of pancreatic esterases and bile acids. Indeed, disorders such as cystic fibrosis or cholestatic liver disease that result in the impairment of biliary secretions result in vitamin E deficiency. These secretions are needed for the micellarization of dietary fats, including vitamin E, and the hydrolysis of triglycerides that release free fatty acids. The micelles are taken up by intestinal enterocytes, then vitamin E is subsequently incorporated into chylomicrons, secreted into the lymphatic system, and finally moves into the plasma (Sokol *et al.*, 1983).

B. REQUIREMENT FOR DIETARY FAT FOR ABSORPTION

Vitamin E absorption is also dependent on the fat content of food consumed with the vitamin E such that little absorption occurs from vitamin E supplements in the absence of dietary fat (Borel *et al.*, 2001; Bruno *et al.*, 2006b; Hayes *et al.*, 2001).

No studies have been carried out on the absorption of vitamin E naturally occurring in food, but some data are available examining foods fortified with vitamin E. As might be expected from a low fat food, α-tocopherol absorption from apples fortified with d_6-α-tocopheryl acetate was minimal, but increased from 10 to 33% with added dietary fat (from 0- to 11-g fat)

(Bruno *et al.*, 2006b). Remarkably, vitamin E added in an emulsifier onto the surface of breakfast cereal was highly bioavailable, despite the low fat content of the food, suggesting that the micellarization of the vitamin E is critical for its absorption (Leonard *et al.*, 2004).

Importantly, the various vitamin E forms, such as α- and γ-tocopherols (Meydani *et al.*, 1989; Traber and Kayden, 1989), or *RRR*- and *SRR*-α-tocopherols (Traber *et al.*, 1990a, 1992), have similar apparent efficiencies of intestinal absorption and secretion in chylomicrons. Thus, no discrimination exists between the various vitamin E forms during absorption.

C. REQUIREMENT FOR CHYLOMICRON SYNTHESIS

Early observations from studies examining human vitamin E requirements indicated that patients with abetalipoproteinemia, a genetic defect in the lipidation of apolipoprotein B (Wetterau *et al.*, 1992), developed symptoms characteristic of vitamin E deficiency. Interestingly, the few particles that are lipidated do contain detectable amounts of α-tocopherol and may allow some absorption and delivery of α-tocopherol to tissues (Aguie *et al.*, 1995).

Circulating chylomicrons undergo triglyceride lipolysis by lipoprotein lipase (LPL). Studies *in vitro* demonstrated that LPL bound to the cell surface could transfer vitamin E to cells (Traber *et al.*, 1985). Subsequent studies in mice overexpressing muscle LPL demonstrated that lipase effectively delivers vitamin E to muscle (Sattler *et al.*, 1996). Furthermore, some of the newly absorbed vitamin E is transferred to circulating lipoproteins, for example high-density lipoproteins (HDL), while some remains with the chylomicron remnants. Thus, during the process of triglyceride lipolysis by LPL, some tocopherols are likely transferred to other lipoproteins and/or taken up by peripheral tissues. The relative proportion of the vitamin E distributed between chylomicrons and HDL has not been identified.

D. ROLE OF EXCHANGE PROTEINS

Vitamin E is readily transferred between HDL and other lipoproteins. Thus, some of the intestinally absorbed vitamin E is distributed to all of the circulating lipoproteins. Vitamin E exchange between lipoproteins was shown to be catalyzed by the phospholipid transfer protein (PLTP) (Kostner *et al.*, 1995). Indeed, PLTP has been demonstrated to be critical for enriching circulating lipoproteins with vitamin E (Jiang *et al.*, 2002).

Remarkably, PLTP also appears to be critical for intratissue distribution of vitamin E. PLTP was required for the normal distribution of brain vitamin E (Desrumaux *et al.*, 2005). This PLTP function for vitamin E also extended to spermatozoa (Drouineaud *et al.*, 2006). Alternatively, PLTP may be necessary for the proper membrane distribution of vitamin E because mice lacking PLTP had modified phospholipid distribution that could be normalized by the addition of vitamin E (Klein *et al.*, 2006).

E. SECRETION OF α-TOCOPHEROL BY THE LIVER

The liver takes up the chylomicron remnants and repackages the dietary fats into very low density lipoproteins (VLDL) for secretion into the plasma (Havel, 1994). Studies using deuterated tocopherols determined that *RRR*-α-tocopherol is preferentially secreted into the plasma for distribution to the peripheral tissues. Until recently, it was thought that incorporation of α-tocopherol into VLDL was required for the delivery of α-tocopherol to the peripheral tissues. However, studies in mice lacking the ability to produce VLDL demonstrated that this lipoprotein is not critical for distribution of α-tocopherol to the peripheral tissues (Minehira-Castelli *et al.*, 2006). The percentage of plasma α-tocopherol in the HDL fraction in mice lacking VLDL increased significantly as compared to wild-type mice suggesting that HDL, at least in mice, may suffice for delivery to peripheral tissues.

IV. THE α-TOCOPHEROL TRANSFER PROTEIN

The preferential secretion of *RRR*-α-tocopherol by the liver is under the control of the α-tocopherol transfer protein (α-TTP) as observed in patients with genetic α-TTP defects (Traber *et al.*, 1990b, 1993) and in α-TTP knockout mice ($Ttpa^{-/-}$) (Leonard *et al.*, 2002; Terasawa *et al.*, 2000). α-TTP selectively binds *RRR*-α-tocopherol, as compared to other vitamin E forms, including the 2*S*-α-tocopherols, and facilitates its secretion from the liver into the plasma for distribution to the tissues (Brigelius-Flohé and Traber, 1999; Hosomi *et al.*, 1997; Panagabko *et al.*, 2002). The mechanism by which α-TTP facilitates the transfer of α-tocopherol to the plasma membrane for incorporation into VLDL, and/or HDL, has yet to be elucidated.

A. STRUCTURE AND LOCALIZATION

α-TTP has been purified from rat (Sato *et al.*, 1991; Yoshida *et al.*, 1992) and human livers (Arita *et al.*, 1995; Kuhlenkamp *et al.*, 1993) and the amino acid sequences reported (Arita *et al.*, 1995; Sato *et al.*, 1993). α-TTP is predominantly described as a cytosolic liver protein, but it has also been found in rat brain (Hosomi *et al.*, 1998) and pregnant mouse uterus (Jishage *et al.*, 2001; Kaempf-Rotzoll *et al.*, 2002). The biological function of α-TTP in these extrahepatic tissues has yet to be determined.

TABLE I. Human Vitamin E Kinetics

	Fractional disappearance rates (pools per day)	Half-life (h)	References
RRR-α-tocopherol	0.3 ± 0.1	57 ± 19	Leonard *et al.*, 2005; Traber *et al.*, 1994
SRR-α-tocopherol	1.2 ± 0.6	16 ± 6	Traber *et al.*, 1994
γ-Tocopherol	1.4 ± 0.4	13 ± 4	Leonard *et al.*, 2005
α-Tocotrienol	4.0 ± 0.9	∼4	Yap *et al.*, 2001

The crystal structure of α-TTP has been described both with and without *RRR*-α-tocopherol present in the ligand-binding pocket (Meier *et al.*, 2003; Min *et al.*, 2003). The structural features required for recognition and optimal binding within the ligand pocket are as follows: (1) a fully methylated chroman ring provides optimal hydrophobic interactions explaining the binding affinities of the four naturally occurring tocopherols, α- > β- > γ- > δ-tocopherol; (2) the ability of the vitamin E tail to fold back in a U-turn significantly increases ligand-binding affinity, a feature that would only be expected of the tocopherol phytyl tail and not the unsaturated tail of the tocotrienols; and (3) the stereochemical position of the methyl group in the 2*R*-position (Panagabko *et al.*, 2003). This third requirement makes α-TTP selective for the 2*R*-isomers of synthetic α-tocopherol, but even then, the binding affinity of *SRR*-α-tocopherol is only 11% that of *RRR*-α-tocopherol (Hosomi *et al.*, 1997). The preferential binding of *RRR*-α-tocopherol by α-TTP is reflected in the threefold greater half-life of *RRR*-α-tocopherol (57 ± 19 h) compared to the half-lives of other vitamin E forms (Table I).

Thus, only natural α-tocopherol (*RRR*-α-tocopherol) and 2*R*-α-tocopherol in synthetic all-*rac*-α-tocopherol, not the other vitamin E forms, are maintained in human plasma and tissues. Moreover, it is the liver, not the intestine, which discriminates between the various forms of vitamin E for distribution to other tissues (Traber *et al.*, 2005a).

B. CRAL-TRIO FAMILY

The CRAL-TRIO family is a small group of lipid-binding proteins, including the cellular retinaldehyde-binding protein (CRALBP), α-TTP, yeast phosphatidylinositol transfer protein (Sec14p), and supernatant protein factor (SPF) (Panagabko *et al.*, 2003). SPF was isolated from bovine liver cytosol and identified as an α-tocopherol-binding protein and renamed tocopherol-associated protein (TAP) (Stocker *et al.*, 1999; Zimmer *et al.*, 2000). However, both heterologous and homologous competition experiments demonstrated that SPF/TAP exhibits only a very weak, nonselective-binding affinity for tocopherols, 25-fold less than that of α-TTP for α-tocopherol, and that SPF binds phosphatidylinositol with greater affinity than α-tocopherol (Manor and Atkinson, 2003; Panagabko *et al.*, 2003). Thus, α-TTP is the only known protein, to date, with a high affinity for tocopherols and, importantly, specificity for α-tocopherol.

V. REGULATION OF VITAMIN E METABOLISM AND EXCRETION

Although human dietary intake of γ-tocopherol is significantly higher than that of α-tocopherol, particularly in the United States, human plasma and tissue levels of α-tocopherol are several-fold higher than those of γ-tocopherol. The enrichment of plasma and tissues with α-tocopherol is mediated at two levels: (1) selectivity of the hepatic α-TTP and (2) the regulation of hepatic vitamin E metabolism and excretion.

A. WHAT IS CEHC?

"Simon metabolites" were the first compounds identified as vitamin E metabolites (Eisengart *et al.*, 1956), but modern techniques have since demonstrated that they are produced largely by *in vitro* sample oxidation (Schultz *et al.*, 1995). In fact, the biologically relevant vitamin E metabolites are the 2′-carboxyethyl-6-hydroxychroman (CEHC) products of the respective forms of vitamin E, that is α-, β-, γ-, and δ-CEHCs. CEHCs were first described in rats injected with δ-tocopherol (Chiku *et al.*, 1984).

Unlike other fat-soluble vitamins, and despite the activity of the α-TTP to facilitate the selective enrichment of plasma and tissues with α-tocopherol, no vitamin E form accumulates to "toxic" levels, suggesting that mechanisms, that is metabolism and/or excretion, prevent excess accumulation (Schultz *et al.*, 1995). To evaluate metabolism, humans ingested a supplement containing 50 mg each d_6-α- and d_2-γ-tocopherols with their breakfast (Leonard *et al.*, 2005). Within 3-h plasma d_6-α- and d_2-γ-tocopherols were present at similar concentrations, indicating similar absorption of these vitamin E forms. However, by 24-h postdose plasma d_6-α-tocopherol concentrations were nearly tenfold greater than d_2-γ-tocopherol concentrations, demonstrating the selective enrichment of plasma with α-tocopherol. In addition, while d_6-α-CEHC (the metabolic end product of d_6-α-tocopherol) was undetectable in either plasma or urine over the course of the experiment (72 h), d_2-γ-CEHC was detectable within a few hours, indicating increased metabolism of γ-tocopherol but not α-tocopherol (Leonard *et al.*, 2005). Furthermore, the percent d_2 enrichment of both γ-tocopherol and γ-CEHC were identical, as were their half-lives, suggesting that metabolism of γ-tocopherol to γ-CEHC is in rapid equilibrium with the plasma γ-tocopherol pool (Leonard *et al.*, 2005).

Similarly, participants provided a single dose of 100-mg d_2-γ-tocopherol had plasma d_2-γ-tocopherol concentrations that rapidly increased simultaneously with a corresponding increase in both plasma and urinary d_2-γ-CEHC concentrations and returned to baseline by 72-h postdose (Galli *et al.*, 2003). Although administration of 100-mg d_2-γ-tocopherol was determined to have no marked effect on the metabolism of α-tocopherol (Galli *et al.*, 2003), in rats supplemented with a single oral dose of 10 mg each α- and γ-tocopherols, urinary and biliary levels of γ-CEHC were increased compared to rats provided with 10-mg γ-tocopherol alone (Kiyose *et al.*, 2001). These data suggest that supplementation with α-tocopherol increases the metabolism of γ-tocopherol, but not the reverse.

In another study, human subjects consumed a single capsule containing 150 mg each of d_3-*RRR*-α- and d_6-all-*rac*-α-tocopherols (Traber *et al.*, 1998). By 6-h postdose plasma concentrations of d_3-α- and d_6-α-tocopherols were similar. However, plasma d_3-α-tocopherol at 12 and 24 h was approximately twofold greater than d_6-α-tocopherol and plasma γ-tocopherol concentrations decreased to less than 50% of baseline levels. Importantly, excretion of d_6-α-CEHC was nearly threefold that of d_3-α-CEHC. The data from the above studies suggest that *RRR*-α-tocopherol regulates the metabolism and, possibly excretion, of other natural and synthetic forms of vitamin E, thus maintaining itself as the predominant form of vitamin E in plasma and tissues.

B. METABOLISM OF VITAMIN E

The proposed metabolic pathway of α- and γ-tocopherols to their respective CEHCs (shown for α-tocopherol in Fig. 2) is based mainly on data from *in vitro* studies in which intermediate metabolites were isolated and identified from HepG2 cells and rat liver subcellular fractions incubated with various forms of vitamin E (Birringer *et al.*, 2001, 2002; Sontag and Parker, 2002). Initially, tocopherols undergo ω-oxidation of the side chain to produce 13′-OH-tocopherol metabolites (Fig. 2). The formation of 13′-OH-tocopherol is hypothesized to be followed by several steps of β-oxidation, leading to the formation of 5′-CMBHC (2,5,7,8-tetramethyl-2-(4′-carboxy-4′-methylbutyl)-6-hydroxychroman) and finally the respective CEHC (Fig. 2). The current dogma is that the initial oxidation step is catalyzed by the cytochrome P450 (CYP) system of xenobiotic metabolizing enzymes. Studies in cell culture have shown that ketoconozole and sesamin, known inhibitors of CYP activity, prevent tocopherol metabolism, while treatment with rifampicin, an inducer of CYP3A, increased tocopherol metabolism (Birringer *et al.*, 2001, 2002; Ikeda *et al.*, 2002; Parker *et al.*, 2000). Thus, the hypothesis that CYP enzymes are required for tocopherol metabolism is supported and suggests that CYP3A4 plays a role in tocopherol metabolism. However, in insect cells

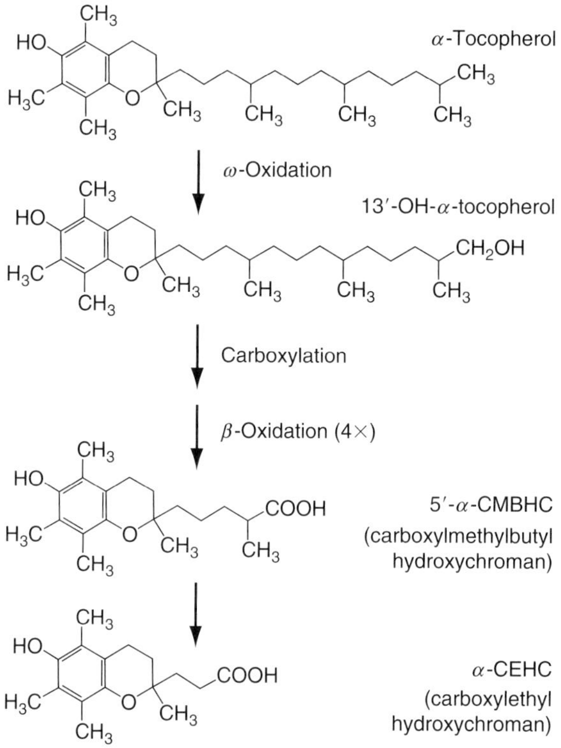

FIGURE 2. Scheme of the metabolism of tocopherols using α-tocopherol as the model compound.

expressing individual CYP enzymes, CYP4F2 has been shown to be the tocopherol hydrolase involved in ω-oxidation of the side chain (Sontag and Parker, 2002).

To evaluate the role of metabolism in protection against excess accumulation of vitamin E, hepatic levels of intermediate α-tocopherol metabolites, 13′-OH-α-tocopherol and 5′-α-CMBHC, were measured in rats given daily subcutaneous injections of α-tocopherol (100-mg/kg body weight) (Mustacich *et al.*, 2006). By the third day, the following were observed: (1) hepatic α-tocopherol concentrations increased 40- to 75-fold, (2) hepatic α-CEHC increased ~100-fold, (3) hepatic 13′-OH-α-tocopherol increased 20-fold, and (4) hepatic 5′-α-CMBHC, which was undetectable prior to α-tocopherol supplementation, increased to 1.0-nmol/g tissue. These data indicate that high-level supplementation with α-tocopherol increases α-tocopherol metabolism. In addition, hepatic protein levels of CYP3A, CYP2B, and CYP2C, but not CYP4F, were increased as much as twofold in rats following 3 days of α-tocopherol injections and remained increased over the 18 days of daily α-tocopherol injections despite a decrease in hepatic α-tocopherol levels after day 9 (Mustacich *et al.*, 2006; Fig. 3). Others have also demonstrated that CYP2C protein levels increase following dietary α-tocopherol supplementation (Murray, 1991).

Although all forms of vitamin E are absorbed, the liver preferentially secretes α-, but not γ-tocopherol, into plasma. Therefore, to assess the bioactivities of γ-tocopherol, mice that do or do not express TTP (*Ttpa*$^{-/-}$, *Ttpa*$^{+/-}$, and *Ttpa*$^{+/+}$ mice) were fed for 5 weeks diets containing either γ-tocopherol (550-mg γ-tocopherol/kg diet or 60-mg γ-tocopherol/kg diet), a vitamin E-deficient diet or a control diet (Traber *et al.*, 2005b). The two γ-tocopherol diets, as well as the control diet, contained α-tocopherol. Plasma and tissues from mice fed γ-T550 diets were found to contain similar γ- and α-tocopherol

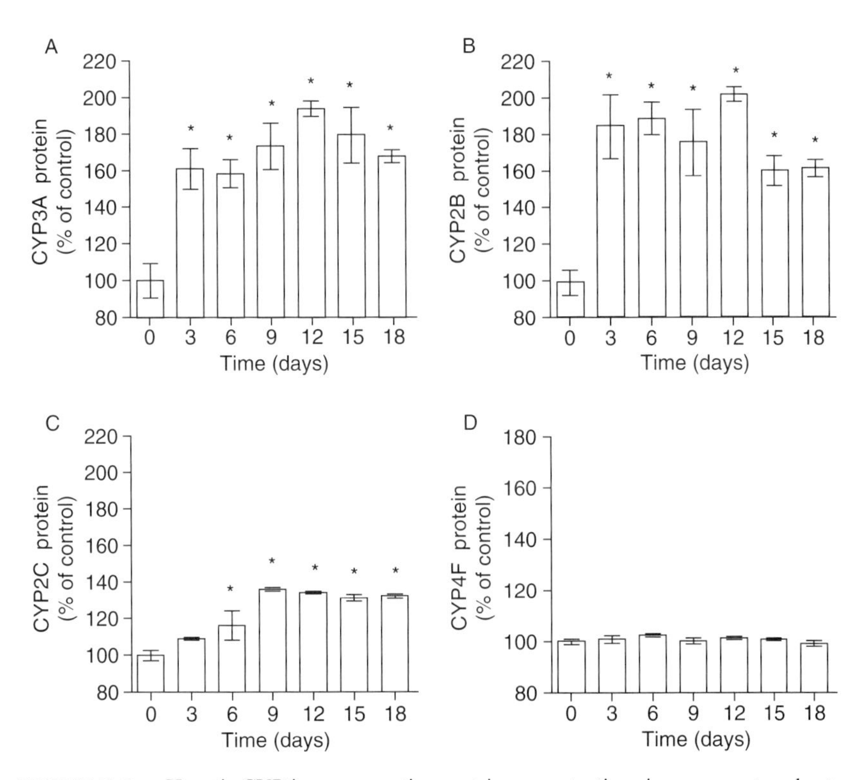

FIGURE 3. Hepatic CYP immunoreactive protein concentrations in response to subcutaneous (SC) α-tocopherol injections. Quantification of the indicated CYP protein by densitometry: (A) CYP3A2, (B) CYP2B, (C) CYP2C, and (D) CYP4F protein levels. Data are expressed as mean ±SE of four rats, * = $p < 0.001$ as compared with their respective concentrations in day 0 rats. Insect microsomes expressing the individual rat CYP protein of interest were utilized as controls to confirm the identification of bands on each blot. Adapted from Mustacich *et al.* (2006).

concentrations despite the high dietary γ-tocopherol content; nervous tissues contained almost no γ-tocopherol, irrespective of diet or genotype. Hepatic Cyp3a protein levels correlated with hepatic α-tocopherol levels and not γ-tocopherol levels. Cyp4F protein levels did not vary between any of the dietary groups. (Note: In rats and humans, CYP protein nomenclature is capitalized, while in mice only the first letter is capitalized, for example CYP3A in humans and rats and Cyp3a in mice.)

Similarly, hepatic Cyp3a mRNA increased in mice fed 20-mg/kg α-tocopherol for 3 months compared to mice fed a 2-mg/kg diet for the same duration (Kluth *et al.*, 2005). In addition, 24-h urinary α-CEHC levels were increased in mice fed a 200-mg/kg diet as compared to mice fed the 20-mg/kg diet (Kluth *et al.*, 2005). Data for Cyps other than Cyp3a were not reported in this last study. The above *in vivo* data support the hypothesis that α-tocopherol modulates a subset of CYP enzymes, thus increasing its own metabolism and preventing toxic accumulation of vitamin E.

Contrary to the *in vivo* results above, which suggest that CYP3A plays a role in increased tocopherol metabolism, studies using insect microsomes expressing individual recombinant human CYP enzymes found that CYP3A4 did not metabolize either α- or γ-tocopherol (Sontag and Parker, 2002). Rather, CYP4F2 was determined to be the only CYP to metabolize γ-tocopherol, and to a much lesser extent α-tocopherol, to their respective 13′-OH-tocopherol metabolites. Further studies are needed to elucidate the role of CYP enzymes in the *in vivo* metabolism of α-tocopherol, as well as metabolism of other forms of vitamin E.

As stated above, following ω-oxidation of the side chain, vitamin Es are thought to undergo several steps of β-oxidation. However, the *in vivo* subcellular localization of the ω- and β-oxidation steps and the enzymes involved in vitamin E β-oxidation have not been identified. In fact, the proposed intermediate products of β-oxidation, other than 5′-α-CMBHC, have not, to date, been isolated *in vivo*.

C. CONJUGATION OF CEHCs

Prior to excretion in either the bile or the urine, most of the CEHCs undergo enzymatic conjugation catalyzed by either sulfotransferases (SULTs) or uridine diphosphate (UDP) glucuronosyl transferases (UGTs), which produce sulfate ester and glucuronide conjugation products, respectively. Although most reports agree that nearly all of the CEHCs found in the bile and urine are conjugated, the percentage of the two conjugation products has yet to be definitively determined and may differ between species and for the various forms of vitamin E (Lodge *et al.*, 2000; Schultz *et al.*, 1995; Stahl *et al.*, 1999; Swanson *et al.*, 1999). The mechanism by which CEHCs are excreted into either the bile or the urine is not known.

D. BILIARY EXCRETION OF α-TOCOPHEROL

In addition to α-CEHC, α-tocopherol itself is excreted into bile. In rats treated with piperonyl butoxide, a P450 inhibitor, the biliary excretion of α-tocopherol increased by nearly 60% as compared to controls (Mustacich *et al.*, 1998). This biliary excretion of α-tocopherol was blocked by pretreatment with verapamil, an inhibitor of MDR1 and MDR3, two ATP-binding cassette (ABC) transporters located in the canalicular membranes of hepatocytes and known to transport lipophilic compounds into the bile (Mustacich *et al.*, 1998). In addition, the biliary excretion of α-tocopherol in Mdr2 knockout mice was only 25% that of wild-type mice fed the same standard chow diet. Thus, the Mdr2 transporter is required for biliary excretion of α-tocopherol under basal conditions (Mustacich *et al.*, 1998). (Note: In rats and humans, ABC transport protein nomenclature is capitalized while in mice only the first letter is capitalized. Additionally, the mouse ABC transport protein designated Mdr2 in mice is designated MDR3 in humans and rats.)

Multidrug resistance protein 1 (MDR1, also ABCB1 or *p*-glycoprotein), one of the ABC transporter proteins located in the canalicular membranes of hepatocytes, shares many of the same substrates with CYP3A (Christians, 2004). Compounds that modulate CYP3A have been observed to similarly modulate MDR1 (Christians, 2004). Importantly, hepatic MDR1 protein levels of rats increased with high-dose α-tocopherol supplementation and this increase in MDR1 protein corresponded with a decrease in hepatic α-tocopherol levels despite continued high-dose supplementation with α-tocopherol (Mustacich *et al.*, 2006). These data indicate that, in addition to the previously demonstrated role of Mdr2 (MDR3 in humans and rats) in the basal biliary excretion of α-tocopherol in nonsupplemented mice (Mustacich *et al.*, 1998), MDR1 may also play a role in the biliary excretion of α-tocopherol, particularly under conditions of high-dose supplementation. The ability of α-tocopherol to modulate the *in vivo* expression of additional biliary transporters has not been reported.

E. IMPLICATIONS OF ALTERED XENOBIOTIC METABOLISM

In addition to metabolizing vitamin E, the hepatic CYP enzyme system is responsible for the metabolism of numerous endogenous and exogenous compounds, including the majority of pharmaceutical drugs (Guengerich, 2003). The human genome encodes ~57 CYP proteins that make up 18 families and 43 subfamilies; however, members of the CYP3A, CYP2B, and CYP2C subfamilies are responsible for the metabolism of most pharmaceutical drugs (Guengerich, 2003). The CYP3A subfamily alone is responsible for the metabolism of ~50% of all pharmaceutical drugs (Kliewer *et al.*, 2002). Thus, dietary supplements that alter one or more of these three CYP subfamilies may potentially alter drug efficacy.

Hepatic CYP3A, CYP2B, and CYP2C protein levels in rats increased with high-dose α-tocopherol supplementation (Mustacich *et al.*, 2006). In addition, 43 days of α-tocopherol supplementation (75 IU) in *Cebus* monkeys increased the plasma clearance and decreased the elimination half-life of antipyrine, a probe compound for assessing hepatic CYP3A, CYP2B, and CYP2C activities (Meydani and Greenblatt, 1990). These data suggest supplemental vitamin E might alter human drug metabolism and disposition. Moreover, reports from clinical trials suggest possible nutrient–drug interactions between α-tocopherol and pharmaceutical drugs metabolized by CYP enzymes, that is cholesterol-lowering agents, for example simvastatin (Brown *et al.*, 2001; Cheung *et al.*, 2001).

Finally, MDR1 is responsible for the biliary excretion of numerous pharmaceutical drugs and their metabolites. Thus, altered expression of MDR1 by high-dose supplementation with α-tocopherol, as was discussed in the aforementioned rat studies, may alter the bioavailability and efficacy of drugs excreted from the body that utilize this transporter.

VI. IMPLICATIONS FOR HUMANS SUPPLEMENTING WITH VITAMIN E

Several studies have reported that vitamin E is associated with decreased chronic disease risk. The Women's Health Study, a 10-year prevention trial in otherwise healthy women, found that 600-IU vitamin E decreased cardiovascular mortality by 24% and in women over 65 by 49% (Lee *et al.*, 2005). Antioxidant treatment with vitamins E and C slowed atherosclerotic progression in intimal thickness of coronary and carotid arteries in hypercholesterolemic (Salonen *et al.*, 2003) and in heart transplant patients (Fang *et al.*, 2002). The Cache County Study reported that antioxidant use (vitamin E >400 IU and vitamin C >500 mg) was associated with reduced prevalence of Alzheimer's disease in the elderly (Zandi *et al.*, 2004). Regular vitamin E supplement use for 10 years or more was associated with a lower risk of mortality from the neurodegenerative disease, amyotrophic lateral sclerosis (Lou Gehrig's disease) (Ascherio *et al.*, 2005). Therefore, it is not surprising that vitamin E supplements are used daily by >35 million people in the United States (Ford *et al.*, 2005).

Although the vitamin E recommended dietary allowance (RDA) is 15 mg of *RRR*-α-tocopherol, vitamin E supplements are commercially available in doses of 100–1000 IU. As many as 15 million Americans consume dietary supplements concurrently with prescription medications (Peng *et al.*, 2004). Drug–nutrient interactions are an area of increased concern because dietary supplements can have potent adverse effects. In 2000, the FDA warned that St. John's Wort (hypericum perforatum), an herbal product, and indinavir, a protease inhibitor used to treat HIV infection, should not be consumed together because the herbal supplement increased the catabolic breakdown of the protease inhibitor (http://www.fda.gov/cder/drug/advisory/stjwort.htm). Hyperforin, the active constituent of St. John's wort, upregulates CYP3A4 as evidenced by increased 6-β-hydroxycortisol excretion (Bauer *et al.*, 2002). Increases in hepatic α-tocopherol concentrations in mice were correlated with increased Cyp3a protein (Traber *et al.*, 2005b) and mRNA (Kluth *et al.*, 2005). Moreover, in rats, increased hepatic α-tocopherol upregulates a constellation of xenobiotic pathways (Mustacich *et al.*, 2006). Thus, given that CYP3A4 isoforms metabolize >50% of therapeutic drugs (Kliewer *et al.*, 2002), it is quite possible that high-dose α-tocopherol may alter drug metabolism.

Reports of adverse effects of vitamin E supplements in humans are sufficiently rare that the Food and Nutrition Board set the upper tolerance level (UL) for α-tocopherol at 1000 mg [1100-IU synthetic (all-*rac*), 1500-IU natural (*RRR*)] per day using data from studies in rats (Food and Nutrition Board and Institute of Medicine, 2000). However, in the last 5 years, findings from two vitamin intervention trials have suggested adverse vitamin effects with therapeutic drugs. One study was a 3-year, double-blind trial of antioxidants (vitamins E and C, β-carotene, and selenium) or placebos in 160 subjects taking both simvastatin and niacin (Brown *et al.*, 2001; Cheung *et al.*, 2001). Simvastatin is a 3-hydroxy-3-methylglutaryl coenzyme A (HMG-CoA) reductase inhibitor (this class of drugs is referred to as statins) that is widely utilized in the treatment of hypercholesterolemia. In subjects taking antioxidants, there was less benefit of the drugs in raising HDL cholesterol than was expected (Cheung *et al.*, 2001), and there was an increase in adverse clinical endpoints (Brown *et al.*, 2001). The other study was the Women's Angiographic Vitamin and Estrogen (WAVE) trial, a randomized, double-blind trial of 423 postmenopausal women with at least one coronary stenosis at baseline coronary angiography. In postmenopausal women on hormone replacement therapy—all-cause mortality was increased in women assigned to antioxidant vitamins compared with placebo (HR, 2.8; 95% CI, 1.1–7.2; $p = 0.047$) (Waters *et al.*, 2002). Although plasma simvastatin and estrogen-concentrations were not measured in these trials, both simvastatin (Williams and Feely, 2002) and estrogen (Lee *et al.*, 2001) are metabolized by CYP3A4, lending support to the hypothesis that pharmacological doses of α-tocopherol stimulated drug metabolism and potentially decreased drug concentrations in these trials.

Vitamin E drug interactions at this point are speculative but represent a potential explanation for the reported adverse affects in some intervention studies. It should be noted that in all of these studies, vitamin E was taken in combination with one or more additional antioxidants, thus it is not clear that any adverse effects are due to vitamin E alone. Finally, vitamin E metabolism and excretion appear to limit the hepatic accumulation of vitamin E and thus, its potential adverse effects.

VII. CONCLUSION

The goal of this chapter was to provide an overview of α-tocopherol, as it relates to its bioavailability and biodistribution. Numerous studies have assisted in explaining the mechanisms by which α-tocopherol is delivered to peripheral tissues or metabolized and excreted in the bile and/or urine. Further elucidation of the ability of α-tocopherol to modulate metabolism of both endogenous and exogenous compounds will increase our ability to utilize this vitamin to its full potential for improved human health and healthcare, while avoiding possible adverse drug interactions.

ACKNOWLEDGMENTS

Grant support was provided by the National Institutes of Health to MGT (NIH DK59576 and DK067930).

REFERENCES

Aguie, G. A., Rader, D. J., Clavery, V., Traber, M. G., Torpier, G., Kayden, H. J., Fruchart, J. C., Brewer, H. B., and Castro, G. (1995). Lipoproteins containing apolipoprotein B in abetalipoproteinemia and homozygous hypobetalipoproteinemia. Identification and characterization. *Atherosclerosis* **118,** 183–191.

Arita, M., Sato, Y., Miyata, A., Tanabe, T., Takahashi, E., Kayden, H., Arai, H., and Inoue, K. (1995). Human α-tocopherol transfer protein: cDNA cloning, expression and chromosomal localization. *Biochem. J.* **306,** 437–443.

Ascherio, A., Weisskopf, M. G., O'Reilly, E. J., Jacobs, E. J., McCullough, M. L., Calle, E. E., Cudkowicz, M., and Thun, M. J. (2005). Vitamin E intake and risk of amyotrophic lateral sclerosis. *Ann. Neurol.* **57,** 104–110.

Bauer, S., Stormer, E., Kerb, R., Johne, A., Brockmoller, J., and Roots, I. (2002). Differential effects of Saint John's Wort (hypericum perforatum) on the urinary excretion of D-glucaric acid and 6β-hydroxycortisol in healthy volunteers. *Eur. J. Clin. Pharmacol.* **58,** 581–585.

Birringer, M., Drogan, D., and Brigelius-Flohe, R. (2001). Tocopherols are metabolized in HepG2 cells by side chain omega-oxidation and consecutive β-oxidation. *Free Radic. Biol. Med.* **31,** 226–232.

Birringer, M., Pfluger, P., Kluth, D., Landes, N., and Brigelius-Flohe, R. (2002). Identities and differences in the metabolism of tocotrienols and tocopherols in HepG2 cells. *J. Nutr.* **132,** 3113–3118.

Borel, P., Pasquier, B., Armand, M., Tyssandier, V., Grolier, P., Alexandre-Gouabau, M. C., Andre, M., Senft, M., Peyrot, J., Jaussan, V., Lairon, D., and Azais-Braesco, V. (2001). Processing of vitamin A and E in the human gastrointestinal tract. *Am. J. Physiol. Gastrointest. Liver Physiol.* **280,** G95–G103.

Brigelius-Flohé, R., and Traber, M. G. (1999). Vitamin E: Function and metabolism. *FASEB J.* **13,** 1145–1155.

Brown, B. G., Zhao, X. Q., Chait, A., Fisher, L. D., Cheung, M. C., Morse, J. S., Dowdy, A. A., Marino, E. K., Bolson, E. L., Alaupovic, P., Frohlich, J., and Albers, J. J. (2001). Simvastatin and niacin, antioxidant vitamins, or the combination for the prevention of coronary disease. *N. Engl. J. Med.* **345,** 1583–1592.

Bruno, R. S., Ramakrishnan, R., Montine, T. J., Bray, T. M., and Traber, M. G. (2005). α-Tocopherol disappearance is faster in cigarette smokers and is inversely related to their ascorbic acid status. *Am. J. Clin. Nutr.* **81,** 95–103.

Bruno, R. S., Leonard, S. W., Atkinson, J. K., Montine, T. J., Ramakrishnan, R., Bray, T. M., and Traber, M. G. (2006a). Faster vitamin E disappearance in smokers is normalized by vitamin C supplementation. *Free Radic. Biol. Med.* **40,** 689–697.

Bruno, R. S., Leonard, S. W., Park, S.-I., Zhao, Y., and Traber, M. G. (2006b). Human vitamin E requirements assessed with the use of apples fortified with deuterium-labeled α-tocopheryl acetate. *Am. J. Clin. Nutr.* **83,** 299–304.

Buettner, G. R. (1993). The pecking order of free radicals and antioxidants: Lipid peroxidation, α-tocopherol, and ascorbate. *Arch. Biochem. Biophys.* **300,** 535–543.

Burton, G. W., Cheeseman, K. H., Doba, T., Ingold, K. U., and Slater, T. F. (1983). Vitamin E as an antioxidant *in vitro* and *in vivo*. *Ciba Found. Symp.* **101,** 4–18.

Burton, G. W., Doba, T., Gabe, E. J., Hughes, L., Lee, F. L., Prasad, L., and Ingold, K. U. (1985). Autoxidation of biological molecules. 4. Maximizing the antioxidant activity of phenols. *J. Am. Chem. Soc.* **107,** 7053–7065.

Cheung, M. C., Zhao, X. Q., Chait, A., Albers, J. J., and Brown, B. G. (2001). Antioxidant supplements block the response of HDL to simvastatin-niacin therapy in patients with coronary artery disease and low HDL. *Arterioscler. Thromb. Vasc. Biol.* **21,** 1320–1326.

Chiku, S., Hamamura, K., and Nakamura, T. (1984). Novel urinary metabolite of d-δ-tocopherol in rats. *J. Lipid Res.* **25,** 40–48.

Christians, U. (2004). Transport proteins and intestinal metabolism: P-glycoprotein and cytochrome P4503A. *Ther. Drug Monit.* **26,** 104–106.

Desrumaux, C., Risold, P. Y., Schroeder, H., Deckert, V., Masson, D., Athias, A., Laplanche, H., Guern, N. L., Blache, D., Jiang, X. C., Tall, A., Desor, D., *et al.* (2005). Phospholipid transfer protein (PLTP) deficiency reduces brain vitamin E content and increases anxiety in mice. *FASEB J.* **19,** 296–297.

Drouineaud, V., Lagrost, L., Klein, A., Desrumaux, C., Le Guern, N., Athias, A., Menetrier, F., Moiroux, P., Sagot, P., Jimenez, C., Masson, D., and Deckert, V. (2006). Phospholipid transfer protein (PLTP) deficiency reduces sperm motility and impairs fertility of mouse males. *FASEB J.* **20**(6), 794–796.

Eisengart, A., Milhorat, A. T., Simon, E. J., and Sundheim, L. (1956). The metabolism of vitamin E. II. Purification and characterization of urinary metabolites of α-tocopherol. *J. Biol. Chem.* **221,** 807–817.

Evans, H. M., and Bishop, K. S. (1922). On the existence of a hitherto unrecognized dietary factor essential for reproduction. *Science* **56,** 650–651.

Fang, J. C., Kinlay, S., Beltrame, J., Hikiti, H., Wainstein, M., Behrendt, D., Suh, J., Frei, B., Mudge, G. H., Selwyn, A. P., and Ganz, P. (2002). Effect of vitamins C and E on progression of transplant-associated arteriosclerosis: A randomised trial. *Lancet* **359,** 1108–1113.

Food and Nutrition Board, and Institute of Medicine (2000). *In* "Dietary Reference Intakes for Vitamin C, Vitamin E, Selenium, and Carotenoids." National Academy Press, Washington.

Ford, E. S., Ajani, U. A., and Mokdad, A. H. (2005). Brief communication: The prevalence of high intake of vitamin E from the use of supplements among U. S. adults. *Ann. Intern. Med.* **143,** 116–120.

Galli, F., Lee, R., Atkinson, J., Floridi, A., and Kelly, F. J. (2003). γ-Tocopherol biokinetics and transformation in humans. *Free Radic. Res.* **37,** 1225–1233.

Guengerich, F. P. (2003). Cytochromes P450, drugs, and diseases. *Mol. Interv.* **3,** 194–204.

Havel, R. (1994). McCollum Award Lecture, 1993: Triglyceride-rich lipoproteins and atherosclerosis—new perspectives. *Am. J. Clin. Nutr.* **59,** 795–799.

Hayes, K. C., Pronczuk, A., and Perlman, D. (2001). Vitamin E in fortified cow milk uniquely enriches human plasma lipoproteins. *Am. J. Clin. Nutr.* **74,** 211–218.

Hosomi, A., Arita, M., Sato, Y., Kiyose, C., Ueda, T., Igarashi, O., Arai, H., and Inoue, K. (1997). Affinity for α-tocopherol transfer protein as a determinant of the biological activities of vitamin E analogs. *FEBS Lett.* **409,** 105–108.

Hosomi, A., Goto, K., Kondo, H., Iwatsubo, T., Yokota, T., Ogawa, M., Arita, M., Aoki, J., Arai, H., and Inoue, K. (1998). Localization of α-tocopherol transfer protein in rat brain. *Neurosci. Lett.* **256,** 159–162.

Ikeda, S., Tohyama, T., and Yamashita, K. (2002). Dietary sesame seed and its lignans inhibit 2,7,8-trimethyl-2(2′-carboxyethyl)-6-hydroxychroman excretion into urine of rats fed γ-tocopherol. *J. Nutr.* **132,** 961–966.

Jiang, X. C., Tall, A. R., Qin, S., Lin, M., Schneider, M., Lalanne, F., Deckert, V., Desrumaux, C., Athias, A., Witztum, J. L., and Lagrost, L. (2002). Phospholipid transfer protein deficiency protects circulating lipoproteins from oxidation due to the enhanced accumulation of vitamin E. *J. Biol. Chem.* **277,** 31850–31856.

Jishage, K., Arita, M., Igarashi, K., Iwata, T., Watanabe, M., Ogawa, M., Ueda, O., Kamada, N., Inoue, K., Arai, H., and Suzuki, H. (2001). α-Tocopherol transfer protein is important for the normal development of placental labyrinthine trophoblasts in mice. *J. Biol. Chem.* **273,** 1669–1672.

Kaempf-Rotzoll, D. E., Igarashi, K., Aoki, J., Jishage, K., Suzuki, H., Tamai, H., Linderkamp, O., and Arai, H. (2002). α-Tocopherol transfer protein is specifically localized at the implantation site of pregnant mouse uterus. *Biol. Reprod.* **67,** 599–604.

Kiyose, C., Saito, H., Kaneko, K., Hamamura, K., Tomioka, M., Ueda, T., and Igarashi, O. (2001). α-Tocopherol affects the urinary and biliary excretion of 2,7,8-trimethyl-2 (2′-carboxyethyl)-6-hydroxychroman, γ-tocopherol metabolite, in rats. *Lipids* **36,** 467–472.

Klein, A., Deckert, V., Schneider, M., Dutrillaux, F., Hammann, A., Athias, A., Le Guern, N., Pais de Barros, J. P., Desrumaux, C., Masson, D., Jiang, X. C., and Lagrost, L. (2006). α-Tocopherol modulates phosphatidylserine externalization in erythrocytes. Relevance in phospholipid transfer protein-deficient mice. *Arterioscler. Thromb. Vasc. Biol.* **26,** 2160–2167.

Kliewer, S. A., Goodwin, B., and Willson, T. M. (2002). The nuclear pregnane X receptor: A key regulator of xenobiotic metabolism. *Endocr. Rev.* **23,** 687–702.

Kluth, D., Landes, N., Pfluger, P., Muller-Schmehl, K., Weiss, K., Bumke-Vogt, C., Ristow, M., and Brigelius-Flohe, R. (2005). Modulation of Cyp3a11 mRNA expression by α-tocopherol but not γ-tocotrienol in mice. *Free Radic. Biol. Med.* **38,** 507–514.

Kostner, G. M., Oettl, K., Jauhiainen, M., Ehnholm, C., Esterbauer, H., and Dieplinger, H. (1995). Human plasma phospholipid transfer protein accelerates exchange/transfer of α-tocopherol between lipoproteins and cells. *Biochem. J.* **305,** 659–667.

Kuhlenkamp, J., Ronk, M., Yusin, M., Stolz, A., and Kaplowitz, N. (1993). Identification and purification of a human liver cytosolic tocopherol binding protein. *Protein Expr. Purif.* **4,** 382–389.

Lee, A. J., Kosh, J. W., Conney, A. H., and Zhu, B. T. (2001). Characterization of the NADPH-dependent metabolism of 17β-estradiol to multiple metabolites by human liver microsomes and selectively expressed human cytochrome P450 3A4 and 3A5. *J. Pharmacol. Exp. Ther.* **298,** 420–432.

Lee, I. M., Cook, N. R., Gaziano, J. M., Gordon, D., Ridker, P. M., Manson, J. E., Hennekens, C. H., and Buring, J. E. (2005). Vitamin E in the primary prevention of cardiovascular disease and cancer: The Women's Health Study: A randomized controlled trial. *JAMA* **294,** 56–65.

Leonard, S. W., Terasawa, Y., Farese, R. V., Jr., and Traber, M. G. (2002). Incorporation of deuterated *RRR*- or *all rac* α-tocopherol into plasma and tissues of α-tocopherol transfer protein null mice. *Am. J. Clin. Nutr.* **75,** 555–560.
Leonard, S. W., Good, C. K., Gugger, E. T., and Traber, M. G. (2004). Vitamin E bioavailability from fortified breakfast cereal is greater than that from encapsulated supplements. *Am. J. Clin. Nutr.* **79,** 86–92.
Leonard, S. W., Paterson, E., Atkinson, J. K., Ramakrishnan, R., Cross, C. E., and Traber, M. G. (2005). Studies in humans using deuterium-labeled α- and γ-tocopherol demonstrate faster plasma γ-tocopherol disappearance and greater γ-metabolite production. *Free Radic. Biol. Med.* **38,** 857–866.
Lodge, J. K., Traber, M. G., Elsner, A., and Brigelius-Flohe, R. (2000). A rapid method for the extraction and determination of vitamin E metabolites in human urine. *J. Lipid Res.* **41,** 148–154.
Manor, D., and Atkinson, J. (2003). Is tocopherol associated protein a misnomer? *J. Nutr. Biochem.* **14,** 421–422; author reply 423.
Meier, R., Tomizaki, T., Schulze-Briese, C., Baumann, U., and Stocker, A. (2003). The molecular basis of vitamin E retention: Structure of human α-tocopherol transfer protein. *J. Mol. Biol.* **331,** 725–734.
Meydani, M., and Greenblatt, D. J. (1990). Influence of vitamin E supplementation on antipyrine clearance in the cebus monkey. *Nutr. Res.* **10,** 1045–1051.
Meydani, M., Cohn, J. S., Macauley, J. B., McNamara, J. R., Blumberg, J. B., and Schaefer, E. J. (1989). Postprandial changes in the plasma concentration of α- and γ-tocopherol in human subjects fed a fat-rich meal supplemented with fat-soluble vitamins. *J. Nutr.* **119,** 1252–1258.
Min, K. C., Kovall, R. A., and Hendrickson, W. A. (2003). Crystal structure of human α-tocopherol transfer protein bound to its ligand: Implications for ataxia with vitamin E deficiency. *Proc. Natl. Acad. Sci. USA* **100,** 14713–14718.
Minehira-Castelli, K., Leonard, S. W., Walker, Q. M., Traber, M. G., and Young, S. G. (2006). Absence of VLDL secretion does not affect α-tocopherol content in peripheral tissues. *J. Lipid Res.* **47,** 1773–1778.
Murray, M. (1991). *In vitro* and *in vivo* studies of the effect of vitamin E on microsomal cytochrome P450 in rat liver. *Biochem. Pharmacol.* **42,** 2107–2114.
Mustacich, D. J., Shields, J., Horton, R. A., Brown, M. K., and Reed, D. J. (1998). Biliary secretion of α-tocopherol and the role of the mdr2 P-glycoprotein in rats and mice. *Arch. Biochem. Biophys.* **350,** 183–192.
Mustacich, D. J., Leonard, S. W., Devereaux, M. W., Sokol, R. J., and Traber, M. G. (2006). α-Tocopherol regulation of hepatic cytochrome P450s and ABC transporters in rats. *Free Radic. Biol. Med.* **41,** 1069–1078.
Panagabko, C., Morley, S., Neely, S., Lei, H., Manor, D., and Atkinson, J. (2002). Expression and refolding of recombinant human α-tocopherol transfer protein capable of specific α-tocopherol binding. *Protein Expr. Purif.* **24,** 395–403.
Panagabko, C., Morley, S., Hernandez, M., Cassolato, P., Gordon, H., Parsons, R., Manor, D., and Atkinson, J. (2003). Ligand specificity in the CRAL-TRIO protein family. *Biochemistry* **42,** 6467–6474.
Parker, R. S., Sontag, T. J., and Swanson, J. E. (2000). Cytochrome P4503A-dependent metabolism of tocopherols and inhibition by sesamin. *Biochem. Biophys. Res. Commun.* **277,** 531–534.
Peng, C. C., Glassman, P. A., Trilli, L. E., Hayes-Hunter, J., and Good, C. B. (2004). Incidence and severity of potential drug-dietary supplement interactions in primary care patients: An exploratory study of 2 outpatient practices. *Arch. Intern. Med.* **164,** 630–636.
Salonen, R. M., Nyyssonen, K., Kaikkonen, J., Porkkala-Sarataho, E., Voutilainen, S., Rissanen, T. H., Tuomainen, T. P., Valkonen, V. P., Ristonmaa, U., Lakka, H. M., Vanharanta, M., Salonen, J. T., *et al.* (2003). Six-year effect of combined vitamin C and E supplementation on atherosclerotic progression: The Antioxidant Supplementation in Atherosclerosis Prevention (ASAP) Study. *Circulation* **107,** 947–953.

Sato, Y., Hagiwara, K., Arai, H., and Inoue, K. (1991). Purification and characterization of the α-tocopherol transfer protein from rat liver. *FEBS Lett.* **288,** 41–45.

Sato, Y., Arai, H., Miyata, A., Tokita, S., Yamamoto, K., Tanabe, T., and Inoue, K. (1993). Primary structure of α-tocopherol transfer protein from rat liver. Homology with cellular retinaldehyde-binding protein. *J. Biol. Chem.* **268,** 17705–17710.

Sattler, W., Levak-Frank, S., Radner, H., Kostner, G., and Zechner, R. (1996). Muscle-specific overexpression of lipoprotein lipase in transgenic mice results in increased α-tocopherol levels in skeletal muscle. *Biochem. J.* **318,** 15–19.

Schultz, M., Leist, M., Petrzika, M., Gassmann, B., and Brigelius-Flohé, R. (1995). Novel urinary metabolite of α-tocopherol, 2,5,7,8-tetramethyl-2(2′-carboxyethyl)-6-hydroxychroman, as an indicator of an adequate vitamin E supply? *Am. J. Clin. Nutr.* **62**(Suppl.), 1527S–1534S.

Sokol, R. J., Heubi, J. E., Iannaccone, S., Bove, K. E., and Balistreri, W. F. (1983). Mechanism causing vitamin E deficiency during chronic childhood cholestasis. *Gastroenterology* **85,** 1172–1182.

Sontag, T. J., and Parker, R. S. (2002). Cytochrome P450 omega-hydroxylase pathway of tocopherol catabolism: Novel mechanism of regulation of vitamin E status. *J. Biol. Chem.* **277,** 25290–25296.

Stahl, W., Graf, P., Brigelius-Flohe, R., Wechter, W., and Sies, H. (1999). Quantification of the α- and γ-tocopherol metabolites 2,5,7,8-tetramethyl-2-(2′-carboxyethyl)-6-hydroxychroman and 2,7,8-trimethyl-2-(2′-carboxyethyl)-6-hydroxychroman in human serum. *Anal. Biochem.* **275,** 254–259.

Stocker, A., Zimmer, S., Spycher, S. E., and Azzi, A. (1999). Identification of a novel cytosolic tocopherol-binding protein: Structure, specificity, and tissue distribution. *IUBMB Life* **48,** 49–55.

Swanson, J. E., Ben, R. N., Burton, G. W., and Parker, R. S. (1999). Urinary excretion of 2,7, 8-trimethyl-2-(β-carboxyethyl)-6-hydroxychroman is a major route of elimination of γ-tocopherol in humans. *J. Lipid Res.* **40,** 665–671.

Terasawa, Y., Ladha, Z., Leonard, S. W., Morrow, J. D., Newland, D., Sanan, D., Packer, L., Traber, M. G., and Farese, R. V., Jr. (2000). Increased atherosclerosis in hyperlipidemic mice deficient in α-tocopherol transfer protein and vitamin E. *Proc. Natl. Acad. Sci. USA* **97,** 13830–13834.

Traber, M. G., and Kayden, H. J. (1989). Preferential incorporation of α-tocopherol vs γ-tocopherol in human lipoproteins. *Am. J. Clin. Nutr.* **49,** 517–526.

Traber, M. G., Olivecrona, T., and Kayden, H. J. (1985). Bovine milk lipoprotein lipase transfers tocopherol to human fibroblasts during triglyceride hydrolysis *in vitro*. *J. Clin. Invest.* **75,** 1729–1734.

Traber, M. G., Burton, G. W., Ingold, K. U., and Kayden, H. J. (1990a). RRR- and SRR-α-tocopherols are secreted without discrimination in human chylomicrons, but RRR-α-tocopherol is preferentially secreted in very low density lipoproteins. *J. Lipid Res.* **31,** 675–685.

Traber, M. G., Sokol, R. J., Burton, G. W., Ingold, K. U., Papas, A. M., Huffaker, J. E., and Kayden, H. J. (1990b). Impaired ability of patients with familial isolated vitamin E deficiency to incorporate α-tocopherol into lipoproteins secreted by the liver. *J. Clin. Invest.* **85,** 397–407.

Traber, M. G., Burton, G. W., Hughes, L., Ingold, K. U., Hidaka, H., Malloy, M., Kane, J., Hyams, J., and Kayden, H. J. (1992). Discrimination between forms of vitamin E by humans with and without genetic abnormalities of lipoprotein metabolism. *J. Lipid Res.* **33,** 1171–1182.

Traber, M. G., Sokol, R. J., Kohlschütter, A., Yokota, T., Muller, D. P. R., Dufour, R., and Kayden, H. J. (1993). Impaired discrimination between stereoisomers of α-tocopherol in patients with familial isolated vitamin E deficiency. *J. Lipid Res.* **34,** 201–210.

Traber, M. G., Ramakrishnan, R., and Kayden, H. J. (1994). Human plasma vitamin E kinetics demonstrate rapid recycling of plasma *RRR*-α-tocopherol. *Proc. Natl. Acad. Sci. USA* **91,** 10005–10008.

Traber, M. G., Elsner, A., and Brigelius-Flohe, R. (1998). Synthetic as compared with natural vitamin E is preferentially excreted as α-CEHC in human urine: Studies using deuterated α-tocopheryl acetates. *FEBS Lett.* **437,** 145–148.

Traber, M. G., Burton, G. W., and Hamilton, R. L. (2005a). Vitamin E trafficking. *Ann. NY Acad. Sci.* **1031,** 1–12.

Traber, M. G., Siddens, L. K., Leonard, S. W., Schock, B., Gohil, K., Krueger, S. K., Cross, C. E., and Williams, D. E. (2005b). α-Tocopherol modulates Cyp3a expression, increases γ-CEHC production and limits tissue γ-tocopherol accumulation in mice fed high γ-tocopherol diets. *Free Radic. Biol. Med.* **38,** 773–785.

Waters, D. D., Alderman, E. L., Hsia, J., Howard, B. V., Cobb, F. R., Rogers, W. J., Ouyang, P., Thompson, P., Tardif, J. C., Higginson, L., Bittner, V., Steffes, M., *et al.* (2002). Effects of hormone replacement therapy and antioxidant vitamin supplements on coronary atherosclerosis in postmenopausal women: A randomized controlled trial. *JAMA* **288,** 2432–2440.

Wetterau, J. R., Aggerbeck, L. P., Bouma, M. E., Eisenberg, C., Munck, A., Hermier, M., Schmitz, J., Gay, G., Rader, D. J., and Gregg, R. E. (1992). Absence of microsomal triglyceride transfer protein in individuals with abetalipoproteinemia. *Science* **258,** 999–1001.

Williams, D., and Feely, J. (2002). Pharmacokinetic-pharmacodynamic drug interactions with HMG-CoA reductase inhibitors. *Clin. Pharmacokinet.* **41,** 343–370.

Yap, S. P., Yuen, K. H., and Wong, J. W. (2001). Pharmacokinetics and bioavailability of α-, γ- and δ-tocotrienols under different food status. *J. Pharm. Pharmacol.* **53,** 67–71.

Yoshida, H., Yusin, M., Ren, I., Kuhlenkamp, J., Hirano, T., Stolz, A., and Kaplowitz, N. (1992). Identification, purification, and immunochemical characterization of a tocopherol-binding protein in rat liver cytosol. *J. Lipid Res.* **33,** 343–350.

Zandi, P. P., Anthony, J. C., Khachaturian, A. S., Stone, S. V., Gustafson, D., Tschanz, J. T., Norton, M. C., Welsh-Bohmer, K. A., and Breitner, J. C. (2004). Reduced risk of Alzheimer disease in users of antioxidant vitamin supplements: The Cache County Study. *Arch. Neurol.* **61,** 82–88.

Zimmer, S., Stocker, A., Sarbolouki, M. N., Spycher, S. E., Sassoon, J., and Azzi, A. (2000). A novel human tocopherol-associated protein: Cloning, *in-vitro* expression and characterization. *J. Biol. Chem.* **275,** 25672–25680.

2

Structure and Function of α-Tocopherol Transfer Protein: Implications for Vitamin E Metabolism and AVED

K. Christopher Min

Department of Neurology, Columbia University, New York, New York 10032

Human α-tocopherol transfer protein (α-TTP) plays a central role in vitamin E homeostasis: mutations in the protein are a cause of a progressive neurodegenerative disorder known as ataxia with vitamin E deficiency (AVED). Despite normal dietary intake of vitamin E, affected individuals suffer from a relative deficiency of this essential lipophilic antioxidant. Disease-associated mutations in α-TTP impair its ability

Vitamins and Hormones, Volume 76

0083-6729/07 $35.00
DOI: 10.1016/S0083-6729(07)76002-8

to prevent the degradation and excretion of α-T. Recently, we and others solved the crystal structures of α-TTP bound to a molecule of (2*R*, 4′*R*, 8′*R*)-α-T, which has led to a better understanding of the molecular basis of its biochemical activity. Surprisingly, the ligand was found buried in the hydrophobic core of the protein, completely sequestered from the aqueous milieu. In this chapter, the implications of the structure of α-TTP bound to its ligand regarding the mechanism of α-T retention are discussed. A comparison to a crystal structure of the *apo* form of α-TTP indicates a possible specific conformational change that allows the entry and exit of the ligand. The effect of known disease-associated point mutations is examined in light of the crystal structure as well as recent biochemical studies. Despite the knowledge gained from these studies, the exact molecular mechanism by which α-TTP retains α-T remains enigmatic and will likely prove a fruitful area for future research.

I. Introduction

First recognized as a nutrient present in leafy vegetables that was required for fertility in rats, vitamin E is thought to be most important as the major lipophilic antioxidant, and has been proposed to be protective in diseases with oxidative stress, including cardiovascular disease, cancer, and neurodegenerative disease (Brigelius-Flohe and Traber, 1999; Evans and Bishop, 1922). There is also evidence that α-tocopherol (α-T) influences cell signaling pathways involving protein kinase C and arachidonic acid (Zingg and Azzi, 2004). Vitamin E is actually composed of eight naturally occurring forms, four tocopherols and four tocotrienols, and all forms are composed of a chromanol ring and an aliphatic side chain (Fig. 1). The tocopherols and tocotrienols differ in the level of saturation in the side chain, with tocotrienols containing three double bonds and the tocopherols containing a phytyl tail. The four tocopherols and tocotrienols, α, β, γ, and δ, differ in the number and position of methyl groups on the chromanol ring: α-T is methylated at the C5- and C7-positions, whereas β-T does not contain the C7-position methyl group, γ-T does not contain the C5-position methyl group, and δ-T contains neither the C5-position nor the C7-position methyl groups. In addition, there are three stereocenters in the tocopherols at the C2-position of the chroman ring and at two positions along the phytyl chain at the C4′- and C8′-positions. α- and γ-T are the most abundant vitamin E forms in natural sources, and in particular, α-T plays a special role in the human body as mechanisms exist to specifically retain this form of vitamin E.

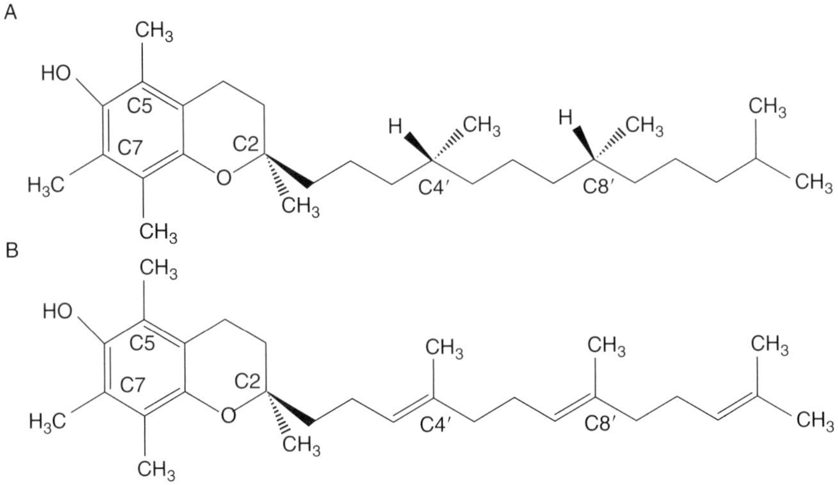

FIGURE 1. Structure of vitamin E. (A) A Natta projection representation of *RRR*-α-T. On the left side is the bicyclic chroman ring, and the C2-, C5-, and C7-positions are labeled. There are three stereocenters at C2 on the chroman ring and on the phytyl tail at C4′ and C8′. Other tocopherols differ in the presence of various methyl groups: β-T does not have a methyl group at C7, γ-T does not have a methyl group at C5, and δ-T is lacking methyl groups at both positions. (B) A Natta projection representation of *R*-α-tocotrienol. The tocotrienols are analogous to the tocopherols except that the aliphatic tail has three double bonds. As a result there is only one stereocenter in the tocotrienols.

α-Tocopherol transfer protein (α-TTP) was first described as a 31-kDa soluble protein from rat liver extracts that bound specifically to α-T (Catignani, 1975; Catignani and Bieri, 1977). The cDNA was eventually isolated by expression screening using an antibody from rat and subsequently cloned from human (Sato *et al.*, 1993). The gene is expressed most highly in the liver, although it is present at lower levels in the brain, kidney, lung, and spleen (Hosomi *et al.*, 1998). Expression of α-TTP in hepatocyte cell lines conferred the ability to secrete α-T (Arita *et al.*, 1997). When the human gene was cloned, it was found to reside in the 8q13.1–13.3 region of chromosome 8 (Arita *et al.*, 1995). Up to this point, the research in α-TTP, although motivated by the possibility of its involvement in vitamin E homeostasis, was limited to the observations of the specificity of binding to α-T and other *in vitro* measures of its function.

The central role of α-TTP in the homeostasis of vitamin E was confirmed when mutations in the gene-encoding α-TTP were found to be causative for an autosomal recessive disease that is often phenotypically indistinguishable from Friedreich's ataxia (Gotoda *et al.*, 1995). An inherited deficiency of vitamin E in the absence of abnormalities of gastrointestinal absorption was initially found in extremely rare sporadic cases and referred to as familial isolated vitamin E (FIVE) deficiency (Burck *et al.*, 1981; Harding *et al.*, 1985; Krendel *et al.*, 1987; Laplante *et al.*, 1984; Sokol *et al.*, 1988; Yokota *et al.*, 1987). Affected individuals had a progressive neurological disorder comprising ataxia, areflexia, marked proprioceptive loss, as well as retinitis pigmentosa late in life. Onset was typically in the second decade of life. The discovery of two large pedigrees in Tunisia led to progress in identifying the causative gene (Ben Hamida *et al.*, 1993b). In these original families, the clinical presentation was similar to severe Friedreich's ataxia, and another designation, ataxia with vitamin E deficiency (AVED), was used to describe these cases. Genetic studies of the affected families showed that AVED mapped to a region of chromosome 8 (Ben Hamida *et al.*, 1993a). The cloning of α-TTP led to the sequencing of the α-TTP gene in these families, and a number of mutations have been discovered that were responsible for both FIVE and AVED (Table I) (Cavalier *et al.*, 1998; Gotoda *et al.*, 1995; Hentati *et al.*, 1996; Hoshino *et al.*, 1999; Ouahchi *et al.*, 1995; Shimohata

TABLE I. Mutations Associated with AVED

Nucleotide change	Effect on coding sequence	Disease onset	α-T binding K_d (nM)[b]	Transfer rate $t_{1/2}$ (min)[b]
None	Wild type		36.1 ± 5	5.9 ± 1.1
T2C	Abolish start codon	Early		
C175T	R59W	Early	123.2 ± 11	11.8 ± 0.6
A191G	D64G	Early		
G205(−1)C	Aberrant splicing	Early		
T303G	H101Q	Late	63.4 ± 3.5	6.2 ± 1.2
A306G	Aberrant splicing	NR[a]		
G358A	A120T	Late	70.0 ± 3.8	6.0 ± 0.9
C400T	A134X (truncation)	Early		
G421A	E141K	Early	76.4 ± 8	15.7 ± 3.8
486delT	Frameshift and truncation	Early		
513insTT	Frameshift and truncation	Early		
530insGTA/533insT	Frameshift and truncation	Early		
T548C	L183P	NR		
G552A	Aberrant splicing	Early		
G575A	R192H	Late	40.9 ± 3.7	6.2 ± 0.5
C661T	R221W	Early	86.1 ± 11	17.3 ± 1.6
744delA	Frameshift and truncation	Early		

[a]NR—not reported.
[b]Reprinted with permission from Morley *et al.* (2004). © 2004 American Chemical Society.

et al., 1998; Usuki and Maruyama, 2000; Yokota *et al.*, 1997). Many of the mutations are splice site or truncation mutations, resulting in a defective protein. Several missense mutations have also been found in families, affecting both alleles either as a homozygous defect or in combination with one of the other mutations. In the absence of α-TTP, ingested α-T is absorbed normally from the intestine in association with chylomicrons, but serum levels associated with lipoprotein particles drop below normal quickly (Traber *et al.*, 1990). α-TTP exerts its action by preventing the excretion of α-T and facilitating the release of α-T in association with VLDL particles from hepatocytes. Treatment of affected individuals consists of supplementation of the diet with large doses of tocopherol to overcome the inefficient retention of vitamin E and has been reported to stabilize symptoms or even ameliorate the conditions of patients who have been treated.

α-TTP is a member of the family of lipid-binding proteins containing the CRAL-TRIO N-terminal (residues 11–83) and C-terminal (residues 89–275) domains (PF03765 and PF00650, respectively), so-called because the domains common to this family were first recognized in cellular retinaldehyde-binding protein (CRALBP) and the triple function domain of the Trio protein (Bateman *et al.*, 2002; Sha *et al.*, 1998). Other notable family members that include both domains include Sec14p and supernatant protein factor (SPF), whereas a much larger group of proteins contain the C-terminal ligand-binding domain. CRALBP is expressed in the pigmented epithelium, appeared to have some functional similarities to α-TTP, binding specifically to the 11-*cis* form of retinaldehyde or retinol, and playing a role in the transfer of this lipophilic molecule between the pigmented epithelial cell and the photoreceptor cell (Saari and Crabb, 2005). Structures of two of the other family members have been determined: Sec14p, a yeast phosphatidyl choline transfer protein essential for the secretory pathway (Sha *et al.*, 1998), and human SPF that stimulates the epoxidation of squalene through an incompletely understood mechanism (Stocker and Baumann, 2003). A sequence alignment of these proteins with CRALBP reveals a number of conserved residues throughout the length of the protein (Fig. 2). Interestingly, another family member of this family of proteins, caytaxin, has been reported to be a cause of autosomal recessive ataxia, but the ligand of this protein is not known (Bomar *et al.*, 2003).

We decided to determine the structure of α-TTP in order to better understand the molecular defects that lead to α-TTP (Min *et al.*, 2003). We also felt that a structure of α-TTP might lead to a better understanding of how these proteins function, from relatively closely related family members like CRALBP and Sec14p to the much larger family of proteins that contain only the C-terminal domain. The purpose of this chapter is to focus on the structure and function of α-TTP, which despite intensive study remains enigmatic in its molecular details.

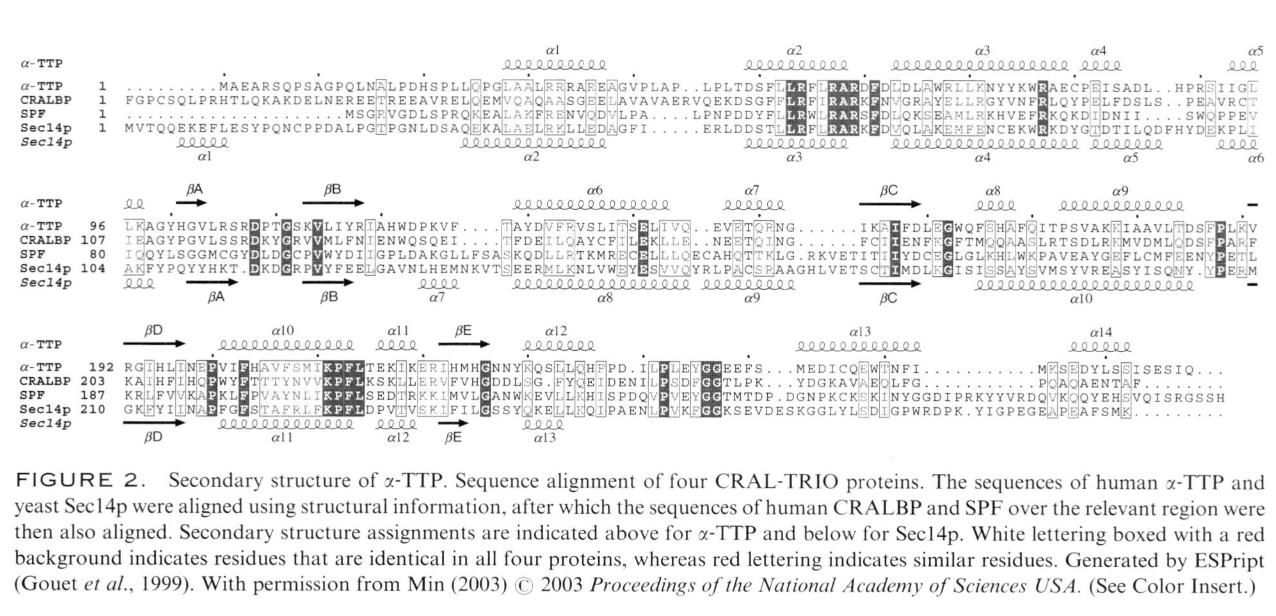

FIGURE 2. Secondary structure of α-TTP. Sequence alignment of four CRAL-TRIO proteins. The sequences of human α-TTP and yeast Sec14p were aligned using structural information, after which the sequences of human CRALBP and SPF over the relevant region were then also aligned. Secondary structure assignments are indicated above for α-TTP and below for Sec14p. White lettering boxed with a red background indicates residues that are identical in all four proteins, whereas red lettering indicates similar residues. Generated by ESPript (Gouet *et al.*, 1999). With permission from Min (2003) © 2003 *Proceedings of the National Academy of Sciences USA*. (See Color Insert.)

II. STRUCTURE OF α-TTP

A. CRYSTALLIZATION

Human α-TTP was expressed as a hexahistidine-tagged protein in *Escherichia coli.* It was observed that α-TTP when isolated without its ligand tended to aggregate over time. To avoid this problem, α-T was present from the time of lysis of the bacteria and at each purification step. Initial attempts at crystallization resulted in small crystals using PEG4000 as precipitant. Subsequent trials failed at reproducing these initial crystals, in spite of seeding trials using crystals from the initial drop. Analysis by MALDI-TOF mass spectrometry indicated that a cleavage event had occurred in the polypeptide, but N-terminal sequencing of protein from the initial crystals failed. On the basis of susceptibility of the N-terminus to proteolysis, a construct was made in which the first 20 residues were removed. This N-terminal deletion mutant was readily crystallized using PEG4000 as precipitant in a hanging drop vapor diffusion arrangement. Selenomethionine substitution was utilized in order to determine phases, and fortuitously, the crystals for theselenomethionine-substituted protein were larger than those obtained for native protein. The crystals belonged to space group $P2_12_12_1$, with unit cell a = 40.1 Å, b = 77.2 Å, and c = 85.4 Å, and diffracted to 1.5-Å spacings at beam line X4A of NSLS synchrotron.

B. OVERALL FOLD

The crystal structure of α-TTP with (2R, 4′P, 8′P)-α-T was solved by initial phases to 2.1 Å using three-wavelength MAD data from a single crystal of selenomethionyl proteins, and subsequently refined to 1.5-Å resolution by using a single-wavelength dataset from another crystal of selenomethionyl proteins. The final model contained residues 25–275, a single molecule of *RRR*-α-T and 253 water molecules. As predicted, α-TTP is composed of two domains, an N-terminal all-helical domain (residues 11–83) and a C-terminal domain (residues 89–275), which at its core is composed of a *βαβαβαβ*-fold and forms the binding pocket for α-T (Fig. 3). The interactions between the domains are primarily derived from the packing of helix 3 against helix 9 and mostly involve hydrophobic contacts. However, there is one hydrogen bond between the side chain of Asn72 and both the side chain and carbonyl oxygen atoms of Asp185, and an apparent covalent bond between Cys80 and Lys178 involving the terminal sulfur atom and ε amino group. There was no evidence for the presence of this bond in a sample of purified protein subjected to tryptic digestion and mass spectroscopic analysis. The formation of this bond may be specific to the milieu of the protein crystals, and it is unlikely to play a role in the function of α-TTP.

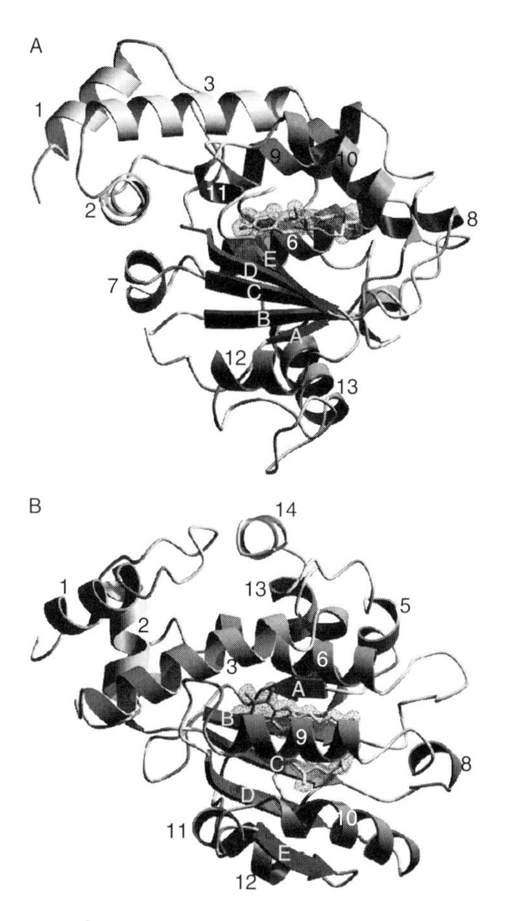

FIGURE 3. Structure of α-TTP. (A and B) A ribbon diagram of α-TTP (PDB:1R5L) with α-T colored in yellow. The N-terminal domain helices are indicated in green, and the C-terminal domain helices in blue and strands in red. The C-terminal domain forms a binding pocket for α-T in the form of a cage, with a floor formed by the β sheet and a ceiling formed by α-helices. Helices are indicated by numerals and strands by letters, in the same order as in Fig. 2. The $2f_o$-f_c electron density contoured at 1σ is drawn as a blue mesh over α-T. The two views are related by a 90° rotation along the horizontal axis. Ribbon diagram generated with POVScript and rendered with POVRay (Fenn *et al.*, 2003). With permission from Min (2003) © 2003 *Proceedings of the National Academy of Sciences USA*. (See Color Insert.)

C. LIGAND BINDING BY α-TTP

The first striking feature of the structure of α-TTP bound to the ligand is the location of α-T in the hydrophobic core of the C-terminal domain. The β sheet forms the floor of a cage, which is covered by three α-helices (Figs. 3 and 4A). Several van der Waals interactions in the pocket with α-T are composed of side chain contributions from strands βC (Ile154, Phe158) and

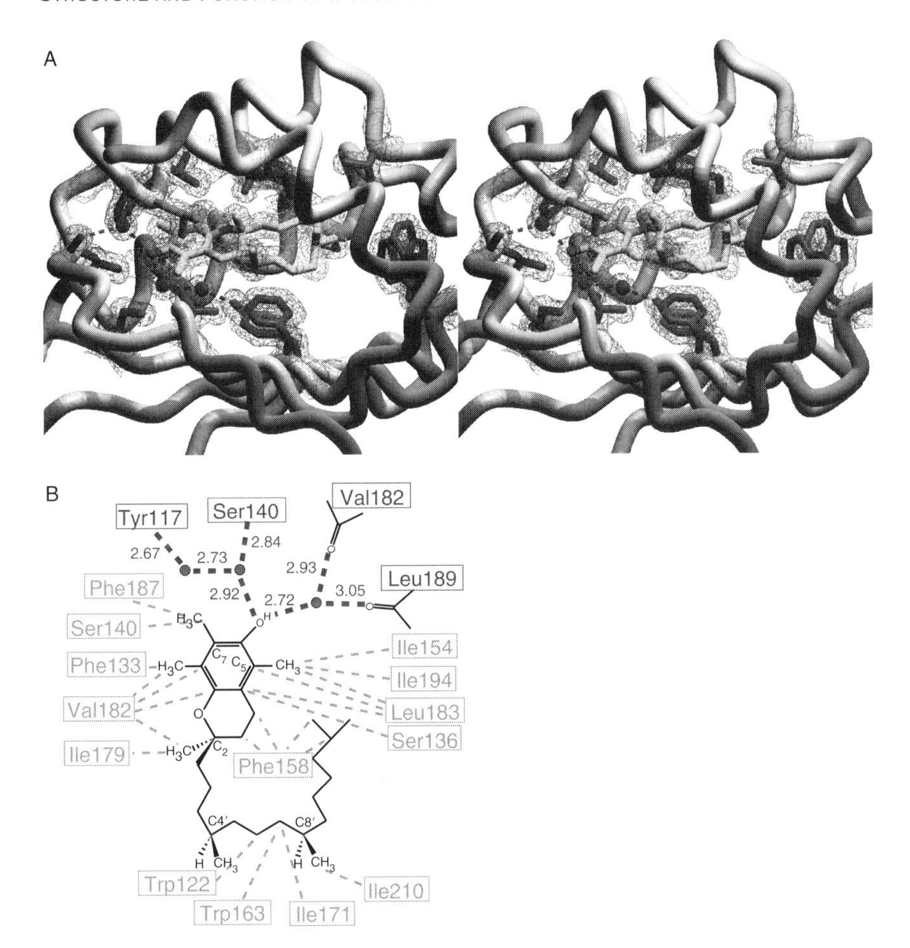

FIGURE 4. Ligand-binding pocket of α-TTP. (A) Residues that form van der Waals contacts with α-T or form hydrogen-bonding networks with ordered water molecules in the ligand-binding pocket are depicted in this stereo view from PDB:1R5L. The ligand is depicted in yellow in a stick representation with a $2f_o$-f_c map contoured at 1σ drawn in yellow. Side chains are also depicted in stick representation but in green, except for Leu183 in cyan, with a $2f_o$-f_c map contoured at 1σ drawn in blue. Three well-ordered water molecules located in the binding pocket are represented as red spheres. Dashed red lines indicate hydrogen bonds. The protein backbone of the C-terminal domain is shown in light gray, excluding residues 216–220 and 249–275 for clarity. Figure generated in POVScript and rendered in POVRay. (B) A schematic representation of the ligand-binding pocket. (2*R*, 4′*R*, 8′*R*)-α-T are shown with the C2, C4′, and C8′ stereocenters labeled, respectively. The C5- and C7-positions on the chroman ring, where there are differences in methylation with β-, γ-, and δ-T, are also labeled. Side chains are indicated by three-letter codes and the position of the residue in the sequence. The carbonyl oxygen atoms of Val182 and Leu189 are drawn to illustrate the hydrogen bond interactions with one of the water molecules. Three well-ordered water molecules located in the binding pocket are depicted as blue circles. Dashed red lines indicate hydrogen bonds, and the lengths of the bonds are indicated. Dashed green lines indicate van der Waals interactions. With permission from Min (2003) © 2003 *Proceedings of the National Academy of Sciences USA*. (See Color Insert.)

βD (Ile194) and from helices α6 (Phe133, Ser136, Ser140), α9 (Ile179, Val182, Leu183), and α10 (Ile210) (Fig. 4). In addition, there are contributions from several of the connecting loops (Trp122, Trp163, Ile171, Phe187). There are four well-ordered water molecules that are also enclosed with α-T in the binding pocket. Two water molecules are hydrogen bonded to the hydroxyl group of the chroman ring. One of these water molecules is in turn hydrogen bonded to the carbonyl oxygen atoms of Val182 and Leu189 and the other is hydrogen bonded to the side chain hydroxyl group of Ser140 as well as the third water molecule. The fourth water molecule is near the phytyl tail but is not closely associated with α-T. The surface interactions of α-TTP with α-T were extensive, and the combined buried surface area was calculated to be 1200 $Å^2$ using GRASP.

The observed interactions between α-TTP and α-T are most extensive with the chroman ring. In a study in which the binding affinity of α-TTP for tocopherols and related molecules was directly measured, the degree of methylation on the chroman ring determined the affinity of α-TTP for ligand, with binding affinities in the order α- > β- > γ- > δ-T. The difference in affinity between the β- and γ-T was 2-fold, and between α- and δ-T was 20-fold. The extensive van der Waals contacts between the methyl groups and protein likely explain the observed affinity differences, which are primarily due to the energy of burying the hydrophobic methyl groups. The C7-position methyl group that is absent from β- and δ-T interacts with Ser140 of helix α6 and Phe187 from the loop between helix α9 and strand βD. The C5-position methyl group that is missing from γ- and δ-T interacts with Ile154 of strand βC, Leu183 of helix α9, and Ile194 of strand βD. Furthermore, the nature of the binding pocket demonstrates why α-TTP is selective for the *R*-stereoisomer at the C2-position. Several residues, Phe133, Ile179, Val182, and Leu183 form a binding pocket for the methyl group at this position. A switch to the *S*-isomer at the C2-position would require a major rearrangement of the ligand-binding pocket in this area of α-TTP. In addition, modification of the hydroxyl group of the phenol ring affects the binding affinity of α-T analogues. Given the hydrogen-bonding network involving the phenol group of the chroman ring, it is not surprising that substituents at the C6-position have dramatic effects on ligand affinity. Finally, the remarkable way in which the phytyl chain is bent within the ligand-binding pocket would not be as well supported by the unsaturated side chains of the tocotrienols, thus explaining the lower affinity of α-TTP for these potential ligands.

D. AN OPEN CONFORMATION OF α-TTP

In our experience, α-TTP obtained from *E. coli* in the absence of α-T was aggregated over time. Others have also determined the structure of α-TTP bound to α-T but by employing a different purification strategy, using a nonionic detergent to stabilize the *apo* form of α-TTP (Meier *et al.*, 2003). The purified α-TTP was then combined with α-T and crystallized. Despite the presence of the entire N-terminus in their construct, the crystallographic model again was missing the N-terminal 25 residues, implying that this region is disordered in the ligand-bound state. As would be expected, a comparison of the Cα-positions of the two α-TTP structures bound to α-T agrees closely with a root mean square deviation (rmsd) of 0.33 Å over 251 residues. In addition, an examination of the ligand pocket is essentially identical with the same observed van der Waals contacts as well as hydrogen-bonded water molecules. The one difference is an additional water molecule observed in our structure, which forms hydrogen bonds with Tyr117, and another water molecule seen in both models.

In the same report, Meier *et al.* also solved the structure of an *apo* form of α-TTP utilizing a nonionic detergent to stabilize the protein. In this case, a dimer of α-TTP was crystallized with densities observed for three Triton X-100 detergent molecules per α-TTP in the ligand-binding pocket (PDB:1OIZ). The significance of the dimeric nature observed in the crystal is uncertain; a similar dimer with a twofold axis of symmetry coinciding with one of the crystal axes was also observed in the structure of Sec14p. In the *apo* structure of α-TTP, more of the N-terminus was ordered, and the models of the dimer encompassed residues 9–273 (chain A) and 11–274 (chain B). A comparison of the *apo* form of α-TTP in chain A with our α-T-bound structure (PDB:1R5L) showed little difference over most of the molecule, with an rmsd of 0.61 Å for the Cα-positions over 237 residues, with nearly identical results for chain B from the dimer. The major differences were the additional residues in the model at the N-terminus and a stretch of residues in helix 10 displaced by ~10 Å from the ligand-bound form of the protein (Figs. 5 and 6). In addition to occupying the pocket for α-T, the detergent molecules were found to be in the new crevice between helices α9 and α10.

Although artificially derived through the use of nonionic detergent, it is attractive to consider this second conformation of α-TTP as functionally relevant. The ligand-bound state of α-TTP must clearly undergo a conformational change in order to release α-T for transport out of the hepatocyte for delivery to peripheral tissues. A similar open conformation was also observed for the distantly related protein Sec14p. This yeast protein is necessary for the function of the protein secretion pathway and has been characterized as a phosphatidyl choline exchange factor between the Golgi network and the endoplasmic reticulum. Structural studies of this protein have been limited to a β-octylglucoside-bound form; attempts to crystallize Sec14p with *bona fide* ligands have failed to date. The crystal structure of Sec14p appears to be similar to the *apo* form of α-TTP, again with the equivalent helices open and exposing the ligand-binding pocket. This open conformation could be relevant either for the binding or for release, or possibly for both actions.

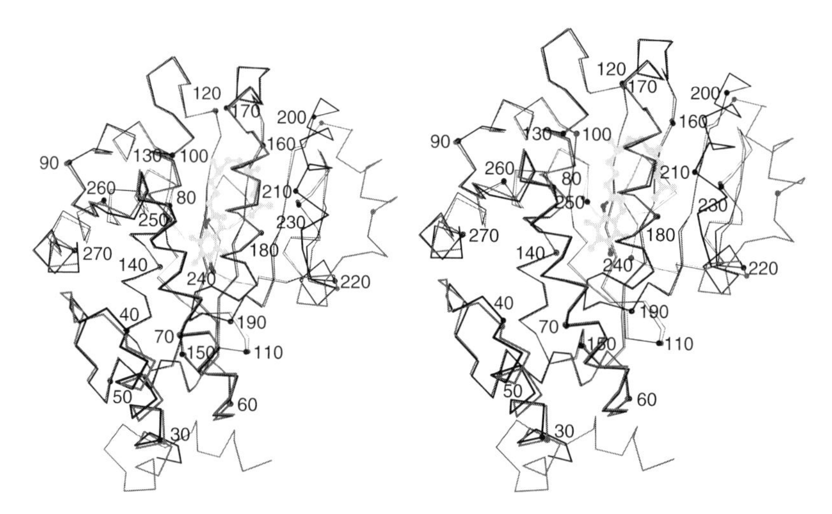

FIGURE 5. Comparison of two conformations of α-TTP. A stereo diagram of the Cα trace of α-TTP (PDB:1R5L) is drawn in black, with α-T shown in ball-and-stick representation in yellow. Small spheres and a corresponding numeral indicate positions of every 10th Cα atom. The *apo* form of α-TTP (PDB:1OIZ:chain A) is superimposed and colored in red. The diagrams were generated using BOBScript (Esnouf, 1999). (See Color Insert.)

Closer examination of the binding pocket in the *apo* form of α-TTP seems to indicate that no significant changes occur relative to the ligand-binding pocket itself other than the movement of helix α10. After superposition of the Cα-positions, the rmsd of all the atoms in the 15 residues that have interactions with α-T is 0.74 or 0.86 Å for models A and B; Ile210 was excluded since it lies on helix α10. This is similar to the variance seen between the two ligand-bound structures, 1R5L and 1OIP; the same calculation resulted in an rmsd of 0.62 Å for all atoms of the same residues. Therefore, the differences seem to lie within that expected for experimental errors of the structural determinations, and the release or entry of α-T into the binding pocket may be determined solely by the rigid body movement of helix α10.

III. MUTATIONS ASSOCIATED WITH AVED

A. STRUCTURAL CONSIDERATIONS

Several mutations in the gene-encoding α-TTP have been described, which result in a late onset spinocerebellar ataxia (Table I, Fig. 7). As the phenotype is recessive, the simplest explanation is that the disease results in a loss of function of α-TTP. Knockout mouse models in which the murine analogue of the α-TTP gene was deleted confirmed the importance of the function of α-TTP for vitamin E homeostasis and replicated the AVED phenotype in

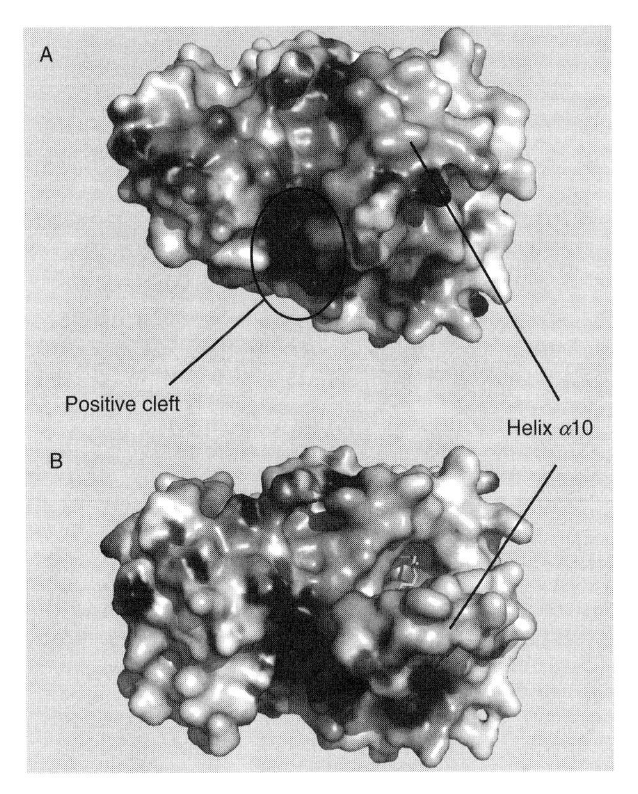

FIGURE 6. A positively charged cleft located near the proposed conformationally sensitive region of α-TTP. (A) Potential surface map of ligand-bound α-TTP (PDB:1R5L) is shown, colored from red (−) to dark blue (+) (>+15 kT). A strongly positively charged cleft is apparent near helix α10 that changes position in the *apo* form of α-TTP. (B) Potential surface map of the *apo* form of α-TTP (PDB:1OIZ:chain A) is shown, calculated in the same fashion as for A. Here the open view of the ligand-binding pocket is apparent from the movement of helixα10. A stick representation of α-T in yellow is shown in the position it occupies in the ligand-bound structure. Molecular surface and surface potential map calculated with GRASP and rendered with PyMOL (DeLano, 2002; Nicholls *et al.*, 1991). (See Color Insert.)

mice (Leonard *et al.*, 2002; Terasawa *et al.*, 2000; Yokota *et al.*, 2001). Although many of the mutations result either in truncations or to prevent the translation of any protein, there are several missense mutations that lead to dysfunction of α-TTP; these mutations may provide a deeper understanding of how this protein functions in α-T trafficking within the cell.

Of the eight missense mutations reported, only one involves a change in the ligand-binding pocket itself, Leu183Pro. This side chain, in combination with those of Val191 and Ile194, forms a hydrophobic pocket for the methyl group on the C5-position of the chroman ring of α-T (Figs. 4 and 7). Leu183 is located in the middle of helix α9, which contains a number of residues

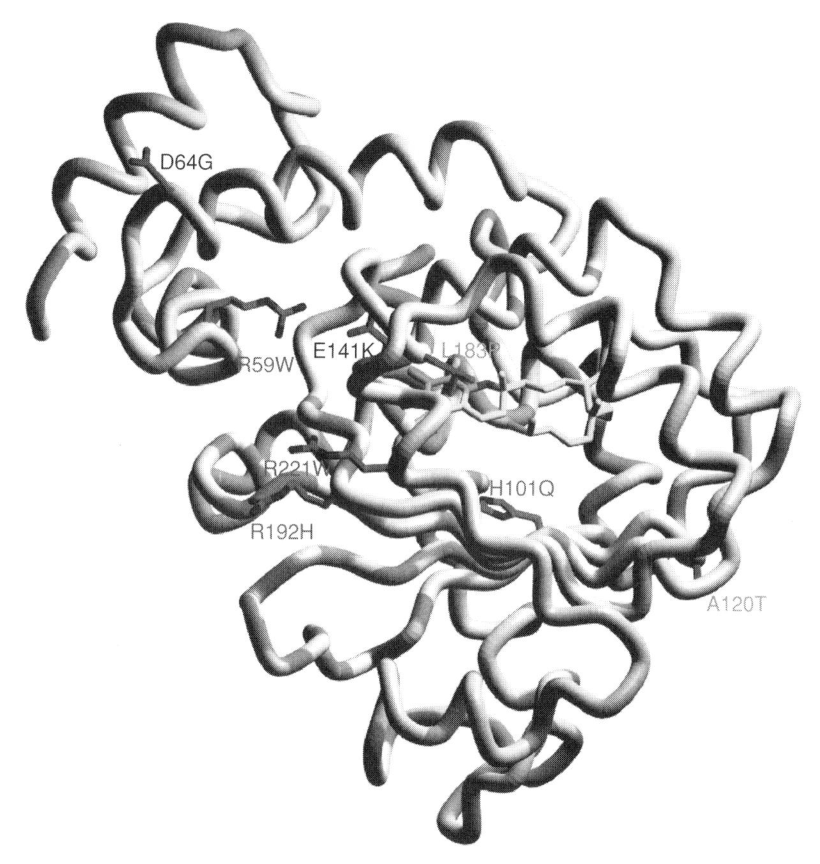

FIGURE 7. Missense mutations associated with AVED. A coil representation of α-TTP (PDB:1R5L) is presented, with α-T indicated in yellow. The side chains of residues that are mutated in AVED are shown as stick models with the carbon atoms colored green and numbered. Table I contains a list of all known mutations associated with AVED. Drawing generated in POVScript and rendered with POVRay. With permission from Min (2003) © 2003 *Proceedings of the National Academy of Sciences USA*. (See Color Insert.)

that form significant contacts with the ligand. As a mutation to proline might interrupt the secondary structure of the helix, this mutation may disrupt several interactions between α-TTP and α-T, resulting in a much lower binding affinity to the ligand.

Two other missense mutations are located toward the interior of α-TTP, Glu141Lys and His101Gln. The side chain of Glu141 forms a hydrogen bond with Tyr73, but also may serve to partially neutralize the relatively buried positive charge from Arg77, which together with Glu141 is strictly conserved in this group of proteins (Fig. 2). It is likely the introduction of yet another positive charge in the form of a lysine residue at this position would serve to destabilize the protein. The clinical phenotype of this mutation has been reported to be severe, with affected individuals having disease onset before the age of 10.

In contrast, the relatively conservative mutation of histidine to glutamine at position 101 results in a relatively mild phenotype. Both nitrogen atoms of the imidazole ring participate in hydrogen bonds. One is a donor to the carbonyl oxygen of Ile116 and the other is an acceptor of the hydroxyl group of Thr139. A glutamine residue at this position would not be capable of forming both hydrogen bonds, and may lead to some destabilization of the protein.

The rest of the disease-associated mutations map to the surface. Of these, three of the mutations involve arginine residues, and all three side chains contribute to a positively charged cleft in α-TTP (Fig. 6). Two of the mutations, at residues 59 and 221, involve a change to tryptophan and are associated with severe phenotypes. The third change, at residue 192, is from arginine to histidine, a relatively conservative mutation, and results in a milder phenotype. Arg59 is strictly conserved among the CRAL-TRIO proteins as shown in Fig. 2 and forms a salt bridge with Asp185. The positive charge at position 221 is conserved within this family, and at position 192, three of four members have a positively charged residue at this position.

In Sec14p, it is possible that the conserved charge at position equivalent to residue 221 in α-TTP plays a role in binding to some ligands. A variant in which both Lys239 (the equivalent of Arg221 in α-TTP) and Lys66 (equivalent to Asp60) are changed to alanine can no longer bind to phosphatidyl inositol (Phillips *et al.*, 1999). These residues were studied because they appeared to stabilize a hydrogen bond network with the head group of one of the two β-OG detergent molecules found in the binding pocket in the crystal structure. When crystallized, the mutant still had two β-OG molecules in the putative binding pocket although with some slight alterations in configuration. In addition, it was found that the Lys66Ala/Lys239A mutant of Sec14p was able to complement the lethal phenotype of the deletion of this gene in yeast, calling into question the importance of phosphatidyl inositol binding for the function of Sec14p.

In another related protein, mutations of CRALBP have been implicated in a number of hereditary retinopathies in humans, Arg150Gln, Met225Lys, and Arg233Trp (Maw *et al.*, 1997). The most interesting CRALBP mutant was Arg233Trp, which corresponds to the Arg192His mutation in α-TTP. This mutation did not disrupt binding of 11-*cis*-retinaldehyde but actually increased the affinity of CRALBP for ligand approximately twofold (Golovleva *et al.*, 2003). In order to test whether this mutant interfered with the transfer activity of this protein, the ability of the protein to present 11-*cis*-retinol to retinal dehydrogenase (RDH) was determined. In that assay, the apparent K_m of RDH was approximately sevenfold higher for the mutant than wild type, leading to the conclusion that part of the phenotype of

Arg233Trp was a defect in presenting the ligand to RDH. CRALBP would be expected to have a similar positively charged cleft as was seen with α-TTP, with arginine residues at 102 and 262, which are equivalent to Arg59 and Arg221 in α-TTP, respectively. Furthermore, other positively charged residues found in the positively charged cleft observed in α-TTP are conserved for CRALBP. It is tempting to envision this positively charged surface of α-TTP and CRALBP as a region that contributes to protein–protein interactions in a way that coordinates release and entry of the ligand in a regulated fashion.

B. BIOCHEMICAL CHARACTERIZATION OF AVED-ASSOCIATED MUTANTS

Recent work has begun to address the specific biochemical defects of disease-associated mutations of α-TTP. In general, three types of assays have been used to characterize the function of α-TTP *in vitro*. First is an assay for the affinity of α-TTP for α-T and other related ligands. Second is a transfer assay in which the ability of α-TTP to catalyze the transfer of α-T between artificial liposomes and mitochondrial membranes as a function of time is measured. Third is an assay in which cells, which have been transfected with the α-TTP gene, are examined for the ability to secrete α-T into the media.

All of the disease-associated missense mutations have been studied with regard to affinity for α-T except for the Leu183Pro and Asp64Gly mutations (Morley *et al.*, 2004). Whereas wild-type α-TTP has been found to have high affinity for α-T ($K_d \sim 30$ nM), the measured K_d of mutants varied from 40 to 123 nM, with most mutations resulting in a measured K_d of ~70 nM (Table I). Interestingly, mutations associated with an early onset of disease appeared to have a rough correlation with the lowest measured affinity for α-T, with the Arg59Trp having the lowest measured affinity for α-T.

In another assay, the ability of recombinant α-TTP to catalyze the exchange of α-T between liposomes and mitochondrial membranes was measured for the same mutants (Morley *et al.*, 2004). Again there was a degree of correlation between the age at onset and the decrease in rate of transfer of the various mutations. Early onset mutations such as Arg59Trp, Glu141Lys, and Arg221Trp had rates that were two to three times slower than wild type, whereas the other mutations were quite similar to wild type in this transfer assay (Table I). Another group previously reported that the His101Gln mutation was completely inactive in a similar assay, and it is not clear what caused this discrepancy (Gotoda *et al.*, 1995). Although these two assays did seem to show some relevance to the observed phenotypes, but they did not seem to explain the nearly complete lack of function often observed in individuals carrying the equivalent mutations with a complete lack of α-T in the serum.

Perhaps the best *in vitro* assay thus far is one in which the ability of α-TTP to confer the ability to secrete α-T in cell culture is measured (Arita *et al.*, 1997). Three mutants of α-TTP—Ala120Thr, Arg221Trp, and Arg59Trp—have been examined in such a system (Qian *et al.*, 2006). All three were found to be defective in secretion of α-T, with Ala120Thr having 67% of wild-type activity, Arg221Trp 40%, and Arg59Trp 17%. In affected individuals, the Ala120Thr mutation is associated with a later onset form of the disease, whereas both Arg221Trp and Arg59Trp are associated with an early onset phenotype. Studies of the remaining mutants will be needed to see whether this assay will correlate the genotype to the phenotype of the disease.

In addition to these biochemical assays, several groups have reported on the cellular distribution of α-TTP using immunofluorescence. In the case of Horiguchi *et al.* (2003), α-TTP was found diffusely in the cytoplasm of hepatocytes unless the cells were treated with chloroquine, after which α-TTP was observed in a punctate pattern. In the chloroquine-treated cells, α-TTP was found to colocalize with markers from late endosome membranes (Horiguchi *et al.*, 2003). This activity was found to be dependent on a short segment in the N-terminal domain, in residues 21–51. In contrast, others have reported that α-TTP is constitutively associated with late endosome membranes: one possible explanation for the difference is that the construct in the latter results contains a hemagglutinin tag on the 5′ end of α-TTP (Qian *et al.*, 2005). The three mutants studied for α-T secretion—Ala120T, Arg221Trp, and Arg59Trp—were also found to be associated with late endosome membranes, indistinguishable from wild type. Although the functional significance of the association of α-TTP with the late endosome is unclear, it is thought that it may be the site where it picks up α-T that has entered the cell by endocytosis in association with chylomicrons.

IV. SUMMARY

The combination of structural, genetic, and biochemical studies has elucidated the central role that α-TTP has in α-T homeostasis, but the precise molecular mechanisms by which it routes α-T away from excretion to secretion remain elusive. The two conformations of α-TTP known through X-ray crystallography show a closed ligand-bound state as well as a putative open state ready for entry of ligand. In the closed state, α-TTP likely acts as a chaperone for α-T, ferrying the lipophilic molecule from the endocytic pathway through the cytoplasm for eventual secretion. It seems likely that the first step is the uptake of α-T from late endosome membranes. Presumably, α-T-bound α-TTP would then dissociate from the late endosome and release its cargo in an appropriate location for secretion. It has been shown that the secretion of α-T by hepatocytes depends on the transporter ABCA1, and it has been proposed that the transporter itself may directly interact with α-TTP to release α-T from the binding pocket (Oram *et al.*, 2001; Qian *et al.*, 2006).

Thus far, several mutations responsible for AVED have been studied in detail *in vitro*, and minor alterations in affinity for α-T as well as in transfer rates of α-T between liposomes and membranes have been described. However, these perturbations do not appear to adequately explain the observed deficiencies in affected individuals or cells in terms of α-TTP function. In the future, more work will need to be done to understand how binding and release of α-T from α-TTP is regulated. Questions that will need to be addressed include: What proteins interact with α-TTP as it carries out its functions? Can it be shown that α-TTP directly interacts with the ABCA1 transporter? How does α-TTP bind to late endosome membranes—does it bind to a protein in the membrane or does it recognize a specific lipid component particular to these vesicles? It is interesting to note that others have found that α-TTP appeared to bind specifically to phosphoinositide species in a proteomic screen (Krugmann *et al.*, 2002). Once the structural basis of these questions is understood, we anticipate gaining a precise understanding of the molecular basis of AVED. Moreover, we would expect that the lessons learned with α-TTP might apply broadly to the entire family of CRAL-TRIO proteins.

REFERENCES

Arita, M., Sato, Y., Miyata, A., Tanabe, T., Takahashi, E., Kayden, H. J., Arai, H., and Inoue, K. (1995). Human α-tocopherol transfer protein: cDNA cloning, expression and chromosomal localization. *Biochem. J.* **306**(Pt. 2), 437–443.

Arita, M., Nomura, K., Arai, H., and Inoue, K. (1997). α-Tocopherol transfer protein stimulates the secretion of α-tocopherol from a cultured liver cell line through a brefeldin A-insensitive pathway. *Proc. Natl. Acad. Sci. USA* **94**(23), 12437–12441.

Bateman, A., Birney, E., Cerruti, L., Durbin, R., Etwiller, L., Eddy, S. R., Griffiths-Jones, S., Howe, K. L., Marshall, M., and Sonnhammer, E. L. (2002). The Pfam protein families database. *Nucleic Acids Res.* **30**(1), 276–280.

Ben Hamida, C., Doerfligner, N., Belal, S., Linder, L., Reutenauer, C., Dib, C., Gyapay, A., Vignal, D., Le Paslier, D., Cohen, M., Pandolfo, V., Mokini, G., *et al.* (1993a). Localization of Friedreich ataxia phenotype with selective vitamin E deficiency to chromosome 8q by homozygosity mapping. *Nat. Genet.* **5**(2), 195–200.

Ben Hamida, M., Belal, S., Sirugo, G., Ben Hamida, C., Panayides, K., Ionannou, P., Beckmann, J., Mandel, J. L., Hentati, F., Koenig, M., *et al.* (1993b). Friedreich's ataxia phenotype not linked to chromosome 9 and associated with selective autosomal recessive vitamin E deficiency in two inbred Tunisian families. *Neurology* **43**(11), 2179–2183.

Bomar, J. M., Benke, P. J., Slattery, E. L., Puttagunta, R., Taylor, L. P., Seong, E., Nystuen, A., Chen, W., Albin, R. L., Patel, P. D., Kittles, R. A., Sheffield, V. C., *et al.* (2003). Mutations in a novel gene encoding a CRAL-TRIO domain cause human Cayman ataxia and ataxia/dystonia in the jittery mouse. *Nat. Genet.* **35**(3), 264–269.

Brigelius-Flohe, R., and Traber, M. G. (1999). Vitamin E: Function and metabolism. *FASEB J.* **13**(10), 1145–1155.

Burck, U., Goebel, H. H., Kuhlendahl, H. D., Meier, C., and Goebel, K. M. (1981). Neuromyopathy and vitamin E deficiency in man. *Neuropediatrics* **12**(3), 267–278.

Catignani, G. L. (1975). An α-tocopherol binding protein in rat liver cytoplasm. *Biochem. Biophys. Res. Commun.* **67**(1), 66–72.

Catignani, G. L., and Bieri, J. G. (1977). Rat liver α-tocopherol binding protein. *Biochim. Biophys. Acta* **497**(2), 349–357.

Cavalier, L., Ouahchi, K., Kayden, H. J., Di Donato, S., Reutenauer, L., Mandel, J. L., and Koenig, M. (1998). Ataxia with isolated vitamin E deficiency: Heterogeneity of mutations and phenotypic variability in a large number of families. *Am. J. Hum. Genet.* **62**(2), 301–310.

DeLano (2002). The PyMOL Molecular Graphics System DeLano Scientific, San Carlos.

Esnouf, R. M. (1999). Further additions to MolScript version 1.4, including reading and contouring of electron-density maps. *Acta Crystallogr. D Biol. Crystallogr.* **55**(Pt. 4), 938–940.

Evans, H. E., and Bishop, K. S. (1922). On the existence of a hitherto unrecognized dietary factor essential for reproduction. *Science* **56**(1458), 650–651.

Fenn, T. D., Ringe, D., and Petsko, G. A. (2003). POVScript+: A program for model and data visualization using persistence of vision ray-tracing. *J. Appl. Crystallogr.* **36,** 944–947.

Golovleva, I., Bhattacharya, S., Wu, Z., Shaw, N., Yang, Y., Andrabi, K., West, K. A., Burstedt, M. S., Forsman, K., Holmgren, G., Sandgren, O., Noy, N., *et al.* (2003). Disease-causing mutations in the cellular retinaldehyde binding protein tighten and abolish ligand interactions. *J. Biol. Chem.* **278**(14), 12397–12402.

Gotoda, T., Arita, M., Arai, H., Inoue, K., Yokota, T., Fukuo, Y., Yazaki, Y., and Yamada, N. (1995). Adult-onset spinocerebellar dysfunction caused by a mutation in the gene for the α-tocopherol-transfer protein. *N. Engl. J. Med.* **333**(20), 1313–1318.

Gouet, P., Courcelle, E., Stuart, D. I., and Metoz, F. (1999). ESPript: Analysis of multiple sequence alignments in PostScript. *Bioinformatics* **15**(4), 305–308.

Harding, A. E., Matthews, S., Jones, S., Ellis, C. J., Booth, I. W., and Muller, D. P. (1985). Spinocerebellar degeneration associated with a selective defect of vitamin E absorption. *N. Engl. J. Med.* **313**(1), 32–35.

Hentati, A., Deng, H. X., Hung, W. Y., Nayer, M., Ahmed, M. S., He, X., Tim, R., Stumpf, D. A., and Siddique, T. (1996). Human α-tocopherol transfer protein: Gene structure and mutations in familial vitamin E deficiency. *Ann. Neurol.* **39**(3), 295–300.

Horiguchi, M., Arita, M., Kaempf-Rotzoll, D. E., Tsujimoto, M., Inoue, K., and Arai, H. (2003). pH-dependent translocation of α-tocopherol transfer protein (α-TTP) between hepatic cytosol and late endosomes. *Genes Cells* **8**(10), 789–800.

Hoshino, M., Masuda, N., Ito, Y., Murata, M., Goto, J., Sakurai, M., and Kanazawa, I. (1999). Ataxia with isolated vitamin E deficiency: A Japanese family carrying a novel mutation in the α-tocopherol transfer protein gene. *Ann. Neurol.* **45**(6), 809–812.

Hosomi, A., Goto, K., Kondo, H., Iwatsubo, T., Yokota, T., Ogawa, M., Arita, M., Aoki, J., Arai, H., and Inoue, K. (1998). Localization of α-tocopherol transfer protein in rat brain. *Neurosci. Lett.* **256**(3), 159–162.

Krendel, D. A., Gilchrist, J. M., Johnson, A. O., and Bossen, E. H.. (1987). Isolated deficiency of vitamin E with progressive neurologic deterioration. *Neurology* **37**(3), 538–540.

Krugmann, S., Anderson, K. E., Ridley, S. H., Risso, N., McGregor, A., Coadwell, J., Davidson, K., Eguinoa, A., Ellson, C. D., Lipp, P., Manifava, M., Ktistakis, N., *et al.* (2002). Identification of ARAP3, a novel PI3K effector regulating both Arf and Rho GTPases, by selective capture on phosphoinositide affinity matrices. *Mol. Cell* **9**(1), 95–108.

Laplante, P., Vanasse, M., Michaud, J., Geoffroy, G., and Brochu, P. (1984). A progressive neurological syndrome associated with an isolated vitamin E deficiency. *Can. J. Neurol. Sci.* **11**(4 Suppl.), 561–564.

Leonard, S. W., Terasawa, Y., Farese, R. V., Jr., and Traber, M. G. (2002). Incorporation of deuterated RRR- or all-rac-α-tocopherol in plasma and tissues of α-tocopherol transfer protein–null mice. *Am. J. Clin. Nutr.* **75**(3), 555–560.

Maw, M. A., Kennedy, B., Knight, A., Bridges, R., Roth, K. E., Mani, E. J., Mukkadan, J. K., Nancarrow, D., Crabb, J. W., and Denton, M. J. (1997). Mutation of the gene encoding cellular retinaldehyde-binding protein in autosomal recessive retinitis pigmentosa. *Nat. Genet.* **17**(2), 198–200.

Meier, R., Tomizaki, T., Schulze-Briese, C., Baumann, U., and Stocker, A. (2003). The molecular basis of vitamin E retention: Structure of human α-tocopherol transfer protein. *J. Mol. Biol.* **331**(3), 725–734.

Min, K. C., Kovall, R. A., and Hendrickson, W. A. (2003). Crystal structure of human α-tocopherol transfer protein bound to its ligand: Implications for ataxia with vitamin E deficiency. *Proc. Natl. Acad. Sci. USA* **100**(25), 14713–14718.

Morley, S., Panagabko, C., Shineman, D., Mani, B., Stocker, A., Atkinson, J., and Manor, D. (2004). Molecular determinants of heritable vitamin E deficiency. *Biochemistry* **43**(14), 4143–4149.

Nicholls, A., Sharp, K. A., and Honig, B. (1991). Protein folding and association: Insights from the interfacial and thermodynamic properties of hydrocarbons. *Proteins* **11**(4), 281–296.

Oram, J. F., Vaughan, A. M., and Stocker, R. (2001). ATP-binding cassette transporter A1 mediates cellular secretion of α-tocopherol. *J. Biol. Chem.* **276**(43), 39898–39902.

Ouahchi, K., Arita, M., Kayden, H., Hentati, F., Ben Hamida, M., Sokol, R., Arai, H., Inoue, K., Mandel, J. L., and Koenig, M. (1995). Ataxia with isolated vitamin E deficiency is caused by mutations in the α-tocopherol transfer protein. *Nat. Genet.* **9**(2), 141–145.

Phillips, S. E., Sha, B., Topalof, L., Xie, Z., Alb, J. G., Klenchin, V. A., Swigart, P., Cockcroft, S., Martin, T. F., Luo, M., and Bankaitis, V. A. (1999). Yeast Sec14p deficient in phosphatidylinositol transfer activity is functional *in vivo*. *Mol. Cell* **4**(2), 187–197.

Qian, J., Morley, S., Wilson, K., Nava, P., Atkinson, J., and Manor, D. (2005). Intracellular trafficking of vitamin E in hepatocytes: The role of tocopherol transfer protein. *J. Lipid Res.* **46**(10), 2072–2082.

Qian, J., Atkinson, J., and Manor, D. (2006). Biochemical consequences of heritable mutations in the α-tocopherol transfer protein. *Biochemistry* **45**(27), 8236–8242.

Saari, J. C., and Crabb, J. W. (2005). Focus on molecules: Cellular retinaldehyde-binding protein (CRALBP). *Exp. Eye Res.* **81**(3), 245–246.

Sato, Y., Arai, H., Miyata, A., Tokita, S., Yamamoto, K., Tanabe, T., and Inoue, K. (1993). Primary structure of α-tocopherol transfer protein from rat liver. Homology with cellular retinaldehyde-binding protein. *J. Biol. Chem.* **268**(24), 17705–17710.

Sha, B., Phillips, S. E., Bankaitis, V. A., and Luo, M. (1998). Crystal structure of the Saccharomyces cerevisiae phosphatidylinositol-transfer protein. *Nature* **391**(6666), 506–510.

Shimohata, T., Date, H., Ishiguro, H., Suzuki, T., Takano, H., Tanaka, H., Tsuji, S., and Hirota, K. (1998). Ataxia with isolated vitamin E deficiency and retinitis pigmentosa. *Ann. Neurol.* **43**(2), 273.

Sokol, R. J., Kayden, H. J., Bettis, D. B., Traber, M. G., Neville, H., Ringel, S., Wilson, W. B., and Stumpf, D. A. (1988). Isolated vitamin E deficiency in the absence of fat malabsorption–familial and sporadic cases: Characterization and investigation of causes. *J. Lab. Clin. Med.* **111**(5), 548–559.

Stocker, A., and Baumann, U. (2003). Supernatant protein factor in complex with RRR-α-tocopherylquinone: A link between oxidized vitamin E and cholesterol biosynthesis. *J. Mol. Biol.* **332**(4), 759–765.

Terasawa, Y., Ladha, Z., Leonard, S. W., Morrow, J. D., Newland, D., Sanan, D., Packer, L., Traber, M. G., and Farese, R. V., Jr. (2000). Increased atherosclerosis in hyperlipidemic mice deficient in α-tocopherol transfer protein and vitamin E. *Proc. Natl. Acad. Sci. USA* **97** (25), 13830–13834.

Traber, M. G., Sokol, R. J., Burton, G. W., Ingold, K. U., Papas, A. M., Huffaker, J. E., and Kayden, H. J. (1990). Impaired ability of patients with familial isolated vitamin E deficiency to incorporate α-tocopherol into lipoproteins secreted by the liver. *J. Clin. Invest.* **85**(2), 397–407.

Usuki, F., and Maruyama, K. (2000). Ataxia caused by mutations in the α-tocopherol transfer protein gene. *J. Neurol. Neurosurg. Psychiatry* **69**(2), 254–256.

Yokota, T., Wada, Y., Furukawa, T., Tsukagoshi, H., Uchihara, T., and Watabiki, S. (1987). Adult-onset spinocerebellar syndrome with idiopathic vitamin E deficiency. *Ann. Neurol.* **22**(1), 84–87.

Yokota, T., Shiojiri, T., Gotoda, T., Arita, M., Arai, H., Ohga, T., Kanda, T., Suzuki, J., Imai, T., Matsumoto, H., Harino, S., Kiyosawa, M., *et al.* (1997). Friedreich-like ataxia with retinitis pigmentosa caused by the His101Gln mutation of the α-tocopherol transfer protein gene. *Ann. Neurol.* **41**(6), 826–832.

Yokota, T., Igarashi, K., Uchihara, T., Jishage, K., Tomita, H., Inaba, A., Li, Y., Arita, M., Suzuki, H., Mizusawa, H., and Arai, H. (2001). Delayed-onset ataxia in mice lacking α-tocopherol transfer protein: Model for neuronal degeneration caused by chronic oxidative stress. *Proc. Natl. Acad. Sci. USA* **98**(26), 15185–15190.

Zingg, J. M., and Azzi, A. (2004). Non-antioxidant activities of vitamin E. *Curr. Med. Chem.* **11**(9), 1113–1133.

3

THE α-TOCOPHEROL TRANSFER PROTEIN

D. MANOR*,† AND S. MORLEY*

**Division of Nutritional Sciences, Cornell University, Ithaca, New York 14853*
†Department of Nutrition, Case School of Medicine, Case Western Reserve University, Cleveland, Ohio 44106

Almost a century ago, plant extracts were documented to be critical for the fertility of rodents. This activity was later ascribed to vitamin E, a term comprising a number of structurally related plant lipids that function as fat soluble antioxidants. The α-tocopherol transfer protein

0083-6729/07 $35.00
DOI: 10.1016/S0083-6729(07)76003-X

(TTP) is a critical regulator of vitamin E status that stimulates the movement of vitamin E between membrane vesicles *in vitro* and facilitates the secretion of tocopherol from hepatocytes. Heritable mutations in the *ttpA* gene cause ataxia with vitamin E deficiency (AVED), an autosomal recessive disorder characterized by low plasma vitamin E levels and progressive neurodegeneration. This chapter summarizes recent advances in our understanding of the molecular and physiological aspects of TTP activity.

I. INTRODUCTION

In 1922, Evans and Bishop discovered that some plant oils reduced the occurrence of fetus resorption in diet-restricted rats (Evans and Bishop, 1922). This "fertility factor" was later isolated and characterized as tocopherol (*Gr*. to bear offspring, denoted herein as TOH), a member of a family of naturally occurring plant lipids composed of a phytyl side chain and a chromanol ring that have come to be commonly called vitamin E:

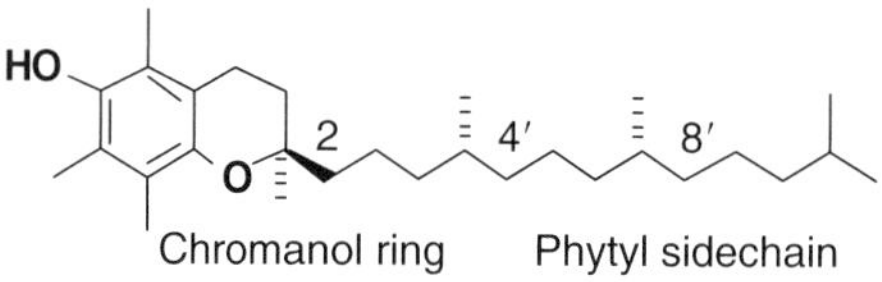

HO O 2 4′ 8′

Chromanol ring Phytyl sidechain

(2*R*,4′*R*,8′*R*)-α-tocopherol

The α-, β-, γ-, and δ-vitamers differ in the number and position of methyl substitutions on the chromanol ring, whereas tocopherols differ from tocotrienols in the saturation of the phytyl chain. Naturally occurring tocols possess the *R* configuration at each of three chiral centers in the phytyl side chain (i.e., *RRR*), whereas synthetic tocopherols are racemic mixtures of both *R* and *S* configurations. The ability of vitamin E to quench free radicals, and its pronounced hydrophobicity, have led to its common definition as the single most important lipid-soluble antioxidant (cf., Burton, 1994). Indeed, a number of epidemiological studies have shown an inverse correlation between vitamin E intake and risk for pathologies thought to involve oxidative damage such as cardiovascular disease and inflammation. These observations led to the current widespread supplementation with this vitamin (Stampfer *et al.*, 1993; Stephens *et al.*, 1996).

The extreme hydrophobicity of tocopherol poses a major thermodynamic barrier to its distribution and transport through the aqueous milieu of the cytosol and circulation. As in the case of other small lipids (e.g., cholesterol, retinoids, fatty acids), this barrier appears to have been overcome by the

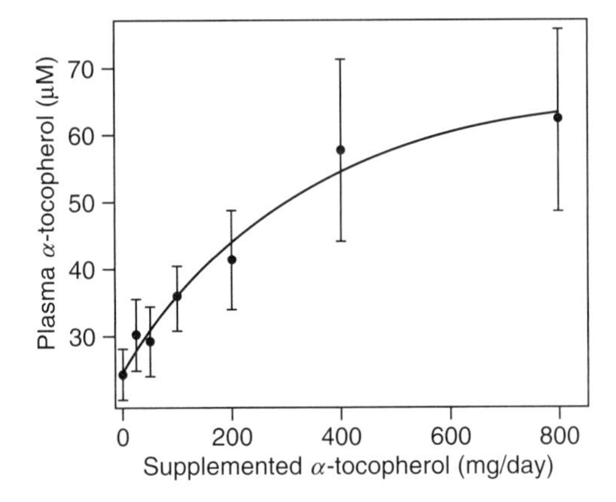

FIGURE 1. Response of plasma tocopherol levels to vitamin E supplementation in women. Data courtesy of Hans Princen (Gaubius Laboratory, Leiden, The Netherlands). See Princen *et al.* (1995) for details.

evolution of soluble-binding proteins that mediate the transport of hydrophobic compounds and regulate their availability. The existence of specific protein(s) that regulate vitamin E levels is also suggested by reports that tocopherol levels in the body are *not* in equilibrium with dietary tocopherols. Thus, *RRR*-α-TOH is selectively retained over other forms of vitamin E despite the dietary prevalence of the latter. Furthermore, plasma levels of vitamin E do not respond linearly to changes in intake. For example, an increase in vitamin E supplementation from 400 to 800 mg/day is accompanied by only a minor (<10%) increase in plasma tocopherol (Fig. 1), suggesting that factors other than intake regulate and limit vitamin E levels, and that these factors become saturated at certain intake levels. Moreover, inter- and intra-individual plasma tocopherol levels in humans are remarkably constant, suggesting that genetically encoded factor(s) determine vitamin E concentrations *in vivo* (Kelly *et al.*, 2004; Roxborough *et al.*, 2000).

II. IDENTIFICATION OF TTP

The identification of hepatic α-tocopherol transfer protein (TTP) as a critical mediator of vitamin E function came about from two independent lines of experimental research: the biochemical isolation of TOH binding and transfer activities, and, independently, from the study of the molecular basis for heritable primary vitamin E deficiency in humans.

Catignani (1975) first described a 31-kDa [^{3}H]-TOH-binding activity in rat liver cytosol that exhibited preferential binding of α-TOH over γ-TOH (Catignani and Bieri, 1977; Kuhlenkamp *et al.*, 1993). Due to the hydrophobic nature of vitamin E and its "site of action" in lipid bilayers, considerable efforts were devoted to the identification of soluble factors that facilitate vitamin E transfer between membranes. Murphy and Mavis (1981) and Mowri *et al.* (1981) characterized a protein fraction that stimulated the transfer of radiolabeled TOH from liposomes to microsomes. Sato *et al.* (1991) purified TTP to homogeneity from rat liver cytosol, and later cloned the *ttpA* transcript from rat liver (Sato *et al.*, 1993). The human transcript encoding TTP was then cloned and mapped to the q13 region of human chromosome 8 by Arita *et al.* (1995). The 278-residue TTP protein shows high sequence homology to a family of small lipid-binding proteins, all sharing a region that has come to be known as the CRAL_TRIO domain (PFAM 00650). In summary, biochemical studies established TTP as a liver-enriched, ~32-kDa protein that binds TOH with high affinity and catalyzes its transfer between lipid membranes.

III. TTP AND VITAMIN E STATUS: ATAXIA WITH VITAMIN E DEFICIENCY

In the early 1980s, a number of independent clinical reports established the occurrence of primary familial vitamin E deficiency in humans, presently termed ataxia with vitamin E deficiency (AVED) (OMIM 277640; Sokol, 1988; Ouahchi *et al.*, 1995 for references). AVED patients display autosomal recessive inheritance of progressive neurodegenerative symptoms (e.g., ataxia, dysarthria, loss of deep tendon reflexes), coupled with low plasma vitamin E levels ($\leq$3 μM). Plasma and tissue levels of other lipids (e.g., cholesterol, triglycerides) are typically unaffected. In a unique and insightful collaboration between biochemists, geneticists, and clinicians, it was established that the molecular defects at the root of the AVED pathology are mutations in the gene-encoding TTP (Ouahchi *et al.*, 1995, and references therein). The functional linkage between tocopherol, TTP, and specific neuropathologies has markedly advanced our understanding of the molecular mechanisms by which vitamin E status is regulated.

Known mutations in the *ttpA* gene that are associated with the AVED syndrome are shown in Table I. Several mutations arise from nucleotide insertions or deletions that cause incorrect initiation or termination of TTP, presumably leading to a misfolded protein. In other cases, AVED mutations cause single amino acid changes that vary from conserved (e.g., R192H) to drastically different (e.g., E141K) substitutions. The type of mutation harbored by the affected patient appears to correlate with the severity of symptoms and age of onset of the syndrome. Truncations, frame-shift mutations and nonconserved substitutions (e.g., 744delA, 513insTT, 485delT, 421G-> A) result in a severe, early onset form of the AVED pathology, and a drastic reduction in plasma TOH levels (Amiel *et al.*, 1995; Hentati *et al.*, 1996; Krendel *et al.*, 1987; Mariotti *et al.*, 2004; Roubertie *et al.*, 2003;

TABLE I. Mutations in the ***ttpA*** Gene Associated with AVED

Nucleotide change, references	Effect on coding sequence	Clinical phenotype	Stereoselectivity	Corresponding residue in CRALBP, MEG2, Sec14p, and SPF
−1C → T; Usuki and Maruyama, 2000	Decrease in TTP levels	Severe	ND	–
2T → C; Hoshino *et al.*, 1999	Disruption of initiation	ND[a]	ND	M,M,M,M
175C → T; Cavalier *et al.*, 1998	R59W	Severe	ND	R,R,R,R
191A → G; Usuki and Maruyama, 2000	D64G	Severe	ND	G,L,Q,Q
205G → C; Cavalier *et al.*, 1998	Mis-splicing?	ND	ND	–
219insAT; Mariotti *et al.*, 2004	Premature termination	Severe	ND	–
303T → G; Gotoda *et al.*, 1995; Pang *et al.*, 2001; Yokota *et al.*, 1996, 1997, 2000	H101Q	Mild	Discriminator	P,F,P,S
306A → G; Cavalier *et al.*, 1998	Mis-splicing?	Mild	ND	G,T,Q,G
358G → A; Cavalier *et al.*, 1998	A120T	Mild	ND	E,R,G,G
400C → T; Cavalier *et al.*, 1998; Cellini *et al.*, 2002	Premature termination	Severe	Nondiscriminator	Q,Q,K,R
421G → T; Schuelke *et al.*, 2000	Premature termination	ND	ND	E,D,E,E
421G → A; Cavalier *et al.*, 1998	E141K	Severe	ND	E,D,E,E
485delT; Hentati *et al.*, 1996	Frameshift	Severe	ND	–
486delT; Cavalier *et al.*, 1998; Roubertie *et al.*, 2003	Frameshift	Severe	ND	–

(*Continues*)

TABLE I. *(Continued)*

Nucleotide change; references	Effect on coding sequence	Clinical phenotype	Stereoselectivity	Corresponding residue in CRALBP, MEG2, Sec14p, and SPF
513insTT; Cavalier *et al.*, 1998; Cellini *et al.*, 2002; Schuelke *et al.*, 2000; Hentati *et al.*, 1996; Mariotti *et al.*, 2004; Angelini *et al.*, 2002; Martinello *et al.*, 1998	Frameshift	Severe	Nondiscriminator	–
530AG → GTAAGT; Schuelke *et al.*, 2000; Ouahchi *et al.*, 1995; Traber *et al.*, 1993; Burck *et al.*, 1981	Frameshift	Severe	Nondiscriminator	–
552G → A; Schuelke *et al.*, 1999, 2000; Tamaru *et al.*, 1997	Exon 3 skipping	Severe	ND	Q,K,Q,E
566T → C; Shimohata *et al.*, 1998	L183P	Severe	ND	L,L,S,F
574G → A; Hentati *et al.*, 1996	R192H	Mild	Discriminator	K,K,G,K
661C → T; Cavalier *et al.*, 1998	R221W	Severe	ND	R,R,K,K
736G → C; Mariotti *et al.*, 2004	G246D	Mild	ND	G,G,G,G
744delA; Angelini *et al.*, 2002; Aparicio *et al.*, 2001; Benomar *et al.*, 2002; Cavalier *et al.*, 1998; Mariotti *et al.*, 2004	Frameshift	Severe	Nondiscriminator	–

[a] ND—not determined.

Schuelke *et al.*, 1999; Stumpf *et al.*, 1987). These mutations likely cause large perturbations to the TTP structure, possibly leading to weakened affinity for vitamin E. AVED mutations that result in conserved substitutions, on the other hand (e.g., H101Q, R192H), result in milder clinical symptoms and later onset of the disease (Cavalier *et al.*, 1998). The correlation between clinical phenotype and the functional integrity of TTP is also manifested in the ability of AVED patients to discriminate between tocopherol stereoisomers. Thus, in all tested cases, an inability to discriminate between the *RRR* and *SRR* configurations of α-TOH is presented by patients harboring severe mutations in the *ttpA* gene (Traber *et al.*, 1993).

IV. TTP AND VITAMIN E STATUS: $TTP^{-/-}$ MICE

The causal relationship between vitamin E status, normal health, and TTP integrity was unequivocally demonstrated in mouse models in which the expression of TTP was disrupted (Jishage *et al.*, 2001, 2005; Schock *et al.*, 2004; Terasawa *et al.*, 2000; Yokota *et al.*, 2001). While the homozygous $TTP^{-/-}$ mice are normal in many respects, they manifest a number of pathologies that underscore the critical role of TTP in vitamin E function. Plasma tocopherol levels are very low in $TTP^{-/-}$ mice, and intermediate in the heterozygous ($TTP^{+/-}$) mice (Jishage *et al.*, 2001), indicating that TTP levels directly influence overall vitamin E status. Female $TTP^{-/-}$ mice are infertile, a defect attributed to improper placental development (in accordance with the original identification of vitamin E as a "fertility factor"). As $TTP^{-/-}$ mice age, they display the neurological symptoms associated with AVED (Yokota *et al.*, 2001). Removal of vitamin E from the diet exacerbates the neurological symptoms, while tocopherol supplementation alleviates them. $TTP^{-/-}$ mice also display enhanced lipid peroxidation in the central nervous system, supporting the notion that ataxia stems from the increased oxidative stress that accompanies TOH deficiency (Yokota *et al.*, 2001). Disruption of TTP expression in $ApoE^{-/-}$ mice leads to enhanced aortal lipid peroxidation and increased occurrence of atherogenic plaques (Terasawa *et al.*, 2000), suggesting a causal relationship between vitamin E, oxidative stress, and atherosclerosis (Keaney *et al.*, 1999; Lonn *et al.*, 2005; Pratico *et al.*, 1998; Yusuf *et al.*, 2000). Other studies demonstrated elevated levels of the cytokines TNF-α and MIP-2 in $TTP^{-/-}$ mice, pointing at the important role that vitamin E plays in the regulation of inflammatory responses (Schock *et al.*, 2004).

V. BIOCHEMICAL ACTIVITIES OF TTP

Two *in vitro* biochemical hallmarks were assigned as "signature" activities of TTP.

A. SPECIFIC, HIGH-AFFINITY BINDING OF *RRR*-α-TOH

Incubation of liver extracts or purified TTP with radiolabeled *RRR*-α-TOH, followed by chromatographic separation of protein-bound from free ligand, established the tight interaction between TTP and TOH (Catignani and Bieri, 1977; Kuhlenkamp *et al.*, 1993; Sato *et al.*, 1991; Yoshida *et al.*, 1992). The substrate specificity of TTP was studied in detail, revealing that the affinity of different tocols for TTP mirrors their potency in biological assays (i.e., $\alpha > \beta > \gamma > \delta$; Hosomi *et al.*, 1997; Panagabko *et al.*, 2003). Importantly, binding of α-TOH to TTP is sensitive to the stereochemistry of the phytyl chain; the *RRR* stereoisomer exhibits some 25-fold higher affinity for TTP than its *SRR* counterpart (Panagabko *et al.*, 2003). While the physiological significance of this selectivity is not known, this

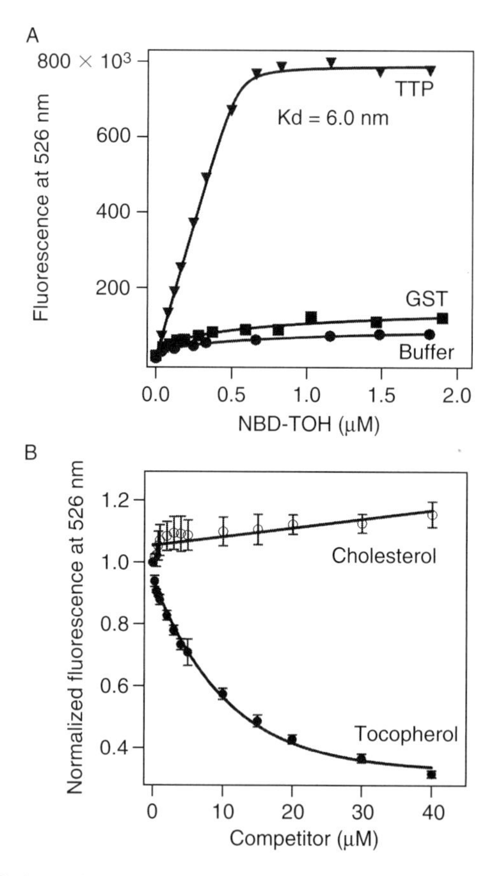

FIGURE 2. Affinity and specificity of the fluorescent vitamin E analogue NBD-TOH for TTP. (A) NBD-TOH was titrated into buffer lacking or containing 1 μM of the indicated protein. (B) Preloaded TTP•NBD-TOH was titrated with the indicated competitor, and fluorescence of NBD-TOH was monitored. Excitation = 466 nm. See Morley *et al.* (2006) for details.

stereochemical preference serves as a convenient and sensitive measure for the integrity of TTP in live organisms (Cheng *et al.*, 1987; Ingold *et al.*, 1987; Parker and McCormick, 2005; Traber *et al.*, 1992, 1993). More accurate measurements of ligand binding became amenable with the advent of novel fluorescent analogues of vitamin E that allow sensitive, real-time monitoring of ligand binding in solution, and verify the high affinity (Kd ~ 10 nM; Fig. 2A) and specificity (Fig. 2B) of the interaction between TTP and TOH (Morley *et al.*, 2006; Nava *et al.*, 2006).

B. FACILITATION OF LIGAND TRANSFER BETWEEN LIPID VESICLES

Dating back to its original identification in liver extracts, multiple reports established that TTP functions as a lipid transfer protein, that is, that it facilitates the transfer of its ligand between 'donor' and 'acceptor' hydrophobic phases. The protein stimulates the transfer of αTOH from liposomes to mitochondria, liposomes to microsomes, liposomes to erythrocytes, and HDL to triglyceride-rich lipoprotein particles (Catignani, 1975; Catignani and Bieri, 1977; Kaplowitz *et al.*, 1989; Massey, 1984; Murphy and Mavis, 1981; Verdon and Blumberg, 1988). Presently, neither the molecular mechanisms by which TTP facilitates ligand transfer nor the role of this activity in TTP's function *in vivo* are clear. It is tempting to speculate that TTP stimulates the distribution of its ligand between different intracellular organelles, as is the case for other lipid transfer proteins such as Sec14p (Bankaitis *et al.*, 1990) or the glycolipid transfer protein (Sasaki, 1990). Interestingly, TTP also facilitates transfer of TOH between synthetic lipid bilayers (Atkinson *et al.*, 2004; Nakagawa *et al.*, 1980) raising the possibility that, at least *in vitro*, no additional proteins are required for this activity.

VI. PHYSIOLOGICAL ACTIVITIES OF TTP

The first clue to TTPs physiological function was reported by Arita *et al.*, who showed that stable expression of TTP in cultured rat hepatoma (McA-RH7777) cells caused facilitation of TOH secretion from the cells into the media (Arita *et al.*, 1997; Horiguchi *et al.*, 2003). Qian *et al.* (2005) demonstrated that induction of TTP expression in human hepatoblastoma cells (HepG2) also lead to stimulation of tocopherol secretion (Fig. 3), and demonstrated that this effect is not due to involvement of TTP in uptake of the vitamin. While the exact cellular mechanisms employed by TTP to stimulate tocopherol secretion are still unknown, recent subcellular localization studies provided important insights into this question. The protein localizes to the endocytic compartment, specifically to endosomes and lysosomes (Horiguchi *et al.*, 2003; Qian *et al.*, 2005, 2006). Studies on the uptake

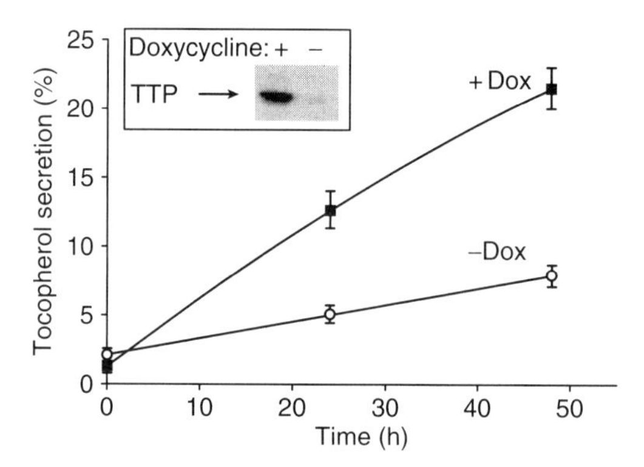

FIGURE 3. TTP-mediated secretion of tocopherol from TetOn-TTP-HepG2 cells. Cells were loaded with [^{14}C]-labeled *RRR*-α-TOH, washed, and the appearance of radioactivity in the media was monitored at the indicated time points. Where indicated, expression of TTP was induced by addition of doxycyclin (1 μg/ml) to the media. Inset: Doxycycline-induced expression of TTP was assessed by immunoblotting. See Qian *et al.* (2005) for details.

of the fluorescent tocopherol analogue NBD-TOH revealed that TTP and its ligand colocalize to the same vesicular compartment after internalization of the vitamin (Qian *et al.*, 2005). A few hours after uptake, NBD-TOH "travels" from the endocytic compartment in small (ca., 100 nm), concentrated structures that later release the vitamin at the plasma membrane (Fig. 4). The regular shape and small size (~100 nm) of the structures with which NBD-TOH is associated, and the sensitivity of tocopherol secretion to drugs that interfere with tubulin polymerization (e.g., colchicine; Bjorneboe *et al.*, 1987), suggest that vitamin E "travels" in microtubule-based transport vesicles. Whether these transport vesicles are derived from the lysosomes, from the endocytic recycling compartment, or from another source, awaits further study. Notably, in cells that express variants of TTP-bearing AVED mutations, intracellular transit of tocopherol is blocked and the vitamin remains "trapped" in endocytic vesicles (Qian *et al.*, 2006; Fig. 4).

Through what portal does vitamin E leave the hepatocyte? Postprandial vitamin E is associated with VLDL shortly after ingestion (Blatt *et al.*, 2001; Traber and Sies, 1996), and it was originally proposed that the vitamin is packaged into these particles as they are being assembled in hepatocytes. This notion was refuted when TTP activity was shown to be resistant to treatment with brefeldin A, an agent that disrupts the integrity of the Golgi compartment and, consequently, blocks VLDL secretion (Arita *et al.*, 1997). Furthermore, Mtp$^{-/-}$ mice, devoid of circulating VLDL, display normal transport and distribution of vitamin E (Minehira-Castelli *et al.*, 2006). Thus, TTP-dependent secretion of vitamin E appears to involve a novel, Golgi-independent pathway that is yet to be fully characterized. An important clue emerged from the observations of Oram *et al.* that treatment of cells with cholesterol

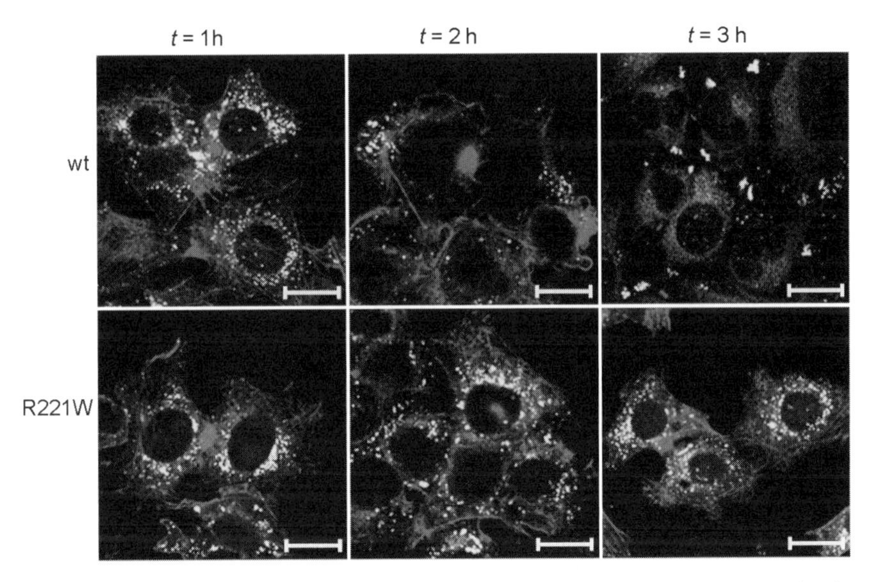

FIGURE 4. TTP-mediated intracellular transport of NBD-TOH. Cells stably expressing the indicated TTP variant were incubated with NBD-TOH and washed. At the indicated times, cells were fixed and imaged under a confocal fluorescence microscope. NBD-TOH fluorescence is shown in green, whereas the actin cytoskeleton is shown in red (Texas Red-conjugated phalloidin stain). Scale bar = 12 μm. See Qian *et al.* (2006) for details. (See Color Insert.)

or 8-Br-cAMP, known to induce the expression of the ABCA1 membrane transporter, caused an increase in TOH secretion. Conversely, TOH secretion was markedly attenuated in cells that lack ABCA1 (Oram *et al.*, 2001). Furthermore, treatment of hepatocytes with glyburide, an inhibitor of ABC proteins, was shown to abolish TTP-dependent TOH secretion (Qian *et al.*, 2005). These observations indicate that ABC-type transporters play an important role in the physiological distribution of vitamin E.

It is interesting to note that some naturally occurring mutations in TTP have no measurable effect on the ability of TTP to facilitate ligand transfer *in vitro*, despite the fact that they cause vitamin E deficiency in human carriers (e.g., the R192H, H101Q substitutions; Morley *et al.*, 2004). This apparent discrepancy may be explained if additional cellular components are required for normal physiological function of TTP. TTP action *in vivo* could involve two separate activities: (1) facilitation of intermembrane transfer of tocopherol and (2) coupling to other cellular proteins that participate in TOH transport. Class I AVED mutations (e.g., R59W) affect ligand binding and transfer, thus abolishing both activities and causing severe vitamin E deficiency in human patients. Class II mutations (e.g., R192H), on the other hand, do not impede ligand binding and transfer, and thus are indistinguishable from the wild-type protein *in vitro*. In intact cells, however, class II mutant TTPs may be unable to interact with the appropriate molecular "partners," and unable to stimulate tocopherol secretion from hepatocytes, leading to vitamin E deficiency (albeit mild) in human carriers. The identification of cellular factors that associate with TTP is a critical undertaking toward achieving a better understanding of the mechanisms of action of the protein.

It is now clear that TTP is also expressed in a number of extrahepatic tissues such as the central nervous system (Copp *et al.*, 1999; Hosomi *et al.*, 1998), the uterus (Kaempf-Rotzoll *et al.*, 2002), and the placenta (Jishage *et al.*, 2001; Kaempf-Rotzoll *et al.*, 2003). How TTP affects vitamin E transport in these (as opposed to other) tissues is the subject of ongoing investigations.

VII. THREE-DIMENSIONAL STRUCTURE OF TTP

The three-dimensional structure of TTP was determined by two groups using X-ray crystallography (Meier *et al.*, 2003; Min *et al.*, 2003), yielding important insights into the molecular basis of its action (Fig. 5A). Like other CRAL_TRIO family members, the defining structural feature of TTP is a lipid-binding domain, composed of an amino terminal three-helix coil (residues 1–86), and a larger C-terminal domain (residues 87–278), composed of

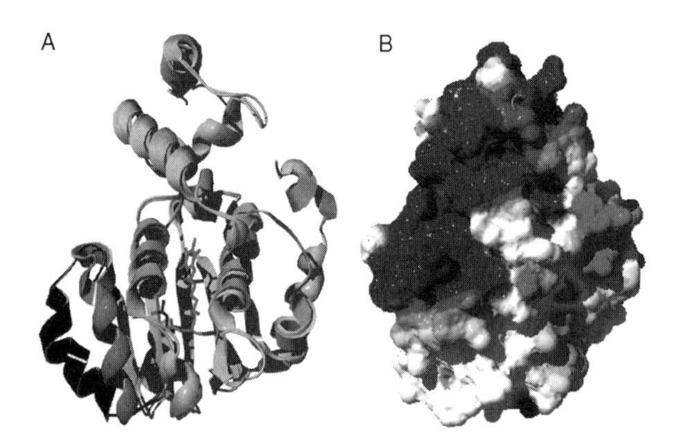

FIGURE 5. Three-dimensional structure and calculated electrostatic surface potential of TTP. (A) Superposition of the apo and holo conformations of TTP. The ligand (*RRR*-α-TOH) is shown in red. The amphipathic lid is colored purple in the holo conformation and black in the apo conformation. (B) Calculated electrostatic surface potential of TTP. Basic residues are shown in blue and acidic residues in red. Drawing and calculations were done using Swiss-PDB software and the 1OIZ and 1OIP coordinate files from the Protein Data Bank (Meier *et al.*, 2003). (See Color Insert.)

alternating α-helices and β-sheets. Together, these elements form a buried, solvent inaccessible binding cavity for α-TOH. The binding pocket is lined with hydrophobic residues that stabilize the bound tocopherol. The ligand's 5-methyl group fits into a hydrophobic "socket" formed by the side chains of Ile^{194}, Val^{191}, Ile^{154}, and Leu^{183}, explaining the preference of TTP for α-TOH over non-5′-methylated vitamers (e.g., γ-TOH; Hosomi *et al.*, 1997; Panagabko *et al.*, 2003). The phenolic group of tocopherol is stabilized within the predominantly hydrophobic core by the polar side chain of Ser140, and by several water molecules. In the protein's binding pocket, α-tocopherol adopts an unusual U-turn conformation, formed by a "kink" in the 4′ and 8′ positions of the phytyl tail. Interestingly, a similar conformation is also adopted by α-tocopheryl quinone when bound to the related protein SPF (Stocker and Baumann, 2003). The "bend" that is induced in the ligand geometry on binding, together with the interaction between the protein and the phytyl chain's 2′ methyl group, explain the pronounced stereoselectivity of the protein toward the *RRR* configuration of its ligand (Hosomi *et al.*, 1997; Panagabko *et al.*, 2003).

An important structural element of TTP is a mobile amphipathic helix (residues 198–221) that adopts different conformations in the apo *versus* holo states of the protein (Fig. 5A). In the absence of tocopherol, the hydrophobic face of this "lid" protrudes away from the binding pocket, thereby allowing ligand entry into the cavity. In the ligand-bound state of TTP, the "lid" closes onto the binding cavity and side chains protruding from its hydrophobic face (Phe^{203}, Val^{206}, Phe^{207}, Ile^{210}, and Leu^{214}) contact the bound ligand. The polar face of the lid, lined with charged residues, points away from the binding pocket and aids in solubilizing the protein in the aqueous milieu.

As the hydrophobic tocopherol is predominantly embedded in biological membranes, it is likely that TTP interacts with lipid bilayers during ligand transport. In support of a physical interaction between TTP and membranes are the observations that, *in vivo*, the protein is localized to endocytic vesicles (Horiguchi *et al.*, 2003; Qian *et al.*, 2005). Early observations that, in cell homogenates, TTP is found in the soluble fraction (100,000-g supernatant; Sato *et al.*, 1991) suggest that the interactions of the protein with membranes are weak, indirect, or both. It is possible that TTP associates with membranes through binding to organelle-specific, resident membrane proteins. However, the observations that TTP facilitates ligand transfer between synthetic lipid vesicles that are devoid of protein (Atkinson *et al.*, 2004; Nakagawa *et al.*, 1980) suggest that protein–lipid interactions play at least some role in this function. Direct interactions with bilayers could be based on electrostatic interactions; TTP's surface potential is markedly polarized, with a pronounced positive "lobe" that arises from high spatial concentration of charged arginine and lysine residues, as shown in Fig. 5B. Perhaps through this region TTP associates with the negatively charged phospholipid headgroups.

VIII. SELECTIVE RETENTION OF *RRR*-α-TOH AND THE EVOLUTIONARY ORIGINS OF TTP

Discrimination among dietary tocols by selectively accumulating *RRR*-α-TOH over other vitamers is not unique to humans. Organisms as diverse as fish (Parazo *et al.*, 1998), turkey (Surai *et al.*, 1999), cow (Hidiroglou *et al.*, 1978), and eel (Ando, 1995) discriminate between different forms of vitamin E. Thus, the molecular mechanisms responsible for this selectivity appear to be evolutionarily conserved.

In humans, discrimination between different vitamin E forms is achieved by two activities. The cytochrome P450 ω-hydroxylase CYP4F2 efficiently catabolizes non-α-tocols to polar, short-chained metabolites that are then excreted in urine (Sontag and Parker, 2002). This activity is the major pathway by which γ- and δ-tocopherols, the predominant dietary forms of vitamin E, are eliminated such that the organism is not exposed to their cytotoxic effects (McCormick and Parker, 2004). The substrate specificity of TTP provides an additional discriminatory tool in that it selectively retains the natural (*RRR*) stereoisomer of α-tocopherol. Together, these mechanisms facilitate the selective retention of α-TOH and the elimination of other vitamers.

Putative orthologues of TTP can be found in the genomes of various organisms. Furthermore, heritable vitamin E deficiency coupled to neurological symptoms have been reported in dogs (McLellan *et al.*, 2002) and sheep (Menzies *et al.*, 2004). Alignment of TTP-homologous sequences from different species and their calculated evolutionary relationship are shown in Fig. 6. It can be seen that TTP arose relatively early in evolution, as evidenced by the presence of a homologous sequence in the pufferfish. An important cautionary note is that significant substrate promiscuity is found among CRAL_TRIO proteins (Manor and Atkinson, 2003; Panagabko *et al.*, 2003) and thus, that in addition to sequence analyses, the presence of TTP in a given organism should be verified using functional criteria, such as discrimination between the different stereoisomers of α-TOH (Parker and McCormick, 2005; Traber *et al.*, 1990). Keeping this caveat in mind, we note that sequence identity between TTP and related CRAL_TRIO proteins that do *not* bind α-TOH with high affinity, such as Sec14p and SPF, is $\leq 30\%$. In contrast, the orthologous sequences shown in Fig. 6 show $>56\%$ sequence identity to the human TTP and therefore likely encode genuine TTP orthologues. It is noteworthy that alignment of the putative TTP orthologue sequences reveals remarkable conservation of the binding pocket architecture. Specifically, Phe^{133}, Leu^{137}, Ile^{154}, Ile^{179}, Val^{182}, Leu^{183}, Phe^{187}, Val^{191}, and Ile^{194}, which stabilize tocopherol in the binding pocket and determine ligand stereoselectivity are highly conserved among the species analyzed here (Fig. 6). Understanding the evolutionary relationship between the emergence of TTP and that of tocopherol ω-hydroxylase is an exciting area of study that will greatly advance our understanding of vitamin E biology.

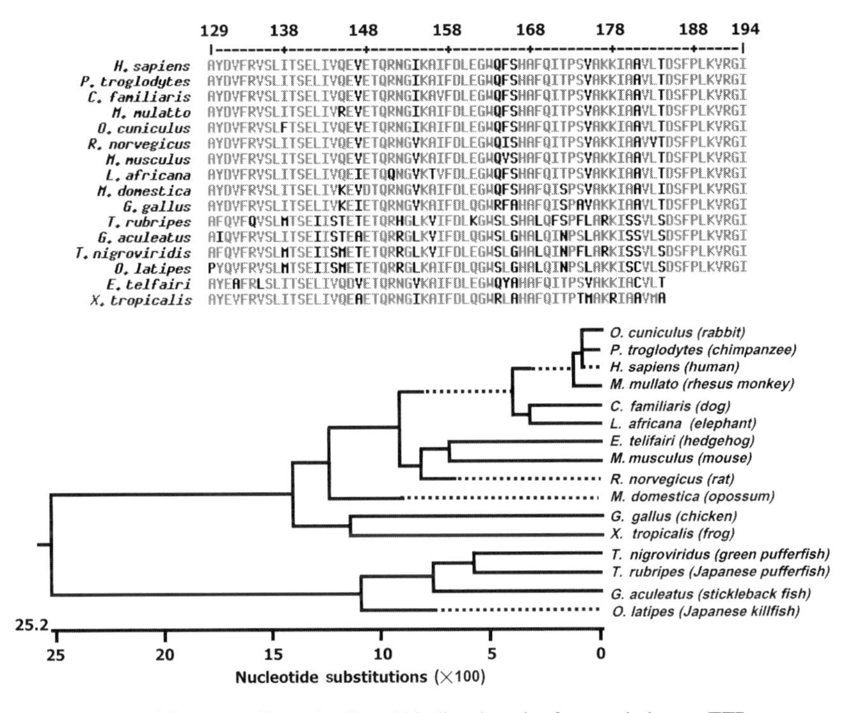

FIGURE 6. Alignment of putative ligand binding domains from orthologous TTP sequences. Numbering is based on the human TTP sequence. Identical and conserved residues are shown in gray. Sequences were obtained from the Ensembl database (www.ensembl.org), using the human sequence (ENSG00000137561) as reference. Alignment generated on the Multalin website (http://bioinfo.genopole-toulouse.prd.fr/multalin/multalin.html; Corpet, 1988).

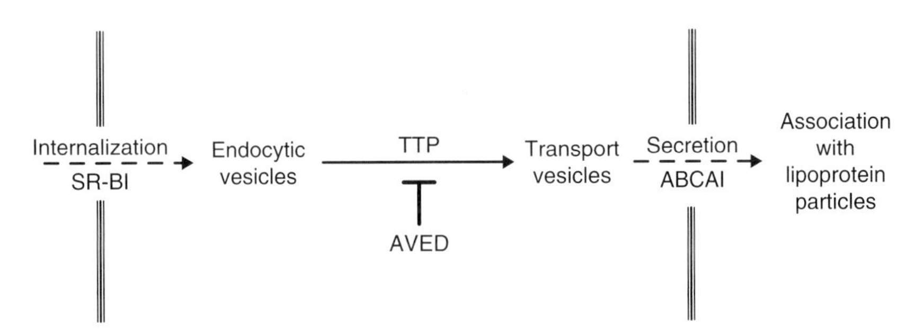

FIGURE 7. TTP and hepatic trafficking of vitamin E. Tocopherol, complexed to lipoprotein particles is internalized in an incompletely understood process that involves the scavenger receptor SR-BI (Cohn *et al.*, 1992; Goti *et al.*, 1998; Mardones *et al.*, 2002; Qian *et al.*, 2005). The vitamin arrives at the endosome/lysosome compartment with which TTP is associated. TTP then facilitates the transfer of its ligand to transport vesicles that travel to the plasma membrane, where the vitamin is secreted from the cell through the ABCA1 transporter. Outside the cell, tocopherol associates with lipoprotein particles that distribute it to target tissues.

IX. EPILOGUE

Numerous studies over the past four decades established that TTP plays critical roles in vitamin E biology, both in health and in disease states. This protein is key for the selective retention of *RRR*-α-TOH. By controlling trafficking of the vitamin in the liver, TTP directly regulates the body-wide status of vitamin E. Figure 7 depicts a simplified model of the main events that comprise vitamin E trafficking in hepatocytes, and the central role that TTP plays in this pathway. As is often the case, as new information is revealed, new questions emerge. Open questions that remain in understanding the complete scope of TTP function include: what is the nature of the "acceptor" complex that transports tocopherol from TTP to the plasma membrane? What accessory proteins assist TTP in its function? What is the exact function of TTP in extrahepatic tissue? How does TTP facilitate the inter-membrane transport of its ligand? With continued cooperation between nutritionists, chemists, cell biologists, and clinicians, future studies will provide insights into these interesting issues.

REFERENCES

Amiel, J., Maziere, J. C., Beucler, I., Koenig, M., Reutenauer, L., Loux, N., Bonnefont, D., Feo, C., and Landrieu, P. (1995). Familial isolated vitamin E deficiency. Extensive study of a large family with a 5-year therapeutic follow-up. *J. Inherit. Metab. Dis.* **18,** 333–340.

Ando, S. (1995). Tocopherol levels in the plasma lipoproteins from Japanese eel Anguilla japonica. *Biosci. Biotechnol. Biochem.* **59,** 2326–2327.

Angelini, L., Erba, A., Mariotti, C., Gellera, C., Ciano, C., and Nardocci, N. (2002). Myoclonic dystonia as unique presentation of isolated vitamin E deficiency in a young patient. *Mov. Disord.* **17,** 612–614.

Aparicio, J. M., Belanger-Quintana, A., Suarez, L., Mayo, D., Benitez, J., Diaz, M., and Escobar, H. (2001). Ataxia with isolated vitamin E deficiency: Case report and review of the literature. *J. Pediatr. Gastroenterol. Nutr.* **33,** 206–210.

Arita, M., Sato, Y., Miyata, A., Tanabe, T., Takahashi, E., Kayden, H. J., Arai, H., and Inoue, K. (1995). Human α-tocopherol transfer protein: cDNA cloning, expression and chromosomal localization. *Biochem. J.* **306,** 437–443.

Arita, M., Nomura, K., Arai, H., and Inoue, K. (1997). α-Tocopherol transfer protein stimulates the secretion of α-tocopherol from a cultured liver cell line through a brefeldin A-insensitive pathway. *Proc. Natl. Acad. Sci. USA* **94,** 12437–12441.

Atkinson, J. K., Nava, P., Frahm, G., Curtis, V., and Manor, D. (2004). Fluorescent tocopherols as probes of inter-vesicular transfer catalyzed by the α-tocopherol transfer protein. *Ann. NY Acad. Sci.* **1031,** 324–327.

Bankaitis, V. A., Aitken, J. R., Cleves, A. E., and Dowhan, W. (1990). An essential role for a phospholipid transfer protein in yeast Golgi function. *Nature* **347,** 561–562.

Benomar, A., Yahyaoui, M., Meggouh, F., Bouhouche, A., Boutchich, M., Bouslam, N., Zaim, A., Schmitt, M., Belaidi, H., Ouazzani, R., Chkili, T., and Koenig, M. (2002). Clinical comparison between AVED patients with 744 del A mutation and Friedreich ataxia with GAA expansion in 15 Moroccan families. *J. Neurol. Sci.* **198,** 25–29.

Bjorneboe, A., Bjorneboe, G. E., Hagen, B. F., Nossen, J. O., and Drevon, C. A. (1987). Secretion of α-tocopherol from cultured rat hepatocytes. *Biochim. Biophys. Acta* **922,** 199–205.

Blatt, D. H., Leonard, S. W., and Traber, M. G. (2001). Vitamin E kinetics and the function of tocopherol regulatory proteins. *Nutrition* **17,** 799–805.

Burck, U., Goebel, H. H., Kuhlendahl, H. D., Meier, C., and Goebel, K. M. (1981). Neuromyopathy and vitamin E deficiency in man. *Neuropediatrics* **12,** 267–278.

Burton, G. W. (1994). Vitamin E: Molecular and biological function. *Proc. Nutr. Soc.* **53,** 251–262.

Catignani, G. L. (1975). An α-tocopherol binding protein in rat liver cytoplasm. *Biochem. Biophys. Res. Commun.* **67,** 66–72.

Catignani, G. L., and Bieri, J. G. (1977). Rat liver α-tocopherol binding protein. *Biochim. Biophys. Acta* **497,** 349–357.

Cavalier, L., Ouahchi, K., Kayden, H. J., Di Donato, S., Reutenauer, L., Mandel, J. L., and Koenig, M. (1998). Ataxia with isolated vitamin E deficiency: Heterogeneity of mutations and phenotypic variability in a large number of families. *Am. J. Hum. Genet.* **62,** 301–310.

Cellini, E., Piacentini, S., Nacmias, B., Forleo, P., Tedde, A., Bagnoli, S., Ciantelli, M., and Sorbi, S. (2002). A family with spinocerebellar ataxia type 8 expansion and vitamin E deficiency ataxia. *Arch. Neurol.* **59,** 1952–1953.

Cheng, S. C., Burton, G. W., Ingold, K. U., and Foster, D. O. (1987). Chiral discrimination in the exchange of α-tocopherol stereoisomers between plasma and red blood cells. *Lipids* **22,** 469–473.

Cohn, W., Goss-Sampson, M. A., Grun, H., and Muller, D. P. (1992). Plasma clearance and net uptake of α-tocopherol and low-density lipoprotein by tissues in WHHL and control rabbits. *Biochem. J.* **287**(Pt. 1), 247–254.

Copp, R. P., Wisniewski, T., Hentati, F., Larnaout, A., Ben Hamida, M., and Kayden, H. J. (1999). Localization of α-tocopherol transfer protein in the brains of patients with ataxia with vitamin E deficiency and other oxidative stress related neurodegenerative disorders. *Brain Res.* **822,** 80–87.

Corpet, F. (1988). Multiple sequence alignment with hierarchical clustering. *Nucleic Acids Res.* **16,** 10881–10890.

Evans, H. M., and Bishop, K. S. (1922). On the Existance of a Hitherto unrecognized dietary factor essential for reproduction. *Science* **56,** 650–651.

Goti, D., Reicher, H., Malle, E., Kostner, G. M., Panzenboeck, U., and Sattler, W. (1998). High-density lipoprotein (HDL3)-associated α-tocopherol is taken up by HepG2 cells via the selective uptake pathway and resecreted with endogenously synthesized apo-lipoprotein B-rich lipoprotein particles. *Biochem. J.* **332,** 57–65.

Gotoda, T., Arita, M., Arai, H., Inoue, K., Yokota, T., Fukuo, Y., Yazaki, Y., and Yamada, N. (1995). Adult-onset spinocerebellar dysfunction caused by a mutation in the gene for the α-tocopherol-transfer protein. *N. Engl. J. Med.* **333,** 1313–1318.

Hentati, A., Deng, H. X., Hung, W. Y., Nayer, M., Ahmed, M. S., He, X., Tim, R., Stumpf, D. A., Siddique, T., and Ahmed, A. (1996). Human α-tocopherol transfer protein: Gene structure and mutations in familial vitamin E deficiency. *Ann. Neurol.* **39,** 295–300.

Hidiroglou, M., Lessard, J. R., and Wauthy, J. M. (1978). Blood serum tocopherol levels in calves born from cows winter fed hay or grass silage. *Can. J. Comp. Med.* **42,** 128–131.

Horiguchi, M., Arita, M., Kaempf-Rotzoll, D. E., Tsujimoto, M., Inoue, K., and Arai, H. (2003). pH-dependent translocation of α-tocopherol transfer protein (α-TTP) between hepatic cytosol and late endosomes. *Genes Cells* **8,** 789–800.

Hoshino, M., Masuda, N., Ito, Y., Murata, M., Goto, J., Sakurai, M., and Kanazawa, I. (1999). Ataxia with isolated vitamin E deficiency: A Japanese family carrying a novel mutation in the α-tocopherol transfer protein gene. *Ann. Neurol.* **45,** 809–812.

Hosomi, A., Arita, M., Sato, Y., Kiyose, C., Ueda, T., Igarashi, O., Arai, H., and Inoue, K. (1997). Affinity for α-tocopherol transfer protein as a determinant of the biological activities of vitamin E analogs. *FEBS Lett.* **409,** 105–108.

Hosomi, A., Goto, K., Kondo, H., Iwatsubo, T., Yokota, T., Ogawa, M., Arita, M., Aoki, J., Arai, H., and Inoue, K. (1998). Localization of α-tocopherol transfer protein in rat brain. *Neurosci. Lett.* **256,** 159–162.

Ingold, K. U., Burton, G. W., Foster, D. O., Hughes, L., Lindsay, D. A., and Webb, A. (1987). Biokinetics of and discrimination between dietary RRR- and SRR-α-tocopherols in the male rat. *Lipids* **22,** 163–172.

Jishage, K., Arita, M., Igarashi, K., Iwata, T., Watanabe, M., Ogawa, M., Ueda, O., Kamada, N., Inoue, K., Arai, H., and Suzuki, H. (2001). α-Tocopherol transfer protein is important for the normal development of placental labyrinthine trophoblasts in mice. *J. Biol. Chem.* **276,** 1669–1672.

Jishage, K., Tachibe, T., Ito, T., Shibata, N., Suzuki, S., Mori, T., Hani, T., Arai, H., and Suzuki, H. (2005). Vitamin E is essential for mouse placentation but not for embryonic development itself. *Biol. Reprod.* **73,** 983–987.

Kaempf-Rotzoll, D. E., Igarashi, K., Aoki, J., Jishage, K., Suzuki, H., Tamai, H., Linderkamp, O., and Arai, H. (2002). α-Tocopherol transfer protein is specifically localized at the implantation site of pregnant mouse uterus. *Biol. Reprod.* **67,** 599–604.

Kaempf-Rotzoll, D. E., Horiguchi, M., Hashiguchi, K., Aoki, J., Tamai, H., Linderkamp, O., and Arai, H. (2003). Human placental trophoblast cells express α-tocopherol transfer protein. *Placenta* **24,** 439–444.

Kaplowitz, N., Yoshida, H., Kuhlenkamp, J., Slitsky, B., Ren, I., and Stolz, A. (1989). Tocopherol-binding proteins of hepatic cytosol. *Ann. NY Acad. Sci.* **570,** 85–94.

Keaney, J. F., Jr., Simon, D. I., and Freedman, J. E. (1999). Vitamin E and vascular homeostasis: Implications for atherosclerosis. *FASEB J.* **13,** 965–975.

Kelly, F. J., Lee, R., and Mudway, I. S. (2004). Inter- and intra-individual vitamin E uptake in healthy subjects is highly repeatable across a wide supplementation dose range. *Ann. NY Acad. Sci.* **1031,** 22–39.

Krendel, D. A., Gilchrist, J. M., Johnson, A. O., and Bossen, E. H. (1987). Isolated deficiency of vitamin E with progressive neurologic deterioration. *Neurology* **37,** 538–540.

Kuhlenkamp, J., Ronk, M., Yusin, M., Stolz, A., and Kaplowitz, N. (1993). Identification and purification of a human liver cytosolic tocopherol binding protein. *Protein Expr. Purif.* **4,** 382–389.

Lonn, E., Bosch, J., Yusuf, S., Sheridan, P., Pogue, J., Arnold, J. M., Ross, C., Arnold, A., Sleight, P., Probstfield, J., and Dagenais, G. R. (2005). Effects of long-term vitamin E supplementation on cardiovascular events and cancer: A randomized controlled trial. *JAMA* **293,** 1338–1347.

Manor, D., and Atkinson, J. (2003). Is tocopherol associated protein a misnomer? *J. Nutr. Biochem.* **14,** 421–422; author reply 423.

Mardones, P., Strobel, P., Miranda, S., Leighton, F., Quinones, V., Amigo, L., Rozowski, J., Krieger, M., and Rigotti, A. (2002). α-Tocopherol metabolism is abnormal in scavenger receptor class B type I (SR-BI)-deficient mice. *J. Nutr.* **132,** 443–449.

Mariotti, C., Gellera, C., Rimoldi, M., Mineri, R., Uziel, G., Zorzi, G., Pareyson, D., Piccolo, G., Gambi, D., Piacentini, S., Squitieri, F., Capra, R., *et al.* (2004). Ataxia with isolated vitamin E deficiency: Neurological phenotype, clinical follow-up and novel mutations in TTPA gene in Italian families. *Neurol. Sci.* **25,** 130–137.

Martinello, F., Fardin, P., Ottina, M., Ricchieri, G. L., Koenig, M., Cavalier, L., and Trevisan, C. P. (1998). Supplemental therapy in isolated vitamin E deficiency improves the peripheral neuropathy and prevents the progression of ataxia. *J. Neurol. Sci.* **156,** 177–179.

Massey, J. B. (1984). Kinetics of transfer of α-tocopherol between model and native plasma lipoproteins. *Biochim. Biophys. Acta* **793,** 387–392.

McCormick, C. C., and Parker, R. S. (2004). The cytotoxicity of vitamin E is both vitamer- and cell-specific and involves a selectable trait. *J. Nutr.* **134,** 3335–3342.

McLellan, G. J., Elks, R., Lybaert, P., Watte, C., Moore, D. L., and Bedford, P. G. (2002). Vitamin E deficiency in dogs with retinal pigment epithelial dystrophy. *Vet. Rec.* **151,** 663–667.

Meier, R., Tomizaki, T., Schulze-Briese, C., Baumann, U., and Stocker, A. (2003). The molecular basis of vitamin E retention: Structure of human α-tocopherol transfer protein. *J. Mol. Biol.* **331,** 725–734.

Menzies, P., Langs, L., Boermans, H., Martin, J., and McNally, J. (2004). Myopathy and hepatic lipidosis in weaned lambs due to vitamin E deficiency. *Can. Vet. J.* **45,** 244–247.

Min, K. C., Kovall, R. A., and Hendrickson, W. A. (2003). Crystal structure of human α-tocopherol transfer protein bound to its ligand: Implications for ataxia with vitamin E deficiency. *Proc. Natl. Acad. Sci. USA* **100,** 14713–14718.

Minehira-Castelli, K., Leonard, S. W., Walker, Q. M., Traber, M. G., and Young, S. G. (2006). Absence of VLDL secretion does not affect α-tocopherol content in peripheral tissues. *J. Lipid Res.* **47,** 1733–1738.

Morley, S., Panagabko, C., Shineman, D., Mani, B., Stocker, A., Atkinson, J., and Manor, D. (2004). Molecular determinants of heritable vitamin E deficiency. *Biochemistry* **43,** 4143–4149.

Morley, S., Cross, V., Cecchini, M., Nava, P., Atkinson, J., and Manor, D. (2006). Utility of a fluorescent vitamin E analogue as a probe for tocopherol transfer protein activity. *Biochemistry* **45,** 1075–1081.

Mowri, H., Nakagawa, Y., Inoue, K., and Nojima, S. (1981). Enhancement of the transfer of α-tocopherol between liposomes and mitochondria by rat-liver protein(s). *Eur. J. Biochem.* **117,** 537–542.

Murphy, D. J., and Mavis, R. D. (1981). Membrane transfer of α-tocopherol. Influence of soluble α-tocopherol-binding factors from the liver, lung, heart, and brain of the rat. *J. Biol. Chem.* **256,** 10464–10468.

Nakagawa, Y., Nojima, S., and Inoue, K. (1980). Transfer of steroids and α-tocopherol between liposomal membranes. *J. Biochem. (Tokyo)* **87,** 497–502.

Nava, P., Cecchini, M., Chirico, S., Gordon, H., Morley, S., Manor, D., and Atkinson, J. (2006). Preparation of fluorescent tocopherols for use in protein binding and localization with the a-tocopherol transfer protein. *Bioorg. Med. Chem.* **14**(11), 3721–3736.

Oram, J. F., Vaughan, A. M., and Stocker, R. (2001). ATP-binding cassette transporter A1 mediates cellular secretion of α-tocopherol. *J. Biol. Chem.* **276,** 39898–39902.

Ouahchi, K., Arita, M., Kayden, H., Hentati, F., Ben Hamida, M., Sokol, R., Arai, H., Inoue, K., Mandel, J. L., and Koenig, M. (1995). Ataxia with isolated vitamin E deficiency is caused by mutations in the α-tocopherol transfer protein. *Nat. Genet.* **9,** 141–145.

Panagabko, C., Morley, S., Hernandez, M., Cassolato, P., Gordon, H., Parsons, R., Manor, D., and Atkinson, J. (2003). Ligand specificity in the CRAL-TRIO protein family. *Biochemistry* **42,** 6467–6474.

Pang, J., Kiyosawa, M., Seko, Y., Yokota, T., Harino, S., and Suzuki, J. (2001). Clinicopathological report of retinitis pigmentosa with vitamin E deficiency caused by mutation of the α-tocopherol transfer protein gene. *Jpn. J. Ophthalmol.* **45,** 672–676.

Parazo, M. P., Lall, S. P., Castell, J. D., and Ackman, R. G. (1998). Distribution of α- and γ-tocopherols in Atlantic salmon (Salmo salar) tissues. *Lipids* **33,** 697–704.

Parker, R. S., and McCormick, C. C. (2005). Selective accumulation of α-tocopherol in Drosophila is associated with cytochrome P450 tocopherol-omega-hydroxylase activity but not α-tocopherol transfer protein. *Biochem. Biophys. Res. Commun.* **338,** 1537–1541.

Pratico, D., Tangirala, R. K., Rader, D. J., Rokach, J., and FitzGerald, G. A. (1998). Vitamin E suppresses isoprostane generation *in vivo* and reduces atherosclerosis in ApoE-deficient mice. *Nat. Med* **4,** 1189–1192.

Princen, H. M., van Duyvenvoorde, W., Buytenhek, R., van der Laarse, A., van Poppel, G., Gevers Leuven, J. A., and van Hinsbergh, V. W. (1995). Supplementation with low doses of vitamin E protects LDL from lipid peroxidation in men and women. *Arterioscler. Thromb. Vasc. Biol.* **15,** 325–333.

Qian, J., Morley, S., Wilson, K., Nava, P., Atkinson, J., and Manor, D. (2005). Intracellular trafficking of vitamin E in hepatocytes: The role of tocopherol transfer protein. *J. Lipid Res.* **46,** 2072–2082.

Qian, J., Atkinson, J., and Manor, D. (2006). Biochemical consequences of heritable mutations in the α-tocopherol transfer protein. *Biochemistry* **45,** 8236–8242.

Roubertie, A., Biolsi, B., Rivier, F., Humbertclaude, V., Cheminal, R., and Echenne, B. (2003). Ataxia with vitamin E deficiency and severe dystonia: Report of a case. *Brain Dev.* **25,** 442–445.

Roxborough, H. E., Burton, G. W., and Kelly, F. J. (2000). Inter- and intra-individual variation in plasma and red blood cell vitamin E after supplementation. *Free Radic. Res.* **33,** 437–445.

Sasaki, T. (1990). Glycolipid transfer protein and intracellular traffic of glucosylceramide. *Experientia* **46,** 611–616.

Sato, Y., Hagiwara, K., Arai, H., and Inoue, K. (1991). Purification and characterization of the α-tocopherol transfer protein from rat liver. *FEBS Lett.* **288,** 41–45.

Sato, Y., Arai, H., Miyata, A., Tokita, S., Yamamoto, K., Tanabe, T., and Inoue, K. (1993). Primary structure of α-tocopherol transfer protein from rat liver. Homology with cellular retinaldehyde-binding protein. *J. Biol. Chem.* **268,** 17705–17710.

Schock, B. C., Van der Vliet, A., Corbacho, A. M., Leonard, S. W., Finkelstein, E., Valacchi, G., Obermueller-Jevic, U., Cross, C. E., and Traber, M. G. (2004). Enhanced inflammatory responses in α-tocopherol transfer protein null mice. *Arch. Biochem. Biophys.* **423,** 162–169.

Schuelke, M., Mayatepek, E., Inter, M., Becker, M., Pfeiffer, E., Speer, A., Hubner, C., and Finckh, B. (1999). Treatment of ataxia in isolated vitamin E deficiency caused by α-tocopherol transfer protein deficiency. *J. Pediatr.* **134,** 240–244.

Schuelke, M., Finckh, B., Sistermans, E. A., Ausems, M. G., Hubner, C., and von Moers, A. (2000). Ataxia with vitamin E deficiency: Biochemical effects of malcompliance with vitamin E therapy. *Neurology* **55,** 1584–1586.

Shimohata, T., Date, H., Ishiguro, H., Suzuki, T., Takano, H., Tanaka, H., Tsuji, S., and Hirota, K. (1998). Ataxia with isolated vitamin E deficiency and retinitis pigmentosa. *Ann. Neurol.* **43,** 273.

Sokol, R. J. (1988). Vitamin E deficiency and neurologic disease. *Annu. Rev. Nutr.* **8,** 351–373.

Sontag, T. J., and Parker, R. S. (2002). Cytochrome P450 omega-hydroxylase pathway of tocopherol catabolism. Novel mechanism of regulation of vitamin E status. *J. Biol. Chem.* **277,** 25290–25296.

Stampfer, M. J., Hennekens, C. H., Manson, J. E., Colditz, G. A., Rosner, B., and Willett, W. C. (1993). Vitamin E consumption and the risk of coronary disease in women. *N. Engl. J. Med.* **328,** 1444–1449.

Stephens, N. G., Parsons, A., Schofield, P. M., Kelly, F., Cheeseman, K., and Mitchinson, M. J. (1996). Randomised controlled trial of vitamin E in patients with coronary disease: Cambridge Heart Antioxidant Study (CHAOS). *Lancet* **347,** 781–786.

Stocker, A., and Baumann, U. (2003). Supernatant protein factor in complex with RRR-α-tocopherylquinone: A link between oxidized Vitamin E and cholesterol biosynthesis. *J. Mol. Biol.* **332,** 759–765.

Stumpf, D. A., Sokol, R., Bettis, D., Neville, H., Ringel, S., Angelini, C., and Bell, R. (1987). Friedreich's disease: V. Variant form with vitamin E deficiency and normal fat absorption. *Neurology* **37,** 68–74.

Surai, P. F., Sparks, N. H., and Noble, R. C. (1999). Antioxidant systems of the avian embryo: Tissue-specific accumulation and distribution of vitamin E in the turkey embryo during development. *Br. Poult. Sci.* **40,** 458–466.

Tamaru, Y., Hirano, M., Kusaka, H., Ito, H., Imai, T., and Ueno, S. (1997). α-Tocopherol transfer protein gene: Exon skipping of all transcripts causes ataxia. *Neurology* **49,** 584–588.

Terasawa, Y., Ladha, Z., Leonard, S. W., Morrow, J. D., Newland, D., Sanan, D., Packer, L., Traber, M. G., and Farese, R. V., Jr. (2000). Increased atherosclerosis in hyperlipidemic mice deficient in α-tocopherol transfer protein and vitamin E. *Proc. Natl. Acad. Sci. USA* **97,** 13830–13834.

Traber, M. G., and Sies, H. (1996). Vitamin E in humans: Demand and delivery. *Annu. Rev. Nutr.* **16,** 321–347.

Traber, M. G., Burton, G. W., Ingold, K. U., and Kayden, H. J. (1990). RRR- and SRR-α-tocopherols are secreted without discrimination in human chylomicrons, but RRR-α-tocopherol is preferentially secreted in very low density lipoproteins. *J. Lipid Res.* **31,** 675–685.

Traber, M. G., Burton, G. W., Hughes, L., Ingold, K. U., Hidaka, H., Malloy, M., Kane, J., Hyams, J., and Kayden, H. J. (1992). Discrimination between forms of vitamin E by humans with and without genetic abnormalities of lipoprotein metabolism. *J. Lipid Res.* **33,** 1171–1182.

Traber, M. G., Sokol, R. J., Kohlschutter, A., Yokota, T., Muller, D. P., Dufour, R., and Kayden, H. J. (1993). Impaired discrimination between stereoisomers of α-tocopherol in patients with familial isolated vitamin E deficiency. *J. Lipid Res.* **34,** 201–210.

Usuki, F., and Maruyama, K. (2000). Ataxia caused by mutations in the α-tocopherol transfer protein gene. *J. Neurol. Neurosurg. Psychiatry* **69,** 254–256.

Verdon, C. P., and Blumberg, J. B. (1988). Influence of dietary vitamin E on the intermembrane transfer of α-tocopherol as mediated by an α-tocopherol binding protein. *Proc. Soc. Exp. Biol. Med.* **189,** 52–60.

Yokota, T., Shiojiri, T., Gotoda, T., and Arai, H. (1996). Retinitis pigmentosa and ataxia caused by a mutation in the gene for the α-tocopherol-transfer protein. *N. Engl. J. Med.* **335,** 1770–1771.

Yokota, T., Shiojiri, T., Gotoda, T., Arita, M., Arai, H., Ohga, T., Kanda, T., Suzuki, J., Imai, T., Matsumoto, H., Harino, S., Kiyosawa, M., *et al.* (1997). Friedreich-like ataxia with retinitis pigmentosa caused by the His101Gln mutation of the α-tocopherol transfer protein gene. *Ann. Neurol.* **41,** 826–832.

Yokota, T., Uchihara, T., Kumagai, J., Shiojiri, T., Pang, J. J., Arita, M., Arai, H., Hayashi, M., Kiyosawa, M., Okeda, R., and Mizusawa, H. (2000). Postmortem study of ataxia with retinitis pigmentosa by mutation of the α-tocopherol transfer protein gene. *J. Neurol. Neurosurg. Psychiatry* **68,** 521–525.

Yokota, T., Igarashi, K., Uchihara, T., Jishage, K., Tomita, H., Inaba, A., Li, Y., Arita, M., Suzuki, H., Mizusawa, H., and Arai, H. (2001). Delayed-onset ataxia in mice lacking α-tocopherol transfer protein: Model for neuronal degeneration caused by chronic oxidative stress. *Proc. Natl. Acad. Sci. USA* **98,** 15185–15190.

Yoshida, H., Yusin, M., Ren, I., Kuhlenkamp, J., Hirano, T., Stolz, A., and Kaplowitz, N. (1992). Identification, purification, and immunochemical characterization of a tocopherol-binding protein in rat liver cytosol. *J. Lipid Res.* **33,** 343–350.

Yusuf, S., Dagenais, G., Pogue, J., Bosch, J., and Sleight, P. (2000). Vitamin E supplementation and cardiovascular events in high-risk patients. The Heart Outcomes Prevention Evaluation Study Investigators. *N. Engl. J. Med.* **342,** 154–160.

4

Molecular Associations of Vitamin E

Peter J. Quinn

Department of Biochemistry, King's College London
London SE2 9NH, United Kingdom

Vitamins and Hormones, Volume 76

0083-6729/07 $35.00
DOI: 10.1016/S0083-6729(07)76004-1

To understand how vitamin E fulfills its functions in membranes and lipoproteins, it is necessary to know how it associates with the lipid components of these structures and the effects its presence has on their structure and stability. Studies of model membrane systems containing vitamin E have proved to be an informative approach to address these questions. A review of the way vitamin E interacts with phospholipid bilayers, how it distributes within the structure, its motional diffusion characteristics, and orientation has been undertaken. The effect of vitamin E on membrane stability and permeability has been described. The tendency of vitamin E to form complexes with certain phospholipids is examined as is the way modulation of protein functions takes place. Finally, recent evidence relevant to the putative role of vitamin E in protecting membranes from free radical attack and the consequences of lipid oxidation in lipoproteins and membranes is examined.

I. INTRODUCTION

Among the broad classification of the vitamins into water soluble and water insoluble, vitamin E falls firmly into the latter category. While vitamin E is a term that refers to a generic group of chemically similar compounds sharing the tocopherol and tocotrienol structures, their physical properties are alike. The fat-soluble property is dictated by the chromanol ring system to which is attached a polyisopentenyl side chain which can be either saturated (tocopherol) or unsaturated (tocotrienol) (Kamal-Eldin and Appelqvist, 1996). A weak polarity is imparted by the presence of a hydroxyl group attached to the aromatic ring, and this serves to produce some surface activity on the molecule. The physical properties thus endowed ensure that vitamin E partitions into tissue lipids whether in the form of storage organelles or into the different subcellular membranes. Because of its weakly amphipathic character, it tends to locate at the interface of aqueous phases and hydrophobic domains of membranes and lipoproteins.

The distribution of vitamin E in the body differs from one tissue to another (Viana *et al.*, 1999) and is altered by, among other factors, oxidative stress (Galinier *et al.*, 2006; Lu *et al.*, 2006) and temperature (Behrens and Madere, 1990). Adipose tissue, liver, and muscle represent the major stores of vitamin E in the body, with about 90% of the vitamin being contained in the adipose tissue (Traber and Kayden, 1987). Of all the subcellular membrane fractions, the greatest concentrations of vitamin E are found in the Golgi membranes and lysosomes (Zhang *et al.*, 1996). The molar ratio of vitamin E:phospholipid in these membranes is of the order of 1:65 phospholipid molecules, which is about an order of magnitude greater than that found in the other subcellular membranes. In all tissues, α-tocopherol, as opposed to other types of the vitamin, was found to be the major form of vitamin E.

Although vitamin E is present in relatively low proportions relative to other membrane lipids, it is believed to play an important part in preserving the integrity of membranes. Foremost among its functions is its ability to protect polyunsaturated lipids of the lipid bilayer matrix of membranes and of serum lipoproteins against oxidation (Singh *et al.*, 2005). Another important function in membranes is the formation of complexes between vitamin E and products of membrane lipid hydrolysis such as lysophospholipids and free fatty acids. The complexes thus formed tend to stabilize membranes and prevent the detergent-like actions of lipid hydrolytic products on the membrane (Kagan, 1989). The formation of complexes between vitamin E and polyunsaturated fatty acids like docosahexaenoic acid has been shown to be particularly favorable (Stillwell, 2000; Stillwell *et al.*, 1992).

Other roles have been ascribed to vitamin E in addition to actions associated with stabilizing membranes and lipoproteins against oxidative stress. The rate of transcription of certain genes such as those coding CD36 (Ozer *et al.*, 2006), α-TTP (Fechner *et al.*, 1998), α-tropomyosin (Aratri *et al.*, 1999), and collagenase (Chojkier *et al.*, 1998) is affected by vitamin E. A range of other effects of vitamin E on gene expression has been reviewed by Azzi *et al.* (2004). Vitamin E also influences events at a posttranslational level (Zingg and Azzi, 2004). Thus, it has been shown to inhibit protein kinase C (Devaraj *et al.*, 1996), regulate lipid second messenger metabolism in cells (Beharka *et al.*, 1997; Tran *et al.*, 1994, 1996), and can directly regulate phospholipase A_2 (Grau and Ortiz, 1998) and phospholipase D (Yamamoto *et al.*, 1994) activity in model membranes. One indirect effect of vitamin E in the inhibition of protein kinase C is manifest in an inhibition of the respiratory burst of superoxide production in monocytes by NADPH-oxidase. The effect has been shown to be due to an interference in the assembly of the enzyme and specifically to an inhibition of phosphorylation and translocation of the cytosolic factor $p47^{phox}$ (Cachia *et al.*, 1998). Vitamin E has also been reported to inhibit cell proliferation (Schindler and Mentlein, 2006), platelet aggregation (Marsh and Coombes, 2006), monocyte adhesion (Murphy *et al.*, 2005), and other cellular functions (Singh *et al.*, 2005). These effects appear to be unrelated to the antioxidant activity of vitamin E, but rather are thought to result from specific interactions of vitamin E with components of the cell such as proteins, enzymes, and membranes.

One of the key factors in understanding how vitamin E fulfills its functions in membranes and lipoproteins is the way it associates with the lipid components of these structures and the effect its presence has on their structure and stability. Studies of model membrane systems containing vitamin E have proved to be an informative approach to address these questions. In this chapter, recent evidence relevant to the putative role of vitamin E in protecting membranes from free radical attack and the consequences of lipid oxidation in lipoproteins and membranes will be examined.

II. PHYSICAL PROPERTIES OF VITAMIN E

Vitamin E is composed of a range of isomeric compounds all of which are derivatives of tocol. The natural stereoisomer of α-tocopherol is (2*R*,4′*R*,8′*R*)-α-tocopherol or *RRR*-α-tocopherol. Vitamin E encompasses a group of eight isomeric molecules which are characterized by a chromanol ring structure and a side chain. The tocopherols possess a 4′,8′,12′-trimethyltridecyl phytol side chain, and the tocotrienols differ by the presence of double bonds at the 3′, 7′, and 11′ positions of the side chain (Kamal-Eldin and Appelqvist, 1996). The α-, β-, γ-, and δ-isomers of tocopherols and tocotrienols differ in the number and position of the methyl subsistents attached to the chromanol ring. Natural α-tocopherol was shown conclusively to have the 2*R*,4′*R*,8′*R* configuration (Mayer *et al.*, 1963).

The melting point of *RRR*-α-tocopherol is 2.5–3.5°C so that, at ambient temperature, the compound has the property of a viscous oil which is soluble in aprotic solvents. The optical rotations of these tocopherols are relatively small and are dependent on the nature of the solvent. The ultraviolet absorption spectra of the tocol and tocotrienol in ethanol show maxima in the range 292–298 nm; infrared absorption spectra show –OH (2.8–3.0 μm) and –CH– (3.4–3.5 μm) stretching and a characteristic band at 8.6 μm. Vitamin E is fluorescent with an emission maximum about 325 nm in hydrophobic environments. The spectral shift associated with changes in environment is a useful method for investigating the interaction of vitamin E with other molecules.

Vitamin E has amphipathic properties because the hydroxyl group attached to the chromanol ring system has affinity for a polar environment while the remainder of the molecule is hydrophobic in character. The overall amphipathic balance favors a hydrophobic affinity because the polarity of the hydroxyl group is weak relative to the remainder of the molecule. This is reflected in the relatively low solubility of vitamin E in water (Castle and Perkins, 1986). The shape of the molecule tends to be an elongated cylinder (Wang and Quinn, 1999a) with the hydroxyl group located at one end of the cylinder, and this serves to orient the molecule at polar–nonpolar interfaces.

III. INTERACTION OF VITAMIN E WITH LIPIDS IN MONOLAYERS

The surface activity of vitamin E has been demonstrated by the ability of the molecule to form a monomolecular film at an air–water interface. This is shown in Fig. 1A where a surface pressure–area/molecule relationship of

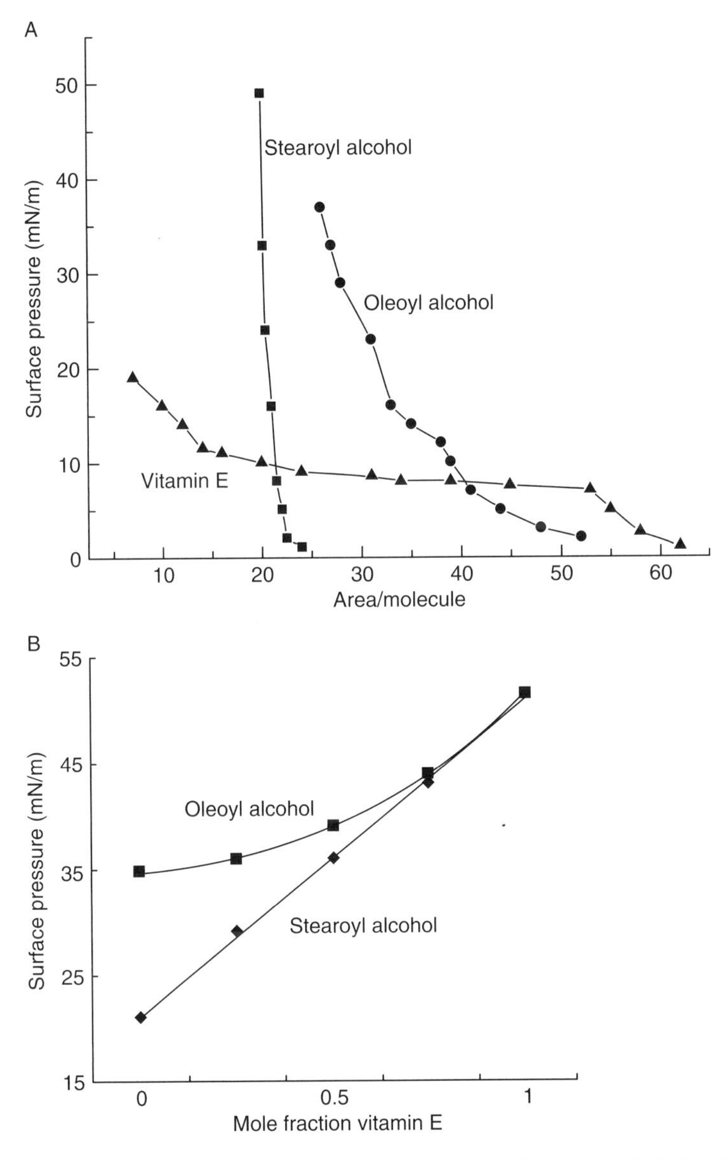

FIGURE 1. (A) Surface pressure–area isotherms of vitamin E, stearoyl alcohol, and oleoyl alcohol spread on aqueous subphases of 0.1-N NaOH, 0.1-M NaCl at 24°C. (B) Average surface areas per molecule at 17 mN/ m of mixed monolayers of vitamin E with stearoyl alcohol or oleoyl alcohol.

vitamin E at an air–water interface is presented. Vitamin E forms a monolayer which has the property of a gaseous film similar to that of oleoyl alcohol, also shown in the figure, but in contrast to stearoyl alcohol which forms condensed monolayers. The surface area of vitamin E is 0.52 nm^2/molecule at 17 mN/m on a subphase of 0.1-N NaOH, 0.1-M NaCl. The properties of vitamin E monolayers depend on the number of isopentane units in the side chain (Maggio *et al.*, 1977). Surface area–pressure isotherms have an increasingly liquid-expanded character such that they have a greater area/molecule for a given surface pressure as the number of isopentane units increases. It was noted that the largest effect was due to the presence of the first three isopentane units attached to the chromanol ring system and thereafter the change was relatively small up to nine isopentane units. Isomers containing four to nine isopentane units occupied similar areas (0.65 nm^2/molecule) at surface pressure of 5 mN/m.

The interaction of vitamin E in mixed monolayers at the air–water interface with other lipids to investigate the nature of the interactions with molecules in their biological environment has been reported. The analysis of the behavior of mixed monolayers is generally undertaken in terms of the additivity rule. Thus, miscibility of two components of high equilibrium spreading pressures in a mixed monolayer at an air–water interface can be established by additivity of the average surface area of each of the individual components. When the components are immiscible, the average surface area of a mixed monolayer, $A_{1,2}$, obeys the additivity rule:

$$A_{1,2} = N_1 A_1 + N_2 A_2$$

where N_1 and N_2 are the mole fractions of the components and A_1 and A_2 the surface area of each component in pure films maintained at specified surface pressure. When the components are miscible, the average surface area deviates from the surface area determined from the additivity equation. Although immiscible films must obey the additivity rule, they must also obey the phase rule which predicts that the collapse pressure of a binary mixture will be independent of the composition of the monolayer. Both these conditions were found to apply for mixed monolayers of vitamin E and stearoyl alcohol but not mixed monolayers of oleoyl alcohol and vitamin E (Patil and Cornwell, 1978). This is illustrated in Fig. 1B, which shows that mixed monolayers of stearoyl alcohol and vitamin E obey a straight-line relationship, whereas the relationship deviates from linearity for mixed monolayers of oleoyl alcohol and vitamin E. Immiscibility of mixed monolayers of vitamin E and stearoyl alcohol was further confirmed by the observation that surface pressure–area isotherms of mixed monolayers had an inflection point near the collapse pressure of pure vitamin E monolayers, and this was independent of the mole fraction of each component in the film. The properties of mixed monolayers of vitamin E and fatty acids have been reported by other researchers (Fukuzawa *et al.*, 1977).

Miscibility of vitamin E and other isomers of α-tocopherol differing in length of isopentane side chain with phosphatidylcholine monolayers at the air–water interface has been investigated (Maggio *et al.*, 1977). In general, vitamin E and its long-chain homologues were found to be immiscible with distearoylphosphatidylcholine reminiscent in the way that it behaves with condensed monolayers like stearoyl alcohol. By contrast, vitamin E was miscible with dioleoyl and arachidonyl molecular species of phosphatidylcholine not only by the criterion of the additivity rule but also due to the fact that an inflection point, presumably due to the vitamin E in the monolayer, progressively increased as the proportion of phospholipid in the monolayer increased.

Interaction of vitamin E with phospholipid monolayers has also been studied by measuring the penetration of vitamin E into phospholipid films formed at pressures that exceed the collapse pressure of vitamin E monolayers as judged by an increase in surface pressure when vitamin E is introduced into the subphase (Maggio *et al.*, 1977). Under these conditions, the extent of penetration of vitamin E homologues into dioleoylphosphatidylcholine monolayers increases as the length of the isopentane side chain increases until the length is similar to the length of the fatty acyl residues of the phospholipid; molecules with longer isopentane side chains penetrate less efficiently. It was noteworthy in this respect that the most effective penetration is observed with the natural homologue of vitamin E, indicating that van der Waals interactions between the two lipids is an important factor in stabilizing the monolayer. Vitamin E was also reported to penetrate more readily into monolayers of phospholipid with unsaturated fatty acid substituents than into monolayers of their saturated counterparts and the presence of relatively small proportions of unsaturated phospholipid in mixed monolayers with saturated molecular species of phospholipid greatly facilitated penetration of vitamin E.

IV. INTERACTION OF VITAMIN E WITH PHOSPHOLIPID BILAYER MEMBRANES

The partitioning of vitamin E into the hydrophobic domain of membranes has been inferred from the fluorescence emission properties of the molecule in environments of differing polarity. Vitamin E has a fluorescence emission maximum at 325 nm when excited at 300 nm and a fluorescence lifetime in the order of 2 ns (Bisby and Ahmed, 1989; Sow and Durocher, 1990). The effect of solvent environment can be seen from the fluorescence emission intensity data of the natural vitamin E with an isopentane side chain of 16 carbon atoms (C_{16}) compared with a shorter chain homologue (C_6) in nonaqueous and aqueous model membrane systems, which is presented in Table I. This shows that the fluorescence intensity is relatively high in the nonpolar solvent, cyclohexane, but is practically abolished by location in water.

TABLE I. Fluorescence Emission Intensity (Arbitary Units) of 20-mM Tocopherol Homologue at 325 nm[a]

System	Vitamin E (C_{16})	Vitamin E (C_6)
Cyclohexane	165	164
Water	23	6
0.04% Triton X-100	246	242
DMPC dispersion	52	183
DMPC + Triton X-100	248	242

[a]Data from Kagan and Quinn (1988).

This effect is unlikely to be the result of reduction of quantum yield in the polar environment but more probably is due to self-quenching of the fluorescence in insoluble aggregates of the molecule. This is consistent with the effect of detergent on fluorescence yield in which case the aggregates are solubilized resulting in a dramatic increase in fluorescence. The biologically interesting observation is the fluorescence intensity recorded from dispersions of dimyristoylphosphatidylcholine (DMPC) containing vitamin E. This is markedly less than in the presence of detergent suggesting that vitamin E does not distribute randomly in the phospholipid bilayer, but exists in domains enriched with vitamin E. There is apparently a dependence on the length of the prenyl chain on the ability to disperse in the bilayer as the tendency of the natural homologue to associate into domains appears to be greater than the shorter chain homologue.

Acylation of vitamin E with succinic acid generates an amphiphilic molecule which by itself can form multilamellar bilayers (Lai *et al.*, 1985). The addition of the succinyl moiety adds a free carboxyl group at the end of a four carbon chain which increases the overall length of the molecule and most likely locates the chromanol ring deeper into the hydrophobic domain of phospholipid bilayers. Acylation of α-tocopheryl with an acetate group reduces the amphilicity of the molecule by adding a nonpolar methyl group and reducing the hydrogen-bonding capacity of the molecule.

V. DISTRIBUTION AND ORIENTATION OF VITAMIN E IN PHOSPHOLIPID MEMBRANES

Three possible locations have been considered for the orientation of vitamin E in phospholipid bilayer membranes (Fukuzawa *et al.*, 1992). The first, based on paramagnetic resonance and other studies of vitamin E incorporated into phospholipid bilayer membranes, was a location in the region of the aqueous interface of the structure (Perly *et al.*, 1985; Srivastava *et al.*, 1983). The second, based on studies reported by Fragata and Bellemare (1980), which indicated a location of the chromanol ring of vitamin E 1 nm from the aqueous interface. Finally, fluorescence quenching of probes located in different domains of the bilayer places the chromanol nucleus at varying depths from the aqueous interface within the hydrophobic interior of the structure (Bisby and Ahmed, 1989).

A systematic study of the location of vitamin E and homologues with side chains of varying lengths using fluorescence-quenching and fluorescence energy transfer methods has been reported by Kagan and Quinn (1988). Because the excitation and fluorescence emission wavelengths of vitamin E are appropriate for fluorescence energy transfer with anthroyloxystearate (AS) probes, it was possible to obtain fairly precise distance measurements between the interacting species. The efficiency of fluorescence energy transfer between vitamin E and different *n*-AS probes differing in the location of the 9-anthroyloxy group in egg phosphatidylcholine bilayers was measured, and it was found that the order of efficiency followed the series 7-AS > 2-AS > 9-AS = 12-AS. This led to the conclusion that the chromanol nucleus of vitamin E is located in the bilayer in a region between the 9-anthroyloxy group attached to carbon 7 and carbon 2, and preferentially nearest to carbon 7. Water-soluble fluorescence-quenching agents were found to have low efficiency in quenching the fluorescence of vitamin E, suggesting that the chromanol nucleus does not protrude extensively into the lipid–water interface.

Fluorescence techniques that measure acyl-chain mobility (Massey, 1998), interfacial polarity (Hutterer *et al.*, 1996; Krasnowska *et al.*, 1998; Parasassi *et al.*, 1994) and surface charge (Fromherz, 1989) have been used to study the arrangement of vitamin E and its derivatives in phospholipid membranes (Massey, 2001). The location of the chromanol nucleus established on the basis of fluorescence energy transfer and quenching studies is supported by Fourier transform infrared spectroscopy measurements (Gomez-Fernandez *et al.*, 1989; Salgado *et al.*, 1993). These studies showed evidence of hydrogen bonding of the phenoxyl hydroxy group to either the carbonyl or the phosphate oxygen of the phospholipid molecules. The possibility of hydrogen bonding to water cannot be excluded at this stage but seems unlikely. Hydrogen bonding of vitamin E to the phospholipid is consistent with bilayer permeability experiments reported by Urano *et al.* (1990), which led them to conclude that the phenoxy group of vitamin E is hydrogen bonded to the carbonyl group of the ester bond of unsaturated fatty acids in the phospholipids arranged in a bilayer configuration. Dynamic spectroscopic studies of α-tocopherol and α-tocotrienol indicate that the molecules rotate about their long axis perpendicular to the plane of the membrane (Fukuzawa *et al.*, 1992).

VI. MOTION OF VITAMIN E IN LIPID ASSEMBLIES

The lateral diffusion of vitamin E in micelles and fluid phospholipid bilayers has been investigated by fluorescence-quenching methods (Gramlich *et al.*, 2004). A fluorescent probe, Fluorazophore-L, which has a similar size and orientation as vitamin E in lipid structures, is efficiently quenched by vitamin E and an analysis of the time-resolved fluorescence in surfactant micelles and multibilayer dispersions of palmitoyloleoylphosphatidylcholine was reported. The quenching efficiency decreased according to the microviscosity of the lipid assembly and was greatest in micelles of sodium dodecyl sulphate and least efficient in the fluid phospholipid bilayers. The results obtained from the phospholipid bilayer membranes containing vitamin E gave a mutual lateral diffusion coefficient of 18 pm^2/s at 27°C. This parameter provides an upper limit for diffusion of the individual molecules in the bilayer; and because the probe and vitamin E are of similar size and presumably have the same lateral diffusion coefficient, the lateral diffusion coefficient of vitamin E alone would be 9 pm^2/sec. This value suggests that vitamin E has a diffusion rate consistent with its size relative to the phospholipid which itself has a lateral diffusion coefficient in the range 4.2–4.7 pm^2/s (Bockmann *et al.*, 2003; Schutz *et al.*, 1997; Vaz *et al.*, 1985). Values of lateral diffusion coefficients of vitamin E in fluid phospholipid bilayers significantly greater than this (48 nm^2/sec) has been reported (Aranda *et al.*, 1989), but the underlying assumptions in the steady state fluorescence-quenching methods employed have been criticized (Rajarathnam *et al.*, 1989).

The diffusion rate of different isomers of vitamin E and related compounds in palmitoleoylphosphatidylcholine bilayers using Fluorazophore-L has also been measured (Sonnen *et al.*, 2005). It was found that the pattern of fluorescence quenching of a fluorescence probe in a solvent of acetonitrile–water was not influenced by the nature of the isoprentane substituent (tocopherol compared with tocotrienol), but was dependent on the methylation pattern of the chromanol ring. The order of quenching efficiency of the different isomers was found to be $\alpha > \beta = \gamma > \delta$ consistent with their reactivities toward peroxyl radicals as well as the phenolic O–H bond dissociation energies. The mutual lateral diffusion coefficient in palmitoleoylphosphatidylcholine liposomes was almost identical for different tocopherols and tocotrienols with a value of 8 pm^2/s similar to the values reported earlier (Gramlich *et al.*, 2004). The rate of lateral diffusion in phospholipid bilayers was greatly influenced by temperature with activation energies in the order of 44 ± 6 kJ/mol. Interestingly, the presence of up to 30 mol% of cholesterol in the fluid bilayers had little effect on the lateral diffusion coefficient of vitamin E.

The diffusion rate of vitamin E from one leaflet of a phospholipid bilayer to the other is significantly more constrained than is diffusion laterally in the plane of the bilayer (Tyurin *et al.*, 1986, 1988). In these studies, vitamin E was incorporated into single bilayer vesicles and fluorescence intensity measured on addition of an impermeant oxidizing agent (potassium ferricyanide) which abolishes the signal from molecules residing in the outer monolayer of the vesicle. It was found that about 54% of the vitamin E incorporated into dioleoylphosphatidylcholine vesicles in a ratio of 5 mol% of phospholipid was located in the outer monolayer, consistent with the volume distribution of molecules in small vesicles. The fluorescence emitted from the molecules located in the inner leaflet remained constant over a period of hours suggesting that migration across the bilayer was a relatively slow process. In other experiments, when vitamin E was added to pure phospholipid vesicles, it becomes incorporated into the outer monolayer where it remains for many hours without measurable migration to the inner leaflet. Similar results were obtained with vesicles composed of both saturated and unsaturated phospholipid molecular species indicating that the constraint on *trans*-bilayer flip-flop is not dependent on the nature of the fatty acyl chains of the phospholipid.

VII. EFFECT OF VITAMIN E ON PHOSPHOLIPID PHASE BEHAVIOR

Many studies have shown that vitamin E influences the phase behavior of phospholipids. Calorimetric studies (Massey *et al.*, 1982) have shown that the presence of vitamin E in aqueous dispersions of dimyristoylphosphatidylcholine decreased the temperature of the gel to liquid crystalline phase transition and reduced the enthalpy of the transition. Similar effects were noted for fully saturated derivatives of other phospholipid classes. It was noted that these effects were similar to those observed in phospholipid bilayers containing cholesterol, but in quantitative terms the effect of vitamin E appeared to be greater on a per mole basis than cholesterol. This was attributed to a greater flexibility of the isopentanoyl side chain compared to the rigid ring structure of the sterol nucleus. The hydroxyl group of the chromanol ring was also shown to be important in modulating the phase transition behavior of phospholipids (Lai *et al.*, 1985). Other studies with unsaturated phospholipids (Fukuzawa *et al.*, 1979), which indicated relatively minor effects of vitamin E even when present in proportions of 10 mol%, are in marked contrast with the effect on transitions from the gel phase in saturated molecular species of phospholipid where relatively large perturbations are observed.

Examination of the thermal properties of mixed aqueous dispersions of dipalmitoylphosphatidylcholine (DPPC) with vitamin E by differential scanning calorimetry showed that increasing proportions of vitamin E cause a progressive broadening of the gel to liquid crystalline phase transition (decrease in cooperativity of the transition) such that the onset temperature decreased but the temperature of completion of the transition was unaltered (De Kruyff *et al.*, 1974; Massey *et al.*, 1982). A report by Villalain *et al.* (1986)

that the temperature of completion of the transition decreases with increasing proportions of vitamin E has not been confirmed in subsequent studies (Quinn, 1995). Proportions of vitamin E greater than about 20 mol% resulted in almost complete loss of transition enthalpy. The effect of increasing proportions of vitamin E on the endothermic transitions in heating thermograms of aqueous dispersions of DPPC is shown in Fig. 2. It can be seen that the pretransition endotherm disappears in the presence of 2.5-mol% vitamin E and the main transition progressively broadens. There is also an inverse relationship between the proportion of vitamin E in the phospholipid and the transition enthalpy. Similar findings have been reported for the dimyristoyl derivative of phosphatidylcholine (McMurchie and McIntosh, 1986).

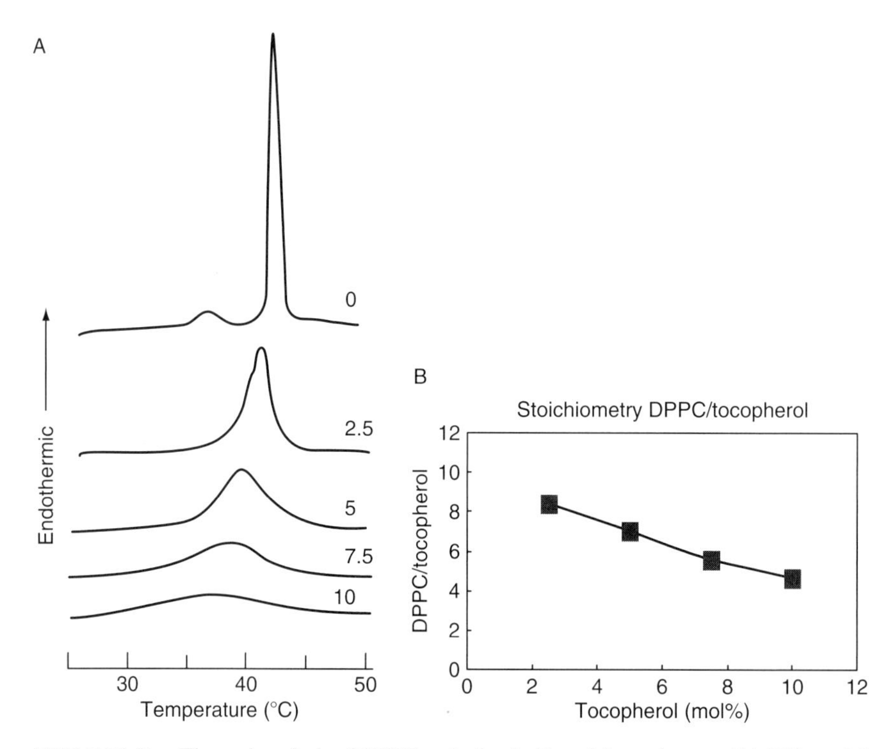

FIGURE 2. Thermal analysis of DPPC and vitamin E model membranes. (A) Differential scanning calorimetric heating curves recorded from aqueous dispersions of DPPC and vitamin E in mol% indicated on the thermogram. The samples contained approximately the same amount of phospholipid. No enthalpy changes were observed over this temperature range in the absence of lipid (data from reference Quinn, 1995). (B) The ratio of phospholipid molecules/vitamin E molecule removed from the phase transition calculated from the thermal data in (A), assuming that the transition enthalpy of the pure phospholipid is 35 kJ/mol, plotted as a function of the mol% vitamin E in the mixture. A linear regression of the form $y = -0.56x + 9.6$ is obtained from the plot. Numerical values are presented in Table II for comparison with the data obtained from X-ray measurements.

It has been consistently observed in the thermal studies of mixtures of vitamin E with saturated phosphatidycholines that the presence of vitamin E apparently eliminates the pretransition enthalpy. This does not mean that the pretransition itself is eliminated by the presence of vitamin E as it is possible that the enthalpies of the two transitions merge in the broad transition produced. Evidence for this comes from synchrotron X-ray diffraction studies of DPPC and vitamin E system which showed that a diffraction band at about 15 nm persisted in mixtures up to at least 5-mol% vitamin E, and signs of an uncoordinated ripple structure were observed at even higher proportions of vitamin E (Quinn, 1995). The same results were observed in systems of vitamin E with dilauroyl-, distearoyl-, and palmitoyl-oleoyl-phosphatidylcholines. This suggests that vitamin E reduces the cooperativity of the gel to liquid crystalline phase transition but does not eliminate the pretransition, that is the transition from gel to ripple conformation is unaffected, at least in those molecules undergoing the transition. A reduction in overall transition enthalpy indicates that an increasing proportion of the phospholipid molecules are removed from the transition.

The effect of vitamin E on phase behavior of heteroacid phosphatidylcholines is complex with evidence of domain formation. Sanchez-Migallon *et al.* (1996) examined the thermal behavior of vitamin E with 1,2-di-18-phosphatidylcholines having 18:0 fatty acids acylated in the *sn*-1 position and 18:0, 18:1, 18:3, or 20:4 fatty acids acylated at the *sn*-2 position of the glycerol. They constructed partial phase diagrams of the mixed acyl dispersions and concluded there was fluid phase immiscibility between the phospholipids and vitamin E. They concluded that lateral phase separations of domains containing different amounts of vitamin E were formed in the fluid phase. The magnitude of the effect of vitamin E on transition enthalpy was found to be dependent on the extent of unsaturation of the hydrocarbon chain located at the *sn*-2 position of the glycerol backbone of the phospholipid. It was suggested that vitamin E served to reduce the differences between gel and fluid states of the phospholipids which, in turn, is related to the molecular shape of the unsaturated phosphatidylcholines.

VIII. Effect of Vitamin E on the Structure of Phospholipid Model Membranes

Measurements of intrinsic fluorescence of vitamin E in bilayers of DPPC suggest that vitamin E is closely associated in particular domains of the phospholipid bilayer (Aranda *et al.*, 1989; Kagan and Quinn, 1988). Information about the forces which may operate between phospholipids and vitamin E has come from Fourier transform infrared spectroscopy of DPPC and vitamin E-mixed aqueous dispersions (Quinn, 1995). It was deduced from measurements of bandwidth and frequencies of the $-CH_2-$ antisymmetric and symmetric stretch vibrations of the phospholipid acyl chains that the presence of vitamin E causes a decrease in the average number of *gauche*$^{\pm}$ chain conformers at temperatures above the gel to liquid crystalline phase transition. Interestingly, the bandwidth and frequency of maximum absorbance do not appear to change in concert with changes in temperature, which were interpreted as evidence for the coexistence of two phases. This conclusion, however, assumes that temperature has identical effects on *trans-gauche*$^{\pm}$ isomerizations on the one hand, and lipid order and mobility on the other.

The effect of vitamin E on the vibrational modes of fatty acyl chains of DPPC in multilamellar dispersions using FTIR and Raman spectroscopy has been reported (Lefevre and Picquart, 1996). Temperature profiles of Raman intensity height ratios I-2860:I-2850 and I-2935:I-2880 indicated that increasing amounts of vitamin E broadens the main gel to liquid crystalline phase transition and decreases the onset temperature of the transition. The results indicated that the presence of vitamin E disrupts packing of the acyl chains of the phospholipid in gel phase resulting in an increase in *gauche*$^{\pm}$ chain rotamers as judged by the intensity of $-CH_2-$ wagging progression modes. The motional constraints on the terminal methyl groups of the acyl chains are also diminished by the presence of vitamin E. It was suggested that an observed perturbation of C=O stretching vibration modes by vitamin E may result from increased hydration of the interfacial region, hydrogen-bond formation between vitamin E and the phospholipid, and/or alterations in the molecular configuration at the lipid–water interface.

Evidence for complex formation between vitamin E and saturated phosphatidylcholines and the coexistence of these stoichiometric complexes with bilayers of pure phospholipid at temperatures below the main chain melting transition temperature have been obtained from synchrotron X-ray diffraction studies (Quinn, 1995). The results of such studies are presented in Fig. 3 which shows small-angle X-ray scattering intensities plotted as a function of reciprocal spacing ($S = 1/d = 2\sin(\theta)\lambda$, where d is the repeat spacing, θ is the diffraction angle, and λ is the X-ray wavelength) recorded from codispersions of DPPC with indicated amounts of vitamin E recorded at 47 °C, a temperature above the main chain melting temperature. The pure phospholipid shows a lamellar gel phase with a single *d*-spacing indexed by the intense first-order lamellar reflection on the left and the smaller second-order reflection at $2/S$. Multiple small-angle diffraction peaks were observed in mixed aqueous dispersions both above and below the gel to liquid crystalline phase transition temperature of the phospholipid when vitamin E was included in the dispersion. The relative intensity of scattering of peaks centered at about 6.4 and 6.7 nm in the gel and liquid crystal phases, respectively, decreases with increasing proportions of vitamin E. There is a corresponding increase in relative intensity of scattering of peaks centered at about 7.5 and 6.2 nm,

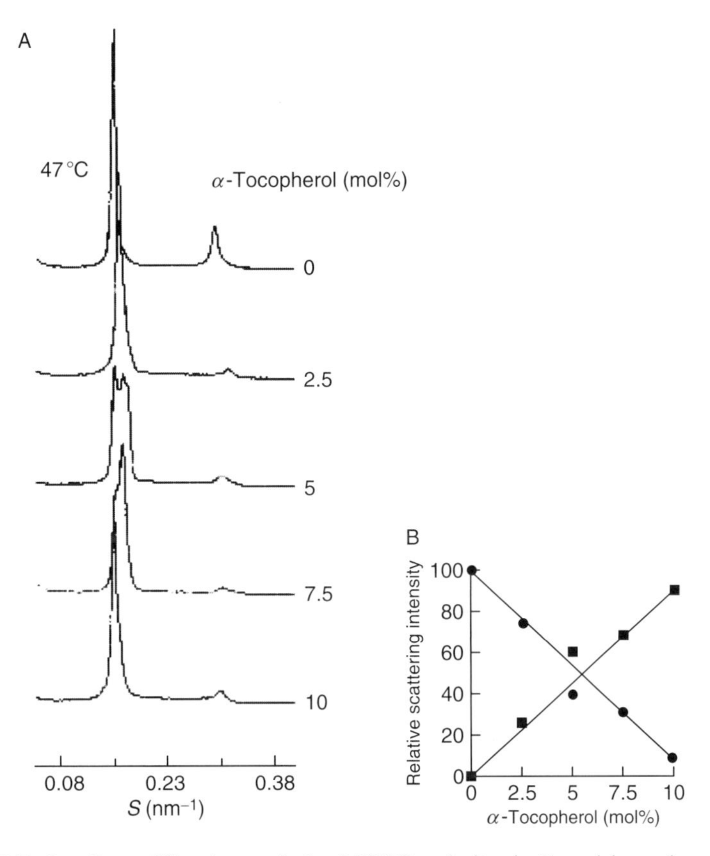

FIGURE 3. X-ray diffraction analysis of DPPC and vitamin E model membranes. (A) Small-angle X-ray scattering intensities versus reciprocal spacing (S) recorded at 47°C from mixed aqueous dispersions of DPPC with indicated amounts of vitamin E (data from reference Wang *et al.*, 2000). (B) Plots of the relative scattering intensities obtained from integrated areas of the peak in the diffraction pattern assigned to a pure phospholipid bilayer (■) and phospholipid bilayer enriched with α-tocopherol (●) versus the mol% of vitamin E in the mixture (data calculated from X-ray scattering intensity patterns recorded at 45°C; reference Quinn, 1995).

respectively, suggesting that these spacings correspond to domains enriched in vitamin E. An increase in relative intensity of a peak centered at 8.4 nm at higher proportions of vitamin E in DPPC in the gel phase may be due to formation of a unit cell containing a higher proportion of vitamin E than that giving rise to a *d*-spacing of 7.5 nm. This could be due either to formation of a new phase containing an increased proportion of vitamin E or to an arrangement of domains created with lower proportions of vitamin E so as to produce a longer *d*-spacing. Since there are only two low-angle peaks in the fluid phase, there is only one structural coordinate of the vitamin E-enriched domain of DPPC. The relative scattering intensity of these two peaks is shown in Fig. 3B.

Estimates of the stoichiometry of the components of the domains enriched with respect to vitamin E were obtained from a comparison of the respective changes in phase transition enthalpy and low-angle X-ray diffraction intensities due to the presence of vitamin E in the bilayer. Calorimetric data showed that the presence of vitamin E caused a loss in cooperativity in the gel to liquid crystal phase transition of DPPC and a decrease in transition enthalpy.

Because of the high degree of correlation between the change in excess specific heat capacity and disordering of the hydrocarbon chains, broadening of the transition with increasing vitamin E concentration indicates that the normal phase transition of the phospholipid is perturbed. It is not possible, on the basis of the present evidence, to conclude that the loss in enthalpy is due to a decrease in molar enthalpy of the phospholipid or removal of a proportion of the total phospholipid from undergoing a gel to liquid crystal phase transition. If it is assumed, however, that domains of pure DPPC contribute normally to transition enthalpy (35 kJ/mol) then the number of molecules of DPPC that do not contribute to enthalpy in mixtures with vitamin E can be calculated from the measured transition enthalpies. When the number of DPPC molecules that do not contribute to transition enthalpy expressed as a ratio of vitamin E molecules in the mixture is plotted as a function of mol/100-mol vitamin E in the mixture (Fig. 2), a linear regression of the form $y = -0.56x + 9.6$ is obtained which indicates that increasing proportions of vitamin E result in a decrease in efficiency in removing DPPC from an endothermic transition. These DSC data are shown in Table II. The intercept with the y-axis yields a stoichiometry of 9.6 DPPC/vitamin E. In order to account for the loss in cooperativity of the gel to liquid crystal phase transition, a large-scale phase separation of pure phospholipid and vitamin E-rich domains of phospholipid would appear to be excluded.

Calculations of stoichiometry of vitamin E and DPPC have also been performed using the ratio of integrated intensities of the first-order lamellar repeat X-ray scattering bands recorded in mixtures in the gel phase (25 °C)

TABLE II. Stoichiometry of Vitamin E/DPPC Complexes

Vitamin E and DPPC (mol%)	DSC	X-ray gel (25 °C)	Fluid (45 °C)
2.5	8.4	10.5	10.2
5.0	7.0	9.1	11.5
7.5	5.6	8.8	8.5
10.0	4.7	8.3	8.2
Consensus value[a]	9.6	9.2	9.6

[a]DSC enthalpy values from Fig. 2b; mean values from X-ray experiments on mixtures containing 2.5, 5.0, 7.5, and 10.0 mol% vitamin E in diplamitoylphosphatidylcholine.

and the fluid phase (50 °C). The results are remarkably similar to those from the thermal analysis. A summary of the enthalpy and X-ray scattering data in determination of the stoichiometry of phospholipid–vitamin E complexes is presented in Table II. The values obtained from integrated X-ray scattering intensities in the gel phase (9.2 $\pm$ 1.1) and fluid phase (9.6 $\pm$ 1.6) suggest that the forces responsible for cohesion of the vitamin E-rich domains of phospholipid are not dependent on the phase state of the bilayer in which they are formed.

The formation of domains enriched in vitamin E in bilayers of egg phosphatidylcholine has been detected by electrochemical impedance spectroscopy (Naumowicz and Figaszewski, 2005). Evidence for formation of complexes with a stoichiometry of about 10:1 phospholipid:vitamin E was obtained. Vitamin E apparently had a condensing effect on the phospholipid as there was a reduction of mean molecular area at the lipid–water interface but the effect was not as pronounced as that observed with cholesterol.

One of the manifestations of phase separation of vitamin E-enriched domains in bilayers of phosphatidylcholine is the creation of ripple or ribbon structures seen using freeze-fracture electron microscopy (Wang and Quinn, 2002a; Wang *et al.*, 2000). This is illustrated in Fig. 4 which shows electron micrographs of freeze-fracture replicas prepared from codispersions of distearoylphosphatidylcholine containing 5- and 10-mol% vitamin E thermally quenched from 25 °C, well below the pretransition temperature of the phospholipid, after heating from low temperature. Ripple structures can be seen which have unusually low periodicities (16 nm) which contrast with a periodicity of 25–30 nm in the pure phospholipid in P_{β} phase observed only in the temperature range 50–54 °C. It can be seen from Fig. 4A that the 16 nm symmetrically rippled areas in the dispersion containing 5-mol% vitamin E are periodically tilted forming another ripple with a large periodicity of $\sim$50–150 nm running perpendicular to the small ripples. In addition to areas with straight ripples, bilayer areas were seen which show a fine, wormlike morphology (Fig. 4B, asterisk), which appear to represent disturbed ripple structure. Such features are also seen in codispersions of vitamin E with dimyristoylphosphatidylcholine (Wang *et al.*, 2000) suggesting that beside low-temperature stabilization of a symmetric ripple phase, a further effect of vitamin E is to disturb the linear order of the symmetric ripples.

Phase separation of vitamin E-rich domains has also been observed in mixtures of vitamin E and phosphatidylethanolamines (Wang and Quinn, 1999b, 2002b; Wang *et al.*, 1999). Figure 5 shows a sequence of small- and wide-angle X-ray diffraction patterns recorded from an aqueous dispersion of dipalmitoylphosphatidylethanolamine containing 5-mol% vitamin E during a heating scan from 39 to 77 °C. A vitamin E-enriched lamellar phase can be seen emerging as a broad diffraction band on the higher angle side of the sharp lamellar spacing originating from the pure phospholipid from which

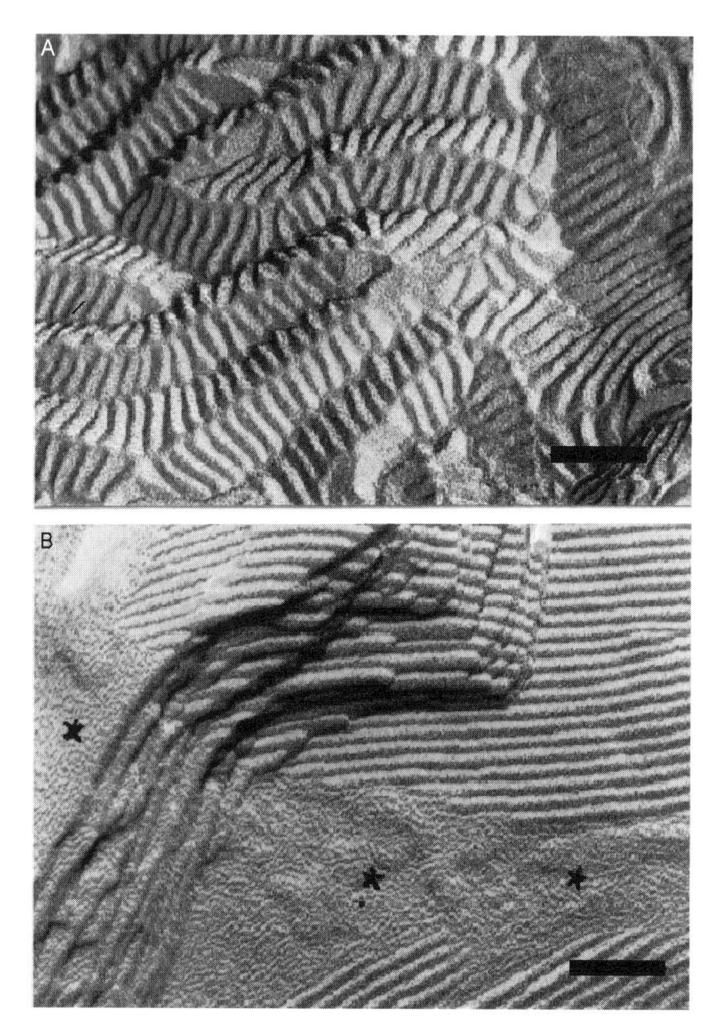

FIGURE 4. Electron micrographs of freeze-fracture replicas prepared from aqueous dispersions of distearoylphosphatidylcholine containing (A) 5-mol% and (B) 10-mol% vitamin E thermally quenched from 25°C. Scale bar = 100 nm (data from reference Wang *et al.*, 2000).

vitamin E had been excluded. An hexagonal-II phase, characterized by a set of diffraction orders in a ratio $1{:}1/\sqrt{3}{:}1/2$, appears at a temperature of about 57°C which arises from the vitamin E-rich domain and is distinct from the bilayer phase of the pure phospholipid. This suggests that vitamin E does not form a stable lamellar phase with the phosphatidylethanolamine but tends to induce hexagonal phase structure.

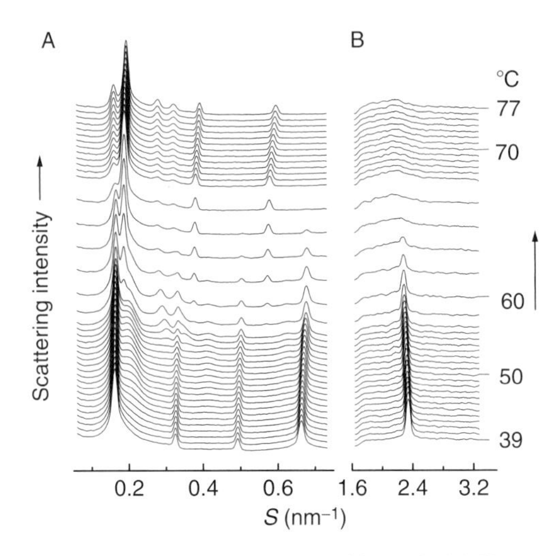

FIGURE 5. A sequence of small-angle (A) and wide-angle (B) X-ray scattering intensities versus reciprocal spacing (*S*) recorded from a codispersion of dipalmitoylphosphatidylethanolamine with 5-mol% vitamin E during a heating scan at 3°C/min (data from reference Wang and Quinn, 1999b).

IX. PHASE SEPARATION OF VITAMIN E IN PHOSPHOLIPID MIXTURES

The effect of vitamin E on the phase behavior of mixed chain diacylphosphatidylcholines has been examined to determine whether any preferential interactions take place. Ortiz *et al.* (1987) have studied the effect of vitamin E on the equimolar mixtures of dipalmitoyl-/distearoylphosphatidylcholine and dimyristoyl-/distearoylphosphatidylcholine. The equimolar dipalmitoyl-/distearoylphosphatidylcholine mixture only showed a single endotherm, which was modified by the presence of vitamin E in a similar manner to the individual phosphatidylcholines. The equimolar dimyristoyl-/distearoylphosphatidylcholine mixture, however, showed monotectic behavior. With increasing proportions of vitamin E, the temperature range of highest temperature endotherm was broadened and shifted toward lower temperatures, while the lowest temperature endotherm was weakened and disappeared in the mixture containing 20-mol% vitamin E. This was interpreted such that vitamin E preferentially affects the lower melting component which corresponds to the dimyristoyl molecular species. Incrcasing proportions of vitamin E in an equimolar mixture of 18:0/18:1 and 18:0/22:6 phosphatidylcholine also result in a two component transition enthalpy (Stillwell *et al.*, 1996). The temperature range of the lower temperature transition is progressively broadened as the vitamin E content of the mixture increases but the higher temperature component is virtually unaffected by the presence of up to 10-mol% vitamin E. This was interpreted as a phase separation of vitamin E within the mixed phospholipid dispersion with a preference for an association with the more unsaturated molecular species of phospholipid.

The inverted hexagonal phase induced by vitamin E in phosphatidylethanolamines can be used as a marker for (phosphatidylethanolamine + vitamin E) domain in phosphatidylcholine/phosphatidylethanolamine mixtures (Wang and Quinn, 2000a,b). Synchrotron X-ray diffraction studies of mixtures of vitamin E with dioleoylphosphatidylcholine and dioleoylphosphatidylethanolamine showed that the effect of vitamin E on unsaturated phospholipids follows a similar pattern to that observed for their saturated counterparts. However, in the X-ray diffraction study of mixed aqueous dispersions of vitamin E with equimolar mixtures of dioleoylphosphatidylethanolamine/dioleoylphosphatidylcholine and dioleoylphosphatidylethanolamine/dimyristoylphosphatidylcholine, it was found that lamellar gel and liquid crystalline phases dominated the phase structure. This is consistent with the fact that vitamin E does not preferentially interact with phosphatidylethanolamine. The changes in the SAXS patterns are similar to those of mixtures of vitamin E and phosphatidylcholines or of equimolar mixtures of phosphatidylethanolamine/phosphatidylcholine (Wang and Quinn, 2000a). This again suggests that vitamin E preferentially interacts with the phosphatidylcholine in the mixture or distribute randomly in domains containing both phosphatidylethanolamine and phosphatidylcholine regardless the saturation and the length of hydrocarbon chains of the two phospholipids.

The effect of vitamin E on equimolar mixtures of saturated phosphatidylethanolamine and phosphatidylcholine has been investigated using differential scanning calorimetry (Ortiz *et al.*, 1987) and X-ray diffraction methods (Wang and Quinn, 2000b). The calorimetric results were interpreted such that vitamin E preferentially partitioned in the most fluid phase irrespective of whether this was phosphatidylethanolamine or phosphatidylcholine. However, the conclusions drawn from the X-ray data appear to be different (Wang and Quinn, 2000b). The change in the wide-angle X-ray scattering intensity of the sharp peak at 0.43 nm has been plotted as a function of temperature in Fig. 6A for equimolar mixtures of DMPC and DPPE containing up to 20-mol% vitamin E. The curves show maxima corresponding approximately to the midpoint of the phase transition of the dimyristoylphosphatidylcholine (Peak 1) and dipalmitoylphosphatidylethanolamine (Peak 2) components of the mixture. The positions of the peaks do not correspond to identical transition temperatures in mixtures containing vitamin E because of the precise conditions that underlie the complicated processes associated with phase separation in a ternary mixture. The ratio in height, Peak 2:Peak 1, is found to increase with increasing vitamin E in the mixture. This is consistent with a preferential partition of vitamin E into dimyristoylphosphatidylcholine so as to reduce the contribution of change in

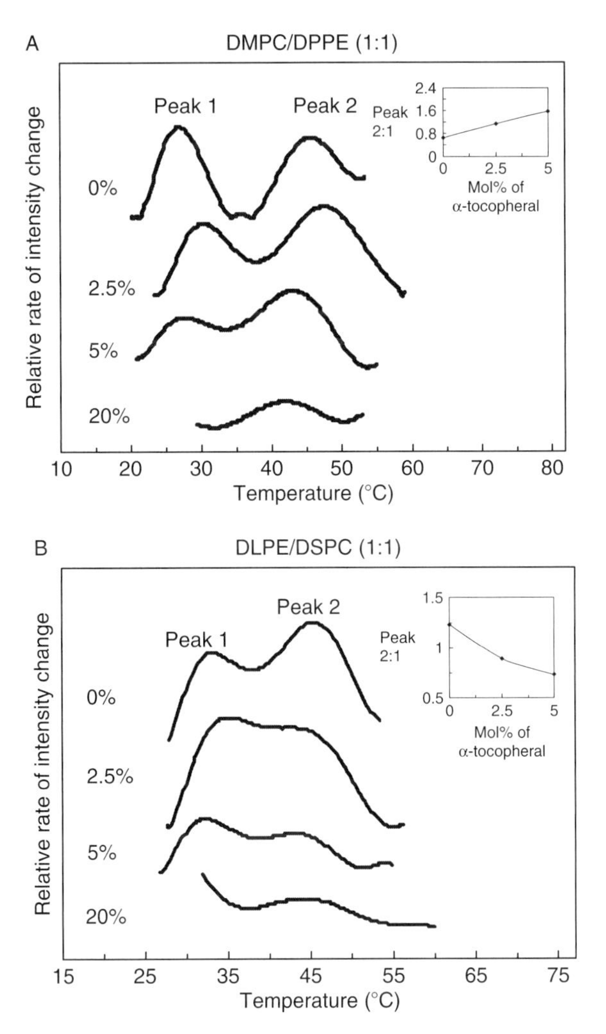

FIGURE 6. Rate of change of chain packing as judged by the wide-angle scattering intensity peaks from aqueous dispersions of equimolar mixtures of (A) dimyristoylphosphatidylcholine/ dipalmitoylphosphatidylethanolamine and (B) dilauroylphosphatidylethanolamine/distearoyl-phosphatidylcholine containing different mol% vitamin E in the phospholipid plotted as a function of temperature. The insets show plots of the relative heights of Peak 2:Peak 1 as a function of the proportion of vitamin E in the mixture (data from reference Wang and Quinn, 2000b). The scattering intensities of the wide-angle peaks originating from the acyl chain packing arrangement were normalized (I) and the relative rate of change of scattering intensity, d_I/d_t, obtained from the data recorded from heating scans at 2°C/min.

the intensity of the wide-angle reflection due to dimyristoylphosphatidylcholine relative to dipalmitoylphosphatidylethanolamine. A similar analysis of the wide-angle X-ray scattering intensity data of mixtures of dilauorylphosphatidylethanolamine and distearoylphosphatidylcholine containing different proportions of vitamin E has been undertaken and the results are presented in Fig. 6B. The inset to the figure shows the relationship between the relative heights of Peaks 2 and 1 and the mol% vitamin E in the mixture. This indicates that as the proportion of vitamin E in the mixture increases, the contribution to the change in scattering intensity of the wide-angle X-ray scattering peak from distearoylphosphatidylcholine decreases relative to that from dilauorylphosphatidylethanolamine. Such data can be interpreted as a preferential partitioning of vitamin E into the high melting point phospholipid component of the mixture, which is again phosphatidylcholine. This is consistent with the tendency of vitamin E to form complexes with phosphatidylcholines in both the fluid and gel phases and the exclusion of vitamin E from bilayer phases of phosphatidylethanolamines.

X. EFFECT OF VITAMIN E ON MEMBRANE PERMEABILITY

The interaction of vitamin E with phospholipids results in decreased permeability of bilayer membranes to small solute molecules. The effect of vitamin E on proton permeability of large unilamellar vesicles (diameter >0.1 μm) composed of unsaturated molecular species of phosphatidylcholine has been investigated using pH-sensitive fluorescence probes (Stillwell *et al.*, 1996). It was found that the presence of vitamin E in amounts up to 30 mol% resulted in a proportionate decease in proton permeability across the phospholipid bilayer, and the effect was not dependent on the degree of unsaturation of the fatty acyl chains. By contrast, the decrease in proton permeability caused by the presence of cholesterol was only observed if the phospholipid was relatively saturated and a condensing effect was manifest.

Somewhat different effects on proton permeability were reported in large unilamellar vesicles formed from digalactosydiacylglyceride extracted from oat seeds but the effect of vitamin E on the permeability to glucose was greater than that of cholesterol (Berglund *et al.*, 1999). Thus, the permeability coefficient of glucose across bilayers of the galactolipid decreased from $P = 7$ pm/s to $P = 5$ pm/s due to the presence of 5 mol% of vitamin E in the membrane. The decrease in glucose permeability was significantly greater than that observed when the galactolipid bilayers contained 10-mol% cholesterol ($P = 6$ pm/s). The permeability of the galactolipid bilayers to protons, however, was reduced from $P = 1$ nm/s to $P = 0.35$ nm/s in the presence of 10-mol% cholesterol but to only $P = 0.9$ nm/s when 5-mol% vitamin E was included in the bilayer. These results indicate that vitamin E interacts with polar lipids in a way that reduces the permeability of membranes to polar solutes and ions but the molecular mechanism may differ from that of sterols like cholesterol.

XI. EFFECT OF VITAMIN E ON MEMBRANE STABILITY

The interaction of vitamin E with other membrane constituents is said to influence the stability of membranes but how this is achieved is still conjectural. There is some evidence to indicate that the formation of complexes with certain membrane components tends to stabilize bilayer structures, while other evidence suggests that tocopherol destabilizes membranes and promotes fusion.

Evidence that supplementation of membranes with vitamin E results in increased stability of membranes has been reported from studies of model membrane systems and from subcellular organelles. Mouse liver microsomal membranes, for example, have been shown to be stabilized by supplementation with vitamin E (Fukuzawa *et al.*, 1977). This is shown in Table III in which the rate of leakage of acid phosphatase from suspensions of lysosomes supplemented with different vitamin E homologues was measured. It can be seen that a concentration-dependent stabilization of the lysosomes results from the presence of vitamin E, and the natural homologue with three isopentane units is the most effective. The effect is much less pronounced with homologues with shorter side chains. Evidence has been reported of the formation of complexes between lysophosphatidylcholine and vitamin E, which was said to explain the protective effect of vitamin E against the action of products of phospholipase A hydrolysis on lysosomal membranes (Mukherjee *et al.*, 1997). It has also been reported that tamoxifen-induced haemolysis of human erythrocytes (Silva *et al.*, 2000), retinol-induced hemolysis of rabbit erythrocytes (Urano *et al.*, 1992), and hemin-induced hemolysis of sheep erythrocytes (Wang *et al.*, 2006) are all inhibited by vitamin E in a

TABLE III. Stabilization of Mouse Liver Lysosomes by Vitamin E and Related Homologues[a]

	Concentration (μM)		
Isopentane units	10	50	100
3 (vitamin E)	14	26	31
2	9	17	25
0	6	14	–

[a]Values are percentage stabilization calculated as the decrease in relative rate of release of acid phosphatase. Data from Fukuzawa *et al.* (1977).

manner that is not related to an antioxidant action of the vitamin. Moreover, it was demonstrated that both the chromanol ring structure and the isopentane side chain were important in stabilizing the membrane.

One of the ways in which vitamin E is believed to stabilize membranes is to form complexes with membrane lipid components that have a tendency to destabilize the bilayer structure, thereby countering their effects and rendering the membrane more stable. Kagan (1989) has used paramagnetic resonance spectroscopy to demonstrate the formation of complexes between free fatty acids and vitamin E in phospholipid bilayer membranes. The incorporation of a fatty acid spin-probe into bilayers of palmitoyloleoylphosphatidylcholine was found to take place when the phospholipid was in a liquid crystalline phase. At temperatures below the gel to liquid crystalline phase transition temperature, the spin-probe could become incorporated into the bilayer when 15-mol% arachidonic acid was added alone but not with 5-mol% vitamin E. This was interpreted as a formation of a complex between the arachidonic acid and vitamin E which did not perturb the phase behavior of the phospholipid to the same extent as the free fatty acid. Similar conclusions were drawn from [^{1}H] NMR studies of DPPC–20-mol% linoleic acid mixed dispersions with and without 5-mol% vitamin E.

The mode of interaction of unsaturated fatty acids with vitamin E has been investigated by Urano *et al.* (1993) using fluorescence and NMR methods. They showed that vitamin E decreased the fluidity of phospholipid liposomes that were perturbed by the presence of free fatty acids with more than one double bond. Examination of spin-lattice relaxation modes of the methyl carbons led to the conclusion that the three methyl groups attached to the aromatic ring rather than the isoprenoid side chain have the strongest affinity for unsaturated lipids.

The rationale for the stabilizing effect of vitamin E is that by formation of a molecular complex with molecules such as free fatty acids and lysophospholipids, the amphipathic balance of the complex is restored to that which approximates a bilayer-forming phospholipid. This is illustrated in Fig. 7 which shows space-filling molecular models of phosphatidylcholine and compares this with complexes of vitamin E and lysophosphatidylcholine and palmitic acid. This again emphasizes the hydrophobic character of vitamin E.

In other experiments, the presence of vitamin E in membranes leads to destabilization of the membrane. The electrical properties of black lipid films have been used to examine the effect of vitamin E on membrane stability. It was found that incorporation of vitamin E into lipid bilayers resulted in destabilization of the films and facilitated the electrogeneration of pores in the membrane (Koronkiewicz *et al.*, 2001). With respect to membrane fusion, there are conflicting reports on the effects of vitamin E. An early observation by Ahkong *et al.* (1972) showed that when α-tocopherol was added to a suspension of avian erythrocytes, the cells were induced to fuse, presumably because vitamin E destabilized the membrane lipid bilayer. Similar studies with chromaffin

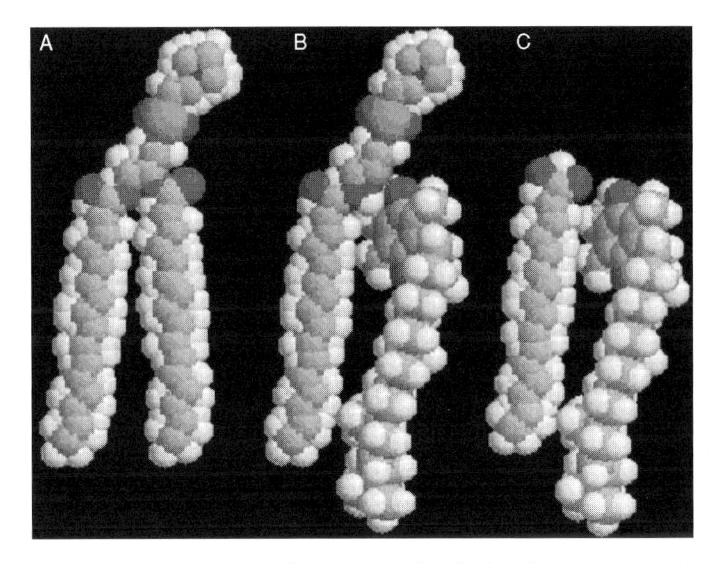

FIGURE 7. Space-filling models of (A) phosphatidylcholine, (B) lysophosphatidylcholine–vitamin E complex, and (C) palmitic acid–vitamin E complex. (See Color Insert.)

granules aggregated with synexin, however, showed that the addition of vitamin E did not induce the granules to fuse (Creutz, 1981). Moreover, in other experiments vitamin E was found to inhibit aggregation of platelets (Anastasi and Steiner, 1976; Steiner and Mower, 1982).

A series of studies in which the stability of model phospholipid bilayer membranes was determined in the presence of different amounts of vitamin E also supported the notion that vitamin E stabilizes lipid bilayers and prevents fusion. For example, the aggregation rate of phosphatidylcholine vesicles containing free oleic acid is decreased (Ortiz and Gomez-Fernandez, 1988), and Ca^{2+}-induced fusion of large unilamellar vesicles of phosphatidylserine (Aranda *et al.*, 1996) is inhibited by vitamin E. One explanation for the inhibitory effect of vitamin E on Ca^{2+}-induced fusion is that vitamin E reduces the binding of Ca^{2+} to phosphatidylserine vesicles (Sanchez-Migallon *et al.*, 1996).

XII. DOMAINS ENRICHED IN VITAMIN E IN MEMBRANES

There is emerging evidence that vitamin E is not randomly distributed throughout the lipid bilayer matrix of biological membranes but instead forms complexes with specific membrane constituents. The suggestion that vitamin E is preferentially located at membrane sites most susceptible to oxidation is not based on direct evidence. The formation of complexes with membrane-destabilizing agents is more convincing, suggesting that although present as a relatively minor membrane component it may exert a significant structural role via a localized effect where the vitamin becomes enriched. A preferential association with phosphatidycholines as opposed to phosphatidylethanolamines may be a consequence of the fact that vitamin E forms a stable bilayer complex with phosphatidylcholine. Studies of vitamin E and phosphatidylethanolamine mixtures indicate that vitamin E is largely excluded from bilayer phases of phosphatidylethanolamine but phase separates into domains of inverted hexagonal phase of the phospholipid which would clearly have a destabilizing effect on the membrane lipid bilayer matrix. Failure to form stable complexes with phosphatidylethanolamine may be explained on the same basis as that used to explain why vitamin E readily forms complexes with free fatty acids and lysophospholipids, namely, the amphipathic balance within the complex is conducive to bilayer formation. The predominantly hydrophobic character of phosphatidylethanolamine is matched by that of vitamin E and combination of the two can only exist in nonbilayer phases.

There is clearly compelling evidence that, although present in relatively minor proportions in biological membranes, vitamin E segregates in membranes and forms complexes with specific lipid constituents. These reactions have the effect of stabilizing the lipid bilayer matrix. The wider significance of these preferential interactions, especially in regard to the putative antioxidant function of vitamin E, remains to be established.

XIII. EFFECT OF VITAMIN E ON MEMBRANE PROTEIN FUNCTION

The activity of a variety of membrane enzymes is known to be modulated by vitamin E without apparent binding of the vitamin directly to the enzyme (Azzi and Stocker, 2000; Ricciarelli *et al.*, 1998; Tasinato *et al.*, 1995). It has been suggested that the effects are exerted indirectly by influencing the lipid milieu surrounding the enzymes. A number of such enzyme activities have been shown to be influenced, for example, by the curvature of the lipid bilayer in which they are embedded (Cornell, 1991; Davies *et al.*, 2001; Drobnies *et al.*, 2002; Hubner *et al.*, 1998). To address these questions, measurements of the effect of vitamin E on the spontaneous curvature and bending modulus of phospholipids have been undertaken (Bradford *et al.*, 2003). In these studies, the lattice dimensions and properties of hydrated dioleoylphosphatidylethanolamine containing vitamin E were determined by X-ray diffraction and osmotic stress methods. It was found that lattice spacing of the hexagonal-II phase of the phospholipid decreases with increasing proportion of vitamin E indicating that the curvature of the mixed monolayers is increased. Calculation of the spontaneous curvature of the lipid monolayers due to the presence of vitamin E gave a value of $R_{op} = -1.37$ nm. This negative curvature is comparable with those observed with diacylglycerols in phospholipid membranes and considerably more negative than the more closely structurally related cholesterol. Osmotic stress measurements performed on vitamin E/dioleoylphosphatidylethanolamine mixtures were used to calculate values for bending modulus and the results showed that vitamin E does not change the bending elasticity of the phospholipid monolayers. The overall conclusion from these studies is that the pronounced effect on membrane curvature imposed by the presence of vitamin E is a plausible mechanism for modulating the activity of membrane-bound enzymes.

XIV. CONCLUSIONS

Vitamin E as a fat-soluble vitamin partitions into hydrophobic lipid assemblies that form structures like membranes and lipoproteins. The vitamin has weak surface-active properties that serve to orient the molecule at the interface between the aqueous phase and the hydrophobic domain of these structures. The vitamin is not randomly distributed within membrane structures and there is evidence that complexes are formed with products of phospholipid hydrolysis that prevents the products from destabilizing bilayer structures. There is also evidence that vitamin E undergoes preferred interactions with particular phospholipid classes irrespective of their acyl chain composition. Thus, stoichiometric complexes are formed with choline phosphatides but stable bilayer structures cannot form with mixtures of vitamin E and phosphatidylethanolamines. The presence of vitamin-enriched domains in phosphatidylcholine bilayers may induce localized changes in membrane properties that could act to modulate the activity of membrane-bound enzymes. The detailed molecular mechanisms responsible for these effects have yet to be fully characterized.

REFERENCES

Ahkong, Q. F., Lucy, J. A., Cramp, F. C., Tampion, W., Fisher, D., and Howell, J. I. (1972). Fusion of hen erythrocytes by nonionic surface-active adjuvants. *Biochem. J.* **130,** 44–45P.

Anastasi, J., and Steiner, M. (1976). Effect of alpha-tocopherol on human platelet-aggregation. *Clin. Res.* **24,** A498–A498.

Aranda, F. J., Coutinho, A., Berberan-Santos, M. N., Prieto, M. J. E., and Gomez-Fernandez, J. C. (1989). Fluorescence study of the location and dynamics of α-tocopherol in phospholipid vesicles. *Biochim. Biophys. Acta* **985,** 26–32.

Aranda, F. J., Sanchez-Migallon, M. P., and Gomez-Fernandez, J. C. (1996). Influence of alpha-tocopherol incorporation on Ca2+-induced fusion of phosphatidylserine vesicles. *Arch. Biochem. Biophys.* **333,** 394–400.

Aratri, E., Spycher, S. E., Breyer, I., and Azzi, A. (1999). Modulation of alpha-tropomyosin expression by alpha-tocopherol in rat vascular smooth muscle cells. *FEBS Lett.* **447,** 91–94.

Azzi, A., and Stocker, A. (2000). Vitamin E: Non-antioxidant roles. *Prog. Lipid Res.* **39,** 231–255.

Azzi, A., Gysin, R., Kempna, P., Munteanu, A., Villacorta, L., Visarius, T., and Zingg, J. M. (2004). Regulation of gene expression by alpha-tocopherol. *Biol. Chem.* **385,** 585–591.

Beharka, A. A., Wu, D., Han, S. N., and Meydani, S. N. (1997). Macrophage prostaglandin production contributes to the age-associated decrease in T cell function which is reversed by the dietary antioxidant vitamin E. *Mech. Ageing Dev.* **93,** 59–77.

Behrens, W. A., and Madere, R. (1990). Kinetics of tissue RRR-alpha-tocopherol depletion and repletion. Effect of cold exposure. *J. Nutr. Biochem.* **1,** 528–532.

Berglund, A. H., Nilsson, R., and Liljenberg, C. (1999). Permeability of large unilamellar digalactosyldiacylglycerol vesicles for protons and glucose—influence of alpha-tocopherol, beta-carotene, zeaxanthin and cholesterol. *Plant Physiol. Biochem.* **37,** 179–186.

Bisby, R. H., and Ahmed, S. (1989). Transverse distribution of alpha-tocopherol in bilayer membranes studied by fluorescence quenching. *Free Radic. Biol. Med.* **6,** 231–239.

Bockmann, R. A., Hac, A., Heimburg, T., and Grubmuller, H. (2003). Effect of sodium chloride on a lipid bilayer. *Biophys. J.* **85,** 1647–1655.

Bradford, A., Atkinson, J., Fuller, N., and Rand, R. P. (2003). The effect of vitamin E on the structure of membrane lipid assemblies. *J. Lipid Res.* **44,** 1940–1945.

Cachia, O., El Benna, J., Pedruzzi, E., Descomps, B., Gougerot-Pocidalo, M. A., and Leger, C. L. (1998). Alpha-tocopherol inhibits the respiratory burst in human monocytes—attenuation of p47(phox) membrane translocation and phosphorylation. *J. Biol. Chem.* **273,** 32801–32805.

Castle, L., and Perkins, M. J. (1986). Inhibition kinetics of chain-breaking phenolic antioxidants in SDS micelles. Evidence that intermicellar diffusion rates may be rate-limiting for hydrophobic inhibitors such as alpha-tocopherol. *J. Am. Chem. Soc.* **108,** 6381–6382.

Chojkier, M., Houglum, K., Lee, K. S., and Buck, M. (1998). Long- and short-term D-alpha-tocopherol supplementation inhibits liver collagen alpha1(I) gene expression. *Am. J. Physiol.* **275,** G1480–G1485.

Cornell, R. B. (1991). Regulation of Ctp-phosphocholine cytidylyltransferase by lipids. 2. Surface curvature, acyl chain-length, and lipid-phase dependence for activation. *Biochemistry* **30,** 5881–5888.

Creutz, C. E. (1981). Cis-unsaturated fatty-acids induce the fusion of chromaffin granules aggregated by synexin. *J. Cell Biol.* **91,** 247–256.

Davies, S. M. A., Epand, R. M., Kraayenhof, R., and Cornell, R. B. (2001). Regulation of CTP: Phosphocholine cytidylyltransferase activity by the physical properties of lipid membranes: An important role for stored curvature strain energy. *Biochemistry* **40,** 10522–10531.

De Kruyff, B., Van Dijck, P. W., Demel, R. A., Schuijff, A., Brants, F., and Van Deene, L. L. (1974). Nonrandom distribution of cholesterol in phosphatidylcholine bilayers. *Biochim. Biophys. Acta* **356,** 1–7.

Devaraj, S., Li, D., and Jialal, I. (1996). The effects of alpha tocopherol supplementation on monocyte function. Decreased lipid oxidation, interleukin 1 beta secretion, and monocyte adhesion to endothelium. *J. Clin. Invest.* **98,** 756–763.

Drobnies, A. E., Davies, S. M. A., Kraayenhof, R., Epand, R. F., Epand, R. M., and Cornell, R. B. (2002). CTP: Phosphocholine cytidylyltransferase and protein kinase C recognize different physical features of membranes: Differential responses to an oxidized phosphatidylcholine. *Biochim. Biophys. Acta Biomembr.* **1564,** 82–90.

Fechner, H., Schlame, M., Guthmann, F., Stevens, P. A., and Rustow, B. (1998). Alpha- and delta-tocopherol induce expression of hepatic alpha-tocopherol-transfer-protein mRNA. *Biochem. J.* **331**(Pt. 2), 577–581.

Fragata, M., and Bellemare, F. (1980). Model of singlet oxygen scavenging by alpha-tocopherol in biomembranes. *Chem. Phys. Lipids* **27,** 93–99.

Fromherz, P. (1989). Lipid coumarin dye as a probe of interfacial electrical potential in biomembranes. *Methods Enzymol.* **171,** 376–387.

Fukuzawa, K., Hayashi, K., and Suzuki, A. (1977). Effects of alpha-tocopherol analogs on lysosome membranes and fatty-acid monolayers. *Chem. Phys. Lipids* **18,** 39–48.

Fukuzawa, K., Ikeno, H., Tokumura, A., and Tsukatani, H. (1979). Effect of alpha-tocopherol incorporation on glucose permeability and phase-transition of lecithin liposomes. *Chem. Phys. Lipids* **23,** 13–22.

Fukuzawa, K., Ikebata, W., Shibata, A., Kumadaki, I., Sakanaka, T., and Urano, S. (1992). Location and dynamics of alpha-tocopherol in model phospholipid membranes with different charges. *Chem. Phys. Lipids* **63,** 69–75.

Galinier, A., Carriere, A., Fernandez, Y., Carpene, C., Andre, M., Caspar-Bauguil, S., Thouvenot, J. P., Periquet, B., Penicaud, L., and Casteilla, L. (2006). Adipose tissue proadipogenic redox changes in obesity. *J. Biol. Chem.* **281,** 12682–12687.

Gomez-Fernandez, J. C., Villalain, J., Aranda, F. J., Ortiz, A., Micol, V., Coutinho, A., Berberan-Santos, M. N., and Prieto, M. J. (1989). Localization of alpha-tocopherol in membranes. *Ann. NY Acad. Sci.* **570,** 109–120.

Gramlich, G., Zhang, J., and Nau, W. M. (2004). Diffusion of alpha-tocopherol in membrane models: Probing the kinetics of vitamin E antioxidant action by fluorescence in real time. *J. Am. Chem. Soc.* **126,** 5482–5492.

Grau, A., and Ortiz, A. (1998). Dissimilar protection of tocopherol isomers against membrane hydrolysis by phospholipase A2. *Chem. Phys. Lipids* **91,** 109–118.

Hubner, S., Couvillon, A. D., Kas, J. A., Bankaitis, V. A., Vegners, R., Carpenter, C. L., and Janmey, P. A. (1998). Enhancement of phosphoinositide 3-kinase (PI 3-kinase) activity by membrane curvature and inositol-phospholipid-binding peptides. *Eur. J. Biochem.* **258,** 846–853.

Hutterer, R., Schneider, F. W., Sprinz, H., and Hof, M. (1996). Binding and relaxation behaviour of prodan and patman in phospholipid vesicles: A fluorescence and 1H NMR study. *Biophys. Chem.* **61,** 151–160.

Kagan, V. E. (1989). Tocopherol stabilizes membrane against phospholipase A, free fatty acids, and lysophospholipids. *Ann. NY Acad. Sci.* **570,** 121–135.

Kagan, V. E., and Quinn, P. J. (1988). The interaction of alpha-tocopherol and homologues with shorter hydrocarbon chains with phospholipid bilayer dispersions. A fluorescence probe study. *Eur. J. Biochem.* **171,** 661–667.

Kamal-Eldin, A., and Appelqvist, L. A. (1996). The chemistry and antioxidant properties of tocopherols and tocotrienols. *Lipids* **31,** 671–701.

Koronkiewicz, S., Kalinowski, S., and Bryl, K. (2001). Changes of structural and dynamic properties of model lipid membranes induced by alpha-tocopherol: Implication to the membrane stabilization under external electric field. *Biochim. Biophys. Acta Biomembr.* **1510,** 300–306.

Krasnowska, E. K., Gratton, E., and Parasassi, T. (1998). Prodan as a membrane surface fluorescence probe: Partitioning between water and phospholipid phases. *Biophys. J.* **74,** 1984–1993.

Lai, M. Z., Duzgunes, N., and Szoka, F. C. (1985). Effects of replacement of the hydroxyl group of cholesterol and tocopherol on the thermotropic behavior of phospholipid membranes. *Biochemistry* **24,** 1646–1653.

Lefevre, T., and Picquart, M. (1996). Vitamin E-phospholipid interactions in model multilayer membranes: A spectroscopic study. *Biospectroscopy* **2,** 391–403.

Lu, L., Hackett, S. F., Mincey, A., Lai, H., and Campochiaro, P. A. (2006). Effects of different types of oxidative stress in RPE cells. *J. Cell. Physiol.* **206,** 119–125.

Maggio, B., Diplock, A. T., and Lucy, J. A. (1977). Interactions of tocopherols and ubiquinones with monolayers of phospholipids. *Biochem. J.* **161,** 111–121.

Marsh, S. A., and Coombes, J. S. (2006). Vitamin E and alpha-lipoic acid supplementation increase bleeding tendency via an intrinsic coagulation pathway. *Clin. Appl. Thromb. Hemost.* **12,** 169–173.

Massey, J. B. (1998). Effect of cholesteryl hemisuccinate on the interfacial properties of phosphatidylcholine bilayers. *Biochim. Biophys. Acta* **1415,** 193–204.

Massey, J. B. (2001). Interfacial properties of phosphatidylcholine bilayers containing vitamin E derivatives. *Chem. Phys. Lipids* **109,** 157–174.

Massey, J. B., She, H. S., and Pownall, H. J. (1982). Interaction of Vitamin-E with saturated phospholipid-bilayers. *Biochem. Biophys. Res. Commun.* **106,** 842–847.

Mayer, H., Schudel, P., Isler, O., and Ruegg, R. (1963). Die absolute konfiguration des naturlichen alpha-tocopherols. 4. *Helv. Chim. Acta* **46,** 963–969.

McMurchie, E. J., and McIntosh, G. H. (1986). Thermotropic interaction of Vitamin-E with dimyristoyl and dipalmitoyl phosphatidylcholine liposomes. *J. Nutr. Sci. Vitaminol.* **32,** 551–558.

Mukherjee, A. K., Ghosal, S. K., and Maity, C. R. (1997). Lysosomal membrane stabilization by alpha-tocopherol against the damaging action of Vipera russelli venom phospholipase A(2). *Cell. Mol. Life Sci.* **53,** 152–155.

Murphy, N., Grimsditch, D. C., Vidgeon-Hart, M., Groot, P. H., Overend, P., Benson, G. M., and Graham, A. (2005). Dietary antioxidants decrease serum soluble adhesion molecule (sVCAM-1, sICAM-1) but not chemokine (JE/MCP-1, KC) concentrations, and reduce atherosclerosis in C57BL but not apoE*3 Leiden mice fed an atherogenic diet. *Dis. Markers* **21,** 181–190.

Naumowicz, M., and Figaszewski, Z. A. (2005). Impedance analysis of phosphatidylcholine/alpha-tocopherol system in bilayer lipid membranes. *J. Membr. Biol.* **205,** 29–36.

Ortiz, A., and Gomez-Fernandez, J. C. (1988). Calcium-induced aggregation of phosphatidylcholine vesicles containing free oleic-acid. *Chem. Phys. Lipids* **46,** 259–266.

Ortiz, A., Aranda, F. J., and Gomez-Fernandez, J. C. (1987). A differential scanning calorimetry study of the interaction of alpha-tocopherol with mixtures of phospholipids. *Biochim. Biophys. Acta* **898,** 214–222.

Ozer, N. K., Negis, Y., Aytan, N., Villacorta, L., Ricciarelli, R., Zingg, J. M., and Azzi, A. (2006). Vitamin E inhibits CD36 scavenger receptor expression in hypercholesterolemic rabbits. *Atherosclerosis* **184,** 15–20.

Parasassi, T., Di Stefano, M., Loiero, M., Ravagnan, G., and Gratton, E. (1994). Cholesterol modifies water concentration and dynamics in phospholipid bilayers: A fluorescence study using Laurdan probe. *Biophys. J.* **66,** 763–768.

Patil, G. S., and Cornwell, D. G. (1978). Interfacial oxidation of alpha-tocopherol and the surface properties of its oxidation products. *J. Lipid Res.* **19,** 416–422.

Perly, B., Smith, I. C., Hughes, L., Burton, G. W., and Ingold, K. U. (1985). Estimation of the location of natural alpha-tocopherol in lipid bilayers by 13C-NMR spectroscopy. *Biochim. Biophys. Acta* **819,** 131–135.

Quinn, P. J. (1995). Characterization of clusters of alpha-tocopherol in gel and fluid phases of dipalmitoylglycerophosphocholine. *Eur. J. Biochem.* **233,** 916–925.

Rajarathnam, K., Hochman, J., Schindler, M., and Ferguson-Miller, S. (1989). Synthesis, location, and lateral mobility of fluorescently labeled ubiquinone 10 in mitochondrial and artificial membranes. *Biochemistry* **28,** 3168–3176.

Ricciarelli, R., Tasinato, A., Clement, S., Ozer, N. K., Boscoboinik, D., and Azzi, A. (1998). Alpha-tocopherol specifically inactivates cellular protein kinase C alpha by changing its phosphorylation state. *Biochem. J.* **334,** 243–249.

Salgado, J., Villalain, J., and Gomez-Fernandez, J. C. (1993). Magic-angle-spinning C-13-Nmr spin-lattice relaxation study of the location and effects of alpha-tocopherol, ubiquinone-10 and ubiquinol-10 in unsonicated model membranes. *Eur. Biophys. J.* **22,** 151–155.

Sanchez-Migallon, M., Aranda, F. J., and Gomez-Fernandez, J. C. (1996). Interaction between alpha-tocopherol and heteroacid phosphatidylcholines with different amounts of unsaturation. *Biochim. Biophys. Acta Biomembr.* **1279,** 251–258.

Schindler, R., and Mentlein, R. (2006). Flavonoids and vitamin E reduce the release of the angiogenic peptide vascular endothelial growth factor from human tumor cells. *J. Nutr.* **136,** 1477–1482.

Schutz, G. J., Schindler, H., and Schmidt, T. (1997). Single-molecule microscopy on model membranes reveals anomalous diffusion. *Biophys. J.* **73,** 1073–1080.

Silva, M. M. C., Madeira, V. M. C., Almeida, L. M., and Custodio, J. B. A. (2000). Hemolysis of human erythrocytes induced by tamoxifen is related to disruption of membrane structure. *Biochim. Biophys. Acta Biomembr.* **1464,** 49–61.

Singh, U., Devaraj, S., and Jialal, I. (2005). Vitamin E, oxidative stress, and inflammation. *Annu. Rev. Nutr.* **25,** 151–174.

Sonnen, A. F., Bakirci, H., Netscher, T., and Nau, W. M. (2005). Effect of temperature, cholesterol content, and antioxidant structure on the mobility of vitamin E constituents in biomembrane models studied by laterally diffusion-controlled fluorescence quenching. *J. Am. Chem. Soc.* **127,** 15575–15584.

Sow, M., and Durocher, G. (1990). Spectroscopic and photophysical properties of some biological antioxidants: Structural and solvent effects. *J. Photochem. Photobiol. A Chem.* **54,** 349–365.

Srivastava, S., Phadke, R. S., Govil, G., and Rao, C. N. R. (1983). Fluidity, permeability and antioxidant behaviour of model membranes incorporated with alpha-tocopherol and vitamin-E acetate. *Biochim. Biophys. Acta* **734,** 353–362.

Steiner, M., and Mower, R. (1982). Mechanism of action of vitamin-E on platelet-function. *Ann. NY Acad. Sci.* **393,** 289–299.

Stillwell, W. (2000). Docosahexaenoic acid and membrane lipid domains. *Curr. Org. Chem.* **4,** 1169–1183.

Stillwell, W., Ehringer, W., and Wassall, S. R. (1992). Interaction of alpha-tocopherol with fatty-acids in membranes and ethanol. *Biochim. Biophys. Acta* **1105,** 237–244.

Stillwell, W., Dallman, T., Dumaual, A. C., Crump, F. T., and Jenski, L. J. (1996). Cholesterol versus alpha-tocopherol: Effects on properties of bilayers made from heteroacid phosphatidylcholines. *Biochemistry* **35,** 13353–13362.

Tasinato, A., Boscoboinik, D., Bartoli, G. M., Maroni, P., and Azzi, A. (1995). D-Alpha-tocopherol inhibition of vascular smooth muscle cell proliferation occurs at physiological concentrations, correlates with protein kinase C inhibition, and is independent of its antioxidant properties. *Proc. Natl. Acad. Sci. USA* **92,** 12190–12194.

Traber, M. G., and Kayden, H. J. (1987). Tocopherol distribution and intracellular localization in human adipose tissue. *Am. J. Clin. Nutr.* **46,** 488–495.

Tran, K., Proulx, P. R., and Chan, A. C. (1994). Vitamin E suppresses diacylglycerol (DAG) level in thrombin-stimulated endothelial cells through an increase of DAG kinase activity. *Biochim. Biophys. Acta* **1212,** 193–202.

Tran, K., Wong, J. T., Lee, E., Chan, A. C., and Choy, P. C. (1996). Vitamin E potentiates arachidonate release and phospholipase A2 activity in rat heart myoblastic cells. *Biochem. J.* **319**(Pt. 2), 385–391.

Tyurin, V. A., Kagan, V. E., Serbinova, E. A., Gorbunov, N. V., Erin, A. N., Prilipko, L. L., and Stoichev, T. S. (1986). Interaction of alpha-tocopherol with phospholipid liposomes—absence of transbilayer mobility. *Bull. Exp. Biol. Med.* **102,** 1677–1680.

Tyurin, V. A., Kagan, V. E., Avrova, N. F., and Prozorovskaya, M. P. (1988). Asymmetry of lipids and alpha-tocopherol distribution in the outer and inner monolayer of bilayer lipid-membranes. *Bull. Exp. Biol. Med.* **105,** 790–793.

Urano, S., Kitahara, M., Kato, Y., Hasegawa, Y., and Matsuo, M. (1990). Membrane stabilizing effect of vitamin-E—existence of a hydrogen-bond between alpha-tocopherol and phospholipids in bilayer liposomes. *J. Nutr. Sci. Vitaminol.* **36,** 513–519.

Urano, S., Inomori, Y., Sugawara, T., Kato, Y., Kitahara, M., Hasegawa, Y., Matsuo, M., and Mukai, K. (1992). Vitamin-E—inhibition of retinol-induced hemolysis and membrane-stabilizing behavior. *J. Biol. Chem.* **267,** 18365–18370.

Urano, S., Matsuo, M., Sakanaka, T., Uemura, I., Koyama, M., Kumadaki, I., and Fukuzawa, K. (1993). Mobility and molecular-orientation of vitamin-E in liposomal membranes as determined by F-19 Nmr and fluorescence polarization techniques. *Arch. Biochem. Biophys.* **303,** 10–14.

Vaz, W. L., Clegg, R. M., and Hallmann, D. (1985). Translational diffusion of lipids in liquid crystalline phase phosphatidylcholine multibilayers. A comparison of experiment with theory. *Biochemistry* **24,** 781–786.

Viana, M., Barbas, C., Castro, M., Herrera, E., and Bonet, B. (1999). Alpha-tocopherol concentration in fetal and maternal tissues of pregnant rats supplemented with alpha-tocopherol. *Ann. Nutr. Metab.* **43,** 107–112.

Villalain, J., Aranda, F. J., and Gomez-Fernandez, J. C. (1986). Calorimetric and infrared spectroscopic studies of the interaction of alpha-tocopherol and alpha-tocopheryl acetate with phospholipid-vesicles. *Eur. J. Biochem.* **158,** 141–147.

Wang, X., and Quinn, P. J. (1999a). Vitamin E and its function in membranes. *Prog. Lipid Res.* **38,** 309–336.

Wang, X., and Quinn, P. J. (1999b). The effect of alpha-tocopherol on the thermotropic phase behaviour of dipalmitoylphosphatidylethanolamine. A synchrotron X-ray diffraction study. *Eur. J. Biochem.* **264,** 1–8.

Wang, X. Y., and Quinn, P. J. (2000a). The distribution of alpha-tocopherol in mixed aqueous dispersions of phosphatidylcholine and phosphatidylethanolamine. *Biochim. Biophys. Acta Biomembr.* **1509,** 361–372.

Wang, X. Y., and Quinn, P. J. (2000b). Preferential interaction of alpha-tocopherol with phosphatidylcholines in mixed aqueous dispersions of phosphatidylcholine and phosphatidylethanolamine. *Eur. J. Biochem.* **267,** 6362–6368.

Wang, X. Y., and Quinn, P. J. (2002a). The interaction of alpha-tocopherol with bilayers of 1-palmitoyl-2-oleoyl-phosphatidylcholine. *Biochim. Biophys. Acta Biomembr.* **1567,** 6–12.

Wang, X. Y., and Quinn, P. J. (2002b). Phase separations of alpha-tocopherol in aqueous dispersions of distearoylphosphatidylethanolamine. *Chem. Phys. Lipids* **114,** 1–9.

Wang, X. Y., Takahashi, H., Hatta, I., and Quinn, P. J. (1999). An X-ray diffraction study of the effect of alpha-tocopherol on the structure and phase behaviour of bilayers of dimyristoylphosphatidylethanolamine. *Biochim. Biophys. Acta Biomembr.* **1418,** 335–343.

Wang, X. Y., Semmler, K., Richter, W., and Quinn, P. J. (2000). Ripple phases induced by alpha-tocopherol in saturated diacylphosphatidylcholines. *Arch. Biochem. Biophys.* **377,** 304–314.

Wang, F., Wang, T. H., Lai, J. H., Li, M., and Zou, C. G. (2006). Vitamin E inhibits hemolysis induced by hemin as a membrane stabilizer. *Biochem. Pharmacol.* **71,** 799–805.

Yamamoto, M. I., Mazumi, T., Asai, T., Handa, T., and Miyajima, K. (1994). Effects of α-tocopherol and its acetate on the hydrolytic activity of phospholipase D in egg-yolk phosphatidylcholine bilayers. *Colloid Polym. Sci.* **272,** 598–603.

Zhang, Y., Turunen, M., and Appelkvist, E. L. (1996). Restricted uptake of dietary coenzyme Q is in contrast to the unrestricted uptake of alpha-tocopherol into rat organs and cells. *J. Nutr.* **126,** 2089–2097.

Zingg, J. M., and Azzi, A. (2004). Non-antioxidant activities of vitamin E. *Curr. Med. Chem.* **11,** 1113–1133.

5

Studies in Vitamin E: Biochemistry and Molecular Biology of Tocopherol Quinones

David G. Cornwell and Jiyan Ma

Department of Molecular and Cellular Biochemistry
The Ohio State University College of Medicine, Columbus, Ohio 43210

Vitamins and Hormones, Volume 76

0083-6729/07 $35.00
DOI: 10.1016/S0083-6729(07)76005-3

Tocopherols and tocotrienols, parent congeners in the vitamin E family, function as phenolic antioxidants. However, there has been little interest in their quinone electrophiles formed as a consequence of oxidation reactions, even though unique biological properties were suggested by early studies conducted immediately after the discovery of vitamin E. Oxidation of tocopherols and tocotrienols produces *para*- and *ortho*-quinones, and quinone methides, while oxidation of their carboxyethyl hydroxychroman derivatives produces quinone lactones. These quinone electrophiles are grouped in two subclasses, the nonarylating fully methylated α-family and the arylating desmethyl β-, γ-, and δ-family. Arylating quinone electrophiles form Michael adducts with thiol nucleophiles, provided by cysteinyl proteins or peptides, which can be identified and quantified by tetramethylammonium hydroxide thermochemolysis. They have striking biological properties which differ significantly from their nonarylating congeners. They are highly cytotoxic, inducing characteristic apoptotic changes in cultured cells. Cytotoxicity is intimately associated with the induction of endoplasmic reticulum stress and a consequent unfolded protein response involving the pancreatic ER kinase (PERK) signaling pathway that commits overstressed cells to apoptosis. The step-function difference between arylating and nonarylating tocopherol quinones is conceivably the basis for distinct biological properties of parent tocopherols, including the epigenetic modification of a histone thiol, the ceramide pathway, natriuresis, and the activity of COX-2, NF-κB, PPARγ, and cyclin. The role of α-tocopherol in the origin and evolution of the western hominin diet, the so-called "Mediterranean" diet, and the prominence of α-tocopherol in colostrum, mother's milk, and infant nutrition are considered. Finally, the discordance introduced into the diet by arylating tocopherol quinone precursors through the wide use of vegetable oils in deep-frying is recognized.

I. INTRODUCTION

The multifaceted biological properties of vitamins are overwhelming. The discovery of their specific and often unanticipated effects needed the developing matrix of systems biology which was not available to the basic scientist and clinician during the early years of vitamin research. For example, "vitamin E," which was first called factor X and is now recognized as a family of tocopherols and tocotrienols, was first described as an accessory food factor that prevented fetal death and resorption in the laboratory rat (Evans and Bishop, 1922). It was studied intensely by nutritionists, physiologists, and chemists in this "Golden Age" of vitamin research, and by the late 1930s its fundamental biological and chemical properties seemed to have been explained. The biological properties of tocopherol quinones, the subject of this chapter, are hidden in this early history, a period which is the subject of several excellent reviews (Evans, 1962; Mason, 1977; Mattill, 1939; Smith, 1940). We were guided to the original literature by Mattill, who is most often remembered for identifying "vitamin E" as an antioxidant (Wolf, 2005).

Refined diets deficient in "vitamin E," as measured by a rat fertility test assay, were used initially in studies on biological effects and these studies naturally focused on sterility. "Vitamin E" is widely distributed in nutrients and deficient diets were difficult to prepare. The problem was apparently solved by a procedure in which "vitamin E" in natural foods was destroyed by oxidation when foods were treated with ferric chloride in ether (Waddell and Steenbock, 1928). This "vitamin E-deficient" diet had biological effects that were, unexpectedly, very different from highly refined deficient diets. The ferric chloride-treated diet was associated with high toxicity in rats (Taylor and Nelson, 1930), lymphoblastoma in chicks (Adamstone, 1936), and "nutritional muscular dystrophy" in guinea pigs and rabbits (Goettsch and Pappenheimer, 1931). Most importantly, these effects depended on the composition of the diet. For example, a dramatic toxic effect was observed when the diet was based on oxidized wheat germ oil, while a lesser effect was observed when the diet was based on oxidized cod liver oil. This observation suggested to the original investigators (Taylor and Nelson, 1930) that different biological effects of the "vitamin E deficiency" diets could not be related solely to the absence of "vitamin E," and ferric chloride caused oxidation might convert "vitamin E" or some as yet unknown agent in the diet to a cytotoxic derivative. Shortly after the effects of a "vitamin E" deficiency through oxidative degradation were first reported, other investigators found that "vitamin E," extracted from wheat germ with ether and then concentrated by evaporation, produced transplantable sarcomas in several rat strains but not in mice or guinea pigs (Rowntree *et al.*, 1937). Oxidation induced by wet ether extraction was implied in this work since neither cold-pressed nor hydrocarbon-extracted wheat germ oil produced sarcomas. The study was immediately repeated by several well-known nutritionists who could not reproduce the observation in experiments that did not replicate exactly the original procedure (Carruthers, 1938; Day *et al.*, 1938; Evans and Emerson, 1939). The question of "vitamin E" oxidation to a carcinogenic derivative was not pursued in part because the chemical structure and properties of "vitamin E" had not been established at this time.

An intense period of chemical studies followed and established the structures and properties of congeners within the "vitamin E family." A number of investigators began to associate the term tocopherol with specific chemical compounds. Three crystalline allophanates, α-, β-, and γ-tocopherols (α-, β-, and γ-T), were isolated from a vitamin E concentrate of wheat germ oil (Evans *et al.*, 1936), an observation that began a long series of studies which showed wide variations in the tocopherol content of foods such as the presence of α-T alone in cod liver oil (Sheppard *et al.*, 1993). A chroman structure was proposed for α-T (Fernholz, 1938), which was subsequently synthesized from trimethylhydroquinone and phytyl bromide by a zinc-catalyzed condensation reaction (Karrer *et al.*, 1938a). The β- and γ-T (Fig. 1) were each shown to contain one less methyl group attached to the chroman ring (Emerson, 1938). The oxidation of tocopherols produced substituted 1,4-benzoquinones that were demonstrated with ferric chloride and other oxidizing agents (John *et al.*, 1938; Karrer *et al.*, 1938b). The conclusions reached in the early literature are sustained in many more recent studies. Thus, even in 1938, sufficient data were available to reexamine and perhaps explain the striking differences in cytotoxicity found between ferric chloride-treated wheat germ and cod liver oils, but the subject was not explored at that time.

We hypothesized that, as an early study had suggested (Taylor and Nelson, 1930), the cytotoxicity of oxidized oils was due not merely to the destruction of "vitamin E" but to the synthesis of cytotoxic oxidation products. Cod liver oil contains only α-T and will yield only α-tocopherol quinone (α-TQ) when it is treated with the oxidizing agent ferric chloride. In contrast, wheat germ oil contains α-, β-, γ-, and δ-T (Sheppard *et al.*, 1993) and will yield α-, β-, γ-, and δ-TQ, when it is oxidized. Therefore, the

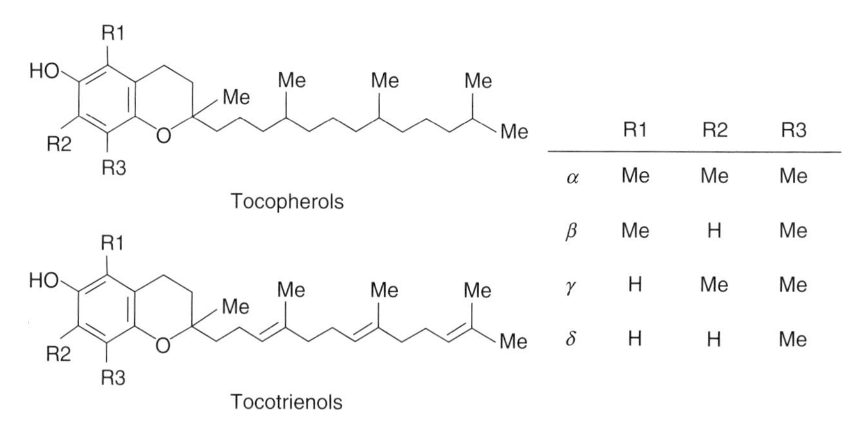

	R1	R2	R3
α	Me	Me	Me
β	Me	H	Me
γ	H	Me	Me
δ	H	H	Me

FIGURE 1. Structures of natural compounds within the "vitamin E" family. Stereoisomers which differ in biological activity are not designated.

differences between refined diets where "vitamin E" is eliminated and oxidized diets where "vitamin E" is destroyed might be explained by the oxidation of tocopherols to their quinone electrophiles. Furthermore, the differences between cod liver and wheat germ oil diets might be explained by the formation of tocopherol quinone electrophiles with very different biological properties. We tested this hypothesis by comparing the effects of purified α-TQ with purified γ-TQ, and found a step-function difference in cytotoxicity between the two quinones (Lindsey *et al.*, 1985). This step-function difference is the subject of our chapter.

II. REDOX CYCLING AND ARYLATING PROPERTIES OF TOCOPHEROL QUINONES

A. REDOX CYCLING

Quinones are redox-cycling agents that generate reactive oxygen species (ROS) and cause oxidative damage to cells (Burton and Ingold, 1986; Liebler, 1993; Niki, 1996; van Acker *et al.*, 1993). However, the biologist Evans and his coworkers (Emerson *et al.*, 1939) proposed that both phenolic precursors and quinone products within the tocopherol family were associated with antioxidant activity, an observation rejected by the organic chemists Karrer and Geiger (1940) but later confirmed in other studies (Gallo-Torres, 1980; Green and McHale, 1965). An early suggestion that α-TQ is converted to a semiquinone (Michaelis and Wollman, 1949) explains, in part, how quinones function as antioxidants. We suggest that quinones are reduced to semiquinones by aqueous superoxide since α-TQ functions as an antioxidant in a model system containing a polyunsaturated fatty acid, aqueous cumene hydroperoxide, and a ferric ion catalyst, but not in a nonaqueous model system that uses azoisobutyronitrile to form cumene hydroperoxide from cumene (Gavino, 1981; Gavino *et al.*, 1981). Such antioxidant activity of tocopherols and tocopherol quinones may account for the early results (Juhasz-Schaffer, 1931; Mason, 1933) that "vitamin E" enhanced cell proliferation, a possible explanation for fetal resorption with a refined diet deficiency. We found that both α-T and α-TQ decreased the population doubling time for cells in culture and these data supported the concept that α-TQ has antioxidant activity (Gavino *et al.*, 1982). Similar antioxidant activity is also associated with other tocopherol quinones. We compared the antioxidant activity of γ and δ TQ by testing their abilities to prevent lipid peroxidation (Lindsey *et al.*, 1985; Thornton *et al.*, 1995). Using the thiobarbituric acid test, we first compared α-T and α-TQ with γ-T and γ-TQ by measuring their effects on polyunsaturated fatty acid oxidation caused by aqueous cumene hydroperoxide (Gavino, 1981; Gavino *et al.*, 1981). All phenols and their quinones were effective antioxidants in this nonbiological model system although a 100-fold increase in γ-TQ concentration enhanced lipid peroxidation (Lindsey *et al.*, 1985). This effect was confirmed with δ-TQ and, furthermore, we showed that all tocopherol quinones inhibited lipid peroxidation in a biological system with confluent (nonproliferating) cultures of smooth muscle cells (SMC) challenged with 120-μM arachidonic acid (Thornton *et al.*, 1995). Therefore, antioxidant activity is a general property for tocopherol quinones and, obviously, cannot explain differences in the biological effects caused by ferric chloride-oxidized wheat germ and cod liver oils.

B. ARYLATION

Since redox cycling is a general property of all tocopherol quinones, it did not appear to be a defining property that would explain the specificity of tocopherol oxidation products in "vitamin E" biology. We then proposed that differences in tocopherol quinone arylation, that is the formation of a Michael adduct with a nucleophile such as the thiol group of a cysteinyl protein, could account for the dramatic difference between ferric chloride-treated wheat germ and cod liver oils. On oxidation, α-T yields a fully substituted 1,4-benzoquinone congener, the nonarylating quinone electrophile α-TQ, as its major oxidation product, and a carboxyethyl hydrochroman (α-CEHC) and its nonarylating quinone electrophile, α-CEHC quinone lactone (α-CEHC-QL). Minor products include a 1,2-benzoquinone tocored which forms a tautomeric arylating quinone methide and an arylating quinone methide, α-TQ methide (α-TQM), formed directly from the oxidation of α-T. The different structures are given in Fig. 2. Arylating quinone methide electrophiles are minor components of the α-T family, but they do occur and may contribute to the biological properties of α-T under certain specific biological conditions.

Partially methylated β-, γ-, and δ-T (Fig. 1) are oxidized to arylating quinone electrophiles as their major oxidation products (structures and arylation sites are described in Fig. 3). The differences between the effects caused by largely nonarylating (Fig. 2) and arylating tocopherol quinones (Fig. 3) are spectacular when their chemical and biological properties are compared.

The synthesis of α-, β-, and γ-TQ in 1938 (John *et al.*, 1938; Karrer *et al.*, 1938b) was followed by many studies with α-TQ, but even as late as 1965 (Green and McHale, 1965), only two papers reported studies with γ-TQ (Henninger *et al.*, 1963; Krogmann and Olivero, 1962). By 1985 (Lindsey *et al.*, 1985), no papers had remarked on the possible biological effect of arylation as an explanation for the difference between nonarylating α-TQ and arylating γ-TQ, even though the arylation of the thiol nucleophile thioglycolic acid with the quinone electrophile benzoquinone had been reported as early as 1888 (Nickerson *et al.*, 1963). Many studies, in addition to

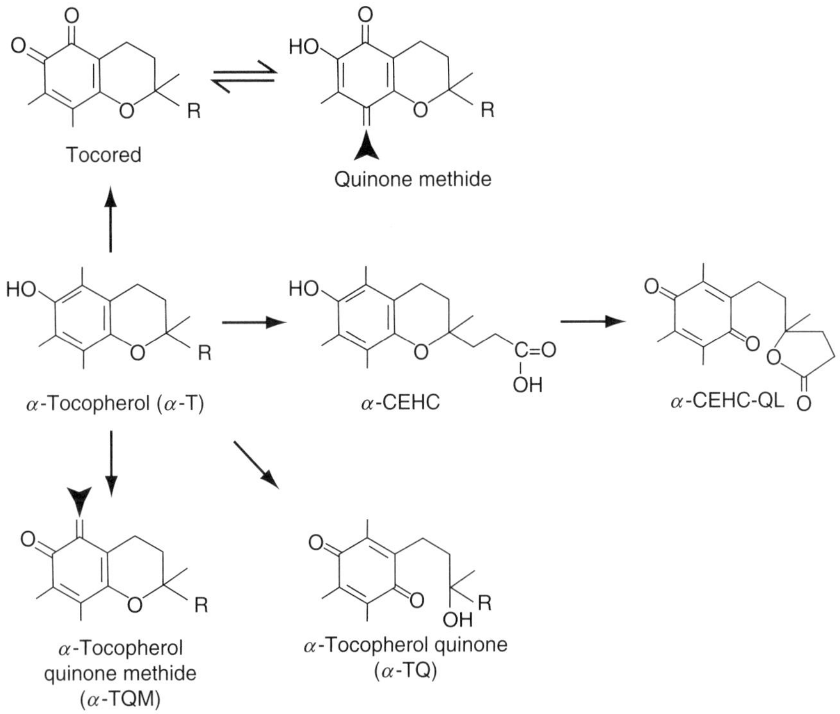

FIGURE 2. Oxidation of α-tocopherol showing positions (➤) available for thiol nucleophile addition.

our own (Mezick *et al.*, 1970), described specific biological effects with arylating quinones (Bolton *et al.*, 2000; Brunmark and Cadenas, 1989; Dennehy *et al.*, 2006; Di Simplicio *et al.*, 2003; Monks and Jones, 2002; Monks and Lau, 1998; O'Brien, 1991).

We explored the striking difference in cytotoxicity between nonarylating α-TQ and arylating γ-TQ in cell culture experiments using SMC and human acute lymphoblastic leukemia cells (CEM) (Thornton *et al.*, 1995). Nonarylating α-TQ had little effect on cell morphology and cells remained viable, retained their growth pattern, and contained normal numbers of mitotic figures. In contrast, arylating γ- and δ-TQ had profound effects on viability and morphology which showed concentration-dependant accumulation of larger vacuoles and frankly pyknotic nuclei. Cytotoxicity was confirmed in suspension cultures of CEM cells by measuring a decrease with nonviable cells in the reduction of MTT (3-(4,5-dimethylthiazol-2-yl)-2,5-diphenyltetrazolium bromide) to blue formazan with dehydrogenases. Again, in contrast to the nonarylating α-TQ, arylating γ- and δ-TQ showed striking concentration-dependent cytotoxicity. The sharp differences between nonarylating α-TQ

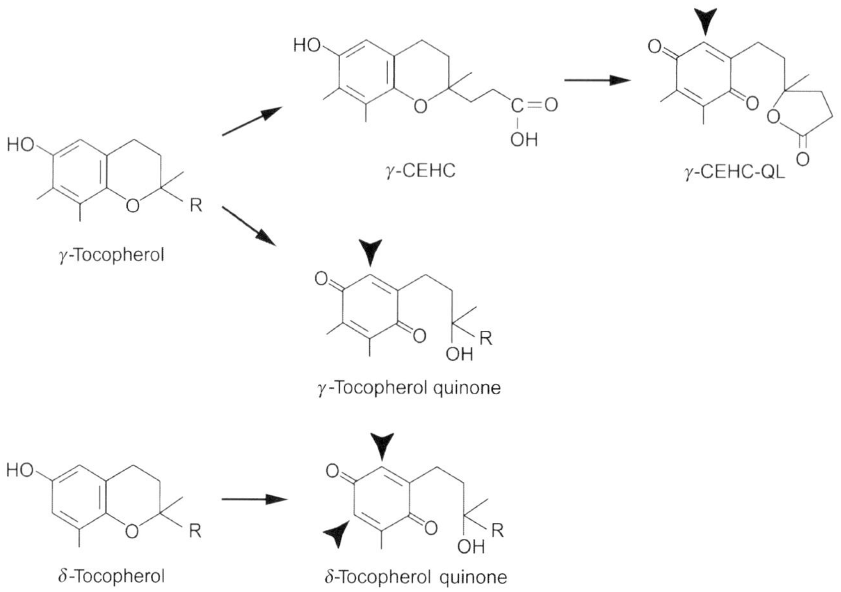

FIGURE 3. Oxidation of γ- and δ-tocopherols showing positions (➤) available for thiol nucleophile addition.

and the arylating tocopherol quinones were confirmed by us in multiple mammalian cell lines derived from different tissues of species ranging from guinea pig to human (Calviello *et al.*, 2003; Cornwell *et al.*, 1998, 2002, 2003; Jones *et al.*, 2002; Lindsey *et al.*, 1985; Sachdeva *et al.*, 2005; Thornton *et al.*, 1995; Wang *et al.*, 2006), clearly establishing that the arylating tocopherol quinine-associated cytotoxicity is ubiquitous.

The mechanism of arylating tocopherol quinone-induced cell death was investigated in a series of experiments. Studies involving transmission electron microscopy (TEM) with rapidly proliferating cells in suspension cultures showed that α-TQ had little effect on morphology while γ-TQ induced progressive changes, including chromatin condensation and margination, fragmentation of the nuclear membrane, and the appearance of apoptotic vesicles (Thornton *et al.*, 1995). These morphological changes suggested that arylating tocopherol quinones stimulated apoptosis in these cells and, indeed, this was confirmed with classic apoptotic markers, including terminal deoxynucleotide transfer-mediated dUTP nick-end labeling (TUNEL), DNA fragmentation, annexin V binding, cytochrome *c* release, the cleavage of poly (ADP-ribose)polymerase (PARP) (Jones *et al.*, 2002), and the activation of caspase-9, caspase-8, followed by caspase-3 (Calviello *et al.*, 2003). These findings, together with a decrease in Bcl-2 and a change in mitochondrial transmembrane potential (Calviello *et al.*, 2003), supported the involvement of the intrinsic mitochondrial apoptotic pathway in arylating tocopherol quinone-induced apoptosis.

The cytotoxic effects of other arylating tocopherol quinone electrophiles produced during tocopherol metabolism were also compared. Besides oxidation to quinones, tocopherols are also metabolized by side chain degradation through a β-oxidation pathway to hydroxyethyl hydrochromans (α- and γ-CEHC) which are further oxidized to their quinone lactones (CEHC-QL) by a pathway (Figs. 2 and 3) first identified for α-T (Eisengart *et al.*, 1956; Gross *et al.*, 1956). This pathway was then described as a general property of isoprenyl compounds and γ-CEHC-QL, the quinone lactone of γ-T, was identified (Gallo-Torres, 1980). Unlike α-CEHC, γ-CEHC is a major tocopherol urinary excretory product (Lodge *et al.*, 2001; Schonfeld *et al.*, 1993; Schultz *et al.*, 1995; Sontag and Parker, 2002; Swanson *et al.*, 1999) with biological properties closely related to its parent γ-T (see below). We reasoned that arylating γ-CEHC-QL would have biological properties similar to arylating γ-TQ. γ-CEHC was synthesized and oxidized to γ-CEHC-QL and the cytotoxicity of each precursor/quinone pair, γ-T/γ-TQ and γ-CEHC/γ-CEHC-QL, was compared in several cell lines (Sachdeva *et al.*, 2005). Both arylating quinones were highly cytotoxic but the cytotoxicity with γ-TQ was tenfold greater than with γ-CEHC-QL. These observations supported a general role for arylation in tocopherol quinone toxicity.

The fundamental role of arylation in determining the biological properties of tocopherol quinones was clearly established by the fact that glutathion-*S*-yl derivatives of arylating tocopherol quinones did not evidence cytotoxicity with CEM cells (Thornton *et al.*, 1995). In addition, the thiol nucleophile *N*-acetylcysteine (NAC) diminished cytotoxicity and buthionine-[*S,R*]-sulfoximine (BSO), which depletes the intracellular glutathione (GSH) pool by inhibiting γ-glutamyl-cysteine synthetase enhanced cytotoxicity in CEM cells (Cornwell *et al.*, 1998). Moreover, CEM cells treated with γ-TQ showed a dose- and time-dependant decrease in intracellular GSH while nonarylating α-TQ had no effect (Jones *et al.*, 2002).

Although cytotoxicity is a general biological effect for arylating tocopherol quinones, chemical structure greatly influences their cytotoxicity with hydrophilic γ-CEHC-QL much less toxic than lipophilic arylating γ-TQ (Sachdeva *et al.*, 2005). Thus, the higher rate of γ-T to γ-CEHC conversion *in vivo* (see below) might reflect a mechanism evolved to synthesize γ-CEHC-QL and avoid the generation of highly toxic-arylating γ-TQ inside the body.

The influence of tocopherol quinone's chemical properties on their biological effects is also reflected by comparison between γ- and δ T and their quinones. Both γ- and δ-T, phenolic precursors for arylating tocopherol quinones, are toxic to cultured cells at high concentrations (50–200 μM, unpublished observation) with δ-T much more toxic than γ-T (unpublished observations). Tocopherol cytotoxicity which decreases in the sequence $\delta > \gamma > \alpha$ is consistent with a mechanism that is predicated on the conversion of tocopherols to their arylating quinones. Since there are two arylation sites on δ-TQ (Fig. 3), it is chemically more active as an arylating agent than γ-TQ with only on arylation site (Fig. 3). However, cytotoxicity is reversed with γ-TQ $>$ δ-TQ when purified quinones, rather than their tocopherol precursors, are added directly to cell cultures (unpublished observations). We attribute this reversal to the high chemical reactivity of δ-TQ, which is unstable at room temperature (unpublished observations) and others found difficult to isolate as an identifiable oxidation product (Green and McHale, 1965). Higher reactivity presumably results in the oxidative degradation of δ-TQ, a reaction which would decreased its effective concentration in cells (Eggitt and Norris, 1956). Therefore, chemical reactivity and cytotoxicity are not in a simple linear relationship. However, the fact that all arylating tocopherol 1,4-quinones and their phenolic precursors are much more toxic than their nonarylating 1,4-quinone congeners (Figs. 2 and 3) clearly establishes a relationship between arylation and cytotoxicity.

III. IDENTIFICATION AND ANALYSIS OF TOCOPHEROLS, QUINONES, AND ADDUCTS

A. TOCOPHEROLS AND THEIR QUINONES

Many elegant techniques have been developed to identify tocopherols and their quinones by HPLC using dual wavelength UV, fluorescence, or multichannel electron capture detector (ECD) systems (described in citations throughout this chapter), GC-MS (Lee *et al.*, 2002), and LC-MS (Lauridsen *et al.*, 2001; Leonard *et al.*, 2005a), but only experimental data, not the standard techniques themselves, are summarized here.

Synthetic arylating tocopherol quinones are powerful biological agents (Section II.B), but unlike the nonarylating quinone α-TQ, there have been very few studies on the natural occurrence of arylating tocopherol quinones. Previous studies (Eggitt and Norris, 1956; Green and McHale, 1965) showed that both α- and γ-T were oxidized *in vitro* to the *ortho*-quinone 2,7,8-trimethyl-2-(4′,8′,12′-trimethyltridecyl)-chroman-5,6-dione (tocored in Fig. 2). This quinone is in equilibrium with its QM tautomer (Fig. 2) that forms adducts with the thiol nucleophile mercaptoethanol and functions as a cytotoxic arylating agent (Cornwell *et al.*, 1998).

Depending on the mode of administration, tocopherols are metabolized to tocopherol quinones (Gallo-Torres, 1980) which vary from a trace (Csallany *et al.*, 1962; Plack and Bieri, 1964) to as much as from 12 to 44% α-TQ in "total" α-T recovered (Alaupovic *et al.*, 1961; Krishnamurthy and Bieri, 1963). Both α- and γ-T are absorbed to a similar extent, but γ-T is metabolized more rapidly than α-T through the CEHC pathway for excretion in the urine (Leonard *et al.*, 2005b; Sontag and Parker, 2002; Traber *et al.*, 1993, 2005). Plasma α-CEHC is 2500-fold lower than plasma α-T, while plasma

γ-CEHC is only 13-fold lower than plasma γ-T (Galli *et al.*, 2002). Even though urine contains much more γ-CEHC than α-CEHC and urine contains α-CEHC-QL, no γ-CEHC-QL was found in urine (Lee *et al.*, 2002), not even when urine containing γ-CEHC is oxidized with ferric chloride (Swanson *et al.*, 1999). A study found similar small amounts of α- and γ-TQ in tissues in spite of the rapid metabolism of γ-T, and only a trace of γ-TQ was recovered in bile (Kiyose *et al.*, 2001). The apparent absence of arylating tocored, γ-TQ, and γ-CEHC-QL from tissues and body fluids suggested to us that the formation of these compounds was masked by arylation which converts them to thiol nucleophile adducts. We devised a strategy to identify their general occurrence as thiol nucleophile adducts.

B. Thiol Nucleophile Adducts

Different strategies are required for adduct identification in tissues and model systems. Tissues contain many thiol nucleophiles that range in complexity from simple peptides such as GSH to large soluble or membrane-bound cysteinyl proteins. A general method is necessary to identify total adduct in these complex mixtures. Strong base is a classic reagent for the cleavage and replacement of a thiol ether (Nickerson *et al.*, 1963) and tetramethylammonium hydroxide (TMAH) is a particularly strong base that cleaves adducts with conjugated macromolecules in humic substances and in plant biopolymers (Del Rio *et al.*, 1998; Hatcher and Clifford, 1994; McKinney *et al.*, 1995). We developed a procedure using thermochemolysis with TMAH to identify adducts between arylating tocopherol quinone electrophiles and thiol nucleophiles (Cornwell *et al.*, 2003; Sachdeva *et al.*, 2005).

The Michael reaction between an arylating quinone electrophile and thiol nucleophile yields a hydroquinone adduct that is reoxidized to its quinone adduct, and reaction products may contain hydroquinone/quinone mixtures which are readily separated by reverse-phase HPLC (Thornton *et al.*, 1995) and identified by ESI-TOF-MS (Sachdeva *et al.*, 2005). TMAH is a reductive methylation agent that reduces quinones to their hydroquinones and then methylates the products forming fully methylated derivatives, which, without the need for silylation, are suitable for GC separation followed by identification using the MS fragmentation pattern. Parent tocopherols and their quinones are identified in this way. TMAH thermochemolysis also cleaves thiol ether Michael adducts forming *S*-methylated products for GC separation and MS fragmentation. A schematic diagram of TMAH derivatives suitable for GC MS is shown in Fig. 4. Mono- and dithiol nucleophile adducts of δ-TQ are also identified but the two stereoisomers of the mono thiol adducts have not been distinguished (Cornwell *et al.*, 2003).

Studies with either pure proteins or tissues showed that Michael adducts were indeed formed and their formation depended on thiol group accessibility. Albumin contains 35 cysteinyl residues which form 17 disulfide bonds leaving one hindered thiol (Sugio *et al.*, 1999) available for adduct formation with the

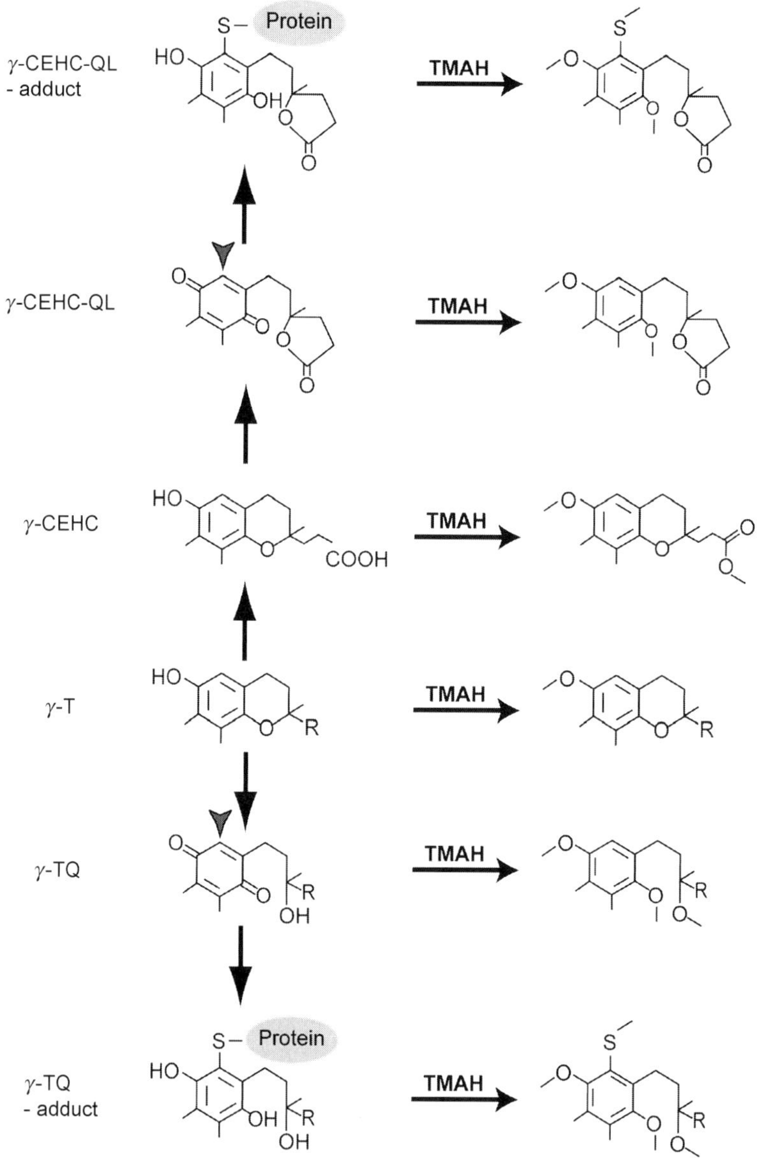

FIGURE 4. TMAH fragmentation/reductive methylation of γ-tocopherol, quinone derivatives, and their thiol nucleophile adducts.

simple arylating quinone electrophile 1,4-benzoquinone (Rappaport *et al.*, 2002). TMAH thermochemolysis showed that the bulky γ-TQ did not form a thiol adduct with albumin nor with thiol-free insulin. Whereas, both γ-TQ and γ-CEHC-QL form adducts with hemoglobin, papain, and histone, proteins which contain accessible thiols (Cornwell *et al.*, 2003; Sachdeva *et al.*, 2005). In addition, TMAH thermochemolysis revealed the formation of adducts when γ-TQ was added to fetal bovine serum (FBS) or to cells in tissue culture. In the latter case, adducts were detected in both culture media and in cells, results that are consistent with cytotoxicity (Section II.B).

The TMAH results are consistent with our hypothesis that the absence of arylating γ-TQ and γ-CEHC-QL in cells and tissues is explained by their conversion to thiol nucleophile adducts. Other studies lend support to our hypothesis that arylation indeed occurs in tissues. We found that cytotoxicity was enhanced when cells treated with γ-TQ were grown in 2% rather than 10% FBS (unpublished results). Thiol adducts are formed when arylating quinone electrophiles are added to FBS (Cornwell *et al.*, 2003) which contains relatively large amounts of antioxidant thiol proteins (Frei *et al.*, 1992). The reduction of FBS from 10 to 2% would increase the effective intracellular γ-TQ arylation and enhance its cytotoxicity. In a second study, cells were treated with γ-TQ and increasing concentrations of NAC. Adduct formation was then quantitated by ESI-TOF-MS and it was shown that cytotoxicity varied inversely with the amount of adduct formed (Wang *et al.*, 2006).

IV. ARYLATING TOCOPHEROL QUINONES AND THE UNFOLDED PROTEIN RESPONSE

The correlation between intracellular adduct formation and arylating γ-TQ caused cytotoxicity implies that the dysfunction of intracellular thiol containing proteins is a cause of apoptotic cell death. Thiol groups are involved in a variety of cellular metabolic activities and one of its major biological functions is the formation of disulfide bonds to maintain a correct protein conformation. In eukaryotic cells, disulfide bonds are formed during the initial protein-folding process within the endoplasmic reticulum (ER), which is catalyzed by the ER thiol disulfide oxidoreductases through a series of disulfide exchange reactions between enzymes and substrates and coordinate disulfide bond formation with the protein-folding process (Ellgaard, 2004; Ellgaard and Ruddock, 2005). Since the protein is not completely folded at these stages and the thiol groups on both substrate and enzyme are involved in exchange reactions, these thiols should be readily accessible for nucleophilic adduction by arylating quinones. Adduct formation will disrupt the proper ER protein folding, and misfolded ER proteins will be recognized by the ER protein folding quality control mechanism, retained in the ER, and retrotranslocated to the cytosol for proteasome degradation (Jorgensen *et al.*, 2003; Ye, 2005; Zhang and Kaufman, 2006).

The accumulation of misfolded proteins causes ER stress (Rutkowski and Kaufman, 2004; Schroder and Kaufman, 2005), inducing a coordinated cellular adaptive program, the unfolded protein response (UPR), which alleviates ER stress by increasing chaperone expression to assist protein folding, inhibiting protein translation to lower the protein load (Harding *et al.*, 2002), and committing overstressed cells to cell death (Oyadomari and Mori, 2004; Schroder and Kaufman, 2005). Three distinct signaling pathways, inositol-requiring enzyme 1 (IRE1), pancreatic ER kinase (PERK), and activating transcription factor 6 (ATF6), are involved in ER stress (Schroder and Kaufman, 2005). The PERK signaling pathway is activated through PERK autophosphorylation, which in turn phosphorylates the translation factor elF2α. Phosphorylated elF2α shuts down protein translation that lowers the ER protein load to alleviate ER stress. Whereas, it allows for the preferential translation of some mRNAs (Harding *et al.*, 2002; Zhang and Kaufman, 2006), including activating transcription factor 4 (ATF4) that activates the expression of CEBP homology protein (CHOP), a proapoptotic transcription factor involved in committing overstressed cells to apoptosis (Marciniak *et al.*, 2004; McCullough *et al.*, 2001; Oyadomari and Mori, 2004).

We analyzed cells treated with arylating γ-TQ and showed that it induced a time-dependent PERK autophosphorylation. In cells treated 24 h with 10-μM α- or γ-TQ, or with their phenolic precursors α- or γ-T, induction of the PERK signaling pathway, including phosphorylated e2Fα and increased expression of ATF4 and CHOP, was only observed with arylating γ-TQ. No effect was found in cells treated with nonarylating α-TQ or parent tocopherols (Wang *et al.*, 2006). Moreover, general stress proteins such as Hsp90 remained the same in cells treated with either tocopherols or tocopherol quinones, indicating that UPR does not simply result from general stress induced by quinones. The intimate association between UPR and arylating tocopherol quinone-caused cytotoxicity was further strengthened by a decrease in CHOP induction in cells where arylating γ-TQ-caused toxicity was alleviated by incubation with the thiol nucleophile NAC (Wang *et al.*, 2006). These observations led us to propose that arylating γ-TQ-causes aberrant disulfide bond formation and protein misfolding. We hypothesize that UPR contributes to the cellular mechanism behind arylating tocopherol quinone-caused cytotoxicity. The fact that UPR plays a central role in a variety of human diseases including Parkinson's disease and diabetes (Wu and Kaufman, 2006) warrants further elucidation of the relationship between tocopherol metabolism and UPR.

V. Tocopherol Quinones and Mutagenesis

Arylating tocopherol quinone-induced cell death explains toxicity in animals fed with a ferric chloride-treated diet, but does not appear to account for tumor formation (Adamstone, 1936; Rowntree *et al.*, 1937; Taylor and Nelson, 1930). These seemingly paradoxical biological effects, cytotoxicity and uncontrolled cell growth, could be explained if cells are exposed to different dosages of arylating tocopherol quinones. With relatively higher dosages, adduct formation destines cells to apoptosis. Whereas, lower dosages induce UPR but do not necessarily overstress cells, and this notion is supported by the observation that the CHOP level increased modestly in cells treated with a nontoxic dosage of γ-TQ (unpublished results). In this case, cells survive and arylating tocopherol quinones may influence cellular metabolism through additional pathways.

Neoplasm growth is generally associated with mutations in the genome that disrupt the control mechanism for cell multiplication (Vogelstein and Kinzler, 1993). Some quinones cause mutagenesis by redox cycling that generates DNA-damaging ROS. The mutagenicity of tocopherol quinones was tested and results showed that neither arylating nor nonarylating tocopherol quinones were genotoxins causing DNA damage in plasmid DNA incubated *in vitro*, or induced mutagenesis monitored in bacteria using the Ames test (Cornwell *et al.*, 2002). However, both arylating γ-TQ and nonarylating α-TQ increased mutation frequencies for the *gpt* gene in mammalian AS52 cells. Arylating γ-TQ was highly cytotoxic, but even at a 10 times lower concentration, it was still more mutagenic than its nonarylating congener α-TQ, suggesting a strong contribution from arylation to tocopherol quinone-caused mutagenesis in mammalian AS52 cells. A major difference between bacteria and eukaryocytes is that DNA is packaged in chromatin, a DNA–histone complex in eukarocytes. Arylating electrophiles have very low reactivity toward amine group nucleophiles (Smithgall *et al.*, 1988) such as those found in free bacterial or plasmid DNA and would not be expected to function as a genotoxin in these systems. However, histones contain a thiol group and we found that γ-TQ forms a Michael adduct with histone (Cornwell *et al.*, 2003). This suggested to us that the Michael reaction might be a new mechanism, in addition to acetylation and methylation (He and Lehming, 2003), to modify the histone code leading to epigenetic signaling mechanisms for mutations. Indeed, a study found that the arylating electrophile menadione formed an adduct with H3 Cys110, suggesting a novel role for this histone site as a nuclear stress sensor (Scott *et al.*, 2005).

VI. Specificity of Phenolic Antioxidant Precursors in Tocopherol Biology

A. The α-T Story

Although its historic functional role as an antioxidant (Evans, 1962; Mason, 1977; Mattill, 1939) has dominated our thinking about vitamin E for many years (Burton and Ingold, 1986; Liebler, 1993; Niki, 1996; van Acker *et al.*, 1993), recent work, much of it generated, collated, and discussed by Azzi and his coworkers (Azzi and Stocker, 2000; Azzi *et al.*, 2004), has raised the question of other signaling functions in the cell biology of vitamin E. Tocopherols all function effectively as antioxidants yet differ widely in their biological effects. We have attributed these differences and the emerging consensus that γ-T is strikingly different from α-T (Giovannucci, 2000; Jiang *et al.*, 2001; McCormick and Parker, 2004; Tan, 2005; Wagner *et al.*, 2004) to the fact that tocopherols are oxidized to quinone electrophiles with unique chemical properties which are exemplified by comparisons between nonarylating α-TQ and α-CEHC-QL and arylating γ-TQ and γ-CEHC-QL. Interestingly, recent studies (Hensley *et al.*, 2004; McCormick and Parker, 2004; Tan, 2005) noted that cytotoxic γ- and δ-T were undermethylated compared to fully methylated α-T but did not connect this difference to the biological properties of arylating tocopherol quinones. The biological properties of tocopherols, which are not always clearly distinguished, may also involve arylation with the highly reactive quinone methide oxidation products of α-T which function as arylating agents (Kohar *et al.*, 1995; Rosenau *et al.*, 2002, 2004; Fig. 2). A purported arylating property for an α-T quinone tautomer that modifies specific cysteinyl groups in key enzymes has been suggested (Dowd and Zheng, 1995). However, tautomerization does not appear to be a general property of α-TQ, which shows little cytotoxicity in most studies (Bolton *et al.*, 1997; Boscoboinik *et al.*, 1995; Van De Water and Pettus, 2002). An alternative explanation for the difference between α-T and γ- or δ-T, which we do not cover in this chapter, may involve a dramatically higher stability of the α-T phenoxonium cation compared to its undermethylated congeners (Wilson *et al.*, 2006).

In 1991, Azzi and his coworkers first reported a highly specific difference between α-T and other tocopherols in that α-T diminished the proliferation of A7r5 SMC (Boscoboinik *et al.*, 1991), which may be a cell type-specific effect since the same group found that other cell lines were specific for γ-T (Galli *et al.*, 2004). The phenomenon in A7r5 cells was studied in great detail (Koya *et al.*, 1997) and a general mechanism that was proposed (Azzi and Stocker, 2000; Azzi *et al.*, 2004) is summarized as follows. Phosphorylated protein kinase Cα (PKCα) promotes cell proliferation and this effect is blocked by dephosphorylation with protein phosphatase-2A (PP2A). The α-T does not act directly on PKCα, instead, it stimulates PKCα dephosphorylation by activating PP2A. Through this indirect effect, α-T inhibits PKCα activity and, consequently, prevents cell proliferation. This mechanism is strongly supported by the *in vitro* experiment with a general phosphatase inhibitor, okadaic acid, which blocked the α-T effect on PKCα (Ricciarelli *et al.*, 1998; Tasinato *et al.*, 1995). A mechanism whereby α-T enhances PP2A activity is not immediately apparent. Neither is it clear whether the α-T effect in A7r5 cells is solely dependent on the PKCα pathway or other cellular mechanisms also contribute to this phenomenon. An alternative mechanism could involve an α-TQM (Fig. 2). The induction and inhibition of cytochrome P450 enzymes is known to vary widely among different cells (Pelkonen *et al.*, 1998). It is possible that, in A7r5 cells, P450-catalyzed α-T oxidation produces both nonarylating α-TQ and arylating α-TQM and that the quinone methide contributes to toxicity through arylation.

The notion that α-TQM may contribute to cytotoxicity is supported by the comparison between α-T and troglitazone (TRO). TRO is an oral antidiabetic drug that contains a chroman moiety and closely resembles α-T except that a side chain with a thiazolidine group replaces the phytyl side chain of α-T. Liver cytochrome P450 enzymes convert TRO to a quinone methide metabolite (Peraza *et al.*, 2006; Yamazaki *et al.*, 1999) that forms adducts with GSH and hepatic proteins (Kassahun *et al.*, 2001; Prabhu *et al.*, 2002) and affects HepG2 cell viability and mitochondrial function (Tirmenstein *et al.*, 2002), properties similar to those of arylating tocopherol 1,4-benzoquinones. Despite the similarity, direct comparison between TRO and α-T in hepatocytes revealed quite different biological properties (Tafazoli *et al.*, 2005), which may reflect that TRO more readily forms an arylating methide than α-T under the intracellular environment of hepatocytes. Moreover, TRO does not alter total cellular PP2A activity (Cho *et al.*, 2006), although PP2A is known to be modulated by specific arylating/alkylating electrophiles (Codreanu *et al.*, 2006). Thus, the α-TQM, if produced in A7r5 cells, might also act through a different cellular mechanism than the PP2A and PKCα pathway.

Quinone methides are highly reactive transitory compounds (Rosenau *et al.*, 2002, 2004; Van De Water and Pettus, 2002) that may only be generated in specific cell types such as A7r5 cells, which is consistent with the fact that the antiproliferation effect by α-T was only observed in certain specific cell lines, as Azzi and his coworkers found. In these cells, a quinone methide would not accumulate as the free methide, but as quinone methide-thiol adducts. The isolation of the glutathion-*S*-yl adduct of a TRO quinone methide in cells suggests to us that it may be possible to identify the glutathion-*S*-yl adduct of α-TQM using TMAH thermochemolysis. Arylating electrophiles of 1,4-quinone electrophiles are also highly reactive, which probably explains, as we suggest, their general absence from cells and tissues. However, these compounds, unlike quinone methides, are sufficiently stable to react at a distance with many thiol nucleophiles.

B. The Multifaceted Effects of γ-T and γ-CEHC

Differences between nonarylating α-TQ and α-CEHC-QL compared to arylating γ-TQ and γ-CEHC-QL are unambiguous and readily interpreted as a function of arylation. Differences between phenolic antioxidant precursors are more nuanced. One precursor α-T has two distinct properties which we attribute to the synthesis of either a nonarylating 1,4-quinone or an arylating quinone methide. There is little evidence for an arylating α-CEHC quinone methide, only the nonarylating α-CEHC-QL. A second precursor γ-T and its γ-CEHC metabolite each have similar properties that were explained through a common arylation pathway used by both quinones in the γ-T family. Interestingly, recent studies (Jiang *et al.*, 2004a,b) found that γ-T inhibited prostate cancer cell proliferation, inducing DNA fragmentation, cytochrome *c* release, PARP cleavage, and caspase-3 activation, all biological effects that were associated with arylating γ-TQ and γ-CEHC-QL. In addition, γ-T interrupted sphingolipid synthesis by an effect enhanced in 2% FBS, similar to our observation that arylating quinone effects are enhanced when the thiol nucleophile content is reduced by substituting 2% for 10% FBS.

These investigators (Jiang *et al.*, 2004a) also found that relatively high levels of thiol nucleophiles such as 1- to 5-mM NAC did not significantly reverse the effect of γ-T and they ruled out the possibility of arylating quinone formation. Notably, relatively high concentrations of NAC are required to reverse γ-TQ caused toxicity when both agents are added concurrently to cells (Wang *et al.*, 2006). Since the assay for the γ-T effects on prostate cancer cell proliferation requires a 3- to 4-day incubation period (Jiang *et al.*, 2004a), it is not clear whether the concentration of NAC remained at a high level during the whole incubation period. In any event, the highly interesting ceramide pathway to cell death (Engedal and Saatcioglu, 2001; Eto *et al.*, 2006; Garzotto *et al.*, 1998; Sumitomo *et al.*, 2002) warrants further study with arylating tocopherol quinones and their precursors.

In some instances, cysteinyl proteins appear to be candidate thiol nucleophiles for the arylating quinone pathway, but most adducts are as yet unproved. Specific biological effects are summarized below and some candidate cysteinyl proteins are identified. One interesting example could involve specific cysteinyl residues in the ErbB family of receptor tyrosine kinases involved in breast cancer, which are inhibited by small molecule Micheal thiol acceptors including HKI-272 (Rabindran, 2004, 2005; Tsou *et al.*, 2005). HKI-272 modulates signal transduction and cell cycle regulatory circuits as does γ-T. It is tempting to speculate that the conversion of γ-T to its γ-TQ Michael acceptor explains some of the unique properties of γ-T (Giovannucci, 2000; Hensley *et al.*, 2004; Jiang *et al.*, 2001; McCormick and Parker, 2004; Wagner *et al.*, 2004). This possibility should be explored.

The role of the γ-T metabolite γ-CEHC as a natriuretic factor is a classical example of the difference between nonarylating quinone precursors in the α-T family and arylating quinone precursors in the γ-T family. Wechter *et al.* (1996) identified γ-CEHC as a natriuretic factor distinct from α-CEHC and speculated that it was derived from γ-T. A number of studies confirmed and extended this work (Hattori *et al.*, 2000a; Kantoci *et al.*, 1997; Murray *et al.*, 1997; Nakamura, 2000), and γ-tocotrienol (γ-T3) was also identified as a natriuretic factor precursor (Hattori *et al.*, 2000b; Saito *et al.*, 2003), an observation readily explained by the metabolism of γ-T3 to γ-CEHC (Birringer *et al.*, 2002; Lodge *et al.*, 2001). Tocopherol congeners that function as natriuretic factors enhance sodium excretion (Uto *et al.*, 2004). These data suggest two mechanisms each involving arylation reactions. The arylating agents TRO and PGJ2 are PPARγ ligands (Peraza *et al.*, 2006), which regulate the expression of renal epithelium sodium channel (Guan *et al.*, 2005; Zhang *et al.*, 2005). Second, renin could be involved through its regulation by a renin-binding protein, which has been identified as an *N*-acetyl-D-glucosamine 2-epimerase. This protein contains several functionally important cysteine residues and is inactivated by thiol-alkylating agents (Takahashi *et al.*, 1988, 2001). We propose that γ-CEHC-QL could act in this way and suggest that adducts between the arylating γ-CEHC-QL and renin-binding protein in model systems should be explored.

Tocopherols and their derivatives function as anti-inflammatory agents, a general effect that has been related to eicosanoid synthesis. Interestingly, both the γ- and α-T families appear to be involved. Several investigators found that γ-T, its hydrophilic precursor γ-tocopherol-*N,N*-dimethylglycinate hydrochloride hydrolyzed *in situ* (Yoshida *et al.*, 2006), and its major metabolite γ-CEHC, but not α-T, diminished proinflammatory PGE2 and LTB4 through the inhibition of COX-2 (Jiang and Ames, 2003; Jiang *et al.*, 2000). Other investigators have reported that α-T also inhibits COX-2 (Egger *et al.*, 2003). It is well known that the expression of COX-2 is regulated by NF-κB (Choi *et al.*, 2005; Lee *et al.*, 2004; St-Germain *et al.*, 2004; Vandoros *et al.*, 2006) and alkylating electrophiles, for example PGJ2 metabolites, form adducts with cysteinyl residues inhibiting NF-κB activity (Straus *et al.*, 2000). Furthermore, cobrotoxin, which is known to bind to cysteinyl residues in biological molecules, downregulates both NF-κB and COX-2 (Park *et al.*, 2005). These data are at least consistent with a hypothesis that the inhibition of COX-2 is closely related to cysteinyl group thiolation reactions with arylating quinones in the γ-T family, and also possibly an arylating quinone methide in the α-T family.

Studies have shown that γ-T and γ-T3 are also involved in the cyclin signaling pathways with the activity of γ-T3 exceeding that of γ-T, and the activity of the γ- family tocopherols greatly exceeding α-family tocopherols (Betti *et al.*, 2006; Boscoboinik *et al.*, 1991; Conte *et al.*, 2004; Gysin *et al.*, 2002). Higher activity of γ-T, the parent tocopherol of arylating quinone electrophiles, may also imply the contribution of arylation. Interestingly, cyclin expression is regulated by PPARγ ligands (Chen and Xu, 2005; He *et al.*, 2004; Huang *et al.*, 2005; Sharma *et al.*, 2004), and as discussed above, arylating agents are ligands for PPARγ. Here again, we see a nexus between Michael reaction ligands and the multifaceted roles of PPARγ in systems biology (Kim *et al.*, 2006; Peraza *et al.*, 2006).

C. SIMILARITIES AND DIFFERENCES BETWEEN TOCOPHEROLS AND TOCOTRIENOLS

Tocotrienols comprise a second major family of phenolic antioxidants, similar in all respects to the tocopherols except that the saturated phytyl side chain in tocopherol is replaced by a geranylgeranyl side chain with three double bonds (Fig. 1). These compounds are minor components of human tissues with plasma ratios that vary from 0.003 for α-T3/α-T to 0.07 for γ-T3/γ-T (Lee *et al.*, 2003). CEHCs, formed by the ω-hydroxylation pathway, are major metabolites of both tocopherols and tocotrienols (You *et al.*, 2005). Shortly after the exploration of signaling pathways unrelated to antioxidant properties of tocopherols, similar studies began with tocotrienols and many investigators quickly established that tocotrienols had many signaling properties that mimicked the tocopherols. These similarities are collated in several reviews and only a few studies are described here.

Early studies showed that tocotrienols were more effective than tocopherols in inhibiting cell proliferation and PKC activity, and causing cell death (Edem, 2002; Kline *et al.*, 2004; Nakagawa *et al.*, 2004; Nesaretnam *et al.*, 2004; Sen *et al.*, 2006). As with tocopherols, biological effects with undermethylated compounds were greater than with methylated compounds in the sequence $\delta > \gamma > \beta > \alpha$ for tocotrienols. One study did show a sequence reversal $\gamma > \delta$ but the effect was not large (Guthrie and Carroll, 1998). A rare tocotrienol isolated from rice bran, 2-desmethyl-T3, has biological effects that are even tenfold greater than δ-T3 (He *et al.*, 1997). Other data centered around apoptosis suggested some differences between the effects of tocotrienols and tocopherols on the mitochondrion (Birringer *et al.*, 2003; Shah and Sylvester, 2004, 2005; Sylvester *et al.*, 2002). Whether tocotrienols act solely through the formation of γ-CEHC-QL or through an arylating quinone mixture that also contains γ-tocotrienol quinone remains an open question.

VII. NATURAL ABUNDANCE OF TOCOPHEROLS AND ITS EFFECTS ON BIOLOGY

A. PLANT SOURCES AND THE ORIGINS OF THE MEDITERRANEAN AND MODERN DIETS

Fully methylated α-T is the principal tocopherol found in animal tissues, whereas the partially methylated phenolic precursors β-, γ-, and δ-T are the principal tocopherols in plants, with γ-T the dominant form in many plants (Firestone, 1999; Gallo-Torres, 1980; Sheppard *et al.*, 1993). In contrast, animals maintain their α-T $\gg$ γ-T distribution in ways that include highly selective α-T transport protein (α-TTP; Stocker, 2004; Traber and Arai, 1999;

Traber *et al.*, 1993, 2004) and the preferential ω-hydroxylation of γ-T to γ-CEHC (Section II.B). The animal diet also contributes to the α-T $\gg$ γ-T distribution in their bodies, and the animal diet, as Bieri and coworkers first noted (Peake and Bieri, 1971), has changed dramatically in the last 150 years leading to evolutionary discordance and its potential effects on health and disease (Boaz, 2002; Gould, 2002). A brief review of the evolutionary biology of the tocopherols is in order here.

The metabolic sequence for tocopherol synthesis in cyanobacteria and photosynthetic eukarocyte plants has been elaborated through many brilliant papers and collated in excellent reviews (Cheng *et al.*, 2003; Della Penna and Pogson, 2006; Eckardt, 2003). Biosynthesis begins with the condensation of homogentisic acid with either phytyl-diphosphate (tocopherol pathway) or geranylgeranyl-diphosphate (tocotrienol pathway) to 2-methl-6-phytyl-1,4-benzoquinone or 2-methyl-6-geranylgeranyl-1,4-benzoquinone (MPBQ/MGBQ). In the tocopherol pathway (the tocotrienol pathway is similar), some MPBQ is methylated forming 2,3-dimethyl-6-phytyl-1,4-benzoquinone (DMPBQ) and both are cyclized yielding δ-T from MPBQ, which is then methylated to β-T, and γ-T from DMPBQ, which is then methylated to α-T. Studies with higher plants and cyanobacteria show that mutants with a cyclase deficiency lack all four tocopherols and accumulate DMPBQ, and, interestingly, the cyclase deficiency has very little effect which may be due in part, to antioxidant functions contributed by ascorbic acid, glutathione, and carotenoids and/or redox cycling with DMPBQ (Porfirova *et al.*, 2002; Sattler *et al.*, 2003). DMPBQ, an arylating quinone electrophile, has been suggested as an agent that disrupts cell signaling (Sattler *et al.*, 2003).

As a general rule, the relative tocopherol content of plants is α-T > γ-T for leaf tissue and γ-T > α-T for seed tissue (Sattler *et al.*, 2003), but composition varies widely between different plants. These variations may have had significant effects on the origin and evolution of the western hominin diet (Cordain *et al.*, 2005; Kennett and Winterhalder, 2006). Herbivores subsisted on leaf tissue rich in α-T. The evolution of α-TTP is not well established but this tocopherol carrier protein would, at some point, have helped to concentrate α-T in the diets of hunter-gatherers who existed on wild plant, animal foods, and their own milk, before the emergence of agriculture with the Natufian culture in the Levant about 13,000 years before the present (BP). The domestication of cereal grains, emmer and einkorn wheat, and barley, began about 10,000–11,000 BP, and the domestication of livestock leading to dairying was around 5500–6100 BP. Both were characteristics of primitive agriculture in the Neolithic period, 5500–10,000 BP. Olive oil then served as the principal dietary source of oil. This diet originating in the Levant has continued as the so-called "Mediterranean diet" to the present time and is characterized by its olive oil and wheat content (Hu, 2003; Simopoulos, 2001; Trichopoulou, 2001).

Analytic tocopherol data (Firestone, 1999) show that the relative α-T content of the Mediterranean diet is striking. These data are frequently summarized as a range in content, and we have used here the midpoint of each range in establishing ratios as an approximation of relative content. Olive oil is unusually rich in α-T ($\alpha/\gamma = 18$), durum wheat a recent variety ($\alpha/\gamma = 4$), and barley ($\alpha/\gamma = 9$). As Bieri and coworkers first noted (Peake and Bieri, 1971), the lower relative α-T content of the modern diet is very different because it contains high concentrations of corn oil ($\alpha/\gamma = 0.2$), soybean oil ($\alpha/\gamma = 0.01$), and rapeseed oil ($\alpha/\gamma = 0.2$). Interestingly, flax seed oil ($\alpha/\gamma = 0.01$), which was known but not used as a food in the ancient world, and borage seed oil ($\alpha/\gamma = 0.2$ and $\alpha/\delta = 0.03$) have also been introduced in the hominin diet because of their fatty acid composition. Is it possible that the historical nutritional advantage of the Mediterranean diet (Owen *et al.*, 2004; Perez-Jimenez, 2005; Visioli and Galli, 2002) is related to the relative $\alpha > \gamma$ content derived from its major constituents olive oil and wheat? The possibility of a negative effect from γ-T, which we discuss in a study (Cornwell *et al.*, 2002), does not seem to have been much considered in the literature. One study that does comment on the Mediterranean diet presents data that γ-T diminishes cell proliferation and promotes apoptosis in colon cancer cell lines without affecting normal colon cell lines (Campbell *et al.*, 2006), This study, interestingly, highlights γ-T which is characteristically deficient in the Mediterranean diet. Olive oil also contains phenolic antioxidants, the catechol hydroxytyrosol and its precursor oleuropein (Covas *et al.*, 2006), but catechols, which are readily oxidized to ortho quinones which are arylating electrophiles (Penning *et al.*, 1999), are beyond the scope of this chapter.

B. THE BREAST MILK/INFANT FORMULA CONUNDRUM

With hominins, the relative plasma and tissue tocopherol content is under partial control due to α-TTP and the ω-hydroxylation of γ-T followed by γ-CEHC excretion. However, there are situations where relative tocopherol content is under assault. The neonatal period is an excellent example of one such situation. Several early studies report high α-T levels in human colostrum which decrease in mother's milk over time, suggesting sequestration by the mammary gland (Chappell *et al.*, 1985; Jain *et al.*, 1996; Jansson *et al.*, 1981; Kiely *et al.*, 1999; Syvaoja *et al.*, 1985; Zheng *et al.*, 1993). A classic study by Mahan (1991), which could only have been accomplished in an animal model, measured the effect of dietary α-T on α- and γ-T levels in the serum, colostrum, and milk of the sow. We found, among his many interesting observations, a dramatic increase in the α-T content of colostrum, and mother's milk compared to serum. This increase is shown by the α-T content (μg/ml). For example, α-T content in serum at 80 days postcoition, 1.02, decreases at parturition, 0.70, and then shows a very high level in colostrum, 7.75. Similarly, the α-T content of mother's milk remains high, 1.76, compared to serum, 1.19, at 8 days. These data suggest that mammals have evolved to provide high levels of α-T to the newborn during a period before α-TTP is expressed (Tamai *et al.*, 1998).

Breast milk and infant formulas have very different tocopherol content. The relative γ-T content of infant formulas varies widely but always exceeds the relative γ-T content of breast milk (Miquel *et al.*, 2004). One study (Tamai *et al.*, 1998) reported the tocopherol content (μg/ml) of breast milk as α-T at 6.70 and γ-T at 0.78, and of infant formula as α-, β-, γ-, and δ-T at 6.50, 0.66, 12.30, and 7.70, respectively, showing that tocopherols were supplied to formula by a vegetable oil and enriched with synthetic α-T acetate (Chavez-Servin *et al.*, 2006). As a consequence, the plasma tocopherol content of breast- and bottle-fed infants showed dramatic differences, with plasma α-T at 7.90 and γ-T at 0.87 ($\alpha/\gamma = 9.1$) in breast-fed infants, and plasma α-T at 6.00 and γ-T at 2.30 ($\alpha/\gamma = 2.6$), and δ-T at 0.63 in bottle-fed infants. The tocopherol content of plasma and the α/γ ratio are similarly affected when formulas for premature infants are enriched with oils, such as borage, that have an unusual tocopherol content (Demmelmair *et al.*, 2001). A consequence of a relative increase in dietary α-T, as would be supplied by breast feeding, is that tissue γ- and δ-T levels are reduced (Handelman *et al.*, 1994; Huang and Appel, 2003). It has been reported that breast feeding is associated with a reduced risk for childhood cancer (Shu *et al.*, 1999), although a meta-analysis suggests that the association is small (Martin *et al.*, 2005). Whether the difference in tocopherol contents between breast milk and formulas contributes to the increased risk of childhood cancer is not clear. If, as seems to be suggested from studies summarized above, there is an evolutionary advantage in maintaining a high α/γ-tocopherol ratio, the tocopherol content of infant formulas should be reassessed and the advantages of breast-feeding in improving health during childhood and adolescence should be reconsidered in terms of an increase in the α/γ tocopherol ratio in the neonatal period (Cornwell *et al.*, 2002).

C. GENERATION OF ARYLATING TOCOPHEROL QUINONES IN VEGETABLE COOKING OILS

Our initial interest in tocopherol quinones arose from a reexamination of the very early literature where profound biological effects were reported when "vitamin E" was destroyed through oxidation with ferric chloride (Adamstone, 1936; Rowntree *et al.*, 1937; Taylor and Nelson, 1930). Similar oxidation rcactions may be recapitulated in our cooking today. Vegetable oils are widely used in food processing that employs high temperatures. For example in 1999, 5.5 billion pounds were used in frying/baking, 2.9 billion pounds in fried snack foods, 2 billion pounds in restaurants (Warner, 1999). Thermal oxidation of α-T results in the formation of 4a,5-epoxy-α-TQ, and 7,8-epoxy-α-TQ, and other epoxy quinones (Liebler, 1993; Verleyen *et al.*, 2001a,b;Yamauchi *et al.*, 2002). There have been many studies on tocopherol, usually α-T depletion and conversion to α-TQ during frying, but not always with the identification of oxidation products (Andrikopoulos *et al.*, 2002; Boskou, 2003; Brenes *et al.*, 2002; Pellegrini *et al.*, 2001; Quiles *et al.*, 2002; Warner, 1999; Yuki and Ishikawa, 1976). Thermal oxidation products of arylating tocopherol quinone precursors, γ- and δ-T, have not been studied in detail and indeed γ- and δ-T were consumed in heated vegetable oils without indentification of their quinones (Rennick and Warner, 2006) which suggests to us their quinone adduct formation in the oils. Thermal oxidation of polyunsaturated fatty acids in vegetable oils is a confounding variable because the genetic modification of fatty acid composition also modifies tocopherol levels (Abidi *et al.*, 1999; Firestone, 1999; McCord *et al.*, 2004; Mounts *et al.*, 1996; Scherder *et al.*, 2006). These interrelationships need to be explored in thermal oxidation studies with simulated vegetable oils containing carefully controlled tocopherol and polyunsaturated fatty acid mixtures. Much work remains before it will be possible to map the products of vegetable oil thermal oxidation onto the biological properties of arylating tocopherol quinones. If our hypothesis is sustained that the oxidation of β-, γ-, and δ-T to arylating tocopherol quinone electrophiles in food preparation is a negative factor in the modern diet, it may be worthwhile to consider the genetic modification of plants with enzymes that redirect the metabolic flux from γ-T to an α-T "Mediterranean-like" diet (Cahoon *et al.*, 2003; Cho *et al.*, 2005; Della Penna and Pogson, 2006; Falk *et al.*, 2005; Kanwischer *et al.*, 2005; Karunanandaa *et al.*, 2005; Shintani and Della Penna, 1998).

ACKNOWLEDGMENTS

We appreciate the many contributions of Patrick Hatcher, Kenneth Jones, Rakesh Sachdeva, Beena Thomas, Xinhe Wang, and Hanfang Zhang to this work, which was supported in part by the Large Interdisciplinary Grant Program in the Office of Research at the Ohio State University.

REFERENCES

Abidi, S. L., List, G. R., and Rennick, K. A. (1999). Effects of genetic modification on the distribution of minor constituents in canola oil. *J. Am. Oil Chem. Soc.* **76,** 463–467.

Adamstone, F. B. (1936). A lymphoblastoma occurring in young chicks reared on a diet treated with ferric chloride to destroy vitamin E. *Am. J. Cancer* **28,** 540–549.

Alaupovic, P., Johnson, B. C., Crider, Q., Bhagavan, H. N., and Johnson, B. J. (1961). Metabolism of α-tocopherol and the isolation of a nontocopherol-reducing substance from animal tissues. *Am. J. Clin. Nutr.* **9**(4, Pt. 2), 76–88.

Andrikopoulos, N. K., Dedoussis, G. V., Falirea, A., Kalogeropoulos, N., and Hatzinikola, H. S. (2002). Deterioration of natural antioxidant species of vegetable edible oils during the domestic deep-frying and pan-frying of potatoes. *Int. J. Food Sci. Nutr.* **53,** 351–363.

Azzi, A., and Stocker, A. (2000). Vitamin E: Non-antioxidant roles. *Prog. Lipid Res.* **39,** 231–255.

Azzi, A., Gysin, R., Kempna, P., Munteanu, A., Negis, Y., Villacorta, L., Visarius, T., and Zingg, J. M. (2004). Vitamin E mediates cell signaling and regulation of gene expression. *Ann. NY Acad. Sci.* **1031,** 86–95.

Betti, M., Minelli, A., Canonico, B., Castaldo, P., Magi, S., Aisa, M. C., Piroddi, M., Di Tomaso, V., and Galli, F. (2006). Antiproliferative effects of tocopherols (vitamin E) on murine glioma C6 cells: Homologue-specific control of PKC/ERK and cyclin signaling. *Free Radic. Biol. Med.* **41,** 464–472.

Birringer, M., Pfluger, P., Kluth, D., Landes, N., and Brigelius-Flohe, R. (2002). Identities and differences in the metabolism of tocotrienols and tocopherols in HepG2 cells. *J. Nutr.* **132,** 3113–3118.

Birringer, M., EyTina, J. H., Salvatore, B. A., and Neuzil, J. (2003). Vitamin E analogues as inducers of apoptosis: Structure-function relation. *Br. J. Cancer* **88,** 1948–1955.

Boaz, N. T. (2002). "The Origins of Illness and How the Modern World is Making Us Sick." Wiley, New York.

Bolton, J. L., Turnipseed, S. B., and Thompson, J. A. (1997). Influence of quinone methide reactivity on the alkylation of thiol and amino groups in proteins: Studies utilizing amino acid and peptide models. *Chem. Biol. Interact.* **107,** 185–200.

Bolton, J. L., Trush, M. A., Penning, T. M., Dryhurst, G., and Monks, T. J. (2000). Role of quinones in toxicology. *Chem. Res. Toxicol.* **13,** 135–160.

Boscoboinik, D., Szewczyk, A., and Azzi, A. (1991). Alpha-tocopherol (vitamin E) regulates vascular smooth muscle cell proliferation and protein kinase C activity. *Arch. Biochem. Biophys.* **286,** 264–269.

Boscoboinik, D., Ozer, N. K., Moser, U., and Azzi, A. (1995). Tocopherols and 6-hydroxy-chroman-2-carbonitrile derivatives inhibit vascular smooth muscle cell proliferation by a nonantioxidant mechanism. *Arch. Biochem. Biophys.* **318,** 241–246.

Boskou, D. (2003). Losses of natural antioxidants and vitamins during deep-fat frying. *Forum Nutr.* **56,** 343–345.

Brenes, M., Garcia, A., Dobarganes, M. C., Velasco, J., and Romero, C. (2002). Influence of thermal treatments simulating cooking processes on the polyphenol content in virgin olive oil. *J. Agric. Food. Chem.* **50,** 5962–5967.

Brunmark, A., and Cadenas, E. (1989). Redox and addition chemistry of quinoid compounds and its biological implications. *Free Radic. Biol. Med.* **7,** 435–477.

Burton, G. W., and Ingold, K. U. (1986). Vitamin E: Application of the principles of physical organic chemistry to the exploration of structure and function. *Acc. Chem. Res.* **19,** 194–201.

Cahoon, E. B., Hall, S. E., Ripp, K. G., Ganzke, T. S., Hitz, W. D., and Coughlan, S. J. (2003). Metabolic redesign of vitamin E biosynthesis in plants for tocotrienol production and increased antioxidant content. *Nat. Biotechnol.* **21,** 1082–1087.

Calviello, G., Di Nicuolo, F., Piccioni, E., Marcocci, M. E., Serini, S., Maggiano, N., Jones, K. H., Cornwell, D. G., and Palozza, P. (2003). Gamma-tocopheryl quinone induces apoptosis in cancer cells via caspase-9 activation and cytochrome c release. *Carcinogenesis* **24,** 427–433.

Campbell, S. E., Stone, W. L., Lee, S., Whaley, S., Yang, H., Qui, M., Goforth, P., Sherman, D., McHaffie, D., and Krishnan, K. (2006). Comparative effects of RRR-α- and RRR-γ-tocopherol on proliferation and apoptosis in human colon cancer cell lines. *BMC Cancer* **6,** 13.

Carruthers, C. (1938). Attempt to produce sarcomas in rats from ingestion of crude wheat germ oil by ether extraction. *Proc. Soc. Exp. Biol. Med.* **40,** 107–108.

Chappell, J. E., Francis, T., and Clandinin, M. T. (1985). Vitamin A and E content of human milk at early stages of lactation. *Early Hum. Dev.* **11,** 157–167.

Chavez-Servin, J. L., Castellote, A. I., and Lopez-Sabater, M. C. (2006). Simultaneous analysis of vitamins A and E in infant milk-based formulae by normal-phase high-performance liquid chromatography-diode array detection using a short narrow-bore column. *J. Chromatogr. A* **1122,** 138–143.

Chen, A., and Xu, J. (2005). Activation of PPAR{γ} by curcumin inhibits Moser cell growth and mediates suppression of gene expression of cyclin D1 and EGFR. *Am. J. Physiol. Gastrointest. Liver Physiol.* **288,** G447–G456.

Cheng, Z., Sattler, S., Maeda, H., Sakuragi, Y., Bryant, D. A., and DellaPenna, D. (2003). Highly divergent methyltransferases catalyze a conserved reaction in tocopherol and plastoquinone synthesis in cyanobacteria and photosynthetic eukaryotes. *Plant Cell* **15,** 2343–2356.

Cho, D. H., Choi, Y. J., Jo, S. A., Ryou, J., Kim, J. Y., Chung, J., and Jo, I. (2006). Troglitazone acutely inhibits protein synthesis in endothelial cells via a novel mechanism involving protein phosphatase 2A-dependent p70 S6 kinase inhibition. *Am. J. Physiol. Cell Physiol.* **291,** C317–C326.

Cho, E. A., Lee, C. A., Kim, Y. S., Baek, S. H., de los Reyes, B. G., and Yun, S. J. (2005). Expression of γ-tocopherol methyltransferase transgene improves tocopherol composition in lettuce (*Latuca sativa* L.). *Mol. Cells* **19,** 16–22.

Choi, J. K., Lee, S. G., Lee, J. Y., Nam, H. Y., Lee, W. K., Lee, K. H., Kim, H. J., and Lim, Y. (2005). Silica induces human cyclooxygenase-2 gene expression through the NF-kB signaling pathway. *J. Environ. Pathol. Toxicol. Oncol.* **24,** 163–174.

Codreanu, S. G., Adams, D. G., Dawson, E. S., Wadzinski, B. E., and Liebler, D. C. (2006). Inhibition of protein phosphatase 2A activity by selective electrophile alkylation damage. *Biochemistry* **45,** 10020–10029.

Conte, C., Floridi, A., Aisa, C., Piroddi, M., and Galli, F. (2004). Gamma-tocotrienol metabolism and antiproliferative effect in prostate cancer cells. *Ann. NY Acad. Sci.* **1031,** 391–394.

Cordain, L., Eaton, S. B., Sebastian, A., Mann, N., Lindeberg, S., Watkins, B. A., O'Keefe, J. H., and Brand-Miller, J. (2005). Origins and evolution of the Western diet: Health implications for the 21st century. *Am. J. Clin. Nutr.* **81,** 341–354.

Cornwell, D. G., Jones, K. H., Jiang, Z., Lantry, L. E., Southwell-Keely, P., Kohar, I., and Thornton, D. E. (1998). Cytotoxicity of tocopherols and their quinones in drug-sensitive and multidrug-resistant leukemia cells. *Lipids* **33,** 295–301.

Cornwell, D. G., Williams, M. V., Wani, A. A., Wani, G., Shen, E., and Jones, K. H. (2002). Mutagenicity of tocopheryl quinones: Evolutionary advantage of selective accumulation of dietary α-tocopherol. *Nutr. Cancer* **43,** 111–118.

Cornwell, D. G., Kim, S., Mazzer, P. A., Jones, K. H., and Hatcher, P. G. (2003). Electrophile tocopheryl quinones in apoptosis and mutagenesis: Thermochemolysis of thiol adducts with proteins and in cells. *Lipids* **38,** 973–979.

Covas, M. I., Ruiz-Gutierrez, V., de la Torre, R., Kafatos, A., Lamuela-Raventos, R. M., Osada, J., Owen, R. W., and Visoli, F. (2006). Minor components of olive oil: Evidence to date of health benefits in humans. *Nutr. Rev.* **64,** S20–S30.

Csallany, A. S., Draper, H. H., and Shah, S. N. (1962). Conversion of d-α-tocopherol-C14 to tocopheryl-p-quinone *in vivo*. *Arch. Biochem. Biophys.* **98,** 142–145.

Day, H. G., Becker, E., and McCollum, E. V. (1938). Effect of ether peroxides in wheat germ oil on production of tumors in rats. *Proc. Soc. Exp. Biol. Med.* **40,** 21–22.

Del Rio, J. C., McKinney, D. E., Knicker, H., Nanny, M. A., Minard, R. D., and Hatcher, P. G. (1998). Structural characterization of bio- and geo-macromolecules by off-line thermochemolysis with tetramethylammonium hydroxide. *J. Chromatogr. A* **823,** 433–448.

Della Penna, D., and Pogson, B. J. (2006). Vitamin synthesis in plants: Tocopherols and carotenoids. *Annu. Rev. Plant Biol.* **57,** 711–738.

Demmelmair, H., Feldl, F., Horvath, I., Niederland, T., Ruszinko, V., Raederstorff, D., De Min, C., Muggli, R., and Koletzko, B. (2001). Influence of formulas with borage oil or borage oil plus fish oil on the arachidonic acid status in premature infants. *Lipids* **36,** 555–566.

Dennehy, M. K., Richards, K. A., Wernke, G. R., Shyr, Y., and Liebler, D. C. (2006). Cytosolic and nuclear protein targets of thiol-reactive electrophiles. *Chem. Res. Toxicol.* **19,** 20–29.

Di Simplicio, P., Franconi, F., Frosali, S., and Di Giuseppe, D. (2003). Thiolation and nitrosation of cysteines in biological fluids and cells. *Amino Acids* **25,** 323–339.

Dowd, P., and Zheng, Z. B. (1995). On the mechanism of the anticlotting action of vitamin E quinone. *Proc. Natl. Acad. Sci. USA* **92,** 8171–8175.

Eckardt, N. A. (2003). Vitamin E—defective mutants of arabidopsis tell tales of convergent evolution. *Plant Cell* **15,** 2233–2235.

Edem, D. O. (2002). Palm oil: Biochemical, physiological, nutritional, hematological, and toxicological aspects: A review. *Human Nutr.* **57,** 319–341.

Egger, T., Schuligoi, R., Wintersperger, A., Amann, R., Malle, E., and Sattler, W. (2003). Vitamin E (α-tocopherol) attenuates cyclo-oxygenase 2 transcription and synthesis in immortalized murine BV-2 microglia. *Biochem. J.* **370,** 459–467.

Eggitt, P. W. R., and Norris, F. W. (1956). Chemical estimation of vitamin E activity in cereal products. IV. ϵ-Tocopherol. *J. Sci. Food. Agric.* **7,** 493–511.

Eisengart, A., Milhorat, A. T., Simon, E. J., and Sundheim, L. (1956). The metabolism of vitamin E. II. Purification and characterization of urinary metabolites of α-tocopherol. *J. Biol. Chem.* **221,** 807–817.

Ellgaard, L. (2004). Catalysis of disulphide bond formation in the endoplasmic reticulum. *Biochem. Soc. Trans.* **32,** 663–667.

Ellgaard, L., and Ruddock, L. W. (2005). The human protein disulphide isomerase family: Substrate interactions and functional properties. *EMBO Rep.* **6,** 28–32.

Emerson, O. H. (1938). The structure of β and γ tocopherols. *J. Am. Chem. Soc.* **60,** 1741–1742.

Emerson, O. H., Emerson, G. A., and Evans, H. M. (1939). The vitamin E activity of α-tocoquinone. *J. Biol. Chem.* **131,** 409–412.

Engedal, N., and Saatcioglu, F. (2001). Ceramide-induced cell death in the prostate cancer cell line LNCaP has both necrotic and apoptotic features. *Prostate* **46,** 289–297.

Eto, M., Bennouna, J., Hunter, O. C., Lotze, M. T., and Amoscato, A. A. (2006). Importance of C16 ceramide accumulation during apoptosis in prostate cancer cells. *Int. J. Urol.* **13,** 148–156.

Evans, H. M. (1962). The pioneer history of vitamin E. *Vitam. Horm.* **20,** 379–387.

Evans, H. M., and Bishop, K. S. (1922). On the existence of a hitherto unrecognized dietary factor essential for reproduction. *Science* **56,** 650–651.

Evans, H. M., and Emerson, G. A. (1939). Failure to produce abdominal neoplasms in rats receiving wheat germ oil extracted in various ways. *Proc. Soc. Exp. Biol. Med.* **41,** 318–320.

Evans, H. M., Emerson, O. H., and Emerson, G. A. (1936). The isolation from wheat germ oil of an alcohol, α-tocopherol, having the properties of vitamin E. *J. Biol. Chem.* **113,** 319–332.

Falk, J., Brosch, M., Schafer, A., Braun, S., and Krupinska, K. (2005). Characterization of transplastomic tobacco plants with a plastid localized barley 4-hydroxyphenylpyruvate dioxygenase. *J. Plant Physiol.* **162,** 738–742.

Fernholz, E. (1938). On the constitution of α-tocopherol. *J. Am. Chem. Soc.* **60,** 700–705.

Firestone, D. (1999). "Physical and Chemical Characteristics of Oils, Fats, and Waxes." AOCS Press, Champaign, Illinois.

Frei, B., Stocker, A., and Ames, B. N. (1992). Small molecule antioxidant defenses in human extracellular fluids. *In* "Molecular Biology of Free Radical Scavenging Systems" (J. G. Scandalios, Ed.), pp. 23–45. Cold Spring Harbor Laboratory Press, Plainview, New York.

Galli, F., Lee, R., Dunster, C., and Kelly, F. J. (2002). Gas chromatography mass spectrometry analysis of carboxyethyl-hydroxychroman metabolites of α- and γ-tocopherol in human plasma. *Free Radic. Biol. Med.* **32,** 333–340.

Galli, F., Stabile, A. M., Betti, M., Conte, C., Pistilli, A., Rende, M., Floridi, A., and Azzi, A. (2004). The effect of α- and γ-tocopherol and their carboxyethyl hydroxychroman metabolites on prostate cancer cell proliferation. *Arch. Biochem. Biophys.* **423,** 97–102.

Gallo-Torres, H. E. (1980). Transport and metabolism. *In* "Vitamin E. A Comprehensive Treatise" (L. J. Machlin, Ed.), pp. 193–267. Dekker, New York.

Garzotto, M., White-Jones, M., Jiang, Y., Ehleiter, D., Liao, W. C., Haimovitz-Friedman, A., Fuks, Z., and Kolesnick, R. (1998). 12-O-tetradecanoylphorbol-13-acetate-induced apoptosis in LNCaP cells is mediated through ceramide synthase. *Cancer Res.* **58,** 2260–2264.

Gavino, V. C. (1981). Polyunsaturated Fatty Acids, Lipid Accumulation, and Oxidant Stress in Cells in Culture, Dissertation Ohio State University, Columbus, Ohio.

Gavino, V. C., Miller, J. S., Ikharebha, S. O., Milo, G. E., and Cornwell, D. G. (1981). Effect of polyunsaturated fatty acids and antioxidants on lipid peroxidation in tissue cultures. *J. Lipid Res.* **22,** 763–769.

Gavino, V. C., Milo, G. E., and Cornwell, D. G. (1982). Image analysis for the automated estimation of clonal growth and its application to the growth of smooth muscle cells. *Cell Tissue Kinet.* **15,** 225–231.

Giovannucci, E. (2000). γ-Tocopherol: A new player in prostate cancer prevention? *J. Natl. Cancer Inst.* **92,** 1966–1967.

Goettsch, M., and Pappenheimer, A. M. (1931). Nutritional muscular dystrophy in the guinea pig and rabbit. *J. Exp. Med.* **54,** 145–165.

Gould, S. J. (2002). "The Structure of Evolutionary Theory." Harvard University Press, Cambridge, Massachusetts.

Green, J., and McHale, D. (1965). Quinones related to vitamin E. *In* "Biochemistry of Quinones" (R. A. Morton, Ed.), pp. 261–285. Academic Press, London.

Gross, C. S., Milhorat, A. T., and Simon, E. J. (1956). The metabolism of vitamin E. I. The absorption and excretion of d-α-tocopheryl-5-methyl-C14-succinate. *J. Biol. Chem.* **221,** 797–805.

Guan, Y., Hao, C., Cha, D. R., Rao, R., Lu, W., Kohan, D. E., Magnuson, M. A., Redha, R., Zhang, Y., and Breyer, M. D. (2005). Thiazolidinediones expand body fluid volume through PPARgamma stimulation of ENaC-mediated renal salt absorption. *Nat. Med.* **11,** 861–866.

Guthrie, N., and Carroll, K. K. (1998). Tocotrienols and cancer. *In* "Biological Oxidants and Antioxidants: Molecular Mechanisms and Health Effects" (L. Packer and A. S. H. Ong, Eds.), pp. 257–264. AOCS Press, Champaign, Illinois.

Gysin, R., Azzi, A., and Visarius, T. (2002). Gamma-tocopherol inhibits human cancer cell cycle progression and cell proliferation by down-regulation of cyclins. *FASEB J.* **16,** 1952–1954.

Handelman, G. J., Epstein, W. L., Peerson, J., Spiegelman, D., Machlin, L. J., and Dratz, E. A. (1994). Human adipose α-tocopherol and γ-tocopherol kinetics during and after 1 y of α-tocopherol supplementation. *Am. J. Clin. Nutr.* **59,** 1025–1032.

Harding, H. P., Calfon, M., Urano, F., Novoa, I., and Ron, D. (2002). Transcriptional and translational control in the mammalian unfolded protein response. *Annu. Rev. Cell Dev. Biol.* **18,** 575–599.

Hatcher, P. G., and Clifford, D. J. (1994). Flash pyrolysis and *in situ* methylation of humic acids from soil. *Org. Geochem.* **21,** 1081–1092.

Hattori, A., Fukushima, T., and Imai, K. (2000a). Occurrence and determination of a natriuretic hormone, 2,7,8-trimethyl-2-(β-carboxyethyl)-6-hydroxy chroman, in rat plasma, urine, and bile. *Anal. Biochem.* **281,** 209–215.

Hattori, A., Fukushima, T., Yoshimura, H., Abe, K., and Imai, K. (2000b). Production of LLU-alpha following an oral administration of γ-tocotrienol or γ-tocopherol to rats. *Biol. Pharm. Bull.* **23,** 1395–1397.

He, G., Thuillier, P., and Fischer, S. M. (2004). Troglitazone inhibits cyclin D1 expression and cell cycling independently of PPARγ in normal mouse skin keratinocytes. *J. Invest. Dermatol.* **123,** 1110–1119.

He, H., and Lehming, N. (2003). Global effects of histone modifications. *Brief. Funct. Genomic. Proteomic.* **2,** 234–243.

He, L., Mo, H., Hadisusilo, S., Qureshi, A. A., and Elson, C. E. (1997). Isoprenoids suppress the growth of murine B16 melanomas *in vitro* and *in vivo*. *J. Nutr.* **127,** 668–674.

Henninger, M. D., Dilley, R. A., and Crane, F. L. (1963). Restoration of ferricyanide reduction in acetone-extracted chloroplasts by β and γ tocopherol quinones. *Biochem. Biophys. Res. Commun.* **10,** 237–242.

Hensley, K., Benaksas, E. J., Bolli, R., Comp, P., Grammas, P., Hamdheydari, L., Mou, S., Pye, Q. N., Stoddard, M. F., Wallis, G., Williamson, K. S., West, M., *et al.* (2004). New perspectives on vitamin E: Gamma-tocopherol and carboxyelthylhydroxychroman metabolites in biology and medicine. *Free Radic. Biol. Med.* **36,** 1–15.

Hu, F. B. (2003). The Mediterranean diet and mortality—olive oil and beyond. *N. Engl. J. Med.* **348,** 2595–2596.
Huang, H. Y., and Appel, L. J. (2003). Supplementation of diets with α-tocopherol reduces serum concentrations of γ- and δ-tocopherol in humans. *J. Nutr.* **133,** 3137–3140.
Huang, J. W., Shiau, C. W., Yang, Y. T., Kulp, S. K., Chen, K. F., Brueggemeier, R. W., Shapiro, C. L., and Chen, C. S. (2005). Peroxisome proliferator-activated receptor γ-independent ablation of cyclin D1 by thiazolidinediones and their derivatives in breast cancer cells. *Mol. Pharmacol.* **67,** 1342–1348.
Jain, S. K., Wise, R., and Bocchini, J. J., Jr. (1996). Vitamin E and vitamin E-quinone levels in red blood cells and plasma of newborn infants and their mothers. *J. Am. Coll. Nutr.* **15,** 44–48.
Jansson, L., Akesson, B., and Holmberg, L. (1981). Vitamin E and fatty acid composition of human milk. *Am. J. Clin. Nutr.* **34,** 8–13.
Jiang, Q., and Ames, B. N. (2003). Gamma-tocopherol, but not α-tocopherol, decreases proinflammatory eicosanoids and inflammation damage in rats. *FASEB J.* **17,** 816–822.
Jiang, Q., Elson-Schwab, I., Courtemanche, C., and Ames, B. N. (2000). Gamma-tocopherol and its major metabolite, in contrast to α-tocopherol, inhibit cyclooxygenase activity in macrophages and epithelial cells. *Proc. Natl. Acad. Sci. USA* **97,** 11494–11499.
Jiang, Q., Christen, S., Shigenaga, M. K., and Ames, B. N. (2001). Gamma-tocopherol, the major form of vitamin E in the US diet, deserves more attention. *Am. J. Clin. Nutr.* **74,** 714–722.
Jiang, Q., Wong, J., Fyrst, H., Saba, J. D., and Ames, B. N. (2004a). Gamma-tocopherol or combinations of vitamin E forms induce cell death in human prostate cancer cells by interrupting sphingolipid synthesis. *Proc. Natl. Acad. Sci. USA* **101,** 17825–17830.
Jiang, Q., Wong, J., and Ames, B. N. (2004b). Gamma-tocopherol induces apoptosis in androgen-responsive LNCaP prostate cancer cells via caspase-dependent and independent mechanisms. *Ann. NY Acad. Sci.* **1031,** 399–400.
John, W., Dietzel, E., Günther, P., and Emte, W. (1938). Zum beweis der chromanstruktur des α-tokopherols. *Naturwissenschaften* **26,** 366–367.
Jones, K. H., Liu, J. J., Roehm, J. S., Eckel, J. J., Eckel, T. T., Stickrath, C. R., Triola, C. A., Jiang, Z., Bartoli, G. M., and Cornwell, D. G. (2002). Gamma-tocopheryl quinone stimulates apoptosis in drug-sensitive and multidrug-resistant cancer cells. *Lipids* **37,** 173–184.
Jorgensen, M. M., Bross, P., and Gregersen, N. (2003). Protein quality control in the endoplasmic reticulum. *APMIS Suppl.* **109,** 86–91.
Juhasz-Schaffer, A. (1931). Wirkung des E-vitamins auf explantate *in vitro*. *Arch. Pathol. Anat. Physiol.* **281,** 35–45.
Kantoci, D., Wechter, W. J., Murray, E. D., Dewind, S. A., Borchardt, D., and Khan, S. I. (1997). Endogenous natriuretic factors 6: The stereochemistry of a natriuretic α-tocopherol metabolite LLU-α. *J. Pharmacol. Exp. Ther.* **282,** 648–656.
Kanwischer, M., Porfirova, S., Bergmuller, E., and Dormann, P. (2005). Alterations in tocopherol cyclase activity in transgenic and mutant plants of Arabidopsis affect tocopherol content, tocopherol composition, and oxidative stress. *Plant Physiol.* **137,** 713–723.
Karrer, P., and Geiger, A. (1940). Über α-tocopherol-chinon. *Helv. Chim. Acta* **23,** 455–459.
Karrer, P., Fritzsche, H., Ringier, B. H., and Salomon, H. (1938a). α-Tocopherol. *Helv. Chim. Acta* **21,** 520–525.
Karrer, P., Escher, R., Fritzsche, H., Keller, H., Ringier, B. H., and Salomon, H. (1938b). Konstitution und Bestimmung des α-Tocopherols und einiger ähnlicher Verbindungen. *Helv. Chim. Acta* **21,** 939–953.
Karunanandaa, B., Qi, Q., Hao, M., Baszis, S. R., Jensen, P. K., Wong, Y. H., Jiang, J., Venkatramesh, M., Gruys, K. J., Moshiri, F., Post-Beittenmiller, D., Weiss, J. D., *et al.* (2005). Metabolically engineered oilseed crops with enhanced seed tocopherol. *Metab. Eng.* **7,** 384–400.
Kassahun, K., Pearson, P. G., Tang, W., McIntosh, I., Leung, K., Elmore, C., Dean, D., Wang, R., Doss, G., and Baillie, T. A. (2001). Studies on the metabolism of troglitazone to

reactive intermediates *in vitro* and *in vivo*. Evidence for novel biotransformation pathways involving quinone methide formation and thiazolidinedione ring scission. *Chem. Res. Toxicol.* **14,** 62–70.

Kennett, D. J., and Winterhalder, B. (2006). "Behavioral Ecology and the Transition to Agriculture." University California Press, Berkeley, California.

Kiely, M., Cogan, P. F., Kearney, P. J., and Morrissey, P. A. (1999). Concentrations of tocopherols and carotenoids in maternal and cord blood plasma. *Eur. J. Clin. Nutr.* **53,** 711–715.

Kim, H. J., Woo, I. S., Kang, E. S., Eun, S. Y., Kim, G. H., Ham, S. A., Kim, H. J., Lee, J. H., Chang, K. C., Kim, J. H., Lee, H. T., and Seo, H. G. (2006). Phorbol ester potentiates the growth inhibitory effects of troglitazone via up-regulation of PPARγ in A549 cells. *Biochem. Biophys. Res. Commun.* **349**(2), 660–667.

Kiyose, C., Saito, H., Ueda, T., and Igarashi, O. (2001). Simultaneous determination of α-, γ-tocopherol and their quinones in rats plasma and tissues using reversed-phase high-performance liquid chromatography. *J. Nutr. Sci. Vitaminol. (Tokyo)* **47,** 102–107.

Kline, K., Yu, W., and Sanders, B. G. (2004). Vitamin E and breast cancer. *J. Nutr.* **134,** 3458S–3462S.

Kohar, I., Baca, M., Suarna, C., Stocker, R., and Southwell-Keely, P. T. (1995). Is α-tocopherol a reservoir for α-tocopheryl hydroquinone? *Free Radic. Biol. Med.* **19,** 197–207.

Koya, D., Lee, I. K., Ishii, H., Kanoh, H., and King, G. L. (1997). Prevention of glomerular dysfunction in diabetic rats by treatment with d-α-tocopherol. *J. Am. Soc. Nephrol.* **8,** 426–435.

Krishnamurthy, S., and Bieri, J. G. (1963). The absorption, storage, and metabolism of α-tocopherol-C14 in the rat and chicken. *J. Lipid Res.* **4,** 330–336.

Krogmann, D. W., and Olivero, E. (1962). The specificity of plastoquinone as a cofactor for photophosphorylation. *J. Biol. Chem.* **237,** 3292–3295.

Lauridsen, C., Leonard, S. W., Griffin, D. A., Liebler, D. C., McClure, T. D., and Traber, M. G. (2001). Quantitative analysis by liquid chromatography-tandem mass spectrometry of deuterium-labeled and unlabeled vitamin E in biological samples. *Anal. Biochem.* **289,** 89–95.

Lee, B. L., New, A. L., and Ong, C. N. (2003). Simultaneous determination of tocotrienols, tocopherols, retinol, and major carotenoids in human plasma. *Clin. Chem.* **49,** 2056–2066.

Lee, K. M., Kang, B. S., Lee, H. L., Son, S. J., Hwang, S. H., Kim, D. S., Park, J. S., and Cho, H. J. (2004). Spinal NF-kB activation induces COX-2 upregulation and contributes to inflammatory pain hypersensitivity. *Eur. J. Neurosci.* **19,** 3375–3381.

Lee, R., Galli, F., and Kelly, F. J. (2002). γ-Tocopherol metabolism and its relationship with α-tocopherol in humans. *In* "The Antioxidant Vitamin C and E" (L. Packer, M. G. Traber, K. Kraemer, and B. Frei, Eds.), pp. 180–194. AOCS Press, Champaign, Illinois.

Leonard, S. W., Gumpricht, E., Devereaux, M. W., Sokol, R. J., and Traber, M. G. (2005a). Quantitation of rat liver vitamin E metabolites by LC-MS during high-dose vitamin E administration. *J. Lipid Res.* **46,** 1068–1075.

Leonard, S. W., Paterson, E., Atkinson, J. K., Ramakrishnan, R., Cross, C. E., and Traber, M. G. (2005b). Studies in humans using deuterium-labeled α- and γ-tocopherols demonstrate faster plasma γ-tocopherol disappearance and greater γ-metabolite production. *Free Radic. Biol. Med.* **38,** 857–866.

Liebler, D. C. (1993). The role of metabolism in the antioxidant function of vitamin E. *Crit. Rev. Toxicol.* **23,** 147–169.

Lindsey, J. A., Zhang, H. F., Kaseki, H., Morisaki, N., Sato, T., and Cornwell, D. G. (1985). Fatty acid metabolism and cell proliferation. VII. Antioxidant effects of tocopherols and their quinones. *Lipids* **20,** 151–157.

Lodge, J. K., Ridlington, J., Leonard, S., Vaule, H., and Traber, M. G. (2001). α- and γ-tocotrienols are metabolized to carboxyethyl-hydroxychroman derivatives and excreted in human urine. *Lipids* **36,** 43–48.

Mahan, D. C. (1991). Assessment of the influence of dietary vitamin E on sows and offspring in three parities: Reproductive performance, tissue tocopherol, and effects on progeny. *J. Anim. Sci.* **69,** 2904–2917.

Marciniak, S. J., Yun, C. Y., Oyadomari, S., Novoa, I., Zhang, Y., Jungreis, R., Nagata, K., Harding, H. P., and Ron, D. (2004). CHOP induces death by promoting protein synthesis and oxidation in the stressed endoplasmic reticulum. *Genes Dev.* **18,** 3066–3077.

Martin, R. M., Gunnell, D., Owen, C. G., and Smith, G. D. (2005). Breast-feeding and childhood cancer: A systematic review with metaanalysis. *Int. J. Cancer* **117,** 1020–1031.

Mason, K. E. (1933). Differences in testis injury and repair after vitamin A-deficiency, vitamin E-deficiency, and inanition. *Am. J. Anat.* **52,** 153–239.

Mason, K. E. (1977). The first two decades of vitamin E. *Fed. Proc.* **36,** 1906–1910.

Mattill, H. A. (1939). Vitamin E. *In* "The Vitamins" (M. Fishbein, Ed.), pp. 575–596. American Medical Association, Chicago, Illinois.

McCord, K. L., Fehr, W. R., Wang, T., Welke, G. A., Cianzio, S. R., and Schenebly, S. T. (2004). Tocopherol content of soybean lines with reduced linolenate in the seed oil. *Crop. Sci.* **44,** 772–776.

McCormick, C. C., and Parker, R. S. (2004). The cytotoxicity of vitamin E is both vitamer- and cell-specific and involves a selectable trait. *J. Nutr.* **134,** 3335–3342.

McCullough, K. D., Martindale, J. L., Klotz, L. O., Aw, T. Y., and Holbrook, N. J. (2001). Gadd153 sensitizes cells to endoplasmic reticulum stress by down-regulating Bcl2 and perturbing the cellular redox state. *Mol. Cell. Biol.* **21,** 1249–1259.

McKinney, D. E., Carson, D. M., Clifford, D. J., Minard, R. D., and Hatcher, P. G. (1995). Off-line thermochemolysis versus flash pyrolysis for the *in situ* methylation of lignin: Is pyrolysis necessary? *J. Anal. Appl. Pyrolysis* **34,** 41–46.

Mezick, J. A., Settlemire, C. T., Brierley, G. P., Barefield, K. P., Jensen, W. N., and Cornwell, D. G. (1970). Erythrocyte membrane interactions with menadione and the mechanism of menadione-induced hemolysis. *Biochim. Biophys. Acta* **219,** 361–371.

Michaelis, L., and Wollman, S. H. (1949). The semiquinone radical of tocopherol. *Science* **109,** 313–314.

Miquel, E., Alegría, A., Barberá, R., Farré, R., and Clemente, G. (2004). Stability of tocopherols in adapted milk-based infant formulas during storage. *Int Dairy J.* **14,** 1003–1011.

Monks, T. J., and Jones, D. C. (2002). The metabolism and toxicity of quinones, quinonimines, quinone methides, and quinone-thioethers. *Curr. Drug Metab.* **3,** 425–438.

Monks, T. J., and Lau, S. S. (1998). The pharmacology and toxicology of polyphenolic-glutathione conjugates. *Annu. Rev. Pharmacol. Toxicol.* **38,** 229–255.

Mounts, T. L., Abidi, S. L., and Rennick, K. A. (1996). Effect of genetic modification on the content and composition of bioactive constituents in soybean oil. *J. Am. Oil Chem. Soc.* **73,** 581–586.

Murray, E. D., Wechter, W. J., Kantoci, D., Wang, W., Pham, T., Quiggle, D. D., Gibson, K. M., Leipold, D., and Anner, B. M. (1997). Endogenous natriuretic factors 7: Biospecificity of a natriuretic α-tocopherol metabolite LLU-α. *J. Pharmacol. Exp. Ther.* **282,** 657–662.

Nakagawa, K., Eitsuka, T., Inokuchi, H., and Miyazawa, T. (2004). DNA chip analysis of comprehensive food function: Inhibition of angiogenesis and telomerase activity with unsaturated vitamin E, tocotrienol. *Biofactors* **21,** 5–10.

Nakamura, T. (2000). Overview of studies on vitamin E metabolism—missing link of vitamin E metabolism. *Bitamin* **74,** 255–261.

Nesaretnam, K., Ambra, R., Selvaduray, K. R., Radhakrishnan, A., Canali, R., and Virgili, F. (2004). Tocotrienol-rich fraction from palm oil and gene expression in human breast cancer cells. *Ann. NY Acad. Sci.* **1031,** 143–157.

Nickerson, W. J., Falcone, G., and Strauss, G. (1963). Studies on quinone-thioethers. I. Mechanism of formation and properties of thiodione. *Biochemistry* **2,** 537–543.

Niki, E. (1996). α-Tocopherol. *In* "Handbook of Antioxidants" (E. Cadenas and L. Packer, Eds.), pp. 3–26. Dekker, New York.

O'Brien, P. J. (1991). Molecular mechanisms of quinone cytotoxicity. *Chem. Biol. Interact.* **80,** 1–41.

Owen, R. W., Haubner, R., Wurtele, G., Hull, E., Spiegelhalder, B., and Bartsch, H. (2004). Olives and olive oil in cancer prevention. *Eur. J. Cancer Prev.* **13,** 319–326.

Oyadomari, S., and Mori, M. (2004). Roles of CHOP/GADD153 in endoplasmic reticulum stress. *Cell Death Differ.* **11,** 381–389.

Park, M. H., Song, H. S., Kim, K. H., Son, D. J., Lee, S. H., Yoon do, Y., Kim, Y., Park, I. Y., Song, S., Hwang, B. Y., Jung, J. K., and Hong, J. T. (2005). Cobrotoxin inhibits NF-kappa B activation and target gene expression through reaction with NF-kappa B signal molecules. *Biochemistry* **44,** 8326–8336.

Peake, I. R., and Bieri, J. G. (1971). Alpha- and γ-tocopherol in the rat: *In vitro* and *in vivo* tissue uptake and metabolism. *J. Nutr.* **101,** 1615–1622.

Pelkonen, O., Maenpaa, J., Taavitsainen, P., Rautio, A., and Raunio, H. (1998). Inhibition and induction of human cytochrome P450 (CYP) enzymes. *Xenobiotica* **28,** 1203–1253.

Pellegrini, N., Visioli, F., Buratti, S., and Brighenti, F. (2001). Direct analysis of total antioxidant activity of olive oil and studies on the influence of heating. *J. Agric. Food Chem.* **49,** 2532–2538.

Penning, T. M., Burczynski, M. A., Hung, C. F., McCoull, K. D., Palackal, N. T., and Tsuruda, L. S. (1999). Dihydrodiol dehydrogenases and polycyclic hydrocarbon activation: Generation of reactive and redox active o-quinones. *Chem. Res. Toxicol.* **12,** 1–18.

Peraza, M. A., Burdick, A. D., Marin, H. E., Gonzalez, F. J., and Peters, J. M. (2006). The toxicology of ligands for peroxisome proliferator-activated receptors (PPAR). *Toxicol. Sci.* **90,** 269–295.

Perez-Jimenez, F. (2005). International conference on the healthy effect of virgin olive oil. *Eur. J. Clin. Invest.* **35,** 421–424.

Plack, P. A., and Bieri, J. G. (1964). Metabolic products of α-tocopherol in the livers of rats given intraperitoneal injections of (14c)-α-tocopherol. *Biochim. Biophys. Acta* **84,** 729–738.

Porfirova, S., Bergmuller, E., Tropf, S., Lemke, R., and Dormann, P. (2002). Isolation of an Arabidopsis mutant lacking vitamin E and identification of a cyclase essential for all tocopherol biosynthesis. *Proc. Natl. Acad. Sci. USA* **99,** 12495–12500.

Prabhu, S., Fackett, A., Lloyd, S., McClellan, H. A., Terrell, C. M., Silber, P. M., and Li, A. P. (2002). Identification of glutathione conjugates of troglitazone in human hepatocytes. *Chem. Biol. Interact.* **142,** 83–97.

Quiles, J. L., Huertas, J. R., Battino, M., Ramirez-Tortosa, M. C., Cassinello, M., Mataix, J., Lopez-Frias, M., and Manas, M. (2002). The intake of fried virgin olive or sunflower oils differentially induces oxidative stress in rat liver microsomes. *Br. J. Nutr.* **88,** 57–65.

Rabindran, S. K. (2005). Antitumor activity of HER-2 inhibitors. *Cancer Lett.* **227,** 9–23.

Rabindran, S. K., Discafani, C. M., Rosfjord, E. C., Baxter, M., Floyd, M. B., Golas, J., Hallett, W. A., Johnson, B. D., Nilakantan, R., Overbeek, E., Reich, M. F., Shen, R., *et al.* (2004). Antitumor activity of HKI-272, an orally active, irreversible inhibitor of the HER-2 tyrosine kinase. *Cancer Res.* **64,** 3958–3965.

Rappaport, S. M., Waidyanatha, S., Qu, Q., Shore, R., Jin, X., Cohen, B., Chen, L. C., Melikian, A. A., Li, G., Yin, S., Yan, H., Xu, B., *et al.* (2002). Albumin adducts of benzene oxide and 1,4-benzoquinone as measures of human benzene metabolism. *Cancer Res.* **62,** 1330–1337.

Rennick, K. A., and Warner, K. (2006). Effect of elevated temperature on development of tocopherolquinones in oils. *J. Agric. Food. Chem.* **54,** 2188–2192.

Ricciarelli, R., Tasinato, A., Clement, S., Ozer, N. K., Boscoboinik, D., and Azzi, A. (1998). Alpha-tocopherol specifically inactivates cellular protein kinase C α by changing its phosphorylation state. *Biochem. J.* **334**(Pt. 1), 243–249.

Rosenau, T., Potthast, A., Elder, T., and Kosma, P. (2002). Stabilization and first direct spectroscopic evidence of the o-quinone methide derived from vitamin E. *Org. Lett.* **4,** 4285–4288.

Rosenau, T., Ebner, G., Stanger, A., Perl, S., and Nuri, L. (2004). From a theoretical concept to biochemical reactions: Strain-induced bond localization (SIBL) in oxidation of vitamin E. *Chemistry* **11,** 280–287.

Rowntree, L. G., Steinberg, A., Dorrance, G. M., and Ciccone, E. F. (1937). Sarcoma in rats from the ingestion of a crude wheat-germ oil made by ether extraction. *Am. J. Cancer* **31,** 359–372.

Rutkowski, D. T., and Kaufman, R. J. (2004). A trip to the ER: Coping with stress. *Trends Cell Biol.* **14,** 20–28.

Sachdeva, R., Thomas, B., Wang, X., Ma, J., Jones, K. H., Hatcher, P. G., and Cornwell, D. G. (2005). Tocopherol metabolism using thermochemolysis: Chemical and biological properties of γ-tocopherol, γ-carboxyethyl-hydroxychroman, and their quinones. *Chem. Res. Toxicol.* **18,** 1018–1025.

Saito, H., Kiyose, C., Yoshimura, H., Ueda, T., Kondo, K., and Igarashi, O. (2003). Gamma-tocotrienol, a vitamin E homolog, is a natriuretic hormone precursor. *J. Lipid Res.* **44,** 1530–1535.

Sattler, S. E., Cahoon, E. B., Coughlan, S. J., and DellaPenna, D. (2003). Characterization of tocopherol cyclases from higher plants and cyanobacteria. Evolutionary implications for tocopherol synthesis and function. *Plant Physiol.* **132,** 2184–2195.

Scherder, C. W., Fehr, W. R., Welke, G. A., and Wang, T. (2006). Tocopherol content and agronomic performance of soybean lines with reduced palmitate. *Crop. Sci.* **46,** 1286–1290.

Schonfeld, A., Schultz, M., Petrizka, M., and Gassmann, B. (1993). A novel metabolite of RRR-α-tocopherol in human urine. *Nahrung* **37,** 498–500.

Schroder, M., and Kaufman, R. J. (2005). ER stress and the unfolded protein response. *Mutat. Res.* **569,** 29–63.

Schultz, M., Leist, M., Petrzika, M., Gassmann, B., and Brigelius-Flohe, R. (1995). Novel urinary metabolite of α-tocopherol, 2,5,7,8-tetramethyl-2(2'-carboxyethyl)-6-hydroxy-chroman, as an indicator of an adequate vitamin E supply? *Am. J. Clin. Nutr.* **62,** 1527S–1534S.

Scott, G. K., Atsriku, C., Kaminker, P., Held, J., Gibson, B., Baldwin, M. A., and Benz, C. C. (2005). Vitamin K3 (menadione)-induced oncosis associated with keratin 8 phosphorylation and histone H3 arylation. *Mol. Pharmacol.* **68,** 606–615.

Sen, C. K., Khanna, S., and Roy, S. (2006). Tocotrienols: Vitamin E beyond tocopherols. *Life Sci.* **78,** 2088–2098.

Shah, S., and Sylvester, P. W. (2004). Tocotrienol-induced caspase-8 activation is unrelated to death receptor apoptotic signaling in neoplastic mammary epithelial cells. *Exp. Biol. Med. (Maywood)* **229,** 745–755.

Shah, S. J., and Sylvester, P. W. (2005). Tocotrienol-induced cytotoxicity is unrelated to mitochondrial stress apoptotic signaling in neoplastic mammary epithelial cells. *Biochem. Cell Biol.* **83,** 86–95.

Sharma, C., Pradeep, A., Wong, L., Rana, A., and Rana, B. (2004). Peroxisome proliferator-activated receptor gamma activation can regulate β-catenin levels via a proteasome-mediated and adenomatous polyposis coli-independent pathway. *J. Biol. Chem.* **279,** 35583–35594.

Sheppard, A. J., Pennington, J. A. T., and Weirauch, J. L. (1993). Analysis and distribution of vitamin E in vegetable oils and foods. *In* "Vitamin E in Health and Disease" (L. Packer and J. Fuchs, Eds.), pp. 9–31. Dekker, New York.

Shintani, D., and Della Penna, D. (1998). Elevating the vitamin E content of plants through metabolic engineering. *Science* **282,** 2098–2100.

Shu, X. O., Linet, M. S., Steinbuch, M., Wen, W. Q., Buckley, J. D., Neglia, J. P., Potter, J. D., Reaman, G. H., and Robison, L. L. (1999). Breast-feeding and risk of childhood acute leukemia. *J. Natl. Cancer Inst.* **91,** 1765–1772.

Simopoulos, A. P. (2001). The Mediterranean diets: What is so special about the diet of Greece? The scientific evidence. *J. Nutr.* **131,** 3065S–3073S.

Smith, L. I. (1940). The chemistry of vitamin E. *Chem. Rev.* **27,** 287–329.

Smithgall, T. E., Harvey, R. G., and Penning, T. M. (1988). Spectroscopic identification of ortho-quinones as the products of polycyclic aromatic trans-dihydrodiol oxidation catalyzed by dihydrodiol dehydrogenase. A potential route of proximate carcinogen metabolism. *J. Biol. Chem.* **263,** 1814–1820.

Sontag, T. J., and Parker, R. S. (2002). Cytochrome P450 omega-hydroxylase pathway of tocopherol catabolism. Novel mechanism of regulation of vitamin E status. *J. Biol. Chem.* **277,** 25290–25296.

St-Germain, M. E., Gagnon, V., Parent, S., and Asselin, E. (2004). Regulation of COX-2 protein expression by Akt in endometrial cancer cells is mediated through NF-kappaB/IkappaB pathway. *Mol. Cancer* **3,** 7.

Stocker, A. (2004). Molecular mechanisms of vitamin E transport. *Ann. N Y Acad. Sci.* **1031,** 44–59.

Straus, D. S., Pascual, G., Li, M., Welch, J. S., Ricote, M., Hsiang, C. H., Sengchanthalangsy, L. L., Ghosh, G., and Glass, C. K. (2000). 15-Deoxy-δ 12,14-prostaglandin J2 inhibits multiple steps in the NF-kappa B signaling pathway. *Proc. Natl. Acad. Sci. USA* **97,** 4844–4849.

Sugio, S., Kashima, A., Mochizuki, S., Noda, M., and Kobayashi, K. (1999). Crystal structure of human serum albumin at 2.5 Å resolution. *Protein Eng.* **12,** 439–446.

Sumitomo, M., Ohba, M., Asakuma, J., Asano, T., Kuroki, T., and Hayakawa, M. (2002). Protein kinase Cdelta amplifies ceramide formation via mitochondrial signaling in prostate cancer cells. *J. Clin. Invest.* **109,** 827–836.

Swanson, J. E., Ben, R. N., Burton, G. W., and Parker, R. S. (1999). Urinary excretion of 2,7,8-trimethyl-2-(β-carboxyethyl)-6-hydroxychroman is a major route of elimination of γ-tocopherol in humans. *J. Lipid Res.* **40,** 665–671.

Sylvester, P. W., Nachnani, A., Shah, S., and Briski, K. P. (2002). Role of GTP-binding proteins in reversing the antiproliferative effects of tocotrienols in preneoplastic mammary epithelial cells. *Asia Pac. J. Clin. Nutr.* **11**(Suppl. 7), S452–S459.

Syvaoja, E. L., Piironen, V., Varo, P., Koivistoinen, P., and Salminen, K. (1985). Tocopherols and tocotrienols in Finnish foods: Human milk and infant formulas. *Int. J. Vitam. Nutr. Res.* **55,** 159–166.

Tafazoli, S., Wright, J. S., and O'Brien, P. J. (2005). Prooxidant and antioxidant activity of vitamin E analogues and troglitazone. *Chem. Res. Toxicol.* **18,** 1567–1574.

Takahashi, S., Miura, R., and Miyake, Y. (1988). Characterization of hog kidney renin-binding protein: Interconversion between monomeric and dimeric forms. *Biochem. Int.* **16,** 1053–1060.

Takahashi, S., Takahashi, K., Kaneko, T., Ogasawara, H., Shindo, S., Saito, K., and Kawamura, Y. (2001). Identification of functionally important cysteine residues of the human renin-binding protein as the enzyme N-acetyl-D-glucosamine 2-epimerase. *J. Biochem. (Tokyo)* **129,** 529–535.

Tamai, H., Kim, H. S., Arai, H., Inoue, K., and Mino, M. (1998). Developmental changes in the expression of α-tocopherol transfer protein during the neonatal period of rat. *Biofactors* **7,** 87–91.

Tan, B. (2005). Appropriate spectrum vitamin E and new perspectives on desmethyl tocopherols and tocotrienols. *JANA* **8,** 35–42.

Tasinato, A., Boscoboinik, D., Bartoli, G. M., Maroni, P., and Azzi, A. (1995). d-Alpha-tocopherol inhibition of vascular smooth muscle cell proliferation occurs at physiological concentrations, correlates with protein kinase C inhibition, and is independent of its antioxidant properties. *Proc. Natl. Acad. Sci. USA* **92,** 12190–12194.

Taylor, M. W., and Nelson, V. E. (1930). Some observations on ferric chloride addition to the diet. *Proc. Soc. Exp. Biol. Med.* **27,** 764–766.

Thornton, D. E., Jones, K. H., Jiang, Z., Zhang, H., Liu, G., and Cornwell, D. G. (1995). Antioxidant and cytotoxic tocopheryl quinones in normal and cancer cells. *Free Radic. Biol. Med.* **18,** 963–976.

Tirmenstein, M. A., Hu, C. X., Gales, T. L., Maleeff, B. E., Narayanan, P. K., Kurali, E., Hart, T. K., Thomas, H. C., and Schwartz, L. W. (2002). Effects of troglitazone on HepG2 viability and mitochondrial function. *Toxicol. Sci.* **69,** 131–138.

Traber, M. G., and Arai, H. (1999). Molecular mechanisms of vitamin E transport. *Annu. Rev. Nutr.* **19,** 343–355.

Traber, M. G., Cohen, W., and Muller, D. P. R. (1993). Absorption, transport, and delivery to tissues. *In* "Vitamin E in Health and Disease" (L. Packer and J. Fuchs, Eds.), pp. 35–52. Dekkers, New York.

Traber, M. G., Burton, G. W., and Hamilton, R. L. (2004). Vitamin E trafficking. *Ann. NY Acad. Sci.* **1031,** 1–12.

Traber, M. G., Siddens, L. K., Leonard, S. W., Schock, B., Gohil, K., Krueger, S. K., Cross, C. E., and Williams, D. E. (2005). Alpha-tocopherol modulates Cyp3a expression, increases γ-CEHC production, and limits tissue γ-tocopherol accumulation in mice fed high γ-tocopherol diets. *Free Radic. Biol. Med.* **38,** 773–785.

Trichopoulou, A. (2001). Mediterranean diet: The past and the present. *Nutr. Metab. Cardiovasc. Dis.* **11,** 1–4.

Tsou, H. R., Overbeek-Klumpers, E. G., Hallett, W. A., Reich, M. F., Floyd, M. B., Johnson, B. D., Michalak, R. S., Nilakantan, R., Discafani, C., Golas, J., Rabindran, S. K., Shen, R., *et al.* (2005). Optimization of 6,7-disubstituted-4-(arylamino) quinoline-3-carbonitriles as orally active, irreversible inhibitors of human epidermal growth factor receptor-2 kinase activity. *J. Med. Chem.* **48,** 1107–1131.

Uto, H., Kiyose, C., Saito, H., Ueda, T., Nakamijra, T., Igarashi, O., and Kondo, K. (2004). Gamma-tocopherol enhances sodium excretion as a natriuretic hormone precursor. *J. Nutr. Sci. Vitaminol. (Tokyo)* **50,** 277–282.

van Acker, S. A., Koymans, L. M., and Bast, A. (1993). Molecular pharmacology of vitamin E: Structural aspects of antioxidant activity. *Free Radic. Biol. Med.* **15,** 311–328.

Van De Water, R. W., and Pettus, T. R. R. (2002). o-Quinone methides: Intermediates underdeveloped and underutilized in organic synthesis. *Tetrahedron* **58,** 5367–5405.

Vandoros, G. P., Konstantinopoulos, P. A., Sotiropoulou-Bonikou, G., Kominea, A., Papachristou, G. I., Karamouzis, M. V., Gkermpesi, M., Varakis, I., and Papavassiliou, A. G. (2006). PPAR-γ is expressed and NF-kB pathway is activated and correlates positively with COX-2 expression in stromal myofibroblasts surrounding colon adenocarcinomas. *J. Cancer Res. Clin. Oncol.* **132,** 76–84.

Verleyen, T., Kamal-Eldin, A., Dobarganes, C., Verhe, R., Dewettinck, K., and Huyghebaert, A. (2001a). Modeling of α-tocopherol loss and oxidation products formed during thermoxidation in triolein and tripalmitin mixtures. *Lipids* **36,** 719–726.

Verleyen, T., Verhe, R., Huyghebaert, A., Dewettinck, K., and De Greyt, W. (2001b). Identification of α-tocopherol oxidation products in triolein at elevated temperatures. *J. Agric. Food. Chem.* **49,** 1508–1511.

Visioli, F., and Galli, C. (2002). Biological properties of olive oil phytochemicals. *Crit. Rev. Food Sci. Nutr.* **42,** 209–221.

Vogelstein, B., and Kinzler, K. W. (1993). The multistep nature of cancer. *Trends Genet.* **9,** 138–141.

Waddell, J., and Steenbock, H. (1928). The destruction of vitamin E in a ration composed of natural and varied foodstuffs. *J. Biol. Chem.* **80,** 431–442.

Wagner, K. H., Kamal-Eldin, A., and Elmadfa, I. (2004). γ-Tocopherol—an underestimated vitamin? *Ann. Nutr. Metab.* **48,** 169–188.

Wang, X., Thomas, B., Sachdeva, R., Arterburn, L., Frye, L., Hatcher, P. G., Cornwell, D. G., and Ma, J. (2006). Mechanism of arylating quinone toxicity involving Michael adduct formation and induction of endoplasmic reticulum stress. *Proc. Natl. Acad. Sci. USA* **103,** 3604–3609.

Warner, K. (1999). Impact of high-temperature food processing on fats and oils. *Adv. Exp. Med. Biol.* **459,** 67–77.

Wechter, W. J., Kantoci, D., Murray, E. D., Jr., D'Amico, D. C., Jung, M. E., and Wang, W. H. (1996). A new endogenous natriuretic factor: LLU-α. *Proc. Natl. Acad. Sci. USA* **93,** 6002–6007.

Wilson, G. J., Lin, C. Y., and Webster, R. D. (2006). Significant differences in the electrochemical behavior of the α-, β-, γ-, and δ-tocopherols (vitamin E). *J. Phys. Chem. B Condens. Matter Mater. Surf. Interfaces Biophys.* **110,** 11540–11548.

Wolf, G. (2005). The discovery of the antioxidant function of vitamin E: The contribution of Henry A. Mattill. *J. Nutr.* **135,** 363–366.

Wu, J., and Kaufman, R. J. (2006). From acute ER stress to physiological roles of the unfolded protein response. *Cell Death Differ.* **13,** 374–384.

Yamauchi, R., Noro, H., Shimoyamada, M., and Kato, K. (2002). Analysis of vitamin E and its oxidation products by HPLC with electrochemical detection. *Lipids* **37,** 515–522.

Yamazaki, H., Shibata, A., Suzuki, M., Nakajima, M., Shimada, N., Guengerich, F. P., and Yokoi, T. (1999). Oxidation of troglitazone to a quinone-type metabolite catalyzed by cytochrome P-450 2C8 and P-450 3A4 in human liver microsomes. *Drug Metab. Dispos.* **27,** 1260–1266.

Ye, Y. (2005). The role of the ubiquitin-proteasome system in ER quality control. *Essays Biochem.* **41,** 99–112.

Yoshida, E., Watanabe, T., Takata, J., Yamazaki, A., Karube, Y., and Kobayashi, S. (2006). Topical application of a novel, hydrophilic γ-tocopherol derivative reduces photo-inflammation in mice skin. *J. Invest. Dermatol.* **126,** 1633–1640.

You, C. S., Sontag, T. J., Swanson, J. E., and Parker, R. S. (2005). Long-chain carboxychromanols are the major metabolites of tocopherols and tocotrienols in A549 lung epithelial cells but not HepG2 cells. *J. Nutr.* **135,** 227–232.

Yuki, E., and Ishikawa, Y. (1976). Tocopherol contents of nine vegetable frying oils, and their changes under simulated deep-fat frying conditions. *J. Am. Oil Chem. Soc.* **53,** 673–676.

Zhang, H., Zhang, A., Kohan, D. E., Nelson, R. D., Gonzalez, F. J., and Yang, T. (2005). Collecting duct-specific deletion of peroxisome proliferator-activated receptor gamma blocks thiazolidinedione-induced fluid retention. *Proc. Natl. Acad. Sci. USA* **102,** 9406–9411.

Zhang, K., and Kaufman, R. J. (2006). The unfolded protein response: A stress signaling pathway critical for health and disease. *Neurology* **66,** S102–S109.

Zheng, M. C., Zhou, L. S., and Zhang, G. F. (1993). Alpha-tocopherol content of breast milk in China.. *J. Nutr. Sci. Vitaminol. (Tokyo)* **39,** 517–520.

6

Vitamin E and NF-κB Activation: A Review

Howard P. Glauert

Graduate Center for Nutritional Sciences, University of Kentucky Lexington, Kentucky 40506

Vitamins and Hormones, Volume 76

0083-6729/07 $35.00
DOI: 10.1016/S0083-6729(07)76006-5

Nuclear factor (NF-κB)[1] is a eukaryotic transcription factor that may be activated by oxidative stress. Because of this hypothesis, the effect of vitamin E on NF-κB activation has been examined in many studies, using both *in vivo* and *in vitro* models. Most of these studies have observed that vitamin E inhibits the activation of NF-κB, with the greatest inhibition seen with the succinate form. Vitamin E may be inhibiting NF-κB by reducing oxidative stress or through one of its nonantioxidant functions; this is not clear at the present time. It also is not known if the inhibition of NF-κB is necessary for any of vitamin E's effects on gene expression and the resulting physiological effects.

I. INTRODUCTION

One of the possible roles of vitamin E and other antioxidants is that they may function in the alteration of gene expression. Vitamin E has both antioxidant and nonantioxidant functions, and both could possibly contribute to changes in the expression of specific genes. One mechanism by which vitamin E might participate in altering gene expression is by altering the activation of specific transcription factors. The activation of several transcription factors may be altered by oxidative stress, including nuclear factor (NF-κB). Numerous studies have examined the effect of vitamin E on the activation of NF-κB. The evidence that vitamin E may influence NF-κB activation, possible mechanisms by which vitamin E may influence NF-κB activation, and whether the alteration of NF-κB activation by vitamin E is an important component of vitamin E's function are discussed in this chapter.

[1]Abbreviations: DEM, diethyl maleate; DSS, dextran sulfate sodium; EGF, epidermal growth factor; EPC-K1, phosphate diester linkage of vitamins E and C; EPU, phosphodiester linkage of vitamin E and uracil; GFP, green fluorescent protein; HCH, γ-hexachlorocyclohexane; HIV, human immunodeficiency virus; IFN, interferon; IRFI-016, 2,3-dihydro-5-acetyloxy-4,6,7-trimethyl-2-benzofuranacetic acid; IRFI-042, butanedioic acid, mono[2-[2-(acetylthio) ethyl]-2,3-dihydro-4,6,7-trimethyl-5-benzofuranyl] ester, (±)-; IKK, IκB kinase; IL, interleukin; LDL, low-density lipoprotein; LPS, lipopolysaccharide; LTR, long terminal repeat; MitoVit E, [2-(3,4-dihydro-6-hydroxy-2,5,7,8-tetramethyl-2H-1-benzopyran-2-yl)ethyl]triphenylphosphonium bromide; NAC, *N*-acetyl cysteine; NF-κB, nuclear factor-κB; PAF, platelet-activating factor; PCB, polychlorinated biphenyl; PCB-77, 3,3′,4,4′-tetrachlorobiphenyl; PGE_2, prostaglandin E_2; PKC, protein kinase C; PMA, phorbol myristate acetate; PMC, 2,2,5,7,8-pentamethyl-6-hydroxychromane; PPAR, peroxisome proliferator-activated receptor; PP2A, protein phosphatase 2A; TNF-α, tumor necrosis factor-α; TRAIL, TNF-related apoptosis-inducing ligand; Trolox, 6-hydroxyl-2,5,7,8-tetramethylchroman-2-carboxylic acid; VLDL, very low density lipoprotein.

II. Nuclear Factor-κB

NF-κB is a eukaryotic transcription factor family consisting of the following proteins: p50 (NF-κB1), p65 (RelA), p52 (NF-κB2), c-Rel, and RelB. It is normally found in the cytoplasm as an inactive dimer, with the most common being the p50–p65 heterodimer, bound to an inhibitory subunit, IκB, which also has several family members, including IκBα, IκBβ, IκBγ, and IκBε (Karin and Lin, 2002). On activation, NF-κB is released from IκB and translocates to the nucleus, where it increases the transcription of specific genes. This process requires the phosphorylation of IκB, followed by the subsequent degradation via an ubiquitin-mediated 26S proteosome pathway (Karin and Lin, 2002). A 900-kDa complex, termed the IκB kinase (IKK) complex, has been identified and consists of two kinase subunits of IKK, IKKα and IKKβ, and a regulatory subunit, IKKγ (Karin and Delhase, 2000; Zandi *et al.*, 1997). These two kinase subunits form homo- or heterodimers that phosphorylate IκB molecules, leading to their degradation. The activation pathway for the p50–p65 heterodimer has been referred to as the classical or canonical NF-κB signaling pathway, and is dependent on the IKKβ and IKKγ subunits of IKK (Karin, 2006). An alternative NF-κB signaling pathway has also been identified in which IKKα is required and results in the activation of the p52–RelB heterodimer (Karin, 2006).

NF-κB has been shown to be important in the activation of genes including several that regulate the immune response, cell proliferation, and apoptosis (Beg *et al.*, 1995; Fitzgerald *et al.*, 1995). Several studies have used genetically modified mice to examine the role of NF-κB subunits in these functions. Knockout mice have been developed for all of the NF-κB subunits (Beg *et al.*, 1995; Franzoso *et al.*, 1998; Kontgen *et al.*, 1995; Sha *et al.*, 1995; Weih *et al.*, 1995); in addition, knockouts for specific tissues, such as the liver, have been developed (Pikarsky *et al.*, 2004; Sakurai *et al.*, 2006). A clear role for NF-κB in inhibiting apoptosis by TNF-α or other apoptosis inducers has been demonstrated in several cell types, in studies in which NF-κB activity has been inhibited by the deletion of one of its subunits, the inhibition of its translocation, or the expression of a dominant negative form of IκB (Beg and Baltimore, 1996; Beg *et al.*, 1995; Schoemaker *et al.*, 2002; Sha *et al.*, 1995; Vanantwerp *et al.*, 1996; Wang *et al.*, 1996; Xu *et al.*, 1998). The deletion of the p65 subunit leads to embryonic lethality at 15–16 days of gestation due to hepatocyte apoptosis (Beg *et al.*, 1995). The deletion of the p50 subunit led to defects in the immune response involving B cells (Sha *et al.*, 1995). Hepatocyte apoptosis is higher in $p50^{-/-}$ mice (Lu *et al.*, 2004; Tharappel *et al.*, 2003), but it is not lethal as in the p65 knockout. However, DNA synthesis and liver regeneration were not affected by the absence of the p50 subunit following partial hepatectomy or carbon tetrachloride treatment; increased levels of the p65 subunit may have compensated for the lack of p50 (Deangelis *et al.*, 2001). Similarly, the hepatic-specific expression of a truncated IκBα superrepressor did not affect DNA synthesis, apoptosis, or liver regeneration following partial hepatectomy, but led to increased apoptosis after treatment with TNF-α (Chaisson *et al.*, 2002). Also, the hepatic inflammatory response after ischemia/reperfusion was not altered in p50$^{-/-}$ mice (Kato *et al.*, 2002). The deletion of p52 leads to defects in humoral immunity and splenic architecture (Franzoso *et al.*, 1998). In RelB$^{-/-}$ mice, multiorgan inflammation, impaired cellular immunity, and hematopoietic abnormalities are observed (Weih *et al.*, 1995). The deletion of c-Rel leads to defects in humoral immunity and in the proliferation of T cells in response to mitogens (Kontgen *et al.*, 1995). In addition, B cells lacking p50, RelB, or c-Rel (but not p52 or p65) have decreased proliferation in response to LPS (Horwitz *et al.*, 1999; Kontgen *et al.*, 1995; Sha *et al.*, 1995; Snapper *et al.*, 1996a,b).

One mechanism by which NF-κB may be activated is by increased oxidative stress. NF-κB can be activated *in vitro* by H_2O_2, and its activation can be inhibited by antioxidants, such as *N*-acetyl cysteine (NAC), or by increased expression of antioxidant enzymes (Meyer *et al.*, 1993; Nilakantan *et al.*, 1998; Schreck *et al.*, 1991, 1992; Staal *et al.*, 1990). In addition, agents that activate NF-κB frequently also increase oxidative stress (Schmidt *et al.*, 1995). However, Hayakawa *et al.* (2003) found that NAC inhibits NF-κB activation independently of its antioxidant function. And as of yet no target for reactive oxygen species has been clearly demonstrated in the NF-κB signal transduction pathway (Bowie and Oneill, 2000; Li and Karin, 1999).

III. *IN VITRO* STUDIES

Vitamin E was first observed to influence the activation of NF-κB in cultured cells. The hypothesis that NF-κB activation could be influenced by antioxidants was first tested by Staal *et al.* (1990), who found that NAC inhibited the activation of NF-κB by TNF-α and PMA (Staal *et al.*, 1990). Israel *et al.* (1992) subsequently tested whether α-tocopherol and other lipid soluble antioxidants (BHA, nordihydroquairetic acid) could also inhibit TNF-α- and PMA-induced NF-κB activation, and found that these antioxidants decreased the transactivation of the long terminal repeat (LTR) of HIV, which contains two NF-κB-binding sites. Shortly afterward, Suzuki and Packer (1993a,b) in two studies examined effects of vitamin E derivatives on TNF-α- or okadaic acid-induced NF-κB activation in human Jurkat cells. When cells were pretreated with one of the vitamin E derivatives, α-tocopheryl acetate, α-tocopheryl succinate, and the vitamin E derivative PMC inhibited TNF-α-induced NF-κB DNA-binding activity, whereas α-tocopherol did not. However, when the derivatives were not added until after nuclear extracts had been prepared only α-tocopheryl succinate inhibited the DNA-binding activity, implying that this vitamin E derivative may be acting late in signal transduction. In addition, only α-tocopheryl succinate inhibited NF-κB DNA-binding activity induced by okadaic acid, an inhibitor of protein phosphatases 1 and 2A.

Following these initial observations, numerous investigators have examined the effect of vitamin E on NF-κB activation *in vitro* in many different cell types. These studies are discussed below.

A. IMMUNE SYSTEM

Many studies have been carried out using primary immune system cells or immune cell lines, including four additional studies in Jurkat cells: three found that α-tocopherol inhibited NF-κB activation by PMA, PMA plus ionomycin, or TNF-related apoptosis-inducing ligand (TRAIL); α-tocopheryl succinate also inhibited TRAIL-induced NF-κB activation (Dalen and Neuzil, 2003; Li-Weber *et al.*, 2002a,b). However, Trolox, a water-soluble derivative of vitamin E, did not affect NF-κB activation by TNF-α in Jurkat cells (Hayakawa *et al.*, 2003). The results in other cell lines have varied, depending on the cell type, NF-κB activator, and form of vitamin E used. In U937 cells, α-tocopherol inhibited the DNA-binding activity induced by glutathione, glycine–glycine and transferrin (Djavaheri-Mergny *et al.*, 2002), and Mito VitE, a derivative of vitamin that is targeted to mitochondria, inhibited NF-κB activation by TNF-α (Hughes *et al.*, 2005). In Thp-1 cells, α-tocopheryl succinate inhibited LPS- and VLDL-induced NF-κB activation, but α-tocopherol did not affect DNA binding in cells treated with LPS (Blanco-Colio *et al.*, 2000; Nakamura *et al.*, 1998). The succinate form was also more active in HL-60 cells: α-tocopheryl succinate inhibited NF-κB activation in the presence of absence of vitamin D_3, whereas α-tocopheryl acetate only was effective in the presence of vitamin D_3 (Sokoloski *et al.*, 1997). In RAW 264.7 macrophage cells, α-tocopheryl succinate inhibited platelet-activating factor (PAF) and H_2O_2-induced NF-κB activation, and the vitamin E derivative PMC inhibited lipoteichoic acid-induced IκB degradation (Choi *et al.*, 2000; Hsiao *et al.*, 2004). However, α-tocopheryl acetate did not affect NF-κB activation by LPS in J774 macrophage cells (Abate *et al.*, 2000). Interestingly, the inhibition of LPS-induced NF-κB activation by oxidized LDL in RAW macrophage cells was prevented by α-tocopherol (Muroya *et al.*, 2003).

In primary cells, vitamin E was also effective at inhibiting NF-κB activation. Vitamin E inhibited NF-κB activation induced by LPS in human monocytes (Hill *et al.*, 1999) and mouse neutrophils (Asehnoune *et al.*, 2004). In addition, the vitamin E derivatives IRFI-042 (Altavilla *et al.*, 2001, 2002) or 6-hydroxy-7-methoxychroman-2-carboxylic acid phenylamide (CP compound) (Rak Min *et al.*, 2005) were found to inhibit NF-κB activation induced by okadaic acid, H_2O_2, PMA, ceramide, or LPS (for IRFI-042) or LPS only (for CP compound) in peritoneal macrophages. In primary dendritic cells, the administration of both α-tocopherol and vitamin C inhibited NF-κB activation induced by the combination of interleukin (IL)-1β, TNF-α, IL-6, LPS, PGE_2, and interferon (IFN)-γ (Tan *et al.*, 2005). In primary splenocytes treated with PMA and ionomycin, α-tocopherol increased the expression of IκBα mRNA (Hsieh *et al.*, 2006).

B. CARDIOVASCULAR

Two cell types from the cardiovascular system have been examined: endothelial cells and cardiomyocytes. Primary endothelial cells were from a variety of sources: human umbilical veins, human coronary arteries, porcine pulmonary arteries, and bovine retinal endothelial cells. Endothelial cell lines have also been used. NF-κB was activated using a variety of substances: IL-1 (Faruqi *et al.*, 1994), oxidized LDL (Li *et al.*, 1999, 2000a,b; Maziere *et al.*, 1996), fatty acids (Hennig *et al.*, 1996, 2000; Toborek *et al.*, 1996), TNF-α (Erl *et al.*, 1997; Neuzil *et al.*, 2001; Theriault *et al.*, 2002; Toborek *et al.*, 1996), phorbol myristate acetate (Roy *et al.*, 1998), lysophosphatidylcholine (Sugiyama *et al.*, 1998), 3,3′,4,4′-tetrachlorobiphenyl (Slim *et al.*, 1999), advanced glycation end products (Mamputu and Renier, 2004), and plasma from women with severe preeclampsia (Takacs *et al.*, 2001). When specified, a number of vitamin E derivatives were used: α-tocopherol, α-tocopheryl acetate or succinate, γ-tocopherol, or α-tocotrienol. In most of the studies, the vitamin E derivative used inhibited NF-κB activation. However, α-tocopherol had no effect on NF-κB DNA-binding activity induced by IL-1(Maziere *et al.*, 1996). α-Tocopheryl succinate but not the acetate form or free α-tocopherol was found to inhibit NF-κB activation induced by TNF-α (Erl *et al.*, 1997; Neuzil *et al.*, 2001); Toborek *et al.* (1996), however, found that α-tocopherol inhibited NF-κB activation by TNF-α.

The other cell type examined from the cardiovascular system was the cardiomyocyte. Hirotani *et al.* (2002) found that vitamin E inhibited the increases in NF-κB-regulated luciferase activity and IκB degradation in primary rat neonatal ventricular myocytes exposed to angiotensin II, endothelin-1, or phenylephrine. Sauer *et al.* (2004) observed that Trolox, a water-soluble derivative of vitamin E, inhibited the nuclear translocation of p65 in mouse cardiomyocytes derived from embryonic stem cells exposed to cardiotropin-1.

C. NEURAL

In HT-22 hippocampal cells, vitamin E inhibited haloperidol-induced NF-κB activation; however, in the absence of an NF-κB activator, α-tocopherol by itself increased NF-κB-related gene expression (Behl, 2000; Post *et al.*, 1998). In both N9 and BV-2 microglial cell lines, α-tocopherol inhibited NF-κB activation induced by LPS (Egger *et al.*, 2003; Li *et al.*, 2001). Vitamin E similarly inhibited IκB degradation induced by H_2O_2 in SH-SY5Y neural cells (Uberti *et al.*, 2004). However, in T98G astrocytoma cells, TNF-α-induced NF-κB activation was not inhibited by vitamin E, but was inhibited by a compound consisting of a phosphodiester linkage of vitamins E and C (Hirano *et al.*, 1998). In primary rat astrocytes, glutamate-induced NF-κB activation was also inhibited by a synthetic vitamin E analogue, IRFI-016 (Caccamo *et al.*, 2005).

D. LIVER

The liver consists of several different cell types, including hepatocytes (which make up the majority of the cells in the liver), Kupffer cells, stellate cells, bile duct cells, and oval cells. Cell lines or primary cells derived from three of these cell types have been used to determine if vitamin E influences NF-κB activation. In hepatoma cell lines derived from transformed hepatocytes, vitamin E inhibited NF-κB activation by the peroxisome proliferator ciprofibrate in H4IIEC3 cells and by TNF-α in Hep G2 cells (Li *et al.*, 2000c; Liu *et al.*, 1995). In primary Kupffer cells, NF-κB activation induced by TNF-α (Hill *et al.*, 1999) or LPS (Bellezzo *et al.*, 1998) was blocked by α-tocopheryl succinate or α-tocopherol, respectively. However, α-tocopheryl succinate had no effect on NF-κB activation in primary Kupffer cells induced by bile duct ligation (Fox *et al.*, 1997). In primary stellate cells, vitamin E inhibited NF-κB activation by either CCl_4 or the use of collagen plate coating (Lee *et al.*, 2001).

E. EPITHELIAL CELLS

Vitamin E was found to inhibit NF-κB activation in all epithelial cell lines studied from the prostate (Gunawardena *et al.*, 2004), mammary gland (Shah and Sylvester, 2005; Sylvester *et al.*, 2005), and lung (Lee *et al.*, 2005). γ-Tocotrienol as well as vitamin E and the combination of vitamins C and E were effective at inhibiting NF-κB activation.

F. FIBROBLASTS

Two studies have been carried out in MRC5 human fibroblasts. In both studies, vitamin E was found to block an inhibitory effect on NF-κB activation by another agent. Vitamin E prevented the inhibitory effect of UVA radiation on the activation of NF-κB by epidermal growth factor (Maziere *et al.*, 2003), and the inhibitory effect of oxidized LDL on the activation of NF-κB by insulin (Maziere *et al.*, 2004).

G. OTHER

Studies in other tissues also have observed an inhibitory effect of vitamin E on NF-κB activation: by crocidolite asbestos in RFL-6 lung fibroblasts (Faux and Howden, 1997), by diethyl maleate/advanced glycation end products in primary renal mesangial cells (Lal *et al.*, 2002), by UVB radiation in 308 epidermal cells (Maalouf *et al.*, 2002), and by hydrogen peroxide in L6 myoblasts (Aoi *et al.*, 2004).

IV. *IN VIVO* STUDIES

As in the *in vitro* studies, most of the *in vivo* studies found that vitamin E inhibited NF-κB activation. The liver has been examined in several studies. The first published *in vivo* study was performed by Liu *et al.* (1995). Vitamin E inhibited both hepatic toxicity and NF-κB activation induced by an acute dose of carbon tetrachloride, raising the possibility that NF-κB activation is important in the hepatic toxicity of CCl_4. In other studies examining the liver, dietary vitamin E was found to inhibit NF-κB activation induced by the environmental agents phenobarbital, γ-hexachlorocyclohexane (HCH), and the peroxisome proliferator ciprofibrate, but did not affect NF-κB activation induced by 3,3′,4,4′-tetrachlorobiphenyl (PCB-77) (Calfee-Mason *et al.*, 2002b, 2004; Glauert *et al.*, 2005; Videla *et al.*, 2004). The induction of a vitamin E deficiency was also sufficient to increase the DNA-binding activity of NF-κB in the liver (Morante *et al.*, 2005).

Two studies examining the lung obtained different results. Choi *et al.* (2000) examined the effect of vitamin E on lethality and NF-κB activation induced by PAF in the lungs of mice. Although lethality was not affected by the injection of vitamin E, NF-κB activation was inhibited. Rocksén *et al.* (2003) examined the effect of vitamin E on the induction of lung injury by LPS. Although vitamin E did prevent lung injury, it did not affect the DNA-binding activity of NF-κB induced by LPS.

Several studies have examined the effect of vitamin E in the cardiovascular system *in vivo*. Rodriguez-Porcel *et al.* (2002) fed an atherogenic diet containing 2% cholesterol and 15% lard to pigs for 12 weeks, which increased the expression of the p65 subunit of NF-κB in the arterial tissue, using both Western blotting and immunohistochemistry. The administration of a combination of vitamins E and C blocked this increase. Horton *et al.* (2001) examined the effect of combined vitamins A, C, and E on the activation of NF-κB in the heart by skin burns. The combined treatment decreased the DNA-binding activity of NF-κB. The combination of vitamins A, C, and E plus zinc was also found to inhibit NF-κB activation in the heart induced by the intratracheal delivery of *Streptococcus pneumoniae* (Carlson *et al.*, 2006).

Studies in other tissues all observed an inhibitory effect of vitamin E. Poynter and Daynes (1998) examined the activation of NF-κB in spleens of peroxisome proliferator-activated receptor-α (PPAR-α)-deficient and wild-type mice. NF-κB was activated both in aged (15-month-old) mice compared to 2-month-old mice and in mice fed vitamin E-deficient diets. In both cases, the DNA-binding activity of NF-κB was decreased after administering α-tocopherol in the diet for 2 weeks. Post *et al.* (2002) examined the effect of vitamin E on the neurotoxicity of the drug haloperidol. Vitamin E inhibited the activation of NF-κB by haloperidol in the hippocampus, and also decreased the phosphorylation of IκB. Aoi *et al.* (2004) found that the activation of NF-κB in the gastrocnemius muscle by exercise was inhibited by vitamin E. In human diabetic patients, dietary vitamin E inhibited the high level of NF-κB activation seen in untreated patients (Cipollone *et al.*, 2005). The activation of NF-κB by dextran sulfate sodium in the colon was found not to be affected by dietary vitamin E (Carrier *et al.*, 2006).

The vitamin E analogues IRFI-042, IRFI-016, and α-tocopheryl polyethylene glycol-100 succinate (α-TPGS) were also examined. The ability of IRFI-042 to prevent myocardial injury induced by ischemia–reperfusion was examined by Altavilla *et al.* (2000). NF-κB activation induced by ischemia–reperfusion was blocked by i.p. injection of IRFI-042. Altavilla *et al.* (2001) examined the effect of IRFI-042 on hepatic NF-κB activation induced by hypovolemic hemorrhagic shock and found that it inhibited NF-κB at those time points where it was induced. In the liver, IRFI-042 also inhibited NF-κB activation induced by LPS (Altavilla *et al.*, 2002), and IRFI-016 inhibited ethanol-induced NF-κB activation (Altavilla *et al.*, 2005). Nguyen *et al.* (2006) observed that the activation of NF-κB during the rejection of cardiac allographs was not affected by α-TPGS.

V. MECHANISMS BY WHICH VITAMIN E MAY INHIBIT NF-κB ACTIVATION

The function of vitamin E as a lipid-soluble antioxidant is well known, but several nonantioxidant functions have also been documented (Brigelius-Flohe and Traber, 1999; Brigelius-Flohe *et al.*, 2002; Rimbach *et al.*, 2002; Zingg and Azzi, 2004). α-Tocopherol has been shown to inhibit protein kinase C (PKC), which is brought about by increasing the activity of protein phosphatase 2A (PP2A), which dephosphorylates PKC (Rimbach *et al.*, 2002; Zingg and Azzi, 2004). Certain forms of vitamin E inhibit the eicosanoid-synthesizing enzymes phospholipase A_2, cyclooxygenase-2, and 5-lipoxygenase (Rimbach *et al.*, 2002; Zingg and Azzi, 2004). Vitamin E has been shown to modify the expression of several specific genes, but the mechanisms by which it does so are not clear (Zingg and Azzi, 2004).

The mechanisms by which vitamin E may inhibit NF-κB activation are likely related to one of these functions. If NF-κB is activated by oxidative stress, which currently is not clear as described earlier, the antioxidant function of vitamin E would likely play a role. Vitamin E may block the formation of lipoxygenase metabolites produced during lipid peroxidation, which may then activate NF-κB (Faux and Howden, 1997). Vitamin E inhibits the oxidation of LDL, which have been shown to be activators of NF-κB in their oxidized form (Li *et al.*, 1999, 2000a,b; Maziere *et al.*, 1996). Vitamin E's nonantioxidant functions are likely to be important. Several PKC isoforms can activate NF-κB (Moscat *et al.*, 2003); since vitamin E can inhibit PKC, it is possible that this could be an important mechanism. In addition to its effect on PKC, PP2A can also influence NF-κB activation by binding to and dephosphorylating the p65 subunit of NF-κB (Yang *et al.*, 2001). Other important targets of vitamin E are the NAD(P)H oxidases, which also may activate NF-κB, although it is not clear if the forms that are inhibited by vitamin E are the same forms that may activate NF-κB (Brar *et al.*, 2003; Zingg and Azzi, 2004).

The particular chemical form of vitamin E that is administered also may be important. Several studies have observed that α-tocopheryl succinate inhibited the activation of NF-κB, whereas free α-tocopherol did not (Dalen and Neuzil, 2003; Erl *et al.*, 1997; Nakamura *et al.*, 1998; Neuzil *et al.*, 2001; Suzuki and Packer, 1993a,b). Numerous other studies, however, have observed that free α-tocopherol or α-tocopheryl acetate inhibited NF-κB activation. The mechanisms of action of α-tocopheryl succinate that have been proposed include the blocking of the binding of NF-κB to DNA (Suzuki and Packer, 1993b) increased access to intracellular compartments (Erl *et al.*, 1997), and increased cleavage of the p65 subunit by caspases (Neuzil *et al.*, 2001). Another possible mechanism relates to the observation that the administration of α-tocopheryl succinate to cultured cells results in the enhanced accumulation of α-tocopherol in mitochondria compared with the administration of α-tocopherol, although using higher concentrations of α-tocopherol can overcome this difference (Fariss *et al.*, 2001; Zhang *et al.*, 2001). These studies imply that the concentration of α-tocopherol in mitochondria is important in NF-κB activation. It is not clear how altered vitamin E concentrations in mitochondria may influence NF-κB activation. Several pathways have been identified in which mitochondria function in NF-κB signaling and which therefore could be a mechanism for the effects of vitamin E on NF-κB. Mitochondrial antiviral signaling protein (MAVS), also known as virus-induced signaling adaptor (VISA) or IFN-β promoter stimulator 1 (IPS-1), activates NF-κB in response to viral infection (Kawai *et al.*, 2005; Seth *et al.*, 2005; Xu *et al.*, 2005). Mitochondrial stress can lead to NF-κB activation via activation of calcineurin, which can then directly inactivate IκBβ (Biswas *et al.*, 2003).

VI. IS THE INHIBITION OF NF-κB ACTIVATION NECESSARY FOR SOME OF THE ACTIVITIES OF VITAMIN E?

Another important question is if vitamin E is exerting some of its effects by blocking the activation of NF-κB. The use of NF-κB knockout models may provide answers to this question. A study has addressed this question by examining if the inhibition of NF-κB by vitamin E is necessary for vitamin E's effects on the induction of cell proliferation by the peroxisome proliferator ciprofibrate and on the inhibition of apoptosis by ciprofibrate (Calfee-Mason *et al.*, 2002a). Wild-type and $p50^{-/-}$ mice were administered ciprofibrate and one of two levels of vitamin E (10 or 250 mg/kg diet). Vitamin E inhibited ciprofibrate-induced cell proliferation only in the $p50^{-/-}$ mice. Dietary vitamin E also increased apoptosis and increased the GSH/GSSG ratio in both wild-type and $p50^{-/-}$ mice. In this study, therefore, vitamin E does not seem to be acting through decreased NF-κB activation, suggesting that vitamin E is acting by other molecular mechanisms. Clearly, other studies using knockout models are necessary to examine whether NF-κB is necessary for other actions of vitamin E.

VII. SUMMARY

Numerous studies, both *in vitro* and *in vivo*, have been published examining the effects of vitamin E on NF-κB activation. Most of these have observed an inhibition with higher levels of vitamin E. Some studies have observed a greater inhibition with the succinate form. At this time, it is not clear if the inhibition of NF-κB activation of vitamin E is from decreased oxidative stress and/or one of vitamin E's nonantioxidant functions. Many more studies are needed to determine the mechanism by which vitamin E inhibits NF-κB activation as well as to determine if the inhibition of NF-κB activation is necessary for at least part of the effects of vitamin E on gene expression.

REFERENCES

Abate, A., Yang, G., Dennery, P. A., Oberle, S., and Schroder, H. (2000). Synergistic inhibition of cyclooxygenase-2 expression by vitamin E and aspirin. *Free Radic. Biol. Med.* **29,** 1135–1142.

Altavilla, D., Deodato, B., Campo, G. M., Arlotta, M., Miano, M., Squadrito, G., Saitta, A., Cucinotta, D., Ceccarelli, S., Ferlito, M., Tringali, M., Minutoli, L., *et al.* (2000). IRFI 042, a novel dual vitamin E-like antioxidant, inhibits activation of nuclear factor-κB and reduces the inflammatory response in myocardial ischemia-reperfusion injury. *Cardiovasc. Res.* **47,** 515–528.

Altavilla, D., Saitta, A., Guarini, S., Galeano, M., Squadrito, G., Cucinotta, D., Santamaria, L. B., Mazzeo, A. T., Campo, G. M., Ferlito, M., Minutoli, L., Bazzani, C., *et al.* (2001). Oxidative stress causes nuclear factor-κB activation in acute hypovolemic hemorrhagic shock. *Free Radic. Biol. Med.* **30,** 1055–1066.

Altavilla, D., Squadrito, G., Minutoli, L., Deodato, B., Bova, A., Sardella, A., Seminara, P., Passaniti, M., Urna, G., Venuti, S. F., Caputi, A. P., and Squadrito, F. (2002). Inhibition of nuclear factor-κB activation by IRFI 042, protects against endotoxin-induced shock. *Cardiovasc. Res.* **54,** 684–693.

Altavilla, D., Marini, H., Seminara, P., Squadrito, G., Minutoli, L., Passaniti, M., Bitto, A., Calapai, G., Calo, M., Caputi, A. P., and Squadrito, F. (2005). Protective effects of antioxidant raxofelast in alcohol-induced liver disease in mice. *Pharmacology* **74,** 6–14.

Aoi, W., Naito, Y., Takanami, Y., Kawai, Y., Sakuma, K., Ichikawa, H., Yoshida, N., and Yoshikawa, T. (2004). Oxidative stress and delayed-onset muscle damage after exercise. *Free Radic. Biol. Med.* **37,** 480–487.

Asehnoune, K., Strassheim, D., Mitra, S., Kim, J. Y., and Abraham, E. (2004). Involvement of reactive oxygen species in Toll-like receptor 4-dependent activation of NF-κB. *J. Immunol.* **172,** 2522–2529.

Beg, A. A., and Baltimore, D. (1996). An essential role for NF-κB in preventing TNF-α-induced cell death. *Science* **274,** 782–784.

Beg, A. A., Sha, W. C., Bronson, R. T., Ghosh, S., and Baltimore, D. (1995). Embryonic lethality and liver degeneration in mice lacking Rel A component of NF-κB. *Nature* **376,** 167–170.

Behl, C. (2000). Vitamin E protects neurons against oxidative cell death *in vitro* more effectively than 17-β estradiol and induces the activity of the transcription factor NF-κB. *J. Neural Transm.* **107,** 393–407.

Bellezzo, J. M., Leingang, K. A., Bulla, G. A., Britton, R. S., Bacon, B. R., and Fox, E. S. (1998). Modulation of lipopolysaccharide-mediated activation in rat Kupffer cells by antioxidants. *J. Lab. Clin. Med.* **131,** 36–44.

Biswas, G., Anandatheerthavarada, H. K., Zaidi, M., and Avadhani, N. G. (2003). Mitochondria to nucleus stress signaling: A distinctive mechanism of NF-κB/Rel activation through calcineurin-mediated inactivation of IκBβ. *J. Cell Biol.* **161,** 507–519.

Blanco-Colio, L. M., Valderrama, M., Alvarez-Sala, L. A., Bustos, C., Ortego, M., Hernandez-Presa, M. A., Cancelas, P., Gomez-Gerique, J., Millan, J., and Egido, J. (2000). Red wine intake prevents nuclear factor-κB activation in peripheral blood mononuclear cells of healthy volunteers during postprandial lipemia. *Circulation* **102,** 1020–1026.

Bowie, A., and Oneill, L. A. J. (2000). Oxidative stress and nuclear factor-κB activation—a reassessment of the evidence in the light of recent discoveries. *Biochem. Pharmacol.* **59,** 13–23.

Brar, S. S., Kennedy, T. P., Quinn, M., and Hoidal, J. R. (2003). Redox signaling of NF-κB by membrane NAD(P)H oxidases in normal and malignant cells. *Protoplasma* **221,** 117–127.

Brigelius-Flohe, R., and Traber, M. G. (1999). Vitamin E: Function and metabolism. *FASEB J.* **13,** 1145–1155.

Brigelius-Flohe, R., Kelly, F. J., Salonen, J. T., Neuzil, J., Zingg, J. M., and Azzi, A. (2002). The European perspective on vitamin E: Current knowledge and future research. *Am. J. Clin. Nutr.* **76,** 703–716.

Caccamo, D., Campisi, A., Marini, H., Adamo, E. B., Li Volti, G., Squadrito, F., and Ientile, R. (2005). Glutamate promotes NF-κB pathway in primary astrocytes: Protective effects of IRFI 016, a synthetic vitamin E analogue. *Exp. Neurol.* **193,** 377–383.

Calfee-Mason, K. G., Lee, E. Y., Glauert, H. P., and Spear, B. T. (2002a). Effects of vitamin E on NF-kB downstream events in p50−/− mice treated with the peroxisome proliferator, ciprofibrate. *Toxicol. Sci.(Suppl.)* **66,** 306.

Calfee-Mason, K. G., Spear, B. T., and Glauert, H. P. (2002b). Vitamin E inhibits hepatic NF-κB activation in rats administered the hepatic tumor promoter, phenobarbital. *J. Nutr.* **132,** 3178–3185.

Calfee-Mason, K. G., Spear, B. T., and Glauert, H. P. (2004). Effects of vitamin E on the NF-kB pathway in rats treated with the peroxisome proliferator, ciprofibrate. *Toxicol. Appl. Pharmacol.* **199,** 1–9.

Carlson, D., Maass, D. L., White, D. J., Tan, J., and Horton, J. W. (2006). Antioxidant vitamin therapy alters sepsis-related apoptotic myocardial activity and inflammatory responses. *Am. J. Physiol. Heart Circ. Physiol.* **291**(6), H2779–H2789.

Carrier, J. C., Aghdassi, E., Jeejeebhoy, K., and Allard, J. P. (2006). Exacerbation of dextran sulfate sodium-induced colitis by dietary iron supplementation: Role of NF-κB. *Int. J. Colorectal Dis.* **21,** 381–387.

Chaisson, M. L., Brooling, J. T., Ladiges, W., Tsai, S., and Fausto, N. (2002). Hepatocyte-specific inhibition of NF-κB leads to apoptosis after TNF treatment, but not after partial hepatectomy. *J. Clin. Invest.* **110,** 193–202.

Choi, J. H., Chung, W. J., Han, S. J., Lee, H. B., Choi, I. W., Lee, H. K., Jang, K. Y., Lee, D. G., Han, S. S., Park, K. H., and Im, S. Y. (2000). Selective involvement of reactive oxygen intermediates in platelet-activating factor-mediated activation of NF-κB. *Inflammation* **24,** 385–398.

Cipollone, F., Chiarelli, F., Iezzi, A., Fazia, M. L., Cuccurullo, C., Pini, B., De Cesare, D., Torello, M., Tumini, S., Cuccurullo, F., and Mezzetti, A. (2005). Relationship between reduced BCL-2 expression in circulating mononuclear cells and early nephropathy in type 1 diabetes. *Int. J. Immunopathol. Pharmacol.* **18,** 625–635.

Dalen, H., and Neuzil, J. (2003). α-Tocopheryl succinate sensitises a T lymphoma cell line to TRAIL-induced apoptosis by suppressing NF-κB activation. *Br. J. Cancer* **88,** 153–158.

Deangelis, R. A., Kovalovich, K., Cressman, D. E., and Taub, R. (2001). Normal liver regeneration in p50/nuclear factor κB1 knockout mice. *Hepatology* **33,** 915–924.

Djavaheri-Mergny, M., Accaoui, M. J., Rouillard, D., and Wietzerbin, J. (2002). γ-Glutamyl transpeptidase activity mediates NF-κB activation through lipid peroxidation in human leukemia U937 cells. *Mol. Cell. Biochem.* **232,** 103–111.

Egger, T., Schuligoi, R., Wintersperger, A., Amann, R., Malle, E., and Sattler, W. (2003). Vitamin E (α-tocopherol) attenuates cyclo-oxygenase 2 transcription and synthesis in immortalized murine BV-2 microglia. *Biochem. J.* **370,** 459–467.

Erl, W., Weber, C., Wardemann, C., and Weber, P. C. (1997). α-Tocopheryl succinate inhibits monocytic cell adhesion to endothelial cells by suppressing NF-κB mobilization. *Am. J. Physiol.* **273,** H634–H640.

Fariss, M. W., Nicholls-Grzemski, F. A., Tirmenstein, M. A., and Zhang, J. G. (2001). Enhanced antioxidant and cytoprotective abilities of vitamin E succinate is associated with a rapid uptake advantage in rat hepatocytes and mitochondria. *Free Radic. Biol. Med.* **31,** 530–541.

Faruqi, R., de la Motte, C., and DiCorleto, P. E. (1994). α-Tocopherol inhibits agonist-induced monocytic cell adhesion to cultured human endothelial cells. *J. Clin. Invest.* **94,** 592–600.

Faux, S. P., and Howden, P. J. (1997). Possible role of lipid peroxidation in the induction of NF-κB and AP-1 in RFL-6 cells by crocidolite asbestos: Evidence following protection by vitamin E. *Environ. Health Perspect.* **105**(Suppl. 5), 1127–1130.

Fitzgerald, M. J., Webber, E. M., Donovan, J. R., and Fausto, N. (1995). Rapid DNA binding by nuclear factor κB in hepatocytes at the start of liver regeneration. *Cell Growth Differ.* **6,** 417–427.

Fox, E. S., Kim, J. C., and Tracy, T. F. (1997). NF-κB activation and modulation in hepatic macrophages during cholestatic injury. *J. Surg. Res.* **72,** 129–134.

Franzoso, G., Carlson, L., Poljak, L., Shores, E. W., Epstein, S., Leonardi, A., Grinberg, A., Tran, T., SchartonKersten, T., Anver, M., Love, P., Brown, K., *et al.* (1998). Mice deficient in nuclear factor (NF)-κB/p52 present with defects in humoral responses, germinal center reactions, and splenic microarchitecture. *J. Exp. Med.* **187,** 147–159.

Glauert, H. P., Kumar, A., Lu, Z., Patel, S., Tharappel, J. C., Stemm, D. N., Bunaciu, R. P., Lee, E. Y., Lehmler, H. J., Robertson, L. W., and Spear, B. T. (2005). Dietary vitamin E does not inhibit the promotion of liver carcinogenesis by polychlorinated biphenyls in rats. *J. Nutr.* **135,** 283–286.

Gunawardena, K., Campbell, L. D., and Meikle, A. W. (2004). Combination therapy with vitamins C plus E inhibits survivin and human prostate cancer cell growth. *Prostate* **59,** 319–327.

Hayakawa, M., Miyashita, H., Sakamoto, I., Kitagawa, M., Tanaka, H., Yasuda, H., Karin, M., and Kikugawa, K. (2003). Evidence that reactive oxygen species do not mediate NF-kB activation. *EMBO J.* **22,** 3356–3366.

Hennig, B., Toborek, M., Joshi-Barve, S., Barger, S. W., Barve, S., Mattson, M. P., and McClain, C. J. (1996). Linoleic acid activates nuclear transcription factor-κB (NF-κB) and induces NF-κB-dependent transcription in cultured endothelial cells. *Am. J. Clin. Nutr.* **63,** 322–328.

Hennig, B., Meerarani, P., Ramadass, P., Watkins, B. A., and Toborek, M. (2000). Fatty acid-mediated activation of vascular endothelial cells. *Metabolism* **49,** 1006–1013.

Hill, D. B., Devalaraja, R., Joshi-Barve, S., Barve, S., and McClain, C. J. (1999). Antioxidants attenuate nuclear factor-κB activation and tumor necrosis factor-α production in alcoholic hepatitis patient monocytes and rat Kupffer cells, *in vitro*. *Clin. Biochem.* **32,** 563–570.

Hirano, F., Tanaka, H., Miura, T., Hirano, Y., Okamoto, K., Makino, Y., and Makino, I. (1998). Inhibition of NF-κB-dependent transcription of human immunodeficiency virus 1 promoter by a phosphodiester compound of vitamin C and vitamin E, EPC-K1. *Immunopharmacology* **39,** 31–38.

Hirotani, S., Otsu, K., Nishida, K., Higuchi, Y., Morita, T., Nakayama, H., Yamaguchi, O., Mano, T., Matsumura, Y., Ueno, H., Tada, M., and Hori, M. (2002). Involvement of nuclear factor-κB and apoptosis signal-regulating kinase 1 in G-protein-coupled receptor agonist-induced cardiomyocyte hypertrophy. *Circulation* **105,** 509–515.

Horton, J. W., White, D. J., Maass, D. L., Hybki, D. P., Haudek, S., and Giroir, B. (2001). Antioxidant vitamin therapy alters burn trauma-mediated cardiac NF-κB activation and cardiomyocyte cytokine secretion. *J. Trauma* **50,** 397–406; discussion 407–398.

Horwitz, B. H., Zelazowski, P., Shen, Y., Wolcott, K. M., Scott, M. L., Baltimore, D., and Snapper, C. M. (1999). The p65 subunit of NF-κB is redundant with p50 during B cell proliferative responses, and is required for germline C-H transcription and class switching to IgG3. *J. Immunol.* **162,** 1941–1946.

Hsiao, G., Huang, H. Y., Fong, T. H., Shen, M. Y., Lin, C. H., Teng, C. M., and Sheu, J. R. (2004). Inhibitory mechanisms of YC-1 and PMC in the induction of iNOS expression by lipoteichoic acid in RAW 264.7 macrophages. *Biochem. Pharmacol.* **67,** 1411–1419.

Hsieh, C.-C., Huang, C.-J., and Lin, B.-F. (2006). Low and high levels of α-tocopherol exert opposite effects on IL-2 possibly through the modulation of PPAR-γ, IkBα, and apoptotic pathway in activated splenocytes. *Nutrition* **22,** 433–440.

Hughes, G., Murphy, M. P., and Ledgerwood, E. C. (2005). Mitochondrial reactive oxygen species regulate the temporal activation of nuclear factor κB to modulate tumour necrosis factor-induced apoptosis: Evidence from mitochondria-targeted antioxidants. *Biochem. J.* **389,** 83–89.

Israel, N., Gougerot-Pocidalo, M. A., Aillet, F., and Virelizier, J. L. (1992). Redox status of cells influences constitutive or induced NF-κB translocation and HIV long terminal repeat activity in human T and monocytic cell lines. *J. Immunol.* **149,** 3386–3393.

Karin, M. (2006). NF-κB and cancer: Mechanisms and targets. *Mol. Carcinog.* **45,** 355–361.

Karin, M., and Delhase, M. (2000). The I κB kinase (IKK) and NF-κB: Key elements of proinflammatory signalling. *Semin. Immunol.* **12,** 85–98.

Karin, M., and Lin, A. (2002). NF-κB at the crossroads of life and death. *Nat. Immunol.* **3,** 221–227.

Kato, A., Edwards, M. J., and Lentsch, A. B. (2002). Gene deletion of NF-κB p50 does not alter the hepatic inflammatory response to ischemia/reperfusion. *J. Hepatol.* **37,** 48–55.

Kawai, T., Takahashi, K., Sato, S., Coban, C., Kumar, H., Kato, H., Ishii, K. J., Takeuchi, O., and Akira, S. (2005). IPS-1, an adaptor triggering RIG-I- and Mda5-mediated type I interferon induction. *Nat. Immunol.* **6,** 981–988.

Kontgen, F., Grumont, R. J., Strasser, A., Metcalf, D., Li, R., Tarlinton, D., and Gerondakis, S. (1995). Mice lacking the c-rel proto-oncogene exhibit defects in lymphocyte proliferation, humoral immunity, and interleukin-2 expression. *Genes Dev.* **9,** 1965–1977.

Lal, M. A., Brismar, H., Eklof, A. C., and Aperia, A. (2002). Role of oxidative stress in advanced glycation end product-induced mesangial cell activation. *Kidney Int.* **61,** 2006–2014.

Lee, C. C., Cheng, Y. W., and Kang, J. J. (2005). Motorcycle exhaust particles induce IL-8 production through NF-κB activation in human airway epithelial cells. *J. Toxicol. Environ. Health A* **68,** 1537–1555.

Lee, K. S., Lee, S. J., Park, H. J., Chung, J. P., Han, K. H., Chon, C. Y., Lee, S. I., and Moon, Y. M. (2001). Oxidative stress effect on the activation of hepatic stellate cells. *Yonsei Med. J.* **42,** 1–8.

Li-Weber, M., Giaisi, M., Treiber, M. K., and Krammer, P. H. (2002a). Vitamin E inhibits IL-4 gene expression in peripheral blood T cells. *Eur. J. Immunol.* **32,** 2401–2408.

Li-Weber, M., Weigand, M. A., Giaisi, M., Suss, D., Treiber, M. K., Baumann, S., Ritsou, E., Breitkreutz, R., and Krammer, P. H. (2002b). Vitamin E inhibits CD95 ligand expression and protects T cells from activation-induced cell death. *J. Clin. Invest.* **110,** 681–690.

Li, D., Saldeen, T., and Mehta, J. L. (1999). γ-Tocopherol decreases ox-LDL-mediated activation of nuclear factor-κB and apoptosis in human coronary artery endothelial cells. *Biochem. Biophys. Res. Commun.* **259,** 157–161.

Li, D., Saldeen, T., and Mehta, J. L. (2000a). Effects of α-Tocopherol on ox-LDL-mediated degradation of IκB and apoptosis in cultured human coronary artery endothelial cells. *J. Cardiovasc. Pharmacol.* **36,** 297–301.

Li, D., Saldeen, T., Romeo, F., and Mehta, J. L. (2000b). Oxidized LDL upregulates angiotensin II type 1 receptor expression in cultured human coronary artery endothelial cells: The potential role of transcription factor NF-κB. *Circulation* **102,** 1970–1976.

Li, N., and Karin, M. (1999). Is NF-κB the sensor of oxidative stress? *FASEB J.* **13,** 1137–1143.

Li, Y., Liu, L., Barger, S. W., Mrak, R. E., and Griffin, W. S. (2001). Vitamin E suppression of microglial activation is neuroprotective. *J. Neurosci. Res.* **66,** 163–170.

Li, Y. X., Glauert, H. P., and Spear, B. T. (2000c). Activation of nuclear factor-κB by the peroxisome proliferator ciprofibrate in H4IIEC3 rat hepatoma cells and its inhibition by the antioxidants N-acetylcysteine and vitamin E. *Biochem. Pharmacol.* **59,** 427–434.

Liu, S. L., Esposti, S. D., Yao, T., Diehl, A. M., and Zern, M. A. (1995). Vitamin E therapy of acute CCl4-induced hepatic injury in mice is associated with inhibition of nuclear factor κB binding. *Hepatology* **22,** 1474–1481.

Lu, Z., Lee, E. Y., Robertson, L. W., Glauert, H. P., and Spear, B. T. (2004). Effect of 2,2′,4,4′,5,5′-hexachlorobiphenyl (PCB-153) on hepatocyte proliferation and apoptosis in mice deficient in the p50 subunit of the transcription factor NF-kB. *Toxicol. Sci.* **81,** 35–42.

Maalouf, S., El-Sabban, M., Darwiche, N., and Gali-Muhtasib, H. (2002). Protective effect of vitamin E on ultraviolet B light-induced damage in keratinocytes. *Mol. Carcinog.* **34,** 121–130.

Mamputu, J. C., and Renier, G. (2004). Signalling pathways involved in retinal endothelial cell proliferation induced by advanced glycation end products: Inhibitory effect of gliclazide. *Diabetes Obes. Metab.* **6,** 95–103.

Maziere, C., Auclair, M., Djavaheri-Mergny, M., Packer, L., and Maziere, J. C. (1996). Oxidized low density lipoprotein induces activation of the transcription factor NF κB in fibroblasts, endothelial and smooth muscle cells. *Biochem. Mol. Biol. Int.* **39,** 1201–1207.

Maziere, C., Floret, S., Santus, R., Morliere, P., Marcheux, V., and Maziere, J. C. (2003). Impairment of the EGF signaling pathway by the oxidative stress generated with UVA. *Free Radic. Biol. Med.* **34,** 629–636.

Maziere, C., Morliere, P., Santus, R., Marcheux, V., Louandre, C., Conte, M. A., and Maziere, J. C. (2004). Inhibition of insulin signaling by oxidized low density lipoprotein. Protective effect of the antioxidant Vitamin E. *Atherosclerosis* **175,** 23–30.

Meyer, M., Schreck, R., and Baeuerle, P. A. (1993). H2O2 and antioxidants have opposite effects on activation of NF-κB and AP-1 in intact cells: AP-1 as secondary antioxidant-responsive factor. *EMBO J.* **12,** 2005–2015.

Morante, M., Sandoval, J., Gomez-Cabrera, M. C., Rodriguez, J. L., Pallardo, F. V., Vina, J. R., Torres, L., and Barber, T. (2005). Vitamin E deficiency induces liver nuclear factor-κB DNA-binding activity and changes in related genes. *Free Radic. Res.* **39,** 1127–1138.

Moscat, J., Diaz-Meco, M. T., and Rennert, P. (2003). NF-κB activation by protein kinase C isoforms and B-cell function. *EMBO Rep.* **4,** 31–36.

Muroya, T., Ihara, Y., Ikeda, S., Yasuoka, C., Miyahara, Y., Urata, Y., Kondo, T., and Kohno, S. (2003). Oxidative modulation of NF-κB signaling by oxidized low-density lipoprotein. *Biochem. Biophys. Res. Commun.* **309,** 900–905.

Nakamura, T., Goto, M., Matsumoto, A., and Tanaka, I. (1998). Inhibition of NF-κB transcriptional activity by α-tocopheryl succinate. *Biofactors* **7,** 21–30.

Neuzil, J., Schroder, A., von Hundelshausen, P., Zernecke, A., Weber, T., Gellert, N., and Weber, C. (2001). Inhibition of inflammatory endothelial responses by a pathway involving caspase activation and p65 cleavage. *Biochemistry (Mosc.)* **40,** 4686–4692.

Nguyen, T. K., Nilakantan, V., Felix, C. C., Khanna, A. K., and Pieper, G. M. (2006). Beneficial effect of α-tocopheryl succinate in rat cardiac transplants. *J. Heart Lung Transplant.* **25,** 707–715.

Nilakantan, V., Spear, B. T., and Glauert, H. P. (1998). Liver-specific catalase expression in transgenic mice inhibits NF-κB activation and DNA synthesis induced by the peroxisome proliferator ciprofibrate. *Carcinogenesis* **19,** 631–637.

Pikarsky, E., Porat, R. M., Stein, I., Abramovitch, R., Amit, S., Kasem, S., Gutkovich-Pyest, E., Urieli-Shoval, S., Galun, E., and Ben-Neriah, Y. (2004). NF-κB functions as a tumour promoter in inflammation-associated cancer. *Nature* **431,** 461–466.

Post, A., Holsboer, F., and Behl, C. (1998). Induction of NF-κB activity during haloperidol-induced oxidative toxicity in clonal hippocampal cells: Suppression of NF-κB and neuroprotection by antioxidants. *J. Neurosci.* **18,** 8236–8246.

Post, A., Rucker, M., Ohl, F., Uhr, M., Holsboer, F., Almeida, O. F., and Michaelidis, T. M. (2002). Mechanisms underlying the protective potential of α-tocopherol (vitamin E) against haloperidol-associated neurotoxicity. *Neuropsychopharmacology* **26,** 397–407.

Poynter, M. E., and Daynes, R. A. (1998). Peroxisome proliferator-activated receptor α activation modulates cellular redox status, represses nuclear factor-κB signaling, and reduces inflammatory cytokine production in aging. *J. Biol. Chem.* **273,** 32833–32841.

Rak Min, K., Lee, H., Hak Kim, B., Chung, E., Min Cho, S., and Kim, Y. (2005). Inhibitory effect of 6-hydroxy-7-methoxychroman-2-carboxylic acid phenylamide on nitric oxide and interleukin-6 production in macrophages. *Life Sci.* **77,** 3242–3257.

Rimbach, G., Minihane, A. M., Majewicz, J., Fischer, A., Pallauf, J., Virgli, F., and Weinberg, P. D. (2002). Regulation of cell signalling by vitamin E. *Proc. Nutr. Soc.* **61,** 415–425.

Rocksén, D., Ekstrand-Hammarstrom, B., Johansson, L., and Bucht, A. (2003). Vitamin E reduces transendothelial migration of neutrophils and prevents lung injury in endotoxin-induced airway inflammation. *Am. J. Respir. Cell Mol. Biol.* **28,** 199–207.

Rodriguez-Porcel, M., Lerman, L. O., Holmes, D. R., Jr., Richardson, D., Napoli, C., and Lerman, A. (2002). Chronic antioxidant supplementation attenuates nuclear factor-κB activation and preserves endothelial function in hypercholesterolemic pigs. *Cardiovasc. Res.* **53,** 1010–1018.

Roy, S., Sen, C. K., Kobuchi, H., and Packer, L. (1998). Antioxidant regulation of phorbol ester-induced adhesion of human Jurkat T-cells to endothelial cells. *Free Radic. Biol. Med.* **25,** 229–241.

Sakurai, T., Maeda, S., Chang, L., and Karin, M. (2006). Loss of hepatic NF-κB activity enhances chemical hepatocarcinogenesis through sustained c-Jun N-terminal kinase 1 activation. *Proc. Natl. Acad. Sci. USA* **103,** 10544–10551.

Sauer, H., Neukirchen, W., Rahimi, G., Grunheck, F., Hescheler, J., and Wartenberg, M. (2004). Involvement of reactive oxygen species in cardiotrophin-1-induced proliferation of cardiomyocytes differentiated from murine embryonic stem cells. *Exp. Cell Res.* **294,** 313–324.

Schmidt, K. N., Amstad, P., Cerutti, P., and Baeuerle, P. A. (1995). The roles of hydrogen peroxide and superoxide as messengers in the activation of transcription factor NF-κB. *Chem. Biol.* **2,** 13–22.

Schoemaker, M. H., Ros, J. E., Homan, M., Trautwein, C., Liston, P., Poelstra, K., vanGoor, H., Jansen, P. L. M., and Moshage, H. (2002). Cytokine regulation of pro- and anti-apoptotic genes in rat hepatocytes: NF-κB-regulated inhibitor of apoptosis protein 2 (CIAP2) prevents apoptosis. *J. Hepatol.* **36,** 742–750.

Schreck, R., Rieber, P., and Baeuerle, P. A. (1991). Reactive oxygen intermediates as apparently widely used messengers in the activation of the NF-κB transcription factor and HIV-1. *EMBO J.* **10,** 2247–2258.

Schreck, R., Albermann, K., and Baeuerle, P. A. (1992). Nuclear factor κB: An oxidative stress-responsive transcription factor of eukaryotic cells (a review). *Free Radic. Res. Commun.* **17,** 221–237.

Seth, R. B., Sun, L., Ea, C. K., and Chen, Z. J. (2005). Identification and characterization of MAVS, a mitochondrial antiviral signaling protein that activates NF-κB and IRF 3. *Cell* **122,** 669–682.

Sha, W. C., Liou, H. C., Tuomanen, E. I., and Baltimore, D. (1995). Targeted disruption of the p50 subunit of NF-kB leads to multifocal defects in immune responses. *Cell* **80,** 321–330.

Shah, S. J., and Sylvester, P. W. (2005). γ-Tocotrienol inhibits neoplastic mammary epithelial cell proliferation by decreasing Akt and nuclear factor κB activity. *Exp. Biol. Med. (Maywood)* **230,** 235–241.

Slim, R., Toborek, M., Robertson, L. W., and Hennig, B. (1999). Antioxidant protection against PCB-mediated endothelial cell activation. *Toxicol. Sci.* **52,** 232–239.

Snapper, C. M., Rosas, F. R., Zelazowski, P., Moorman, M. A., Kehry, M. R., Bravo, R., and Weih, F. (1996a). B cells lacking RelB are defective in proliferative responses, but undergo normal B cell maturation to Ig secretion and Ig class switching. *J. Exp. Med.* **184,** 1537–1541.

Snapper, C. M., Zelazowski, P., Rosas, F. R., Kehry, M. R., Tian, M., Baltimore, D., and Sha, W. C. (1996b). B cells from p50/NF-κB knockout mice have selective defects in proliferation, differentiation, germ-line CH transcription, and Ig class switching. *J. Immunol.* **156,** 183–191.

Sokoloski, J. A., Hodnick, W. F., Mayne, S. T., Cinquina, C., Kim, C. S., and Sartorelli, A. C. (1997). Induction of the differentiation of HL-60 promyelocytic leukemia cells by vitamin E and other antioxidants in combination with low levels of vitamin D3: Possible relationship to NF-κB. *Leukemia* **11,** 1546–1553.

Staal, F. J. T., Roederer, M., Herzenberg, L. A., and Herzenberg, L. A. (1990). Intracellular thiols regulate activation of nuclear factor kB and transcription of human immunodeficiency virus. *Proc. Natl. Acad. Sci. USA* **87,** 9943–9947.

Sugiyama, S., Kugiyama, K., Ogata, N., Doi, H., Ota, Y., Ohgushi, M., Matsumura, T., Oka, H., and Yasue, H. (1998). Biphasic regulation of transcription factor nuclear factor-κB activity in human endothelial cells by lysophosphatidylcholine through protein kinase C-mediated pathway. *Arterioscler. Thromb. Vasc. Biol.* **18,** 568–576.

Suzuki, Y. J., and Packer, L. (1993a). Inhibition of NF-κB activation by vitamin E derivatives. *Biochem. Biophys. Res. Commun.* **193,** 277–283.

Suzuki, Y. J., and Packer, L. (1993b). Inhibition of NF-κB DNA binding activity by α-tocopheryl succinate. *Biochem. Mol. Biol. Int.* **31,** 693–700.

Sylvester, P. W., Shah, S. J., and Samant, G. V. (2005). Intracellular signaling mechanisms mediating the antiproliferative and apoptotic effects of γ-tocotrienol in neoplastic mammary epithelial cells. *J. Plant Physiol.* **162,** 803–810.

Takacs, P., Kauma, S. W., Sholley, M. M., Walsh, S. W., Dinsmoor, M. J., and Green, K. (2001). Increased circulating lipid peroxides in severe preeclampsia activate NF-κB and upregulate ICAM-1 in vascular endothelial cells. *FASEB J.* **15,** 279–281.

Tan, P. H., Sagoo, P., Chan, C., Yates, J. B., Campbell, J., Beutelspacher, S. C., Foxwell, B. M., Lombardi, G., and George, A. J. (2005). Inhibition of NF-kappa B and oxidative pathways in human dendritic cells by antioxidative vitamins generates regulatory T cells. *J. Immunol.* **174,** 7633–7644.

Tharappel, J. C., Nalca, A., Owens, A. B., Ghabrial, L., Konz, E. C., Glauert, H. P., and Spear, B. T. (2003). Cell proliferation and apoptosis are altered in mice deficient in the NF-κB p50 subunit after treatment with the peroxisome proliferator ciprofibrate. *Toxicol. Sci.* **75,** 300–308.

Theriault, A., Chao, J. T., and Gapor, A. (2002). Tocotrienol is the most effective vitamin E for reducing endothelial expression of adhesion molecules and adhesion to monocytes. *Atherosclerosis* **160,** 21–30.

Toborek, M., Barger, S. W., Mattson, M. P., Barve, S., McClain, C. J., and Hennig, B. (1996). Linoleic acid and TNF-α cross-amplify oxidative injury and dysfunction of endothelial cells. *J. Lipid Res.* **37,** 123–135.

Uberti, D., Carsana, T., Francisconi, S., Toninelli, G. F., Canonico, P. L., and Memo, M. (2004). A novel mechanism for pergolide-induced neuroprotection: Inhibition of NF-κB nuclear translocation. *Biochem. Pharmacol.* **67,** 1743–1750.

Vanantwerp, D. J., Martin, S. J., Kafri, T., Green, D. R., and Verma, I. M. (1996). Suppression of TNF-α-induced apoptosis by NF-κB. *Science* **274,** 787–789.

Videla, L. A., Tapia, G., Varela, P., Cornejo, P., Guerrero, J., Israel, Y., and Fernandez, V. (2004). Effects of acute γ-hexachlorocyclohexane intoxication in relation to the redox regulation of nuclear factor-κB, cytokine gene expression, and liver injury in the rat. *Antioxid. Redox Signal.* **6,** 471–480.

Wang, C. Y., Mayo, M. W., and Baldwin, A. S. (1996). TNF- and cancer therapy-induced apoptosis: Potentiation by inhibition of NF-κB. *Science* **274,** 784–787.

Weih, F., Carrasco, D., Durham, S. K., Barton, D. S., Rizzo, C. A., Ryseck, R. P., Lira, S. A., and Bravo, R. (1995). Multiorgan inflammation and hematopoietic abnormalities in mice with a targeted disruption of RelB, a member of the NF-κB/Rel family. *Cell* **80,** 331–340.

Xu, L. G., Wang, Y. Y., Han, K. J., Li, L. Y., Zhai, Z., and Shu, H. B. (2005). VISA is an adapter protein required for virus-triggered IFN-β signaling. *Mol. Cell* **19,** 727–740.

Xu, Y., Bialik, S., Jones, B. E., Iimuro, Y., Kitsis, R. N., Srinivasan, A., Brenner, D. A., and Czaja, M. J. (1998). NF-κB inactivation converts a hepatocyte cell line TNF-α response from proliferation to apoptosis. *Am. J. Physiol. Cell Physiol.* **44,** C1058–C1066.

Yang, J., Fan, G. H., Wadzinski, B. E., Sakurai, H., and Richmond, A. (2001). Protein phosphatase 2A interacts with and directly dephosphorylates RelA. *J. Biol. Chem.* **276,** 47828–47833.

Zandi, E., Rothwarf, D. M., Delhase, M., Hayakawa, M., and Karin, M. (1997). The IkB kinase complex (IKK) contains two kinase subunits, IKK-α and IKK-β, necessary for IkB phosphorylation and NF-kB activation. *Cell* **91,** 243–252.

Zhang, J. G., NichollsGrzemski, F. A., Tirmenstein, M. A., and Fariss, M. W. (2001). Vitamin E succinate protects hepatocytes against the toxic effect of reactive oxygen species generated at mitochondrial complexes I and III by alkylating agents. *Chem. Biol. Interact.* **138,** 267–284.

Zingg, J. M., and Azzi, A. (2004). Non-antioxidant activities of vitamin E. *Curr. Med. Chem.* **11,** 1113–1133.

7

SYNTHESIS OF VITAMIN E

THOMAS NETSCHER

Research and Development, DSM Nutritional Products
CH-4002 Basel, Switzerland

Vitamin E compounds are biologically essential fat-soluble antioxidants derived from 6-chromanol. This chapter covers the representative literature on the preparation of various stereoisomeric forms and homologues of tocopherols and tocotrienols, and of respective starting materials and intermediates. The industrially most relevant (all-*rac*)-α-tocopherol is generally built up by coupling of arenes with aliphatic precursors.

0083-6729/07 $35.00
DOI: 10.1016/S0083-6729(07)76007-7

For the synthesis of optically active vitamin E components, various strategies are compiled. In approaches to chiral chroman and side chain building blocks, a broad variety of methods from the repertoire of asymmetric synthesis were applied, that is optical resolution and use of chiral pool starting materials and chiral auxiliaries in stoichiometric as well as catalytic amounts, including catalysis by metal complexes, microorganisms, and enzymes. Most efforts were directed to (2*R*,4′*R*,8′*R*)-α-tocopherol due to its prominent biological activity. Syntheses of different stereoisomers (and mixtures thereof) and (β-, γ-, δ-) homologues of tocopherols and tocotrienols, as well as methods for their interconversion, are also described.

I. INTRODUCTION

Vitamin E was discovered, isolated from natural material, and structurally elucidated at the beginning of the last century (Evans and Bishop, 1922; Evans *et al.*, 1936; Fernholz, 1938). This essential food ingredient is of great economic importance due to its biological activity (Weimann and Weiser, 1991; Weiser and Vecchi, 1982) and antioxidant properties (Kamal-Eldin and Åppelqvist, 1996; Pongracz *et al.*, 1995). Industrially, it is either prepared by total synthesis or originating from natural sources (vide infra). The predominant amount is used for animal feeding and about a quarter goes to human applications (pharma, food, cosmetics) (Baldenius *et al.*, 1996; Ernst, 1983). The term vitamin E covers tocol and tocotrienol derivatives exhibiting qualitatively the biological activity of α-tocopherol (IUPAC-IUB, 1982). All naturally occurring components of this group hitherto discovered are single-isomer products. The (2*R*,4′*R*,8′*R*) configuration is found in α-, β-, γ-, and δ-tocopherol (*RRR*-**1** to *RRR*-**4**), and α-, β-, γ-, and δ-tocotrienol (**5**–**8**) possess (2*R*,3′*E*,7′*E*) stereochemistry (Fig. 1).

All-racemic-α-tocopherol [(all-*rac*)-**1**, Fig. 2], an equimolar mixture of all eight stereoisomers, is industrially the most important product, manufactured in about 35,000 tons per year worldwide, mainly applied as its acetate derivative (all-*rac*)-**9**.

II. SYNTHESIS OF (all-*rac*)-α-TOCOPHEROL

The large-scale industrial synthesis of (all-*rac*)-α-tocopherol (synthetic vitamin E) consists of three major parts which have been reviewed, for example, by Baldenius *et al.* (1996), Mayer and Isler (1971), Schudel *et al.* (1972), and Bonrath and Netscher (2005): the preparation of the aromatic

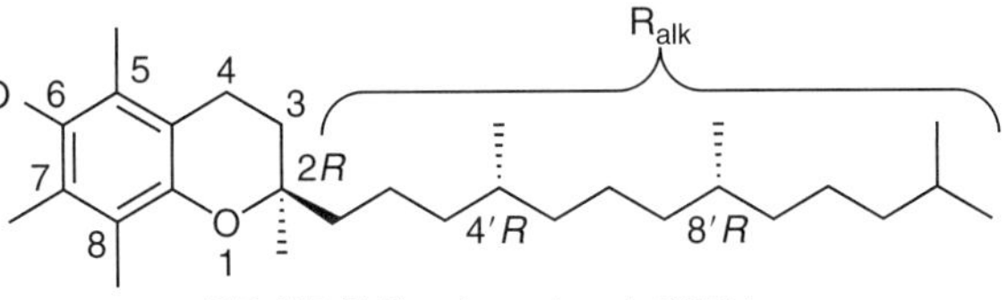

(2*R*,4′*R*,8′*R*)-α-tocopherol *RRR*-**1**

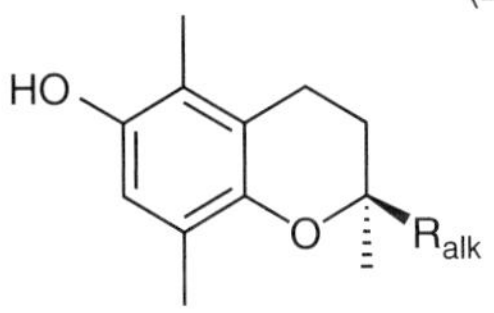

(2*R*,4′*R*,8′*R*)-β-tocopherol *RRR*-**2**

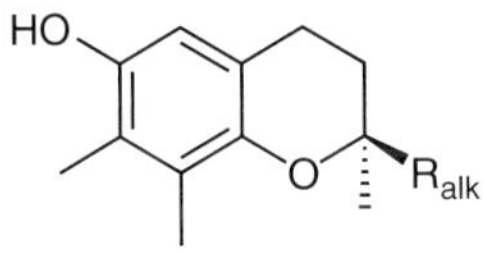

(2*R*,4′*R*,8′*R*)-γ-tocopherol *RRR*-**3**

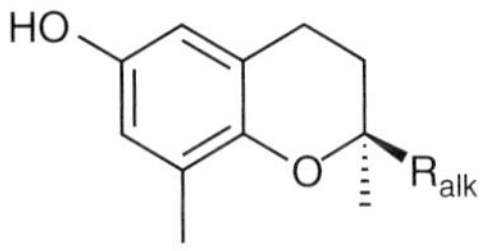

(2*R*,4′*R*,8′*R*)-δ-tocopherol *RRR*-**4**

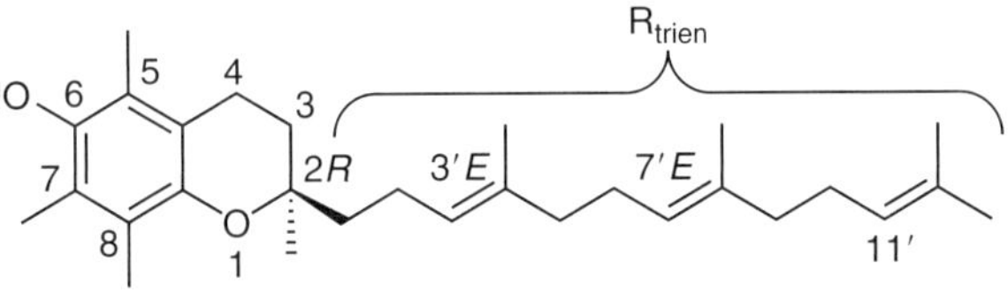

(2*R*,3′*E*,7′*E*)-α-tocotrienol *REE*-**5**

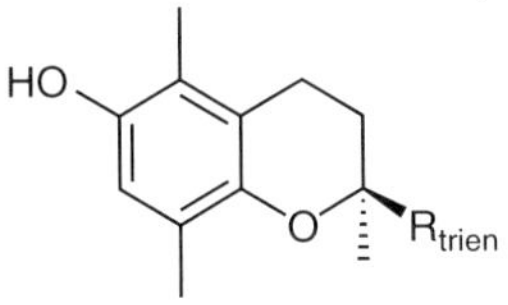

(2*R*,3′*E*,7′*E*)-β-tocotrienol *REE*-**6**

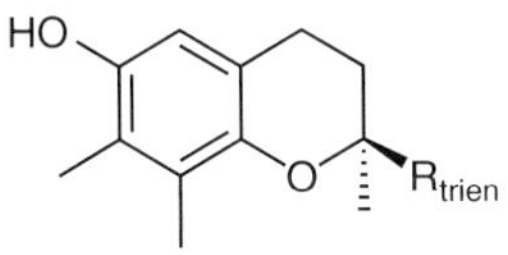

(2*R*,3′*E*,7′*E*)-γ-tocotrienol *REE*-**7**

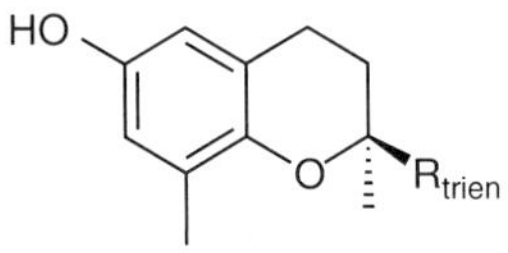

(2*R*,3′*E*,7′*E*)-δ-tocotrienol *REE*-**8**

FIGURE 1. Naturally occurring tocopherols and tocotrienols.

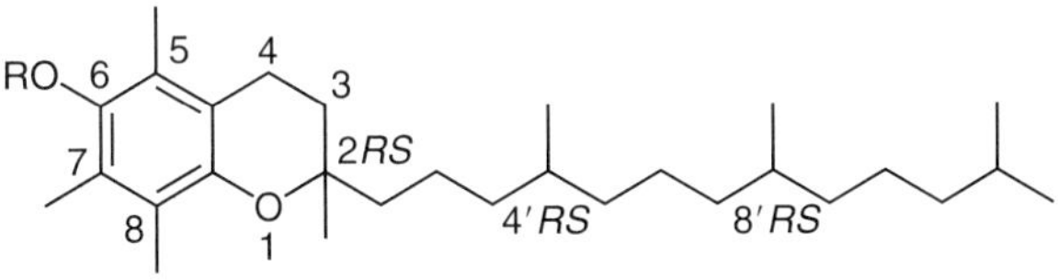

R = H (2*RS*,4′*RS*,8′*RS*)-α-tocopherol (all-*rac*)-**1**

R = Ac (2*RS*,4′*RS*,8′*RS*)-α-tocopheryl acetate (all-*rac*)-**9**

FIGURE 2. (all-*rac*)-α-Tocopherol and α-tocopheryl acetate.

building block (trimethylhydroquinone, **10**), the production of the side chain component [(all-*rac*)-isophytol, (all-*rac*)-**11**, or a corresponding C_{20} derivative], and the condensation reaction (Fig. 3).

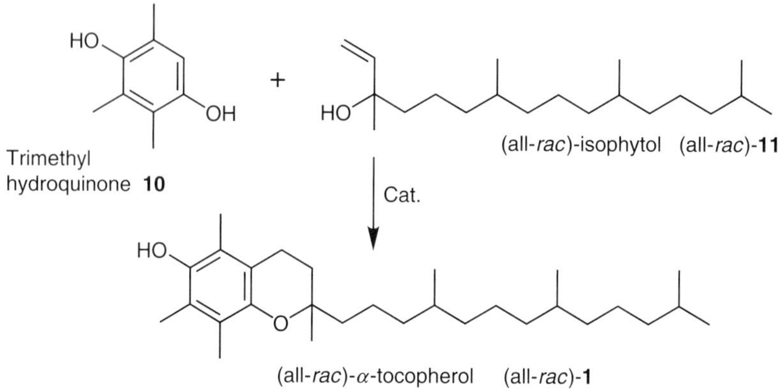

Trimethyl hydroquinone **10**

(all-*rac*)-isophytol (all-*rac*)-**11**

Cat.

(all-*rac*)-α-tocopherol (all-*rac*)-**1**

FIGURE 3. Synthesis of (all-*rac*)-α-tocopherol (synthetic vitamin E).

A. BUILDING BLOCKS FOR THE ARYL CONTAINING CHROMAN MOIETY

Selected routes to 2,3,5-trimethylhydroquinone (**10**) are shown in Fig. 4. m-Cresol (**12**) is catalytically methylated to trimethylphenol **13** which is transformed by oxidation to quinone **15** and subsequently reduced to hydroquinone **10**. Alternative processes start from mesitol (**14**, oxidation and rearrangement), isophorones (**16, 17**, oxidation/hydrogenation/isomerization sequences), or diethyl ketone (**18**, condensation reaction with methyl vinyl ketone or crotonaldehyde) (Baldenius *et al.*, 1996; Bonrath and Netscher, 2005; Ernst, 1983).

The stereoselective preparation of trimethylhydroquinone-1-monoacetate (**22**) opens the direct way to α-tocopheryl acetate (all-*rac*)-**9** by a condensation reaction of **22** with (all-*rac*)-**11** (vide infra), without the need of going via tocopherol **1** (Fig. 5). The diacetate **20**, obtained from α-isophorone (**17**) via β-isophorone (**19**) and ketoisophorone (**16**) (Baldenius *et al.*, 1996; Bonrath and Netscher, 2005; Bonrath and Schneider, 2003; Ernst, 1983; Laufer *et al.*, 2005; Schneider *et al.*, 2001), can be hydrolyzed with complete regioselectivity (no formation of **10** or **21**) by a lipase (thermomyces lanoginosus lipase, TLL)-catalyzed hydrolysis (Bonrath *et al.*, 2002a,b).

B. SIDE CHAIN BUILDING BLOCKS

For the preparation of (all-*rac*)-isophytol [(all-*rac*)-**11**] preferentially used as the side chain building block, various synthetic strategies are applied (Baldenius *et al.*, 1996; Bonrath and Netscher, 2005; Ernst, 1983). Representative pathways are sketched in Fig. 6. A $C_3 + C_2$ scheme starts from acetone and acetylene or the vinyl Grignard compound. For a C_3 elongation of the isoprenoic chain, treatment of the allyl alcohol with methyl acetoacetate or isopropenyl methyl

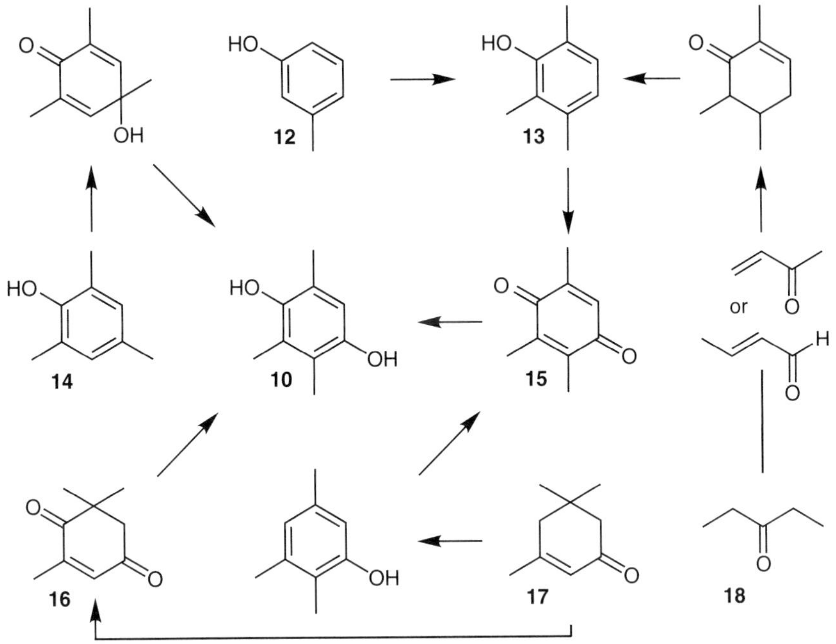

FIGURE 4. Selected routes to 2,3,5-trimethylhydroquinone (**10**).

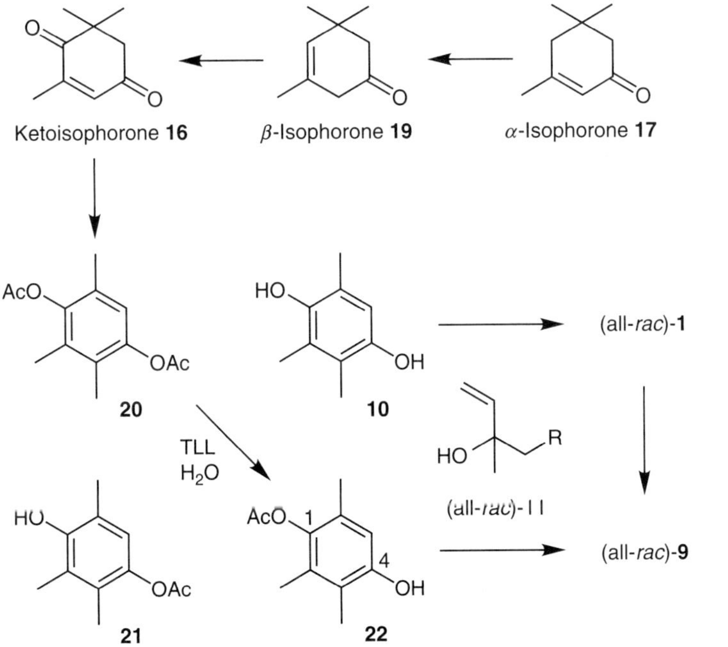

FIGURE 5. Preparation and use of trimethylhydroquinone-1-monoacetate (**22**).

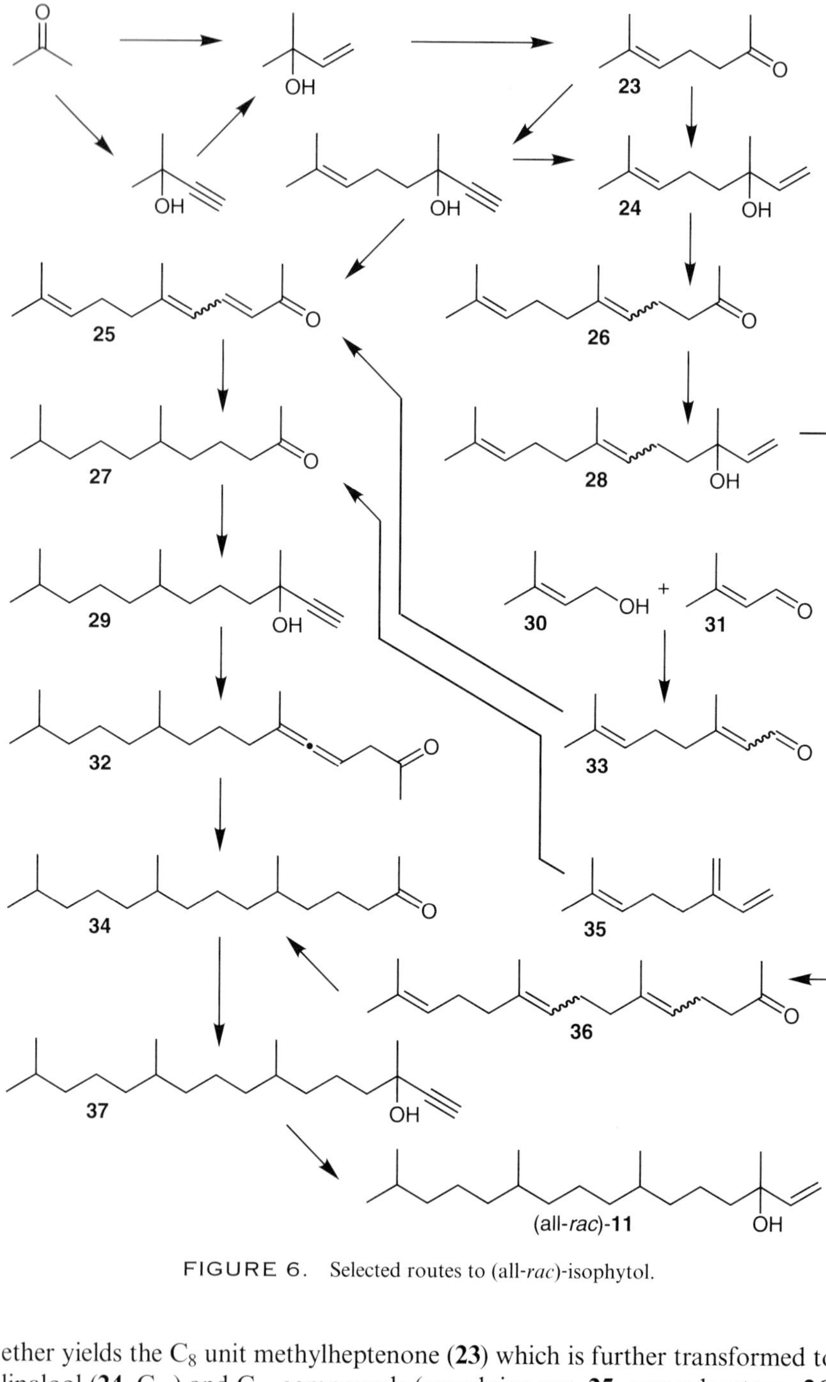

FIGURE 6. Selected routes to (all-*rac*)-isophytol.

ether yields the C_8 unit methylheptenone (**23**) which is further transformed to linalool (**24**, C_{10}) and C_{13} compounds (pseudoionone, **25**; geranylacetone, **26**; hexahydropseudoionone, **27**). Subsequent chain elongations in combination with hydrogenation/rearrangement reactions lead to C_{15} (nerolidol, **28; 29**) and C_{18} intermediates (alleneketone **32**/farnesylacetone **36** → **34**) which yield finally isophytol [(all-*rac*)-**11**] via acetylenecarbinol **37**.

A different approach is the prenol/prenal route (**30/31**) (Chauvel *et al.*, 1994; Mimoun, 1996) starting from inexpensive isobutene (C_4) and formaldehyde (C_1); citral (**33**, C_{10}) is further processed as shown in Fig. 6. Myrcene (**35**, C_{10}) from natural sources can be alkylated by an Rh-catalyzed process using a water-soluble phosphine ligand (TPPTS; Mercier and Chabardes, 1994; Mercier *et al.*, 1991). Hexahydropseudoionone (**27**) is obtained after subsequent decarboxylation and hydrogenation. Special features of this area of isoprenoid chemistry, like the ethynylation of ketones, the methodology for C_3 elongation by Saucy-Marbet and Carroll reactions and Aldol condensation, the Prins reaction (preparation of isoprenol), the total and Lindlar-type hydrogenation (transformation of propargylic to allylic alcohols), and various types of rearrangement reactions of allylic and propargylic alcohols (or derivatives), were reviewed in more detail by Bonrath and Netscher (2005).

C. SYNTHESIS OF (all-*rac*)-α-TOCOPHEROL

The first synthesis of (all-*rac*)-α-tocopherol [(all-*rac*)-**1**] by condensation reaction of **10** with (all-*rac*)-**11** or a C_{20} equivalent (Fig. 3) was described by Karrer and Isler (1946a,b) and Karrer *et al.* (1938), which resulted in the first vitamin E production at F. Hoffmann–La Roche in Basel in the early 1950s. Several classical Lewis and Brønsted acids, or combinations thereof, work well in this reaction. Typical examples are $ZnCl_2/HCl$, BF_3, or $AlCl_3$, applied in various organic solvents. For large-scale production, however, corrosion problems and contamination of wastewater, in particular with zinc and halide ions, are major drawbacks of such procedures. Further disadvantages are the high, often near stoichiometric amounts of catalysts, and the excess of (expensive) isophytol necessary for obtaining satisfying results.

During the last years, several approaches to find environmental friendly procedures were investigated, aimed at high yield and selectivity, as well as reusability of the catalysts. The main focus was on reactions in alternative reaction media (e.g., two- or multiphase solvent systems or supercritical fluids), and the use of new types of (Lewis/Brønsted acid) catalysts, including "superacidic" and supported catalysts, in order to substitute mineral acids or traditional Lewis acids. It would be beyond the scope of this chapter to completely cover all work on this topic. Instead, selected publications dealing with these new developments may be cited here (Baldenius *et al.*, 1996; Bonrath and Netscher, 2005; Bonrath *et al.*, 2002c, 2007; Hasegawa *et al.*, 2003; Ishihara *et al.*, 1996; Kokubo *et al.*, 2005; Matsui, 1996; Matsui *et al.*, 1995; Odinokov *et al.*, 2003; Wildermann *et al.*, 2007). Examples of novel catalysts are clays, ion exchange resins, rare earth and indium metal halides and triflates, heteropolytungsten acids, various polyfluorinated compounds (imides, methides), and boron and phosphorous compounds.

Remarkable features of these systems are not only the high chemical yield, but in particular the extremely high selectivity of the overall condensation reaction of trimethylhydroquinone (**10**) with (all-*rac*)-isophytol [(all-*rac*)-**11**] to (all-*rac*)-α-tocopherol [(all-*rac*)-**1**, Fig. 3], thus avoiding the formation of isomeric products, for example benzofuran compounds **38** (Fig. 7). Instead of the tertiary alcohol (all-*rac*)-**11**, also the primary alcohol phytol (*EZ*, all-*rac*)-**41**, or a corresponding ester, halide, or similar derivative can be used in this acid-catalyzed condensation reaction. Also, 1,3-dienes have been applied as side chain components (Matsui and Yamamoto, 1995). It is generally accepted that the overall condensation reaction proceeds via a two-step mechanism, that is a Friedel-Crafts C-alkylation [→(*EZ*,all-*rac*)-**39**], rather than an O-alkylation [→(*EZ*,all-*rac*)-**40**], followed by a cyclization reaction (Bonrath *et al.*, 2003; Shchegolev *et al.*, 1983).

Recently, alternative routes to (all-*rac*)-α-tocopheryl acetate [(all-*rac*)-**9**] have been worked out (Fig. 8). Reaction of monoacetate derivative **22** with (all-*rac*)-**11** leads directly to (all-*rac*)-**9** by using rare earth metal triflates (e.g., Gd, Sc, Hf triflates) or $Ga(OTf)_3$ and $Bi(OTf)_3$ in yields of up to 94% (Bonrath *et al.*, 2006). Remarkably, practically no saponification (less than 0.3%) of the ester derivatives (starting material and product) was observed under optimized reaction conditions. The stereoselective O → C rearrangement was described by Bonrath *et al.* (2002d, 2004). By application of ruthenium-catalyzed olefin cross-metathesis, the C- and O-alkylated intermediates (all-*rac*)-**43** and (all-*rac*)-**44** (cf. compounds **39** and **40** in Fig. 7) were obtained from olefins **42** (Malaisé *et al.*, 2006; Netscher, 2006; Netscher *et al.*, 2006).

The base-catalyzed condensation reaction of aryl methyl ketone **45**, derived from trimethylhydroquinone (**10**) with farnesylacetone **46**, followed by reduction of **47** and elimination, delivers the dehydro-tocotrienyl derivative **48**,

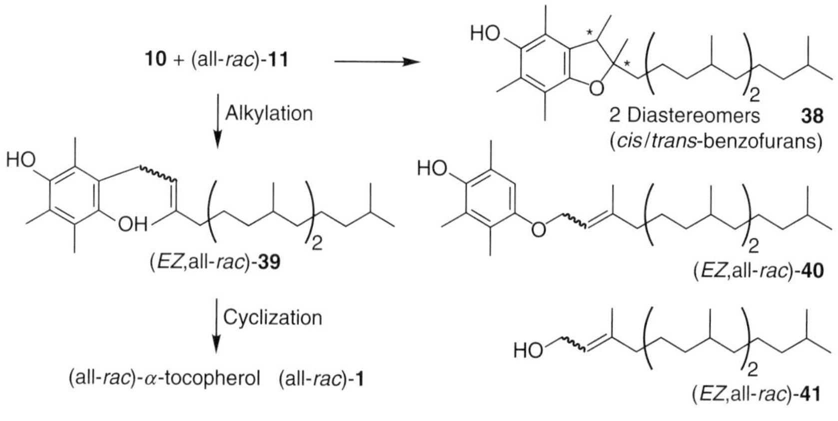

FIGURE 7. Overall condensation reaction to (all-*rac*)-α-tocopherol.

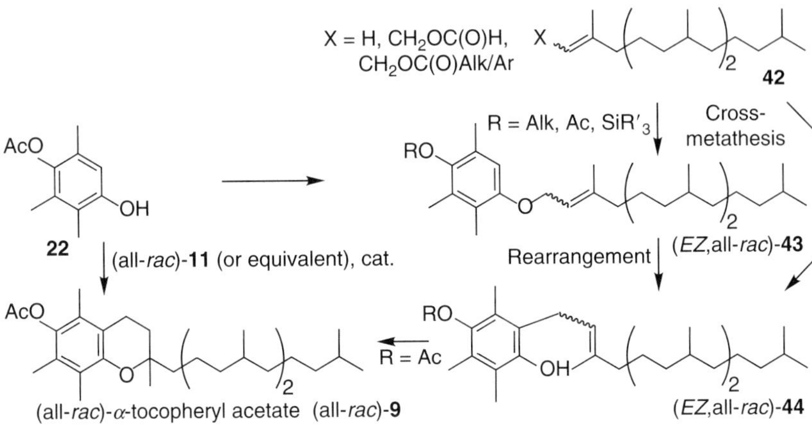

FIGURE 8. Alternative routes to (all-*rac*)-α-tocopheryl acetate, (all-*rac*)-**9**.

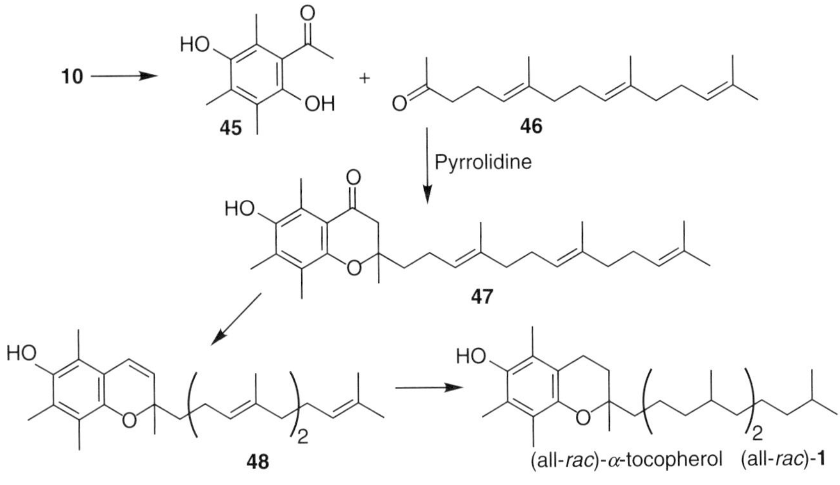

FIGURE 9. Synthesis of (all-*rac*)-**1** by Kabbe and Heitzer.

which can be transformed to (all-*rac*)-**1** by catalytic hydrogenation (Kabbe and Heitzer, 1978; Fig. 9).

The biologically derived diterpene alcohol (all-*E*)-geranylgeraniol [(all-*E*)-**49**] was employed by the sequence sketched in Fig. 10. The highly stereoselective monoepoxidation, followed by hydrogenation, yielded the monoepoxide **50**. Deoxygenation with MTO/triphenylphosphane delivered a ca. 1:3 mixture of phytol and isophytol [(*EZ*,all-*rac*)-**41** + (all-*rac*)-**11**]. Alternatively, **50** was reduced to the phytandiol **51**. Condensation of

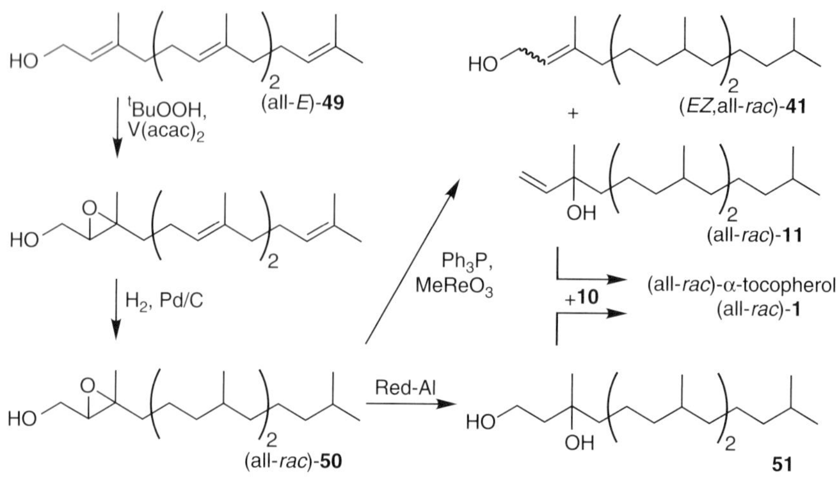

FIGURE 10. Synthesis of (all-*rac*)-**1** from biologically produced geranylgeraniol.

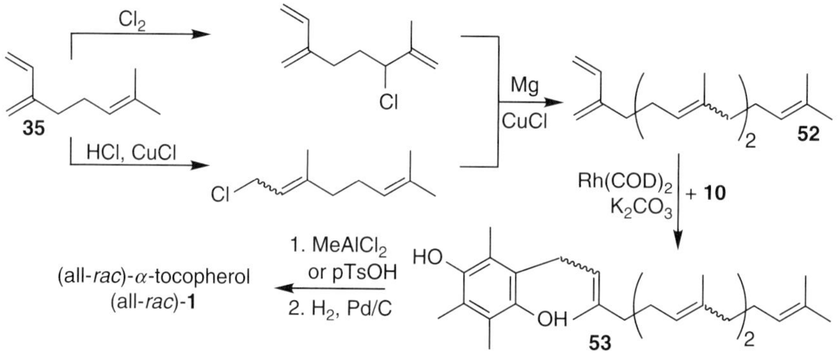

FIGURE 11. Synthesis of (all-*rac*)-**1** by rhodium-catalyzed alkylation.

trimethylhydroquinone (**10**) with either **11**/**41** or **51** with zinc chloride gave (all-*rac*)-**1** (Hyatt *et al.*, 2002). Further methods which use side chain compounds from biological origin are compiled by Millis *et al.* (2000a,b, 2001).

The rhodium-catalyzed alkylation of the conjugated diene **52** is another interesting approach (Fig. 11). The main difference to conventional routes is the use of the natural source starting material myrcene (**35**), which is dimerized to the C_{20} building block **52** via functionalization by chlorine and HCl/CuCl and further transformed to C-alkylation product **53** (Bienayme *et al.*, 2000).

III. Preparation of Optically Active Tocopherols

About 10% of the total amount of vitamin E industrially produced is isomerically pure (2*R*,4′*R*,8′*R*)-α-tocopherol (*RRR*-**1**) prepared by semisynthesis. This product is used almost exclusively for pharma (human) applications. Vegetable oils refined on large scale are the major sources of vitamin E compounds (Gupta, 1993; Walsh *et al.*, 1998; Young *et al.*, 1986). Soya deodorizer distillates (SDD), a waste stream from that vegetable oils production, are applied as starting materials. The SDD contain up to 10% of the four homologous tocopherols (*RRR*-**1** to *RRR*-**4**) which are isolated by a combination of several separation methods. To increase the value of the vitamin E concentrate of mixed tocopherols obtained from SDD, the lower β-, γ-, and δ-homologues (*RRR*-**2** to *RRR*-**4**, content about 90%) have to be converted to the biologically more active α-tocopherol (*RRR*-**1**, only ca. 5% in the original mixture) by permethylation reactions. This is performed by chloro-, amino-, or hydroxymethylation reactions yielding functionalized alkylated intermediates, which are further converted to *RRR*-**1** (Fig. 12) (Netscher, 1999; Netscher *et al.*, 2004b; Riegl *et al.*, 2000). The stannous chloride-mediated alkylation reduction protocol has also been used for the preparation of deuterium-labeled tocopherols (Hughes *et al.*, 1990).

This semisynthetic approach for the synthesis of *RRR*-α-tocopherol has still the general problem of a (given) limited availability of starting material (SDD) from natural sources which prevents an increasing demand of *RRR*-**1** from being satisfied over a certain level. The possibility to improve the α-tocopherol content in agricultural crops by genetically manipulating the vitamin E biosynthetic pathway has been reviewed by DellaPenna and Last (2006) and DellaPenna and Pogson (2006). Additional amounts of this product could also be made available by an economic total synthesis. Therefore, considerable efforts have been directed to the development of stereoselective syntheses of *RRR*-**1** and of corresponding building blocks during the last four decades. Most of them are compiled in the literature (Coffen *et al.*, 1994; Netscher, 1996; Netscher *et al.*, 2004a; Saucy and Cohen, 1981; Schudel *et al.*, 1972). General routes use classical optical resolution, biocatalysis (by microorganisms and isolated enzymes), chiral-pool starting materials, the application of stoichiometric and catalytic amounts of chiral auxiliaries, and asymmetric catalysis (Fig. 13). For large-scale applications, many of those methods suffer from complexity, limited space–time yield, and formation of excessive amounts of waste material. The aim, an economic industrial total synthesis of *RRR*-**1**, could not be reached until today by any of the methods described. It is not the intention of this chapter to draw a complete picture from the vast amount of detailed information on this topic. Instead, only a selection of representative approaches will be described

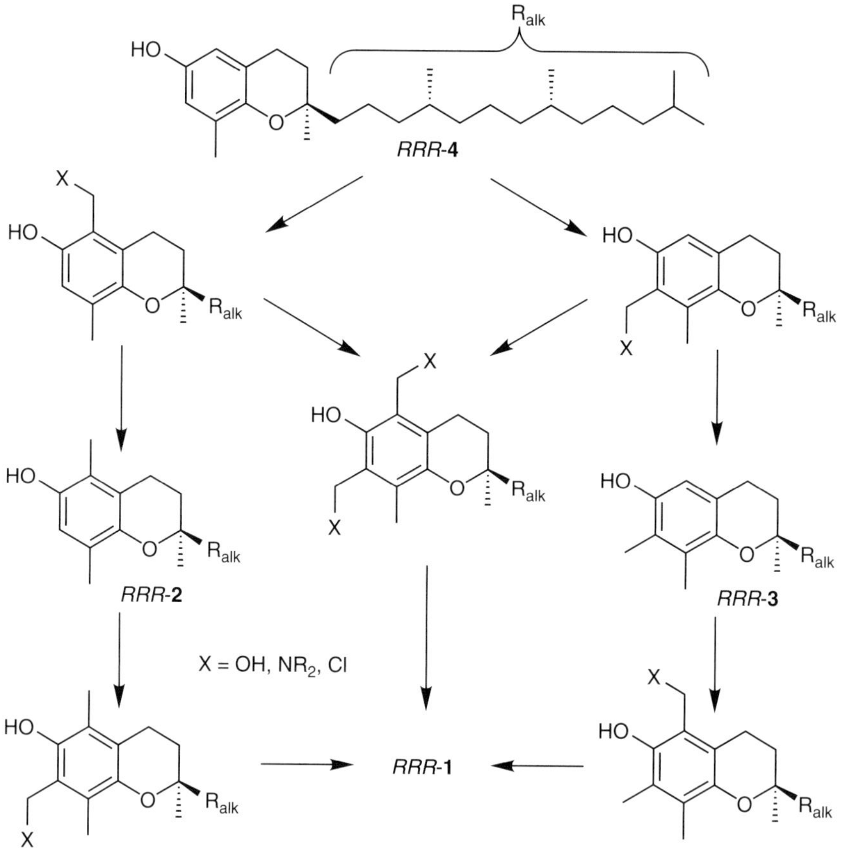

R_{alk}

RRR-**4**

RRR-**2**

RRR-**3**

X = OH, NR_2, Cl

RRR-**1**

FIGURE 12. Permethylation of non-α-tocopherols to *RRR*-**1**.

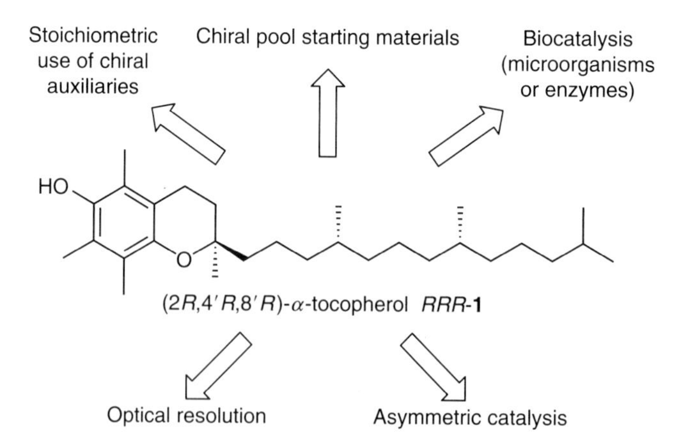

FIGURE 13. General strategies for the synthesis of (2*R*,4′*R*,8′*R*)-α-tocopherol.

here, also reflecting the limited value of many contributions regarding synthetically useful (in particular large-scale) applications.

Concerning the overall strategy, the three major problems of the total synthesis of *RRR*-**1** have been addressed separately in most literature reports. One being the formation of the chiral chroman ring, the second the introduction of the two chiral centers in the aliphatic side chain, and the third coupling of chroman and side chain building blocks. In the following two paragraphs, the first two issues will be illustrated by selected examples for the synthesis of chroman and side chain components. Interestingly, some synthetic strategies use a different approach like combining the solution of two problems, for exapmple C–C (C–O) coupling under concomitant creation of one chiral center (vide infra, Figs. 19, 30, and 37).

A. SYNTHESIS OF CHIRAL CHROMAN COMPOUNDS

In Table I, a selection of typical chiral precursors or building blocks for construction of the 2*R*-chroman moiety of vitamin E compounds are compiled. A high (>98%) enantiomeric excess (ee) value is crucial in general to ensure the method of being useful in the context of single-isomer syntheses. The strategies as categorized in Fig. 13 are illustrated by some details depicted in Figs. 14–19.

Optical resolution was the first method which was employed in the synthesis of building blocks (e.g., **54**–**59**). For example, carboxylic acids *S*- and *R*-**54** were resolved with quinine and used in the first synthesis of *RRR*- and *SRR*-α-tocopherol by Mayer *et al.* (1963a) (cf. Fig. 24, vide infra).

Important contributions came from the area of biocatalysis. Lipases were used successfully in both hydrolysis and esterification reactions to deliver chiral chroman building blocks (e.g., **60**–**66**). As sketched in Fig. 14, this approach was by far more efficient in the preparation of functionalized tocopherol precursors (e.g., **64**–**66**) than in resolution of tocopherol itself. Hydrolysis of (all-*rac*)-α-tocopheryl acetate [(all-*rac*)-**9**] or other ester derivatives (e.g., →**63**) was of limited success. Only the corresponding unnatural parent compound tocol (i.e., the sterically less hindered tris-desmethyl-aryl derivative) could be resolved with excellent selectivity (99%) (Mizuguchi and Achiwa, 1993). Chiral auxiliaries and chiral pool starting materials (e.g., **67**–**70**; Fig. 15 for the application of D-proline; Yoda and Takabe, 1989) were also, but less commonly, used.

A more important and, from the view of synthetic efficiency, more promising approach uses enantioselective catalysis. Here, the Sharpless epoxidation (Fig. 16) was one of the first methods used, for example, for the synthesis of **71** (Imfeld, 1986; Takabe *et al.*, 1985) and **72** (Mizuguchi and Achiwa, 1995). A chroman ring closure reaction with rhodium or palladium catalysis by using chiral ligands was verified by Knierzinger and Scalone

TABLE I. Examples for Preparation of Chiral Chroman Compounds and Precursors

Compound number	Method used and structure of product	References
	Optical resolution	
S-**54**	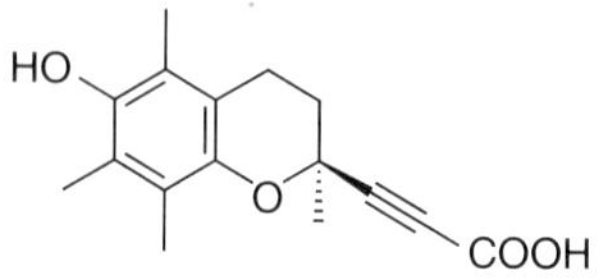 HO, COOH	Mayer *et al.*, 1963a
55	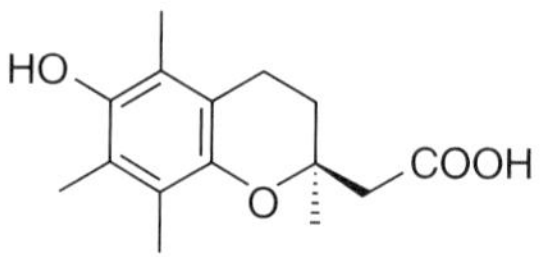 HO, COOH, O	Cohen *et al.*, 1982; Scott *et al.*, 1976
56	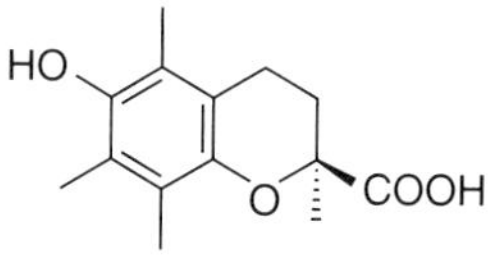 HO, COOH, O	Walther *et al.*, 1991
57	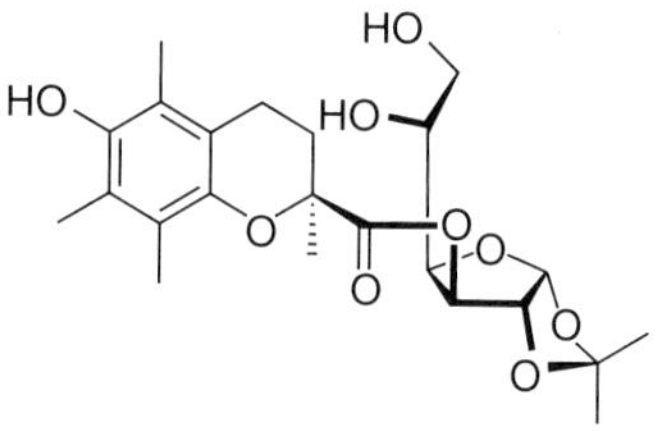 HO, HO, HO, O, O, O, O, O	Netscher and Gautschi, 1992
58	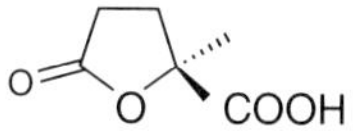 O, O, COOH	Cohen *et al.*, 1979
59	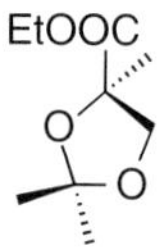 EtOOC, O, O	Barner and Schmid, 1979
	Biocatalysis	
60	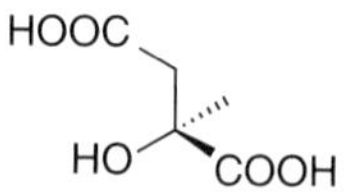 HOOC, HO, COOH	Barner and Schmid, 1979
61	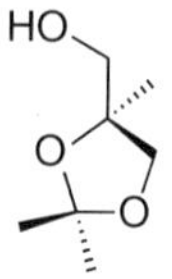 HO, O, O	Wirz *et al.*, 1993

TABLE I. (*Continued*)

Compound number	Method used and structure of product	References
62	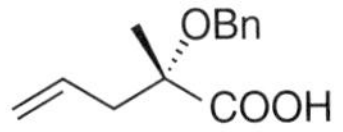 OBn; COOH	Sugai *et al.*, 1991
63	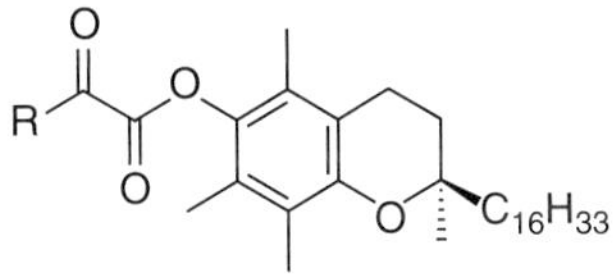 O; R; O; O; O; $C_{16}H_{33}$	Mizuguchi and Achiwa, 1993
64	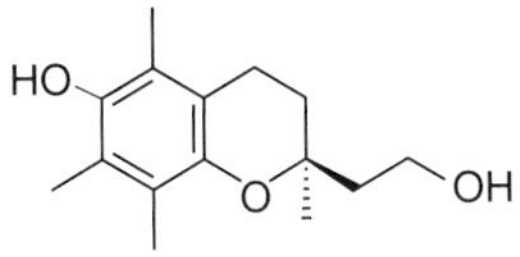 HO; O; OH	Mizuguchi *et al.*, 1994
65	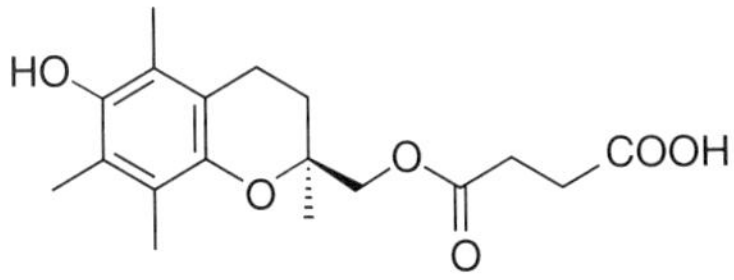 HO; O; O; COOH; O	Hyatt and Skelton, 1997
66	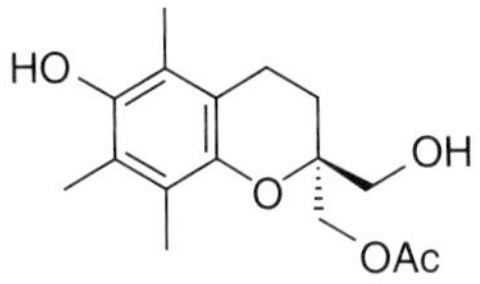 HO; OH; O; OAc	Chênevert and Courchesne, 2002; Chênevert *et al.*, 2006
	Chiral auxiliaries/chiral pool	
67	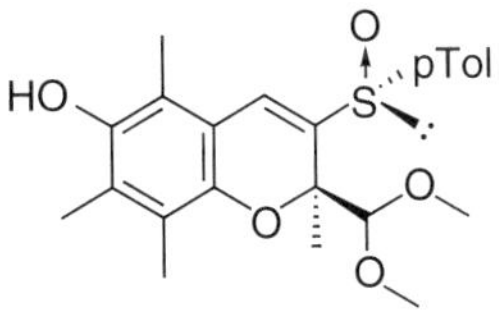 O; pTol; HO; S; O; O; O	Solladie and Moine, 1984
68	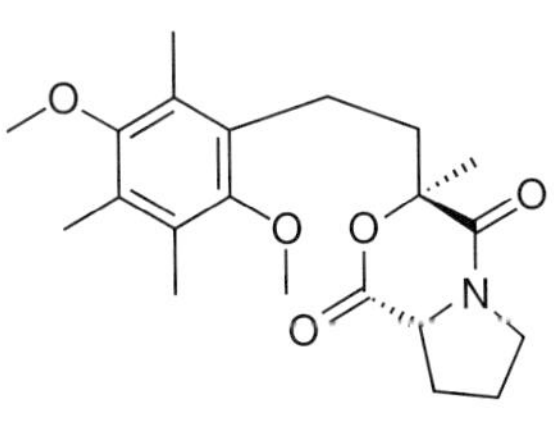 O; O; O; O; N; O	Yoda and Takabe, 1989
69	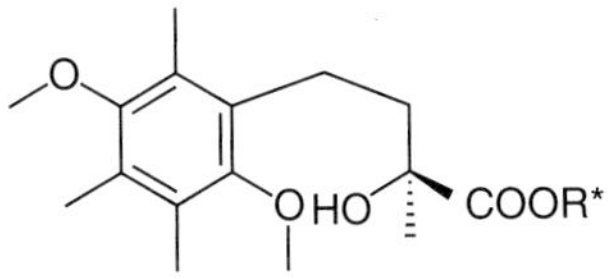 O; O; HO; COOR*	Hübscher and Barner, 1990

(*Continues*)

TABLE I. (*Continued*)

Compound number	Method used and structure of product	References
70	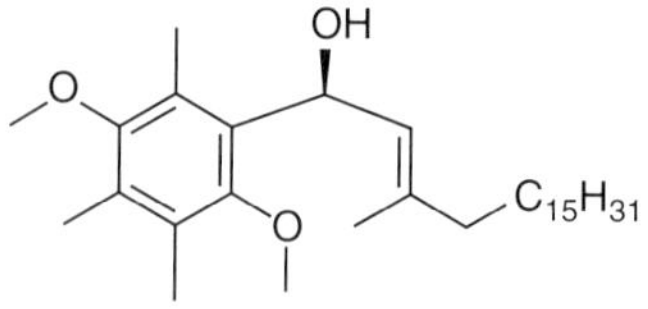	Hübscher and Barner, 1990
	Enantioselective catalysis	
71	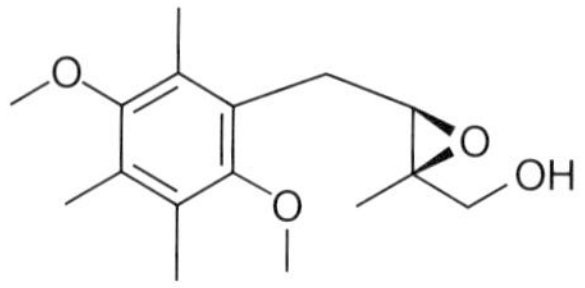	Imfeld, 1986; Takabe *et al.*, 1985
72	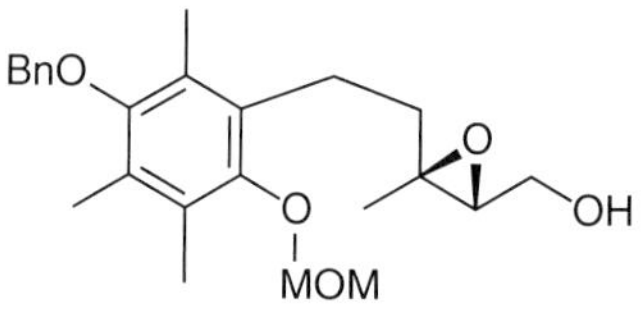	Mizuguchi and Achiwa, 1995
73	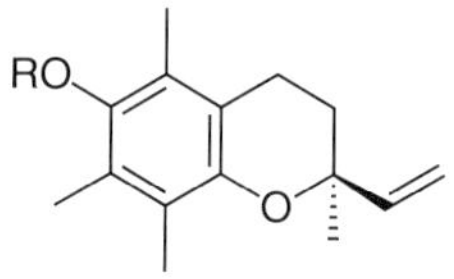	Knierzinger and Scalone, 1990, 1992; Trost and Asakawa, 1999; Trost *et al.*, 2004
74	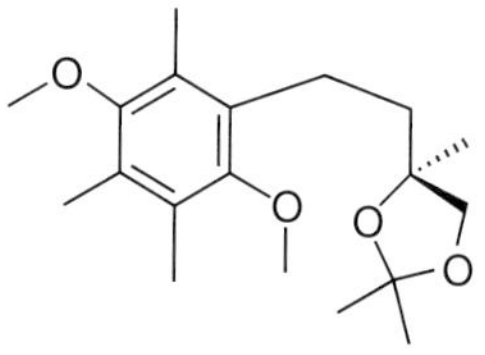	Tietze and Görlitzer, 1998; Tietze *et al.*, 1999
75	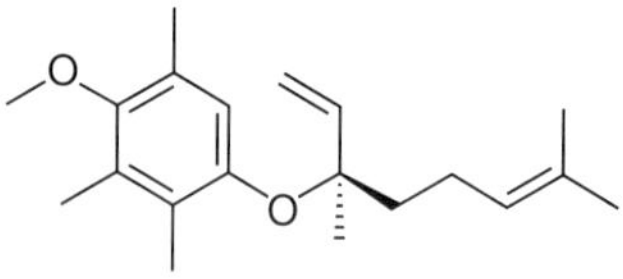	Trost and Toste, 1998; Trost, 2004
76	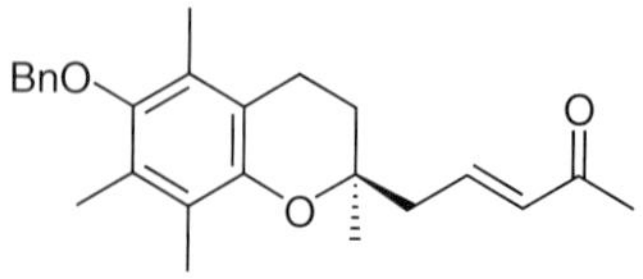	Tietze *et al.*, 2005, 2006

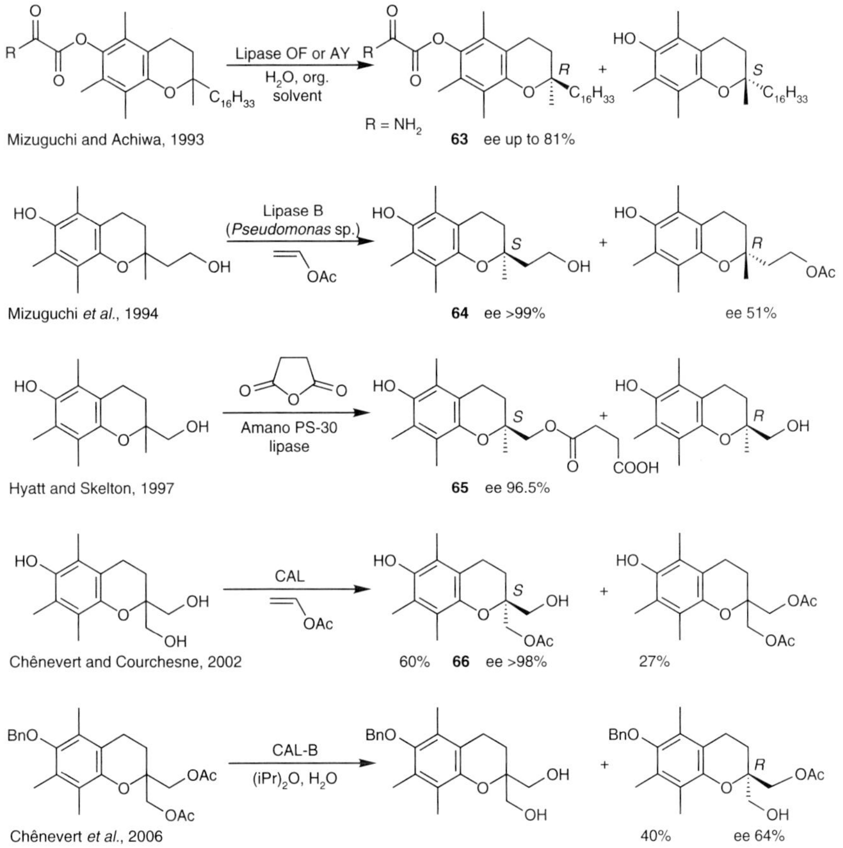

FIGURE 14. Lipase-catalyzed kinetic resolution for the preparation of optically active chroman compounds.

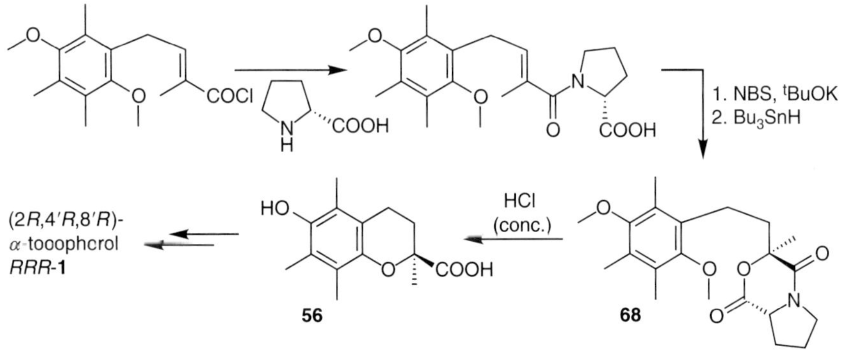

FIGURE 15. Synthesis of a chiral chroman compound by use of D-proline.

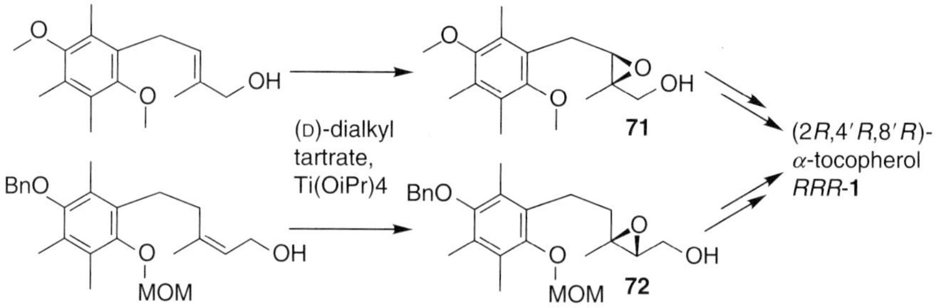

FIGURE 16. Synthesis of chiral chroman precursors by Sharpless epoxidation.

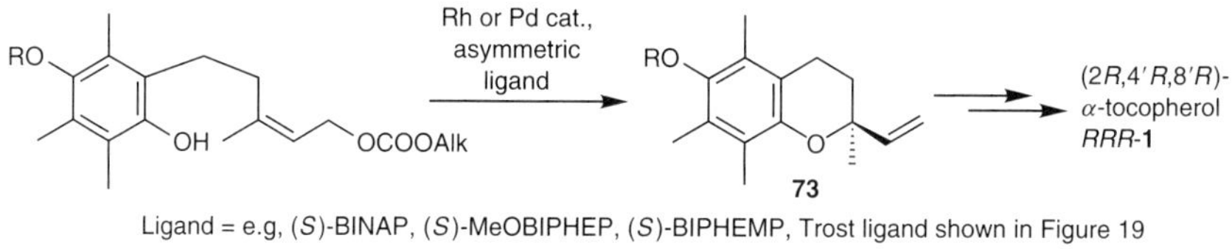

Ligand = e.g, (*S*)-BINAP, (*S*)-MeOBIPHEP, (*S*)-BIPHEMP, Trost ligand shown in Figure 19

FIGURE 17. Intramolecular allylic substitution to chiral vinyl chroman compounds.

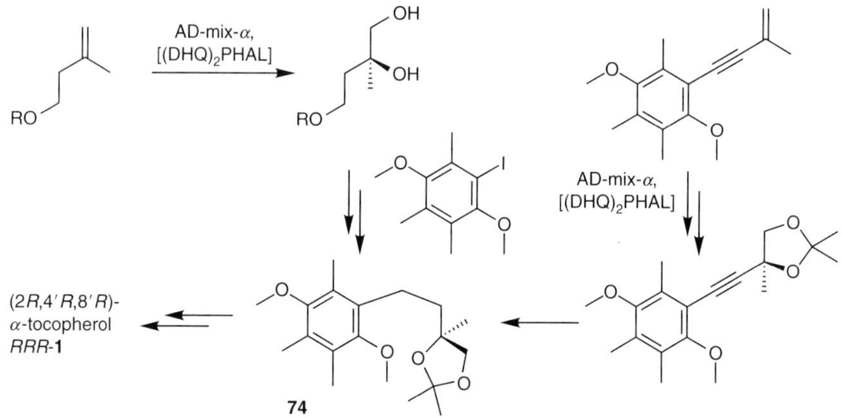

FIGURE 18. Synthesis of chiral chroman precursors by Sharpless bishydroxylation.

(1990, 1992) with moderate ee, and later on somewhat improved by the Trost *et al.* (Trost, 2004; Trost and Asakawa, 1999) with an ee of up to 87% of vinyl chroman compounds **73** (Fig. 17).

Chiral chroman precursor **74** was synthesized by Tietze *et al.* (Tietze and Görlitzer, 1998; Tietze *et al.*, 1999). The chirality was introduced by employing the Sharpless bishydroxylation methodology, combined with aryl-sp^3 [Zn/Cu, $PdCl_2(PPh_3)_2$] or Sonogashira coupling [$PdCl_2(PPh_3)_2$, CuI, $NHEt_2$] (Fig. 18).

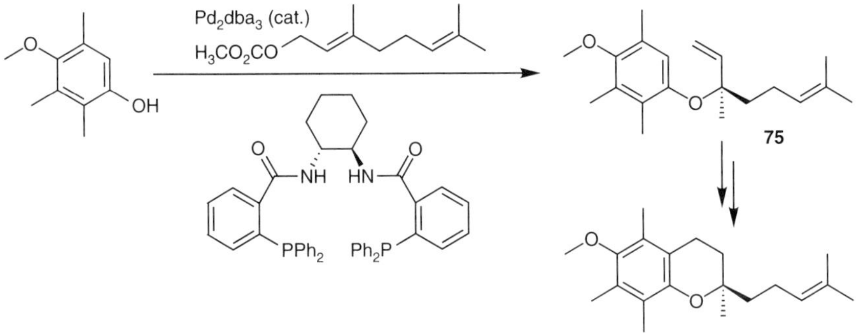

FIGURE 19. Synthesis of a chiral chroman compound by asymmetric allylic substitution.

An intermolecular palladium-catalyzed asymmetric allylic substitution reaction delivered chroman precursor **75** in up to 77% ee and a regioselectivity of 98:2 (tertiary to primary ether) (Trost, 2004; Trost and Toste, 1998; Fig. 19). An enantioselective chroman cyclization was claimed to proceed by the use of catalytic antibodies with an ee of up to 42% (Tietze *et al.*, 2000).

B. SYNTHESIS OF CHIRAL SIDE CHAIN COMPONENTS

Table II contains some representative examples for the preparation of optically active C_4 to C_{20} side chain components, which were applied for syntheses of (2*R*,4′*R*,8′*R*)-α-tocopherol (*RRR*-**1**), according to Fig. 13.

As in the case of chroman building blocks (cf. Section III.A), optical resolution was used in early years. For example, diastereomeric salt formation of a hemiphthalate with (*R*)-α-methylbenzylamine delivered alcohol **77** which was further transformed to aldehyde **78** by highly stereoselective sigmatropic rearrangement reactions (Chan and Saucy, 1977; Chan *et al.*, 1976, 1978).

Microorganisms and enzymes (Leuenberger and Wirz, 1993) were used for the efficient preparation of intermediates **79**–**84** (Fig. 20). Building block **79** was found to be readily available by bacterial hydroxylation of isobutyric acid with *Pseudomonas putida* (Cohen *et al.*, 1976; Goodhue and Schaeffer, 1971). Further biocatalytic transformations include, for example, stereoselective reduction reactions with yeast, esterification, and ester hydrolysis reactions.

(*E,R,R*)-Phytol (*ERR*-**41**) was used as a chiral pool starting material for the transformation to C_{18} ketone *RR*-**34** and C_{15} alcohol *RR*-**85** (Mayer and Isler, 1971; Mayer *et al.*, 1963a), and (*R*)-(+)-pulegone (C_{10}) was transformed to *RR*-**86** (C_{10}) (Chan *et al.*, 1976; Chen *et al.*, 1996; Fujisawa *et al.*, 1981) and to *RR*-**85** (C_{15}) by chirality transfer in a Carroll reaction (Koreeda and Brown, 1983).

Several breakthroughs in the field of enantioselective catalysis resulted in highly efficient procedures for the construction of chiral isoprenoid side chain building blocks. In particular, enantioselective catalytic allylamine → enamine

TABLE II. Examples for Preparation of Chiral Side Chain Compounds

Compound number	Method used and structure of product	References
	Optical resolution	
77	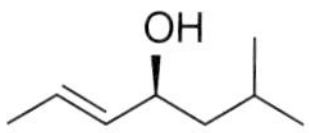 OH	Chan and Saucy, 1977; Chan *et al.*, 1976, 1978
78	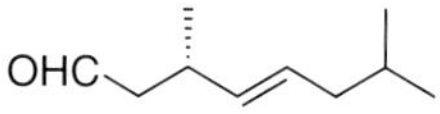 OHC	Chan and Saucy, 1977; Chan *et al.*, 1976, 1978
	Biocatalysis	
79	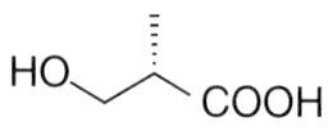 HO COOH	Cohen *et al.*, 1976; Goodhue and Schaeffer, 1971
80	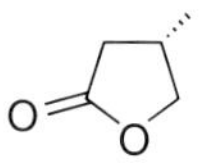 O O	Leuenberger *et al.*, 1979
81	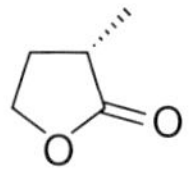 O O	Leuenberger *et al.*, 1979
R-**82**	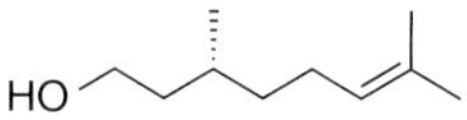 HO	Gramatica *et al.*, 1986, 1987; Oda *et al.*, 2000
83	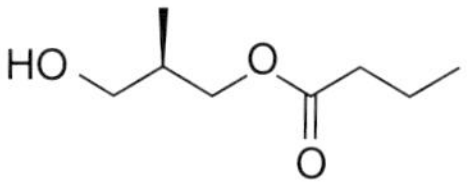 HO O O	Wirz *et al.*, 1990
84	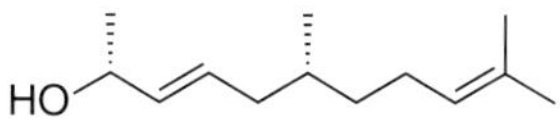 HO	Coffen *et al.*, 1994
	Chiral auxiliaries/chiral pool	
RR-**85**	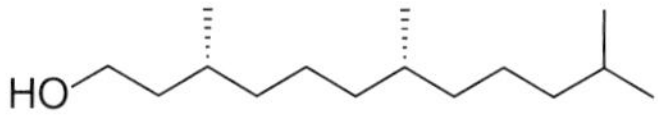 HO	Koreeda and Brown, 1983; Mayer and Isler, 1971; Mayer *et al.*, 1963a
R-**86**	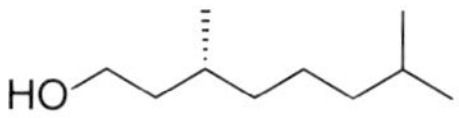 HO	Chan *et al.*, 1976; Chen *et al.*, 1996; Fujisawa *et al.*, 1981
	Enantioselective catalysis	
87	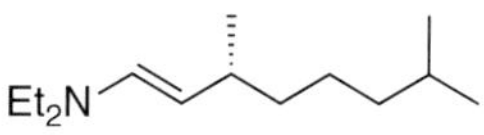 Et_2N	Schmid *et al.*, 1988, 1991; Tani *et al.*, 1984

TABLE II. (*Continued*)

Compound number	Method used and structure of product	References
88	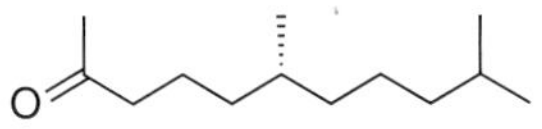	Broger and Müller, 1993; Schmid *et al.*, 1996; Wüstenberg, 2003
RR-**85**	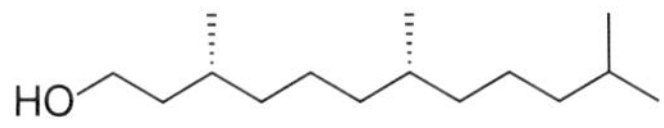 HO	Heiser *et al.*, 1991; Netscher *et al.*, 2004a; Takaya *et al.*, 1987; Wüstenberg, 2003
89	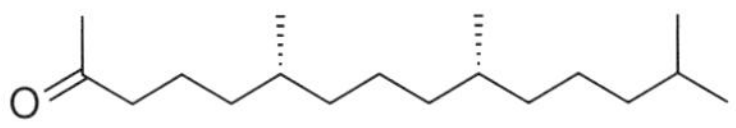	Broger and Müller, 1993; Schmid and Broger, 1994
90	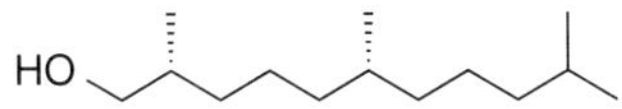 HO	Huo and Negishi, 2001
RR-**50**	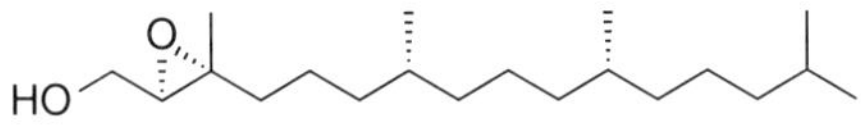 HO O	Inoue *et al.*, 1987; Takano *et al.*, 1990, 1994
91	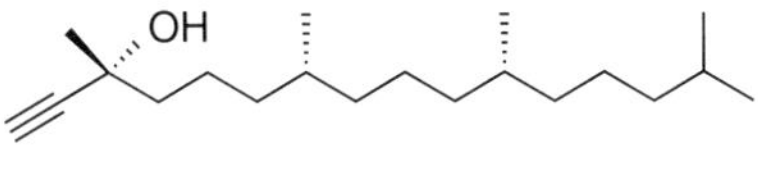 OH	Takano *et al.*, 1990, 1994

isomerization (Tani *et al.*, 1984) and hydrogenation reactions represent milestones in the area of terpenoid chemistry. C_2 symmetric diphosphane ligands like BINAP (Akutagawa, 1995; Noyori and Takaya, 1990), BIPHEMP (Schmid *et al.*, 1988), and MeOBIPHEP (Schmid *et al.*, 1991) showed exceptionally high chemo- and stereoselectivity in ruthenium-catalyzed hydrogenations of allylic alcohols (Heiser *et al.*, 1991). Representative examples (cf. Figs. 21 and 22) are the preparation of highly stereoisomerically enriched (up to 99%) (*R*/*S*)-citronellol (*R*-**82**/*S*-**82**) from geraniol/nerol (*E*-**92**/*Z*-**92**) (Takaya *et al.*, 1987), and ketone **88** and C_{15} alcohol *RR*-**85** via intermediates **93**, *R*-**86**, *E*/*Z*-**94**, and *E*-**95**. The relevant work on this topic has been reviewed in detail (Netscher *et al.*, 2004a). It has to be mentioned that also the analytical procedures for reliable determination of high (>95%) ee/de values are in place (Knierzinger *et al.*, 1990).

Huo and Negishi (2001) performed a zirconium-catalyzed asymmetric carboalumination to C_{14} side chain alcohol **90** (Fig. 23). Although the stereoselectivity in both catalytic steps was insufficient (resulting in an *R*:*S* ratio of 87:13), the isomeric purity could be upgraded via crystallization of stereoisomeric *p*-phenylenebisurethans to >99% *R* at C-2 and 97% *R* at C-6. C_{14} alcohol **90** was further transformed to *RRR*-**1** by known chemistry.

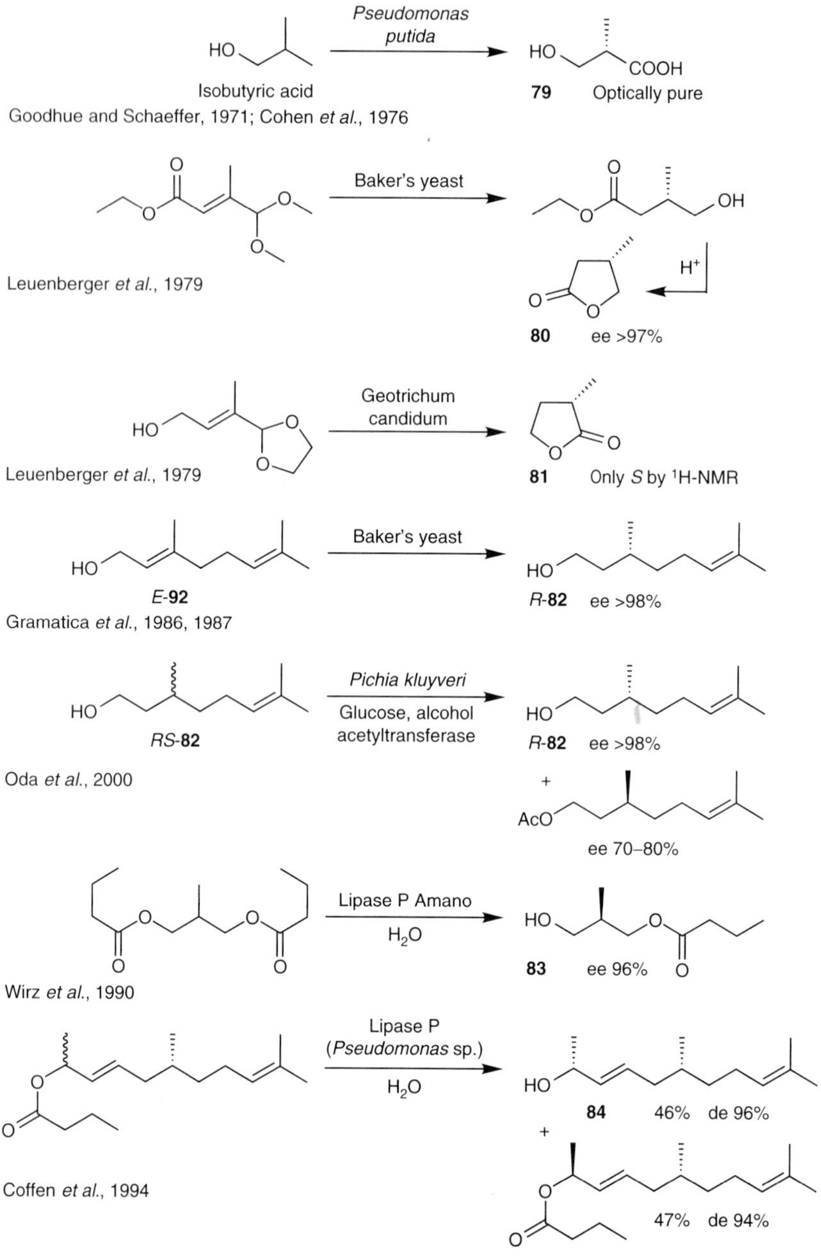

Pseudomonas putida
HO COOH
Isobutyric acid
79 Optically pure
Goodhue and Schaeffer, 1971; Cohen *et al.*, 1976

Baker's yeast
O O O OH
H^+
Leuenberger *et al.*, 1979
80 ee >97%

Geotrichum candidum
HO O O O
Leuenberger *et al.*, 1979
81 Only *S* by ^{1}H-NMR

Baker's yeast
HO HO
E-**92**
R-**82** ee >98%
Gramatica *et al.*, 1986, 1987

Pichia kluyveri
Glucose, alcohol acetyltransferase
HO HO
RS-**82**
R-**82** ee >98%
Oda *et al.*, 2000
+
AcO
ee 70–80%

Lipase P Amano
H_2O
HO
83 ee 96%
Wirz *et al.*, 1990

Lipase P (*Pseudomonas* sp.)
H_2O
HO
84 46% de 96%
+
47% de 94%
Coffen *et al.*, 1994

FIGURE 20. Biotransformations leading to chiral side chain building blocks.

Asymmetric catalysis has also been used for the generation of the (2*R*) chiral chroman center to yield C_{20} intermediates already fully equipped with the chiral (*R*,*R*) side chain moiety. Natural phytol (*ERR*-**41**) was transformed

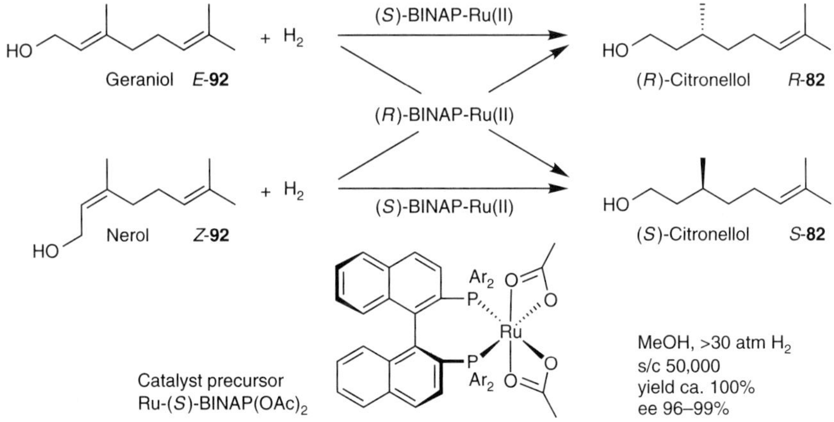

FIGURE 21. Ru-BINAP enantioselective hydrogenation of geraniol and nerol.

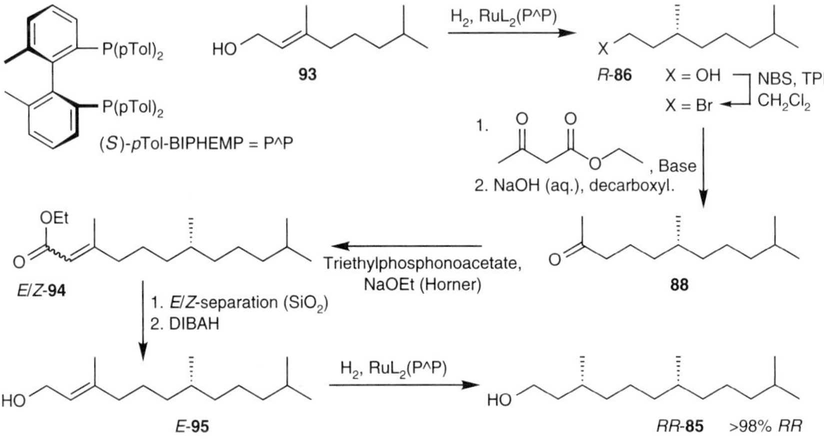

FIGURE 22. Laboratory scale synthesis of C_{15} side chain building block *RR*-**85**.

to (2*R*,3*R*)-2,3-epoxyphytol (*RR*-**50**) by Sharpless epoxidation (Inoue *et al.*, 1987) and to **91** by Sharpless bishydroxylation (Takano *et al.*, 1990, 1994; cf. Fig. 25, vide infra), showing similarities to the shorter-chain examples depicted in Figs. 16 and 18.

C. SYNTHESIS OF *(R,R,R)*-TOCOPHEROLS

In general, it has to be mentioned that efforts regarding this topic were originally almost exclusively directed to the synthesis of α-homologue *RRR*-**1** due to its high biological relevance. The first synthesis of *RRR*-**1** and *SRR*-**1** was published by Mayer *et al.* (1962, 1963a). The enantiomeric aldehydes **96**/*ent*-**96** (derived from trimethylhydroquinone **10** via an optical resolution

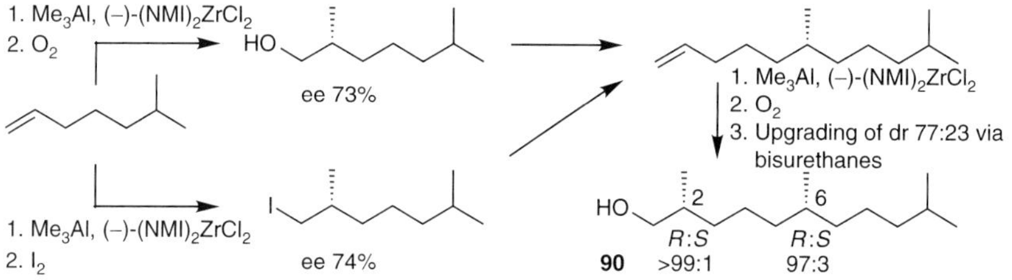

FIGURE 23. Chiral isoprenoic building blocks by asymmetric carboalumination.

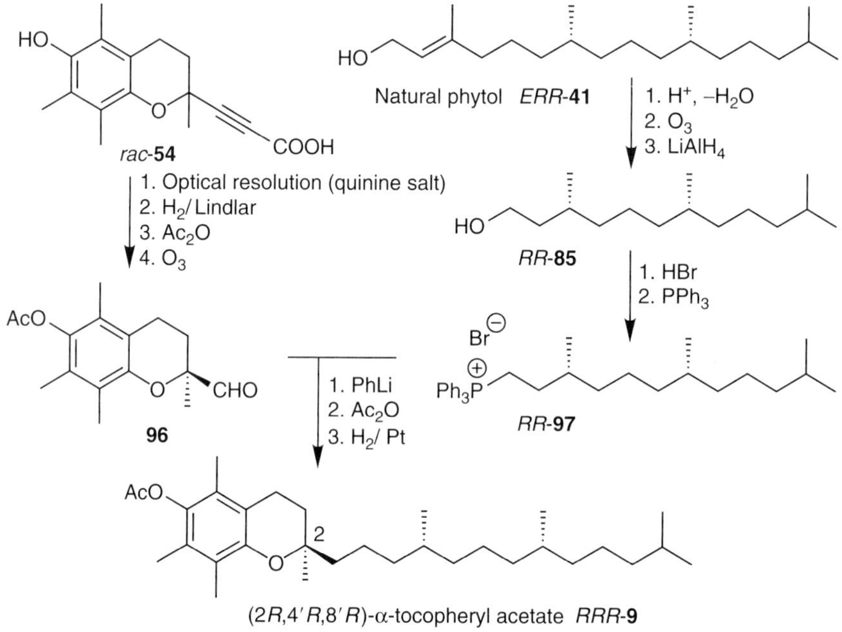

FIGURE 24. First synthesis of (2*R*,4′*R*,8′*R*)-α-tocopherol (acetate derivative **9**).

of quinine salts) were coupled with the C_{15} phosphonium salt *RR*-**97** (obtained by degradation of natural phytol, *ERR*-**41**) by a Wittig reaction (Fig. 24). In combination with further investigations, including the inversion of the configuration at C-2 of the chroman ring (Schudel *et al.*, 1963a), the absolute configuration of natural α-tocopherol could be elucidated (Mayer *et al.*, 1963b).

Later on, various routes and combinations of (chiral) building blocks and methods for their preparation and coupling have been described during the last four decades. In particular, transition metal-mediated C–C coupling methods have been applied to link the side chain with aryl or chroman

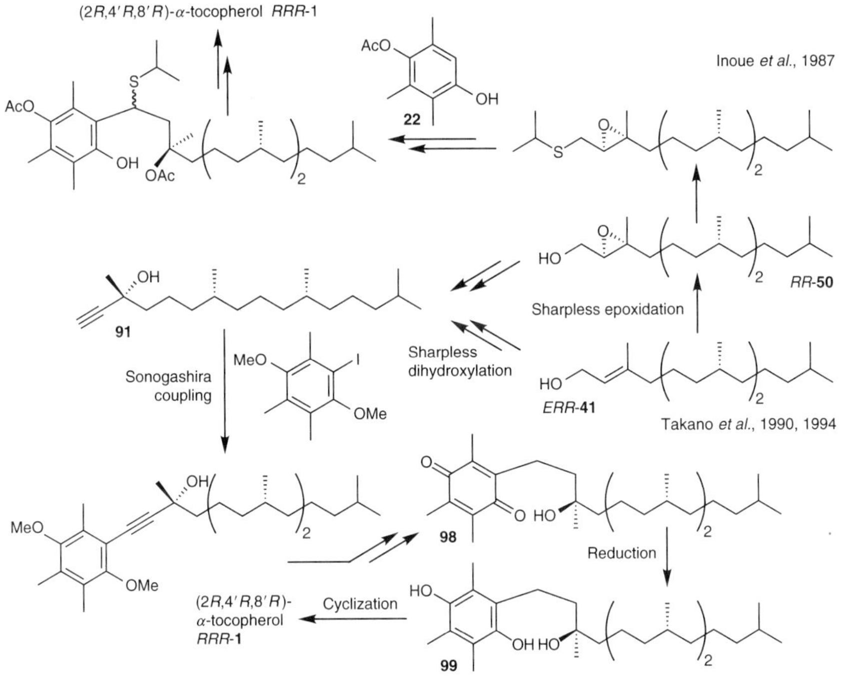

FIGURE 25. Synthesis of *RRR*-**1** via functionalization of *ERR*-phytol.

intermediates, for example, using the Sonogashira-type coupling worked out by Inoue *et al.* (1987) and Takano *et al.* (1990, 1994) sketched in Fig. 25 (cf. also Fig. 18).

A crucial question in several routes was the stereoselectivity of the chemical chroman ring closure reaction **99** → *RRR*-**1** (under retention of configuration) which was investigated in detail (Cohen *et al.*, 1981a; Mayer *et al.*, 1967a; Schudel *et al.*, 1963a); the cyclization under inversion was also used for the preparation of the epimeric *SRR*-**1**, see Fig. 36 infra. A Fouquet-Schlosser coupling of C_{14} side chain Grignard reagent **101** with a chroman-C_2 unit **100** to benzyl ether *RRR*-**102** (or analogous compounds) using copper catalysis was described as an efficient sp^3–sp^3 coupling method (Cohen *et al.*, 1976, Ernst *et al.*, 1984; Hughes *et al.*, 1990; Fig. 26).

Palladium-catalyzed coupling reactions (Heck, Sonogashira) were also used as key steps in the preparation of optically active intermediates **103**–**105** and **110** in which the chirality for transfer to the chroman ring was generated by enzymatic hydrolysis or Sharpless epoxidation or obtained by chromatographic separation of diastereomeric esters (Outten *et al.*, 1999a,b; Fig. 27). As an alternative to Wittig olefination used by Mayer *et al.* (1963a) in the first synthesis of *RRR*-α-tocopheryl acetate (*RRR*-**9**, Fig. 24), a copper-catalyzed coupling between triflate **108** and Grignard reagent **109**

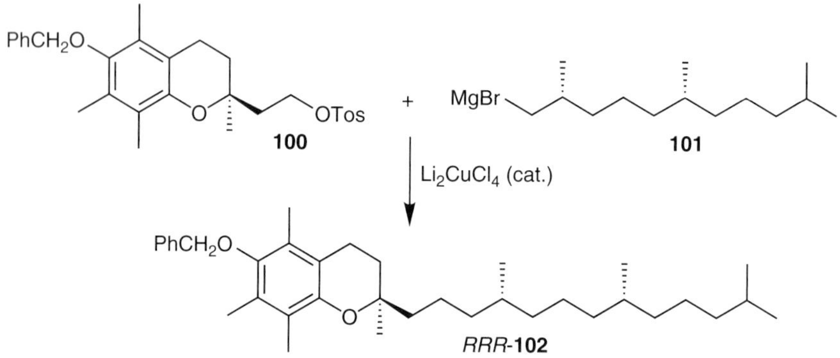

FIGURE 26. Chroman-C_2 + alkyl-C_{14} strategy for the synthesis of tocopherols.

was applied, resulting in a (chroman-C_1 + alkyl-C_{15}) approach involving a rather unusual S_N2-type substitution at an oxa-neopentylic center (Netscher, 2003; Netscher and Bohrer, 1993).

The mechanism of enzymatic chromanol ring closure reaction known from the biosynthetic pathway of tocopherols has been investigated in the group of Woggon (Stocker *et al.*, 1993, 1996). The cyclization reaction of tocopherol precursor *E*-**111** with tocopherol cyclase isolated from *Anabaena variabilis* (blue-green algae) in the presence of D_2O yielded stereospecifically labeled γ-tocopherol derivative **112**, while the corresponding *Z*-isomer (*Z*-**111**) did not react. The methyl ether derivative **113** could be shown to be identical with the product chemically synthesized by two independent routes; in one of them a highly stereoselective, although low-yield S_N2 substitution reaction of secondary, oxa-neopentyl-type sulfonates **114** was a key step (Stocker *et al.*, 1994; Fig. 28).

The first biomimetic chromanol cyclization has been described recently by using proline-derived compound **115** obtained by an aminomethylation route. *RRR*-**1** was obtained with 70% de after acid treatment and deprotection (Grütter *et al.*, 2006; Fig. 29).

A completely novel scheme is shown in Fig. 30 in a slightly modified version of the synthesis of *RRR*-**1** presented (Rein and Breit, 2005). Key steps of this new strategy to build up the two chiral centers of the aliphatic side chain are a rhodium-catalyzed hydroformylation and a copper-mediated allylic substitution reaction, both based on the use of the reagent directing o-DBBP group (Breit and Demel, 2001).

The first total syntheses of the lower γ- and δ-tocopherol homologues *RRR*-**3** and *RRR*-**4** were achieved by using the copper-mediated coupling methodology already mentioned in Fig. 27. Chiral chroman building blocks were obtained by chromatographic separation of diastereomeric glucosyl esters (cf. compound **57** in Table I, Netscher and Gautschi, 1992), and coupled with C_{15} Grignard reagent **109** (Fig. 31; Netscher, 1996, 2003).

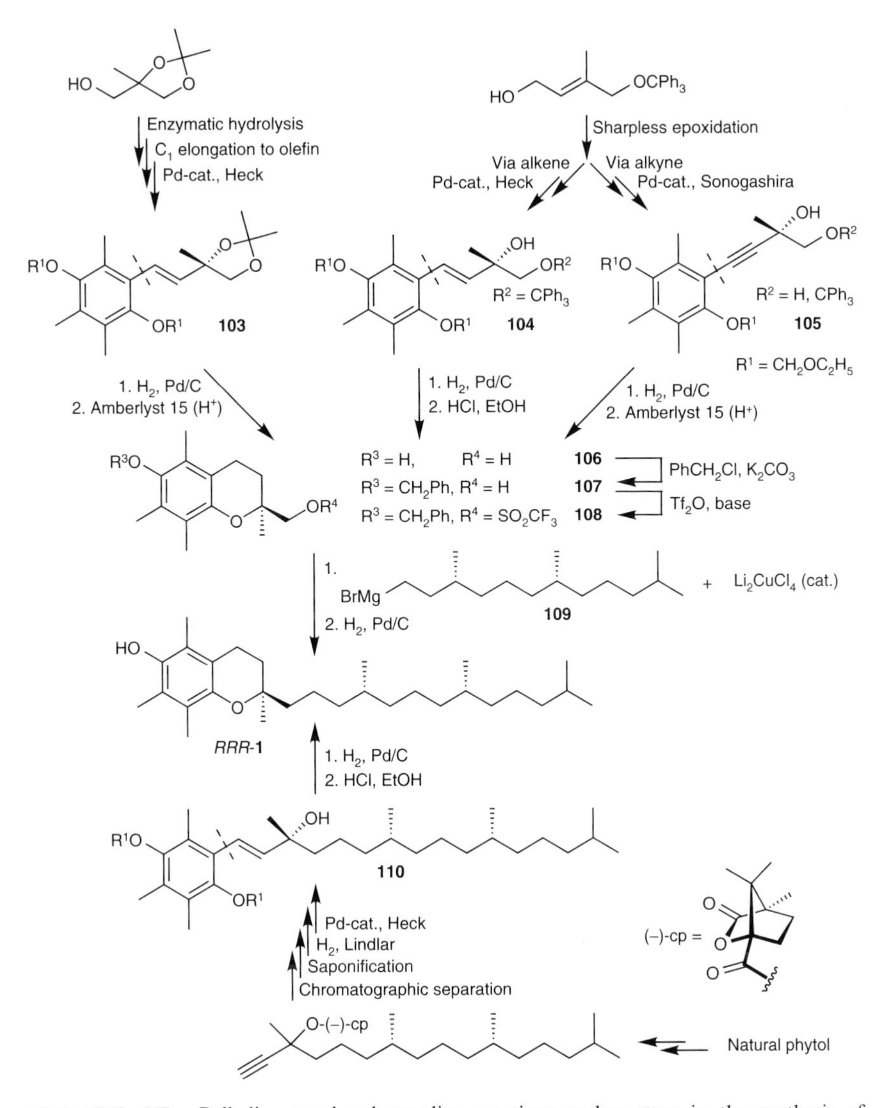

FIGURE 27. Palladium-catalyzed coupling reactions as key steps in the synthesis of tocopherols.

An aminomethylation reduction sequence by using natural δ-tocopherol (*RRR*-**4**) as a starting material resulted in an easy and rather efficient synthesis of β-tocopherol *RRR*-**2**, difficult to obtain by other routes (Netscher *et al.*, 2004b, 2007; Fig. 32). The remarkably high regioselectivity of the C-5 over C-7 alkylation, followed by purification of the crystalline hydrochloride, represents the first practical synthesis of this vitamin E compound.

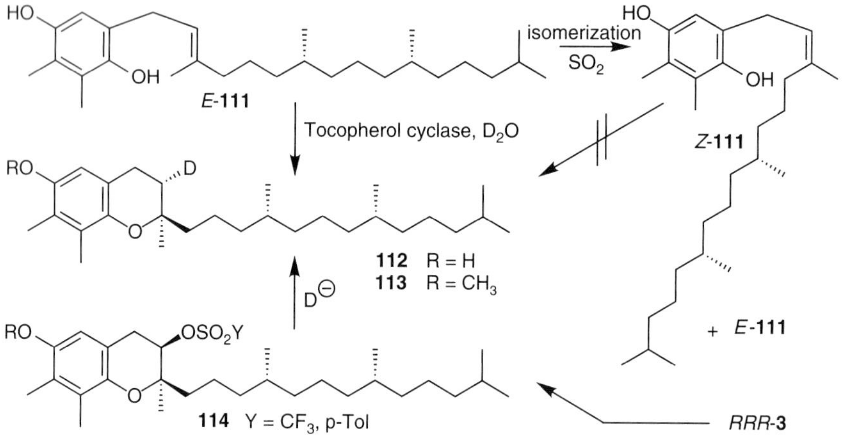

FIGURE 28. Enzymatic chromanol ring formation to (2R,4′R,8′R)-γ-tocopherol.

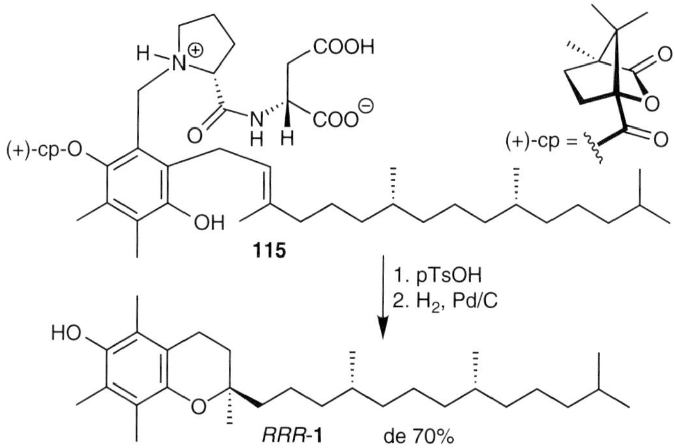

FIGURE 29. Biomimetic diastereoselective chromanol ring formation toward (2R,4′R,8′R)-α-tocopherol.

The (unusual) demethylation of α-tocopherol (**1**) to γ-tocopherol (**3**) is an example of an exceptionally easy decarboxylation reaction (170°C, 3 h) under total retention of chirality when applied to enantiopure starting material (Mazzini *et al.*, 2005; Fig. 33). In summary, there are methods around to interconvert tocopherol homologues into each other, for example β-, γ-, and δ-tocopherol (**2**,**3**,**4**) to α-tocopherol (**1**), δ- to β-tocopherol (**4** → **2**), and α- to γ-tocopherol (**1** → **3**).

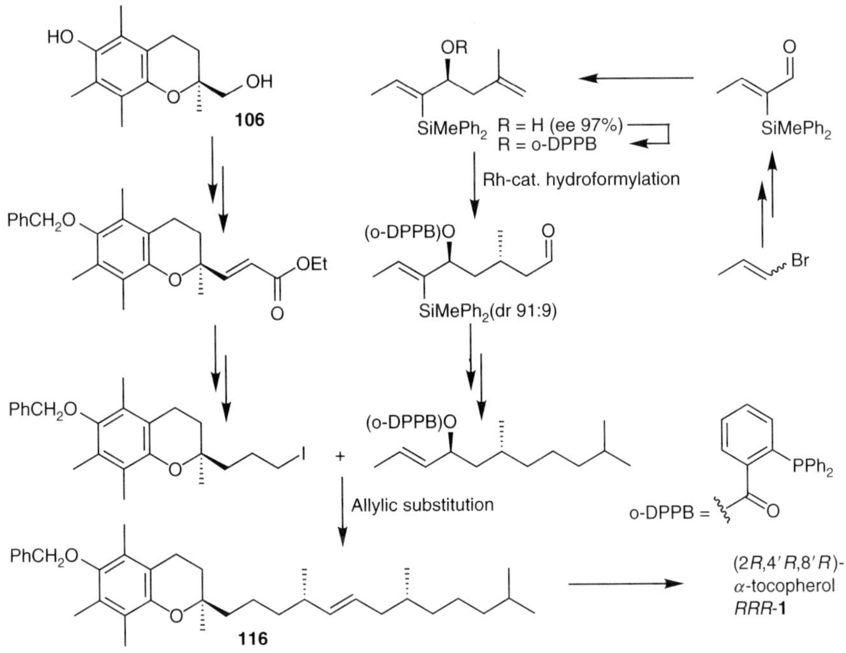

FIGURE 30. Synthesis of $(2R,4'R,8'R)$-α-tocopherol via hydroformylation and allylic substitution.

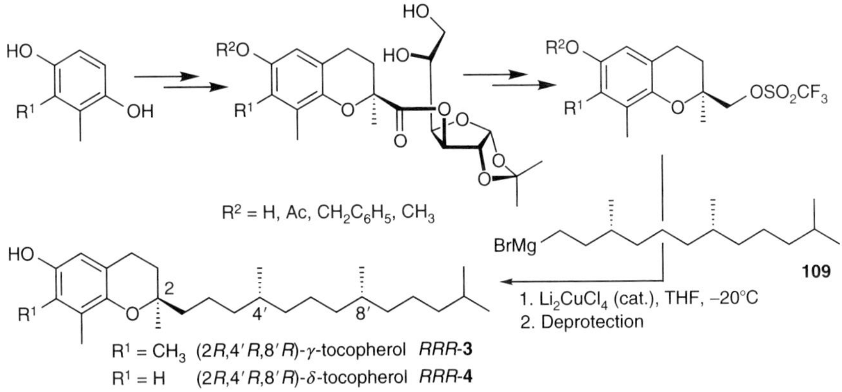

FIGURE 31. First total syntheses of $(2R,4'R,8'R)$-γ- and $(2R,4'R,8'R)$-δ-tocopherol.

Another milestone of vitamin E chemistry was reached by asymmetric catalysis. $(2R,4'R,8'R)$-γ-tocopheryl acetate (*RRR*-**119**) was obtained in a highly stereoselective asymmetric iridium-catalyzed hydrogenation of tocotrienyl derivative *REE*-**117** using the chiral P,N ligand **118**. This is the first

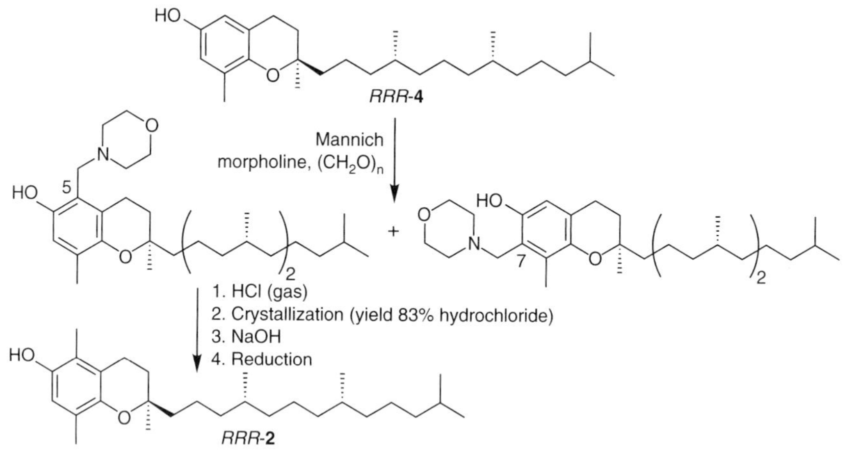

FIGURE 32. Synthesis of (2*R*,4′*R*,8′*R*)-β-tocopherol from (2*R*,4′*R*,8′*R*)-δ-tocopherol.

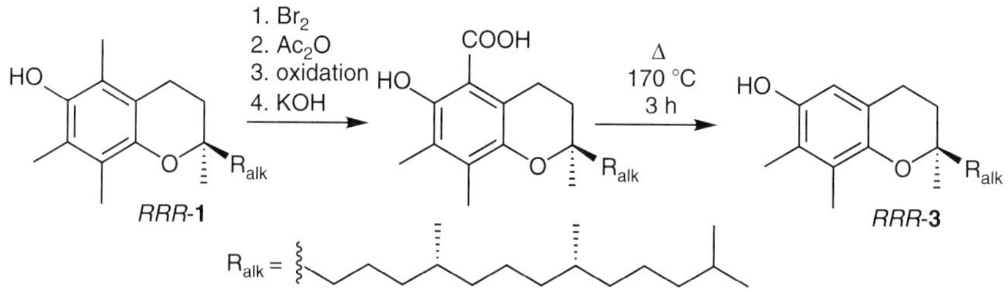

FIGURE 33. Synthesis of *RRR*-**3** from *RRR*-**1** via aryl demethylation.

example in which both stereogenic centers of the side chain could be generated from two unfunctionalized, purely alkyl-substituted olefinic moieties in a highly efficient manner in a single step (Bell *et al.*, 2006; Wüstenberg, 2003; Fig. 34).

D. SYNTHESIS OF STEREOISOMERS AND HOMOLOGUES OTHER THAN *(R,R,R)-α*-TOCOPHEROL

(2*RS*,4′*R*,8′*R*)-α-tocopherol [(2-*ambo*)-**1**] has been synthesized by acid-catalyzed condensation of trimethylhydroquinone (TMHQ, **10**) with (natural) phytol (*ERR*-**41**) or the corresponding bromide by Karrer and Isler (1946a,b; Fig. 35). Alternative chromenylation of **10** with ethylmagnesium bromide and (*R,R*)-phytal dimethylacetal (**120**) gave 3,4-dehydro-α-tocopherol in low (26%) yield, which was further transformed by hydrogenation to (2*RS*, 4′*R*,8′*R*)-α-tocopherol (Asgill *et al.*, 1978). Cohen *et al.* (1992) worked out a procedure for coupling of 2-chlorochroman **121** with C_{16} Grignard reagent

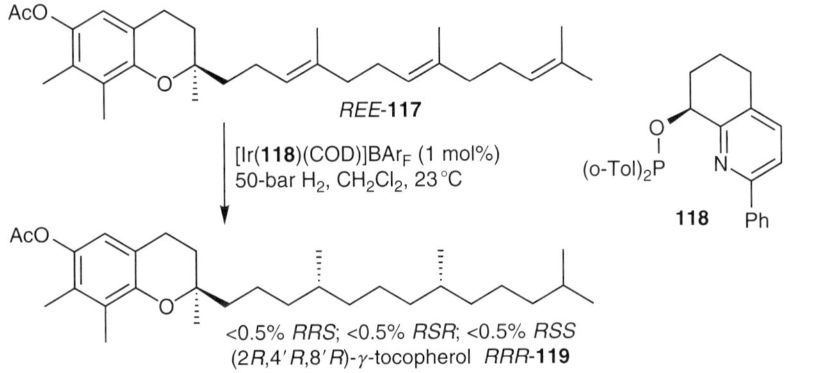

FIGURE 34. Highly stereoselective asymmetric hydrogenation of (all-*E*)-γ-tocotrienyl acetate to (2*R*,4′*R*,8′*R*)-γ-tocopheryl acetate *RRR*-**119**.

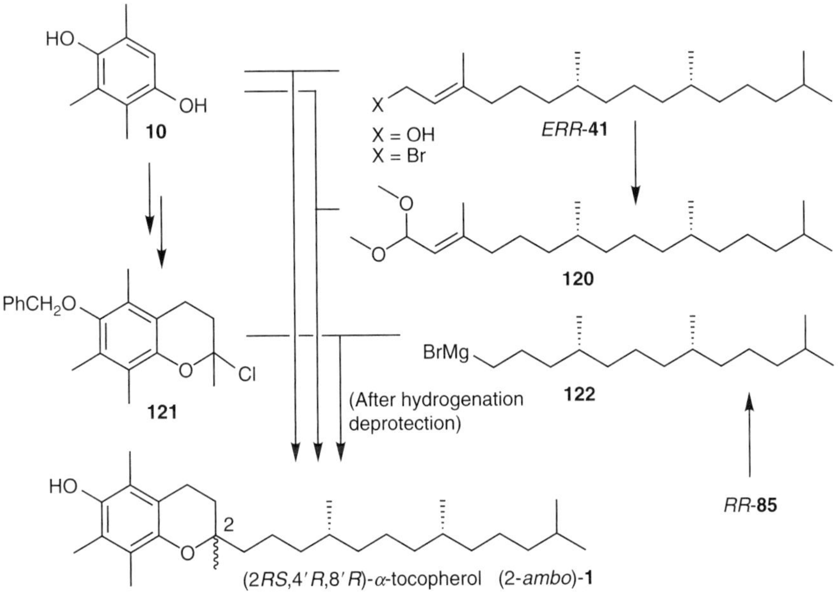

FIGURE 35. Synthesis of (2*RS*,4′*R*,8′*R*)-α-tocopherol [(2-*ambo*)-**1**].

122 (obtained from *RR*-**85**), yielding 44% of the benzyl ether of (2-*ambo*)-**1**. Prolonged heating of *RRR*-**1** with acid causes epimerization at the C-2 chroman center to yield (2-*ambo*)-**1**. Its acetate derivative (2*RS*,4′*R*,8′*R*)-**9** is also available via ruthenium-catalyzed olefin cross-metathesis routes (Malaisé *et al.*, 2006; Netscher, 2006; Netscher *et al.*, 2006). Starting materials *RR*-**42** derived from natural phytol (*ERR*-**41**) deliver intermediates *RR*-**43** or *RR*-**44** (cf. Fig. 8), which can be cyclized to (2-*ambo*)-**9**.

The (2*S*) epimeric α-tocopherol (*SRR*-**1**) has been synthesized for the first time by the Wittig route depicted in Fig. 24, starting from the enantiomeric aldehyde *ent*-**96** (Mayer *et al.*, 1963a), and could also be prepared selectively from *RRR*-**1** by $ZnCl_2$-mediated chroman cyclization of hydroquinone **99** (obtained via quinone **98**) under (predominant; selectivity 8:2) inversion of configuration (Cohen *et al.*, 1981a; Schudel *et al.*, 1963a; Fig. 36).

All eight individual stereoisomers of α-tocopheryl acetate were prepared by Cohen *et al.* (1981b). Also the analytical methods for their separation could be established (Vecchi *et al.*, 1990; Walther and Netscher, 1996). Samples of various stereoisomeric composition were obtained by asymmetric hydrogenation of tocotrienols (Bonrath *et al.*, 2006; Wüstenberg, 2003). During this work, the highly selective preparation of *RRR*-**119** from *REE*-**117** (cf. Bell *et al.*, 2006; Fig. 34) and *SSS*-**119** from *SEE*-**117** (Wüstenberg, 2003) could be achieved. By using the synthetic scheme for the preparation of the *RRR* isomers of α-, γ-, and δ-tocopherols (**1, 3, 4**) (cf. Figs. 27 and 31), also various unnatural (different from *RRR*) stereoisomers have been prepared. On the basis of the assignment by using those synthetic reference materials, analytical (HPLC, GC) procedures for the separation and quantification of all 24 isomers/homologues of these 3 tocopherols are now in place.

Tietze *et al.* (2005, 2006) described a domino Wacker–Heck sequence (Fig. 37) in which the chiral chroman cyclization was achieved with high ee (up to 97%). A method for introducing the chirality at C-4′ is still lacking. So, overall this scheme represents a stereoselective synthesis of (4′-*ambo*)-α-tocopherol.

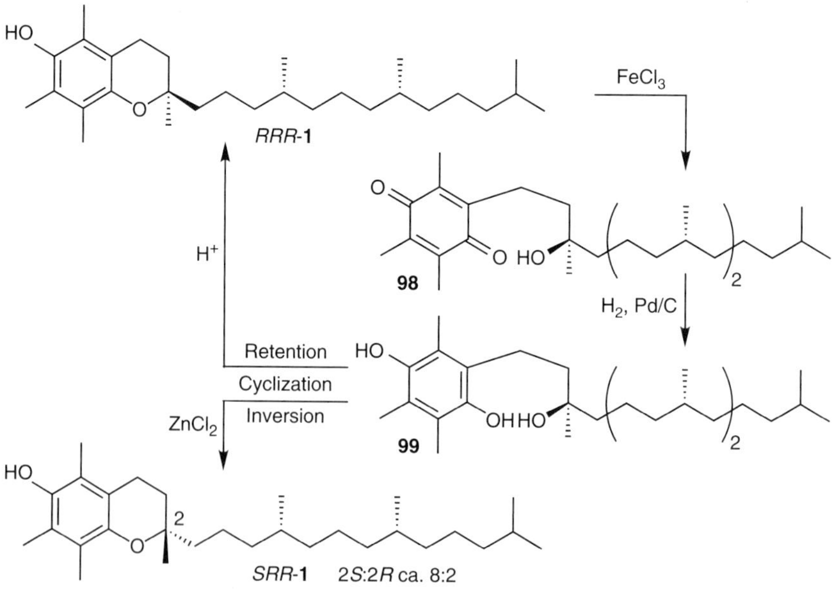

FIGURE 36. Stereoselective inversion at the C-2 chiral chroman center.

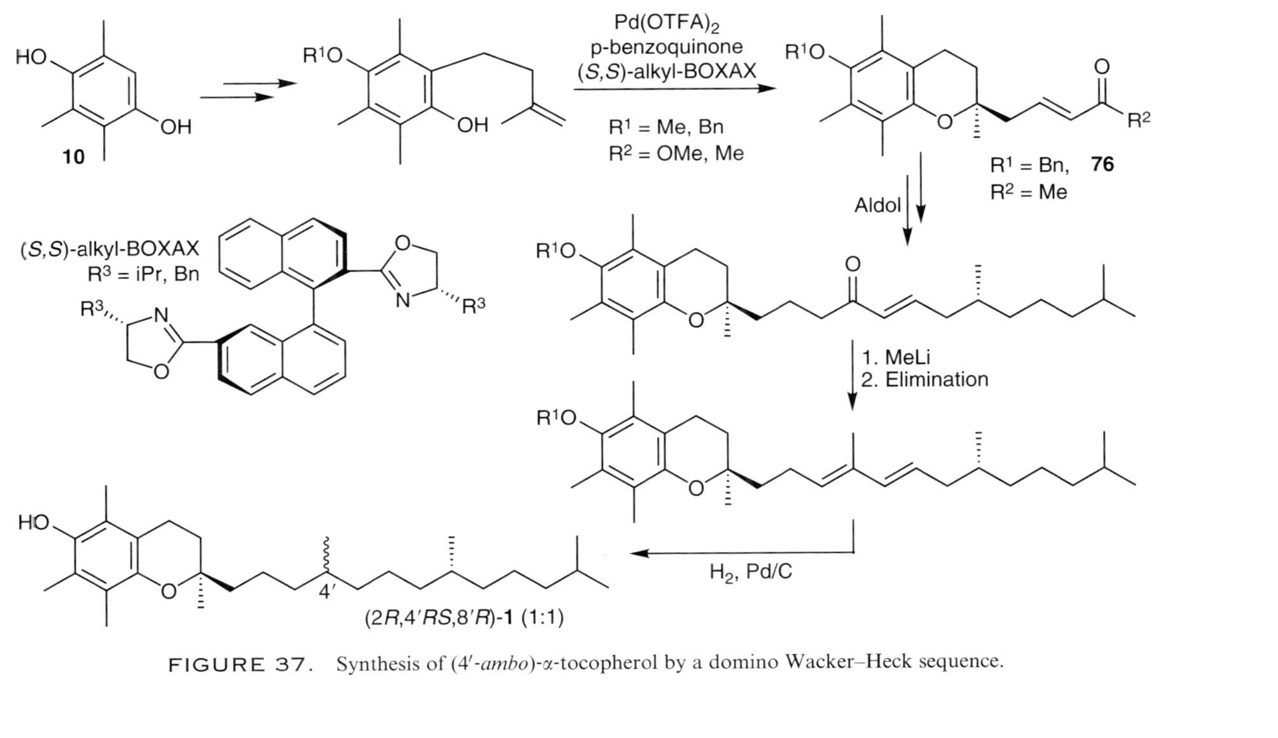

FIGURE 37. Synthesis of (4′-*ambo*)-α-tocopherol by a domino Wacker–Heck sequence.

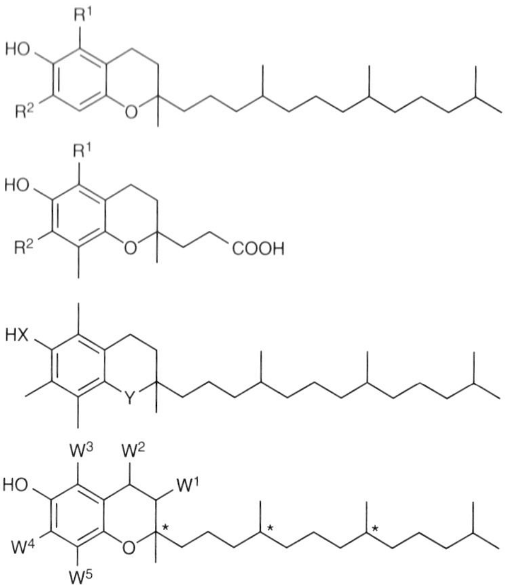

123 $R^1,R^2 = CH_3,CH_3$ or CH_3,H or H,CH_3

124 $R^1,R^2 = CH_3,CH_3$ (α-) or H,CH_3 (γ-) or H,H (δ-)

125 X,Y = O, S, NH

126 W^1,W^2 = H, D
$W^3,W^4,W^5 = CH_3, CH_2D, CHD_2, CD_3$
* = *R* or *S* or *RS* (C = $^{12/13/14}$C)

FIGURE 38. Additional derivatives and analogues of vitamin E compounds.

Several other stereoisomeric mixtures of various tocopherols have been prepared and were compiled by Mayer and Isler (1971). Examples are (2*R*,4′*RS*, 8′*RS*)-α-tocopherol or corresponding mixtures of lower homologues, including (unnatural) mono- and dimethyltocols (**123**, Fig. 38). Data on further compounds, which are mainly of biological interest in the field of vitamin E, can be found in the same reference. Examples are products of metabolic pathways, like α-, γ-, δ-CEHC (**124**), aza- and thia-analogues (**125**), and isotopically labeled tocopherols (**126**). Here, it has to be mentioned that later developments in this field are not covered by the present chapter, that is work on metabolites (cf. Mazzini *et al.*, 2004, 2006) and labeled vitamin E compounds (cf. Mazzini *et al.*, 2003) are omitted.

IV. SYNTHESIS OF TOCOTRIENOLS

The first successful synthesis of (*rac*,all-*E*)-α-tocotrienol [(*rac,E,E*)-**5**] from (*rac*,all-*E*)-geranyllinalool (**127**) was described by Schudel *et al.* (1963b; Fig. 39). The alkylation step proceeds in low to moderate yields only. Problems occur, in general, by (additional) cyclization reactions of the unsaturated side chain (Karrer and Rentschler, 1944; McHale *et al.*, 1963). Nevertheless, a yield of 66% is reported for the transformation of **127** with trimethylbenzoquinone (**15**) in formic acid with 2% copper–zinc powder at 90°C (Kajiwara *et al.*, 1980). In analogy to the first synthesis of (*rac,E,E*)-**5**,

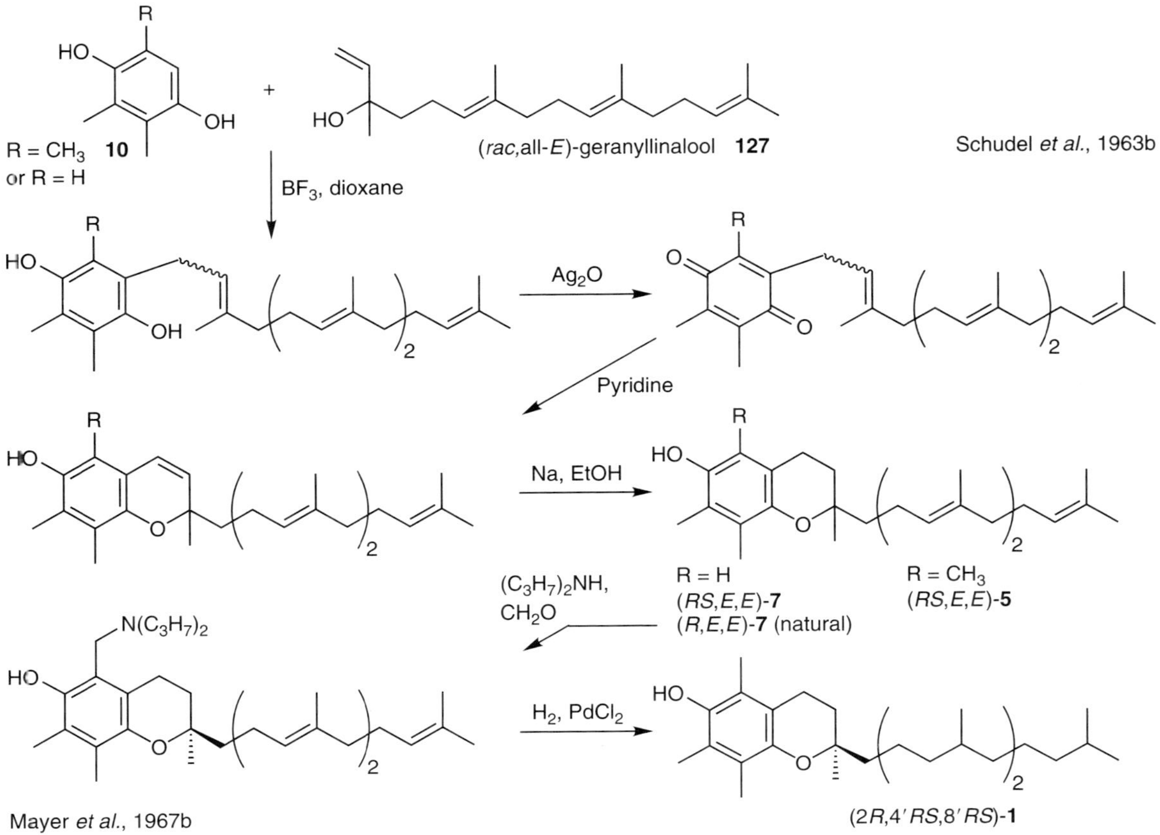

FIGURE 39. Synthesis of (*rac*,*E*,*E*)-tocotrienols via Friedel-Crafts alkylation.

Mayer *et al.* (1967b) synthesized the γ-homologue (R = H). In the same work, aminomethylation of (natural) γ-tocotrienol is described, which was further transformed to (2*R*,4′*RS*,8′*RS*)-**1**. The alternative synthesis of (all-*rac*)-**1** via 3,4-dehydro-α-tocotrienyl derivative **48** was already shown in Fig. 9 (vide supra). A modified version has been applied to the synthesis of various trienyl chromanols by Pearce *et al.* (1994). The reduction step of the intermediate oxochromanol in the original work of Kabbe and Heitzer (1978) has been improved by Baldenius *et al.* (2000). The route of Bienayme *et al.* (2000) goes also through 3,4-dehydro-α-tocotrienol (Fig. 11).

Urano *et al.* (1983) performed a selective side chain functionalization of **128** with selenium dioxide, followed by a C_{10} elongation with geranyl bromide (Fig. 40). It has been reported, however, that this synthesis is delivering a mixture of side chain olefin isomers according to NMR and HPLC data (Pearce *et al.*, 1992).

The first preparation of naturally identical α-tocotrienol (*REE*-**5**, Fig. 41) was achieved by Scott *et al.* (1976). Again, the chirality of the chroman part (**129**) was generated by optical resolution. (2*R*,3′*E*,7′*E*)-α-Tocotrienol (*REE*-**5**) was obtained in a linear multistep synthesis in an overall yield of 3.44% from trimethylhydroquinone (**10**). Sato *et al.* (1988) have claimed a stereoselective synthesis of *REE*-**5** by using a chiral intermediate obtained by Sharpless epoxidation of geranylgeraniol [(all-*E*)-**49**], according to the route applied for the preparation of (2*R*,4′*R*,8′*R*)-α-tocopherol (*RRR*-**1**; cf. Fig. 25).

The chiral chroman compound **66**, obtained by enzymatic resolution, was used as a starting material for both enantiomers of α-tocotrienol (Chênevert and Courchesne, 2002; Chênevert *et al.*, 2006; Fig. 42). For deprotonation of sulfone **130** with *n*-butyllithium, the presence of hexamethylphosphoramide (HMPA) is necessary to get acceptable yields (60%) of a mixture of

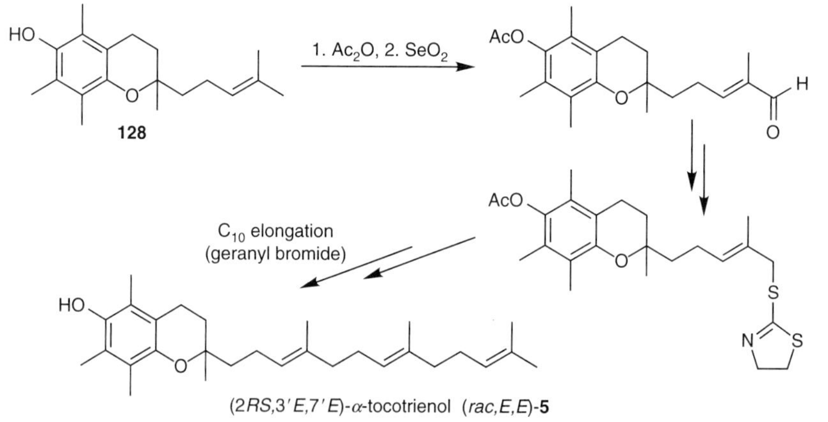

FIGURE 40. Synthesis of (*rac*,*E*,*E*)-α-tocotrienol via SeO_2 oxidation and C_{10} side chain elongation.

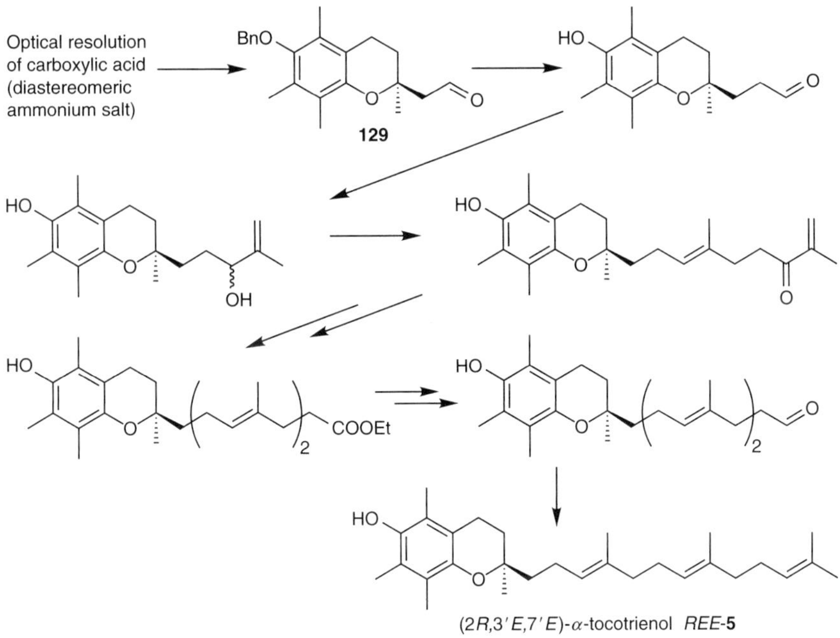

Optical resolution of carboxylic acid (diastereomeric ammonium salt)

BnO HO

129

HO OH

COOEt

2

O

(2R,3′E,7′E)-α-tocotrienol REE-**5**

FIGURE 41. First total synthesis of (2R,3′E,7′E)-α-tocotrienol.

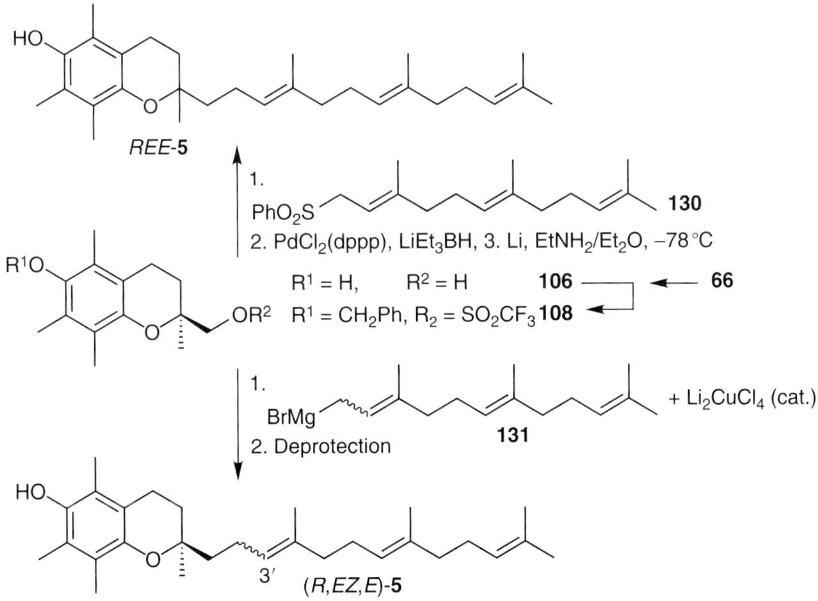

HO

REE-**5**

1. PhO_2S **130**

2. $PdCl_2$(dppp), $LiEt_3BH$, 3. Li, $EtNH_2/Et_2O$, −78 °C

R^1O OR^2

$R^1 = H$, $R^2 = H$ **106** ← **66**

$R^1 = CH_2Ph$, $R_2 = SO_2CF_3$ **108**

1. BrMg **131** + Li_2CuCl_4 (cat.)

2. Deprotection

3′ (R,EZ,E)-**5**

FIGURE 42. Alternative syntheses of (R)-α-tocotrienol.

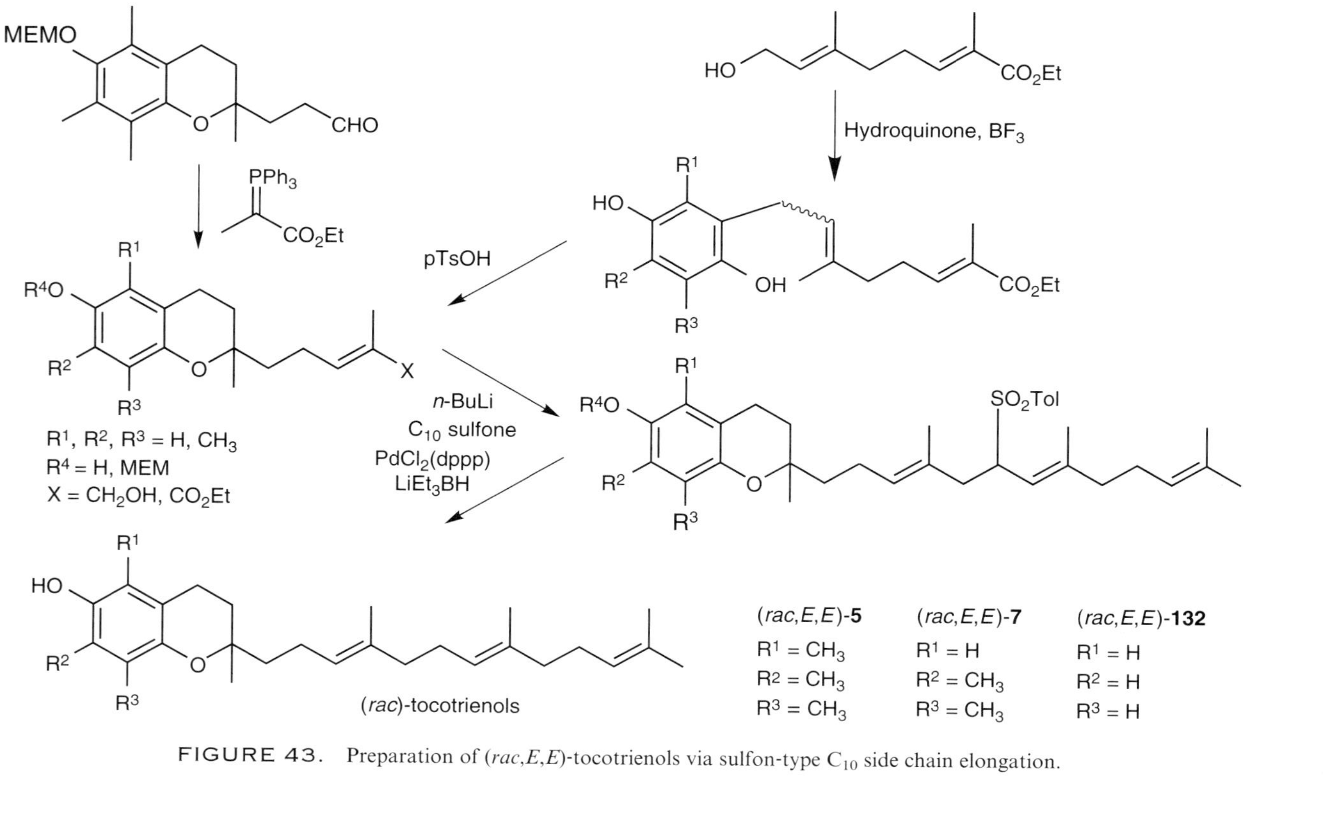

FIGURE 43. Preparation of (*rac*,*E*,*E*)-tocotrienols via sulfon-type C_{10} side chain elongation.

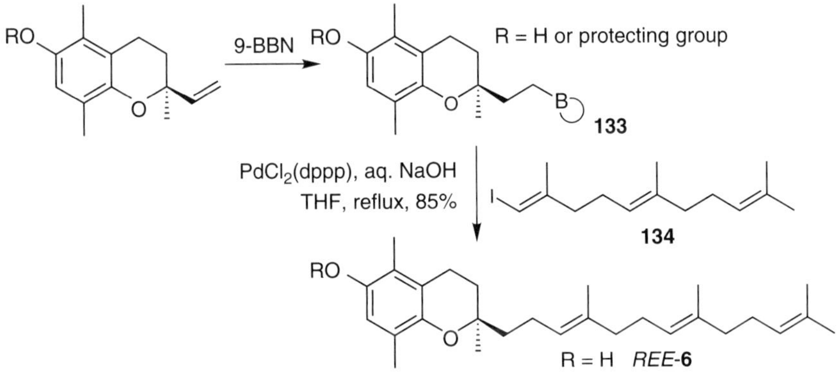

FIGURE 44. Synthesis of (R,E,E)-β-tocotrienol via a hydroboration-coupling protocol.

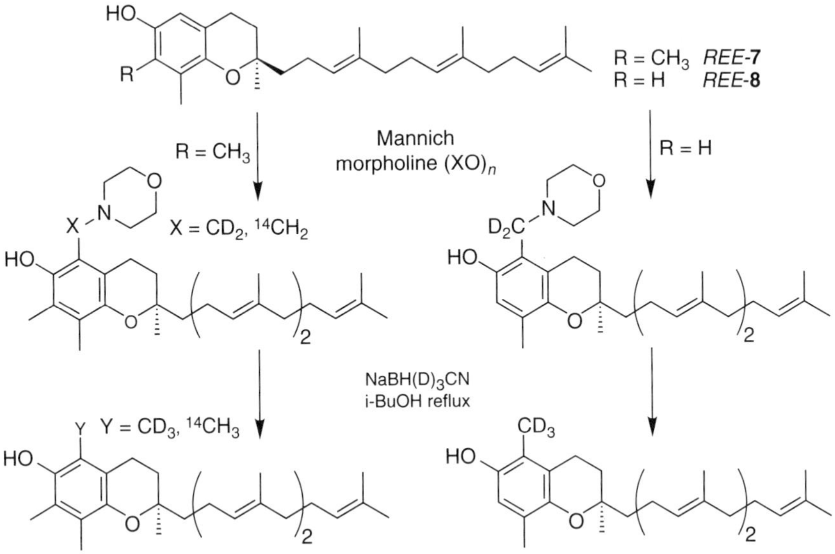

FIGURE 45. Synthesis of labeled $(2R,3'E,7'E)$-α- and $(2R,3'E,7'E)$-β-tocotrienols.

diastereomers, which is desulfonylated with superhydride in 76%, and further deprotected to *REE*-**5**. Applying the copper-mediated chroman-C_1 + alkyl-C_{15} coupling methodology as developed for tocopherol syntheses (cf. Figs. 27 and 31) delivered a (ca. 2:1) (3′*E*/*Z*) mixture of **5** (Netscher, 2003), since the Grignard reagent **131** can be prepared of from (all-*E*)-farnesyl bromide only as a mixture of the allylic isomers.

Two routes for the synthesis of racemic (*E,E*)-tocotrienols (*rac,E,E*)-**5**, -**7**, and -**132** are depicted in Fig. 43 (Pearce *et al.*, 1992). Multistep functionalization sequences are described in a patent application of Sato *et al.* (1989) who

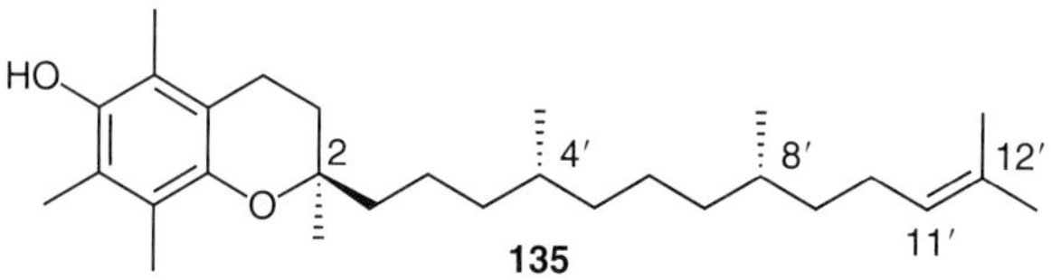

FIGURE 46. (2*R*,4′*R*,8′*S*)-11′,12′-dehydro-α-tocotrienol (**135**).

used similar sulfon-type coupling reactions. By starting from optically active educts (obtained again via Sharpless epoxidation, cf. Fig. 25), C_{10} elongation of the side chain led to nature-identical *REE*-**5**.

A hydroboration coupling protocol has been applied to the synthesis of (2*R*,3′*E*,7′*E*)-β-tocotrienol (*REE*-**6**; Fig. 44). Suzuki-type cross-coupling of organoborane **133**, obtained from the chiral vinylchroman compound, with vinyl iodide **134** gave the tocotrienol in ca. 85% yield (Couladouros *et al.*, 2005).

The work of Gu *et al.* (2006) by using a selective aminomethylation–reduction sequence (Fig. 45) may be also mentioned here, although (formally) only syntheses of labeled materials are described there. The preparation of CD_3- and $^{14}CH_3$-labeled α-tocotrienols belongs, however, to the first examples of preparation of *REE*-**5** from a lower homologue (γ-tocotrienol, *REE*-**7**).

While tocopherols and tocotrienols are well-known naturally occurring vitamin E compounds, also the structurally related tocomonoenol **135** [(2*R*,4′*R*,8′*S*)-11′,12′-dehydro-α-tocotrienol] has been isolated from plant material (Matsumoto *et al.*, 1995; Fig. 46), and prepared by total synthesis (Schiefer *et al.*, 1999), using the (chroman-C_1 + alkyl-C_{15}) Grignard coupling protocol sketched in Figs. 27 and 31 for tocopherol syntheses.

REFERENCES

Akutagawa, S. (1995). Asymmetric synthesis by metal BINAP catalysts. *Appl. Catal. A: General* **128,** 171–207.

Asgill, J. O., Crombie, L., and Whiting, D. A. (1978). Chromenylation of 2-naphthol and alkylhydroquinones: Short syntheses of (2RS,4′R,8′R)-α-tocopherol (vitamin E) and (2RS,4′R,8′R)-β-tocopherol. *J. Chem. Soc., Chem. Commun.* 59–60.

Baldenius, K. U., von dem Bussche-Hünnefeld, L., Hilgemann, E., Hoppe, P., and Stürmer, R. (1996). Vitamin E (tocopherols, tocotrienols). *In* "Ullmann's Encyclopedia of Industrial Chemistry," and 594–597. Vol. A27, pp. 478–488. VCH Verlagsgesellschaft, Weinheim.

Baldenius, K. U., Bockstiegel, B., Jaedicke, H., Ruff, D., Siedenbiedel, C., and Stürmer, R. (2000). Verfahren zur Herstellung von Chromanol-Derivaten. EP 0 989 126 A1.

Barner, R., and Schmid, M. (1979). Totalsynthese von natürlichem α-Tocopherol. 4. Mitteilung. Aufbau des Chromanringsystems aus Trimethylhydrochinon und einem optisch aktiven C_4-bzw. C_5-Synthon. *Helv. Chim. Acta* **62,** 2384–2399.

Bell, S., Wüstenberg, B., Kaiser, S., Menges, F., Netscher, T., and Pfaltz, A. (2006). Asymmetric hydrogenation of unfunctionalized, purely alkyl-substituted olefins. *Science* **311,** 642–644.

Bienayme, H., Ancel, J.-E., Meilland, P., and Simonato, J.-P. (2000). Rhodium(I)-catalyzed addition of phenols to dienes. A new convergent synthesis of vitamin E. *Tetrahedron Lett.* **41,** 3339–3343.

Bonrath, W., and Netscher, T. (2005). Catalytic processes in vitamins synthesis and production. *Appl. Catal. A: General* **280,** 55–73.

Bonrath, W., and Schneider, M. (2003). Manufacture of trimethylhydroquinone diacylates. WO 2003/051812 A1.

Bonrath, W., Karge, R., and Netscher, T. (2002a). Lipase-catalyzed transformations as key-steps in the large-scale preparation of vitamins. *J. Mol. Catal. B: Enzym.* **19–20,** 67–72.

Bonrath, W., Eisenkrätzer, D., Enjolras, V., Karge, R., Netscher, T., and Schneider, M. (2002b). Process for the manufacture of a vitamin E intermediate. EP 1 239 045 A1.

Bonrath, W., Haas, A., Hoppmann, E., Netscher, T., Pauling, H., Schager, F., and Wildermann, A. (2002c). Synthesis of (all-rac)-α-tocopherol using fluorinated NH-acidic catalysts. *Adv. Synth. Catal.* **344,** 37–39.

Bonrath, W., Dittel, C., Pabst, T., Netscher, T., Schmid, R., and Zwartjes, I. (2002d). Stereoselective O → C allylic rearrangement and ring-closure reaction: A novel route to tocopherols, XXXV. Jahrestreffen Deutscher Katalytiker, 20–22 March 2002, Weimar (Germany), Poster P 135, book of abstracts, pp. 312–313.

Bonrath, W., Burdick, D. C., Netscher, T., Schager, F., and Thomas, D. (2003). Manufacture of (all-rac)-α-tocopherol via acid-catalyzed ring closure. WO 2003/037883 A1.

Bonrath, W., Dittel, C., Netscher, T., Pabst, T., and Schmid, R. (2004). Manufacture of tocopheryl acetate. WO 2004/046126 A1.

Bonrath, W., Menges, F., Netscher, T., Pfaltz, A., and Wüstenberg, B. (2006). Asymmetric hydrogenation of alkenes using chiral iridium complexes. WO 2006/066863 A1.

Bonrath, W., Dittel, C., Giraudi, L., Netscher, T., and Pabst, T. (2007). Rare earth triflate catalysts in the synthesis of vitamin E and its derivatives. *Catal. Today* **121,** 65–70.

Breit, B., and Demel, P. (2001). Copper-mediated and -catalyzed *o*-DPPB-directed allylic substitution. *Adv. Synth. Catal.* **343,** 429–432.

Broger, E. A., and Müller, R. K. (1993). Verfahren zur Herstellung von Isoprenderivaten. EP 0 565 975 A2.

Chan, K.-K., and Saucy, G. (1977). Transfer of chirality in the [2,3] sigmatropic rearrangement of allylic alcohols to β,γ-unsaturated amides. Preparation of optically active nine- and fourteen-carbon saturated isoprenoid synthons. *J. Org. Chem.* **42,** 3828–3832.

Chan, K.-K., Cohen, N., De Noble, J. P., Specian, A. C., Jr., and Saucy, G. (1976). Synthetic studies on (2R,4′R,8′R)-α-tocopherol. Facile syntheses of optically active, saturated, acyclic isoprenoids via stereospecific [3,3] sigmatropic rearrangements. *J. Org. Chem.* **41,** 3497–3505.

Chan, K.-K., Specian, A. C., Jr., and Saucy, G. (1978). Synthesis of (2R,4′R,8′R)-α-tocopheryl acetate (vitamin E acetate) using [3,3] sigmatropic rearrangement. *J. Org. Chem.* **43,** 3435–3440.

Chauvel, A., Delmon, B., and Hölderich, W. F. (1994). New catalytic processes developed in Europe during the 1980's. *Appl. Catal. A: General* **115,** 173–217.

Chen, C. Y., Nagumo, S., and Akita, H. (1996). A synthesis of (2R,4′R,8′R)-α-tocopherol (vitamin E) side chain. *Chem. Pharm. Bull.* **44,** 2153–2156.

Chênevert, R., and Courchesne, G. (2002). Synthesis of (S)-α-tocotrienol via an enzymatic desymmetrization of an achiral chroman derivative. *Tetrahedron Lett.* **43,** 7971–7973.

Chênevert, R., Courchesne, G., and Pelchat, N. (2006). Chemoenzymatic synthesis of both enantiomers of α-tocotrienol. *Bioorg. Med. Chem.* **14,** 5389–5396.

Coffen, D. L., Cohen, N., Pico, A. M., Schmid, R., Sebastian, M. J., and Wong, F. (1994). A microbial lipase based stereoselective synthesis of (d)-α-tocopherol from (R)-citronellal and (S)-(6-hydroxy-2,5,7,8-tetramethylchroman-2-yl)acetic acid. *Heterocycles* **39,** 527–552.

Cohen, N., Eichel, W. F., Lopresti, R. J., Neukom, C., and Saucy, G. (1976). Synthetic studies on (2R,4′R,8′R)-α-tocopherol. An approach utilizing side chain synthons of microbiological origin. *J. Org. Chem.* **41,** 3505–3511.

Cohen, N., Lopresti, R. J., and Saucy, G. (1979). A novel total synthesis of (2R,4′R,8′R)-α-tocopherol (vitamin E). Construction of chiral chromans from an optically active, nonaromatic precursor. *J. Am. Chem. Soc.* **101,** 6710–6716.

Cohen, N., Lopresti, R. J., and Neukom, C. (1981a). Studies on the total synthesis of (2R,4′R,8′R)-α-tocopherol (vitamin E). Stereospecific cyclizations leading to optically active chromans. *J. Org. Chem.* **46,** 2445–2450.

Cohen, N., Scott, C. G., Neukom, C., Lopresti, R. J., Weber, G., and Saucy, G. (1981b). Total synthesis of all eight stereoisomers of α-tocopheryl acetate. Determination of their diastereoisomeric and enantiomeric purity by gas chromatography. *Helv. Chim. Acta* **64,** 1158–1173.

Cohen, N., Banner, B. L., and Neukom, C. (1982). Improvements in the synthesis of a key α-tocopherol intermediate—(S)-(−)-3,4-dihydro-6-hydroxy-2,5,7,8-tetramethyl-2H-1-benzopyran-2-acetic acid. *Synth. Commun.* **12,** 57–65.

Cohen, N., Schaer, B., and Scalone, M. (1992). Synthesis of (2RS,4′R,8′R)-α-tocopherol and related compounds via a 2-chlorochroman. *J. Org. Chem.* **57,** 5783–5785.

Couladouros, E. A., Papas, A. M., Moutsos, V. L., and Lampropoulou, M. (2005). Process for synthesizing d-tocotrienols from 2-vinylchromane compound. WO 2005/035491 A2; US 2005/0124688 A1.

DellaPenna, D., and Last, R. L. (2006). Progress in the dissection and manipulation of plant vitamin E biosynthesis. *Physiologica Plantarum* **126,** 356–368.

DellaPenna, D., and Pogson, B. J. (2006). Vitamin synthesis in plants: Tocopherols and carotenoids. *Annu. Rev. Plant Biol.* **57,** 711–738.

Ernst, H. G. (1983). Vitamin E (Tocopherole, Tocotrienole). *In* "Ullmanns Encyclopädie der technischen Chemie," Bd. 23, pp. 643–649 and 717–718. Verlag Chemie, Weinheim.

Ernst, H., Gehrken, H.-P., and Paust, J. (1984). Verbessertes Verfahren zur Herstellung von α-Tocopherol. DE 33 09 158 A1.

Evans, H. M., and Bishop, K. S. (1922). On the existence of a hitherto unrecognized dietary factor essential for reproduction. *Science* **56,** 650–651.

Evans, H. M., Emerson, O. H., and Emerson, G. A. (1936). The isolation from wheat germ oil of an alcohol, α-tocopherol, having the properties of vitamin E. *J. Biol. Chem.* **113,** 319–332.

Fernholz, E. (1938). On the constitution of α-tocopherol. *J. Am. Chem. Soc.* **60,** 700–705.

Fujisawa, T., Sato, T., Kawara, T., and Ohashi, K. (1981). A stereocontrolled total synthesis of optically active (R,R)-phytol. *Tetrahedron Lett.* **22,** 4823–4826.

Goodhue, C. T., and Schaeffer, J. R. (1971). Preparation of L(+) β-hydroxyisobutyric acid by bacterial oxidation of isobutyric acid. *Biotechnol. Bioeng.* **13,** 203–214.

Gramatica, P., Manitto, P., Monti, D., and Speranza, G. (1986). New syntheses of optically active vitamin E side chain by chemoenzymatic approach. *Tetrahedron* **42,** 6687–6692.

Gramatica, P., Manitto, P., Monti, D., and Speranza, G. (1987). Stereoselective total synthesis of natural phytol via double bond reductions by baker's yeast. *Tetrahedron* **43,** 4481–4486.

Grütter, C., Alonso, E., Chougnet, A., and Woggon, W.-D. (2006). A biomimetic chromanol cyclization leading to α-tocopherol. *Angew. Chem. Int. Ed.* **45,** 1126–1130.

Gu, F., Netscher, T., and Atkinson, J. (2006). 5-trideuteromethyl-α-tocotrienol and 5-$^{14}CH_3$-α-tocotrienol as biological tracers of tocotrienols. *J. Labelled Comp. Radiopharm.* **49,** 733–743.

Gupta, M. K. (1993). Processing to improve soybean oil quality. *INFORM* **4**(no. 11), 1267–1272.

Hasegawa, A., Ishihara, K., and Yamamoto, H. (2003). Trimethylsilyl pentafluorophenylbis (trifluoromethanesulfonyl)methide as a super Lewis acid catalyst for the condensation of trimethylhydroquinone with isophytol. *Angew. Chem. Int. Ed.* **42,** 5731–5733.

Heiser, B., Broger, E. A., and Crameri, Y. (1991). New efficient methods for the synthesis and *in-situ* preparation of ruthenium(II) complexes of atropisomeric diphosphines and their application in asymmetric catalytic hydrogenations. *Tetrahedron: Asymmetry* **2,** 51–62.

Hübscher, J., and Barner, R. (1990). Totalsynthese von natürlichem α-Tocopherol. 5. Mitteilung. Asymmetrische Alkylierung und asymmetrische Epoxidierung als Methoden zur Einführung der (R)-Konfiguration an C(2) des Chroman-Systems. *Helv. Chim. Acta* **73,** 1068–1086.

Hughes, L., Slaby, M., Burton, G. W., and Ingold, K. U. (1990). Syntheses of α- and γ-tocopherols selectively labelled with deuterium. *J. Labelled. Comp. Radiopharm.* **28,** 1049–1057.

Huo, S., and Negishi, E. (2001). A convenient and asymmetric protocol for the synthesis of natural products containing chiral alkyl chains via Zr-catalyzed asymmetric carboalumination of alkenes. Synthesis of phytol and vitamins E and K. *Org. Lett.* **3,** 3253–3256.

Hyatt, J. A., and Skelton, C. (1997). A kinetic resolution route to the (S)-chromanmethanol intermediate for synthesis of the natural tocols. *Tetrahedron: Asymmetry* **8,** 523–526.

Hyatt, J. A., Kottas, G. S., and Effler, J. (2002). Development of synthetic routes to D,L-α-tocopherol (vitamin E) from biologically produced geranylgeraniol. *Org. Process Res. Dev.* **6,** 782–787.

Imfeld, M. (1986). Verfahren zur Herstellung von Hydrochinonderivaten EP 0 183 042 A2.

Inoue, S., Ikeda, H., Sato, S., Horie, K., Ota, T., Miyamoto, O., and Sato, K. (1987). Improved general method of ortho-alkylation of phenols using alkyl isopropyl sulfide, sulfuryl chloride, and triethylamine. An expedient synthesis of representative oxygen heterocycles and (2R,4′R,8′R)-α-tocopherol. *J. Org. Chem.* **52,** 5495–5497.

Ishihara, K., Kubota, M., and Yamamoto, H. (1996). Practical synthesis of (±)-α-tocopherol. Trifluoromethanesulfonimide as an extremely active Broensted acid catalyst for the condensation of trimethylhydroquinone with isophytol. *Synlett* 1045–1046.

IUPAC-IUB (1982). Nomenclature of tocopherols and related compounds. *Eur. J. Biochem.* **123,** 473–475; *Pure Appl. Chem.*, **54**, 1507–1510.

Kabbe, H. J., and Heitzer, H. (1978). Eine neue Synthese von 3,4-Dehydro-α-tocotrienol und Vitamin-E. *Synthesis* 888–889.

Kajiwara, M., Sakamoto, O., and Ohta, S. (1980). Studies on tocopherols II. Convenient synthesis of tocopherols. *Heterocycles* **14,** 1995–1998.

Kamal-Eldin, A., and Åppelqvist, L.-A. (1996). The chemistry and antioxidant properties of tocopherols and tocotrienols. *Lipids* **31,** 671–701. Remark: This review provides an excellent overview, but it contains several errors of nomenclature and wrong structures drawn from literature. References with the correct nomenclature and the revised structures are also cited in that review.

Karrer, P., and Isler, O. (1946a). dl-Tocopherols and process for the manufacture of same. US 2,411,967.

Karrer, P., and Isler, O. (1946b). Process for the preparation of synthetic dl-tocopherols. US 2,411,968.

Karrer, P., and Rentschler, H. (1944). Ein Tocol mit bicyclischer Seitenkette. *Helv. Chim. Acta* **27,** 1297–1300.

Karrer, P., Fritzsche, H., Ringier, B. H., and Salomon, H. (1938). α-Tocopherol. *Helv. Chim. Acta* **21,** 520–525; Synthese des α-Tocopherols. *Helv. Chim. Acta* **21,** 820–825.

Knierzinger, A., and Scalone, M. (1990). Verfahren zur Herstellung von Vinylverbindungen. EP 0 392 389 A2.

Knierzinger, A., and Scalone, M. (1992). Tocopherol synthesis: Chromane cyclization and catalysis. US 5,110,955.

Knierzinger, A., Walther, W., Weber, B., Müller, R. K., and Netscher, T. (1990). Eine neue Methode zur stereochemischen Analyse offenkettiger terpenoider Carbonyl-Verbindungen. *Helv. Chim. Acta* **73,** 1087–1107.

Kokubo, Y., Hasegawa, A., Kuwata, S., Ishihara, K., Yamamoto, H., and Ikariya, T. (2005). Synthesis of (all-rac)-α-tocopherol in supercritical carbon dioxide: Tuning of the product selectivity in batch and continuous-flow reactors. *Adv. Synth. Catal.* **347,** 220–224.

Koreeda, M., and Brown, L. (1983). Chirality transfer in stereoselective synthesis. A highly stereoselective synthesis of optically active vitamin E side chain. *J. Org. Chem.* **48,** 2122–2124.

Laufer, M. C., Bonrath, W., and Hoelderich, W. F. (2005). Synthesis of (all-rac)-α-tocopherol using Nafion resin/silica nanocomposite materials as catalysts. *Catal. Lett.* **100,** 101–103.

Leuenberger, H. G. W., and Wirz, B. (1993). Biotransformations leading to optically active synthons for the preparation of fine chemicals. *Chimia* **47,** 82–85.

Leuenberger, H. G. W., Boguth, W., Barner, R., Schmid, M., and Zell, R. (1979). Herstellung bifunktioneller, optisch aktiver Synthesebausteine für die Seitenkette mit Hilfe mikrobiologischer Umwandlungen. *Helv. Chim. Acta* **62,** 455–463.

Malaisé, G., Bonrath, W., Breuninger, M., and Netscher, T. (2006). A new route to vitamin E key-intermediates by olefin cross-metathesis. *Helv. Chim. Acta* **89,** 797–812.

Matsui, M. (1996). Synthesis of α-tocopherol for the next century. *Nihon Yukagakkaishi. J. Jpn. Oil Chem. Soc.* **45,** 821–831; *Chem. Abstr.* **125**, 301241.

Matsui, M., and Yamamoto, H. (1995). Direct construction of the chroman structure from 1,3-diene. Regioselective protonation of acyclic polyene. *Bull. Chem. Soc. Jpn.* **68,** 2657–2661.

Matsui, M., Karibe, N., Hayashi, K., and Yamamoto, H. (1995). Synthesis of α-tocopherol: Scandium(III) trifluoromethanesulfonate as an efficient catalyst in the reaction of hydroquinone with allylic alcohol. *Bull. Chem. Soc. Jpn.* **68,** 3569–3571.

Matsumoto, A., Takahashi, S., Nakano, K., and Kijima, S. (1995). Identification of a new form of vitamin E in plant oil. *Yukagaku* **44,** 593–597; *Chem. Abstr.* **123**, 226445.

Mayer, H., and Isler, O. (1971). Synthesis of vitamins E. *Methods Enzymol.* **18,** 241–348.

Mayer, H., Schudel, P., Rüegg, R., and Isler, O. (1962). Über eine neue Vitamin-E-Synthese. *Chimia* **16,** 367–368.

Mayer, H., Schudel, P., Rüegg, R., and Isler, O. (1963a). Die Totalsynthese von (2*R*,4′*R*,8′*R*)- und (2*S*,4′*R*,8′*R*)-α-Tocopherol. *Helv. Chim. Acta* **46,** 650–671.

Mayer, H., Schudel, P., Rüegg, R., and Isler, O. (1963b). Die absolute Konfiguration des natürlichen α-Tocopherols. *Helv. Chim. Acta* **46,** 963–982.

Mayer, H., Vetter, W., Metzger, J., Rüegg, R., and Isler, O. (1967a). Die stereospezifische Cyclisierung von (3′R,7′R,11′R)-α-Tocopherolchinon. *Helv. Chim. Acta* **50,** 1168–1178.

Mayer, H., Metzger, J., and Isler, O. (1967b). Die Stereochemie von natürlichem γ-Tocotrienol (Plastochromanol-3), Plastochromanol-8 und Plastochromenol-8. *Helv. Chim. Acta* **50,** 1376–1393.

Mazzini, F., Alpi, E., Salvadori, P., and Netscher, T. (2003). First synthesis of (8-$^2H_3$)-(all-rac)-δ-tocopherol. *Eur. J. Org. Chem.* 2840–2844.

Mazzini, F., Alpi, E., Salvadori, P., and Netscher, T. (2004). First synthesis of rac-(5-$^2H_3$)-α-CEHC, a labeled analogue of a major vitamin E metabolite. *J. Org. Chem.* **69,** 9303–9306.

Mazzini, F., Netscher, T., and Salvadori, P. (2005). Easy route to labeled and unlabeled R,R, R-γ-tocopherol by aryl demethylation of α-homologues. *Tetrahedron* **61,** 813–817.

Mazzini, F., Galli, F., and Salvadori, P. (2006). Vitamin E metabolites: Synthesis of [D_2]- and [D_3]-γ-CEHC. *Eur. J. Org. Chem.* 5588–5593.

McHale, D., Green, J., Marcinkiewicz, S., Feeney, J., and Sutcliffe, L. H. (1963). Tocopherols. Part VIII. Structural and synthetic studies of ε-tocopherol. *J. Chem. Soc.* 784–791.

Mercier, C., and Chabardes, P. (1994). Organometallic chemistry in industrial vitamin A and vitamin E synthesis. *Pure Appl. Chem.* **66,** 1509–1518.

Mercier, C., Mignani, G., Aufrand, M., and Allmang, G. (1991). Efficient synthesis of pseodoionone by oxidative decarboxylation of allyl β-ketoesters. *Tetrahedron Lett.* **32,** 1433–1436.

Millis, J. R., Saucy, G. G., Maurina-Brunker, J., McMullin, T. W., and Hyatt, J. A. (2000a). Method of vitamin production. WO 00/01649.

Millis, J. R., Saucy, G. G., Maurina-Brunker, J., and McMullin, T. W. (2000b). Method of vitamin production. WO 00/01650.

Millis, J. R., Saucy, G. G., Maurina-Brunker, J., McMullin, T. W., and Hyatt, J. A. (2001). Method of vitamin production. US 6,242,227 B1.

Mimoun, H. (1996). Catalytic opportunities in the flavor and fragrance industry. *Chimia* **50,** 620–625.

Mizuguchi, E., and Achiwa, K. (1993). Oxalate as an activated ester group in lipase-catalyzed enantioselective hydrolysis: A versatile approach to d-α-tocopherol. *Tetrahedron: Asymmetry* **4,** 2303–2306. Remark: The designations "enantioselectivity" and "enantiomeric purity" have to be changed to "diastereoselectivity" and "diastereoisomeric purity", as well as "ee" (enantiomeric excess) has to be replaced by "de" (diasteroisomeric excess), due to the presence of additional two chiral centers in the side chain of both starting materials and products.

Mizuguchi, E., and Achiwa, K. (1995). Asymmetric synthesis of (S)-chromanmethanol as a useful synthetic unit of vitamin E analogs. *Synlett* 1255–1256.

Mizuguchi, E., Suzuki, T., and Achiwa, K. (1994). Lipase-catalyzed synthesis of optically active N,N,N-triethylethanaminium α-tocopherol analog MDL-73404. *Synlett* 929–930.

Netscher, T. (1996). Stereoisomers of tocopherols—syntheses and analytics. *Chimia* **50,** 563–567.

Netscher, T. (1999). Synthesis and production of vitamin E. *In* "Lipid Synthesis and Manufacture" (F. D. Gunstone, Ed.), pp. 250–267. Sheffield Academic Press Ltd., Sheffield, UK.

Netscher, T. (2003). Sulfonate leaving groups for nucleophilic substitution reactions—improved structures and procedures. *Recent Res. Dev. Org. Chem.* **7,** 71–83.

Netscher, T. (2006). Preparation of trialkyl-substituted olefins by ruthenium catalyzed cross-metathesis. *J. Organomet. Chem.* **691,** 5155–5162.

Netscher, T., and Bohrer, P. (1993). Nucleophilic substitution reaction at the (oxo-) neopentylic center in the synthesis of tocopherols and analogues. *Chimia* **47,** 295. Fall Meeting of the New Swiss Chemical Society, 22 October 1993, Basel, Switzerland, Abstract Organische Chemie, Poster.

Netscher, T., and Gautschi, I. (1992). Präparative Enantiomerentrennung einer Chroman-2-Carbonsäure durch Chromatographie von Glucose-Derivaten. *Liebigs Ann. Chem.* 543–546.

Netscher, T., Scalone, M., and Schmid, R. (2004a). Enantioselective hydrogenation: Towards a large-scale total synthesis of (R,R,R)-α-tocopherol. *In* "Asymmetric Catalysis on Industrial Scale" (H.-U. Blaser and E. Schmidt, Eds.), pp. 71–89. Wiley-VCH, Weinheim, Germany.

Netscher, T., Müller, R. K., Schneider, J., Schneider, H., Bohrer, P., and Jestin, R. (2004b). Aminomethylation of vitamin E compounds. *In* "International Electronic Conferences on Synthetic Organic Chemistry, 5th, 6th, Sept. 1–30, 2001 and 2002, and 7th, 8th, Nov. 1–30, 2003 and 2004" (J. A. Seijas, Ed.). Molecular Diversity Preservation International, Basel, Switzerland. http://www.mdpi.net/ecsoc/ecsoc-5/Papers/c0007/c0007.htm

Netscher, T., Malaisé, G., Bonrath, W., and Breuninger, M. (2006). Olefin cross-metathesis in natural product synthesis: Preparation of trisubstituted olefins on the way to vitamin E. *Actual. Chim.* **293,** 21–23.

Netscher, T., Mazzini, F., and Jestin, R. (2007). Tocopherols by hydride reduction of dialkylamino derivatives. *Eur. J. Org. Chem.* 1176–1183.

Noyori, R., and Takaya, H. (1990). BINAP: An efficient chiral element for asymmetric catalysis. *Acc. Chem. Res.* **23,** 345–350.

Oda, S., Sugai, T., and Ohta, H. (2000). Synthesis of optically active citronellol, citronellal, and citronellic acid by microbial oxidation and double coupling system in an interface bioreactor. *Bull. Chem. Soc. Jpn.* **73,** 2819–2823.

Odinokov, V. N., Spivak, A. Y., Emelyanova, G. A., Mallyabaeva, M. I., Nazarova, O. V., and Dzhemilev, U. M. (2003). Synthesis of α-tocopherol (vitamin E), vitamin K_1-chromanol, and their analogs in the presence of aluminosilicate catalysts Tseokar-10 and Pentasil. *ARKIVOC* **13,** 101–118.

Outten, R. A., Bohrer, P., Müller, R. K., Schneider, H., Rüttimann, A., and Netscher, T. (1999a). Routes to vitamin E: Palladium-mediated arylation of optically active acetylene and ethylene building blocks. *Chem. Listy. Symposia* **93,** S49; presented by T.N. at 18th Conference on Isoprenoids; 10–16 September 1999, Prachatice, Czech Republic.

Outten, R. A., Bohrer, P., Müller, R. K., Schneider, H., Rüttimann, A., and Netscher, T. (1999b). Palladium-catalyzed coupling reactions: Key-steps in stereoselective syntheses of tocopherols; presented by T.N. at 11th European Symposium on Organic Chemistry, Göteborg, Sweden, July 23–28, 1999, Poster.

Pearce, B. C., Parker, R. A., Deason, M. E., Qureshi, A. A., and Kim Wright, J. J. (1992). Hypocholesterolemic activity of synthetic and natural tocotrienols. *J. Med. Chem.* **35,** 3595–3606.

Pearce, B. C., Parker, R. A., Deason, M. E., Dischino, D. D., Gillespie, E., Qureshi, A. A., Volk, K., and Kim Wright, J. J. (1994). Inhibitors of cholesterol biosynthesis. 2. Hypocholesterolemic and antioxidant activities of benzopyran and tetrahydronaphthalene analogues of the tocotrienols. *J. Med. Chem.* **37,** 526–541.

Pongracz, G., Weiser, H., and Matzinger, D. (1995). Tocopherole—Antioxidantien der Natur. *Fat Sci. Technol.* **97,** 90–104.

Rein, C., and Breit, B. (2005). Total synthesis of (2R,4′R,8′R′)-α-tocopherol. 9th International SFB Symposium, p. P184. Aachen, Germany.

Riegl, J., Brüggemann, K., and Netscher, T. (2000). Efficient preparation of (2R,4′R,8′R)-α-tocopherol by hydroxymethylation. *In* "Proceedings of ECSOC-3, and Proceedings of ECSOC-4, Sept. 1–30, 1999 and 2000" (E. Pombo-Villar, Ed.). Molecular Diversity Preservation International, Basel, Switzerland. http://www.mdpi.org/ecsoc-4.htm,http://pages.unibas.ch/mdpi/ecsoc-4/c0035/c0035.htm

Sato, K., Miyamoto, O., Inoue, S., and Sato, S. (1988). Preparation of d-α-tocotrienol from geranylgeraniol. JP 63–63674; *Chem. Abstr.* **110,** 193145.

Sato, K., Inoue, S., and Murayama, T. (1989). Preparation of optically active α-tocotrienol. JP 1–233278; *Chem. Abstr.* **112,** 139621.

Saucy, G., and Cohen, N. (1981). Asymmetric reactions applied to the synthesis of d-α-tocopherol. *In* "New Synthetic Methodology and Biologically Active Substances" (Z. Yoshida, Ed.), pp. 155–175. Elsevier, Amsterdam.

Schiefer, G., Müller, R. K., Schneider, H., Umiker, M., and Netscher, T. (1999). Dehydrotocopherols—new vitamin E components. *Chimia* **53,** 359. Fall Meeting of the New Swiss Chemical Society, 12 October 1999, Basel, Switzerland, Abstract Organic Chemistry 135.

Schmid, R., and Broger, E. A. (1994). Asymmetric hydrogenation in process research of pharmaceuticals, vitamins and fine chemicals. *Proceedings of the Chiral Europe '94 Symposium* 19–20 Sept. 1994, Nice, France, pp. 79–86.

Schmid, R., Cereghetti, M., Heiser, B., Schönholzer, P., and Hansen, H. J. (1988). Axially dissymmetric bis(triaryl)phosphines in the biphenyl series: Synthesis of (6,6′-dimethylbiphenyl-2,2′-diyl)bis(diphenylphosphine) ('BIPHEMP') and analogues, and their use in rhodium(I)-catalyzed asymmetric isomerizations of N,N-diethylnerylamine. *Helv. Chim. Acta* **71,** 897–929.

Schmid, R., Foricher, J., Cereghetti, M., and Schönholzer, P. (1991). Axially dissymmetric diphosphines in the biphenyl series: Synthesis of (6,6′-dimethoxybiphenyl-2,2′-diyl)bis(diphenylphosphine) ('MeOBIPHEP') and analogues via an ortho-lithiation/iodination Ullmann-reaction approach. *Helv. Chim. Acta* **74,** 370–389.

Schmid, R., Broger, E. A., Cereghetti, M., Crameri, Y., Foricher, J., Lalonde, M., Müller, R. K., Scalone, M., Schoettel, G., and Zutter, U. (1996). New developments in enantioselective hydrogenation. *Pure Appl. Chem.* **68,** 131–138.

Schneider, M., Zimmermann, K., Aquino, F., and Bonrath, W. (2001). Industrial application of Nafion-systems in rearrangement-aromatisation, transesterification, alkylation, and ring-closure reactions. *Appl. Catal. A: General* **220,** 51–58.

Schudel, P., Mayer, H., Metzger, J., Rüegg, R., and Isler, O. (1963a). Die Umkehrung der Konfiguration am Kohlenstoffatom 2 von natürlichem (2R,4′R,8′R)-α-Tocopherol. *Helv. Chim. Acta* **46,** 333–343.

Schudel, P., Mayer, H., Metzger, J., Rüegg, R., and Isler, O. (1963b). Über die Chemie des Vitamins E—die Synthese von rac.all-trans-ζ_1- und ε-Tocopherol. *Helv. Chim. Acta* **46,** 2517–2526.

Schudel, P., Mayer, H., and Isler, O. (1972). 16. Tocopherols, II. Chemistry. *In* "The Vitamins" (W. H. Sebrell and R. S. Harris, Eds.), Vol. V, pp. 168–218. Academic Press, New York.

Scott, J. W., Bizzarro, F. T., Parrish, D. R., and Saucy, G. (1976). Syntheses of (2R,4′R,8′R)-α-tocopherol and (2R,3′E,7′E)-α-tocotrienol. *Helv. Chim. Acta* **59,** 290–306.

Shchegolev, A. A., Sarycheva, I. K., Kochetova, E. V., Mosolova, O. V., Kulish, M. A., and Evstigneeva, R. P. (1983). Synthesis of vitamin E acetate. *Khimiko-farmatsevticheskii Zhurnal* **17,** 92–95; *Chem. Abstr.* **98** 160971.

Solladie, G., and Moine, G. (1984). Application of chiral sulfoxides in asymmetric synthesis: The enantiospecific synthesis of the chroman ring of α-tocopherol (vitamin E). *J. Am. Chem. Soc.* **106,** 6097–6098.

Stocker, A., Rüttimann, A., and Woggon, W.-D. (1993). Identification of tocopherol-cyclase in the blue-green algae Anabaena variabilis Kützing (cyanobacteria). *Helv. Chim. Acta* **76,** 1729–1738.

Stocker, A., Netscher, T., Rüttimann, A., Müller, R. K., Schneider, H., Todaro, L. J., Derungs, G., and Woggon, W.-D. (1994). The reaction mechanism of chromanol-ring formation catalyzed by tocopherol cyclase from Anabaena variabilis Kützing (cyanobacteria). *Helv. Chim. Acta* **77,** 1721–1737.

Stocker, A., Fretz, H., Frick, H., Rüttimann, A., and Woggon, W.-D. (1996). The substrate specificity of tocopherol cyclase. *Bioorg. Med. Chem.* **4,** 1129–1134.

Sugai, T., Watanabe, N., and Ohta, H. (1991). A synthesis of natural α-tocopherol intermediate. *Tetrahedron: Asymmetry* **2,** 371–376.

Takabe, K., Okisaka, K., Uchiyama, Y., Katagiri, T., and Yoda, H. (1985). An efficient synthesis of (S)-chromanmethanol, a vitamin E precursor. *Chem. Lett.* 561–562.

Takano, S., Sugihara, T., and Ogasawara, K. (1990). An efficient stereoselective preparation of vitamin E (α-tocopherol) from phytol. *Synlett* 451–452.

Takano, S., Yoshimizu, T., and Ogasawara, K. (1994). Use of the Sharpless asymmetric dihydroxylation as a substitute for the Katsuki-Sharpless asymmetric epoxidation: An alternative route to vitamin E and prostaglandin intermediates. *Synlett* 119–120.

Takaya, H., Ohta, T., Sayo, N., Kumobayashi, H., Akutagawa, S., Inoue, S., Kasahara, I., and Noyori, R. (1987). Enantioselective hydrogenation of allylic and homoallylic alcohols. *J. Am. Chem. Soc.* **109,** 1596–1597.

Tani, K., Yamagata, T., Akutagawa, S., Kumobayashi, H., Taketomi, T., Takaya, H., Miyashita, A., Noyori, R., and Otsuka, S. (1984). Metal-assisted terpenoid synthesis. 7. Highly enantioselective isomerization of prochiral allylamines catalyzed by chiral diphosphine rhodium(I) complexes. Preparation of optically active enamines. *J. Am. Chem. Soc.* **106,** 5208–5217.

Tietze, L. F., and Görlitzer, J. (1998). Preparation of chiral building blocks for a highly convergent vitamin E synthesis. Systematic investigations on the enantioselectivity of the Sharpless bishydroxylation. *Synthesis* 873–878.

Tietze, L. F., Görlitzer, J., Schuffenhauer, A., and Hübner, M. (1999). Enantioselective synthesis of the chroman moiety of vitamin E. *Eur. J. Org. Chem.* 1075–1084.

Tietze, L. F., Peters, H. J., Djalali Bazzaz, F., and Seibel, J. (2000). Katalytische Antikörper und ein Verfahren zur Herstellung von Chromanen über Zyklisierung mit diesen Antikörpern. DE 198 52 903 A1.

Tietze, L. F., Sommer, K. M., Zinngrebe, J., and Stecker, F. (2005). Palladium-catalyzed enantioselective domino reaction for the efficient synthesis of vitamin E. *Angew. Chem. Int. Ed.* **44,** 257–259.

Tietze, L. F., Stecker, F., Zinngrebe, J., and Sommer, K. M. (2006). Enantioselective palladium-catalyzed total synthesis of vitamin E by employing a domino Wacker-Heck reaction. *Chem. Eur. J.* **12,** 8770–8776.

Trost, B. M. (2004). Asymmetric allylic alkylation, an enabling methodology. *J. Org. Chem.* **69,** 5813–5837.

Trost, B. M., and Asakawa, N. (1999). An asymmetric synthesis of the vitamin E core by Pd catalyzed discrimination of enantiotopic alkene faces. *Synthesis* 1491–1494.

Trost, B. M., and Toste, F. D. (1998). A catalytic enantioselective approach to chromans and chromanols. A total synthesis of (−)-calanolides A and B and the vitamin E nucleus. *J. Am. Chem. Soc.* **120,** 9074–9075.

Trost, B. M., Shen, H. C., Dong, L., Surivet, J.-P., and Sylvain, C. (2004). Synthesis of chiral chromans by Pd-catalyzed asymmetric allylic alkylation (AAA): Scope, mechanism, and applications. *J. Am. Chem. Soc.* **126,** 11966–11983.

Urano, S., Nakano, S.-I., and Matsuo, M. (1983). Synthesis of dl-α-tocopherol and dl-α-tocotrienol. *Chem. Pharm. Bull.* **31,** 4341–4345.

Vecchi, M., Walther, W., Glinz, E., Netscher, T., Schmid, R., Lalonde, M., and Vetter, W. (1990). Chromatographische Trennung und quantitative Bestimmung aller acht Stereoisomeren von α-Tocopherol. *Helv. Chim. Acta* **73,** 782–789.

Walsh, L., Winters, R. L., and Gonzalez, R. G. (1998). Optimizing deodorizer distillate tocopherol yields. *INFORM* **9**(no. 1), 78–83.

Walther, W., and Netscher, T. (1996). Design and development of chiral reagents for the chromatographic e.e. determination of chiral alcohols. *Chirality* **8,** 397–401.

Walther, W., Vetter, W., Vecchi, M., Schneider, H., Müller, R. K., and Netscher, T. (1991). (S)-Trolox™ methyl ether: A powerful derivatizing reagent for the GC determination of the enantiomers of aliphatic alcohols. *Chimia* **45,** 121–123.

Weimann, B. J., and Weiser, H. (1991). Function of vitamin E in reproduction and in prostacyclin and immunoglobulin synthesis in rats. *Am. J. Clin. Nutr.* **53,** 1056S–1060S.

Weiser, H., and Vecchi, M. (1982). Stereoisomers of α-tocopheryl acetate. II. Biopotencies of all eight stereoisomers, individually or in mixtures, as determined by rat resorption-gestation tests. *Internat. J. Vit. Nutr. Res.* **52,** 351–370.

Wildermann, A., Foricher, Y., Netscher, T., and Bonrath, W. (2007). New application of indium catalysts: A novel and green concept in fine chemical industry. *Pure Appl. Chem.* in press.

Wirz, B., Schmid, R., and Walther, W. (1990). Enzymatic preparation of (R)-3-hydroxy-2-methylpropyl butyrate by asymmetric hydrolysis. *Biocatalysis* **3,** 159–167.

Wirz, B., Barner, R., and Hübscher, J. (1993). Facile chemoenzymatic preparation of enantiomerically pure 2-methylglycerol derivatives as versatile trifunctional C4-synthons. *J. Org. Chem.* **58,** 3980–3984.

Wüstenberg, B. (2003). Iridium-katalysierte enantioselektive Hydrierung: Neue Substrate und Anwendungen. Ph.D Thesis, University of Basel, Switzerland.

Yoda, H., and Takabe, K. (1989). Novel synthesis of (S)-(−)-chroman-2-carboxylic acid, a vitamin E precursor. *Chem. Lett.* 465–466.

Young, F. V. K., Poot, C., Biernoth, E., Krog, N., O'Neill, L. A., and Davidson, N. G. J. (1986). Processing of fats and oils. *In* "The Lipid Handbook" (F. D. Gunstone, J. L. Harwood, and F. B. Padley, Eds.), pp. 181–247. Chapman and Hall, London.

8

TOCOTRIENOLS: THE EMERGING FACE OF NATURAL VITAMIN E

CHANDAN K. SEN, SAVITA KHANNA, CAMERON RINK, AND SASHWATI ROY

Laboratory of Molecular Medicine, Department of Surgery
Davis Heart and Lung Research Institute
The Ohio State University Medical Center, Columbus, Ohio 43210

Natural vitamin E includes eight chemically distinct molecules: α-, β-, γ-, and δ-tocopherols and α-, β-, γ-, and δ-tocotrienols. More than 95% of all studies on vitamin E are directed toward the specific study of α-tocopherol. The other forms of natural vitamin E remain poorly understood. The abundance of α-tocopherol in the human body and the comparable efficiency of all vitamin E molecules as antioxidants led

Vitamins and Hormones, Volume 76

0083-6729/07 $35.00
DOI: 10.1016/S0083-6729(07)76008-9

biologists to neglect the non-tocopherol vitamin E molecules as topics for basic and clinical research. Recent developments warrant a serious reconsideration of this conventional wisdom. The tocotrienol subfamily of natural vitamin E possesses powerful neuroprotective, anticancer, and cholesterol-lowering properties that are often not exhibited by tocopherols. Current developments in vitamin E research clearly indicate that members of the vitamin E family are not redundant with respect to their biological functions. α-Tocotrienol, γ-tocopherol, and δ-tocotrienol have emerged as vitamin E molecules with functions in health and disease that are clearly distinct from that of α-tocopherol. At nanomolar concentration, α-tocotrienol, not α-tocopherol, prevents neurodegeneration. On a concentration basis, this finding represents the most potent of all biological functions exhibited by any natural vitamin E molecule. Recently, it has been suggested that the safe dose of various tocotrienols for human consumption is 200–1000/day. A rapidly expanding body of evidence supports that members of the vitamin E family are functionally unique. In recognition of this fact, title claims in publications should be limited to the specific form of vitamin E studied. For example, evidence for toxicity of a specific form of tocopherol in excess may not be used to conclude that high-dosage "vitamin E" supplementation may increase all-cause mortality. Such conclusion incorrectly implies that tocotrienols are toxic as well under conditions where tocotrienols were not even considered. The current state of knowledge warrants strategic investment into the lesser known forms of vitamin E. This will enable prudent selection of the appropriate vitamin E molecule for studies addressing a specific health need.

I. Historical Developments and the Vitamin E Family

That certain foods are vital to maintaining healthy life was recognized long before the first vitamins were actually identified. In ancient times, the famous Greek physician Hippocrates not only described night blindness, a disease now known to be caused by a vitamin A deficiency, but recommended the eating of "ox liver dipped in honey" as a cure. In the centuries that followed, observers continued to report that certain diseases appeared to be nutritionally related. By and large, they attributed the problem to some unknown "toxic substance" in various foodstuffs. In 1747 when the Scottish physician James Lind proved he could cure scurvy by feeding citrus fruits to stricken sailors, his fellow physicians continued to ignore his work and to search for the "toxin" responsible for the illness. The unknown "toxins" were never found. In the last quarter of the nineteenth century, scientific thinking began to change. In 1886, Christiaan Eijkman, a physician working in the Dutch East Indies, began a serious investigation into *Beriberi*, a thiamine deficiency disease. Eijkman's studies indicated that beriberi in animals was caused by diets excessively high in polished rice and that it could be cured by substituting unpolished rice. In 1901, a younger colleague, Gerrit Grijns, determined that polished rice lacked an essential "anti-beriberi" substance that could be found in rice hulls and a number of other foods. Contemporary Englishman William Fletcher determined that if special factors (vitamins) were removed from food disease ensued. Fletcher was researching the causes of the disease beriberi when he discovered that eating unpolished rice prevented *Beriberi* and eating polished rice did not. William Fletcher believed that there were special nutrients contained in the husk of the rice. Next year, English biochemist Sir Frederick Gowland Hopkins also discovered that certain "accessory food factors" were important to health. In 1912, Polish scientist Cashmir Funk named the special nutritional parts of food as a "vitamine" after "vita" meaning life and "amine" from compounds found in the thiamine he isolated from rice husks. Vitamine was later shortened to vitamin when it was discovered that not all of the vitamins contain nitrogen, and, therefore, not all are amines. Together, Hopkins and Funk formulated the vitamin hypothesis of deficiency disease—that a lack of vitamins could make people sick. At this point of time, the notion of fat-soluble vitamins was yet to be conceived.

Fat-soluble vitamins have their root in the 1913 discovery by Elmer V. McCollum, Thomas B. Osborne, and Lafayette B. Mendel who isolated a growth-producing substance from egg yolks. The *substance* appeared quite different from the water-soluble vitamins already discovered. In 1916, McCollum went on to show that at least two factors were responsible for the normal growth of rats, factors he named fat-soluble A and water-soluble B. McCollum therefore is credited with initiating the custom of labeling vitamins by letters. Vitamin E was discovered in 1922 in green leafy vegetables by University of California researchers, Herbert Evans and Katherine Bishop (Evans and Bishop, 1922). In 1924, Sure named it vitamin E. Because E supported fertility, it was scientifically named tocopherol. This comes from the Greek word *tokos* meaning childbirth, and *phero* meaning to bring forth, and the *ol* ending was added to indicate the alcohol properties of this molecule. In 1936, it was discovered that vitamin E was abundant in wheat germ oil. Two years later, it was chemically synthesized for the first time. The U.S. National Research Council sponsored studies on deficiencies of vitamin E, and based on the results E was designated an essential vitamin. Vitamin E emerged as an essential, fat-soluble nutrient that functions as an antioxidant in the human body. It is essential, because it is required to sustain life, and the body cannot manufacture its own vitamin E and foods and supplements must provide it. Since the elucidation of the chemical structure of vitamin E in 1938 by Fenholz and the synthesis of DL-α-tocopherol by Karrer in the same year, specific focus was directed on the chemical class of natural compounds that qualify to be vitamin E. Vitamin E was rediscovered as factor 2 antioxidant in 1965 (Schwarz, 1965). α-Tocopherol drew most attention as the first natural form of vitamin E identified while its sisters remained under veil. At present, vitamin E represents a generic term for four tocopherols and four tocotrienols (Bruno and Traber, 2006). In nature, eight substances have been found to have vitamin E activity: α-, β-, γ-, and δ-tocopherols and α-, β-, γ-, and δ-tocotrienols. Although it has been claimed that tocotrienol may be metabolized to tocopherol in the human tissue (Qureshi *et al.*, 2001c, 2002), the concept has not gained wide acceptance and the hypothesis remains open for additional considerations.

II. BIOSYNTHESIS OF TOCOPHEROLS AND TOCOTRIENOLS

Tocopherols consist of a chromanol ring and a 15-carbon tail derived from homogentisate (HGA) and phytyl diphosphate, respectively (Fig. 1). Condensation of HGA and phytyl diphosphate, the committed step in

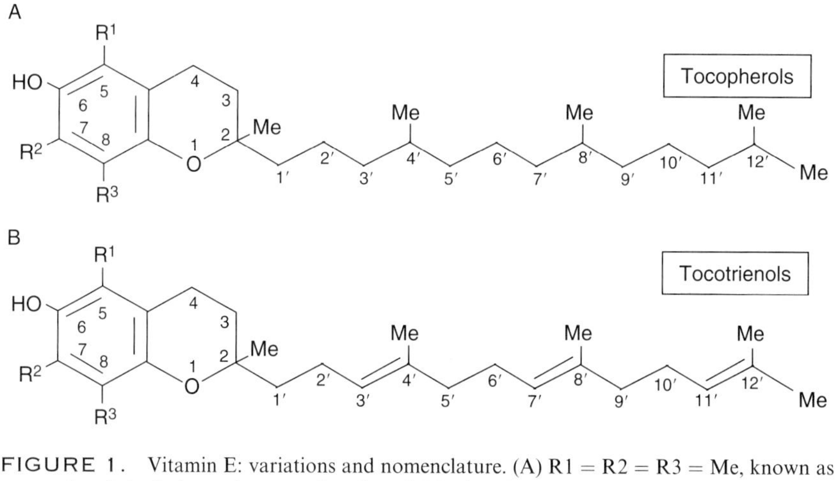

FIGURE 1. Vitamin E: variations and nomenclature. (A) R1 = R2 = R3 = Me, known as α-tocopherol, is designated α-tocopherol or 5,7,8-trimethyltocol; R1 = R3 = Me; R2 = H, known as, β-tocopherol, is designated, β-tocopherol or 5,8-dimethyltocol; R1 = H; R2 = R3 = Me, known as γ-tocopherol, is designated γ-tocopherol or 7,8-dimethyltocol; R1 = R2 = H; R3 = Me, known as δ-tocopherol, is designated δ-tocopherol or 8-methyltocol. (B) R1 = R2 = R3 = H, 2-methyl-2-(4,8,12-trimethyltrideca-3,7,11-trienyl)chroman-6-ol, is designated tocotrienol; R1 = R2 = R3 = Me, formerly known as ζ1- or ζ2-tocopherol, is designated 5,7, 8-trimethyltocotrienol or α-tocotrienol. The name tocochromanol-3 has also been used; R1 = R3 = Me; R2 = H, formerly known as ε-tocopherol, is designated 5,8-dimethyltocotrienol or β-tocotrienol; R1 = H; R2 = R3 = Me, formerly known as η-tocopherol, is designated 7,8-dimethyltocotrienol or γ-tocotrienol. The name plastochromanol-3 has also been used; R1 = R2 = H; R3 = Me is designated 8-methyltocotrienol or δ-tocotrienol.

tocopherol biosynthesis, is catalyzed by HGA phytyltransferase (HPT) (Venkatesh *et al.*, 2006). Tocopherol helps maintain optimal photosynthesis rate under high-light stress (Porfirova *et al.*, 2002). Tocotrienols differ structurally from tocopherols by the presence of three *trans* double bonds in the hydrocarbon tail. Because of these unsaturations in the isoprenoid side chain, tocotrienols are thought to assume a unique conformation (Atkinson, 2006). α-Tocotrienol seems to be very likely much more flexible in the side chain and that it puts a greater curvature stress on phospholipid membranes. This has been confirmed in scanning calorimetry data (Dr. Jeffrey Atkinson, unpublished personal communication).

Tocotrienols are the primary form of vitamin E in the seed endosperm of most monocots, including agronomically important cereal grains such as wheat, rice, and barley. Palm oil contains significant quantities of tocotrienol (Sundram *et al.*, 2003). Tocotrienols are also found in the seed endosperm of a limited number of dicots, including *Apiaceae* species and certain *Solanaeceae* species, such as tobacco. These molecules are found only rarely in vegetative tissues of plants. Crude palm oil extracted from the fruits of *Elaeis guineensis* particularly contains a high amount of tocotrienols (up to 800 mg/kg), mainly consisting of γ- and α-tocotrienols. Compared to tocopherols, tocotrienols are considerably less widespread in the plant kingdom (Horvath *et al.*, 2006). In 80 different plant species studied, 24 were found to contain significant amounts of tocotrienols. No taxonomic relation was apparent among the 16 dicotyledonous species that were found to contain tocotrienol. Monocotyledonous species (eight species) belonged either to the *Poaceae* (six species) or to the *Aracaceae* (two species). A more detailed analysis of tocotrienol accumulation revealed the presence of this natural vitamin E in several nonphotosynthetic tissues and organs, that is seeds, fruits, and latex. No tocotrienols could be detected in mature photosynthetic tissues. Transient accumulation of low levels of tocotrienols is found in the young coleoptiles of plant species whose seeds contained tocotrienols. No measurable tocotrienol biosynthesis was apparent in coleoptiles or in chloroplasts isolated from such coleoptiles. Tocotrienol accumulation in coleoptiles was not associated with chloroplasts. Tocotrienols seem to be transiently present in photosynthetically active tissues; however, it remains to be proven whether they are biosynthesized in such tissues or imported from elsewhere in the plant (Horvath *et al.*, 2006).

In contrast to tocotrienols, tocopherols occur ubiquitously in plant tissues and are the exclusive form of vitamin E in leaves of plants and seeds of most dicots. Transgenic expression of the barley homogentisic acid transferase (HGGT, which catalyzes the committed step of tocotrienol biosynthesis) in *Arabidopsis thaliana* leaves resulted in accumulation of tocotrienols, which were absent from leaves of nontransformed plants, and a 10- to 15-fold increase in total vitamin E antioxidants (tocotrienols plus tocopherols). Overexpression of the barley HGGT in corn seeds resulted in an increase in tocotrienol and tocopherol content of as much as sixfold. These results provide insight into the genetic basis for tocotrienol biosynthesis in plants and demonstrate the ability to enhance the antioxidant content of crops by introduction of an enzyme that redirects metabolic flux (Cahoon *et al.*, 2003). Another strategy involving genetic engineering of metabolic pathways in plants has proved to be efficient in bolstering tocotrienol biosynthesis (Rippert *et al.*, 2004). In plants, phenylalanine is the precursor of a myriad of secondary compounds termed phenylpropanoids. In contrast, much less carbon is incorporated into tyrosine that provides *p*-hydroxyphenylpyruvate and HGA, the aromatic precursors of vitamin E. The flux of these two compounds has been upregulated by deriving their synthesis directly at the level of prephenate. This was achieved by the expression of the yeast prephenate dehydrogenase gene in tobacco plants that already overexpress the *Arabidopsis* *p*-hydroxyphenylpyruvate dioxygenase coding sequence. Massive accumulation of tocotrienols was observed in leaves. These molecules, which were undetectable in wild-type leaves, became the major forms of vitamin E in the leaves of the transgenic lines. An increased resistance of the transgenic plants toward the herbicidal *p*-hydroxyphenylpyruvate dioxygenase inhibitor diketonitril was also observed. Thus, the synthesis of *p*-hydroxyphenylpyruvate is a limiting step for the accumulation of vitamin E in plants (Rippert *et al.*, 2004).

III. CHANGING TRENDS IN VITAMIN E RESEARCH

A striking asymmetry in our understanding of the eight-member natural vitamin E tocol family has deprived us of the full complement of benefits offered by the natural vitamin E molecules (Fig. 2). Approximately, only 1% of the entire literature on vitamin E addresses tocotrienols. A review of the NIH CRISP database shows that funding for tocotrienol research represents less than 1% of all vitamin E research during the last 30+ years. Within the tocopherol literature, the non-α forms remain poorly studied (Dietrich *et al.*, 2006; Hensley *et al.*, 2004; O'Byrne *et al.*, 2000). This represents a major void in vitamin E research. It is important that conclusions drawn about the usefulness of vitamin E as a whole to human health be drawn in this light. At present, conclusions drawn about vitamin E as whole in light of results from α-tocopherol studies alone (Friedrich, 2004; Gorman, 2005; Greenberg, 2005; Hathcock *et al.*, 2005; Miller *et al.*, 2005) can be misleading. It is important to recognize the gaping voids in our understanding of the non-α-tocopherol forms of vitamin E and develop a more symmetrical understanding which would enable us to pick the right form of vitamin E for specific health indications. In this context we need to be cognizant of the fact the biological functions of the different homologues of natural vitamin E are not identical. Evidence supporting the unique biological significance of vitamin E family members is provided by current results derived from α-tocotrienol research.

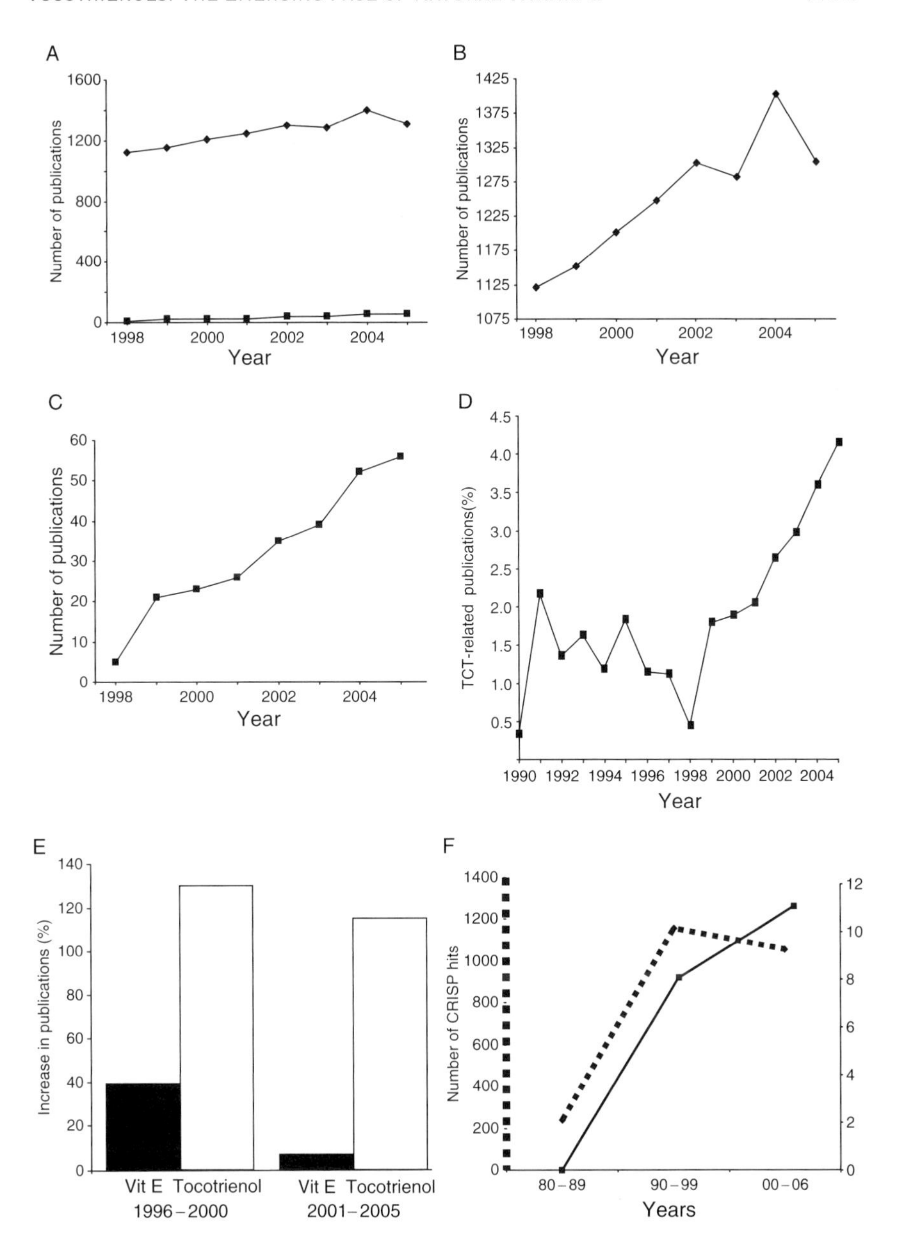

FIGURE 2. Trends in tocotrienol research and in vitamin E research as a whole. Publication data are based on PubMed entries. (A) Comparison of volume of all vitamin E (diamond) research and tocotrienol research (square), (B) time-dependent changes in the volume of vitamin E research as a whole, (C) time-dependent changes in the volume of tocotrienol research, (D) tocotrienol (TCT) publications as a percent of total vitamin E publications reported, (E) percent increase in tocotrienol publications and that of vitamin E as a whole over time, and (F) trends in NIH funding for tocotrienol research (solid line) and for vitamin E research as a whole (broken line). On the basis of hits in the Computer Retrieval of Information on Scientific Projects (CRISP) database.

During the last 5 years, tocotrienol research has gained substantial momentum (Fig. 2). More than two-thirds (189/280) of the entire PubMed literature on tocotrienols has been published on or after 2000. This represents a major swing in the direction of vitamin E research.

IV. UNIQUE BIOLOGICAL FUNCTIONS OF TOCOTRIENOLS

All eight tocols in the vitamin E family share close structural similarity (Fig. 1) and hence comparable antioxidant efficacy. Yet, current studies of the biological functions of vitamin E continue to indicate that members of the vitamin E family possess unique biological functions often not shared by other family members. One of the earliest observations suggesting that α-tocopherol may have functions independent of its antioxidant property came from the study of platelet adhesion. α-Tocopherol strongly inhibits platelet adhesion. Doses of 400 IU/day provide greater than 75% inhibition of platelet adhesion to a variety of adhesive proteins when tested at low shear rate in a laminar flow chamber? The antiadhesive effect of α-tocopherol appears to be related to a reduction in the number and size of pseudopodia on platelet activation and this finding led to the hypothesis that within the body vitamin E may exert functions beyond its antioxidant property (Steiner, 1993). That members of the tocopherol family may have functions independent of their antioxidant properties gained more prominence when vitamin E molecules with comparable antioxidant properties exhibited contrasting biological effects (Boscoboinik *et al.*, 1991). At the posttranslational level, α-tocopherol inhibits protein kinase C, 5-lipoxygenase (5-Lox), and phospholipase A2 and activates protein phosphatase 2A and diacylglycerol kinase. Some genes [e.g., scavenger receptors, α-tocopherol transfer protein (α-TTP), α-tropomyosin, matrix metalloproteinase-19, and collagenase] are specifically modulated by α-tocopherol at the transcriptional level. α-Tocopherol also inhibits cell proliferation, platelet aggregation, and monocyte adhesion. These effects have been characterized to be unrelated to the antioxidant activity of vitamin E and possibly reflect specific interactions of α-tocopherol with enzymes, structural proteins, lipids, and transcription factors (Zingg and Azzi, 2004). γ-Tocopherol represents the major form of vitamin E in the diet in the United States, but not in Europe. Desmethyl tocopherols, such as γ-tocopherol and specific tocopherol metabolites, most notably the carboxyethyl-hydroxychroman (CEHC) products, exhibit functions that are not shared by α-tocopherol. The activities of these other tocopherols do not map directly to their chemical antioxidant behavior but rather reflect anti-inflammatory, antineoplastic, and natriuretic functions possibly mediated through specific binding interactions (Hensley *et al.*, 2004). Metabolites of γ-tocopherol (2,7,8-trimethyl-2-(β-carboxyethyl)-6-hydroxychroman), but not that of α-tocopherol, provides natriuretic activity.

Moreover, a nascent body of epidemiological data suggests that γ-tocopherol is a better negative risk factor for certain types of cancer and myocardial infarction than is α-tocopherol (Wagner *et al.*, 2004).

α-Tocotrienol possesses numerous functions that are not shared by α-tocopherol (Sen *et al.*, 2006). For example, nanomolar concentrations of α-tocotrienol uniquely prevent inducible neurodegeneration by regulating specific mediators of cell death (Khanna *et al.*, 2003, 2006; Sen *et al.*, 2000). Oral supplementation of tocotrienol protects against stroke (Khanna *et al.*, 2005b). Micromolar amounts of tocotrienol suppress the activity of 3-hydroxy-3-methylglutaryl coenzyme A (HMG-CoA) reductase, the hepatic enzyme responsible for cholesterol synthesis (Pearce *et al.*, 1992, 1994). Tocopherols do not share the cholesterol-lowering properties of tocotrienol (Qureshi *et al.*, 1986, 2002). Sterol-regulated ubiquitination marks HMG-CoA reductase for endoplasmic reticulum (ER)-associated degradation by 26S proteasomes. This degradation, which results from sterol-induced binding of reductase to ER membrane proteins called Insigs, contributes to the complex, multivalent feedback regulation of the enzyme. Recently, it has been demonstrated that δ-tocotrienol stimulates ubiquitination and degradation of reductase and blocks processing of sterol regulatory element-binding proteins (SREBPs), another sterol-mediated action of Insigs. The γ-tocotrienol analogue is more selective in enhancing reductase ubiquitination and degradation than blocking SREBP processing. Other forms of vitamin E neither accelerate reductase degradation nor block SREBP processing (Song and Debose-Boyd, 2006).

Tocotrienol, not tocopherol, administration reduces oxidative protein damage and extends the mean life span of *Caenorhabditis elegans* (Adachi and Ishii, 2000). Tocotrienols are thought to have more potent antioxidant properties than α-tocopherol (Serbinova and Packer, 1994; Serbinova *et al.*, 1991). The unsaturated side chain of tocotrienol allows for more efficient penetration into tissues that have saturated fatty layers such as the brain and liver (Suzuki *et al.*, 1993). Experimental research examining the antioxidant, free-radical-scavenging effects of tocopherol, and tocotrienols revealed that tocotrienols appear superior due to their better distribution in the fatty layers of the cell membrane (Suzuki *et al.*, 1993). Furthermore, tocotrienol but not tocopherol, suppresses growth of human breast cancer cells (Nesaretnam *et al.*, 1995).

In humans, tocotrienol supplementation results in peak blood plasma level of α-tocotrienol that is over an order of magnitude higher than that required to protect neurons against a range of neurotoxic insults (Khanna *et al.*, 2003, 2005a,b, 2006; Khosla *et al.*, 2006; Sen *et al.*, 2000). Despite such promising potential, tocotrienol research accounts for roughly 1% of all vitamin E research published in PubMed. The unique vitamin action of α-tocopherol, combined with its prevalence in the human body and the similar efficiency of tocopherols as chain-breaking antioxidants, led biologists to almost completely discount the "minor" vitamin E molecules as topics for basic and clinical research. Recent discoveries warrant a serious reconsideration of this conventional wisdom.

V. NATURAL SOURCES OF TOCOTRIENOLS

Tocotrienol is synthesized in edible as well as inedible plant products. Rubber latex represents a major nonfood natural source of tocotrienols (Chow and Draper, 1970; Horvath *et al.*, 2006; Whittle *et al.*, 1966). Identification of α-tocotrienol as a cholesterogenesis inhibitory factor in barley (*Hordeum vulgare* L.) represents a landmark early discovery highlighting the unique significance of tocotrienols in health and disease (Qureshi *et al.*, 1986). Purification of an oily, nonpolar fraction of high-protein barley flour by high-pressure liquid chromatography yielded 10 major components. Two of these components were identified as potent inhibitors of cholesterogenesis both *in vivo* as well as *in vitro*. Addition of the purified inhibitor I (2.5–20 ppm) to chick diets significantly decreased hepatic cholesterogenesis and serum total and low-density lipoprotein (LDL) cholesterol and concomitantly increased lipogenic activity. The high-resolution mass spectrometric analysis and measurement of different peaks of inhibitor I gave a molecular ion at *m/e* 424 (C29H44O2) and main peaks at *m/e* 205, 203, and 165 corresponding to C13H17O2, C13H15O2, and C10H13O2 moieties, respectively. On the basis of these results, D-α-tocotrienol was identified as the active principle. This identification was confirmed against synthetic samples (Qureshi *et al.*, 1986).

Palm oil represents one of the most abundant natural sources of tocotrienols (Elson, 1992). The distribution of vitamin E in palm oil is 30% tocopherols and 70% tocotrienols (Sundram *et al.*, 2003). The oil palm (*E. guineensis*) is native to many West African countries, where local populations have used its oil for culinary and other purposes. Large-scale plantations, established principally in tropical regions of Asia, Africa, and Latin America are mostly aimed at the production of oil (Solomons and Orozco, 2003), which is extracted from the fleshy mesocarp of the palm fruit, and endosperm or kernel oil. Palm oil is different from other plant and animal oils in that it contains 50% saturated fatty acids, 40% unsaturated fatty acids, and 10% polyunsaturated fatty acids. Because of its high saturated fat content, palm oil has not been very popular in the United States. Hydrogenated fats contain high levels of *trans*-fatty acids which are now thought to have adverse health effects. The U.S. Food and Drug Administration's final ruling on *trans*-fatty acid labeling issued in 2003 has caused a rapid transformation in the fat and oil industries (Tarrago-Trani *et al.*, 2006). Palm oil is free of *trans*-fatty acid and is rapidly gaining wider acceptance by the food industry in the country. Primary applications include bakery products, breakfast cereals, wafers, and candies.

Rice bran oil (RBO), a by-product of the rice-milling industry, is a major natural source of γ-tocotrienol but a poor source of α-tocotrienol. In addition, RBO provides desmethyl tocotrienols. Two novel tocotrienols were isolated from stabilized and heated rice bran, apart from the known α-, β-, γ-, and δ-tocopherols and tocotrienols. These new tocotrienols are known as desmethyl tocotrienol [3,4-dihydro-2-methyl-2-(4,8,12-trimethyltrideca-3′(E),7′(E),11′-trienyl)-2H-1-benzopyran-6-ol] and didesmethy tocotrienol [3,4-dihydro-2-(4,8,12-trimethyltrideca-3′(E),7′(E),11′-trienyl)-2H-1-benzopyran-6-ol] (Qureshi *et al.*, 2000). Although scientific evidence is relatively limited, RBO is tenaciously believed to be a healthy vegetable oil in Asian countries (Sugano *et al.*, 1999).

Cereals such as oat, rye, and barley contain small amounts of tocotrienol in them. α-Tocotrienol is the predominant form of tocotrienol in oat (*Avena sativa* L.) and barley (56 and 40 mg/kg of dry weight, respectively). β-Tocotrienol is the major form of tocotrienol found in hulled and dehulled wheats (from 33 to 43 mg/kg of dry weight) (Panfili *et al.*, 2003). Steaming and flaking of dehulled oat groats results in moderate losses of tocotrienols but not of tocopherols (Bryngelsson *et al.*, 2002). Autoclaving of grains (including the hulls) increases the levels of all tocopherols and tocotrienols analyzed except β-tocotrienol, which was not affected. Drum drying of steamed rolled oats results in an almost complete loss of tocopherols and tocotrienols (Bryngelsson *et al.*, 2002). Although tocotrienols are present in edible natural products, it is questionable whether these dietary sources could provide sufficient amounts of tocotrienol to humans. Of note, processing of 1000 kg of crude palm oil is necessary to derive 1 kg of the commercial product Tocomin 50% (Carotech, New Jersy). Roughly, one would have to consume 100–200 g of palm/rice bran oil or 1.5–4 kg of wheat germ, barley or oat to achieve doses that have been published to be effective biologically. With this consideration in mind, appropriately configured dietary supplements seem to be a prudent choice.

VI. BIOAVAILABILITY OF ORAL TOCOTRIENOLS

During the last two decades, efforts to understand how dietary vitamin E is transported to the tissues have focused on α-tocopherol transport (Blatt *et al.*, 2001; Kaempf-Rotzoll *et al.*, 2003; Traber and Arai, 1999; Traber *et al.*, 2004). α-Tocopherol transfer protein (TTP) has been identified to mediate α-tocopherol secretion into the plasma while other tocopherol-binding proteins seem to play a less important role (Kaempf-Rotzoll *et al.*, 2003). Tocotrienols have been known for decades but why have they not been studied as well as α-tocopherol? Although there does not seem to be straightforward rational answer to this question, one contributing factor is whether tocotrienol taken orally reaches vital organs of the body. This concern was primarily based on a 1997 finding that the transport system, α-tocopherol transfer protein (TTP), responsible to carry α-tocopherol to vital organs has a poorer efficiency to transport tocotrienols to tissues (Hosomi *et al.*, 1997). The lack of relative specific affinity of TTP for tocotrienols led to the notion that availability of dietary tocotrienol to vital organs is negligible.

TTP is a soluble 32-kDa protein expressed in liver that selectively binds and transports α-tocopherol. TTP maintains the concentration of serum α-tocopherol by facilitating α-tocopherol export from the liver. TTP is required to maintain normal α-tocopherol concentrations in plasma and extrahepatic tissues (Traber *et al.*, 2004). Although TTP is known to bind to α-tocotrienol with 8.5-fold lower affinity than that for α-tocopherol (Hosomi *et al.*, 1997), it has not been clear whether, or to what extent, the delivery of orally supplemented α-tocotrienol to vital organs is dependent on TTP. Previously, it has been reported that TTP-deficient females are infertile presumably because of vitamin E deficiency (Terasawa *et al.*, 2000). This important observation was confirmed in a lineage of TTP-deficient mice. Placenta of pregnant TTP-deficient females were severely impaired with marked reduction of labyrinthine trophoblasts, and the embryos died at mid-gestation even when fertilized eggs of TTP-containing wild-type mice were transferred into TTP-deficient recipients (Jishage *et al.*, 2001). Even in the presence of dietary α-tocopherol, TTP knockout mice are known to suffer from α-tocopherol deficiency (Jishage *et al.*, 2001; Terasawa *et al.*, 2000). It has been noted that oral supplementation of female mice with α-tocotrienol restored fertility of TTP knockout mice, suggesting that tocotrienol was successfully delivered to the relevant tissues and that tocotrienol supported reproductive function under conditions of α-tocopherol deficiency (Khanna *et al.*, 2005a). This observation was consistent with another line of evidence from rats where tocotrienol supplementation spared loss of fertility caused by long-term vitamin E deficiency in the diet (Khanna *et al.*, 2005a). TTP continues to be a key transport mechanism for the deliver of α-tocopherol to tissues. The significance of TTP in the transport of other forms of vitamin E remains unclear at present. It is clear, however, that natural isomers of vitamin E do get transported to vital organs even in the absence of TTP. Identification and characterization of TTP-independent vitamin E transport mechanisms *in vivo* is warranted.

Ten years ago in a study testing ligand specificity of vitamin E isomers for TTP concluded that the affinity of vitamin E analogues for TTP is one of the critical determinants of their biological activity (Hosomi *et al.*, 1997). This conclusion was based on the assumption that the biological function of vitamin E molecules is proportionate to their concentration and that vitamin E isomers have redundant function. Early postulates proposing that tissue concentration and relative biological function of tocopherol and tocotrienol are disparate and possibly unrelated (Hayes *et al.*, 1993). Developments during the last decade taught us that both assumptions are incorrect warranting a revisit of the fundamental principles that guide vitamin E research (Azzi and Stocker, 2000; Azzi *et al.*, 1995; Sen *et al.*, 2004, 2006). Another contemporary study reported that tocotrienols, supplemented to laboratory chow, do not reach the brain (Podda *et al.*, 1996). Taken together, the case for *in vivo* efficacy of oral tocotrienol was seriously weakened by these reports (Hosomi *et al.*, 1997; Podda *et al.*, 1996). Today, however, the scenario has strikingly changed in light of new knowledge. For example, it is now clear that oral tocotrienol not only reaches the brain (Khanna *et al.*, 2005a,b; Roy *et al.*, 2002) but it does so in amounts sufficient to protect against stroke (Khanna *et al.*, 2005b). The standard laboratory chow contains excessive amounts of α-tocopherol (Khosla *et al.*, 2006; van der Worp *et al.*, 1998) but negligible amounts of tocotrienol. Long-term lack of tocotrienol in the diet may repress any putative tocotrienol transport mechanism *in vivo*. Thus, long-term supplementation studies are needed. In light of the knowledge that natural analogues of vitamin E may compete for specific transporting mechanisms (Hosomi *et al.*, 1997), it is important that tocotrienol supplementation be performed under conditions of minimized copresence of tocopherols. Another related consideration is that although incorporation of orally supplemented vitamin E into tissues is a slow and progressive process, rapid incorporation of the supplement into tissues of newborns may occur in response to gavaging of pregnant mother rats (Roy *et al.*, 2002). Thus, an experimental design incorporating long-term tocotrienol supplementation under conditions of minimal dietary copresence of tocopherols and breeding of the supplemented colony would be a valuable approach to generate proof of principle testing whether dietary α-tocotrienol is capable of being transported to vital organs *in vivo*. In a recent study, rats were maintained on vitamin E-deficient diet and gavaged with α-tocotrienol alone, α-tocopherol alone, or in combination. Five generations of rats were studied over 60 weeks (Khanna *et al.*, 2005a). Skin, adipose, heart, lungs, skeletal muscle brain, spinal cord, liver, and blood were studied. Oral tocotrienol was delivered to all vital organs. In some tissues, the level of tocotrienol exceeded that of tocopherols, indicating the presence of an efficient tocotrienol transport system *in vivo*. Baseline levels of α-tocotrienol in the skin of tocopherol-fed rats that never received any tocotrienol supplementation were negligible. Orally supplemented tocotrienol was rapidly taken up by the skin. Already in second generation rats, α-tocotrienol levels in the skin of tocotrienol supplemented rats exceeded twice the α-tocopherol levels in that organ. Of note, the α-tocotrienol level in the skin matched the α-tocotrienol level in the skin of rats fed with a comparable amount of tocopherol. When tocotrienol and tocopherol were cosupplemented, the uptake of α-tocotrienol by the skin was clearly blunted. In this group, α-tocotrienol levels were lower than α-tocotrienol levels in the skin, suggesting a direct competition between orally taken tocotrienol and tocopherol for delivery to the skin. Longer supplementation resulted in a marked increase in the α-tocotrienol levels in the skin of tocotrienol-fed rats, indicating a buildup of α-tocotrienol over time. Interestingly, the levels of α-tocotrienol in the skin of these rats were folds higher than the α-tocopherol level in the skin of tocopherol-fed rats. This observation suggests the presence of an effective transport mechanism delivering α-tocotrienol to the skin and efficient retention of α-tocotrienol in the skin over time. Cosupplementation of tocotrienol and tocopherol demonstrated favorable uptake of α-tocopherol over α-tocotrienol. Adipose tissue serves as storage organ for vitamin E (Adachi *et al.*, 1990). Analysis of adipose tissue vitamin E content of fifth generation rats revealed substantially more accumulation of α-tocotrienol in that tissue than α-tocopherol.

In the case of tocotrienol as well as of tocopherol feeding, results from third and fifth generation rats indicate higher levels of vitamin E in the skin of female compared to that of male rats. This gender-specific effect suggesting better transport of tocotrienol in females than in males was noted as a general trend across all organs studied. Gender-based differences in the transport of dietary vitamins are known to exist in specific cases (Garry *et al.*, 1987). Although the effect of several physiological factors on vitamin E transport has been studied, the gender factor remains to be specifically addressed (Lodge *et al.*, 2004). It has been demonstrated that γ-tocopherol is more rapidly metabolized in women than in men (Leonard *et al.*, 2005). The level of α-tocotrienol in the ovary was over fivefold higher than that in the testes from the corresponding male rats (Khanna *et al.*, 2005a). In the ovary, tocopherol is known to accumulate via a lipoprotein receptor-dependent mechanism (Aten *et al.*, 1994). Whether tocotrienol shares that mechanism remain to be tested.

Vitamin E enters the circulation from the intestine in chylomicrons. The conversion of chylomicrons to remnant particles results in the distribution of newly absorbed vitamin E to all of the circulating lipoproteins and ultimately to tissues. This enrichment of lipoproteins with vitamin E is a key mechanism by which vitamin E is delivered to tissues (Traber *et al.*, 2004). In the liver, newly absorbed dietary lipids are incorporated into nascent very LDLs. The liver is responsible for the control and release of α-tocopherol into blood plasma. In the absence of TTP, α-tocopherol is not secreted back into the plasma. Excess vitamin E is not accumulated in the liver, but is excreted, mostly in bile (Traber *et al.*, 2004). It has been noted that α-tocotrienol levels in the liver of rats and of TTP-deficient mice were much lower than the levels of this vitamin E isoform in most peripheral tissues studied (Khanna *et al.*, 2005a). Such observation argues against a central role of the liver in delivering oral α-tocotrienol to peripheral tissues. TTP has the ability to bind to both α-tocopherol as well as α-tocotrienol. The affinity to bind α-tocopherol is severalfold higher than that for α-tocotrienol (Hosomi *et al.*, 1997). Thus, under conditions of coexistence, α-tocopherol is expected to out-compete α-tocotrienol for binding. Although studies with the TTP-deficient mice (Khanna *et al.*, 2005a) indicate the existence of a TTP-independent mechanisms for the tissue delivery of oral α-tocotrienol, observations in the rat (Khanna *et al.*, 2005a) indicate that the mechanisms for transporting α-tocopherol and α-tocotrienol seem to compete such that transport of α-tocopherol is favored. Thus, cosupplementation of α-tocopherol and α-tocotrienol is likely to compromise tissue delivery of α-tocotrienol (Khanna *et al.*, 2005a).

Few studies have specifically looked at the fate of oral tocotrienol supplementation in humans. In a study investigating the pharmacokinetics and bioavailability of α-, γ-, and δ-tocotrienols under fed and fasted conditions in eight healthy volunteers, subjects were administered a single 300-mg oral dose of mixed tocotrienols under fed or fasted conditions. The peak concentration of α-tocotrienol in the blood plasma was just over 1 μM (Yap *et al.*, 2001). The fed state increased the onset as well as the extent of absorption of tocotrienols by more than twofolds. In addition, the mean apparent elimination half-life of α-, γ-, and δ-tocotrienols was estimated to be 4.4, 4.3, and 2.3 h, respectively, being between 4.5- and 8.7-fold shorter than that reported for α-tocopherol (Yap *et al.*, 2001). In another study, human subjects took tocotrienyl acetate supplements (250 mg/day) for eight weeks, while being on low-fat diet. In response to supplementation, the concentrations of tocotrienol in the mean blood plasma were as follows: α-tocotrienol, 0.98 μM; γ-tocotrienol, 0.54 μM; and δ-tocotrienol 0.09 μM (O'Byrne *et al.*, 2000). Thus, tocotrienyl acetate supplements were observed to be hydrolyzed, absorbed, and detectable in human plasma. A novel formulation for improved absorption of tocotrienols has been developed (Ho *et al.*, 2003). Emulsions are known to increase absorption of fat-soluble drugs. This invention is based on self-emulsifying drug delivery systems (SEDDS) technology (Araya *et al.*, 2006; Gao and Morozowich, 2006; Hong *et al.*, 2006). Soft gelatin capsules (Tocovid Suprabio™) containing tocotrienol have been produced. Once ingested, the tocotrienols form emulsion when the contents are released and mixed with human gastrointestinal fluid. In a recent study using Tocovid Suprabio™, the postabsorptive fate of tocotrienol isomers and their association with lipoprotein subfractions were examined in humans (Khosla *et al.*, 2006). The peak α-tocotrienol concentrations in supplemented individuals averaged ~3 μM in blood plasma, 1.7 μM in LDL, 0.9 μM in triglyceride-rich lipoprotein, and 0.5 μM in HDL. This peak plasma concentration of α-tocotrienol is two to three times more than the peak concentration reported in previous studies using generic supplements not based on SEDDS (O'Byrne *et al.*, 2000; Yap *et al.*, 2001).

VII. Biological Functions

The biological functions of tocotrienol known so far have been listed in Table I. In this section, we discuss work that relate to the neuroprotective, anticancer, and cholesterol-lowering activities of tocotrienol.

TABLE I. Tocotrienols: The Emergent Face of Natural Vitamin E[a]

Neuroprotective	Mouse: At nanomolar concentrations, α-tocotrienol, in contrast with α-tocopherol, protects against glutamate-induced neuronal death by suppressing inducible pp60 c-src kinase activation. α-Tocotrienol provided the most potent neuroprotection among all vitamin E analogues. Reported effects of tocotrienol independent of antioxidant property (Sen *et al.*, 2000)	2000
	Rat: Oral tocotrienol crosses the blood–brain barrier to reach brain tissue; more so for fetal brain while pregnant mother is supplemented with tocotrienol (Roy *et al.*, 2002)	2002
	Mouse: At nanomolar concentrations, α-tocotrienol, in contrast with α-tocopherol, protects against glutamate-induced neuronal death by suppressing inducible 12-lipoxygenase activation (Khanna *et al.*, 2003). 12-Lipoxygenase-deficient mice are protected against stroke (Khanna *et al.*, 2005b)	2003
	Mouse: Injected α-tocotrienol decreased the size of the cerebral infarcts 1 day after stroke; γ- and δ-tocotrienols did not protect (Mishima *et al.*, 2003)	2003
	Human: Tocotrienols induced IKBKAP expression: a possible therapy for familial dysautonomia (Anderson *et al.*, 2003)	2003
	Rat: α-Tocotrienol provided the most potent neuroprotection among vitamin E analogues on cultured striatal neurons (Osakada *et al.*, 2004)	2004
	Human: Administration of tocotrienol to individuals with familial dysautonomia resulted in beneficial changes in their peripheral blood cells (Anderson and Rubin, 2005)	2005
	Rat: Attomole quantity of α-tocotrienol, not α-tocopherol, microinjected to primary neurons protects against glutamate cytotoxicity (Khanna *et al.*, 2005b)	2005
	SHR: α-Tocotrienol protects against stroke *in vivo* (Khanna *et al.*, 2005b)	2005
	Rat: α- and γ-tocotrienols have comparable protective effects on H_2O_2-induced death of astrocytes (Mazlan *et al.*, 2006)	2006
	Rat: At nanomolar concentration, α-tocotrienol protects neurons. Vitamin E analogues play an essential role in neuronal maintenance and survival in the CNS (Numakawa *et al.*, 2006)	2006
	Mouse: The neuroprotective property of α-tocotrienol is antioxidant-independent at nanomolar but antioxidant-dependent a micromolar concentrations (Khanna *et al.*, 2006)	2006
	Mouse: At nanomolar concentration, α-tocotrienol protects against homocysteic acid-induced neurotoxicity (Khanna *et al.*, 2006)	2006

Hypocholesterolemic	Chicken: Three double bonds in the isoprenoid chain essential for the inhibition of cholesterogenesis; tocopherols do not share this property (Qureshi *et al.*, 1986)	1986
	Human: Lowered serum cholesterol in hypercholesterolemics (Qureshi *et al.*, 1991b); lowered both serum total cholesterol (TC) and low-density-lipoprotein cholesterol (Tan *et al.*, 1991)	1991
	Pigs: Reduced plasma cholesterol, apolipoprotein B, thromboxane B2, and platelet factor 4 in pigs with inherited hyperlipidemias (Qureshi *et al.*, 1991a)	
	In vitro: Posttranscriptional suppression of HMG-CoA reductase by a process distinct from other known inhibitors of cholesterol biosynthesis (Pearce *et al.*, 1992)	1992
	Regulate cholesterol production in mammalian cells by posttranscriptional suppression of 3-hydroxy-3-methyl-glutaryl-coenzyme A reductase (Parker *et al.*, 1993)	1993
	HepG2: The farnesyl side chain and the methyl/hydroxy substitution pattern of γ-tocotrienol responsible for HMG-CoA reductase suppression (Pearce *et al.*, 1994)	1994
	Isoprenoid-mediated suppression of mevalonate synthesis depletes tumor tissues of two intermediate products, farnesyl pyrophosphate and geranylgeranyl pyrophosphate, which are incorporated posttranslationally into growth control-associated proteins (Elson and Qureshi, 1995)	1995
	Human: Lowered plasma cholesterol level in hypercholesterolemic subjects (Qureshi *et al.*, 1995)	1995
	Chicken: The effects of a tocotrienol–lovastatin combination were no greater than that of tocotrienol alone, indicating that tocotrienol produced a maximum cholesterol lowering effect (Qureshi and Peterson, 2001)	2001
	Swine: Tocotrienols suppress cholesterogenesis in hereditary hypercholesterolemic swine (Qureshi *et al.*, 2001a)	2001
	Human: Tocotrienol, not tocopherol, hypocholesterolemic in humans; claimed that tocotrienol is converted to tocopherol *in vivo* (Qureshi *et al.*, 2001c)	2001
	Human: Dose-dependent suppression of serum cholesterol by tocotrienol-rich fraction of rice bran in hypercholesterolemic humans (Qureshi *et al.*, 2002)	2002
	Hamster: Tocotrienols lower total cholesterol and low-density lipoprotein plasma levels (Raederstorff *et al.*, 2002)	2002
	Rat: Suppression of hypercholesterolemia in rats by tocotrienol-rich fraction isolated from rice bran oil (Iqbal *et al.*, 2003)	2003

(*Continues*)

TABLE I. (*Continued*)

	Rat: TRF lowered HMG-CoA reductase activity in hyperlipidemics (Minhajuddin *et al.*, 2005)	2005
	Rat: Tocotrienol-rich rice bran oil-containing diet can significantly suppress hyperlipidemic and hyperinsulinemic responses in diabetics (Chen and Cheng, 2006)	2006
	δ- and γ-tocotrienols, but not other forms of vitamin E, cause HMG Co-A reductase ubiquitination and degradation. Results explain hypocholesterolemic effects of tocotrienol noted in humans and animals (Song and Debose-Boyd, 2006)	2006
ApoB level reduction in hypercholesterolemic subjects	Human: In HepG2 cells, tocotrienol (not tocopherol) stimulates apoB degradation possibly as the result of decreased apoB translocation into the endoplasmic reticulum lumen (Theriault *et al.*, 1999)	1999
Antihypertensive	Rat: Depressed (better than α-tocopherol) age-related increase in the systolic blood pressure of spontaneously hypertensive rats (Koba *et al.*, 1992)	1992
Hypocholesterolemic and antioxidant	Rat: Spares plasma tocopherol (Watkins *et al.*, 1993)	1993
Lowering blood pressure; antioxidant	SHR: Supplement of γ-tocotrienol may prevent increased blood pressure, reduce lipid peroxides in plasma and blood vessels and enhance total antioxidant status (Newaz and Nawal, 1999)	1999
Cardioprotective	Rat: TRF protected against ischemia-reperfusion in isolated heart by c-Src inhibition (Das *et al.*, 2005)	2005
Antioxidant	*In vitro*: Better than α-tocopherol (Serbinova *et al.*, 1991)	1991
	In vitro: Facilitates antioxidant recycling (Kagan *et al.*, 1992)	1992
	In vitro: Tocotrienol is better than tocopherol; tocotrienol is located closer to the cell membrane surface (Suzuki *et al.*, 1993)	1993
	Human: Dietary tocotrienols become incorporated into circulating human lipoproteins where they react with peroxyl radicals as efficiently as the corresponding tocopherol isomers (Suarna *et al.*, 1993)	1993
	Rat: Protects brain against oxidative damage (Kamat and Devasagayam, 1995)	1995
	Human: Controls the course of carotid atherosclerosis (Tomeo *et al.*, 1995)	1995
	Human: α-Tocotrienol is more potent than α-tocopherol in protecting against free radical-induced impairment of erythrocyte deformability (Begum and Terao, 2002)	2002

	Rat: Comparable effects of a tocotrienol-rich fraction and tocopherol in aspirin-induced lipid peroxidation mediated gastric lesions (Nafeeza *et al.*, 2002)	2002
	Rat: Antioxidant effects of γ-tocotrienol in spontaneously hypertensive rats (Newaz *et al.*, 2003)	2003
	Tocopherols and tocotrienols have comparable antioxidant properties. Some of the vitamin E formulations tested showed antioxidant activities superior to D-α-tocopherol (Naguib *et al.*, 2003)	2003
	The corresponding tocopherols and tocotrienols exert comparable antioxidant activity; tocotrienols are more readily transferred between the membranes and incorporated into the membranes than tocopherols (Yoshida *et al.*, 2003)	2003
	Human: Topical α-tocotrienol supplementation inhibits lipid peroxidation in human skin (Weber *et al.*, 2003)	2003
	Human: Lack of oxidative stress in a selenium-deficient area in Ivory Coast potential nutritional antioxidant role of crude palm oil (Tiahou *et al.*, 2004)	2004
	Rat: Palm oil tocotrienol mixture better than α-tocopherol acetate in protecting bones against free-radical induced elevation of bone-resorbing cytokines (Soelaiman *et al.*, 2004)	2004
	Mouse: Rice–trienol exerted a protective effect against oxidative damage in diabetes mellitus (Kanaya *et al.*, 2004)	2004
	Antioxidant property of tocols: $\alpha > \beta = \gamma > \delta$; not influenced by the nature of the isoprenoid tail (Sonnen *et al.*, 2005)	2005
	α- and α-tocopherols have comparable antioxidant efficacy (Yamasaki *et al.*, 2005)	2005
	Mouse: Both γ-tocopherol as well as γ-tocotrienol has antioxidant properties *in vivo* (Yoshida *et al.*, 2005)	2005
	Polyunsaturated isoprenoid side chain in tocotrienols has antioxidant properties (Yu *et al.*, 2005)	2006
	Individual tocotrienols display different antioxidant potencies: $\delta > \gamma > \alpha$ (Palozza *et al.*, 2006)	2006
	γ-Tocotrienol > α-tocotrienol > α-tocopherol as antioxidant. Tocotrienol regenerated oxidized carotenes demonstrating synergistic action (Schroeder *et al.*, 2006)	2006
Antiaging/ antioxidant	*Caenorhabditis elegans*: Tocotrienol, not tocopherol, administration reduced the accumulation of protein carbonyl and consequently extended the mean life span but not the maximum life span (Adachi and Ishii, 2000; Collins *et al.*, 2006)	2000

(*Continues*)

TABLE I. (*Continued*)

Anticancer	Mouse: Intraperitoneally injected tocotrienol prevented transplanted tumors (Komiyama *et al.*, 1989)	1989
	Rat: Tocotrienol-rich palm oil prevented chemically induced mammary tumorigenesis (Sundram *et al.*, 1989)	1989
	Rat: Tocotrienol, but not tocopherol, increased tumor latency in mammary tumor model (Gould *et al.*, 1991)	1991
	Rat: Tocotrienol chemopreventive in hepatic tumor model (Ngah *et al.*, 1991)	1991
	Rat: Tocotrienol chemopreventive in hepatic tumor model (Rahmat *et al.*, 1993)	1993
	Human: Suppresses activation of Epstein–Barr virus early antigen expression in PMA-activated lymphoblastoid Raji cells (Goh *et al.*, 1994)	1994
	Human: Tocotrienol, not tocopherol, suppresses growth of a human breast cancer cell line in culture (Nesaretnam *et al.*, 1995)	1995
	Human: Inhibited proliferation of estrogen receptor negative MDA-MB-435 and estrogen receptor positive MCF-7 breast cancer cells (Guthrie *et al.*, 1997)	1997
	Mouse: Isoprenoids suppress the growth of murine B16 melanomas *in vitro* and *in vivo* (He *et al.*, 1997)	
	Human: Inhibit the growth of human breast cancer cells irrespective of estrogen receptor status (Nesaretnam *et al.*, 1998)	1998
	Human: Apoptosis and cell cycle arrest in human and murine tumor cells are initiated by isoprenoids (Mo and Elson, 1999)	1999
	Human: Naturally occurring tocotrienols and RRR-δ-tocopherol are effective apoptotic inducers for human breast cancer cells (Yu *et al.*, 1999)	1999
	Human: Tocotrienols inhibit growth of ZR-75–1 breast cancer cells (Nesaretnam *et al.*, 2000)	2000
	Mouse: Highly potent γ- and δ-tocotrienol isoforms may play a physiological role in modulating normal mammary gland growth, function, and remodeling (McIntyre *et al.*, 2000b)	2000
	Mouse: Highly malignant breast cancer cells were the most sensitive, whereas the preneoplastic cells were the least sensitive to the antiproliferative and apoptotic effects of tocotrienols (McIntyre *et al.*, 2000a)	2000
	Mouse: Tocotrienols are significantly more potent than tocopherols in suppressing EGF-dependent normal mammary epithelial cell growth. The inhibitory effects of specific tocopherol and tocotrienol isoforms on EGF-dependent normal mammary epithelial cell mitogenesis occurs downstream from the EGF receptor and appears to be mediated, at least in part, by a reduction in PKCα activation (Sylvester *et al.*, 2001)	2001

Mouse: Antiproliferative effects of tocotrienols in preneoplastic mammary epithelial cells do not reflect a reduction in EGF-receptor mitogenic responsiveness, but rather, result from an inhibition in early postreceptor events involved in cAMP production upstream from EGF-dependent MAPK and phosphoinositide 3-kinase/Akt mitogenic signaling (Sylvester *et al.*, 2002)	2002
Rat: Suppression of 7,12-dimethylbenz[α]anthracene-induced carcinogenesis by tocotrienol-rich fraction isolated from rice bran oil (Iqbal *et al.*, 2003)	2003
Mouse: Tocotrienol-induced apoptosis in mammary cancer cells is mediated through activation of the caspase-8 signaling pathway and is independent of caspase-9 activation (Shah *et al.*, 2003)	2003
Mouse: Tocotrienol induces caspase-8 activation, unrelated to death receptor apoptotic signaling, in neoplastic mammary epithelial cells (Shah and Sylvester, 2004)	2004
Rat: Tocotrienol induces apoptosis in dRLh-84 hepatoma cells (Sakai *et al.*, 2004)	2004
Rat: Tocotrienol-rich fraction isolated from rice bran oil suppressed diethylnitrosamine and 2-acetylaminofluorene-induced hepatocarcinogenesis (Iqbal *et al.*, 2004)	2004
Human: Tocotrienol disrupts mitochondrial function and causes apoptosis of breast cancer cells (Takahashi and Loo, 2004)	2004
Human: Proapoptotic properties of δ-tocotrienol in breast cancer cells (Shun *et al.*, 2004)	2004
Human: Supplementation of tocotrienol-rich fraction of palm oil significantly and specifically affected MCF-7 cell response after tumor formation *in vivo* by an antioxidant-independent mechanism (Nesaretnam *et al.*, 2004)	2004
Human: Tocotrienol-rich fraction of palm oil activated p53, modulated Bax/Bcl-2 ratio, and induced apoptosis independent of cell cycle association in colorectal cancer RKO cells (Agarwal *et al.*, 2004)	2004
Mouse: Tocotrienol kills liver cancer cells (Har and Keong, 2005)	2005
Human: γ-Tocotrienol induces apoptosis of hepatoma Hep3B cells (Sakai *et al.*, 2005)	2005
Human: A redox-silent analogue of α-tocotrienol, 6-*O*-carboxypropyl-α-tocotrienol, possesses anticancer effects against lung adenocarcinoma showing poor prognosis based on the mutation of ras genes (Yano *et al.*, 2005)	2005
Mouse: γ-Tocotrienol is antineoplastic in mammary epithelial cells (Shah and Sylvester, 2005a,b; Sylvester and Shah, 2005a,b; Sylvester *et al.*, 2005)	2005
Mouse: Tocotrienols have anticancer properties *in vitro* and *in vivo* (Wada *et al.*, 2005)	2005

(*Continues*)

TABLE I. (*Continued*)

	Isoprenoid side chain of tocotrienol, not found in tocopherols, may prevent E2 epoxide induced breast cancer carcinogenesis at the initiation (Yu *et al.*, 2005)	2005
	Mouse: Preferential radiation sensitization of prostate cancer by γ-tocotrienol (Kumar *et al.*, 2006)	2006
	Tocotrienols targeted both pol lambda and angiogenesis as anticancer agents (Mizushina *et al.*, 2006)	2006
	Human: TRF of palm oil inhibited cellular proliferation and accelerated apoptosis (Srivastava and Gupta, 2006)	2006
	Human: The vitamin E succinate selenium-conjugated γ-tocotrienyl-2-phenylselenyl succinate decreased prostate cancer cell viability by stimulating caspase-3-dependent apoptosis (Vraka *et al.*, 2006)	2006
	Human: In contrast to tocopherols, tocotrienol potently inhibited telomerase activity in colorectal adenocarcinoma cells (Eitsuka *et al.*, 2006)	2006
Modulating normal mammary gland growth, function, and remodeling	Mouse: Mammary epithelial cells more easily or preferentially took up tocotrienols as compared to tocopherols (McIntyre *et al.*, 2000b)	2000
Antiangiogenic	Bovine: Tocotrienol, but not tocopherol, inhibited both the proliferation and tube formation of aortic endothelial cells (Inokuchi *et al.*, 2003)	2003
	Human/Chicken: Tocotrienol, not tocopherol, inhibited angiogenesis and telomerase activity (Nakagawa *et al.*, 2004)	2004
	Bovine: Tocotrienol, not tocopherol, limited angiogenic responses *in vitro* (Miyazawa *et al.*, 2004)	2004
	Bovine: Tocotrienols inhibited the proliferation of and formation of tubes by aortic endothelial cells, with δ-tocotrienol having the greatest effect. Tocotrienols targeted both pol lambda and angiogenesis as anticancer agents (Mizushina *et al.*, 2006).	2006
Antiproliferative and apoptotic	Mouse: Preneoplastic and neoplastic mammary epithelial cells α- and γ-tocopherols had no effect on cell proliferation (McIntyre *et al.*, 2000a)	2000
	Cancer cell lines: Not α-tocotrienol but γ-tocotrienol was apoptogenic, and more so when succinylated. Shortening the aliphatic side chain of γ-tocotrienol by one isoprenyl unit increased its activity (Birringer *et al.*, 2003)	2003

Hypocholesterolemic, antioxidant and antitumor	Chicken: The number and position of methyl substituents in tocotrienols affect their hypocholesterolemic, antioxidant, and antitumor properties; tocotrienol better than α-tocopherol (Qureshi *et al.*, 2000)	2000
Antiatherogenic	Mouse: Palm tocotrienols protect ApoE+/− mice from diet-induced atheroma formation (Black *et al.*, 2000)	2001
	Mouse: Tocotrienols inhibit atherosclerotic lesions in ApoE-deficient mice (Qureshi *et al.*, 2001b)	2001
	Rat: TRF supplementation decreased the lipid parameters in a dose-dependent manner in rats fed atherogenic diet (Minhajuddin *et al.*, 2005)	2005
	Human: Daily intake of dietary TRF by type 2 diabetics was beneficial against atherogenesis (Baliarsingh *et al.*, 2005)	2005
Serum lipoproteins; platelet function	Human: In men at risk for cardiovascular disease tocotrienol supplements used had no marked favorable effects (Mensink *et al.*, 1999)	1999
Anti-inflammatory	Human: Tocotrienols inhibit monocyte endothelial cell adhesion (Chao *et al.*, 2002)	2002
	Human: Tocotrienol is the most effective vitamin E for reducing endothelial expression of adhesion molecules and adhesion to monocytes (Theriault *et al.*, 2002)	2002
	Human: The efficacy of tocotrienol for reduction of VCAM-1 expression and adhesion of THP-1 cells to HUVECs was tenfold higher than that of tocopherol (Noguchi *et al.*, 2003)	2003
	Human: Compared to α-tocopherol, tocotrienols more potent displayed a more profound inhibitory effect on adhesion molecule expression and monocytic cell adherence (Naito *et al.*, 2005)	2005
Antifibrotic	Human: α-Tocotrienol, not tocopherol, restricted proliferation of human Tenon's capsule fibroblast (Meyenberg *et al.*, 2005)	2005
Hypolipidemic	Rat: Serum triglycerides lower in tocotrienol fed; higher IgM productivity of spleen lymphocytes and IgA, IgG, and higher IgM productivity mesenteric lymph node lymphocytes (Kaku *et al.*, 1999)	1999
	Human: Daily intake of dietary TRF by type 2 diabetics was beneficial against hyperlipidemia (Baliarsingh *et al.*, 2005)	2005
Immune function	Rats: Feeding affects proliferation and function of spleen and mesenetric lymph node lymphocytes (Gu *et al.*, 1999)	1999
Lymphatic transport	Rat: Preferential absorption of α-tocotrienol compared to γ- and δ-tocotrienols and α-tocopherol (Ikeda *et al.*, 1996)	1996

(Continues)

TABLE I. (*Continued*)

Drug metabolism	Tocotrienols inhibit human glutathione *S*-transferase P1–1 (van Haaften *et al.*, 2002)	2002
	Human: Vitamin E is able to activate gene expression via the pregnane X receptor (PXR), a nuclear receptor regulating a variety of drug-metabolizing enzymes. Tocotrienols more potent than tocopherols (Landes *et al.*, 2003)	2003
	Human: Tocotrienols, not tocopherols, activate the steroid and xenobiotic receptor (SXR) and selectively regulate expression of its target genes (Zhou *et al.*, 2004)	2004
	Mouse: Tocopherol, but not tocotrienol, may induce CYP3A11 and interfere with drug metabolism (Kluth *et al.*, 2005)	2005
Eye	Rat: Preferential uptake of topically applied tocotrienol, over tocopherol, by ocular tissues (Tanito *et al.*, 2004)	2004
Bone	Rat: Tocotrienols are needed for normal bone calcification in growing female rats (Norazlina *et al.*, 2002)	2002
	Rat: Tocotrienol offers better protection than tocopherol from free radical-induced damage of bone (Ahmad *et al.*, 2005)	2005
Obesity and osteoporosis	Rat: Tocotrienol, not tocopherol, has the potential to be utilized as a prophylactic agent in preventing side effects of long-term glucocorticoid use (Ima-Nirwana and Suhaniza, 2004)	2004
Diabetes	Rat: Tocotrienols-rich diet decreased advanced glycosylation end products in nondiabetic rats and improved glycemic control in streptozotocin-induced diabetic rats (Wan Nazaimoon and Khalid, 2002)	2002
Gastric lesion	Rat: Tocopherol, not alone, but in combination with tocotrienol and ubiquinone decreased gastric lesion (Nafeeza and Kang, 2005)	2005
	Rat: Tocotrienol, not tocopherol, prevents stress-induced adverse changes in the gastric acidity and gastrin level (Azlina *et al.*, 2005)	2005
Natriuretic function	Rat: An oral administration of γ-tocotrienol increases plasma concentration of 2,7,8-trimethyl-2-(β-carboxyethyl)-6-hydroxy chroman (LLU-α, γ-CEHC), a natriuretic compound (Hattori *et al.*, 2000)	2000
	Rat: γ-Tocotrienol is a natriuretic hormone precursor (Saito *et al.*, 2003)	2003
Bioavailability	Mouse: Supplemented tocotrienol not detected in the brain (Podda *et al.*, 1996). See 2002* below	1996
	Human: Following supplementation, ~1μM tocotrienol detected in human plasma (O'Byrne *et al.*, 2000)	2000
	Rat: The skin is a unique tissue in respect to its ability to discriminate between various vitamin E analogs; it preferentially uptakes dietary tocotrienols (Ikeda *et al.*, 2000)	2000

Human: Increased absorption of the tocotrienols in the fed versus fasted state; ~1 μM tocotrienol detected in human plasma (Yap *et al.*, 2001)	2001
Human: Tocotrienols, like tocopherols, are metabolized to CEHC; however, the quantities excreted in human urine are small in relation to dose size (Lodge *et al.*, 2001)	2001
Rat: Dietary sesame seeds elevate the tissue concentrations of orally taken tocopherols and tocotrienols (Ikeda *et al.*, 2001)	2001
Rat: Oral tocotrienol crosses the blood–brain barrier to reach brain tissue; more so for fetal brain while pregnant mother is supplemented with tocotrienol (Roy *et al.*, 2002)	2002*
Human: In HepG2 cells, tocotrienols are metabolized essentially like tocopherols, that is, by ω-oxidation followed by β-oxidation of the side chain. Quantitatively, tocotrienols are degraded to a larger extent than tocopherols (Birringer *et al.*, 2002)	2002
Rat: Sesame lignans added to diet increased plasma and tissue concentrations of supplemented tocotrienols (Yamashita *et al.*, 2002)	2002
Rat: In epididymal adipose, renal adipose, subcutaneous adipose, and brown adipose tissues and in the heart, the tocotrienol levels were maintained or increased for 24 h after intragastric administration. In the serum, liver, mesenteric lymph node, spleen, and lungs, the tocotrienol levels were highest 8 h after the administration (Okabe *et al.*, 2002)	2002
Human: Novel formulation of tocotrienol developed to improve bioavailability in humans (Ho *et al.*, 2003)	2003
Rat: Dietary α-tocopherol decreases α-tocotrienol but not γ-tocotrienol concentration in rats (Ikeda *et al.*, 2003)	2003
Tocotrienols are more readily transferred between the membranes and incorporated into the membranes than tocopherols (Yoshida *et al.*, 2003)	2003
Human: α-Tocotrienol accumulate in endothelial cells to levels approximately tenfold greater than that of α-tocopherol (Noguchi *et al.*, 2003)	2003
Rat: Of the three tocotrienols, α-tocotrienol had the highest oral bioavailability, at about $27.7 \pm 9.2\%$, compared with γ- and δ-tocotrienols, which had values of $9.1 \pm 2.4\%$ and $8.5 \pm 3.5\%$, respectively. Tocotrienols were found to be negligibly absorbed when administered intraperitoneally and intramuscularly (Yap *et al.*, 2003)	2003
Human: The $t_{1/2}$ of tocotrienols is short, ranging from 3.8 to 4.4 h for γ- and α-tocotrienols (Schwedhelm *et al.*, 2003)	2003

(*Continues*)

TABLE I. (*Continued*)

	Findings	Year
	Human: Following the intervention with palm vitamin E, tocotrienols are detected in total blood plasma, TRP, LDL and HDL. Tocotrienols appeared in the blood stream at 2 h interval and disappeared within 24 h. Tocotrienols concentration in total plasma plasma, TRP and LDL peaked between 4 and 6 h; in HDL, tocotrienol concentrations peaked at 8 h after supplementation. α-Tocopherol was the major vitamin E detected in plasma. Tocotrienols have a very short duration of absorption and distribution in circulating blood (Fairus *et al.*, 2004).	2004
	Rat: Following topical application of small amounts, the concentration of α-tocotrienol increased markedly in ocular tissues (e.g., crystalline lens, neural retina, and eye cup); however, no significant increase was observed in the case of α-tocopherol (Tanito *et al.*, 2004)	2004
	Human: Tocotrienol uptake by aortic endothelial cells ~25- to 95-fold greater than that of α-tocopherol (Naito *et al.*, 2005)	2005
	Rat: Orally taken tocotrienol reaches all vital organs *in vivo* (Khanna *et al.*, 2005a)	2005
	Mouse: Orally fed tocotrienol can be delivered to vital organs *in vivo* even in TTP-deficient mice (Khanna *et al.*, 2005a). There are mechanisms other than TTP to transport tocotrienol to tissues	2005
	Chicken: Estimated that the safe dose of various tocotrienols for human consumption might be 200–1000 mg/day (Yu *et al.*, 2006)	2006
	Human: Single dose of tocotrienol supplementation results in 3-μM peak plasma concentration; 1.7 μM in LDL, 0.9 μM in triglyceride-rich lipoprotein, and 0.5 μM in HDL. The peak plasma level corresponds to 12- to 30-fold more than the concentration of α-tocotrienol required to completely prevent neurodegeneration. Tocotrienols were detected in the blood plasma and all lipoprotein subfractions studied postprandial (Khosla *et al.*, 2006)	2006

[a] CEHC, carboxyethyl-hydroxychromans; EGF, epidermal growth factor; HDL, high-density lipoprotein; HMG-CoA reductase, 3-hydroxy-3-methylglutaryl-coenzyme A reductase; HUVEC, human umbilical vein (derived) endothelial cells; IKBKAP, gene-encoding IκB kinase complex-associated protein; LDL, low-density lipoprotein; SHR, spontaneously hypertensive rats; TRF, tocotrienol-rich fraction; TRP, triglyceride-rich particles, TTP, tocopherol transfer protein.

A. Neuroprotection

Glutamate toxicity is a major contributor to neurodegeneration. It includes excitotoxicity and an oxidative stress component also known as oxytosis (Schubert and Piasecki, 2001; Tan *et al.*, 2001). Murine HT hippocampal neuronal cells, lacking intrinsic excitotoxicity pathway, have been used as a standard model to characterize the oxidant-dependent component of glutamate toxicity. In 1999, we conducted a side by side comparison of all eight forms of natural vitamin E in a model of glutamate-induced neurodegeneration of HT neural cells. In subsequent experiments, it was observed that the neuroprotective property of tocotrienol applies not only to neural cell lines but also to primary cortical neurons. This line of experimentation led to an observation that eventually turned out to be the most potent function of any natural form of vitamin E on a concentration basis reported. Until then, all biological functions of vitamin E studied *in vitro* were observed at micromolar concentration. Our studies led to the first evidence that α-tocotrienol was the most potent neuroprotective form of vitamin E in glutamate-induced degeneration of HT4 hippocampal neurons (Sen *et al.*, 2000). What was striking in this study was the observation that nanomolar concentrations of α-tocotrienol, not α-tocopherol, provide complete neuroprotection. At such low dose, tocotrienol was not protective against direct oxidant insult, suggesting that the observed neuroprotective effects of nanomolar tocotrienol was not dependent on the widely known antioxidant property of vitamin E. That tocotrienol-dependent neuroprotection includes a significant antioxidant-independent mechanism has been now established (Khanna *et al.*, 2006). The neuroprotective property of tocotrienol holds good not only in response to glutamate challenge but also in response to other insults such as homocysteic acid, glutathione deficiency, and linoleic acid-induced oxidative stress (Khanna *et al.*, 2006; Sen *et al.*, 2000). It is now evident that at micromolar concentrations tocotrienol protects neural cells by virtue of its antioxidant property. At nanomolar concentrations, however, tocotrienol regulates specific neurodegenerative signaling processes.

The major tocotrienol-sensitive signaling pathways which are known to be involved in glutamate-induced neurodegeneration include c-Src and 12-lipoxygenase (12-Lox) (Khanna *et al.*, 2003, 2005b, 2006; Sen *et al.*, 2000, 2004). In our initial search for signaling pathways that are sensitive to tocotrienol and play a decisive role in neurodegeneration we were led to c-Src kinase (Sen *et al.*, 2000). c-Src and the structurally related members of the Src family are nonreceptor tyrosine kinases that reside within the cell associated with cell membranes and appear to transduce signals from transmembrane receptors to the cell interior. SH2 and SH3 domains are known to play a central role in regulating the catalytic activity of Src protein tyrosine kinase. High-resolution crystal structures of human Src, in their repressed state, have provided a structural explanation for how intramolecular interactions of the SH3 and

SH2 domains stabilize the inactive conformation of Src (Thomas and Brugge, 1997).

Our hypothesis that tocotrienol prevents neurodegeneration by regulating specific signaling processes involved in neurotoxicity led to screening for potential tocotrienol-sensitive candidate death pathways in HT4 cells. During such screening studies, inhibitors of the protein tyrosine kinase activity completely prevented glutamate-induced cell death. Herbimycin and geldanamycin potently inhibit c-Src tyrosine kinase activity (Hall *et al.*, 1994; Yoneda *et al.*, 1993), whereas lavendustin A is an inhibitor of extracellular growth factor receptor protein tyrosine kinase activity (Hsu *et al.*, 1991). The observation that herbimycin and geldanamycin, but not lavendustin A prevented glutamate-induced death of HT4 neuronal cells hinted the involvement of c-Src kinase activity in the death pathway. Immunoprecipitation of tyrosine phosphorylated protein from cellular extracts confirmed that protein tyrosine phosphorylation reactions were indeed triggered by exposure of cells to elevated levels of glutamate and that such reactions were inhibited by nanomolar concentrations of α-tocotrienol (Sen *et al.*, 2000). These results, however, did not provide any information regarding the specific kinases involved. The involvement of c-Src kinase activity in the death pathway was verified by experiments involving the overexpression of catalytically active or inactive Src kinase. Indeed, overexpression of catalytically active Src kinase markedly sensitized the cells to HT4-induced death. Tocotrienol treatment completely prevented glutamate-induced death even in active c-Src kinase overexpressing cells, indicating that it either inhibited c-Src kinase activity or regulated one or more events upstream of c-Src kinase activation. Further evidence supporting this contention was provided by results obtained from the determination of c-Src kinase activity in HT4 cells. Glutamate treatment resulted in marked enhancement of c-Src kinase activity, and this change was completely blocked in cells treated with nanomolar amounts of α-tocotrienol. Further evidence establishing that signal transduction processes related to the cell death pathway are involved in glutamate-induced cytotoxicity was obtained from the study of ERK1 and ERK2 activation. Mitogen-activated/extracellular response kinase kinase (MEK) kinase (MEKK) is a serine–threonine kinase that regulates sequential protein phosphorylation pathways, leading to the activation of mitogen-activated protein kinases (MAPKs), including members of the extracellular signal-regulated kinases (ERKs). MEKK selectively regulates signal transduction pathways that contribute to the apoptotic response (Johnson *et al.*, 1996). When activated, p44 and p42 MAPKs (ERK1 and ERK2) are phosphorylated at specific threonine and tyrosine residues. ERK has been implicated in mediating the signaling events that precede apoptosis. ERK2 plays an active role in mediating anti-IgM-induced apoptosis of B cells (Lee and Koretzky, 1998). It has also been shown that H_2O_2 induces the activation of multiple MAPKs in oligodendrocyte progenitors and that the activation of ERK is associated with oxidant-mediated cytotoxicity (Bhat and Zhang, 1999). Our studies showed that ERK1 and ERK2 are sensitive to elevated levels of extracellular glutamate. Rapid activation of ERK, particularly ERK2, was observed in response to glutamate treatment. Such response of ERK was completely inhibited in cells treated with α-tocotrienol, suggesting that α-tocotrienol influences an early event in the glutamate-induced death pathway (Sen *et al.*, 2000). In some cases, Src kinase activity is known to be required for the activation of ERK (Aikawa *et al.*, 1997). Thus, it is likely that tocotrienol inhibits inducible ERK activation by downregulating Src kinase activity (Sen *et al.*, 2000).

c-Src is heavily expressed in the brain (Soriano *et al.*, 1991) and in human neural tissues (Pyper and Bolen, 1989). Differentiating rodent neurons are known to express high levels of c-Src. In neurons and astrocytes, c-Src is present at 15–20 times higher levels than that found in fibroblasts. The specific activity of the c-Src protein from neuronal cultures is 6–12 times higher than that from the astrocyte cultures, suggesting a key function of this protein in neurons (Brugge *et al.*, 1985). Initially, c-Src was identified as being important in growth cone-mediated neurite extension and synaptic plasticity (Maness *et al.*, 1988) and in neuronal differentiation (Ingraham *et al.*, 1989). Targeted disruption of c-Src, however, did not cause any abnormality in the brain (Soriano *et al.*, 1991). Our pursuit for the neuroprotective mechanisms of tocotrienols led to the first evidence demonstrating that rapid c-Src activation (Khanna *et al.*, 2002; Sen *et al.*, 2000) plays a central role in executing neurodegeneration. Consistently, it was demonstrated in a subsequent report that Src deficiency or blockade of Src activity in mice provides cerebral protection following stroke (Paul *et al.*, 2001). Further support of our claim that c-Src is a key player in neurodegeneration is provided by observation that the Src family kinase inhibitor PP2 reduces focal ischemic brain injury (Lennmyr *et al.*, 2004). Our observation that tocotrienol-dependent inhibition of c-Src is beneficial for neuroprotection has now been extended to the heart. A recent study showed that c-Src mediates postischemic cardiac injury and dysfunction. Tocotrienol supplementation inhibited c-Src activation and protected the heart (Das *et al.*, 2005). Many intracellular pathways can be stimulated on Src activation, and a variety of cellular consequences can result. High c-Src is tightly associated with carcinogenesis. c-Src inhibitors are being actively studied for cancer therapy (Alper and Bowden, 2005; Ishizawar and Parsons, 2004; Lau, 2005; Shupnik, 2004). On the basis of the inducible c-Src inhibitory properties of tocotrienol, one may postulate that tocotrienol has anticancer properties. The anticancer properties of tocotrienol have been discussed in a separate section below.

GSH is the major cellular thiol present in mammalian cells and is critical for maintenance of redox homeostasis (Sun *et al.*, 2006). GSH is a key survival factor in cells of the nervous system and lowered [GSH]i is one of the early markers of neurotoxicity induced by a variety of agonists

(Bains and Shaw, 1997; Dringen *et al.*, 2000). We observed that α-tocotrienol clearly protects primary cortical neurons against a number of GSH-lowering neurotoxins (Khanna *et al.*, 2003). Of interest, the neurons survived even in the face of GSH loss. These observations led to the hypothesis that loss of [GSH]i alone is not lethal (Khanna *et al.*, 2003). Given that pro-GSH agents are known to be neuroprotective in a variety of scenarios (Bains and Shaw, 1997; Han *et al.*, 1997; Schulz *et al.*, 2000), it becomes reasonable to hypothesize that glutamate-induced lowering of [GSH]i triggers downstream responses that execute cell death. Our works led to the identification of 12-Lox as a key tocotrienol-sensitive mediator of neurodegeneration (Khanna *et al.*, 2003). Specific inhibition of 12-Lox by BL15 protected neurons from glutamate-induced degeneration, although [GSH]i is compromised by 80%. Similar protective effects of BL15 were noted when BSO, a specific inhibitor of GSH synthesis, was used as the agonist. Importantly, neurons isolated from mice lacking the 12-Lox gene were observed to be resistant to glutamate-induced loss of viability (Khanna *et al.*, 2003). This key piece of evidence established that indeed 12-Lox represents a critical checkpoint in glutamate-induced neurodegeneration.

Understanding of the intracellular regulation of 12-Lox requires knowledge of the distribution of both enzyme protein and its activity. For example, in human erythroleukemia cells, the membrane fraction contains about 90% of the total cellular 12-Lox activity, whereas only 10% of 12-Lox activity resides in the cytosol. However, the majority of cellular 12-Lox protein is found in the cytosol (Hagmann *et al.*, 1993). On activation, 12-Lox may translocate to the membrane (Hagmann *et al.*, 1993). Consistently, we have observed the decreased presence of 12-Lox in the cytosol and increased presence in the membrane of glutamate-treated cells. For 5-Lox, both catalytic function and translocation of the enzyme from the cytosol to the membrane are known to be regulated by tyrosine kinases (Lepley *et al.*, 1996). Recently, we have noted that 12-Lox is subject to rapid tyrosine phosphorylation in neuronal cells challenged with glutamate or GSH-lowering agents. Such rapid phosphorylation coincides with the timeline of c-Src activation (Khanna *et al.*, 2005b). Inhibitors of c-Src abrogated inducible 12-Lox tyrosine phosphorylation, supporting the notion that c-Src may directly phosphorylate 12-Lox in challenged neurons. To test this hypothesis, we utilized genetic approaches of overexpressing kinase-active, kinase-dead, or dominant negative c-Src in neuronal cells. Findings from cell biology studies as well as from the study of c-Src and 12-Lox in cell-free systems indicate that in response to challenge by glutamate or GSH-lowering agents, c-Src is rapidly activated and phosphorylates 12-Lox (Khanna *et al.*, 2005b).

Neurons and the brain are rich in arachidonic acid (AA; 20:4ω-6). Massive amounts of AA are released from the membranes in response to brain ischemia or trauma (Bazan, 1970, 1971a,b, 1976; Bazan and Rakowski, 1970). Subsequent work has established that AA and its metabolites may be neurotoxic. There are three major pathways of AA metabolism: Loxs, cycloxygenases, and cytochrome P450. The cycloxygenase pathway has been preliminarily ruled out from being a contributor to neurodegeneration (Kwon *et al.*, 2005). In the Lox pathway, metabolites of 12-Lox seem to be the major metabolite of arachidonic acid in the brain (Adesuyi *et al.*, 1985; Carlen *et al.*, 1994) as well as in cultured cortical neurons (Ishizaki and Murota, 1991; Miyamoto *et al.*, 1987a,b). Lipoxygenases, mainly 5-, 12-, and 15-Lox, are named for their ability to insert molecular oxygen at the 5-, 12-, or 15-carbon atom of arachidonic acid forming a distinct hydroperoxy-eicosatetraenoic (HPETE) acid (Yamamoto, 1992). 12-Lox produces 12(*S*)-HPETE which is further metabolized into four distinct products: an alcohol [12(*S*)-HETE], a ketone (12-keto-eicosatetraenoic acid), or two epoxy alcohols (hepoxilins A3 and B3). Immunohistochemical studies revealed the occurrence of 12-Lox in neurons; particularly in hippocampus, striatum, olivary nucleus, as well as in glial and in cerebral endothelial cells (Nishiyama *et al.*, 1992, 1993). Using immature cortical neurons and HT cells, it has been shown that a decrease in [GSH]i triggers the activation of neuronal 12-Lox, which leads to the production of peroxides, the influx of Ca^{2+}, and ultimately to cell death (Li *et al.*, 1997; Tan *et al.*, 2001). The 12-Lox metabolite 12-HPETE proved to be capable of causing cell death (Gu *et al.*, 2001). Inhibition of 12-Lox protected cortical neurons from β-amyloid-induced toxicity (Lebeau *et al.*, 2001). Intracellular calcium chelation delayed cell death by Lox-mediated free radicals in mouse cortical cultures (Wie *et al.*, 2001). In sum, 12-Lox poses clear threat to neuronal survival especially under GSH-deficient conditions.

Lipoxygenase activity is sensitive to vitamin E. α-Tocopherol strongly inhibits purified 5-Lox with an IC50 of 5 μM. The inhibition is independent of the antioxidant property of tocopherol. Tryptic digestion and peptide mapping of the 5-Lox–tocopherol complex indicated that tocopherol binds strongly to a single peptide (Reddanna *et al.*, 1985). Another study reported inhibition of 15-Lox by tocopherol *via* specific interaction with the enzyme protein (Grossman and Waksman, 1984). Of interest, inhibitors specific for cycloxygenase or 5-Lox are not effective in protecting neuronal cells against glutamate-induced death, suggesting a specific role of 12-Lox in glutamate-induced death (Khanna *et al.*, 2003, 2005b). Our studies addressing the effect of α-tocotrienol on pure 12-Lox indicate that α-tocotrienol directly interacts with the enzyme to suppress arachidonic acid metabolism. *In silico* studies, examining possible docking sites of α-tocotrienol to 12-Lox supports the presence of a α-tocotrienol-binding solvent cavity close to the active site. Previously, it has been demonstrated in 15-Lox that COOH terminal of arachidonic acid enters this solvent cavity while accessing the catalytic site (Borngraber *et al.*, 1999). It is therefore plausible that the binding position of α-tocotrienol prevents access of the natural substrate arachidonic acid to the active site of 12-Lox (Khanna *et al.*, 2003). Does 12-Lox have a tangible impact on neurodegenerative processes *in vivo*? In 1992, it was reported that a mixed Lox/cyclooxygenase inhibitor SK&F 105809 reduced cerebral edema after closed head injury in rat (Shohami *et al.*, 1992). We noted that 12-Lox, but not 5-Lox (Kitagawa *et al.*, 2004), deficient mice were significantly protected against stroke-related injury of the brain (Khanna *et al.*, 2005b). The case for 12-Lox as an important mediator of neurodegeneration *in vivo* is gaining additional support from independent studies (Musiek *et al.*, 2006). 12-Lox has been also implicated in the pathogenesis of Alzheimer's (Yao *et al.*, 2005). α-Tocotrienol is capable of resisting neurodegeneration *in vivo* by opposing the c-Src and 12-Lox pathways.

B. ANTICANCER

Pure and mixed isoprenoids have potent anticancer activity (Mo and Elson, 1999). As discussed earlier in this work, tocotrienols are isoprenoids but tocopherols are not. Unlike in the case of neuroprotection where α-tocotrienol has emerged to be the most potent isoform (Khanna *et al.*, 2005b, 2006; Sen *et al.*, 2004, 2006), there seems to somewhat of a consensus that γ- and δ-tocotrienols are the most potent anticancer isoform of all natural existing tocotrienols. One of the first studies addressing the role of tocotrienols in neoplastic disorders was reported in 1989 (Komiyama *et al.*, 1989). The effects of intraperitoneally injected α- and γ-tocotrienols, as well as that of α-tocopherol, have been examined. Both tocotrienols were effective against sarcoma 180, Ehrlich carcinoma, and invasive mammary carcinoma. γ-Tocotrienol showed a slight life-prolonging effect in mice with Meth A fibrosarcoma, but the tocotrienols had no antitumor activity against P388 leukemia at doses of 5–40 mg/kg/day (Komiyama *et al.*, 1989). In contrast to tocotrienols, α-tocopherol was not as effective. The antitumor activity of γ-tocotrienol was higher than that of α-tocotrienol. In contrast to α-tocopherol, tocotrienols showed growth inhibition of human and mouse tumor cells when the cells were exposed to these agents for 72 h *in vitro* (Komiyama *et al.*, 1989). In an independent study published in the same year, the anticarcinogenic properties of palm oil, a rich source of tocotrienols, was reported (Sundram *et al.*, 1989). In this study, young female Sprague-Dawley rats were treated with a single dose of 5 mg of 7,12-dimethylbenz[α]anthracene (DMBA) intragastrically. Three days after carcinogen treatment, the rats were put on semisynthetic diets containing 20% by weight of corn oil, soybean oil, crude palm oil (CPO), refined, bleached, deodorized palm oil (RBD PO) and metabisulfite-treated palm oil (MCPO) for 5 months. During the course of experiments, rats fed on different dietary fats had similar rate of growth. Rats fed 20% corn oil or soybean oil diet had marginally higher tumor incidence than rats fed on palm oil diets. At autopsy, rats fed on high corn oil or soybean oil diets had significantly more tumors than rats fed on the three palm oil diets. Palm oil is different from corn oil and soybean oil in many ways. In addition to possessing higher levels of tocotrienol, palm oil has a contrasting fatty acid profile and also much higher levels of tocopherol and carotenes. As such, the favorable anticarcinogenic effects noted in this study cannot be directly associated with tocotrienols (Sundram *et al.*, 1989). The antioxidant or redox property of tocotrienol is not responsible for its anticancer property. Results in support of this hypothesis show that a redox-silent analogue of α-tocotrienol, 6-*O*-carboxypropyl-α-tocotrienol is cytotoxic against A549 cells, a human lung adenocarcinoma cell line (Yano *et al.*, 2005). Although the phenolic antioxidant group in tocotrienol may not be implicated in its anticancer property, it is apparent that the phytyl side chain has some antioxidant property which prevents against carcinogenesis (Yu *et al.*, 2005).

1. Breast Cancer

Among the various forms of cancer, breast cancer has been most extensively studied in cell culture and rodent *in vivo* models for the efficacy of tocotrienols. Tocopherol and tocotrienol have been tested side-by-side for chemopreventive activity in a chemically induced rat mammary tumor model. When mammary tumors were induced by DMBA, only the tocotrienol group showed enhanced tumor latency (Gould *et al.*, 1991). The tocotrienol-rich fraction (TRF) of palm oil is not only rich in tocotrienols but also contains some α-tocopherol. The effects of TRF and α-tocopherol on the proliferation, growth, and plating efficiency of the MDA-MB-435 estrogen receptor negative human breast cancer cells have been examined (Nesaretnam *et al.*, 1995). TRF inhibited the proliferation of these cells with a concentration required to inhibit cell proliferation by 50% of 180 μg/ml, whereas α-tocopherol had no effect at concentrations up to 1000 μg/ml. The effects of TRF and α-tocopherol were also tested in longer-term experiments, using concentrations of 180 and 500 μg/ml. TRF, but not α-tocopherol, inhibited the growth as well as plating efficiency of the cells. These findings point toward the hypothesis that α-tocopherol contained in the TRF does not account for its beneficial effects and that tocotrienols may have been the active principle responsible for the observed effects of TRF (Nesaretnam *et al.*, 1995). It is now known that TRF, α-, γ- and δ-tocotrienols inhibited proliferation of estrogen receptor negative MDA-MB-435 human breast cancer cells with 50% inhibitory concentrations (IC50) of 180, 90, 30, and 90 μg/ml, respectively, whereas α-tocopherol is not effective at concentrations up to 500 μg/ml. Tocotrienols inhibit the proliferation of estrogen receptor positive MCF-7 cells. The IC50s for TRF, α-, γ-, and δ-tocotrienols have been estimated to be 4, 6, 2, and 2 μg/ml, respectively. In sharp contrast, the efficiency of α-tocopherol under comparable conditions is 20–50 times lower with an IC50 of 125 μg/ml (Guthrie *et al.*, 1997). Tamoxifen, a widely used synthetic anti-estrogen, inhibits the growth of MCF-7 cells with an IC50 of 0.04 μg/ml. In the MCF-7 cells, only 1:1 combinations of γ- or δ-tocotrienol with tamoxifen showed a synergistic inhibitory effect on the proliferative rate and growth of the cells. α-Tocopherol did not exhibit this beneficial synergistic effect with tamoxifen (Guthrie *et al.*, 1997). The inhibition by tocotrienols was not overcome by addition of excess estradiol to the culture medium, suggesting that tocotrienols are effective inhibitors of both estrogen receptor negative and positive cells and that combinations with tamoxifen may be useful for breast cancer therapy (Guthrie *et al.*, 1997). Studies to come would strengthen support for the case that tocotrienols are effective against breast cancer *in vitro*. TRF inhibits growth of MCF-7 cells in both the presence and absence of estradiol such that complete suppression of growth is achieved at 8 μg/ml. MDA-MB-231 cells are also inhibited by TRF such that 20-μg/ml TRF is needed for complete growth suppression. The study of the individual component tocotrienols in TRF revealed that all fractions inhibit growth of both estrogen-responsive as well as estrogen-nonresponsive cells and of estrogen-responsive cells in both the presence and absence of estradiol. This estradiol-independent effect of tocotrienols is of clinical interest (Nesaretnam *et al.*, 1998, 2000). γ- and δ-Tocotrienol fractions were most potent inhibitors of breast cancer cell growth. Complete inhibition of MCF-7 cell growth was achieved at 6 μg/ml of γ/δ-tocotrienol in the absence of estradiol and 10 μg/ml of δ-tocotrienol in the presence of estradiol. In contrast, complete suppression of MDA-MB-231 cell growth was not achieved even at concentrations of 10 μg/ml of δ-tocotrienol. Of note, unlike tocotrienols α-tocopherol does not inhibit MCF-7, MDA-MB-231, or ZR-75–1 cell growth in either the presence or the absence of estradiol (Mo and Elson, 1999; Nesaretnam *et al.*, 1998, 2000). Studies examining the mechanisms by which tocotrienols check the growth of breast cancer cells have identified that tocotrienols do not act via an estrogen receptor-mediated pathway and must therefore act differently from estrogen antagonists. Furthermore, tocotrienols did not increase levels of growth inhibitory insulin-like growth factor-binding proteins in MCF-7 cells, implying also a different mechanism from that proposed for retinoic acid inhibition of estrogen-responsive breast cancer cell growth (Nesaretnam *et al.*, 1998).

Unlike α-tocopherol, δ-tocopherol seems to be more promising albeit much less so than the tocotrienols. The apoptosis-inducing properties of RRR-α-, β-, γ-, and δ-tocopherols and α-, γ-, and δ-tocotrienols have been compared in estrogen-responsive MCF-7 and estrogen-nonresponsive MDA-MB-435 human breast cancer cell lines. Vitamin E succinate, a known inducer of apoptosis in several cell lines, including human breast cancer cells, served as a positive control. The estrogen-responsive MCF-7 cells were found to be more susceptible than the estrogen-nonresponsive MDA-MB-435 cells, with concentrations for half-maximal response for tocotrienols (α, γ, and δ) and RRR-δ-tocopherol of 14, 15, 7, and 97 μg/ml, respectively. The tocotrienols (α, γ, and δ) and RRR-δ-tocopherol induced MDA-MB-435 cells to undergo apoptosis, with concentrations for half-maximal response of 176, 28, 13, and

145 μg/ml, respectively. With the exception of RRR-δ-tocopherol, the tocopherols (α, β, and γ) and the acetate derivative of RRR-α-tocopherol (RRR-α-tocopheryl acetate) were ineffective in induction of apoptosis in both cell lines when tested within the range of their solubility, that is 10–200 μg/ml (Yu *et al.*, 1999).

Mammary tissue homeostasis depends on dynamic interactions between the epithelial cells, their microenvironment (including the basement membrane and the stroma), and the tissue architecture, which influence each other reciprocally to regulate growth, death, and differentiation in the gland. The study of normal mammary epithelial cells isolated from midpregnant mice grown in collagen gels and maintained on serum-free media showed that treatment with 0- to 120-μM α- or γ-tocopherol had no effect, whereas 12.5- to 100-μM TRF, 100- to 120-μM δ-tocopherol, 50- to 60-μM α-tocotrienol, and 8- to 14-μM γ- or δ-tocotrienol significantly inhibited cell growth in a dose-responsive manner. In acute studies, 24-h exposure to 0- to 250-μM α-, γ-, and δ-tocopherols had no effect, whereas similar treatment with 100- to 250-μM TRF, 140- to 250-μM α-tocotrienol, 25- to 100-μM γ- or δ-tocotrienol significantly reduced cell viability. The observed growth inhibitory doses of TRF, δ-tocopherol, and α-, γ-, and δ-tocotrienols induced apoptosis in these cells. Mammary epithelial cells preferentially took up tocotrienols as compared to tocopherols, suggesting that at least part of the reason tocotrienols display greater potency than tocopherols is because of greater cellular uptake. These observations suggest that the highly biopotent γ- and δ-tocotrienol isoforms may play a physiological role in modulating normal mammary gland growth, function, and remodeling (McIntyre *et al.*, 2000b). A later study identified that highly malignant cells are specifically more sensitive, whereas the preneoplastic cells are least sensitive to the antiproliferative and apoptotic effects of tocotrienols (McIntyre *et al.*, 2000a). The comparative effects of tocopherols and tocotrienols were examined using preneoplastic (CL-S1), neoplastic (−SA), and highly malignant (+SA) mouse mammary epithelial cells. Over a 5-day culture period, treatment with 0- to 120-μM α- and γ-tocopherols had no effect on cell proliferation, whereas cell growth was inhibited 50% (IC50) as compared with controls by treatment with the following: 13-, 7-, and 6-μM tocotrienol-rich fraction of palm oil (TRF); 55-, 47-, and 23-μM δ-tocopherol; 12-, 7-, and 5-μM α-tocotrienol; 8-, 5-, and 4-μM γ-tocotrienol; or 7-, 4-, and 3-μM δ-tocotrienol in CL-S1, −SA, and +SA cells, respectively. Acute 24-h exposure to 0- to 250-μM α- or γ-tocopherol (CL-S1, −SA, and +SA) or 0- to 250-μM δ-tocopherol (CL-S1) had no effect on cell viability, whereas cell viability was reduced 50% (LD50) as compared with controls by treatment with 166- or 125-μM δ-tocopherol in −SA and +SA cells, respectively. Additional LD50 doses were determined as the following: 50-, 43-, and 38-μM TRF; 27-, 28-, and 23-μM α-tocotrienol; 19-, 17-, and 14-μM γ-tocotrienol; or 16-, 15-, or 12-μM δ-tocotrienol in CL-S1, −SA, and +SA cells, respectively. Treatment-induced cell death resulted from activation of apoptosis. Consistent with previous observations, CL-S1, −SA, and +SA cells preferentially accumulated tocotrienols as compared with tocopherols. Highly malignant +SA cells were the most sensitive, whereas the preneoplastic CL-S1 cells were the least sensitive to the antiproliferative and apoptotic effects of tocotrienols (McIntyre *et al.*, 2000a).

How do tocotrienols induce apoptosis in breast cancer cells? δ-Tocotrienol induces TGF-β receptor II expression and activates TGF-β-, Fas-, and JNK signaling pathways (Shun *et al.*, 2004). Are the caspase-3, -8, -9 pathways involved in tocotrienol-induced death of cancer cells? To respond to this question, highly malignant +SA mouse mammary epithelial cells were grown in culture and maintained on serum-free media. Treatment with TRF or γ-tocotrienol, but not α-tocopherol, induced a dose-dependent decrease in +SA cell viability (Shah *et al.*, 2003). TRF- and γ-tocotrienol-induced cell death resulted from apoptosis. Treatment of cells with TRF or γ-tocotrienol increased intracellular activity and levels of processed caspase-8 and -3 but not caspase-9. Furthermore, treatment with specific caspase-8 or -3 inhibitors, but not caspase-9 inhibitor, completely blocked tocotrienol-induced apoptosis in +SA cells, suggesting that tocotrienol-induced apoptosis in +SA mammary cancer cells is mediated through activation of the caspase-8 signaling pathway and is independent of caspase-9 activation (Shah *et al.*, 2003). Tocotrienol-induced caspase-8 activation is not associated with death receptor apoptotic signaling (Shah and Sylvester, 2004). γ-Tocotrienol significantly decreases the relative intracellular levels of phospho-phosphatidylinositol 3-kinase (PI3K)-dependent kinase 1 (phospho-PDK-1 active), phospho-Akt (active), and phospho-glycogen synthase kinase 3. It also decreases the intracellular levels of FLICE-inhibitory protein (FLIP), an antiapoptotic protein that inhibits caspase-8 activation. Because stimulation of the PI3K/PDK/Akt mitogenic pathway is associated with increased FLIP expression, enhanced cellular proliferation, and survival, these observations suggest that tocotrienol-induced caspase-8 activation and apoptosis in malignant +SA mammary epithelial cells is associated with a suppression in PI3K/PDK-1/Akt mitogenic signaling and subsequent reduction in intracellular FLIP levels (Shah and Sylvester, 2004). It has been reported that the antiproliferative effects of γ-tocotrienol results, at least in part, from a reduction in Akt and NF-κB activity in neoplastic +SA mammary epithelial cells (Shah and Sylvester, 2005a).

α-Tocotrienol (20 μM) seems to share some of the cytotoxic effects on cancer cells by inducing caspase-8 and -3 activity (Sylvester and Shah, 2005a). Combined treatment with specific caspase-8 or -3 inhibitors completely blocked α-tocotrienol-induced apoptosis and caspase-8 or -3 activity, respectively. In contrast, α-tocotrienol treatment had no effect on caspase-9 activation, and combined treatment with a specific caspase-9 inhibitor did not block α-tocotrienol-induced apoptosis in +SA cells. α-Tocotrienol-induced caspase-8 activation and apoptosis is not mediated through death receptor activation in malignant +SA mammary epithelial cells. Tocotrienol-induced caspase-8 activation and apoptosis in malignant +SA mammary epithelial cells is not mediated through the activation of death receptors, but appears to result from the suppression of the PI3K/PDK/Akt mitogenic signaling pathway, and subsequent reduction in intracellular FLIP expression (Sylvester and Shah, 2005a).

Bcl-2 family proteins tightly control apoptosis by regulating the permeabilization of the mitochondrial outer membrane and, hence, the release of cytochrome *c* and other proapoptotic factors. Is tocotrienol-induced apoptosis of cancer cells dependent on mitochondrial pathways? Incubation of MDA-MB-231 cells with γ-tocotrienol causes membrane blebbing, formation of apoptotic bodies, chromatin condensation/fragmentation, and phosphatidylserine externalization (Takahashi and Loo, 2004). These are all hallmarks of apoptosis. In γ-tocotrienol-treated cells, mitochondria were disrupted. Collapse of the mitochondrial membrane potential was followed by the release of mitochondrial cytochrome *c*. However, the expression of Bax and Bcl-2 mRNA and protein did not change. In contrast to other studies reporting that tocotrienol-induced cell death is caspase dependent (Shah and Sylvester, 2004; Shah *et al.*, 2003), it was noted that in this model caspases were not involved in γ-tocotrienol-induced apoptosis (Takahashi and Loo, 2004). In a study of +SA cells, it was noted that although γ-tocotrienol induced apoptosis, it did not disrupt mitochondrial membrane potential or cause the release of mitochondrial cytochrome *c* into the cytoplasm. Tocotrienol-treated apoptotic +SA cells showed a paradoxical decrease in mitochondrial levels of proapoptotic proteins Bid, Bax, and Bad, and a corresponding increase in mitochondrial levels of antiapoptotic proteins, Bcl-2 and Bcl-xL, suggesting that mitochondrial membrane stability and integrity might actually be enhanced for a limited period of time following acute tocotrienol exposure. This significance of this unusual finding remains obscure (Shah and Sylvester, 2005b).

Over the past 30 years, a relatively simple growth factor and its cognate receptor have provided seminal insights into the understanding of the genetic basis of cancer, as well as growth factor signaling. The epidermal growth factor (EGF), its cognate receptor (EGFR), and related family members have been shown to be important in normal as well as the malignant growth of many cell types including breast cancer. EGF is a potent mitogen for normal and neoplastic mammary epithelial cells. Initial events in EGFR mitogenic signaling are G-protein activation, stimulation of adenylyl cyclase and cyclic AMP (cAMP) production. Do the antiproliferative effects of tocotrienols associate with reduced EGF-induced G-protein and cAMP-dependent mitogenic signaling? To answer this question, preneoplastic CL-S1 mouse mammary epithelial cells were grown in culture and maintained on serum-free media, containing 0- to 25-μmol/liter tocotrienol-rich fraction of palm oil and/or different doses of pharmacological agents that alter intracellular cAMP levels. Tocotrienol-induced effects on EGF-receptor levels of tyrosine kinase activity, as well as EGF-dependent MAPK and Akt activation, were examined. It was noted that the antiproliferative effects of tocotrienols in preneoplastic mammary epithelial cells do not reflect a reduction in EGF-receptor mitogenic responsiveness, but rather, result from an inhibition in early postreceptor events involved in cAMP production upstream from EGF-dependent MAPK and phosphoinositide 3-kinase/Akt mitogenic signaling (Sylvester *et al.*, 2002).

DMBA is a potent inducer breast cancer in rats. The antitumor and anticholesterol impacts have been examined in rats treated with the chemical carcinogen DMBA, which is known to induce mammary carcinogenesis and hypercholesterolemia. DMBA induced multiple tumors on mammary glands after 6 months. Feeding of TRF (10 mg/kg body weight/day) for 6 months, isolated from RBO, to DMBA-administered rats, attenuated the severity and extent of neoplastic transformation in the mammary glands. Consistently, plasma and mammary alkaline phosphatase activities increased during carcinogenesis were significantly decreased in TRF-treated rats. TRF treatment to rats maintained low levels of glutathione *S*-transferase activities in liver and mammary glands, which is consistent with the anticarcinogenic properties of TRF (Iqbal *et al.*, 2003). Administration of DMBA also caused a significant increase of 30% in plasma total cholesterol and 111% in LDL cholesterol levels compared with normal control levels. Feeding of TRF to rats caused a significant decline of 30% in total cholesterol and 67% in LDL cholesterol levels compared with the DMBA-administered rats. The experimental hypercholesterolemia caused a significant increase in enzymatic activity (23%) and protein mass (28%) of hepatic 3-hydroxy-3-methylglutaryl coenzyme A (HMG-CoA) reductase. Consistent with TRF-mediated reduction in plasma lipid levels, enzymatic activity and protein mass of HMG-CoA reductase was significantly reduced. These observations support that TRF has potent anticancer and anticholesterol effects in rats (Iqbal *et al.*, 2003).

Tocotrienols act on cell proliferation in a dose-dependent manner and can induce programed cell death in breast cancer cells. To elucidate the molecular basis of the effect of tocotrienols, MCF-7 breast cancer cells were injected into athymic nude mice. Feeding quite large amounts (1 mg/day) of TRF for 20 weeks delayed the onset, incidence, and size of tumors. At autopsy, the tumor tissue was excised and cDNA array analysis was performed. Thirty out of 1176 genes were significantly affected by TRF. Ten genes were downregulated and 20 genes upregulated with respect to untreated animals. The expression of the interferon-inducible transmembrane protein-1 gene was significantly upregulated in tumors excised from TRF-treated animals compared with control mice. Within the group of genes related to the immune system, CD59 glycoprotein precursor gene was upregulated.

Among the functional class of intracellular transducers/effectors/modulators, the c-myc gene was significantly downregulated in tumors in response to TRF treatment. This work on the survey of TRF-sensitive genes in the tumor *in vivo* presented useful insight (Nesaretnam *et al.*, 2004).

2. Prostate

Unlike the literature on breast cancer cells, work on prostate cancer cells investigating the effect of tocotrienol is scant. In a model where prostate cancer was induced by injecting PC-3 cells into nude BALB/c mice, it has been noted that the radiotherapy efficacy of prostate cancer can be increased with γ-tocotrienol and a prooxidant if the kidneys can be shielded (Kumar *et al.*, 2006). When the tumors were about 5 mm in diameter, mice were injected subcutaneously with 400-mg/kg γ-tocotrienol and irradiated 24 h later at the site of the tumor with a dose of 12 Gy (60) Cobalt. The size of the tumors was reduced by almost 40%, but only in tocotrienol-treated and irradiated mice (Kumar *et al.*, 2006). The growth inhibitory and apoptotic effects of TRF have been tested on normal human prostate epithelial cells (PrECs), virally transformed normal human prostate epithelial cells (PZ-HPV-7), and human prostate cancer cells (LNCaP, DU145, and PC-3) (Srivastava and Gupta, 2006). TRF selectively resulted in potent growth inhibition in cancer cells but not normal cells. In response to TRF, cancer cells underwent G0/G1 phase arrest and sub G1 accumulation. Colony formation by all three prostate cancer cell lines studied was clearly arrested by TRF. The IC(50) after 24-h TRF treatment in LNCaP, PC-3, and DU145 cells were in the order 16.5, 17.5, and 22.0 μg/ml. TRF treatment resulted in significant apoptosis of cancer cells but not of normal cells (Srivastava and Gupta, 2006).

3. Immune System

Inhibition of tumor promotion by tocopherols and tocotrienols was examined by an *in vitro* assay utilizing the activation of Epstein–Barr virus early antigen expression in Epstein–Barr virus genome-carrying human lymphoblastoid cells. γ- and δ-tocotrienols derived from palm oil exhibited a strong activity against tumor promotion by inhibiting Epstein–Barr virus early antigen expression in Raji cells induced by 12-*O*-tetradecanoylphorbol-13-acetate. In contrast, the corresponding tocopherols lacked this activity (Goh *et al.*, 1994).

4. Liver

Tocotrienol inhibits the growth of hepatoma cells but not that of hepatocytes from healthy rat liver (Sakai *et al.*, 2004). Consistently, tocotrienol killed murine liver cancer cells but not normal cells (Har and Keong, 2005). Of note, this interesting function of tocotrienol is not shared by tocopherol. Tocotrienol-induced apoptosis of hepatoma cells is mediated by caspase-3 activation.

In addition, tocotrienol induced caspase-8 activity. An inhibitor of caspase-8 suppressed the induction of apoptosis in hepatoma by tocotrienol. Compared to tocopherol, tocotrienol was more quickly taken up by the cancer cells suggesting that this could be one reason why tocotrienol was so effective in killing the hepatoma cells (Har and Keong, 2005; Sakai *et al.*, 2004). γ-Tocotrienol inhibits the proliferation of human hepatoma Hep3B cells at lower concentrations and shorter treatment times than α-tocotrienol. γ-Tocotrienol induces poly(ADP-ribose) polymerase (PARP) cleavage activating caspase-3. In addition, γ-tocotrienol activates caspase-8 and -9 and upregulates Bax and fragments of Bid (Sakai *et al.*, 2005). In human hepatocellular carcinoma HepG2 cells, δ-tocotrienol exerts more significant antiproliferative effect than α-, β-, and γ-tocotrienols. δ-Tocotrienol induced apoptosis, and also tended to induce S phase arrest. The phase I enzyme CYP1A1 was induced by δ-tocotrienol (Wada *et al.*, 2005).

2-Acetylaminofluorene is a potent hepatocarcinogen. Prolonged feeding of rats with 2-acetylaminofluorene causes hepatocellular damage. Such damage is prevented by tocotrienol supplementation (Ngah *et al.*, 1991). 2-Acetylaminofluorene significantly increased the activities of both plasma and liver microsomal γ-glutamyltranspeptidase (GGT) and liver microsomal UDP-glucuronyltransferase (UDP-GT). Tocotrienols administered together with AAF significantly decrease the activities of plasma GGT after 12 and 20 weeks and liver microsomal UDP-GT after 20 weeks, when compared with matched controls (Ngah *et al.*, 1991). In a scenario of stronger chemical carcinogen insult caused by 2-acetylaminofluorene in conjunction with diethylnitrosamine (DEN), the effects of tocotrienol turned out to be more encouraging. In response to challenge by the chemical carcinogens, all ten rats in the group showed the presence of two grayish white nodules in the liver. Rats subjected to long-term administration of tocotrienol were protected. Only one out of six rats studied in this group had the hepatocarcinoma (Rahmat *et al.*, 1993).

The anticancer efficacy of TRF has been evaluated during DEN/2-acetylaminofluorene (AAF)-induced hepatocarcinogenesis in male Sprague-Dawley rats. TRF treatment was carried out for 6 months and was started 2 weeks before the initiation phase of hepatocarcinogenesis. Morphological examination of the livers from DEN/AAF rats showed numerous off-white patches and few small nodules, which were significantly reduced by TRF treatment. DEN/AAF caused a twofold increase in the activity of alkaline phosphatase in the plasma as compared with normal control rats. This increase of the tissue damage marker was prevented significantly by TRF treatment. Hepatic activity of glutathione *S*-transferase was also increased (3.5-fold) during the induction of hepatic carcinogenesis. Lipid peroxidation and LDL oxidation increased threefold following initiation by DEN/AAF as compared with normal control rats. TRF treatment to DEN/AAF-treated rats substantially decreased (62–66%) the above parameters and thus limited the action of DEN/AAF. Thus, TRF exhibited clear protective properties in this model of chemical carcinogenesis (Iqbal *et al.*, 2004).

5. Gastrointestinal Tract

RKO, a poorly differentiated colon carcinoma cell line, represents a commonly used *in vitro* model for human colon carcinoma. RKO cells contain wild-type p53 but lack endogenous human thyroid receptor nuclear receptor (h-TRbeta1). In a dose- and time-dependent manner, TRF inhibited the growth and colony formation of RKO. In addition, TRF induced WAF1/p21 which appeared to be independent of cell cycle regulation and was transcriptionally upregulated in p53 dependent fashion. TRF treatment also resulted in alteration in Bax/Bcl-2 ratio in favor of apoptosis, which was associated with the release of cytochrome *c* and induction of apoptotic protease-activating factor-1. This altered expression of Bcl-2 family members triggered the activation of initiator caspase-9 followed by activation of effector caspase-3. Thus, in RKO cells the pathways involved in TRF-induced apoptosis are fairly well characterized (Agarwal *et al.*, 2004). Since the discovery that telomerase is repressed in most normal human somatic cells but strongly expressed in most human tumors, telomerase emerged as an attractive target for diagnostic, prognostic, and therapeutic purposes to combat human cancer (Shay and Wright, 2006). Tocotrienol has been noted to inhibit telomerase activity of DLD-1 human colorectal adenocarcinoma cells in a time- and dose-dependent manner. δ-Tocotrienol demonstrated the highest inhibitory activity. Tocotrienol inhibited protein kinase C activity, resulting in downregulation of c-myc and human telomerase reverse transcriptase (hTERT) expression, thereby reducing telomerase activity. Of note, tocopherol does not share the potent activity of tocotrienol in this regard (Shay and Wright, 2006).

6. Skin

How much tocotrienol is needed to inhibit the increase in population of murine B16(F10) melanoma cells during a 48-h incubation by 50% (IC50 value)? The IC50 value estimated for farnesol, the side chain analogue of the tocotrienols (50 μmol/liter) falls midway between that of α-tocotrienol (110 μmol/liter) and those estimated for γ- (20 μmol/liter) and δ- (10 μmol/liter) tocotrienol. Experimental diets were fed to weanling C57BL female mice for 10 days prior to and 28 days following the implantation of the aggressively growing and highly metastatic B16(F10) melanoma. The isomolar (116 μmol/kg diet) and the vitamin E equivalent (928 μmol/kg diet) substitution of D-γ-tocotrienol for DL-α-tocopherol in the AIN-76A diet produced 36 and 50% retardations, respectively, in tumor growth. Thus, in this skin melanoma model, both tocotrienol as well as tocopherol were significantly effective (He *et al.*, 1997). The growth suppressive effects of γ-tocotrienol on murine B16(F10) melanoma cells have been independently reproduced (Mo and Elson, 1999).

Recent works have led to the identification of antiangiogenic properties of tocotrienol (Table I). This novel development warrants further research testing the anticancer effects of tocotrienol *in vivo*.

C. CHOLESTEROL LOWERING

That the α-tocotrienol form of natural vitamin E, not tocopherol, may have significant cholesterol-lowering properties represents one of the early findings describing the unique biological properties of tocotrienol that was reported two decades ago (Qureshi *et al.*, 1986). The ER enzyme 3-hydroxy-3-methylglutaryl CoA (HMG-CoA) reductase produces mevalonate, which is converted to sterols and other products. It is proposed that tocotrienols are effective in lowering serum total and LDL cholesterol levels by inhibiting the hepatic enzymic activity of HMG-CoA reductase through a posttranscriptional mechanism. α-Tocopherol, however, has an opposite effect (induces) on this enzyme activity (Qureshi *et al.*, 2002). This contrast is of outstanding significance and requires further characterization. α-Tocotrienol, contained in the oily nonpolar fraction of high protein barley (*H. vulgare* L.) flour, decreased hepatic cholesterogenesis and serum total and LDL cholesterol and concomitantly increased lipogenic activity when added to chick diet. It was suspected that the isoprenoid side chain of tocotrienol was responsible for the observed inhibition of cholesterogenesis (Qureshi *et al.*, 1986). Evidence that TRF may indeed lower plasma cholesterol in mammals came from a study of normolipemic and genetically hypercholesterolemic pigs of defined lipoprotein genotype (Qureshi *et al.*, 1991a). The pigs were fed a standard diet supplemented with 50-μg/g TRF isolated from palm oil. Hypercholesterolemic pigs fed the TRF supplement showed a 44% decrease in total serum cholesterol, a 60% decrease in LDL cholesterol, and significant decreases in levels of apolipoprotein B (26%), thromboxane B2 (41%), and platelet factor 4 (PF4; 29%). It was thus noted that TRF had a marked protective effect on the endothelium and platelet aggregation. The effect of the lipid-lowering diet persisted only in the hypercholesterolemic swine after 8 week feeding of the control diet (Qureshi *et al.*, 1991a). These interesting observations were quickly put to test in humans by means of a double-blind, crossover, 8-week study (Qureshi *et al.*, 1991b). The goal was to compare effects of the tocotrienol-enriched fraction of palm oil (200-mg palmvitee capsules/day) with those of 300-mg corn oil/day on serum lipids of hypercholesterolemic human subjects (serum cholesterol 6.21–8.02 mmol/liter). Concentrations of serum total cholesterol (−15%), LDL cholesterol (−8%), Apo B (−10%), thromboxane (−25%), PF4 (−16%), and glucose (−12%) decreased significantly only in the 15 subjects given palmvitee during the initial 4 weeks. Results from the crossover study established that the noted beneficial effects were indeed caused by palmvitee. A carry over effect of palmvitee was reported. Serum cholesterol concentrations of seven hypercholesterolemic subjects (>7.84 mmol/liter) decreased 31% during a 4-week period in which they were given 200-mg γ-tocotrienol/day. These results suggested that γ-tocotrienol could be the active principle cholesterol inhibitor in palmvitee capsules (Qureshi *et al.*, 1991b). Experimental data from the study of hamsters are in agreement (Raederstorff *et al.*, 2002). What added to the interest in tocotrienol as a cholesterol-lowering nutrient in humans was a concurrent independent study reporting the hypocholesterolemic effects of palmvitee (Tan *et al.*, 1991). Each palmvitee capsule contained ~18, 42, and 240 mg of tocopherols, tocotrienols, and palm olein, respectively. All volunteers took one palmvitee capsule per day for 30 consecutive days. Overnight fasting blood was recorded from each volunteer before and after the experiment. Palmvitee lowered both serum total cholesterol and LDL cholesterol concentrations in all subjects. The magnitude of reduction of serum total cholesterol ranged from 5.0 to 35.9%, whereas the reduction of LDL cholesterol values ranged from 0.9 to 37.0% when compared with their respective baseline values (Tan *et al.*, 1991). In another study, the cholesterol-lowering effects of palmvitee and γ-tocotrienol were examined in hypercholesterolemic subjects after acclimation to the American Heart Association Step 1 dietary regimen for 4–8 weeks, respectively (Qureshi *et al.*, 1995). The 4-week dietary regimen alone elicited a 5% significant decrease in the cholesterol level of the 36 subjects. Subjects continuing on the dietary regimen for a second 4-week period benefited from an additional 2% decrease in their cholesterol levels. The subjects experienced significant palmvitee- and γ-tocotrienol-mediated decreases in plasma cholesterol. The group of subjects acclimated to the dietary regimen for 4 weeks responded to palmvitee with a 10% statistically significant decrease in cholesterol. Of interest, α-tocopherol attenuated the cholesterol-suppressive action of the tocotrienols. This antagonism between tocopherol and tocotrienol warrants further research. The second group of subjects acclimated to the dietary regimen for 8 weeks received 200-mg γ-tocotrienol/day for 4 weeks. The cholesterol-suppressive potency of this α-tocopherol-free preparation was calculated to be equivalent to that of the mixture of tocotrienols (220 mg) used in the prior study. Cholesterol levels of the 16 subjects in the second group were significantly decreased by 13% during the 4-week trial. Plasma apolipoprotein B and *ex vivo* generation of thromboxane B2 were similarly responsive to the tocotrienol preparations, whereas neither preparation had an impact on high-density lipoprotein (HDL) cholesterol and apolipoprotein A1 levels (Qureshi *et al.*, 1995).

Tocotrienol not only of palm oil origin but also isolated from rice bran shows cholesterol-lowering properties (Chen and Cheng, 2006; Qureshi *et al.*, 2001a). A human study with 28 hypercholesterolemic subjects has been executed in five phases of 35 days each. The goal was to check the efficacy of a TRF preparation from rice bran alone and in combination with lovastatin. After placing subjects on the American Heart Association (AHA)

Step-1 diet (phase II), the subjects were divided into two groups, A and B. The AHA Step-1 diet was continued in combination with other treatments during phases III–V. Group A subjects were given 10-mg lovastatin, 10-mg lovastatin plus 50-mg TRF, 10-mg lovastatin plus 50-mg α-tocopherol per day, in the third, fourth, and fifth phases, respectively. Group B subjects were treated exactly according to the same protocol except that in the third phase, they were given 50-mg TRF instead of lovastatin. The TRF or lovastatin plus AHA Step-1 diet effectively lowered serum total cholesterol (14%, 13%) and LDL cholesterol (18%, 15%), respectively. The combination of TRF and lovastatin plus AHA Step-1 diet significantly reduced the lipid parameters by 20–25%. Especially significant were the increase in the HDL/LDL ratio to 46% in group A and 53% in group B. None of the subjects reported any side-effects throughout the study of 25 weeks (Qureshi *et al.*, 2001c). Consistent results were obtained using rice bran derived TRF in another human study (Qureshi *et al.*, 2002). A dose of 100 mg/day of TRF decreased the level of serum total cholesterol, LDL cholesterol, apolipoprotein B and triglycerides compared with the baseline values. The work led to the suggestion that a dose of 100 mg/day TRF plus AHA Step-1 diet could control the risk of coronary heart disease in hypercholesterolemic humans (Qureshi *et al.*, 2002).

Mechanistic evidence supporting the cholesterol-lowering properties of tocotrienol is considerable. Tocotrienols cause posttranscriptional suppression of HMG-CoA reductase by a process distinct from other known inhibitors of cholesterol biosynthesis (Pearce *et al.*, 1992). In addition, γ-tocotrienol may stimulate cholesterol catabolism (Chen and Cheng, 2006). *In vitro*, γ-tocotrienol possesses 30-fold greater activity toward cholesterol biosynthesis inhibition compared to α-tocotrienol. The synthetic (racemic) and natural (chiral) tocotrienols exhibited nearly identical cholesterol biosynthesis inhibition and HMG-CoA reductase suppression properties (Pearce *et al.*, 1992). Incubation of several cell types with γ-tocotrienol inhibits the rate of [14C] acetate but not [3H] mevalonate incorporation into cholesterol in a concentration- and time-dependent manner, with 50% inhibition at $\sim$2 μM and maximum $\sim$80% inhibition observed within 6 h in HepG2 cells (Parker *et al.*, 1993). Both HMG-CoA reductase activity and protein expression are sensitive to tocotrienol. *In vivo* studies lend support to that *in vitro* observation (Iqbal *et al.*, 2003). Tocotrienols influence the mevalonate pathway in mammalian cells by posttranscriptional suppression of HMG-CoA reductase, and specifically modulate the intracellular mechanism for controlled degradation of the reductase protein, an activity that mirrors the actions of the putative nonsterol isoprenoid regulators derived from mevalonate (Parker *et al.*, 1993). It is suggested that the farnesyl side chain and the methyl/hydroxy substitution pattern of γ-tocotrienol deliver a high level of HMG-CoA reductase suppression, unsurpassed by synthetic analogues studied (Pearce *et al.*, 1994). HMG-CoA reductase activity in tumor tissues differs from that of liver in being resistant to sterol feedback regulation. Tumor reductase activity retains sensitivity to the posttranscriptional regulation. As a consequence, tocotrienol is effective in suppressing mevalonate synthesis. By doing so, tocotrienol can deplete tumor tissues of two intermediate products, farnesyl pyrophosphate and geranylgeranyl pyrophosphate, which are incorporated posttranslationally into growth control-associated proteins (Elson and Qureshi, 1995).

Ubiquitination followed by rapid degradation by 26S proteasomes represents a key mechanism to silence HMG-CoA reductase. This pathway is activated when sterols and nonsterol end products of mevalonate metabolism accumulate in cells. Sterol-accelerated ubiquitination of HMG-CoA reductase requires Insig-1 and Insig-2, membrane-bound proteins of the ER (Sever *et al.*, 2003). Recently, it has been elegantly demonstrated that δ-tocotrienol stimulates the ubiquitination and degradation of HMG-CoA reductase and blocks processing of SREBPs, another sterol-mediated action of Insigs. The γ-tocotrienol analogue was noted to be more selective in enhancing reductase ubiquitination and degradation than blocking the processing of SREBPs. Interestingly, other forms of vitamin E neither accelerate reductase degradation nor block the processing of SREBPs. δ- and γ-tocotrienols trigger reductase ubiquitination directly and do not require further metabolism for their activity (Song and Debose-Boyd, 2006).

VIII. Conclusion

Often, the term vitamin E is synonymously used with α-tocopherol. While the expression is correct, it is incomplete and may be often misleading. D-α-Tocopherol (RRR-α-tocopherol) has the highest bioavailability and is the standard against which all the others must be compared. However, it is only one out of eight natural forms of vitamin E. The rapidly expanding body of evidence indicating that members of the vitamin E family are functionally unique calls for a revisit of the current practices in vitamin E research and consumption. Research claims should be limited to the specific form of vitamin E studied. For example, evidence for toxicity of a specific form of tocopherol in excess may not be used to conclude that high-dosage vitamin E supplementation may increase all-cause mortality (Miller *et al.*, 2005). Along these lines, it may not be prudent to express frustrations about the net yield of vitamin E research as a whole (Greenberg, 2005) when all that has been tested for efficacy on a limited basis in clinical trials is α-tocopherol—just one out of eight forms. It has been suggested that the safe dose of various tocotrienols for human consumption is 200–1000 mg/day (Yu *et al.*, 2006). Vitamin E represents one of the most fascinating natural resources that have the potential to influence a broad range of mechanisms underlying human health and disease. Yet, clinical outcomes studies have failed to meet expectations (Friedrich, 2004; Greenberg, 2005). The current state of knowledge warrants strategic investment into the lesser known forms of vitamin E with emphasis on uncovering the specific conditions that govern the function of vitamin E molecules *in vivo*. Outcome studies designed in light of such information would yield lucrative returns.

ACKNOWLEDGMENTS

Tocotrienol research in the laboratory is supported by NIH RO1NS42617.

REFERENCES

Adachi, H., and Ishii, N. (2000). Effects of tocotrienols on life span and protein carbonylation in Caenorhabditis elegans. *J. Gerontol. A Biol. Sci. Med. Sci.* **55,** B280–B285.

Adachi, K., Miki, M., Tamai, H., Tokuda, M., and Mino, M. (1990). Adipose tissues and vitamin E. *J. Nutr. Sci. Vitaminol. (Tokyo)* **36,** 327–337.

Adesuyi, S. A., Cockrell, C. S., Gamache, D. A., and Ellis, E. F. (1985). Lipoxygenase metabolism of arachidonic acid in brain. *J. Neurochem.* **45,** 770–776.

Agarwal, M. K., Agarwal, M. L., Athar, M., and Gupta, S. (2004). Tocotrienol-rich fraction of palm oil activates p53, modulates Bax/Bcl2 ratio and induces apoptosis independent of cell cycle association. *Cell Cycle* **3,** 205–211.

Ahmad, N. S., Khalid, B. A., Luke, D. A., and Ima Nirwana, S. (2005). Tocotrienol offers better protection than tocopherol from free radical-induced damage of rat bone. *Clin. Exp. Pharmacol. Physiol.* **32,** 761–770.

Aikawa, R., Komuro, I., Yamazaki, T., Zou, Y., Kudoh, S., Tanaka, M., Shiojima, I., Hiroi, Y., and Yazaki, Y. (1997). Oxidative stress activates extracellular signal-regulated kinases through Src and Ras in cultured cardiac myocytes of neonatal rats. *J. Clin. Invest.* **100,** 1813–1821.

Alper, O., and Bowden, E. T. (2005). Novel insights into c-Src. *Curr. Pharm. Des.* **11,** 1119–1130.

Anderson, S. L., and Rubin, B. Y. (2005). Tocotrienols reverse IKAP and monoamine oxidase deficiencies in familial dysautonomia. *Biochem. Biophys. Res. Commun.* **336,** 150–156.

Anderson, S. L., Qiu, J., and Rubin, B. Y. (2003). Tocotrienols induce IKBKAP expression: A possible therapy for familial dysautonomia. *Biochem. Biophys. Res. Commun.* **306,** 303–309.

Araya, H., Tomita, M., and Hayashi, M. (2006). The novel formulation design of self-emulsifying drug delivery systems (SEDDS) type O/W microemulsion III: The permeation mechanism of a poorly water soluble drug entrapped O/W microemulsion in rat isolated intestinal membrane by the Ussing chamber method. *Drug Metab. Pharmacokinet.* **21,** 45–53.

Aten, R. F., Kolodecik, T. R., and Behrman, H. R. (1994). Ovarian vitamin E accumulation: Evidence for a role of lipoproteins. *Endocrinology* **135,** 533–539.

Atkinson, J. (2006). Chemical investigations of tocotrienols: Isotope substitution, fluorophores and a curious curve. *In* "6th COSTAM/SFRR (ASEAN/Malaysia) International Workshop on Micronutrients, Oxidative Stress, and the Environment" (K. Nesaretnam, Ed.), p. 22. Kuching, Malaysia. COSTAM.

Azlina, M. F., Nafeeza, M. I., and Khalid, B. A. (2005). A comparison between tocopherol and tocotrienol effects on gastric parameters in rats exposed to stress. *Asia Pac. J. Clin. Nutr.* **14,** 358–365.

Azzi, A., and Stocker, A. (2000). Vitamin E: Non-antioxidant roles. *Prog. Lipid Res.* **39,** 231–255.

Azzi, A., Boscoboinik, D., Marilley, D., Ozer, N. K., Stauble, B., and Tasinato, A. (1995). Vitamin E: A sensor and an information transducer of the cell oxidation state. *Am. J. Clin. Nutr.* **62,** 1337S–1346S.

Bains, J. S., and Shaw, C. A. (1997). Neurodegenerative disorders in humans: The role of glutathione in oxidative stress-mediated neuronal death. *Brain Res. Brain Res. Rev.* **25,** 335–358.

Baliarsingh, S., Beg, Z. H., and Ahmad, J. (2005). The therapeutic impacts of tocotrienols in type 2 diabetic patients with hyperlipidemia. *Atherosclerosis* **182,** 367–374.

Bazan, N. G., Jr. (1970). Effects of ischemia and electroconvulsive shock on free fatty acid pool in the brain. *Biochim. Biophys. Acta* **218,** 1–10.

Bazan, N. G., Jr. (1971a). Changes in free fatty acids of brain by drug-induced convulsions, electroshock and anaesthesia. *J. Neurochem.* **18,** 1379–1385.

Bazan, N. G., Jr. (1971b). Phospholipases A 1 and A 2 in brain subcellular fractions. *Acta Physiol. Lat. Am.* **21,** 101–106.

Bazan, N. G. (1976). Free arachidonic acid and other lipids in the nervous system during early ischemia and after electroshock. *Adv. Exp. Med. Biol.* **72,** 317–335.

Bazan, N. G., Jr., and Rakowski, H. (1970). Increased levels of brain free fatty acids after electroconvulsive shock. *Life Sci.* **9,** 501–507.

Begum, A. N., and Terao, J. (2002). Protective effect of α-tocotrienol against free radical-induced impairment of erythrocyte deformability. *Biosci. Biotechnol. Biochem.* **66,** 398–403.

Bhat, N. R., and Zhang, P. (1999). Hydrogen peroxide activation of multiple mitogen-activated protein kinases in an oligodendrocyte cell line: Role of extracellular signal-regulated kinase in hydrogen peroxide-induced cell death. *J. Neurochem.* **72,** 112–119.

Birringer, M., Pfluger, P., Kluth, D., Landes, N., and Brigelius-Flohe, R. (2002). Identities and differences in the metabolism of tocotrienols and tocopherols in HepG2 cells. *J. Nutr.* **132,** 3113–3118.

Birringer, M., EyTina, J. H., Salvatore, B. A., and Neuzil, J. (2003). Vitamin E analogues as inducers of apoptosis: Structure-function relation. *Br. J. Cancer* **88,** 1948–1955.

Black, T. M., Wang, P., Maeda, N., and Coleman, R. A. (2000). Palm tocotrienols protect ApoE+/– mice from diet-induced atheroma formation. *J. Nutr.* **130,** 2420–2426.

Blatt, D. H., Leonard, S. W., and Traber, M. G. (2001). Vitamin E kinetics and the function of tocopherol regulatory proteins. *Nutrition* **17,** 799–805.

Borngraber, S., Browner, M., Gillmor, S., Gerth, C., Anton, M., Fletterick, R., and Kuhn, H. (1999). Shape and specificity in mammalian 15-lipoxygenase active site. The functional interplay of sequence determinants for the reaction specificity. *J. Biol. Chem.* **274,** 37345–37350.

Boscoboinik, D., Szewczyk, A., Hensey, C., and Azzi, A. (1991). Inhibition of cell proliferation by alpha-tocopherol. Role of protein kinase C. *J. Biol. Chem.* **266,** 6188–6194.

Brugge, J. S., Cotton, P. C., Queral, A. E., Barrett, J. N., Nonner, D., and Keane, R. W. (1985). Neurones express high levels of a structurally modified, activated form of pp60c-src. *Nature* **316,** 554–557.

Bruno, R. S., and Traber, M. G. (2006). Vitamin E biokinetics, oxidative stress and cigarette smoking. *Pathophysiology* **13,** 143–149.

Bryngelsson, S., Dimberg, L. H., and Kamal-Eldin, A. (2002). Effects of commercial processing on levels of antioxidants in oats (Avena sativa L.). *J. Agric. Food Chem.* **50,** 1890–1896.

Cahoon, E. B., Hall, S. E., Ripp, K. G., Ganzke, T. S., Hitz, W. D., and Coughlan, S. J. (2003). Metabolic redesign of vitamin E biosynthesis in plants for tocotrienol production and increased antioxidant content. *Nat. Biotechnol.* **21,** 1082–1087.

Carlen, P. L., Gurevich, N., Zhang, L., Wu, P. H., Reynaud, D., and Pace-Asciak, C. R. (1994). Formation and electrophysiological actions of the arachidonic acid metabolites, hepoxilins, at nanomolar concentrations in rat hippocampal slices. *Neuroscience* **58,** 493–502.

Chao, J. T., Gapor, A., and Theriault, A. (2002). Inhibitory effect of δ-tocotrienol, a HMG CoA reductase inhibitor, on monocyte-endothelial cell adhesion. *J. Nutr. Sci. Vitaminol. (Tokyo)* **48,** 332–337.

Chen, C. W., and Cheng, H. H. (2006). A rice bran oil diet increases LDL-receptor and HMG-CoA reductase mRNA expressions and insulin sensitivity in rats with streptozotocin/nicotinamide-induced type 2 diabetes. *J. Nutr.* **136,** 1472–1476.

Chow, C. K., and Draper, H. H. (1970). Isolation of γ-tocotrienol dimers from Hevea latex. *Biochemistry* **9,** 445–450.

Collins, J. J., Evason, K., and Kornfeld, K. (2006). Pharmacology of delayed aging and extended lifespan of Caenorhabditis elegans. *Exp. Gerontol.* **41**(10), 1032–1039.

Das, S., Powell, S. R., Wang, P., Divald, A., Nesaretnam, K., Tosaki, A., Cordis, G. A., Maulik, N., and Das, D. K. (2005). Cardioprotection with palm tocotrienol: Antioxidant activity of tocotrienol is linked with its ability to stabilize proteasomes. *Am. J. Physiol. Heart Circ. Physiol.* **289,** H361–H367.

Dietrich, M., Traber, M. G., Jacques, P. F., Cross, C. E., Hu, Y., and Block, G. (2006). Does γ-tocopherol play a role in the primary prevention of heart disease and cancer? A review. *J. Am. Coll. Nutr.* **25,** 292–299.

Dringen, R., Gutterer, J. M., and Hirrlinger, J. (2000). Glutathione metabolism in brain metabolic interaction between astrocytes and neurons in the defense against reactive oxygen species. *Eur. J. Biochem.* **267,** 4912–4916.

Eitsuka, T., Nakagawa, K., and Miyazawa, T. (2006). Down-regulation of telomerase activity in DLD-1 human colorectal adenocarcinoma cells by tocotrienol. *Biochem. Biophys. Res. Commun.* **348**(1), 170–175.

Elson, C. E. (1992). Tropical oils: Nutritional and scientific issues. *Crit. Rev. Food Sci. Nutr.* **31,** 79–102.

Elson, C. E., and Qureshi, A. A. (1995). Coupling the cholesterol- and tumor-suppressive actions of palm oil to the impact of its minor constituents on 3-hydroxy-3-methylglutaryl coenzyme A reductase activity. *Prostaglandins Leukot. Essent. Fatty Acids* **52,** 205–207.

Evans, H. M., and Bishop, K. S. (1922). On the existence of a hitherto unrecognized dietary factor essential for reproduction. *Science* **56,** 650–651.

Fairus, S., Rosnah, M. N., Cheng, H. M., and Sundram, K. (2004). Metabolic fate of palm tocotrienols in human postprandial plasma model. *Asia Pac. J. Clin. Nutr.* **13,** S77.

Friedrich, M. J. (2004). To "E" or not to "E," vitamin E's role in health and disease is the question. *JAMA* **292,** 671–673.

Gao, P., and Morozowich, W. (2006). Development of supersaturatable self-emulsifying drug delivery system formulations for improving the oral absorption of poorly soluble drugs. *Expert Opin. Drug Deliv.* **3,** 97–110.

Garry, P. J., Hunt, W. C., Bandrofchak, J. L., VanderJagt, D., and Goodwin, J. S. (1987). Vitamin A intake and plasma retinol levels in healthy elderly men and women. *Am. J. Clin. Nutr.* **46,** 989–994.

Goh, S. H., Hew, N. F., Norhanom, A. W., and Yadav, M. (1994). Inhibition of tumour promotion by various palm-oil tocotrienols. *Int. J. Cancer* **57,** 529–531.

Gorman, C. (2005). Vitamin E-gads. *Time* **165,** 73.

Gould, M. N., Haag, J. D., Kennan, W. S., Tanner, M. A., and Elson, C. E. (1991). A comparison of tocopherol and tocotrienol for the chemoprevention of chemically induced rat mammary tumors. *Am. J. Clin. Nutr.* **53,** 1068S–1070S.

Greenberg, E. R. (2005). Vitamin E supplements: Good in theory, but is the theory good? *Ann. Intern. Med.* **142,** 75–76.

Grossman, S., and Waksman, E. G. (1984). New aspects of the inhibition of soybean lipoxygenase by α-tocopherol. Evidence for the existence of a specific complex. *Int. J. Biochem.* **16,** 281–289.

Gu, J., Liu, Y., Wen, Y., Natarajan, R., Lanting, L., and Nadler, J. L. (2001). Evidence that increased 12-lipoxygenase activity induces apoptosis in fibroblasts. *J. Cell. Physiol.* **186,** 357–365.

Gu, J. Y., Wakizono, Y., Sunada, Y., Hung, P., Nonaka, M., Sugano, M., and Yamada, K. (1999). Dietary effect of tocopherols and tocotrienols on the immune function of spleen and mesenteric lymph node lymphocytes in Brown Norway rats. *Biosci. Biotechnol. Biochem.* **63,** 1697–1702.

Guthrie, N., Gapor, A., Chambers, A. F., and Carroll, K. K. (1997). Inhibition of proliferation of estrogen receptor-negative MDA-MB-435 and -positive MCF-7 human breast cancer cells by palm oil tocotrienols and tamoxifen, alone and in combination. *J. Nutr.* **127,** 544S–548S.

Hagmann, W., Kagawa, D., Renaud, C., and Honn, K. V. (1993). Activity and protein distribution of 12-lipoxygenase in HEL cells: Induction of membrane-association by phorbol ester TPA, modulation of activity by glutathione and 13-HPODE, and Ca(2+)-dependent translocation to membranes. *Prostaglandins* **46,** 471–477.

Hall, T. J., Schaeublin, M., and Missbach, M. (1994). Evidence that c-src is involved in the process of osteoclastic bone resorption. *Biochem. Biophys. Res. Commun.* **199,** 1237–1244.

Han, D., Sen, C. K., Roy, S., Kobayashi, M. S., Tritschler, H. J., and Packer, L. (1997). Protection against glutamate-induced cytotoxicity in C6 glial cells by thiol antioxidants. *Am. J. Physiol.* **273,** R1771–R1778.

Har, C. H., and Keong, C. K. (2005). Effects of tocotrienols on cell viability and apoptosis in normal murine liver cells (BNL CL.2) and liver cancer cells (BNL 1ME A.7R.1), *in vitro*. *Asia Pac. J. Clin. Nutr.* **14,** 374–380.

Hathcock, J. N., Azzi, A., Blumberg, J., Bray, T., Dickinson, A., Frei, B., Jialal, I., Johnston, C. S., Kelly, F. J., Kraemer, K., Packer, L., Parthasarathy, S., *et al.* (2005). Vitamins E and C are safe across a broad range of intakes. *Am. J. Clin. Nutr.* **81,** 736–745.

Hattori, A., Fukushima, T., Yoshimura, H., Abe, K., and Imai, K. (2000). Production of LLU-α following an oral administration of γ-tocotrienol or γ-tocopherol to rats. *Biol. Pharm. Bull.* **23,** 1395–1397.

Hayes, K. C., Pronczuk, A., and Liang, J. S. (1993). Differences in the plasma transport and tissue concentrations of tocopherols and tocotrienols: Observations in humans and hamsters. *Proc. Soc. Exp. Biol. Med.* **202,** 353–359.

He, L., Mo, H., Hadisusilo, S., Qureshi, A. A., and Elson, C. E. (1997). Isoprenoids suppress the growth of murine B16 melanomas *in vitro* and *in vivo*. *J. Nutr.* **127,** 668–674.

Hensley, K., Benaksas, E. J., Bolli, R., Comp, P., Grammas, P., Hamdheydari, L., Mou, S., Pye, Q. N., Stoddard, M. F., Wallis, G., Williamson, K. S., West, M., *et al.* (2004). New perspectives on vitamin E: γ-Tocopherol and carboxyelthylhydroxychroman metabolites in biology and medicine. *Free Radic. Biol. Med.* **36,** 1–15.

Ho, D., Yuen, K. H., and Yap, S. P. (2003). Drug delivery system: Formulation for fat-soluble drugs. *US patent* 6, 596, 306.

Hong, J. Y., Kim, J. K., Song, Y. K., Park, J. S., and Kim, C. K. (2006). A new self-emulsifying formulation of itraconazole with improved dissolution and oral absorption. *J. Control. Release* **110,** 332–338.

Horvath, G., Wessjohann, L., Bigirimana, J., Jansen, M., Guisez, Y., Caubergs, R., and Horemans, N. (2006). Differential distribution of tocopherols and tocotrienols in photosynthetic and non-photosynthetic tissues. *Phytochemistry* **67,** 1185–1195.

Hosomi, A., Arita, M., Sato, Y., Kiyose, C., Ueda, T., Igarashi, O., Arai, H., and Inoue, K. (1997). Affinity for α-tocopherol transfer protein as a determinant of the biological activities of vitamin E analogs. *FEBS Lett.* **409,** 105–108.

Hsu, C. Y., Persons, P. E., Spada, A. P., Bednar, R. A., Levitzki, A., and Zilberstein, A. (1991). Kinetic analysis of the inhibition of the epidermal growth factor receptor tyrosine kinase by Lavendustin-A and its analogue. *J. Biol. Chem.* **266,** 21105–21112.

Ikeda, I., Imasato, Y., Sasaki, E., and Sugano, M. (1996). Lymphatic transport of α-, γ- and δ-tocotrienols and α-tocopherol in rats. *Int. J. Vitam. Nutr. Res.* **66,** 217–221.

Ikeda, S., Niwa, T., and Yamashita, K. (2000). Selective uptake of dietary tocotrienols into rat skin. *J. Nutr. Sci. Vitaminol. (Tokyo)* **46,** 141–143.

Ikeda, S., Toyoshima, K., and Yamashita, K. (2001). Dietary sesame seeds elevate α- and γ-tocotrienol concentrations in skin and adipose tissue of rats fed the tocotrienol-rich fraction extracted from palm oil. *J. Nutr.* **131,** 2892–2897.

Ikeda, S., Tohyama, T., Yoshimura, H., Hamamura, K., Abe, K., and Yamashita, K. (2003). Dietary α-tocopherol decreases α-tocotrienol but not γ-tocotrienol concentration in rats. *J. Nutr.* **133,** 428–434.

Ima-Nirwana, S., and Suhaniza, S. (2004). Effects of tocopherols and tocotrienols on body composition and bone calcium content in adrenalectomized rats replaced with dexamethasone. *J. Med. Food* **7,** 45–51.

Ingraham, C. A., Cox, M. E., Ward, D. C., Fults, D. W., and Maness, P. F. (1989). c-src and other proto-oncogenes implicated in neuronal differentiation. *Mol. Chem. Neuropathol.* **10,** 1–14.

Inokuchi, H., Hirokane, H., Tsuzuki, T., Nakagawa, K., Igarashi, M., and Miyazawa, T. (2003). Anti-angiogenic activity of tocotrienol. *Biosci. Biotechnol. Biochem.* **67,** 1623–1627.

Iqbal, J., Minhajuddin, M., and Beg, Z. H. (2003). Suppression of 7,12-dimethylbenz[α]anthracene-induced carcinogenesis and hypercholesterolaemia in rats by tocotrienol-rich fraction isolated from rice bran oil. *Eur. J. Cancer Prev.* **12,** 447–453.

Iqbal, J., Minhajuddin, M., and Beg, Z. H. (2004). Suppression of diethylnitrosamine and 2-acetylaminofluorene-induced hepatocarcinogenesis in rats by tocotrienol-rich fraction isolated from rice bran oil. *Eur. J. Cancer Prev.* **13,** 515–520.

Ishizaki, Y., and Murota, S. (1991). Arachidonic acid metabolism in cultured astrocytes: Presence of 12-lipoxygenase activity in the intact cells. *Neurosci. Lett.* **131,** 149–152.

Ishizawar, R., and Parsons, S. J. (2004). c-Src and cooperating partners in human cancer. *Cancer Cell* **6,** 209–214.

Jishage, K., Arita, M., Igarashi, K., Iwata, T., Watanabe, M., Ogawa, M., Ueda, O., Kamada, N., Inoue, K., Arai, H., and Suzuki, H. (2001). Alpha-tocopherol transfer protein is important for the normal development of placental labyrinthine trophoblasts in mice. *J. Biol. Chem.* **276,** 1669–1672.

Johnson, N. L., Gardner, A. M., Diener, K. M., Lange-Carter, C. A., Gleavy, J., Jarpe, M. B., Minden, A., Karin, M., Zon, L. I., and Johnson, G. L. (1996). Signal transduction pathways regulated by mitogen-activated/extracellular response kinase kinase kinase induce cell death. *J. Biol. Chem.* **271,** 3229–3237.

Kaempf-Rotzoll, D. E., Traber, M. G., and Arai, H. (2003). Vitamin E and transfer proteins. *Curr. Opin. Lipidol.* **14,** 249–254.

Kagan, V. E., Serbinova, E. A., Forte, T., Scita, G., and Packer, L. (1992). Recycling of vitamin E in human low density lipoproteins. *J. Lipid Res.* **33,** 385–397.

Kaku, S., Yunoki, S., Mori, M., Ohkura, K., Nonaka, M., Sugano, M., and Yamada, K. (1999). Effect of dietary antioxidants on serum lipid contents and immunoglobulin productivity of lymphocytes in Sprague-Dawley rats. *Biosci. Biotechnol. Biochem.* **63,** 575–576.

Kamat, J. P., and Devasagayam, T. P. (1995). Tocotrienols from palm oil as potent inhibitors of lipid peroxidation and protein oxidation in rat brain mitochondria. *Neurosci. Lett.* **195,** 179–182.

Kanaya, Y., Doi, T., Sasaki, H., Fujita, A., Matsuno, S., Okamoto, K., Nakano, Y., Tsujiwaki, S., Furuta, H., Nishi, M., Tsuno, T., Taniguchi, H., *et al.* (2004). Rice bran extract prevents the elevation of plasma peroxylipid in KKAy diabetic mice. *Diabetes Res. Clin. Pract.* **66**(Suppl. 1), S157–S160.

Khanna, S., Venojarvi, M., Roy, S., and Sen, C. K. (2002). Glutamate-induced c-Src activation in neuronal cells. *Methods Enzymol.* **352,** 191–198.

Khanna, S., Roy, S., Ryu, H., Bahadduri, P., Swaan, P. W., Ratan, R. R., and Sen, C. K. (2003). Molecular basis of vitamin E action: Tocotrienol modulates 12-lipoxygenase, a key mediator of glutamate-induced neurodegeneration. *J. Biol. Chem.* **278,** 43508–43515.

Khanna, S., Patel, V., Rink, C., Roy, S., and Sen, C. K. (2005a). Delivery of orally supplemented α-tocotrienol to vital organs of rats and tocopherol-transport protein deficient mice. *Free Radic. Biol. Med.* **39,** 1310–1319.

Khanna, S., Roy, S., Slivka, A., Craft, T. K., Chaki, S., Rink, C., Notestine, M. A., DeVries, A. C., Parinandi, N. L., and Sen, C. K. (2005b). Neuroprotective properties of the natural vitamin E α-tocotrienol. *Stroke* **36,** 2258–2264.

Khanna, S., Roy, S., Parinandi, N. L., Maurer, M., and Sen, C. K. (2006). Characterization of the potent neuroprotective properties of the natural vitamin E α-tocotrienol. *J. Neurochem.* **98**(5), 1474–1486.

Khosla, P., Patel, V., Whinter, J. M., Khanna, S., Rakhkovskaya, M., Roy, S., and Sen, C. K. (2006). Postprandial levels of the natural vitamin E tocotrienol in human circulation. *Antioxid. Redox Signal.* **8,** 1059–1068.

Kitagawa, K., Matsumoto, M., and Hori, M. (2004). Cerebral ischemia in 5-lipoxygenase knockout mice. *Brain Res.* **1004,** 198–202.

Kluth, D., Landes, N., Pfluger, P., Muller-Schmehl, K., Weiss, K., Bumke-Vogt, C., Ristow, M., and Brigelius-Flohe, R. (2005). Modulation of Cyp3a11 mRNA expression by α-tocopherol but not γ-tocotrienol in mice. *Free Radic. Biol. Med.* **38,** 507–514.

Koba, K., Abe, K., Ikeda, I., and Sugano, M. (1992). Effects of α-tocopherol and tocotrienols on blood pressure and linoleic acid metabolism in the spontaneously hypertensive rat (SHR). *Biosci. Biotechnol. Biochem.* **56,** 1420–1423.

Komiyama, K., Iizuka, K., Yamaoka, M., Watanabe, H., Tsuchiya, N., and Umezawa, I. (1989). Studies on the biological activity of tocotrienols. *Chem. Pharm. Bull.* **37,** 1369–1371.

Kumar, K. S., Raghavan, M., Hieber, K., Ege, C., Mog, S., Parra, N., Hildabrand, A., Singh, V., Srinivasan, V., Toles, R., Karikari, P., Petrovics, G., *et al.* (2006). Preferential radiation sensitization of prostate cancer in nude mice by nutraceutical antioxidant γ-tocotrienol. *Life Sci.* **78,** 2099–2104.

Kwon, K. J., Jung, Y. S., Lee, S. H., Moon, C. H., and Baik, E. J. (2005). Arachidonic acid induces neuronal death through lipoxygenase and cytochrome P450 rather than cyclooxygenase. *J. Neurosci. Res.* **81,** 73–84.

Landes, N., Pfluger, P., Kluth, D., Birringer, M., Ruhl, R., Bol, G. F., Glatt, H., and Brigelius-Flohe, R. (2003). Vitamin E activates gene expression via the pregnane X receptor. *Biochem. Pharmacol.* **65,** 269–273.

Lau, A. F. (2005). c-Src: Bridging the gap between phosphorylation- and acidification-induced gap junction channel closure. *Sci. STKE* **2005,** pe33.

Lebeau, A., Esclaire, F., Rostene, W., and Pelaprat, D. (2001). Baicalein protects cortical neurons from β-amyloid (25–35) induced toxicity. *Neuroreport* **12,** 2199–2202.

Lee, J. R., and Koretzky, G. A. (1998). Extracellular signal-regulated kinase-2, but not c-Jun NH2-terminal kinase, activation correlates with surface IgM-mediated apoptosis in the WEHI 231 B cell line. *J. Immunol.* **161,** 1637–1644.

Lennmyr, F., Ericsson, A., Gerwins, P., Akterin, S., Ahlstrom, H., and Terent, A. (2004). Src family kinase-inhibitor PP2 reduces focal ischemic brain injury. *Acta Neurol. Scand.* **110,** 175–179.

Leonard, S. W., Paterson, E., Atkinson, J. K., Ramakrishnan, R., Cross, C. E., and Traber, M. G. (2005). Studies in humans using deuterium-labeled α- and γ-tocopherols demonstrate faster plasma γ-tocopherol disappearance and greater γ-metabolite production. *Free Radic. Biol. Med.* **38,** 857–866.

Lepley, R. A., Muskardin, D. T., and Fitzpatrick, F. A. (1996). Tyrosine kinase activity modulates catalysis and translocation of cellular 5-lipoxygenase. *J. Biol. Chem.* **271,** 6179–6184.

Li, Y., Maher, P., and Schubert, D. (1997). A role for 12-lipoxygenase in nerve cell death caused by glutathione depletion. *Neuron* **19,** 453–463.

Lodge, J. K., Ridlington, J., Leonard, S., Vaule, H., and Traber, M. G. (2001). Alpha- and γ-tocotrienols are metabolized to carboxyethyl-hydroxychroman derivatives and excreted in human urine. *Lipids* **36,** 43–48.

Lodge, J. K., Hall, W. L., Jeanes, Y. M., and Proteggente, A. R. (2004). Physiological factors influencing vitamin e biokinetics. *Ann. NY Acad. Sci.* **1031,** 60–73.

Maness, P. F., Aubry, M., Shores, C. G., Frame, L., and Pfenninger, K. H. (1988). c-src gene product in developing rat brain is enriched in nerve growth cone membranes. *Proc. Natl. Acad. Sci. USA* **85,** 5001–5005.

Mazlan, M., Sue Mian, T., Mat Top, G., and Zurinah Wan Ngah, W. (2006). Comparative effects of α-tocopherol and γ-tocotrienol against hydrogen peroxide induced apoptosis on primary-cultured astrocytes. *J. Neurol. Sci.* **243,** 5–12.

McIntyre, B. S., Briski, K. P., Gapor, A., and Sylvester, P. W. (2000a). Antiproliferative and apoptotic effects of tocopherols and tocotrienols on preneoplastic and neoplastic mouse mammary epithelial cells. *Proc. Soc. Exp. Biol. Med.* **224,** 292–301.

McIntyre, B. S., Briski, K. P., Tirmenstein, M. A., Fariss, M. W., Gapor, A., and Sylvester, P. W. (2000b). Antiproliferative and apoptotic effects of tocopherols and tocotrienols on normal mouse mammary epithelial cells. *Lipids* **35,** 171–180.

Mensink, R. P., van Houwelingen, A. C., Kromhout, D., and Hornstra, G. (1999). A vitamin E concentrate rich in tocotrienols had no effect on serum lipids, lipoproteins, or platelet function in men with mildly elevated serum lipid concentrations. *Am. J. Clin. Nutr.* **69,** 213–219.

Meyenberg, A., Goldblum, D., Zingg, J. M., Azzi, A., Nesaretnam, K., Kilchenmann, M., and Frueh, B. E. (2005). Tocotrienol inhibits proliferation of human Tenon's fibroblasts *in vitro*: A comparative study with vitamin E forms and mitomycin C. *Graefes Arch. Clin. Exp. Ophthalmol.* **243,** 1263–1271.

Miller, E. R., III, Pastor-Barriuso, R., Dalal, D., Riemersma, R. A., Appel, L. J., and Guallar, E. (2005). Meta-analysis: High-dosage vitamin E supplementation may increase all-cause mortality. *Ann. Intern. Med.* **142,** 37–46.

Minhajuddin, M., Beg, Z. H., and Iqbal, J. (2005). Hypolipidemic and antioxidant properties of tocotrienol rich fraction isolated from rice bran oil in experimentally induced hyperlipidemic rats. *Food Chem. Toxicol.* **43,** 747–753.

Mishima, K., Tanaka, T., Pu, F., Egashira, N., Iwasaki, K., Hidaka, R., Matsunaga, K., Takata, J., Karube, Y., and Fujiwara, M. (2003). Vitamin E isoforms α-tocotrienol and γ-tocopherol prevent cerebral infarction in mice. *Neurosci. Lett.* **337,** 56–60.

Miyamoto, T., Lindgren, J. A., Hokfelt, T., and Samuelsson, B. (1987a). Formation of lipoxygenase products in the rat brain. *Adv. Prostaglandin Thromboxane Leukot. Res.* **17B,** 929–933.

Miyamoto, T., Lindgren, J. A., Hokfelt, T., and Samuelsson, B. (1987b). Regional distribution of leukotriene and mono-hydroxyeicosatetraenoic acid production in the rat brain. Highest leukotriene C4 formation in the hypothalamus. *FEBS Lett.* **216,** 123–127.

Miyazawa, T., Inokuchi, H., Hirokane, H., Tsuzuki, T., Nakagawa, K., and Igarashi, M. (2004). Anti-angiogenic potential of tocotrienol *in vitro*. *Biochemistry (Mosc.)* **69,** 67–69.

Mizushina, Y., Nakagawa, K., Shibata, A., Awata, Y., Kuriyama, I., Shimazaki, N., Koiwai, O., Uchiyama, Y., Sakaguchi, K., Miyazawa, T., and Yoshida, H. (2006). Inhibitory effect of tocotrienol on eukaryotic DNA polymerase lambda and angiogenesis. *Biochem. Biophys. Res. Commun.* **339,** 949–955.

Mo, H., and Elson, C. E. (1999). Apoptosis and cell-cycle arrest in human and murine tumor cells are initiated by isoprenoids. *J. Nutr.* **129,** 804–813.

Musiek, E. S., Breeding, R. S., Milne, G. L., Zanoni, G., Morrow, J. D., and McLaughlin, B. (2006). Cyclopentenone isoprostanes are novel bioactive products of lipid oxidation which enhance neurodegeneration. *J. Neurochem.* **97,** 1301–1313.

Nafeeza, M. I., and Kang, T. T. (2005). Synergistic effects of tocopherol, tocotrienol, and ubiquinone in indomethacin-induced experimental gastric lesions. *Int. J. Vitam. Nutr. Res.* **75,** 149–155.

Nafeeza, M. I., Fauzee, A. M., Kamsiah, J., and Gapor, M. T. (2002). Comparative effects of a tocotrienol-rich fraction and tocopherol in aspirin-induced gastric lesions in rats. *Asia Pac. J. Clin. Nutr.* **11,** 309–313.

Naguib, Y., Hari, S. P., Passwater, R., Jr., and Huang, D. (2003). Antioxidant activities of natural vitamin E formulations. *J. Nutr. Sci. Vitaminol. (Tokyo)* **49,** 217–220.

Naito, Y., Shimozawa, M., Kuroda, M., Nakabe, N., Manabe, H., Katada, K., Kokura, S., Ichikawa, H., Yoshida, N., Noguchi, N., and Yoshikawa, T. (2005). Tocotrienols reduce 25-hydroxycholesterol-induced monocyte-endothelial cell interaction by inhibiting the surface expression of adhesion molecules. *Atherosclerosis* **180,** 19–25.

Nakagawa, K., Eitsuka, T., Inokuchi, H., and Miyazawa, T. (2004). DNA chip analysis of comprehensive food function: Inhibition of angiogenesis and telomerase activity with unsaturated vitamin E, tocotrienol. *Biofactors* **21,** 5–10.

Nesaretnam, K., Guthrie, N., Chambers, A. F., and Carroll, K. K. (1995). Effect of tocotrienols on the growth of a human breast cancer cell line in culture. *Lipids* **30,** 1139–1143.

Nesaretnam, K., Stephen, R., Dils, R., and Darbre, P. (1998). Tocotrienols inhibit the growth of human breast cancer cells irrespective of estrogen receptor status. *Lipids* **33,** 461–469.

Nesaretnam, K., Dorasamy, S., and Darbre, P. D. (2000). Tocotrienols inhibit growth of ZR-75–1 breast cancer cells. *Int. J. Food Sci. Nutr.* **51**(Suppl.), S95–S103.

Nesaretnam, K., Ambra, R., Selvaduray, K. R., Radhakrishnan, A., Reimann, K., Razak, G., and Virgili, F. (2004). Tocotrienol-rich fraction from palm oil affects gene expression in tumors resulting from MCF-7 cell inoculation in athymic mice. *Lipids* **39,** 459–467.

Newaz, M. A., and Nawal, N. N. (1999). Effect of γ-tocotrienol on blood pressure, lipid peroxidation and total antioxidant status in spontaneously hypertensive rats (SHR). *Clin. Exp. Hypertens. (New York)* **21,** 1297–1313.

Newaz, M. A., Yousefipour, Z., Nawal, N., and Adeeb, N. (2003). Nitric oxide synthase activity in blood vessels of spontaneously hypertensive rats: Antioxidant protection by γ-tocotrienol. *J. Physiol. Pharmacol.* **54,** 319–327.

Ngah, W. Z., Jarien, Z., San, M. M., Marzuki, A., Top, G. M., Shamaan, N. A., and Kadir, K. A. (1991). Effect of tocotrienols on hepatocarcinogenesis induced by 2-acetylaminofluorene in rats. *Am. J. Clin. Nutr.* **53,** 1076S–1081S.

Nishiyama, M., Okamoto, H., Watanabe, T., Hori, T., Hada, T., Ueda, N., Yamamoto, S., Tsukamoto, H., Watanabe, K., and Kirino, T. (1992). Localization of arachidonate 12-lipoxygenase in canine brain tissues. *J. Neurochem.* **58,** 1395–1400.

Nishiyama, M., Watanabe, T., Ueda, N., Tsukamoto, H., and Watanabe, K. (1993). Arachidonate 12-lipoxygenase is localized in neurons, glial cells, and endothelial cells of the canine brain. *J. Histochem. Cytochem.* **41,** 111–117.

Noguchi, N., Hanyu, R., Nonaka, A., Okimoto, Y., and Kodama, T. (2003). Inhibition of THP-1 cell adhesion to endothelial cells by α-tocopherol and α-tocotrienol is dependent on intracellular concentration of the antioxidants. *Free Radic. Biol. Med.* **34,** 1614–1620.

Norazlina, M., Ima-Nirwana, S., Abul Gapor, M. T., and Abdul Kadir Khalid, B. (2002). Tocotrienols are needed for normal bone calcification in growing female rats. *Asia Pac. J. Clin. Nutr.* **11,** 194–199.

Numakawa, Y., Numakawa, T., Matsumoto, T., Yagasaki, Y., Kumamaru, E., Kunugi, H., Taguchi, T., and Niki, E. (2006). Vitamin E protected cultured cortical neurons from oxidative stress-induced cell death through the activation of mitogen-activated protein kinase and phosphatidylinositol 3-kinase. *J. Neurochem.* **97,** 1191–1202.

O'Byrne, D., Grundy, S., Packer, L., Devaraj, S., Baldenius, K., Hoppe, P. P., Kraemer, K., Jialal, I., and Traber, M. G. (2000). Studies of LDL oxidation following α-, γ-, or δ-tocotrienyl acetate supplementation of hypercholesterolemic humans. *Free Radic. Biol. Med.* **29,** 834–845.

Okabe, M., Oji, M., Ikeda, I., Tachibana, H., and Yamada, K. (2002). Tocotrienol levels in various tissues of Sprague-Dawley rats after intragastric administration of tocotrienols. *Biosci. Biotechnol. Biochem.* **66,** 1768–1771.

Osakada, F., Hashino, A., Kume, T., Katsuki, H., Kaneko, S., and Akaike, A. (2004). α-Tocotrienol provides the most potent neuroprotection among vitamin E analogs on cultured striatal neurons. *Neuropharmacology* **47,** 904–915.

Palozza, P., Verdecchia, S., Avanzi, L., Vertuani, S., Serini, S., Iannone, A., and Manfredini, S. (2006). Comparative antioxidant activity of tocotrienols and the novel chromanyl-polyisoprenyl molecule FeAox-6 in isolated membranes and intact cells. *Mol. Cell. Biochem.* **287,** 21–32.

Panfili, G., Fratianni, A., and Irano, M. (2003). Normal phase high-performance liquid chromatography method for the determination of tocopherols and tocotrienols in cereals. *J. Agric. Food Chem.* **51,** 3940–3944.

Parker, R. A., Pearce, B. C., Clark, R. W., Gordon, D. A., and Wright, J. J. (1993). Tocotrienols regulate cholesterol production in mammalian cells by post-transcriptional suppression of 3-hydroxy-3-methylglutaryl-coenzyme A reductase. *J. Biol. Chem.* **268,** 11230–11238.

Paul, R., Zhang, Z. G., Eliceiri, B. P., Jiang, Q., Boccia, A. D., Zhang, R. L., Chopp, M., and Cheresh, D. A. (2001). Src deficiency or blockade of Src activity in mice provides cerebral protection following stroke. *Nat. Med.* **7,** 222–227.

Pearce, B. C., Parker, R. A., Deason, M. E., Qureshi, A. A., and Wright, J. J. (1992). Hypocholesterolemic activity of synthetic and natural tocotrienols. *J. Med. Chem.* **35,** 3595–3606.

Pearce, B. C., Parker, R. A., Deason, M. E., Dischino, D. D., Gillespie, E., Qureshi, A. A., Volk, K., and Wright, J. J. (1994). Inhibitors of cholesterol biosynthesis. 2. Hypocholesterolemic and antioxidant activities of benzopyran and tetrahydronaphthalene analogues of the tocotrienols. *J. Med. Chem.* **37,** 526–541.

Podda, M., Weber, C., Traber, M. G., and Packer, L. (1996). Simultaneous determination of tissue tocopherols, tocotrienols, ubiquinols, and ubiquinones. *J. Lipid Res.* **37,** 893–901.

Porfirova, S., Bergmuller, E., Tropf, S., Lemke, R., and Dormann, P. (2002). Isolation of an Arabidopsis mutant lacking vitamin E and identification of a cyclase essential for all tocopherol biosynthesis. *Proc. Natl. Acad. Sci. USA* **99,** 12495–12500.

Pyper, J. M., and Bolen, J. B. (1989). Neuron-specific splicing of C-SRC RNA in human brain. *J. Neurosci. Res.* **24,** 89–96.

Qureshi, A. A., and Peterson, D. M. (2001). The combined effects of novel tocotrienols and lovastatin on lipid metabolism in chickens. *Atherosclerosis* **156,** 39–47.

Qureshi, A. A., Burger, W. C., Peterson, D. M., and Elson, C. E. (1986). The structure of an inhibitor of cholesterol biosynthesis isolated from barley. *J. Biol. Chem.* **261,** 10544–10550.

Qureshi, A. A., Qureshi, N., Hasler-Rapacz, J. O., Weber, F. E., Chaudhary, V., Crenshaw, T. D., Gapor, A., Ong, A. S., Chong, Y. H., and Peterson, D. (1991a). Dietary tocotrienols reduce concentrations of plasma cholesterol, apolipoprotein B, thromboxane B2, and platelet factor 4 in pigs with inherited hyperlipidemias. *Am. J. Clin. Nutr.* **53,** 1042S–1046S.

Qureshi, A. A., Qureshi, N., Wright, J. J., Shen, Z., Kramer, G., Gapor, A., Chong, Y. H., DeWitt, G., Ong, A., and Peterson, D. M. (1991b). Lowering of serum cholesterol in hypercholesterolemic humans by tocotrienols (palmvitee). *Am. J. Clin. Nutr.* **53,** 1021S–1026S.

Qureshi, A. A., Bradlow, B. A., Brace, L., Manganello, J., Peterson, D. M., Pearce, B. C., Wright, J. J., Gapor, A., and Elson, C. E. (1995). Response of hypercholesterolemic subjects to administration of tocotrienols. *Lipids* **30,** 1171–1177.

Qureshi, A. A., Mo, H., Packer, L., and Peterson, D. M. (2000). Isolation and identification of novel tocotrienols from rice bran with hypocholesterolemic, antioxidant, and antitumor properties. *J. Agric. Food Chem.* **48,** 3130–3140.

Qureshi, A. A., Peterson, D. M., Hasler-Rapacz, J. O., and Rapacz, J. (2001a). Novel tocotrienols of rice bran suppress cholesterogenesis in hereditary hypercholesterolemic swine. *J. Nutr.* **131,** 223–230.

Qureshi, A. A., Salser, W. A., Parmar, R., and Emeson, E. E. (2001b). Novel tocotrienols of rice bran inhibit atherosclerotic lesions in C57BL/6 ApoE-deficient mice. *J. Nutr.* **131,** 2606–2618.

Qureshi, A. A., Sami, S. A., Salser, W. A., and Khan, F. A. (2001c). Synergistic effect of tocotrienol-rich fraction (TRF(25)) of rice bran and lovastatin on lipid parameters in hypercholesterolemic humans. *J. Nutr. Biochem.* **12,** 318–329.

Qureshi, A. A., Sami, S. A., Salser, W. A., and Khan, F. A. (2002). Dose-dependent suppression of serum cholesterol by tocotrienol-rich fraction (TRF25) of rice bran in hypercholesterolemic humans. *Atherosclerosis* **161,** 199–207.

Raederstorff, D., Elste, V., Aebischer, C., and Weber, P. (2002). Effect of either γ-tocotrienol or a tocotrienol mixture on the plasma lipid profile in hamsters. *Ann. Nutr. Metab.* **46,** 17–23.

Rahmat, A., Ngah, W. Z., Shamaan, N. A., Gapor, A., and Abdul Kadir, K. (1993). Long-term administration of tocotrienols and tumor-marker enzyme activities during hepatocarcinogenesis in rats. *Nutrition* **9,** 229–232.

Reddanna, P., Rao, M. K., and Reddy, C. C. (1985). Inhibition of 5-lipoxygenase by vitamin E. *FEBS Lett.* **193,** 39–43.

Rippert, P., Scimemi, C., Dubald, M., and Matringe, M. (2004). Engineering plant shikimate pathway for production of tocotrienol and improving herbicide resistance. *Plant Physiol.* **134,** 92–100.

Roy, S., Lado, B. H., Khanna, S., and Sen, C. K. (2002). Vitamin E sensitive genes in the developing rat fetal brain: A high-density oligonucleotide microarray analysis. *FEBS Lett.* **530,** 17–23.

Saito, H., Kiyose, C., Yoshimura, H., Ueda, T., Kondo, K., and Igarashi, O. (2003). γ-Tocotrienol, a vitamin E homolog, is a natriuretic hormone precursor. *J. Lipid Res.* **44,** 1530–1535.

Sakai, M., Okabe, M., Yamasaki, M., Tachibana, H., and Yamada, K. (2004). Induction of apoptosis by tocotrienol in rat hepatoma dRLh-84 cells. *Anticancer Res.* **24,** 1683–1688.

Sakai, M., Okabe, M., Tachibana, H., and Yamada, K. (2005). Apoptosis induction by gamma-tocotrienol in human hepatoma Hep3B cells. *J. Nutr. Biochem.* **17**(10), 672–676.

Schroeder, M. T., Becker, E. M., and Skibsted, L. H. (2006). Molecular mechanism of antioxidant synergism of tocotrienols and carotenoids in palm oil. *J. Agric. Food Chem.* **54,** 3445–3453.

Schubert, D., and Piasecki, D. (2001). Oxidative glutamate toxicity can be a component of the excitotoxicity cascade. *J. Neurosci.* **21,** 7455–7462.

Schulz, J. B., Lindenau, J., Seyfried, J., and Dichgans, J. (2000). Glutathione, oxidative stress and neurodegeneration. *Eur. J. Biochem.* **267,** 4904–4911.

Schwarz, K. (1965). Role of vitamin E, selenium, and related factors in experimental nutritional liver disease. *Fed. Proc.* **24,** 58–67.

Schwedhelm, E., Maas, R., Troost, R., and Boger, R. H. (2003). Clinical pharmacokinetics of antioxidants and their impact on systemic oxidative stress. *Clin. Pharmacokinet.* **42,** 437–459.

Sen, C. K., Khanna, S., Roy, S., and Packer, L. (2000). Molecular basis of vitamin E action. Tocotrienol potently inhibits glutamate-induced pp60(c-Src) kinase activation and death of HT4 neuronal cells. *J. Biol. Chem.* **275,** 13049–13055.

Sen, C. K., Khanna, S., and Roy, S. (2004). Tocotrienol: The natural vitamin E to defend the nervous system? *Ann. NY Acad. Sci.* **1031,** 127–142.

Sen, C. K., Khanna, S., and Roy, S. (2006). Tocotrienols: Vitamin E beyond tocopherols. *Life Sci.* **78,** 2088–2098.

Serbinova, E. A., and Packer, L. (1994). Antioxidant properties of α-tocopherol and α-tocotrienol. *Methods Enzymol.* **234,** 354–366.

Serbinova, E., Kagan, V., Han, D., and Packer, L. (1991). Free radical recycling and intramembrane mobility in the antioxidant properties of α-tocopherol and α-tocotrienol. *Free Radic. Biol. Med.* **10,** 263–275.

Sever, N., Song, B. L., Yabe, D., Goldstein, J. L., Brown, M. S., and DeBose-Boyd, R. A. (2003). Insig-dependent ubiquitination and degradation of mammalian 3-hydroxy-3-methylglutaryl-CoA reductase stimulated by sterols and geranylgeraniol. *J. Biol. Chem.* **278,** 52479–52490.

Shah, S., and Sylvester, P. W. (2004). Tocotrienol-induced caspase-8 activation is unrelated to death receptor apoptotic signaling in neoplastic mammary epithelial cells. *Exp. Biol. Med. (Maywood)* **229,** 745–755.

Shah, S. J., and Sylvester, P. W. (2005a). Gamma-tocotrienol inhibits neoplastic mammary epithelial cell proliferation by decreasing Akt and nuclear factor kappaB activity. *Exp. Biol. Med. (Maywood)* **230,** 235–241.

Shah, S. J., and Sylvester, P. W. (2005b). Tocotrienol-induced cytotoxicity is unrelated to mitochondrial stress apoptotic signaling in neoplastic mammary epithelial cells. *Biochem. Cell Biol.* **83,** 86–95.

Shah, S., Gapor, A., and Sylvester, P. W. (2003). Role of caspase-8 activation in mediating vitamin E-induced apoptosis in murine mammary cancer cells. *Nutr. Cancer* **45,** 236–246.

Shay, J. W., and Wright, W. E. (2006). Telomerase therapeutics for cancer: Challenges and new directions. *Nat. Rev. Drug Discov.* **5,** 577–584.

Shohami, E., Glantz, L., Nates, J., and Feuerstein, G. (1992). The mixed lipoxygenase/cyclooxygenase inhibitor SK&F 105809 reduces cerebral edema after closed head injury in rat. *J. Basic. Clin. Physiol. Pharmacol.* **3,** 99–107.

Shun, M. C., Yu, W., Gapor, A., Parsons, R., Atkinson, J., Sanders, B. G., and Kline, K. (2004). Pro-apoptotic mechanisms of action of a novel vitamin E analog (α-TEA) and a naturally occurring form of vitamin E (δ-tocotrienol) in MDA-MB-435 human breast cancer cells. *Nutr. Cancer* **48,** 95–105.

Shupnik, M. A. (2004). Crosstalk between steroid receptors and the c-Src-receptor tyrosine kinase pathways: Implications for cell proliferation. *Oncogene* **23,** 7979–7989.

Soelaiman, I. N., Ahmad, N. S., and Khalid, B. A. (2004). Palm oil tocotrienol mixture is better than α-tocopherol acetate in protecting bones against free-radical induced elevation of bone-resorbing cytokines. *Asia Pac. J. Clin. Nutr.* **13,** S111.

Solomons, N. W., and Orozco, M. (2003). Alleviation of vitamin A deficiency with palm fruit and its products. *Asia Pac. J. Clin. Nutr.* **12,** 373–384.

Song, B. L., and Debose-Boyd, R. A. (2006). Insig-dependent ubiquitination and degradation of 3-hydroxy-3-methylglutaryl coenzyme A reductase stimulated by δ- and γ-tocotrienols. *J. Biol. Chem.* **281,** 25054–25061.

Sonnen, A. F., Bakirci, H., Netscher, T., and Nau, W. M. (2005). Effect of temperature, cholesterol content, and antioxidant structure on the mobility of vitamin E constituents in biomembrane models studied by laterally diffusion-controlled fluorescence quenching. *J. Am. Chem. Soc.* **127,** 15575–15584.

Soriano, P., Montgomery, C., Geske, R., and Bradley, A. (1991). Targeted disruption of the c-src proto-oncogene leads to osteopetrosis in mice. *Cell* **64,** 693–702.

Srivastava, J. K., and Gupta, S. (2006). Tocotrienol-rich fraction of palm oil induces cell cycle arrest and apoptosis selectively in human prostate cancer cells. *Biochem. Biophys. Res. Commun.* **346,** 447–453.

Steiner, M. (1993). Vitamin E: More than an antioxidant. *Clin. Cardiol.* **16,** I16–I18.

Suarna, C., Hood, R. L., Dean, R. T., and Stocker, R. (1993). Comparative antioxidant activity of tocotrienols and other natural lipid-soluble antioxidants in a homogeneous system, and in rat and human lipoproteins. *Biochim. Biophys. Acta* **1166,** 163–170.

Sugano, M., Koba, K., and Tsuji, E. (1999). Health benefits of rice bran oil. *Anticancer Res.* **19,** 3651–3657.

Sun, X., Shih, A. Y., Johannssen, H. C., Erb, H., Li, P., and Murphy, T. H. (2006). Two-photon imaging of glutathione levels in intact brain indicates enhanced redox buffering in developing neurons and cells at the cerebrospinal fluid and blood-brain interface. *J. Biol. Chem.* **281,** 17420–17431.

Sundram, K., Khor, H. T., Ong, A. S., and Pathmanathan, R. (1989). Effect of dietary palm oils on mammary carcinogenesis in female rats induced by 7,12-dimethylbenz(a)anthracene. *Cancer Res.* **49,** 1447–1451.

Sundram, K., Sambanthamurthi, R., and Tan, Y. A. (2003). Palm fruit chemistry and nutrition. *Asia Pac. J. Clin. Nutr.* **12,** 355–362.

Suzuki, Y. J., Tsuchiya, M., Wassall, S. R., Choo, Y. M., Govil, G., Kagan, V. E., and Packer, L. (1993). Structural and dynamic membrane properties of α-tocopherol and α-tocotrienol: Implication to the molecular mechanism of their antioxidant potency. *Biochemistry* **32,** 10692–10699.

Sylvester, P. W., and Shah, S. (2005a). Intracellular mechanisms mediating tocotrienol-induced apoptosis in neoplastic mammary epithelial cells. *Asia Pac. J. Clin. Nutr.* **14,** 366–373.

Sylvester, P. W., and Shah, S. J. (2005b). Mechanisms mediating the antiproliferative and apoptotic effects of vitamin E in mammary cancer cells. *Front Biosci.* **10,** 699–709.

Sylvester, P. W., McIntyre, B. S., Gapor, A., and Briski, K. P. (2001). Vitamin E inhibition of normal mammary epithelial cell growth is associated with a reduction in protein kinase C(α) activation. *Cell Prolif.* **34,** 347–357.

Sylvester, P. W., Nachnani, A., Shah, S., and Briski, K. P. (2002). Role of GTP-binding proteins in reversing the antiproliferative effects of tocotrienols in preneoplastic mammary epithelial cells. *Asia Pac. J. Clin. Nutr.* **11**(Suppl. 7), S452–S459.

Sylvester, P. W., Shah, S. J., and Samant, G. V. (2005). Intracellular signaling mechanisms mediating the antiproliferative and apoptotic effects of γ-tocotrienol in neoplastic mammary epithelial cells. *J. Plant Physiol.* **162,** 803–810.

Takahashi, K., and Loo, G. (2004). Disruption of mitochondria during tocotrienol-induced apoptosis in MDA-MB-231 human breast cancer cells. *Biochem. Pharmacol.* **67,** 315–324.

Tan, D. T., Khor, H. T., Low, W. H., Ali, A., and Gapor, A. (1991). Effect of a palm-oil-vitamin E concentrate on the serum and lipoprotein lipids in humans. *Am. J. Clin. Nutr.* **53,** 1027S–1030S.

Tan, S., Schubert, D., and Maher, P. (2001). Oxytosis: A novel form of programmed cell death. *Curr. Top. Med. Chem.* **1,** 497–506.

Tanito, M., Itoh, N., Yoshida, Y., Hayakawa, M., Ohira, A., and Niki, E. (2004). Distribution of tocopherols and tocotrienols to rat ocular tissues after topical ophthalmic administration. *Lipids* **39,** 469–474.

Tarrago-Trani, M. T., Phillips, K. M., Lemar, L. E., and Holden, J. M. (2006). New and existing oils and fats used in products with reduced trans-fatty acid content. *J. Am. Diet. Assoc.* **106,** 867–880.

Terasawa, Y., Ladha, Z., Leonard, S. W., Morrow, J. D., Newland, D., Sanan, D., Packer, L., Traber, M. G., and Farese, R. V., Jr. (2000). Increased atherosclerosis in hyperlipidemic mice deficient in α-tocopherol transfer protein and vitamin E. *Proc. Natl. Acad. Sci. USA* **97,** 13830–13834.

Theriault, A., Wang, Q., Gapor, A., and Adeli, K. (1999). Effects of γ-tocotrienol on ApoB synthesis, degradation, and secretion in HepG2 cells. *Arterioscler. Thromb. Vasc. Biol.* **19,** 704–712.

Theriault, A., Chao, J. T., and Gapor, A. (2002). Tocotrienol is the most effective vitamin E for reducing endothelial expression of adhesion molecules and adhesion to monocytes. *Atherosclerosis* **160,** 21–30.

Thomas, S. M., and Brugge, J. S. (1997). Cellular functions regulated by Src family kinases. *Annu. Rev. Cell Dev. Biol.* **13,** 513–609.

Tiahou, G., Maire, B., Dupuy, A., Delage, M., Vernet, M. H., Mathieu-Daude, J. C., Michel, F., Sess, E. D., and Cristol, J. P. (2004). Lack of oxidative stress in a selenium deficient area in Ivory Coast Potential nutritional antioxidant role of crude palm oil. *Eur. J. Nutr.* **43,** 367–374.

Tomeo, A. C., Geller, M., Watkins, T. R., Gapor, A., and Bierenbaum, M. L. (1995). Antioxidant effects of tocotrienols in patients with hyperlipidemia and carotid stenosis. *Lipids* **30,** 1179–1183.

Traber, M. G., and Arai, H. (1999). Molecular mechanisms of vitamin E transport. *Annu. Rev. Nutr.* **19,** 343–355.

Traber, M. G., Burton, G. W., and Hamilton, R. L. (2004). Vitamin E trafficking. *Ann. NY Acad. Sci.* **1031,** 1–12.

van der Worp, H. B., Bar, P. R., Kappelle, L. J., and de Wildt, D. J. (1998). Dietary vitamin E levels affect outcome of permanent focal cerebral ischemia in rats. *Stroke* **29,** 1002–1005; discussion 1005–1006.

van Haaften, R. I., Haenen, G. R., Evelo, C. T., and Bast, A. (2002). Tocotrienols inhibit human glutathione S-transferase P1–1. *IUBMB Life* **54,** 81–84.

Venkatesh, T. V., Karunanandaa, B., Free, D. L., Rottnek, J. M., Baszis, S. R., and Valentin, H. E. (2006). Identification and characterization of an Arabidopsis homogentisate phytyltransferase paralog. *Planta* **223,** 1134–1144.

Vraka, P. S., Drouza, C., Rikkou, M. P., Odysseos, A. D., and Keramidas, A. D. (2006). Synthesis and study of the cancer cell growth inhibitory properties of α-, γ-tocopheryl and γ-tocotrienyl 2-phenylselenyl succinates. *Bioorg. Med. Chem.* **14,** 2684–2696.

Wada, S., Satomi, Y., Murakoshi, M., Noguchi, N., Yoshikawa, T., and Nishino, H. (2005). Tumor suppressive effects of tocotrienol *in vivo* and *in vitro*. *Cancer Lett.* **229,** 181–191.

Wagner, K. H., Kamal-Eldin, A., and Elmadfa, I. (2004). Gamma-tocopherol—an underestimated vitamin? *Ann. Nutr. Metab.* **48,** 169–188.

Wan Nazaimoon, W. M., and Khalid, B. A. (2002). Tocotrienols-rich diet decreases advanced glycosylation end-products in non-diabetic rats and improves glycemic control in streptozotocin-induced diabetic rats. *Malays J. Pathol.* **24,** 77–82.

Watkins, T., Lenz, P., Gapor, A., Struck, M., Tomeo, A., and Bierenbaum, M. (1993). γ-Tocotrienol as a hypocholesterolemic and antioxidant agent in rats fed atherogenic diets. *Lipids* **28,** 1113–1118.

Weber, S. U., Thiele, J. J., Han, N., Luu, C., Valacchi, G., Weber, S., and Packer, L. (2003). Topical α-tocotrienol supplementation inhibits lipid peroxidation but fails to mitigate increased transepidermal water loss after benzoyl peroxide treatment of human skin. *Free Radic. Biol. Med.* **34,** 170–176.

Whittle, K. J., Dunphy, P. J., and Pennock, J. F. (1966). The isolation and properties of δ-tocotrienol from Hevea latex. *Biochem. J.* **100,** 138–145.

Wie, M. B., Koh, J. Y., Won, M. H., Lee, J. C., Shin, T. K., Moon, C. J., Ha, H. J., Park, S. M., and Kim, H. C. (2001). BAPTA/AM, an intracellular calcium chelator, induces delayed necrosis by lipoxygenase-mediated free radicals in mouse cortical cultures. *Prog. Neuropsychopharmacol. Biol. Psychiatry* **25,** 1641–1659.

Yamamoto, S. (1992). Mammalian lipoxygenases: Molecular structures and functions. *Biochim. Biophys. Acta* **1128,** 117–131.

Yamasaki, M., Nishida, E., Nou, S., Tachibana, H., and Yamada, K. (2005). Cytotoxity of the trans10,cis12 isomer of conjugated linoleic acid on rat hepatoma and its modulation by other fatty acids, tocopherol, and tocotrienol. *In Vitro Cell. Dev. Biol. Anim.* **41,** 239–244.

Yamashita, K., Ikeda, S., Iizuka, Y., and Ikeda, I. (2002). Effect of sesaminol on plasma and tissue α-tocopherol and α-tocotrienol concentrations in rats fed a vitamin E concentrate rich in tocotrienols. *Lipids* **37,** 351–358.

Yano, Y., Satoh, H., Fukumoto, K., Kumadaki, I., Ichikawa, T., Yamada, K., Hagiwara, K., and Yano, T. (2005). Induction of cytotoxicity in human lung adenocarcinoma cells by 6-O-carboxypropyl-α-tocotrienol, a redox-silent derivative of α-tocotrienol. *Int. J. Cancer* **115,** 839–846.

Yao, Y., Clark, C. M., Trojanowski, J. Q., Lee, V. M., and Pratico, D. (2005). Elevation of 12/15 lipoxygenase products in AD and mild cognitive impairment. *Ann. Neurol.* **58,** 623–626.

Yap, S. P., Yuen, K. H., and Wong, J. W. (2001). Pharmacokinetics and bioavailability of α-, γ- and δ-tocotrienols under different food status. *J. Pharm. Pharmacol.* **53,** 67–71.

Yap, S. P., Yuen, K. H., and Lim, A. B. (2003). Influence of route of administration on the absorption and disposition of α-, γ- and delta-tocotrienols in rats. *J. Pharm. Pharmacol.* **55,** 53–58.

Yoneda, T., Lowe, C., Lee, C. H., Gutierrez, G., Niewolna, M., Williams, P. J., Izbicka, E., Uehara, Y., and Mundy, G. R. (1993). Herbimycin A, a pp60c-src tyrosine kinase inhibitor, inhibits osteoclastic bone resorption *in vitro* and hypercalcemia *in vivo*. *J. Clin. Invest.* **91,** 2791–2795.

Yoshida, Y., Niki, E., and Noguchi, N. (2003). Comparative study on the action of tocopherols and tocotrienols as antioxidant: Chemical and physical effects. *Chem. Phys. Lipids* **123,** 63–75.

Yoshida, Y., Itoh, N., Hayakawa, M., Piga, R., Cynshi, O., Jishage, K., and Niki, E. (2005). Lipid peroxidation induced by carbon tetrachloride and its inhibition by antioxidant as evaluated by an oxidative stress marker, HODE. *Toxicol. Appl. Pharmacol.* **208,** 87–97.

Yu, F. L., Gapor, A., and Bender, W. (2005). Evidence for the preventive effect of the polyunsaturated phytol side chain in tocotrienols on 17β-estradiol epoxidation. *Cancer Detect. Prev.* **29,** 383–388.

Yu, S. G., Thomas, A. M., Gapor, A., Tan, B., Qureshi, N., and Qureshi, A. A. (2006). Dose-response impact of various tocotrienols on serum lipid parameters in 5-week-old female chickens. *Lipids* **41,** 453–461.

Yu, W., Simmons-Menchaca, M., Gapor, A., Sanders, B. G., and Kline, K. (1999). Induction of apoptosis in human breast cancer cells by tocopherols and tocotrienols. *Nutr. Cancer* **33,** 26–32.

Zhou, C., Tabb, M. M., Sadatrafiei, A., Grun, F., and Blumberg, B. (2004). Tocotrienols activate the steroid and xenobiotic receptor, SXR, and selectively regulate expression of its target genes. *Drug Metab. Dispos.* **32,** 1075–1082.

Zingg, J. M., and Azzi, A. (2004). Non-antioxidant activities of vitamin E. *Curr. Med. Chem.* **11,** 1113–1133.

9

Vitamin E Biotransformation in Humans

Francesco Galli,*[,1] M. Cristina Polidori,[†,1] Wilhelm Stahl,[†] Patrizia Mecocci,[‡] and Frank J. Kelly[§]

**Department of Internal Medicine*
Section of Applied Biochemistry and Nutritional Sciences
University of Perugia, Italy
[†]Institute of Biochemistry and Molecular Biology I
Heinrich-Heine University, Düsseldorf, Germany
[‡]Institute of Gerontology and Geriatrics
University of Perugia, Italy
[§]Pharmaceutical Science Division
School of Biomedical and Health Sciences
King's College London, London, United Kingdom

The presence and activity of vitamin E in the organism as well as its role in disease prevention depend, as for any other microconstituent in food, on a number of factors related to its release from the food matrix,

[1]Equal contributors.

Vitamins and Hormones, Volume 76

0083-6729/07 $35.00
DOI: 10.1016/S0083-6729(07)76009-0

extent of absorption, and metabolic fate. Biotransformation can be defined as the sum of processes in which vitamin E compounds are altered by the body. It involves the bioactivation and production of reactive metabolites, a series of processes generally referred to as "vitamin E metabolism." This chapter will provide an overview of the known and less known steps of vitamin E biotransformation in humans. Due to recent advances related to the biological activities and metabolic processing of vitamin E compounds, particular attention will be given to the description of the formation, identification, and functions of vitamin E metabolites. The hypothesis of a transformation-dependent bioactivation of vitamin E represents an intriguing and emerging aspect of research that deserves further investigation.

I. INTRODUCTION

The presence and activity of vitamin E in the organism as well as its role in disease prevention depend, as for any other microconstituent in food, on a number of factors related to release from food, extent of absorption, and metabolic fate. These phases have been referred to as bioaccessibility, bioavailability, and biokinetics (Stahl *et al.*, 2002). Biotransformation can be defined as the sum of the processes by which vitamin E compounds—similarly to other micronutrients or foreign chemicals introduced into the organism—are altered by the body. This chapter will provide an overview of the known and less known steps of vitamin E biotransformation in humans. Due to recent advances related to the biological activities and metabolic processing of vitamin E isomers, particular attention will be given to the description of the formation, identification, and functions of the known vitamin E metabolites.

II. THE FATE OF VITAMIN E FROM INGESTION TO EXCRETION

Once a vitamin E-containing food has been ingested, the vitamer has first to be released from the food matrix. This step varies and is dependent on the dietary source, as shown by the observation that release of vitamin E from fats and oils occurs at a higher rate than release from nuts and seeds (Stahl *et al.*, 2002). The absorption of tocopherols and tocotrienols is generally assumed to follow that of lipids. It is, therefore, dependent on processes also necessary for fat absorption. These include the formation of mixed micelles with the aid of bile acids, activity of pancreatic lipases and esterases for hydrolysis of triglycerides, and cleavage of tocopheryl esters, as well as synthesis of vitamin E-containing chylomicrons in the intestinal mucosa with subsequent chylomicrons secretion into the lymph (Drevon, 1991).

Following chylomicron catabolism, some of the newly absorbed vitamin E is released and transferred to other lipoproteins and tissues, and some remains with the chylomicron remnants. After uptake by the liver, *RRR*-α-tocopherol and the other 2*R*-stereoisomers are differentiated from different forms of vitamin E and secreted by the liver via very low density lipoproteins (VLDL)[1]. The α-tocopherol transfer protein (α-TTP) has the function to discriminate between different tocopherols for selective enrichment of nascent VLDL with the 2*R*-stereoisomers of α-tocopherol. Subsequently, about half of the VLDL is transformed to low-density lipoproteins (LDL), with LDL being the major carrier for α-tocopherol in blood. The affinities of α-TTP for β-, γ-, and δ-tocopherols and for α-tocotrienol are 38, 9, 2, and 11% of that for α-tocopherol, respectively (reviewed in Polidori *et al.*, 2006), demonstrating its high selectivity for α-tocopherol. Cellular mechanisms leading to vitamin E assembly into VLDL are, however, still incompletely understood. The enrichment of plasma in *RRR*-α-tocopherol does not seem to correspond linearly with the enrichment of nascent VLDL precursors from either the endoplasmic reticulum or the Golgi apparatus, indicating that VLDL enrichment with α-tocopherol may occur as a post-VLDL secretory process (Traber *et al.*, 2004). It might therefore be possible that α-TTP facilitates the trafficking of α-tocopherol to the hepatocyte plasmalemma—by largely unknown mechanisms—where newly secreted, nascent VLDLs could acquire both α-tocopherol and unesterified cholesterol within the space of Disse.

A primary pathway of vitamin E delivery to tissues appears to be the direct transfer following chylomicron catabolism, with the fastest uptake occurring in lungs, liver, spleen, kidney, and erythrocytes (reviewed in Polidori *et al.*, 2006; Stahl *et al.*, 2002). Recent experiments suggest that VLDL-related delivery and storage of α-tocopherol to tissues may have a secondary role as compared to other less well-understood mechanisms.

[1]Abbreviations: AAPH, 2,2′-azo-bis(2-amidinopropane); ABC-type transport system, ATP-binding cassette-type transport system; $ABTS^+$, 2,2′-azino-bis(3-ethylbenzthiazoline-6-sulfonic acid; α-CEHC, 2,5,7,8-tetramethyl-2(2′-carboxyethyl)-6-hydroxychroman; α-TQ, α-tocopheryl quinone; α-TTP, α-tocopherol transfer protein; COX-2, cyclooxygenase-2; CYP, cytochrome P450-dependent monooxygenases; γ-CMBHC, 2,7,8-trimethyl-2-(δ-carboxymethylbutyl) 6 hydroxychroman; γ-CEHC, 2,7,8-trimethyl-2-(2′-carboxyethyl)-6-hydroxychroman; GC–MS, gas chromatography–mass spectrometry; HDL, high-density lipoproteins; HPLC, high-performance liquid chromatography; IL-1β, interleukin-1β; LDL, low-density lipoproteins; LLU-α, Loma Linda University-α; MTTP, microsomal triglyceride transfer protein; ORAC, oxygen radical absorbance capacity; SR-BI, scavenger receptor class B type I; TEAC, trolox equivalent antioxidant capacity; TNF-α, tumor necrosis factor-α; VLDL, very low density lipoproteins.

Mice lacking microsomal triglyceride transfer protein (MTTP) expression in the liver show impaired VLDL secretion as well as decreased plasma levels and increased liver stores of lipids and α-tocopherol with otherwise peripheral α-tocopherol levels nearly identical to those of control mice (Minehira-Castelli *et al.*, 2006). Accordingly, the incorporation of deuterated α-tocopherol into plasma lipoproteins and various peripheral tissues was identical in MTTP knockout and wild-type mice (Minehira-Castelli *et al.*, 2006). HDL may play a role in the liver reuptake of tocopherols as shown by experiments on scavenger receptor class B type I (SR-BI)-deficient mice (Mardones *et al.*, 2002). This pathway may recycle vitamin E together with other lipids from peripheral tissues and blood, providing a quota of vitamin E to reenter the hepatic pool for excretion or tissue redistribution depending on the liver expression of genes that are involved in its metabolism and trafficking (Mustacich *et al.*, 2006). As reviewed in Mardones and Rigotti (2004), several studies strongly suggest that HDL, by means of SR-BI-mediated selective lipid uptake, represents a significant and specific source of α-tocopherol for peripheral tissues and cells such as the endothelium and the central nervous system. SR-BI-deficient mice showed a marked impairment in α-tocopherol homeostasis with increased plasma α-tocopherol levels mainly in the HDL fraction, decreased biliary excretion of α-tocopherol, and decreased α-tocopherol levels in selected tissues such as brain, lung, and gonads (Mardones and Rigotti, 2004). Studies provide some evidence that SR-BI function may provide a mechanism for higher recycling and metabolic rate of tocopherols, particularly γ-tocopherol, in humans (Leonard *et al.*, 2005b). This type of HDL-related vitamin E delivery pathway, downstream to SR-BI, also seems to involve the function of α-TTP- and ATP-binding cassette (ABC)-type transporter system as investigated by means of a fluorescent vitamin E analogue in hepatocytes (Qian *et al.*, 2005). Mustacich *et al.* (2006) showed that the ABC-type transporter system is upregulated together with selected cytochrome P450 (CYP).

Ingested vitamin E, which is not absorbed and distributed to the tissues, has been estimated to be between 30% and 70% of the initial amount, and is directly excreted in the feces. An alternative route of elimination is the urinary or biliary excretion of tocopherol and tocotrienol metabolic products.

III. BIOTRANSFORMATION AND METABOLISM OF VITAMIN E AS BIOACTIVATION PROCESSES

Vitamin E compounds undergo extensive metabolic processing mainly in the liver before excretion, with a high rate of vitamin E metabolism/excretion being important in preventing liver (and overall tissue) accumulation which could possibly result in adverse effects (Traber, 2004).

The first type of tocopherol metabolites described was reported to arise as a consequence of the antioxidant action of α-tocopherol (Simon *et al.*, 1956). Figure 1 shows the main metabolites resulting from the two-electron oxidation of α-tocopherol. α-Tocopheryl quinone (TQ), 8α-hydroxytocopherones or alkyldioxy-tocopherones, and a series of epoxide derivatives recognized to be early products of vitamin E oxidation (Fig. 1A). Under appropriate reducing conditions, TQ can coexist with its hydroquinone derivative (THQ) to form a redox couple of possible physiological relevance

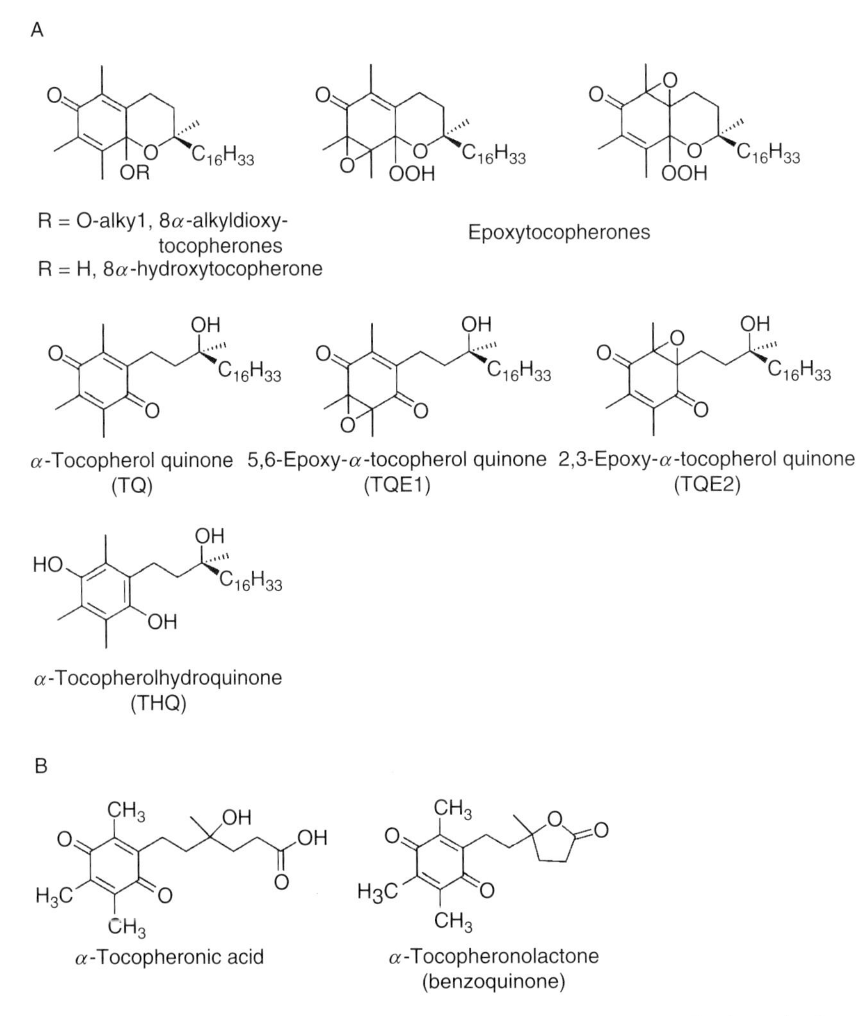

FIGURE 1. Main oxidation products of vitamin E (as α-tocopherol) (A) and urinary (Simon's) metabolites (B). In the case of tocopheryl quinone, it is also shown as the corresponding hydroquinone derivative.

in cell membranes. Two major urinary metabolites derived via this pathway are α-tocopheronic acid and α-tocopheronolactone—the so-called Simon's metabolites (Fig. 1B)—which are excreted in the urine as glucuronides or sulfates. Simon's metabolites have been proposed to be, at least in part, oxidation artifacts generated during sample preparation—as their formation occurs when oxygenation is not avoided.

In the early 1980s, another class of metabolic products belonging to the family of 2-(2′-carboxyethyl)-6-hydroxychromans (CEHCs) (Fig. 2) was identified in the urine of rats given δ-tocopherol intravenously (Chiku *et al.*, 1984). After the original discovery of δ-CEHC formation, it was shown that α-tocopherol metabolism in humans results in the formation and urinary excretion of 2,5,7,8-tetramethyl-2(2′-carboxyethyl)-6-hydroxychroman (α-CEHC) (Schultz *et al.*, 1995). At this time, Wechter *et al.* (1996), while searching for a natriuretic hormone, described the isolation of a compound from the urine of human uremic subjects, which was subsequently identified as 2,7,8-trimethyl-2-(2′-carboxyethyl)-6-hydroxychroman (γ-CEHC) and named Loma Linda University-α (LLU-α) (Wechter *et al.*, 1996). This class of metabolic product deriving from both tocopherols and tocotrienols consists of an intact chroman structure and a shortened side chain (Fig. 2). The intact chroman structure indicates that CEHCs arise from vitamer precursors not having reacted through the Simon's pathway, that is, without previous chromanoxyl radical and quinone formation. The assessment of urinary CEHC metabolites has been

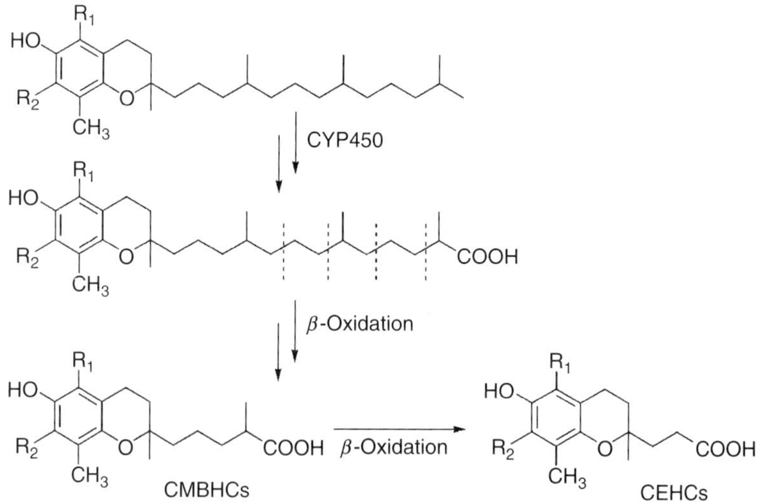

FIGURE 2. Scheme of vitamer processing to form 2-(2′-carboxyethyl)-6-hydroxychromans (CEHCs) and 2-(2′-carboxymethylbutyl)-6-hydroxychroman intermediate (CMBHC). Further details are reported in the text.

recognized to be the reference strategy to monitor vitamin E metabolism, and it has been extensively applied to human and animal studies (Galli *et al.*, 2003; Hattori *et al.*, 2000a,b; Traber *et al.*, 1998). Further advancements in this strategy have been achieved when α- and γ-CEHCs were first measured in nanomolar amounts in human blood, first by means of high-performance liquid chromatography (HPLC) (Stahl *et al.*, 1999) and then by gas chromatography–mass spectrometry (GC–MS) (Galli *et al.*, 2002). These analytical procedures have since been applied to cells and culture media and to solid tissues (Traber *et al.*, 2005). Typical HPLC and GC–MS runs of human serum and cell samples are shown in Figs. 3–5. Moreover, a new method for the analysis of conjugated metabolites in human urine by tandem MS has been described (Pope *et al.*, 2002), which may overcome the possible disadvantages caused by deconjugation.

The formation of CEHCs occurs after initial ω-hydroxylation followed by β-oxidation as well as subsequently to the formation of several intermediates of tocopherols and tocotrienols, all of which have been identified (Birringer *et al.*, 2001, 2002; Schuelke *et al.*, 2000; Sontag and Parker, 2002; Swanson *et al.*, 1999; Fig. 2). As known for the degradation of unsaturated fatty acids, auxiliary enzymes are required for the degradation of tocotrienols which do not possess a saturated side chain. As mentioned above, the β-oxidation pathway of the side chain requires an initial ω-hydroxylation step with subsequent oxidation of the hydroxyl group. Further insight into this phase of vitamin E metabolism have been obtained by a large spectrum of enzyme inhibition and induction studies which confirm the involvement of the a xenobiotic-metabolising system: a CYP-mediated ω-hydroxylation of the side chain (Sontag and Parker, 2002). The first study, investigating the metabolism of γ-tocopherol to carboxychroman metabolites in cell culture, used HepG2 cells incubated with a medium containing fetal bovine serum enriched with *RRR*-γ-tocopherol (Parker and Swanson, 2000). The analysis of γ-tocopherol metabolites released into the cell culture medium by GC–MS revealed the cellular secretion of γ-CEHC as well as of the 5′-carboxychroman analogue, 2,7,8-trimethyl-2-(δ-carboxymethylbutyl)-6-hydroxychroman (Parker and Swanson, 2000), now called γ-CMBHC (Fig. 2). All further intermediates of vitamin E metabolism have been identified in HepG2 cells.

According to early experiments on cancer cell lines such as PC-3 prostate carcinoma cells (Conte *et al.*, 2004) or human lung epithelial cancer cells A549 (You *et al.*, 2005), it is possible to assume that vitamin E metabolism may take place also in extrahepatic tissues, as suggested by the observation that supplementation of PC-3 cells with increasing concentrations of the γ-forms of tocopherols and tocotrienols leads to the simultaneous increase of intracellular concentrations of γ-CEHC and cellular/extracellular concentrations of the parent compounds (Conte *et al.*, 2004). Studies have shown that the cellular supplementation of α-homologues results in a much

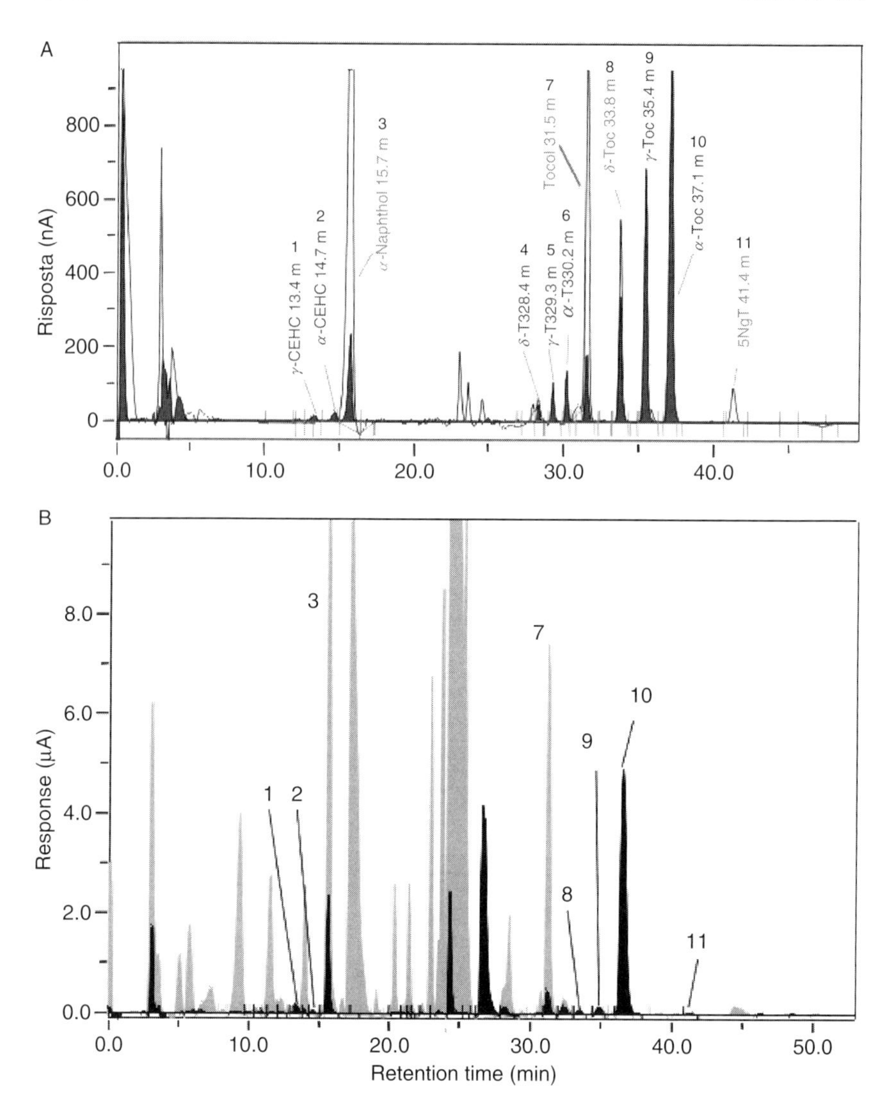

FIGURE 3. Simultaneous analysis by HPLC-ESA of vitamers and main metabolites of vitamin E. Panel A shows two chromatographic profiles obtained while assessing a standard mixture of vitamers and metabolites with the ESA 8-channel coularray detector set at the cathodic voltages of +250 (black or white profile) and +750 mV (gray profile). In Panel B, the same analysis is shown with a plasma sample from one of the authors (GF) who was not taking any supplemental form of vitamin E. With this approach, it is possible to resolve a mixture containing α- and γ-CEHC metabolites (peaks 1 and 2 in Panel B), δ, γ, and α-forms of tocotrienols (peaks 4, 5, and 6, respectively) and tocopherols (peaks 8, 9, and 10, respectively), and the nitration product 5′-nitro-γ-tocopherol (peak 11). α-Naphthol and tocol were used as internal standards (peaks 3 and 7, respectively). Individual concentrations of the test compounds in the standard mixture were between 0.5 and 30 μM. A good linearity in the detector response was obtained for all the test compounds in the concentration range 0.005–100 μM, with LDC

less effective metabolism than when γ-homologues are employed and that supplementation with the tocotrienol homologue produces the highest metabolic response (Conte *et al.*, 2004; You *et al.*, 2005). Interestingly, tocotrienol supplementation also influences the transformation of endogenous tocopherols (Conte *et al.*, 2004). This, together with results reported above, indicates that nonhepatic cells also have the ability to process all forms of vitamin E taken up in excess.

The preferential uptake of γ-tocopherol rather than α-tocopherol found in different cancer cell lines (Betti *et al.*, 2006; Gao *et al.*, 2002; Tran and Chan, 1992) might be pivotal in the explanation of the differential extent of transformation and thus of the amounts of α- and γ-CEHC formed in the cell, although this process appears, to date, to be cell line dependent.

Once formed, α- and γ-CEHC are conjugated with glucuronic acid or sulphate, transferred into the circulation, and excreted in the urine. As shown in rats, parts of these are then directly eliminated via biliary excretion (Hattori *et al.*, 2000b). Only 1–3% of the ingested *RRR*-α-tocopherol appears in the human urine as α-CEHC (Schuelke *et al.*, 2000). In contrast, up to 50% of γ-tocopherol is degraded and eliminated as γ-CEHC (Swanson *et al.*, 1999). Up to 2% and 6% of α- and γ-tocotrienols administered orally have been found as urinary α- and γ-CEHCs, respectively (Lodge *et al.*, 2001). The estimation of α- and γ-tocopherol kinetics in humans by means of oral administration of 50 mg each of an equimolar ratio of deuterium-labeled α- and γ-tocopherols showed that γ-tocopherol fractional disappearance rates are more than three times greater than those of α-tocopherol (Leonard *et al.*, 2005b). In this study, the evidence of a fast kinetic of disappearance of

---

values between 0.001 and 0.005 μM. The chromatographic analysis was carried out using a stationary phase C8, 100 Å, 5 μm, 250 × 4.6 mm^2 (Kromasil, Eka Nobel, Nobel Industries, Sweden), protected with a precolumn filled with the same phase (or C18 phase). A discontinuous gradient was developed at a flow rate of 1 ml/min using as mobile phases, $LiClO_4$ 0.2% in water with final pH 3 (phase A) and acetonitrile 98%, 2% HPLC grade water with 0.2% final concentration of $LiClO_4$, pH 3 (phase B). The percentage of B was maintained at 10% for the first 10 min and then it increased linearly to 100% in 15 min. It remained at 100% for 20 min and then it returned to 10% in 5 min. To allow a simultaneous extraction of CEHC metabolite in the free (or unesterified) form and vitamers, 1 ml of sample was treated with 25 μl of 10-mg/ml ascorbic acid and 1500 IU of *E. coli* β-glucoronidase (Sigma Chemicals) in 10-mM potassium buffer (pH 6.8) to obtain a final reaction volume of 2 ml. The mixture was incubated at 37°C for 30 min. After acidification to pH 1–2 with acetic acid, the samples were extracted twice by extensive vortexing with 2 ml of *tert*-butyl methyl ether/hexane/acetonitrile (2/2/1 v/v/v) containing 10 μl of 100-mg/ml butylated hydroxytoluene (BHT) in hexane. At each extraction step, the organic layer was collected and evaporated under a stream of nitrogen. The dried residue was dissolved by extensive vortexing in 200 μl of phase B. The injection volume was 50 μl. The chromatogram shows that the main tocopherol vitamers are present in micromolar concentrations in human plasma, while α- and γ-CEHC metabolites are present at low nanomolar concentrations. The nitration product 5′-nitro-γ-tocopherol is present in only trace amounts.

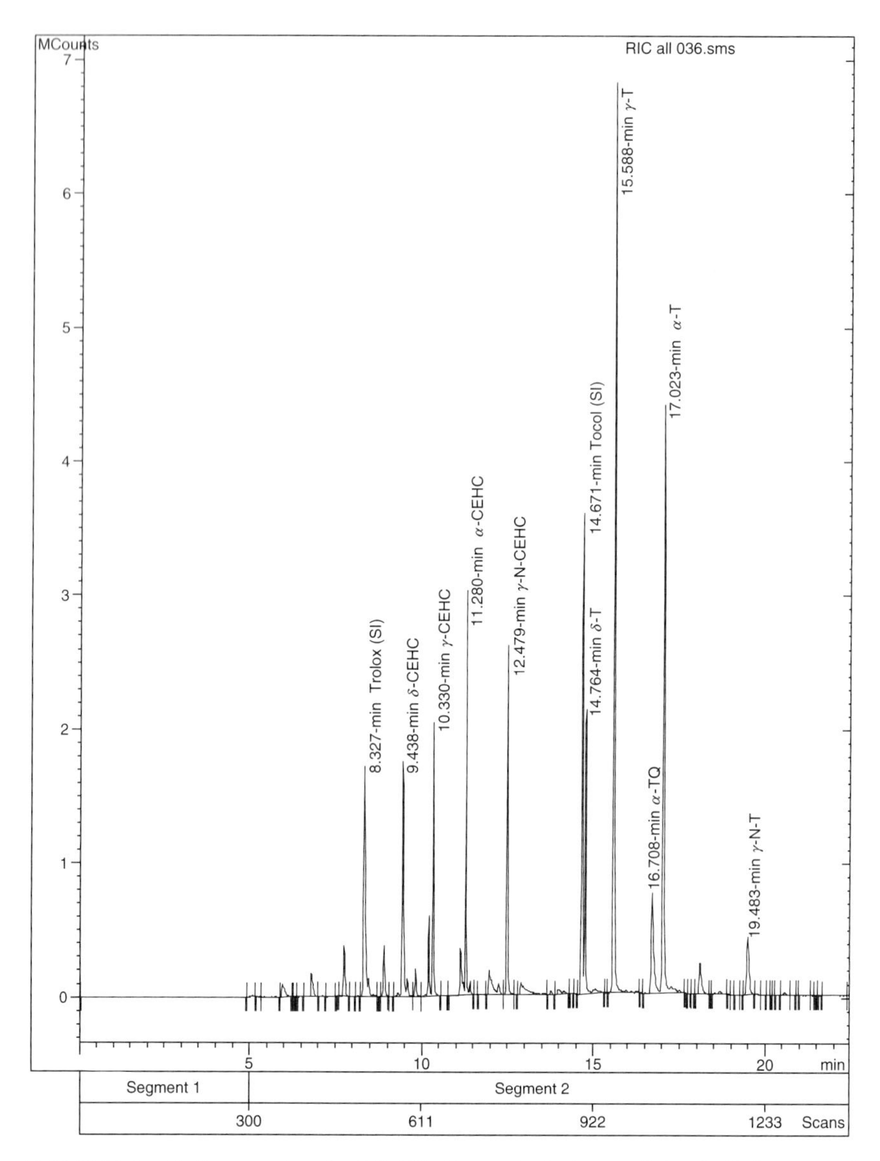

FIGURE 4. Full scan analysis by GC–MS of a standard mixture of vitamin E vitamers and metabolites. The same pattern of components as found in Fig. 3A are shown with the exception of the tocotrienols, but with the addition of α-tocopheryl quinone (α-TQ). The analysis was carried out according to the procedure described in Galli *et al.* (2002) using a GC apparatus Varian CP 3800 coupled with an EI mass spectrometer Sturn 2000 MS. A factor four 5-ms low-polarity column (30-m (Ø, 0.25-mm DF) was used, with helium gas as carrier at the flow 1.0 ml/min. The injector temperature was 260 °C and the column temperature was programed as previously described (Galli *et al.*, 2002).

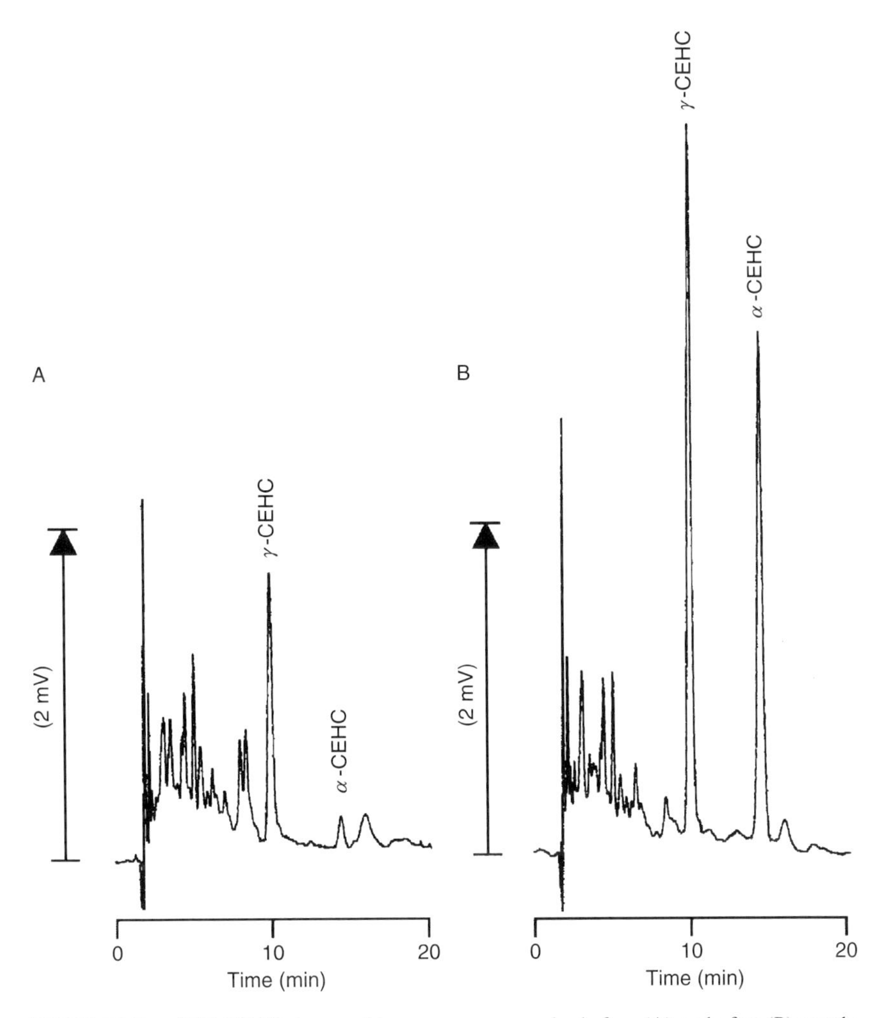

FIGURE 5. HPLC-ECD traces of human serum samples before (A) and after (B) supplementation with a single dose of vitamin E from a natural source (306 mg of *RRR*-α-tocopherol and 1.77 mg of γ-tocopherol). After supplementation, α- and γ-CEHC concentrations increase significantly.

γ-CEHC originally shown in the study of the Kelly's group (Galli *et al.*, 2003) was strengthened, as labeled plus unlabeled γ-CEHC were shown to reach a peak 6–12 h after oral ingestion of a single dose of 100-mg deuterium-labeled γ-tocopherol acetate (Galli *et al.*, 2003). On the contrary, neither plasma nor urinary deuterium-labeled α-CEHC was detectable (Galli *et al.*, 2003). It should be mentioned, however, that an increased metabolism of α-tocopherol to α-CEHC was shown in a rat model of enriched liver α-tocopherol levels (Leonard *et al.*, 2005a) and that mice fed for 5 weeks with high doses of

γ-tocopherol reached similar plasma and liver levels of α- and γ-tocopherols (Traber *et al.*, 2005). In the latter study, γ-CEHC reached levels between 0.1 and 0.8 nmol/g while α-CEHC was undetectable, confirming the rapid clearance of α-CEHC metabolites from hepatic tissue.

The mean apparent elimination half-life of α-tocotrienol was found to be twice that of δ-tocotrienol in humans (Yap *et al.*, 2001). These and other studies (Kaneko *et al.*, 2000) show that different forms of vitamin E are metabolized at different metabolic rates and that urinary excretion of CEHC appears to be more relevant for tocopherols and tocotrienols distinct from *RRR*-α-tocopherol.

In 1995, α-CEHC was described as the major urinary metabolite of α-tocopherol appearing in human urine after supplementation of *RRR*-α-tocopherol dosages ranging from 0 to 800 mg/day (Schultz *et al.*, 1995). HPLC and GC–MS analyses revealed that α-CEHC was only excreted when a plasma threshold of 30- to 50-μM α-tocopherol was exceeded. This concentration was obtained by a daily intake of ~50- to 150-mg α-tocopherol, suggesting that α-CEHC excretion may be considered a marker of (super) optimal vitamin E intake. However, this may only hold true if α-TTP is working properly. Subsequently, it was shown that synthetic compared with natural vitamin E is preferentially metabolized to α-CEHC and excreted in human urine (for review, see Traber and Arai, 1999), and that tocotrienols, like tocopherols, are metabolized to respective CEHCs—although the amounts excreted in human urine are small in relation to dose (Lodge *et al.*, 2001).

The formation of γ-CEHC was identified as a major route of urinary elimination of γ-tocopherol in humans (Swanson *et al.*, 1999). However, Parker and Swanson (2000) reported the consistent presence of the γ-CEHC precursor γ-CMBHC in human urine samples; both metabolites increased in the urine after oral supplementation with purified *RRR*-γ-tocopherol, although the concentration of γ-CMBHC was consistently and substantially less than that of γ-CEHC. The same has been reported for α-CEHC and α-CMBHC (Schuelke *et al.*, 2000).

When the time course of serum levels of α- and γ-CEHCs after a single dose of vitamin E from a natural source (306 mg of *RRR*-α-tocopherol and 1.77 mg of γ-tocopherol) was investigated, maximum levels for both compounds were measured 12 h after application (33.3 ± 11.1 μmol α-tocopherol per liter and 42.4 ± 18.3 nmol α-CEHC per liter); baseline values were reached again after 72 h (Radosavac *et al.*, 2002). While γ-tocopherol levels decreased during the study period, the increase in γ-CEHC remained. It is remarkable that the concentration of α-CEHC in serum is about 1000-fold lower than the parent compound. Even on the ingestion of high amounts of α-tocopherol (306 mg) and the resulting increase of α-tocopherol serum levels by roughly 10 μmol/liter, α-CEHC concentrations increased by only about 30 nmol/liter. Although the dose of γ-tocopherol was 170-fold lower than α-tocopherol, the levels of γ-CEHC increased to about the same extent as α-CEHC. The concentrations of γ-tocopherol, however, decreased (Radosavac *et al.*, 2002), indicating that in humans also, γ-tocopherol is more prone to metabolism than α-tocopherol. Galli *et al.* (2003) showed that supplementation of healthy volunteers with 100 mg of γ-tocopheryl acetate leads to time course and concentration-dependent increases of plasma γ-tocopherol similar to those reported by Radosavac *et al.* (2002), ultimately confirming a sustained transformation of this vitamer to CEHC (Leonard *et al.*, 2005b). Importantly, high-dose administration of α-tocopherol has been reported to increase γ-tocopherol metabolism, which might be the reason for the often observed decrease of plasma γ-tocopherol levels after the intake of high amounts of α-tocopherol (Morinobu *et al.*, 2003).

According to data on vitamin E levels, urinary and plasma levels of CEHC metabolites also show large interindividual variability in healthy subjects (Galli *et al.*, 2003; Kelly *et al.*, 2004; Lee and Kelly, 1999). Moreover, work has led to the hypothesis that women may have an altered metabolic setting which ensures increased vitamin E transformation to CEHC, particularly from the γ-form (Leonard *et al.*, 2005b). Together these findings suggest that tocopherol uptake and transformation are part of stable individual phenotype which is under genetic regulation.

Vitamin E intake *per se,* other dietary factors and drug regimens can increase this variability, affecting either the intake of vitamers or the expression of ω-hydroxylase-dependent metabolic activity in the liver (reviewed in Frank, 2005; Traber, 2004). In addition, particular pathological conditions may affect vitamin E biotransformation in humans. When studying the effect of kidney damage on CEHC levels in humans, Galli *et al.* (2004a) observed that the progressive deterioration of the kidney function produced an exponential increase of both α- and γ-CEHCs in plasma. Compared with healthy controls, CEHCs reached approximately threefold levels in hemodialytic patients (α-CEHC: 20.1 nmol/liter in controls versus 77.3 nmol/liter in patients; γ-CEHC: 231 nmol/liter in controls versus 637 nmol/liter in patients) (Galli *et al.*, 2004a). These findings are in agreement with previous reports of markedly increased levels of α- and γ-CEHCs after vitamin E supplementation in uremic patients, which was taken as an indication that α-tocopherol metabolites may contribute to an anti-inflammatory effect observed after vitamin E administration (Smith *et al.*, 2003).

The reported differences in bioactivity and biotransformation between γ- and α-CEHCs led to the hypothesis that γ-tocopherol has a physiological role which might be mediated by its specific transformation to the corresponding CEHC metabolite. In other words, metabolite production might not simply represent a mechanism for disposal of unnecessary vitamin E forms but also a bioactivation process. The biological functions of CEHC metabolites explored in cell cultures have shown that γ-CEHC can inhibit cyclooxygenase-2 (COX-2) activity (Jiang *et al.*, 2001), that α- and γ-CEHCs inhibit microglial prostaglandin E_2 production, and that α-CEHC suppresses tumor necrosis factor (TNF-α)-stimulated or bacterial lipopolysaccharide-stimulated nitrite production in rat aortic endothelial cells and mouse microglial cultures (Grammas *et al.*, 2004). Experiments in animals have demonstrated protective effects of γ-CEHC against metal-induced nephrotoxicity, and this has been discussed in context of the possible antioxidant properties of the metabolite (Appenroth *et al.*, 2001). In PC-3 prostate carcinoma cells and HTB-82 rhabdomyosarcoma cells, γ-CEHC has been observed to exert a much stronger antiproliferative activity than α-CEHC. CEHCs were as effective as their vitamer precursors to produce this effect, which was shown to occur through a specific inhibition of the cyclin signaling with subsequent arrest of the cell cycle in G0 during the G1-S phase transition (Galli *et al.*, 2004b). Furthermore, hypomethylated forms of tocotrienols, which are more extensively metabolized to CEHCs, are more effective antiproliferative agents than tocopherols in different cancer cell lines (Betti *et al.*, 2006; Conte *et al.*, 2004), further suggesting a relationship between metabolic transformation and extent of the biological effect produced by the individual vitamers.

The antioxidant activity of α-CEHC has been investigated *in vitro* using several methods, which utilize different pro-oxidant challenges. α-CEHC has been reported to exhibit antioxidant properties similar to those of trolox (Betancor-Fernandez *et al.*, 2002). This was shown both using a fluorescent protein acting as a marker for peroxyl radical-induced oxidative damage [oxygen radical absorbance capacity (ORAC) assay] with the trolox equivalent antioxidant capacity (TEAC) assay [in which a stable free radical, 2,2′-azino-bis(3-ethylbenzthiazoline-6-sulfonic acid) ($ABTS^+$) is reduced directly by antioxidants], and by means of the measurement of scavenging properties against reactive nitrogen species (through the evaluation of the modulation of peroxynitrite-induced tyrosine nitration) (Betancor-Fernandez *et al.*, 2002). Further experiments have confirmed these findings showing that α- and γ-CEHCs exhibit similar antioxidant activities against lipid peroxidation in organic solution as the corresponding parent compound (Yoshida and Niki, 2002). Indeed, when exposed to hydroperoxyl radicals, CEHCs generate ESR signals which confirm the formation of chromanoxyl radicals corresponding to the tocopheryl radicals generated by the corresponding precursors (Galli *et al.*, 2004c). α- and γ-CEHCs have been shown to scavenge aqueous radicals more efficiently to inhibit membrane lipid peroxidation less efficiently than the parent compounds and to exert a minimal pro-oxidant effect in the presence of cupric ion (Yoshida and Niki, 2002). Rather, CEHCs were observed to exert a protective effect *in vitro* against copper- or macrophage-induced LDL oxidation (Galli *et al.*, 2004c). These data together provide evidence to support the hypothesis that local (either extracellular or intracellular) sustained production/accumulation of CEHCs might represent a biologically relevant event to control important responses associated with the control of cell signaling and gene expression. The hypothesis of a transformation-dependent bioactivation of vitamin E represents an intriguing and emerging aspect of research that deserves further investigation.

Acknowledgments

We are indebted with Dr. Francesco Mazzini (Department of Chemistry and Industrial Chemistry, University of Pisa, Italy) for his precious assistance in the preparation of artworks.

References

Appenroth, D., Karge, E., Kiessling, G., Wechter, W. J., Winnefeld, K., and Fleck, C. (2001). LLU-alpha, an endogenous metabolite of gamma-tocopherol, is more effective against metal nephrotoxicity in rats than gamma-tocopherol. *Toxicol. Lett.* **122,** 255–265.

Betancor-Fernandez, A., Sies, H., Stahl, W., and Polidori, M. C. (2002). *In vitro* antioxidant activity of 2,5,7,8-tetramethyl-2-(2′-carboxyethyl)-6-hydroxychroman (alpha-CEHC), a vitamin E metabolite. *Free Radic. Res.* **36,** 915–921.

Betti, M., Minelli, A., Canonico, B., Castaldo, P., Magi, S., Aisa, M. C., Piroddi, M., Di Tomaso, V., and Galli, F. (2006). Antiproliferative effects of tocopherols (vitamin E) on murine glioma C6 cells: Homologue-specific control of PKC/ERK and cyclin signaling. *Free Radic. Biol. Med.* **41,** 464–472.

Birringer, M., Drogan, D., and Brigelius-Flohé, R. (2001). Tocopherols are metabolized in HepG2 cells by side chain ω-oxidation and consecutive β-oxidation. *Free Radic. Biol. Med.* **31,** 226–232.

Birringer, M., Pfluger, P., Kluth, D., Landes, N., and Brigelius-Flohé, R. (2002). Identities and differences in the metabolism of tocotrienols and tocopherols in HepG2 cells. *J. Nutr.* **132,** 3113–3118.

Chiku, S., Hamamura, K., and Nakamura, T. (1984). Novel urinary metabolite of α-tocopherol in rats. *J. Lipid Res.* **25,** 40–48.

Conte, C., Floridi, A., Aisa, C., Piroddi, M., Floridi, A., and Galli, F. (2004). Gamma-tocotrienol metabolism and antiproliferative effect in prostate cancer cells. *Ann. NY Acad. Sci.* **1031,** 391–394.

Drevon, C. A. (1991). Absorption, transport and metabolism of vitamin E. *Free Radic. Res. Commun.* **4,** 229–246.

Frank, J. (2005). Beyond vitamin E supplementation: An alternative strategy to improve vitamin E status. *J. Plant Physiol.* **162,** 834–843.

Galli, F., Lee, R., Dunster, C., and Kelly, F. J. (2002). Gas chromatography mass spectrometry analysis of carboxyethyl-hydroxychroman metabolites of alpha- and gamma-tocopherol in human plasma. *Free Radic. Biol. Med.* **32,** 333–340.

Galli, F., Lee, R., Atkinson, J., Floridi, A., and Kelly, F. J. (2003). Gamma-tocopherol biokinetics and transformation in humans. *Free Radic. Res.* **37,** 1225–1233.

Galli, F., Floridi, A. G., Floridi, A., and Buoncristiani, U. (2004a). Accumulation of vitamin E metabolites in the blood of renal failure patients. *Clin. Nutr.* **23,** 205–212.

Galli, F., Stabile, A. M., Betti, M., Conte, C., Pistilli, A., Rende, M., Floridi, A., and Azzi, A. (2004b). The effect of alpha- and gamma-tocopherol and their carboxyethyl hydroxychroman metabolites on prostate cancer cell proliferation. *Arch. Biochem. Biophys.* **423,** 97–102.

Galli, F., Piroddi, M., Lannone, A., Pagliarani, S., Tomasi, A., and Floridi, A. (2004c). A comparison between the antioxidant and peroxynitrite-scavenging functions of the vitamin E metabolites alpha- and gamma-carboxyethyl-6-hydroxychromans. *Int. J. Vitam. Nutr. Res.* **74,** 362–373.

Gao, R., Stone, W. L., Huang, T., Papas, A. M., and Qui, M. (2002). The uptake of tocopherols by RAW 264.7 macrophages. *Nutr. J.* **1,** 2–8.

Grammas, P., Hamdheydari, L., Benaksas, E. J., Mou, S., Pye, Q. N., Wechter, W. J., Floyd, R. A., Stewart, C., and Hensley, K. (2004). Anti-inflammatory effects of tocopherol metabolites. *Biochem. Biophys. Res. Commun.* **319,** 1047–1052.

Hattori, A., Fukushima, T., and Imai, K. (2000a). Occurrence and determination of a natriuretic hormone, 2,7,8-trimethyl-2-(beta-carboxyethyl)-6-hydroxychroman, in rat plasma, urine, and bile. *Anal. Biochem.* **281,** 209–215.

Hattori, A., Fukushima, T., Yoshimura, H., Abe, K., and Imai, K. (2000b). Production of LLU-alpha following an oral administration of gamma-tocotrienol or gamma-tocopherol to rats. *Biol. Pharm. Bull.* **23,** 1395–1397.

Jiang, Q., Christen, S., Shigenaga, M. K., and Ames, B. N. (2001). Gamma-tocopherol, the major form of vitamin E in the US diet, deserves more attention. *Am. J. Clin. Nutr.* **74,** 714–722.

Kaneko, K., Kiyose, C., Ueda, T., Ichikawa, H., and Igarashi, O. (2000). Studies of the metabolism of alpha-tocopherol stereoisomers in rats using [5-methyl-(14)C]SRR- and RRR-alpha-tocopherol. *J. Lipid Res.* **41,** 357–367.

Kelly, F. J., Lee, R., and Mudway, I. S. (2004). Inter- and intra-individual vitamin E uptake in healthy subjects is highly repeatable across a wide supplementation dose range. *Ann. NY Acad. Sci.* **1031,** 22–39.

Lee, R., and Kelly, F. J. (1999). Quantification of urinary metabolites of α-tocopherol and γ-tocopherol in normal European subjects. *Free Radic. Biol. Med.* **27**(Suppl. 1), S38–S39.

Leonard, S. W., Gumpricht, E., Devereaux, M. W., Sokol, R. J., and Traber, M. (2005a). Quantitation of rat liver vitamin E metabolites by LC-MS during high-dose vitamin E administration. *J. Lipid Res.* **46,** 1068–1075.

Leonard, S. W., Paterson, E., Atkinson, J. K., Ramakrishnan, R., Cross, C. E., and Traber, M. G. (2005b). Studies in humans using deuterium-labeled α- and γ-tocopherols demonstrate faster plasma γ-tocopherol disappearance and greater γ-metabolite production. *Free Radic. Biol. Med.* **38,** 857–866.

Lodge, J. K., Ridlington, J., Leonard, S., Vaule, H., and Traber, M. G. (2001). α- and γ-Tocotrienols are metabolized to carboxyethyl-hydroxychroman derivatives and excreted in human urine. *Lipids* **36,** 43–48.

Mardones, P., and Rigotti, A. (2004). Cellular mechanisms of vitamin E uptake: Relevance in alpha-tocopherol metabolism and potential implications for disease. *J. Nutr. Biochem.* **15,** 252–260.

Mardones, P., Strobel, P., Miranda, S., Leighton, F., Quinones, V., Amigo, L., Rozowski, J., Krieger, M., and Rigotti, A. (2002). Alpha-tocopherol metabolism is abnormal in scavenger receptor class B type I (SR-BI)-deficient mice. *J. Nutr.* **132,** 443–449.

Minehira-Castelli, K., Leonard, S. W., Walker, Q. M., Traber, M. G., and Young, S. G. (2006). Absence of VLDL secretion does not affect α-tocopherol content in peripheral tissues. *J. Lipid Res.* **47,** 1733–1738.

Morinobu, T., Yoshikawa, S., Hamamura, K., and Tamai, H. (2003). Measurement of vitamin E metabolites by high-performance liquid chromatography during high-dose administration of alpha-tocopherol. *Eur. J. Clin. Nutr.* **57,** 410–414.

Mustacich, D. J., Leonard, S. W., Devereaux, M. W., Sokol, R. J., and Traber, M. G. (2006). Alpha-tocopherol regulation of hepatic cytochrome P450s and ABC transporters in rats. *Free Radic. Biol. Med.* **41,** 1069–1078.

Parker, R. S., and Swanson, J. E. (2000). A novel 5′-carboxychroman metabolite of gamma-tocopherol secreted by HepG2 cells and excreted in human urine. *Biochem. Biophys. Res. Commun.* **269,** 580–583.

Polidori, M. C., Stahl, W., Sies, H., and Brigelius-Flohé, R. (2006). Vitamin E, metabolism and biological activity of metabolic products. *In* "Encyclopedia of Vitamin E" (Preedy and Watson, Eds.).

Pope, S. A., Burtin, G. E., Clayton, P. T., Madge, D. J., and Muller, D. P. (2002). Synthesis and analysis of conjugates of the major vitamin E metabolite, alpha-CEHC. *Free Radic. Biol. Med.* **33,** 807–817.

Qian, J., Morley, S., Wilson, K., Nava, P., Atkinson, J., and Manor, D. (2005). Intracellular trafficking of vitamin E in hepatocytes: The role of tocopherol transfer protein. *J. Lipid Res.* **46,** 2072–2082.

Radosavac, D., Graf, P., Polidori, M. C., Sies, H., and Stahl, W. (2002). Tocopherol metabolites 2,5,7,8-tetramethyl-2-(2′-carboxyethyl)-6-hydroxychroman (alpha-CEHC) and 2,7,8-trimethyl-2-(2′-carboxyethyl)-6-hydroxychroman (gamma-CEHC) in human serum after a single dose of natural vitamin E. *Eur. J. Nutr.* **41,** 119–124.

Schuelke, M., Elsner, A., Finckh, B., Kohlschütter, A., Hübner, C., and Brigelius-Flohé, R. (2000). Urinary alpha-tocopherol metabolites in alpha-tocopherol transfer protein-deficient patients. *J. Lipid Res.* **41,** 1543–1551.

Schultz, M., Leist, M., Petrzika, M., Gassmann, B., and Brigelius-Flohé, R. (1995). Novel urinary metabolite of alpha-tocopherol, 2,5,7,8-tetramethyl-2(2′-carboxyethyl)-6-hydroxychroman, as an indicator of an adequate vitamin E supply? *Am. J. Clin. Nutr.* **62**(Suppl. 6), 1527S–1534S.

Simon, E. J., Eisengart, A., Sundneim, L., and Milorat, A. T. (1956). The metabolism of vitamin E. II. Purification and characterization of urinary metabolites of alpha-tocopherol. *J. Biol. Chem.* **221,** 807–817.

Smith, K. S., Lee, C. L., Ridlington, J. W., Leonard, S. W., Devaraj, S., and Traber, M. G. (2003). Vitamin E supplementation increases circulating vitamin E metabolites tenfold in end-stage renal disease patients. *Lipids* **38,** 813–819.

Sontag, T. J., and Parker, R. S. (2002). Cytochrome P450 ω-hydroxylase pathway of tocopherol catabolism. *J. Biol. Chem.* **277,** 25290–25296.

Stahl, W., Graf, P., Brigelius-Flohé, R., Wechter, W., and Sies, H. (1999). Quantification of the alpha- and gamma-tocopherol metabolites 2,5,7,8-tetramethyl-2-(2′-carboxyethyl)-6-hydroxychroman and 2,7,8-trimethyl-2-(2′-carboxyethyl)-6-hydroxychroman in human serum. *Anal. Biochem.* **275,** 254–259.

Stahl, W., van den Berg, H., Arthur, J., Bast, A., Dainty, J., Faulks, R. M., Gärtner, C., Haenen, G., Hollman, P., Holst, B., Kelly, F. J., Polidori, M. C., *et al.* (2002). European research on the functional effects of dietary antioxidants (EUROFEDA): Bioavailability and metabolism. *Mol. Aspects Med.* **23,** 39–100.

Swanson, J. E., Ben, R. N., Burton, G. W., and Parker, R. S. (1999). Urinary excretion of 2,7,8-trimethyl-2-(beta-carboxyethyl)-6-hydroxychroman is a major route of elimination of gamma-tocopherol in humans. *J. Lipid Res.* **40,** 665–671.

Traber, M. G. (2004). Vitamin E, nuclear receptors and xenobiotic metabolism. *Arch. Biochem. Biophys.* **423,** 6–11.

Traber, M. G., and Arai, H. (1999). Molecular mechanisms of vitamin E transport. *Annu. Rev. Nutr.* **19,** 343–355.

Traber, M. G., Elsner, A., and Brigelius-Flohe, R. (1998). Synthetic as compared with natural vitamin E is preferentially excreted as alpha-CEHC in human urine: Studies using deuterated alpha-tocopheryl acetates. *FEBS Lett.* **437,** 145–148.

Traber, M. G., Burton, G. W., and Hamilton, R. L. (2004). Vitamin E trafficking. *Ann. NY Acad. Sci.* **1031,** 1–12.

Traber, M. G., Siddens, L. K., Leonard, S. W., Schock, B., Gohil, K., Krueger, S. K., Cross, C. E., and Williams, D. E. (2005). Alpha-tocopherol modulates Cyp3a expression, increases gamma-CEHC production, and limits tissue gamma-tocopherol accumulation in mice fed high gamma-tocopherol diets. *Free Radic. Biol. Med.* **38,** 773–785.

Tran, K., and Chan, A. C. (1992). Comparative uptake of alpha- and gamma-tocopherol by human endothelial cells. *Lipids* **27,** 38–41.

Wechter, J. W., Kantoci, D., Murray, E. D., D'Amico, D. C., Jung, M. E., and Wang, W.-H. (1996). A new endogenous natriuretic factor: LLU-α. *Proc. Natl. Acad. Sci. USA* **93,** 6002–6007.

Yap, S. P., Yuen, K. H., and Wong, J. W. (2001). Pharmacokinetics and bioavailability of alpha-, gamma- and delta-tocotrienols under different food status. *J. Pharm. Pharmacol.* **53,** 67–71.

Yoshida, Y., and Niki, E. (2002). Antioxidant effects of alpha- and gamma-carboxyethyl-6-hydroxychromans. *Biofactors* **16,** 93–103.

You, C. S., Sontag, T. J., Swanson, J. E., and Parker, R. S. (2005). Long-chain carboxychromanols are the major metabolites of tocopherols and tocotrienols in A549 lung epithelial cells but not HepG2 cells. *J. Nutr.* **135,** 227–232.

10

α-TOCOPHEROL STEREOISOMERS

SØREN KROGH JENSEN AND
CHARLOTTE LAURIDSEN

Department of Animal Health, Welfare and Nutrition, Faculty of Agricultural Sciences
University of Aarhus, DK-8830 Tjele, Denmark

Vitamins and Hormones, Volume 76

0083-6729/07 $35.00
DOI: 10.1016/S0083-6729(07)76010-7

Vitamin E comprises a group of compounds possessing vitamin E activity. α-Tocopherol is the compound demonstrating the highest vitamin E activity, which is available both in its natural form as *RRR*-α-tocopherol isolated from plant sources, but more common as synthetically manufactured all-*rac*-α-tocopherol. Synthetic all-*rac*-α-tocopherol consists of a racemic mixture of all eight possible stereoisomers. Assessing the correct biological activity in form of bioavailability and biopotency has been a great challenge during many years as it is difficult to measure clinical endpoints in larger animals than rats and poultry. Thus, the biological effects in focus are resorption of fetuses, testicular degeneration, muscle dystrophy, anemia, encephalomalacia, and in recent years the influence of vitamin E on the immune system are the most important clinical markers of interest. For humans and animals, only different biomarkers or surrogate markers of bioactivity have been measured. In studies with rats, a good consistency between the classical resorption–gestation test and the bioavailability of the individual stereoisomers in fluids and tissues has been shown. For humans and other animals, only different biomarkers or surrogate markers of bioactivity have been measured, and due to the lack of good biological markers for bioactivities, bioavailability is often used as one of the surrogate markers for bioactivities with those limitations this must give. Therefore, a relatively simple analytical method, which allows analysis of the individual stereoisomers of α-tocopherol, is an important tool in order to quantify relative bioavailability of the individual stereoisomers. The analytical method presented here allows the quantification of total tocopherol content and composition by normal phase HPLC and subsequent separation of the stereoisomers of α-tocopherol as methyl ethers by chiral HPLC. Using this method, the α-tocopherol stereoisomers are separated into five peaks. The first peak consists of the four 2*S* isomers (*SSS*-, *SSR*-, *SRR*-, *SRS*-), the second peak consists of *RSS*-, the third peak consists of *RRS*-, the fourth peak consists of *RRR*-, and the fifth peak consists of *RSR*-α-tocopherol. The discussion on the bioavailability of *RRR*- and all-*rac*-α-tocopheryl acetate has primarily been based on human and animal studies using deuterium-labeled forms, whereby a higher biopotency of 2:1 (of *RRR*: all-*rac*) has been demonstrated, differing from the accepted biopotency ratio of 1.36:1. In agreement with previous studies, the 2*S*-forms exert very little importance for the vitamin E activity due to their limited bioavailability. We find notable differences between animal species with regard to the biodiscrimination between the 2*R*-forms. Especially, cows preferentially transfer *RRR*-α-tocopherol into the milk and blood system. The distribution of the stereoisomer forms varies from tissue to tissue, and in some cases, higher levels of the synthetic 2*R*-forms than of the *RRR*-form are obtained, for example, for rats. However, the biodiscrimination of the stereoisomers forms is influenced by other factors such as age, dietary levels, and time after dosage. More focus should be given on the bioactivity of the individual 2*R*-forms rather than just the comparison between *RRR*- and all-rac-α-tocopheryl acetate.

I. INTRODUCTION

Vitamin E is the exception to the paradigm that natural and synthetic vitamins are equivalent because their molecular structures are identical. Natural α-tocopherol (*RRR*-α-tocopherol) is a single stereoisomer. Plants and other oxygenic, photosynthetic organisms are the only organisms able to synthesize tocopherols (DellaPenna, 2005), and since this synthesis is facilitated by stereo-specific enzymes, the resulting tocopherols always posses the same stereochemical structure, namely the *RRR*-structure (Fig. 1).

Synthetic α-tocopherol (all-*rac*-α-tocopherol) is produced commercially by a chemical reaction of tetramethylhydroquinone (TMHQ) with racemic isophytol (VERIS Research summary, 1999). Racemic isophytol is synthesized from isoprenoid units and since isophytol has three chiral centres the resulting α-tocopherol has 2^3 possible conformations and thus yields a racemic mixture of all eight possible stereoisomers.

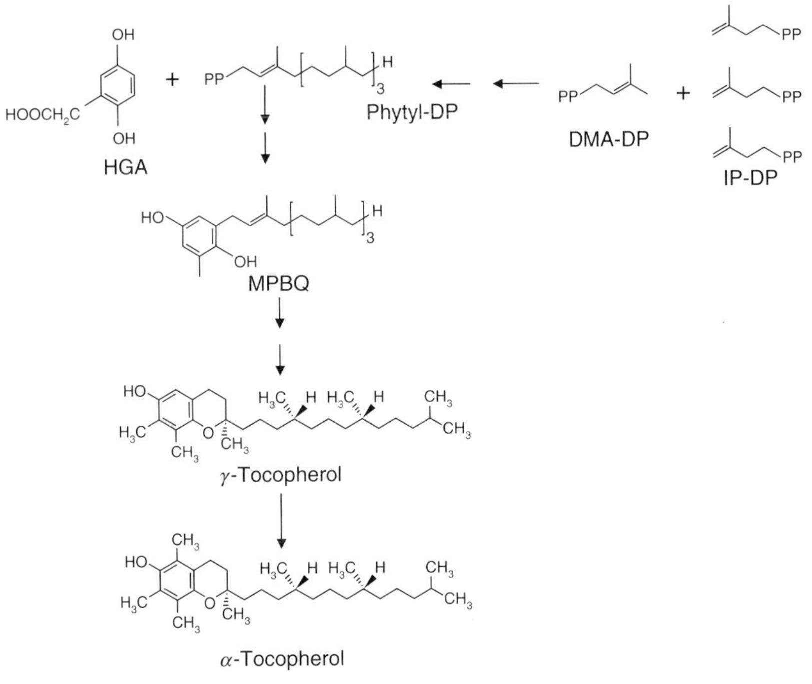

FIGURE 1. Synthesis of α-tocopherol in higher plants. (DMA-DP—dimethylallyl diphosphate; IPDP—isopentenyl diphosphate; HGA—homogentisic acid; Phytyl DP—phytyl diphosphate; MPBQ—2-methyl-6-phytyl-1,4-benzoquinone) (modified after DellaPenna, 2005).

II. SOURCES OF TOCOPHEROL, NOMENCLATURE, AND BIOACTIVITY

A. PRESENCE IN FOOD/FEED INGREDIENTS

Commercial vitamin E supplements can be classified into several distinct categories: fully synthetic vitamin E (all-*rac*-α-tocopherol), the most inexpensive, most commonly sold supplement forms usually as the acetate ester.

Most natural vitamin E is derived during refining of vegetable oils, mainly soybean oil, sunflower oil, and canola/rapeseed oil.

The natural sources of vitamin E can be divided into truly natural *RRR*-α-tocopherol, where *RRR*-α-tocopherol is extracted and isolated directly from a vegetable source without any modifications. Semisynthetic natural source α-tocopherol is *RRR*-α-tocopherol produced from a vegetable source, but where the manufacturer has converted the common natural β, γ, and δ-tocopherol isomers into esters using acetic or succinic acid and eventually added methyl groups to yield *RRR*-α-tocopheryl esters such as *RRR*-α-tocopheryl acetate or *RRR*-α-tocopheryl succinate.

The most abundant sources of vitamin E are vegetable oils such as wheat germ oil, sunflower, canola/rapeseed oil, maize/corn, soybean, palm oil, and olive oil. Nuts, sunflower seeds, sea buckthorn berries, and wheat germ are also good sources. Other sources of vitamin E are whole grains, fish, peanut butter, and green leafy vegetables. In general, green photosynthesizing parts of the plants contain the highest concentration of α-tocopherol (DellaPenna, 2005) and the green parts of the plants are the major source of natural vitamin E for domestic animals.

B. NOMENCLATURE

Vitamin E was discovered by Evans and Bishop, 1923, who observed that its deficiency caused fetal resorption in the rat. An active substance was isolated from wheat germ oil in 1936 and named "tocopherol" from the Greek words *tokos* (childbirth) and *pherein* (to carry) plus the *-ol* suffix designating phenol or alcohol (Packer, 1994). In 1938, the structure of natural α-tocopherol was elucidated and the same time the first synthesis of a biologically active product, consisting of a mixture of the natural diastereoisomer and its two-epimer, was reported. Studies in the following years revealed the existence in nature of a whole family of structurally related compounds with qualitatively identical biological action (IUPAC-IUB, 1981).

Recommendations for the nomenclature of the vitamins, including the tocopherols, were published in 1966 and in 1981 IUPAC-IUB recommended the following nomenclature. The term tocol is the trivial designation for 2-methyl-2-(4,8,12-trimethyldecyl)chroman-6-ol. The term tocopherol(s) should be used as a generic descriptor for the methylated tocols, and although the term tocopherol is not equivalent to the term vitamin E, it include all mono- (δ-tocopherol), di- (β- and γ-tocopherol), and trimethyltocols (α-tocopherol).

Chemical species are said to possess the property of chirality if a molecule and its mirror image of that molecule are distinguishable. In such situation, the molecule and its mirror image are described as enantiomers or stereoisomers. The fact that isolated enantiomers rotate the plane of polarized light has given rise to the expression optical isomers.

The only naturally occurring stereoisomer of α-tocopherol found in nature has the configuration 2*R*,4′*R*,8′*R* according to the sequence rule system. Its semisystematic name is therefore (2*R*,4′*R*,8′*R*)-α-tocopherol and also known as D-α-tocopherol. Today the short term *RRR*-α-tocopherol has been well adopted and the same system is used for all other individual stereoisomers of tocopherols. It is important to notice that today the *RS* system with respect to stereochemical configuration has replaced the dl system.

The diastereoisomer of *RRR*-α-tocopherol, formerly known as L-α-tocopherol, being the epimer of *RRR*-α-tocopherol at C2 with the configuration *SRR* should be called 2-*epi*-α-tocopherol.

As mentioned above the first synthesis of α-tocopherol was an equimolar mixture of *RRR*- and *SRR*-α-tocopherol, this compound is called 2-*ambo*-α-tocopherol (*R*/*SRR*-α-tocopherol) and formerly known as DL-α-tocopherol until the optical activity of phytol was recognized, when DL-α-tocopherol was restricted to all-*rac*-α-tocopherol. The acetate of 2-*ambo*-α-tocopheryl acetate was the former international standard vitamin E activity.

All-*rac*-α-tocopheryl acetate is the international standard vitamin E compound today with an activity of 1 International Unit (IU)/mg (Blatt *et al.*, 2004; VERIS Research summary, 1999). However, since all-*rac*-α-tocopheryl acetate consists of an equimolar mixture of all eight possible stereoisomers (*RRR, RRS, RSS, RSR, SRR, SSR, SRS*, and *SSS*), each of them has its own biological activity.

C. BIOACTIVITY AND BIOAVAILABILITY

"Biologic activity," "bioactivity," or "biopotency" are expressions describing the biological effect of a nutrient on the living organism. Assays of biological activity measure effects on clinical endpoints, biomarkers, or surrogate markers of clinical endpoints (Derendorf *et al.*, 2000). For substances with vitamin E activity, the biological effects in focus are resorption of fetuses, testicular degeneration, muscle dystrophy, anemia, encephalomalacia, and in recent years the influence of vitamin E on the immune system has become more and more important (Scherf *et al.*, 1996). Measures of bioavailability, however, are plasma or tissue concentrations of tocopherols and their metabolites, which are not measures of activity. According to the pharmacology, bioavailability is defined as the plasma concentration of a water-soluble drug given orally compared with the concentration when the drug is given intravenously (Traber, 2000). However, for fat-soluble nutrients, bioavailability is difficult to assess because the nutrient cannot be given intravenously, so relative bioavailability is often measured (Traber, 2000). From a nutritional point of view, the bioavailability may be defined as the proportion of vitamin E ingested that undergoes intestinal absorption and utilization by the body, and this definition therefore encompasses the processes of vitamin E absorption, transport, distribution, and metabolism (Bramley *et al.*, 2000). The relative bioavailability of the different forms of vitamin E varies between tissues as well as with dose, time after dosing, and duration of dosing as demonstrated in an excellent review by Blatt *et al.* (2004). This nonconstant relative bioavailability predicts nonconstant relative activity (Blatt *et al.*, 2004), and preferably, both measures of bioavailability and bioactivity should be taken into account when assessing the biopotency of the different vitamin E forms.

Ideally, assays of bioactivity should measure clinical endpoints like resorption of fetuses, testicular degeneration, muscle dystrophy, anemia, encephalomalacia, and so on. However, these experiments are difficult and costly to conduct—like they often involve ethical considerations (Blatt *et al.*, 2004; Scherf *et al.*, 1996) and the tests are not applicable to larger animals or human subjects (Jensen *et al.*, 2006). Experiments involving clinical endpoints have only been performed with rats, chicks, and a single experiment with rabbits back in 1947 (Scherf *et al.*, 1996), and the relative bioactivity of the various vitamin E compounds used for all animals has been established from these experiments.

Clinical endpoints of vitamin E activity by definition involve only the prevention or resolution of vitamin E deficiency. There are no valid clinical assays of the relative vitamin E activity of the individual stereoisomers of α-tocopherol in humans and animals except rats and chickens because no clinical trials show their relative dose–effect relationships in preventing or treating vitamin E deficiency. For humans and other animals, only different biomarkers or surrogate markers of bioactivity have been measured (Blatt *et al.*, 2004).

It is important to distinguish between bioavailability and bioactivity as bioavailability only gives information about availability of the compound, but nothing about its bioactivity. Bioavailability is a prerequisite for bioactivity, and due to lack of satisfactory biomarkers to compensate for the lack of measurements of clinical endpoint, bioavailability is an often reported biomarker (Jensen *et al.*, 2006; Scherf *et al.*, 1996). In studies with rats, a good consistency between the classical resorption–gestation test and the bioavailability of the individual stereoisomers in fluids and tissues has been shown (Scherf *et al.*, 1996; Weiser *et al.*, 1996). However, this may have been a coincidence because the relative bioavailability of the stereoisomers is not constant in humans and animals because their relative concentrations vary between tissues and vary with time after dosing, duration of dosing, and amount of each dose (Blatt *et al.*, 2004).

Harris and Ludwig (1949) established in a series of resorption–gestation test with rats a relative bioactivity of *RRR*-α-tocopheryl acetate against 2-*ambo*-α-tocopherylacetate of 1.36, which later has been accepted as the official bioactivity ratio between natural and synthetic α-tocopherol and esters thereof. This ratio has later been confirmed in several other experiments with rats, chickens, and a single experiment with rabbits. These experiments has comprised experiments where clinical endpoints as resorption–gestation, red blood cell hemolysis, and curative myopathy with rats. In chick, the experiments has comprised studies both involving clinical endpoints as muscular dystrophy, encephalomalacia, as well as different biomarkers as liver storage and plasma levels. With rabbits, one experiment with creatin–urea as biomarker has been performed. On average, these experiments have shown a bioactivity of *RRR*-α-tocopheryl acetate against 2-*ambo*-α-tocopheryl acetate (*R*/*SRR*-α-tocopheryl acetate) of 1.41 with variations from 1.22 to 1.68 (Scherf *et al.*, 1996).

The relative bioactivity ratio between *RRR*-α-tocopheryl acetate, 2-*ambo*-α-tocopheryl acetate, and all-*rac*-α-tocopheryl acetate was studied by Weiser and Vecchi (1981) in resorption–gestation tests with rats. These experiments showed a small discrepancy between the old standard 2-*ambo*-α-tocopheryl acetate and all-*rac*-α-tocopheryl acetate. Thus, in this experiment, the bioactivity of 2-*ambo*-α-tocopheryl acetate was on average 1.09 compared to all-*rac*-α-tocopheryl acetate. This discrepancy is likely caused by the fact that 2-*ambo*-α-tocopherol consist of 50% *RRR*-α-tocopherol and 50% SRR-α-tocopherol, whereas all-*rac*-α-tocopherol only consist of 12.5% α-tocopherol, 37.5% *RR*/*SR*/*S*-α-tocopherol and 50% *SR*/*SR*/*S*-α-tocopherol. Accordingly, the bioactivity of *RRR*-α-tocopheryl acetate compared to 2-*ambo*-α-tocopheryl acetate and all-*rac*-α-tocopheryl acetate was 1.32 and 1.50, respectively. The relative bioactivity of different stereoisomers of α-tocopheryl esters and mixtures thereof determined on basis of the rat resorption gestation test are shown in Table I.

There is also a need to clarify the relative bioavailability of natural and synthetic vitamin E, which is currently a subject of some controversy (Lodge, 2005). Previous studies using a competitive uptake approach have found bioavailability ratios of natural: synthetic vitamin E of 2:1, differing from the accepted biopotency ratio of 1.36:1. However, when using a noncompetitive uptake approach to compare the plasma biokinetics of *RRR*- and all-*rac*-α-tocopheryl acetate in smokers and nonsmokers, the relative bioavailability ratio was close to the currently accepted ratio of 1.36:1 (Lodge, 2005). However, the current debate on the bioavailability focuses mostly on the comparison of *RRR*- versus all-*rac*-; however, analysis of the individual stereoisomers will be of greater importance for the bioavailability subject.

TABLE I. Relative Biological Activities of Different Stereoisomers of α-Tocopherol and Mixtures Thereof Found Resorption–Gestation Test with Rats

Compounds	References	Relative bioactivity
S/*RRR*-α-tocopheryl acetate relative to *RRR*-α-tocopheryl acetate	Harris and Ludwig, 1949	0.73
S/*RRR*-α-tocopheryl acetate relative to *RRR*-α-tocopheryl acetate	Weiser and Vecchi, 1981	0.76
All-*rac*-α-tocopheryl acetate relative to *RRR*-α-tocopheryl acetate	Weiser and Vecchi, 1981	0.67
S/*RRR*-α-tocopheryl acetate relative to all-*rac*-α-tocopheryl acetate	Weiser and Vecchi, 1981	1.09
SSS-α-tocopheryl acetate relative to *RRR*-α-tocopheryl acetate	Ames *et al.*, 1963	0.21
All-*rac*-α-tocopheryl acetate relative to *RRR*-α-tocopheryl acetate	Weiser and Vecchi, 1982	0.77
RR/*SR*/*S*-α-tocopheryl acetate relative to *RRR*-α-tocopheryl acetate	Weiser and Vecchi, 1982	0.92
RRS-α-tocopheryl acetate relative to *RRR*-α-tocopheryl acetate	Weiser and Vecchi, 1982	0.90
RSS-α-tocopheryl acetate relative to *RRR*-α-tocopheryl acetate	Weiser and Vecchi, 1982	0.73
RSR-α-tocopheryl acetate relative to *RRR*-α-tocopheryl acetate	Weiser and Vecchi, 1982	0.57
SRR-α-tocopheryl acetate relative to *RRR*-α-tocopheryl acetate	Weiser and Vecchi, 1982	0.31
SSR-α-tocopheryl acetate relative to *RRR*-α-tocopheryl acetate	Weiser and Vecchi, 1982	0.21
SRS-α-tocopheryl acetate relative to *RRR*-α-tocopheryl acetate	Weiser and Vecchi, 1982	0.37
SSS-α-tocopheryl acetate relative to *RRR*-α-tocopheryl acetate	Weiser and Vecchi, 1982	0.60

III. ANALYTICAL METHODS FOR SEPARATION OF α-TOCOPHEROL STEREOISOMERS

Analysis of the individual stereoisomers of α-tocopherol is an important tool in order to quantify relative bioavailability of the individual stereoisomers (Jensen *et al.*, 2006). Separations of the eight stereoisomers of α-tocopherol are a great challenge and until now no single method allow the separation of all eight stereoisomers in one single chromatographic run (Nelis *et al.*, 2000). The use of deuterium labeled α-tocopherol in conjunction with GC-MS (Ingold *et al.*, 1987) or HPLC-MS (Lauridsen *et al.*, 2001b) is another important method to study bioavailability of different preparations of natural and synthetic vitamin E.

A. GC AND LC METHODS

Neither GC or LC methods is alone capable of distinguishing between all eight α-tocopherol stereoisomers, but the two methods are complementary and hence, can be applied consecutively (Nelis *et al.*, 2000). The first reported chromatographic method that allowed separation of α-tocopherolstereoisomers was a GC method which allowed separation of the four pairs of diastereoisomers (*RSS*/*SRR*, *RSR*/*SRS*, *RRR*/*SSS*, and *RRS*/*SSR*) on a 115-m SP 2340 capillary column of trimethylsilyl (TMS) ethers of α-tocopherol stereoisomers. Weiser and Vecchi (1981) performed the same separation but used methyl ethers of α-tocopherol stereoisomers instead and Cohen *et al.* (1981) performed synthesis of all eight stereoisomers of α-tocopheryl acetate and determined their diastereisomeric and enantomeric purity by GC. These initial methods for separation of the stereoisomers were based on the principle that diastereoisomers possess different physicochemical properties and can be separated on conventional columns.

Chiral HPLC has proven to be one of the best methods for direct separation and analysis of enantiomers. It is more versatile than chiral GC because the compounds do not need to be volatile. Current chiral HPLC methods are either direct, which utilizes chiral stationary phases and chiral additives in the mobile phase, or indirect, which involves derivatization of samples.

The chiral HPLC methods published for separation of α-tocopherol stereoisomers until today is all based on chiral stationary phases and in the first published methods the columns were prepared in the research laboratories. Thus, Ueda *et al.* (1993) developed a HPLC method based on separation of stereoisomers of α-tocopheryl acetates on a Chiralpak OP(+) column (250 × 4.6 mm^2) with methanol–water (96:4 v/v) as mobile phase at a flow rate of 0.3 ml/min. This column consists of poly(diphenyl-2-pyridylmethyl methacrylate) and is capable of separating compounds possessing an aromatic group. This method allowed separation of stereoisomers of α-tocopheryl acetates into four distinct peaks (peak 1: consisted of the four 2*R*-stereoisomers, *RRR*/*RRS*/*RSS*/*RSR*; peak 2 consisted of *SSS*/*SSR*; peak 3 consisted of *SRR* and peak four consisted of *SRS*).

Weiser *et al.* (1996) further developed the method published by Vecchi *et al.* (1990) based on preparative chiral HPLC of stereoisomers of α-tocopheryl acetates followed by GC of the corresponding methyl ethers resulting in separation of all eight stereoisomers. The preparative chiral HPLC was performed on a Nucleosil 1000–5 Machery-Nagel 250 × 4.6 mm^2 HPLC column coated with (+)poly(triphenylmethylmetacrylate), whereby four pairs of diastereoisomers were separated (*RSR*/*RSS*; *RRR*/*RRS*; *SSS*/*SSR*, and *SRS*/*SRR*). These four peaks were preparative collected two and two, thus the first fraction contained the four 2*R*-stereoisomers and the second fraction contained the four 2*S*-stereoisomers. The four 2*R*-stereoisomers and the four 2*S*-stereoisomers were then transferred into their corresponding methyl ethers and subsequently separated individually on a 100-m glass capillary GC column with an internal diameter of 0.3 mm, dynamically coated with Silar 10C held isothermally at 165°C.

Chiral HPLC columns are made by immobilizing single enantiomers onto the stationary phase. Resolution relies on the formation of transient diastereoisomers on the surface of the column packing. The compound, which forms the most stable diastereoisomer, will be most retained, whereas the opposite enantiomer will form a less stable diastereoisomer and will elute first. To achieve discrimination between enantiomers, a minimum of three points of interaction to achieve chiral recognition are needed.

The fast development of commercial HPLC columns with chiral stationary phases mainly for the pharmaceutical industry has provided new challenges and possibilities for the separation of α-tocopherol stereoisomers.

One of the most widespread chiral HPLC column today is based on modified cellulose coated onto silica. The cellulose is further modified with different aromatic compounds. For separation of stereoisomers of tocopherols tris(3,5-dimethylphenylcarbamate)-cellulose from the Japanese manufacturer Daicel has proven to be excellent. The separations occur due to a multimode mechanism involving hydrogen bonding, π–π interactions, dipole stacking, and inclusion complexes. The phases contain chiral cavities or ravines, which have a high affinity for aromatic groups. For separation to occur, there must be a tight fit of part of the solute into the cavity, and at least one of the substituents on the chiral center needs to interact with the steric environment just outside the cavity. The forces that lead to this interaction are very weak and require careful optimization by adjustment of the mobile phase and temperature to maximize selectivity. Typically, a free energy of interaction difference of only 0.03 kJ/mol between the enantiomers and the stationary phase will lead to resolution. Obviously, the highly polar hydroxyl group of tocopherols give rise to nonstereoselctive binding between the tocopherol molecule and the carbamate site of the stationary phase. Thus, the derivatization into methyl ethers is essential and much more efficient than acetylation in order to block these unspecific associations and provides the tocopherol molecules with groups capable of interacting with the chiral cellulose stationary phase (Drotleff and Ternes, 2001). Another advantage of the methyl ethers over acetylated tocopherol is their considerable fluorescence properties of the methyl ethers that allow a very specific and sensitive detection.

The method published by Lauridsen and Jensen (2005) and Jensen *et al.* (2006) offers an easy and simple combination of analysis of both total tocopherol content and stereoisomer composition of α-tocopherol in biological samples. This method allows the quantification of total tocopherol

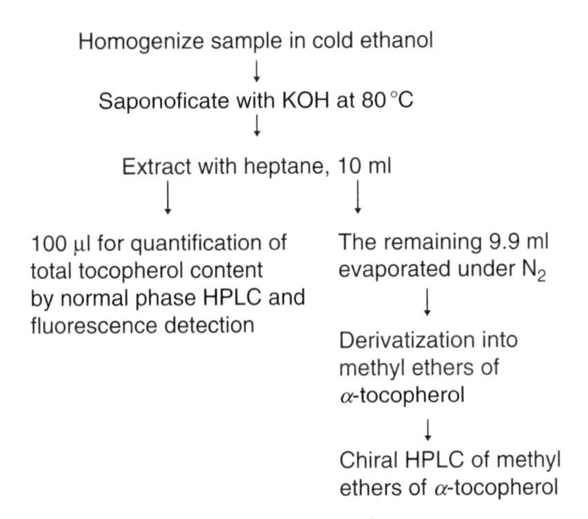

FIGURE 2. Scheme showing the clean up procedure, derivatization, and analytical steps for analysis of total tocopherol content and stereoisomer composition of α-tocopherol (Jensen *et al.*, 2006).

content and composition by normal phase HPLC after saponification and extraction into heptane (Jensen *et al.*, 1999) and subsequent separation of the stereoisomers of α-tocopherol as methyl ethers by chiral HPLC as described by Drotleff and Ternes (2001). By this method, the α-tocopherol stereoisomers are separated into five peaks. The first peak consists of the four 2*S* isomers (*SSS*-, *SSR*-, *SRR*-, *SRS*-), the second peak consists of *RSS*-, the third peak consists of *RRS*-, the fourth peak consists of *RRR*-, and the fifth peak consists of *RSR*-α-tocopherol. The analytical procedure is summarized in Fig. 2 and a HPLC chromatogram of the methyl ethers of α-tocopherol is shown in Fig. 3.

B. Deuterium-Labeling and Mass Spectrometry

The use of stable deuterium-labeled α-tocopherol has been described for the quantification of labeled tocopherols in blood and tissue after administration of deuterium-labeled vitamin E. The evaluation of samples by GC-MS enabled the investigators to distinguish tissue and blood bioavailability of all-*rac*- versus *RRR*-α-tocopherol (Burton *et al.*, 1998). The use of this technique provides several advantages. First, unlabeled endogenous as well as labeled exogenous vitamin E can be determined by single-ion monitoring by MS. Second, as a result of substituting different amounts of deuterium on the chromanol ring, it is possible to administer different forms of vitamin E simultaneously, allowing the subject to act as his/her own control. Hereby the necessity of a crossover protocol is eliminated and newly absorbed vitamin E

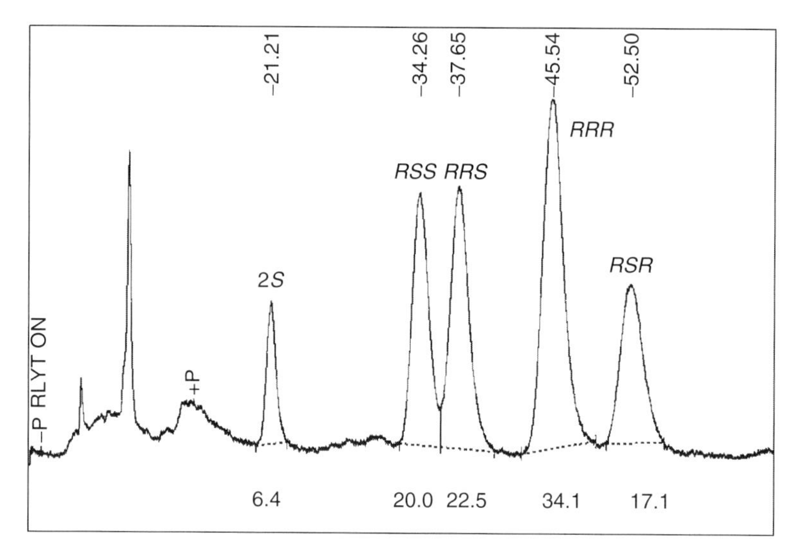

FIGURE 3. HPLC chromatogram of methyl ethers of α-tocopherol from sow plasma separated on a Chiralcel OD-H column. 2SR/SR/S denotes the mixture *SSS*-, *SRR*-, *SSR*-, *SRS*- α-tocopherol. *RSS*, *RRS*, *RRR*, and *RSR* denote *RSS*-, *RRS*-, *RRR*-, and *RSR*-α-tocopherol, respectively. The numbers below the baseline denote the relative distribution of the stereoisomers.

can be distinguished from the unlabeled vitamin E already present in the body. In addition, the deuterated vitamin E technique allows direct comparison of two distinctly labeled vitamin E forms in the same animal, whereby most of the variation attributable to differences between individuals, and factors that change with time can be eliminated (Lauridsen *et al.*, 2002a).

The deuterated technique is based on specific labeling of α-tocopherol with trimethyl deuterium (CD_3). α-Tocopherol can be labeled in either position 5, 7, or 8 or any combination thereof (Fig. 4). Hereby the tocopherol molecules are labeled with 3, 6, or 9 deuterium atoms located in the nonlabile aromatic methyl positions, making it possible to distinguish the silated ethers by GC-MS due to different parent ion masses. For labeling *RRR*-α-tocopherol, either γ- or δ-tocopherol are methylated with CD_3 labeled in position 5 or 5 and 7 on the chromanol ring yielding *RRR*-α-5-(CD_3)-tocopherol (d_3) and *RRR*-α-5,7-$(CD_3)_2$-tocopherol (d_6), respectively (Burton and Traber, 1990; Fig. 4).

Although being very sensitive and specific, the GC-MS method of deuterated tocopherols is time-consuming and costly. The introduction of liquid chromatography tandem mass spectrometry (LC-MS/MS) is a convenient and sensitive technique for the analysis of vitamin E in extracts from biological samples (Lauridsen *et al.*, 2001b). The use of liquid chromatography instead of gas chromatography decreases preparation time and cost, as

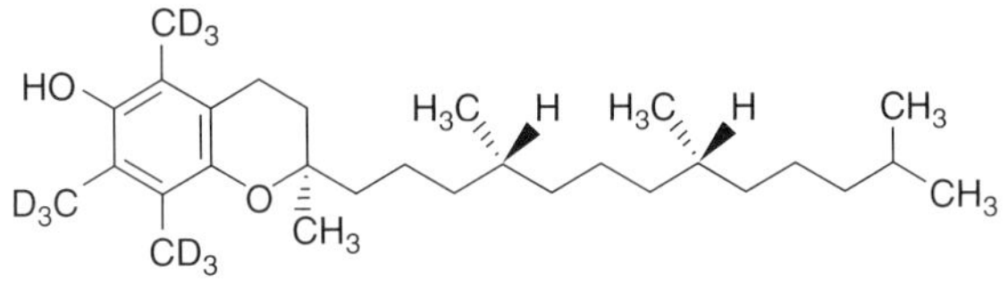

FIGURE 4. Possible positions of deuterium in deuterated α-tocopherol (modified after Burton and Traber, 1990).

the derivitization step is unneeded. Thus, the LC-MS/MS analysis has simplified the clean up procedure and shortened the analysis time when compared to the GC-MS measurement.

IV. BIOAVAILABILITY AND SECRETION INTO MILK

As discussed above, bioavailability is an important part of the term bioactivity and is quantitative measurable in blood, tissue, and excreta. In addition, the lack of good biological markers for bioactivities, bioavailability is often used as one of the surrogate markers for bioactivities with those limitations this must give (Blatt *et al.*, 2004; Jensen *et al.*, 2006). Therefore, results of α-tocopherol stereoisomers in different animal species and humans are presented with the focus on bioavailability rather than bioactivity.

A. RATS

Bioavailability of stereoisomers of α-tocopherol in rats has been studied both with the deuterium-labeling technique (Blatt *et al.*, 2004; Burton and Traber 1990; Ingold *et al.*, 1987) as well as chiral GC and LC techniques (Blatt *et al.*, 2004; Jensen *et al.*, 2006; Weiser *et al.*, 1996).

Ingold *et al.* (1987) studied the uptake and biodiscrimination of deuterated *RRR*-α-tocopherol versus SRR-α-tocopherol in male rats fed a diet containing 36-mg *RRR*-α-tocopheryl acetate/kg diet for 4 weeks and then 18-mg d_6*RRR*- and 18-mg d_3SRR-α-tocopheryl acetate/kg diet over a 5-month period. This experiment demonstrated the variation in bioavailability of different stereoisomers in blood and various tissues as well as changes in bioavailability over time. Thus, as shown in Fig. 5, there was a 1.4-fold enrichment of d_6*RRR*-α-tocopherol in plasma after the first day, and this enrichment increased to 2.5 over the 5-month duration of this experiment. In the liver, d_3*SRR*-α-tocopherol was retained twice as much as

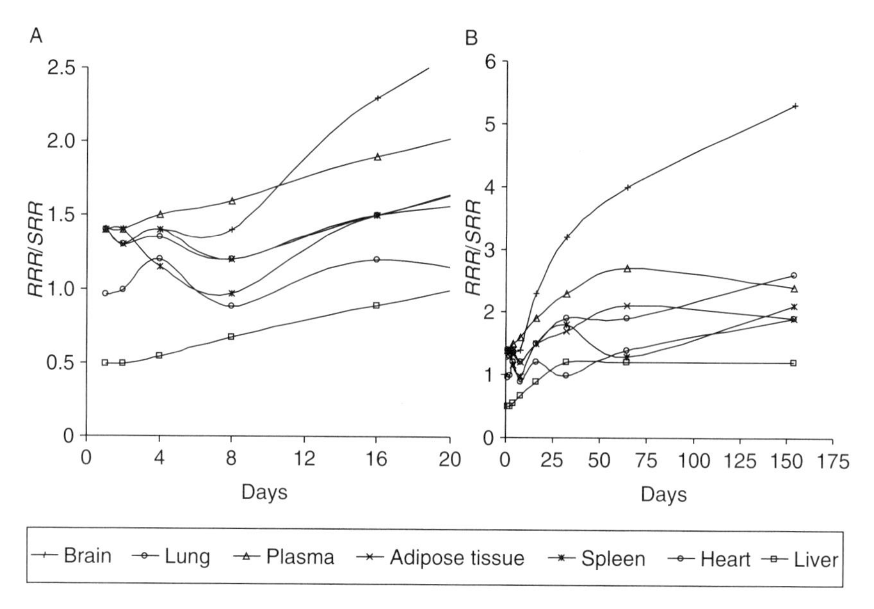

FIGURE 5. Time dependence of the discrimination in net uptake of deuterated *RRR*- and *SRR*-α-tocopherols in plasma and tissue of rats (after Ingold *et al.*, 1987). (A) Day 1–16 and (B) day 1–154.

d_6*RRR*- α-tocopherol during the first period and after 32 days the ratio between d_6*RRR*- and d_3SRR-α-tocopherols approached 1. The greatest discrimination in favor of d_6*RRR*-α-tocopherol was observed in the brain.

Weiser *et al.* (1996) studied the relative distribution of all eight stereoisomers of α-tocopherol in plasma and tissue from rats fed a daily dose of 0.82-mg all-*rac*-α-tocopheryl acetate for 90 days by a combination of chiral HPLC and GC, some of the results are summarized in Fig. 6. Generally, the sum of all 2*R*-α-tocopherol in the analyzed plasma and tissues comprised 74–88% of all the α-tocopherols with exception of the liver, which at day 8 comprised equal amount of the 2*R*- and 2*S*-α-tocopherols. At day 90, 70% of the α-tocopherol in the liver was 2*R*-α-tocopherol. Keeping in mind that *RRR*-α-tocopherol has the highest biological activity, it is interesting to notice that the three synthetic 2*R*-forms generally occurred in higher concentration than *RRR*-α-tocopherol. Among the 2*S* forms, *SRS*-α-tocopherol showed the highest relative concentration in plasma as well as in the analyzed tissues.

However, according to Blatt *et al.* (2004) only exist a weak and variable coherence between relative bioactivity assessed on the basis of the rat resorption–gestation test as used by Weiser and Vecchi (1982) and bioavailability assessed using comparable conditions in Weiser *et al.* (1996).

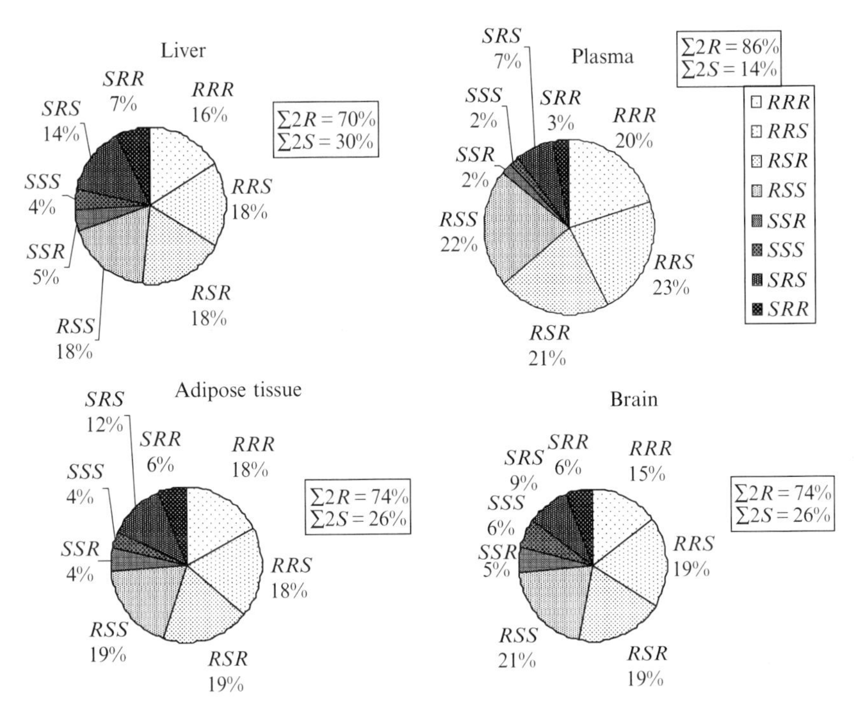

FIGURE 6. Relative distribution of all eight stereoisomers in liver, adipose tissue, brain, and plasma from rats fed at a daily dose of 0.82-mg all-*rac*-α-tocopheryl acetate for 90 days (after Weiser *et al.*, 1996).

Jensen *et al.* (2006) studied the effect of different dietary doses varying from 25 to 200 mg/kg diet of all-*rac*-α-tocopheryl acetate for 10 days on the relative distribution of stereoisomers in plasma and tissues. With increasing dietary content of all-*rac*-α-tocopheryl acetate in the feed, the proportion of *RRR*-α-tocopherol decreased in plasma (22–18%), liver (20–13%), lung (21–19%), and adipose tissue (25–20%). The three synthetic 2*R*-stereoisomers remained fairly constant within each of the tissues. *RRS*-α-tocopherol constituted the highest proportion of the four 2*R*-stereoisomers with exception of the liver, where the sum of the four 2*S*-stereoisomers constituted the highest proportion (29–33%). Thus, the relative bioavailability of *RRR*-α-tocopherol compared with all-*rac*-α-tocopherol varied from 1.07 to 2.02, *RRS*-α-tocopherol varied from 1.40 to 2.12, *RSS*-α-tocopherol varied from 1.35 to 2.00, *RSR*-α-tocopherol varied from 1.40 to 1.94, and the relative bioavailability of the sum of the four 2*S*-forms varied from 0.16 to 0.66. In summary for rats on the basis of the referred literature, a clear preference for the 2*R*-stereoisomers over the 2*S*-stereoisomers has been observed; however, between the four 2*R*-stereoisomers no significant biodiscrimination has been noticed.

B. PIGS

There are three critical phases with regard to vitamin E status in pigs, and these are during reproduction, at birth, and after weaning. In this paragraph, we will focus on the α-tocopherol stereoisomers in relation to the periods in which vitamin E is essential for the development of growth and health. The studies with stereoisomers of α-tocopherol in pigs have been performed both with the stable isotope-labeled technique and chiral HPLC.

1. At Birth

Sows discriminate between *RRR*- and all-*rac*-α-tocopheryl acetate with a preference for *RRR*-α-tocopherol (Lauridsen *et al.*, 2002a,b). When using the stable isotope-labeled α-tocopherols, the natural and synthetic forms can be ingested simultaneously for intraindividual purposes. Pregnant sows were provided 150 mg each of d_3-*RRR*-α- and d_6-all-*rac*-α-tocopheryl acetate daily from 7 days before to 7 days after giving birth, and the α-tocopherol delivery to fetuses and to suckling piglets were followed (Lauridsen *et al.*, 2002a). Despite elevated sow plasma-deuterated α-tocopherol concentrations, no labeled α-tocopherol was detected in piglet plasma or tissues. Sow plasma and milk d_3-α- to d_6-α tocopherol were 2:1, leading to a 2:1 ratio in suckling piglet plasma and tissues. At day 7 after birth, most tissues of the piglets contained a tenfold increase in total α-tocopherol concentration compared with piglets at birth. The highest concentrations of deuterated vitamin E were in the liver, followed by the lung, heart, kidney, muscle, intestine, and brain (Lauridsen *et al.*, 2002a). It was concluded that the bioactivity of synthetic all-*rac*-α-tocopherol is roughly one-half of that of natural *RRR*-α-tocopherol during pregnancy and lactation in sows, resulting in a 2:1 ratio of the natural and the synthetic vitamin E forms in milk and the suckling progeny.

In a later experiment with sows, we studied the concentration of α-tocopherol and its stereoisomer distribution by using the HPLC-method, which separates the stereoisomers of α-tocopherol. Sows were provided with either 75 IU of all-*rac*-α-tocopheryl acetate/kg feed (synthetic) or 75 IU of *RRR*-α-tocopheryl acetate/kg feed (natural) 1 week prior to parturition and during lactation for 28 days. Milk samples were obtained from the sows during lactation (day 2, 16, and 28 after parturition), and plasma samples were obtained from the sows at 1 week before parturition (day 108 of gestation) and at day 2 and 16 of lactation. In addition, plasma samples were obtained from the suckling piglets at day 4 and 16 after birth. The results of the concentration of α-tocopherol and its composition of the stereoisomers in the milk and plasma samples are given in detail in Lauridsen and Jensen (2006a). In brief, the most predominant form of stereoisomer forms of α-tocopherol was the *RRR*-form irrespective of feeding natural or synthetic vitamin E, whereas the 2*S*-forms were only present in very limited proportions.

Among the sows fed the synthetic vitamin E, the *RRR*-form contributed with between 32 and 36% in milk, followed by 24–26% of the *RRS*-, 18–20% of the *RSR*-, and 17–18% of the *RSS*-, and 6–8% of the *SR*/*SR*/*S*- during the period from 2 to 28 days of lactation. For comparison, sows fed the natural form secreted between 69 and 91% of the *RRR*-form into the milk, followed by 2–12% of the *RRS*-, 5–10% of the *RSR*-, and 2–7% of the *RSS*-, and 2–3% of the *SR*/*SR*/*S*- with a clear trend of a higher proportion of the *RRR*- with increasing lactation time on the expense of the other 2*R*-isomers. The stereoisomer composition of α-tocopherol in the plasma of sows fed the synthetic vitamin E and their piglets followed the same pattern as for the milk composition, except for a higher contribution (~17%) of the *SR*/*SR*/*S*- in plasma of sows at day 108 of gestation on expense of all four the 2*R*-forms. Thus, these results confirmed the data obtained by Lauridsen *et al.* (2002a,b) that sows discriminated between the stereoisomer forms of α-tocopherols, which resulted in higher milk and plasma α-concentrations arising from the *RRR*-form.

2. At Weaning

The sterosiomer composition of α-tocopherol in plasma and tissues of suckling piglets nicely reflects the stereoisomer profile of the milk, and the reported results above do not seem to indicate a further biodiscrimination by the suckling piglet than carried out by the lactating sow. However, when considering the piglets after weaning from the sow, one may speculate when the piglets are getting mature to be able to discriminate between stereoisomer form α-tocopherols. The role of α-TTP during the biodiscrimination is described below, and the expression of the protein by piglets is therefore reserved for this paragraph. In a recent study, we investigated the stereoisomer profile of α-tocopherol in plasma of piglets throughout the postweaning period, and the piglets were weaned from sows fed 70-IU all-*rac*-α-tocopheryl acetate on day 28 of age. Piglets were throughout the experiment (day 28–49 of age) given 70 IU of vitamin E either in the form all-*rac*-α-tocopherol or in the form of all-*rac*-α-tocopheryl acetate. Regarding the stereoisomer composition of α-tocopherol in plasma of the piglets, *RRR*-α-tocopherol contributed with the highest proportion (32%) in plasma, followed by *RRS*- (27%), *RSR*- (22%), *RSS*- (17%), while the four 2*S*-forms together only provided 1.7% of the α-tocopherol (Lauridsen and Jensen, 2006b).

The age of the piglets (day 35, 42, and 49) had no influence on the stereoisomer composition of α-tocopherol. By using the quantitative method in this study, we were not able to distinguish the "old pool" of vitamin E (provided with the sow milk) from the "newly absorbed pool" of vitamin E (provided with the weaning feed), and as such it is not possible to conclude regarding the eventual discrimination capability of pigs during this age period.

Besides the stereoisomer form of vitamin E, another important issue of the molecule when addressing pigs after weaning is the alcohol versus the acetate form. The vitamin E in milk of animals or humans is provided in the form of alcohol, that is free vitamin E. However, supplementation of vitamin E to animal feed is normally with the acetate-bound form, α-tocopheryl acetate, which may not be well absorbed by young infants having an immature digestive tract. In the study referred to above (Lauridsen and Jensen, 2006b), the α-tocopherol concentration in plasma and tissues of piglets fed the alcohol form of vitamin E (all-*rac*-α-tocopherol) was higher than the acetate (although no statistical differences were obtained).

Although several lines of evidence illustrate that plasma enrichment with *RRR*-α-tocopherol depends on the hepatic α-TTP, and that no discrimination of stereoisomers takes place during the absorption process of the small intestine (Kiyose *et al.*, 1995), it may be speculated that the steroisomer configuration of the dietary α-tocopherol plays a role for the hydrolysis process. The principal enzyme-hydrolysing tocopheryl acetate (Müller *et al.*, 1976) is carboxyl ester hydrolase (CEH), and the presence of bile salts is a prerequisite both as emulsifier and as activators of this enzyme (Lauridsen *et al.*, 2001a).

A study (Knarreborg *et al.*, 2004) provided evidence *in vivo* that the bioavailability of α-tocopheryl acetate is highly dependent on an adequate amount of bile salts to generate enzymatic hydrolysis of α-tocopheryl acetate and subsequent absorption of α-tocopherol to the blood plasma. Besides their role as detergents and CEH-activators, bile acids furthermore modulate CEH's chiral selectivity (Moore *et al.*, 1994). In addition to the chiral properties of the tocopherol molecule, the rate of hydrolysis of the α-tocopheryl esters will depend on the affinity of the CEH toward the ester, at this may depend on the affinity of the CEH toward the ester, and this may depend on the presence of other dietary molecules, for example retinol (Lauridsen *et al.*, 2001a) or tocopherol derivates.

Pigs may have an insufficient secretion of pancreatic enzymes after weaning, and it cannot be excluded that a discrimination of the esters (the *RRR*- versus the all-*rac*-α-tocopheryl acetate) could be caused by the enzyme's specific substrate recognition.

3. During Reproduction

Vitamin E deficiency has been shown to affect reproduction in several animal species, resulting in fetal death and resorption (Nielsen *et al.*, 1979). Consequently, the vitamin E standards for reproducing swine have increased over the past 25 years from 10- to 44-IU/kg diet (Mahan *et al.*, 2000; NRC, 1998). Given the differences in the biological activities of the stereoisomer forms of α-tocopherol, dietary sources of natural (D-α-tocopheryl acetate) or synthetic (DL-α-tocopheryl acetate) vitamin E have called for attention in reproducing sows for several parities (Mahan *et al.*, 2000). No effect was

TABLE II. Concentration of α-Tocopherol in Feed (mg/kg), Plasma (mg/liter) and Fetuses (μg/Fetus), and Composition of Stereoisomers (%) in Gilts at Day 28 of Gestation (Mean and SD in Parentheses)

		α-Tocopherol stereoisomers				
	Total	*RRR-*	*RRS-*	*RSS-*	*RSR-*	*SR/SR/S-*
Feed	57.6	28.9	11.7	9.8	8.4	41.2
Plasma	1.49 (0.35)	43.2 (2.9)	20.9 (2.5)	16.9 (4.1)	16.4 (1.6)	2.2 (1.1)
Fetus	1.07 (0.82)	42.3 (3.1)	20.3 (1.7)	16.5 (2.6)	16.5 (1.4)	2.1 (0.52)

obtained for vitamin E source or level (30 versus 60 IU/kg) on the various sow reproductive measurements, litter size, or incidences of mastitis–metritis–agalactia (MMA) or fluid discharges from the vagina, but serum and liver α-tocopherol contents in 21-day-old-nursing pigs were higher when the sow had been fed the natural compared with the synthetic source or when the 60-IU level had been fed (Mahan *et al.*, 2000).

In a preliminary study, the stereoisomer composition of plasma and fetus obtained from gilts at day 28 of gestation was investigated (personal communication Charlotte Lauridsen). Homogenates of fetus obtained from each gilt ($n = 24$) were prepared and analyzed for the content of α-tocopherol. The gilts were provided a feed supplemented with 32 mg/kg of vitamin E in the form of all-*rac*-α-tocopheryl acetate. Table II shows the stereoisomer distribution of α-tocopherol in the feed, plasma, and fetus of the gilts.

The data confirm other animal assays providing evidence that the chiral center at the position 2 is the major and possibly sole determinant of the biological differences between α-tocopherol stereoisomers. Obviously, the stereoisomer composition of α-tocopherol in the plasma and fetus of the gilts showed the same proportions, thus indicating that no further biodiscrimination by the fetus or the placenta at this stage of reproduction. Interestingly, when compared with plasma of lactating sows (Lauridsen and Jensen, 2006a), the *RRR*-form contributed with a higher proportion on expense of the other stereosiomers in plasma of the gilts at day 28 of gestation. This result might indicate a trend toward a higher biopotency of the *RRR*-form during the critical phase of reproduction, which should be investigated further in the future.

C. HUMANS

With regard to the α-tocopherol stereoisomers in humans, relatively little information is available on the bioavailability and the secretion into milk, and the studies performed on pigs as described above may therefore provide important information applicable to human nutrition. Investigations on the comparison of natural and synthetic vitamin E in humans have mainly been performed using the deuterium-labeled vitamin E, and these results have indicated overall that the natural vitamin E (*RRR*-α-tocopheryl acetate) has roughly twice the availability of synthetic vitamin (all-*rac*-α-tocopheryl acetate) (Burton *et al.*, 1998).

However, the focus has been given specifically to the ratio of *RRR*:*rac*, and very little information is available on the distribution of the single stereoisomer forms of α-tocopherol in humans after consumption of a dose of all-*rac*-α-tocopheryl acetate. In 1993, Ueda *et al.* developed a new chiral HPLC method to separate the eight stereosisomers of α-tocopherol into four peaks—the first peak was composed of the four 2*R*-isomers, the second peak of the *SSS*- and *SSR*-α-tocopherol, the third peak of *SRR*-α-tocopherol, and the four peak of *SRS*-α-tocopherol (Ueda *et al.*, 1993).

Kiyose *et al.* (1997) studied the single 2*S*-stereoisomer forms and the sum of the 2*R*-forms in women, who had received oral administration of natural and synthetic α-tocopheryl acetate. After oral administration, the concentration of 2*S*-isomers increased gradually but was significantly lower than that of 2*R*-isomers. The 2*S*-isomers contributed with around 4% of all-*rac*-α-tocopherol in LDL on the first day of oral administration, and on day 28 the 2*S*-isomer proportion had increased to ~20%. In HDL, the concentration of 2*S*-isomers was lower than that of HDL on days 1, 7, and 14 after administration. The concentrations of 2*S*-isomers were in the order (*SSS* = *SSR*) > *SRS* > *SRR* in HDL and LDL. This study by Kiyose *et al.* (1997) confirmed that the bioavailability of *RRR*-α-tocopherol was greater than that of all-*rac*-α-tocopheryl acetate because the 2*R*-isomers were preferentially incorporated in the serum lipoproteins of humans administered all-*rac*-α-tocopherol.

In addition, several studies suggest that the relative bioavailability of all-*rac*- and *RRR*-α-tocopherols varies with dosage in humans ingesting multiple doses sufficient to achieve steady-state conditions (Blatt *et al.*, 2004), and it would be interesting to know to which extent the actual proportion of the 2*R*-isomers varies when different dosages of the synthetic vitamin E form when administered to humans.

Our data (using our recent developed chiral HPLC method) showing that the 2*S*-forms of α-tocopherol contributed with <3% in human milk (Lauridsen and Jensen, 2006a), and that the ranking of the 2*R*-forms seemed to be similar to the data obtained on sow milk, lead to some suggestions for the future research: that focus should be given on the biological significance of the 2*R*-forms of α-tocopherol, rather than just differentiating between the *RRR*-form versus the other seven forms. Furthermore, that the relative bioavailability of the stereoisomer forms of α-tocopherol varies between tissues and with dosage, and even between animal species.

D. MINK

Mink (*Mustella vision*) is a carnivore belonging to the marten family (*mustelidae*) and is domesticated because of its excellent fur. A major part of mink feed is different fish by-products, as well as slaughter offal. Due to this high intake of by-products with a high proportion of polyunsaturated fatty acids, mink has been shown to have a high vitamin E requirement (Børsting *et al.,* 1998).

In this connection, we have been looking at the biodiscrimination of stereoisomers of α-tocopherol in lactating mink and suckling mink kits, where the dams were fed a diet containing 100-mg all-*rac*-α-tocopheryl acetate per 10 MJ. In Table III, the relative distribution of stereoisomers of α-tocopherol in feed, plasma, and milk of the lactating mink at day 28 and plasma from mink kit also at day 28 are shown. These data clearly demonstrate the efficient exclusion of the 2*S*-stereoisomers in plasma, resulting in a complete absence in tissues from mink kits. It is interesting to notice that in contrast to rats and pigs, mink kit liver does not contain 2*S*-stereoisomers. Mink heart contain 59% *RRR*-α-tocopherol.

E. POULTRY

Although quite a few experiments where the bioactivity of various vitamin E compounds has been tested in poultry, only one experiment dealing with the distribution of stereoisomers of α-tocopherol has been performed (Cortinas *et al.*, 2004). In contrast to the other animal species investigated, chickens fed 100-, 200-, or 400-mg/kg diet all-*rac*-α-tocopheryl acetate the

TABLE III. Total Content and Relative Distribution of Stereoisomers of α-Tocopherol in Feed, Plasma, and Milk of Lactating Mink at Day 28 and Plasma, Liver, Heart, and White Adipose Tissue of Mink Kit at Day 28

		α-Tocopherol stereoisomer				
	α-Tocopherol (mg/kg)	*RRR*-	*RRS*-	*RSS*-	*RSR*-	*SR/SR/S*-
Feed	56	16.2	11.0	9.5	13.0	50.4
Lactating mink plasma	16.8	39.0	21.2	16.3	16.4	7.2
Milk	7.1	37.5	18.3	14.7	16.7	12.8
Mink kit plasma	23.6	46.2	25.3	13.5	13.6	1.5
Liver	78	37.8	24.8	15.3	22.2	0
Heart	15	59.0	14.2	12.0	14.8	0
White adipose tissue	12	44.4	27.0	14.2	14.4	0

distribution of α-tocopherol stereoisomers in liver and thigh was very similar. In general, both tissues predominantly accumulated the 2*R*-stereoisomers (52% in the diet, 74% in liver, and 70% in thigh).

Further, this experiment showed that increased oxidation pressure due to a higher inclusion of polyunsaturated fatty acids in the diet increased the proportion of 2*R*-stereoisomers on the expense of the 2*S*-stereoisomers in both liver and thigh muscle. Within the 2*R*-stereoisomers, the difference between liver and thigh muscle was negligible. Thus, the average distribution of α-tocopherol stereoisomers in liver and thigh of chickens fed synthetic vitamin E was of 20% *RRR*-, 21% RRS-, 17% RSS, 15% RSR-, and 28% of the four 2S-α-tocopherols.

F. RUMINANTS

Studies with cattle and sheep have indicated that the natural form of vitamin E resulted in higher serum α-tocopherol concentrations than did the synthetic form Hidiroglou *et al.* (1992). However, until the work of Jensen *et al.* (2005) and Meglia *et al.* (2006) was published, no information existed about the distribution of the stereoisomers in ruminants. Meglia *et al.* (2006) fed cows daily with 917-mg all-*rac*-α-tocopheryl acetate in addition to the 300–525-mg *RRR*-α-tocopherol occurring naturally in the basal ration for 5 weeks around parturition and analyzed the distribution of stereoisomers in plasma and milk.

Likewise, Jensen *et al.* (2005) fed cows a daily dose of 3000-mg all-*rac*-α-tocopheryl acetate in addition to the 450-mg *RRR*-α-tocopherol occurring naturally in the basal ration for 16 days and analyzed the composition of the stereoisomers in plasma. These results are listed in Table IV together with

TABLE IV. Relative Distribution of α-Tocopherol Stereoisomers in Plasma and Milk from Ruminants Fed all-*rac*-α-Tocopheryl Acetate

		α-Tocopherol stereoisomer				
		RRR-	*RRS*-	*RSS*-	*RSR*-	*SR/SR/S*-
Cow plasma	Meglia *et al.*, 2006	88.1	4.7	4.3	2.3	0.6
Cow milk		87.4	2.8	2.9	5.1	1.8
Cow plasma	Jensen *et al.*, 2005	84.4	5.5	5.0	3.7	1.4
Calf plasma[a]	Milk replacer, 250 mg/day	34.1	22.5	20.0	17.1	6.3
Lamb plasma[a]	200 mg/day	42.3	20.7	15.9	14.5	6.7

[a]Unpublished data.

unpublished results from calves fed milk replacer with 250-mg all-*rac*-α-tocopheryl acetate per kilogram DM and lambs fed a daily dose of 200-mg all-*rac*-α-tocopheryl acetate per day for 3 months.

From Table IV it is clear that cows even though they are fed a higher proportion of the natural stereoisomer has a preferential accumulation of *RRR*-α-tocopherol in plasma as well as secretion into milk. It is important to notice that the 2*S*-forms are almost absent from plasma and milk and that the three synthetic 2*R*-stereoisomers are present in a considerable lower concentration than *RRR*-α-tocopherol.

Calves fed solely on milk replacer with synthetic vitamin E, which means they have been provided with 12.5% of each of the eight stereoisomers shows also a clear biodiscrimination in favor of *RRR*-α-tocopherol as it constitutes 34% of total α-tocopherol in plasma. The proportion of 2*S*-stereoisomers in both calves and lambs were very low. The three synthetic 2*R*-stereoisomers in both calves and lambs were represented in the order *RRS* > *RSS* > *RSR*. Generally, the stereoisomer patterns in these young animals were very similar to the distribution reported in pigs.

V. α-Tocopherol-Binding Protein (α-TTP)

In contrast to the relatively well-investigated binding proteins for vitamins A and D, proteins that bind and transport vitamin E have only been identified in the past decade and many of their specific biological roles remain elusive. The term "tocopherol-associated proteins" has been used to distinguish a molecularly defined family of proteins that are capable of binding α-tocopherol (Zimmer *et al.*, 2000) with a higher affinity than other tocopherols (Yamauchi *et al.*, 2001) and are also capable of binding phospholipids.

The α-tocopherol transfer protein (α-TTP) is of 32-kilodalton (kDa) that acts primarily in the liver to specifically select α-tocopherol with the side chain linked to the chromanol ring in the *R*-configuration at position 2 for incorporation into nascent very low-density lipoprotein (VLDL) (Hosomi *et al.*, 1997). Thus, α-TTP preferentially binds α-tocopherol over the γ-homologue (or others) as it can discriminate between the number and position of methyl groups on the chromanol ring. The predominant homologue *in vivo* is therefore α-tocopherol because most of the γ-tocopherol is excreted into the bile while the α-tocopherol is preferentially retained.

The overall role of α-TTP is the maintenance of normal plasma tocopherol concentrations, which is a conclusion that has been supported by the existence of two human diseases, "FIVE" (familial-isolated vitamin E deficiency) and "AVED" (ataxia with vitamin E deficiency) apparently caused by a genetic defect of α-TTP (Ben Hamida *et al.*, 1993; Traber *et al.*, 1990). In patients, the genetic defect causes impaired incorporation of α-tocopherol into nascent VLDL, leading to very low levels of plasma vitamin E. Absorption and transport of vitamin E in chylomicrons, however, is normal in these diseases.

α-TTP is predominantly expressed in liver but low concentrations have also been detected in brain, spleen, lung, and kidney (Hosomi *et al.*, 1998). The ability of hepatic α-TTP to discriminate between tocopherols based on the chemistry of the chromanol ring and the position C2 appears to be the main reason for the predominance of *RRR*-α-tocopherol *in vivo*. As well as discriminating between tocopherol homologues, α-TTP can also differentiate between different tocopherol stereoisomers.

This indication has mainly been attributed during the study of Burton *et al.* (1998). In that study, the bioavailability of synthetic α-tocopherol in humans was determined to be roughly the half of that of natural *RRR*-α-tocopherol, and the 2*R*-stereoisomers were preferentially retained and 2*S*-stereoisomers being eliminated. At present, it is therefore generally accepted in the literature that the tocopherol-binding and transport proteins discovered hitherto selectively select *RRR*-α-tocopherol, leaving β-, γ-, and δ-tocopherols and the synthetic stereoisomers of racemic α-tocopherol to degradation.

Our results on the stereoisomer composition of fluids in different animal species show, however, that the other 2*R*-forms also contribute with a considerable proportion of the α-tocopherol, and it would be interesting to know to which extent the biodiscrimination between these stereoisomer forms is dependent on the expression of α-TTP. As shown for rats, the *RRS*-α-tocopherol constituted the highest proportion of the four 2*R*-stereoisomers in plasma and all measured tissues with exception of the liver (Jensen *et al.*, 2006).

Interestingly, cows have a much stronger preference for the *RRR*-form than other animal species or even calves, and to which extent the above mentioned results between animals species is exclusively related to the discrimination by α-TTP remains to be investigated. How many other α-tocopherol-binding proteins exist apart from α-TTP is still unclear (Brigelius-Flohé *et al.*, 2002).

Dietary vitamin E influences hepatic α-TTP concentrations (Fechner *et al.*, 1998), however, when vitamin E-depleted rats were fed a diet containing α- or δ-tocopherol or both, α-TTP messenger RNA increased (Fechner *et al.*, 1998). This was not expected, because the biological activity of δ-tocopherol, which does not bind α-TTP (Hosomi *et al.*, 1997), differs strongly from that of α-tocopherol. This shows that other tocopherols may have similar effects on gene expression that still await detection. Other authors (Kim *et al.*, 1998) have found that vitamin E depletion of rats strongly increased α-TTP messenger RNA concentrations, which then decreased after refeeding of α-tocopherol to below the concentrations in the control rats. Thus, the biodiscrimination appears to depend on dietary tocopherols, and as for example shown for rats (Jensen *et al.*, 2006), the proportion of *RRR*-α-tocopherol decreased in plasma and tissues with increasing dietary content of all-*rac*-α-tocopheryl acetate, whereas the three synthetic 2*R*-isomers remained fairly constant within these tissues, and only small changes were observed.

Obviously, α-TTP may reach a saturation limit with regard to the preference toward the *RRR*-form, or the general biodiscrimination among the 2*R*-forms are actually not notable. Unpublished results on piglets fed increasing levels of all-*rac*-α-tocopheryl acetate during the period after weaning showed no influence on the expression of hepatic α-TTP (C. Lauridsen, personal communication). Interestingly, we observed a higher concentration *RRR*-α-tocopherol on the expense of the other 2*R*-forms in immune cells of pigs supplemented with vitamin C at a level of 500 mg/kg during the period after weaning (Lauridsen and Jensen, 2005), which may reflect a regeneration of the *RRR*-form during absorption and transportation by this antioxidant supplementation. However, a direct influence of the antioxidant supplementation on the regulation and concentration of binding proteins may also be speculated as seen for vitamin A- and D-binding proteins (Baker *et al.*, 1984; Ong, 1985).

VI. CONCLUSIONS

The discussion on the bioavailability of *RRR*- and all-*rac*-α-tocopheryl acetate has primarily been based on human and animal studies using deuterium-labeled forms, whereby a higher biopotency of 2:1 (of *RRR*: all-*rac*) has been demonstrated, differing from the accepted biopotency ratio of 1.36:1. However, the quantitative separation of the individual stereoisomers of the 2*R*-forms allow us to get a more detailed picture of the bioavailability of natural and synthetic vitamin E forms.

In agreement with previous studies, the 2*S*-forms exert very little importance for the vitamin E activity due to their limited bioavailability. We find notable differences between animal species with regard to the biodiscrimination between the 2*R*-forms, and with regard to cows, the preference of *RRR*-form is by far the most predominant one.

The distribution of the stereoisomer forms varies from tissue to tissue, and in some cases, higher levels of the other 2*R*-forms than the *RRR*-form are obtained, for example, for rats. However, the biodiscrimination of the stereoisomers forms is influenced by other factors such as age, dietary levels, and time after dosage. More focus should be given on the bioactivity of the individual 2*R*-forms rather than just the comparison between *RRR*- and all-*rac*-α-tocopheryl acetate.

REFERENCES

Ames, S. R., Ludwig, M. I., Nelam, D. R., and Robeson, C. D. (1963). Biological activity of an l-Epimer of d-α-tocopheryl acetate. *Biochem.* **2,** 188–190.

Baker, L. R. I., Clark, M. L., Fairclough, P. D., and Goble, H. (1984). Identification of two intestinal vitamin D-dependent calcium-binding proteins in the X-linked hypophosphataemic mouse. *Biochem. Biophys. Res. Commun.* **119,** 850–853.

Ben Hamida, C., Doerflinger, N., Belal, S., Linder, C., Reutenauer, L., Dib, C., Gypay, G., Vignal, A., Le Pasilier, D., Cohen, D., Pandolfo, M., Mokini, V., *et al.* (1993). Localization of Friedreich ataxia phenotype with selective vitamin E deficiency to chromosome 8q by homozygosity mapping. *Nat. Genet.* **5,** 195–200.

Blatt, D. H., Pryor, W. A., Mata, J. E., and Rodriguez-Proteau, R. (2004). Re-evaluation of the relative potency of synthetic and natural α-tocopherol: Experimental and clinical observations. *J. Nutr. Biochem.* **15,** 380–395.

Bramley, P. M., Elmadfa, I., Kafatos, A., Kelly, F. J., Manios, Y., Roxborough, H. E., Schuch, W., Sheehy, P. J. A., and Wagner, K.-H. (2000). Review. Vitamin E. *J. Sci. Food Agric.* **80,** 913–938.

Brigelius-Flohé, R., Kelly, F. J., Salonen, J. T., Neuzil, J., Zingg, J.-M., and Azzi, A. (2002). The European perspective on vitamin E: Current knowledge and future research. *Am. J. Clin. Nutr.* **76,** 703–716.

Burton, G. W., and Traber, M. G. (1990). Vitamin E: Antioxidant activity, biokinetics, and bioavailability. *In* "Annual Review of Nutrition" (R. E. Olson, D. M. Brier, and D. B. McCormick, Eds.). Annual Reviews Inc., Palo Alto, CA.

Burton, G. W., Traber, M. G., Acuff, R. V., Walters, D. N., Kayden, H., Hughes, L., and Ingold, K. U. (1998). Human plasma and tissue alpha-tocopherol concentrations in respons to supplementation with deuterated natural and synthetic vitamin E. *Am. J. Clin. Nutr.* **67,** 669–684.

Børsting, C. F., Engberg, R. M., Jensen, S. K., and Damgaard, B. M. (1998). Effects of high amounts of dietary fish oil of different oxidative quality on performance and health of growing-furring male mink (*Mustela vision*) and nursing female mink. *J. Anim. Physiol. Anim. Nutr.* **79,** 210–223.

Cohen, N., Scott, C. G., Neukom, C., Lopresti, R. J., Weber, G., and Saucy, G. (1981). Total synthesis of all eight stereoisomers of α-tocopheryl acetate. Determination of their diastereoisomeric and enantiomeric purity by gas chromatography. *Helv. Chim. Acta* **64,** 1158–1173.

Cortinas, L., Baroeta, A. C., Galobart, J., and Jensen, S. K. (2004). Distribution of α-tocopherol stereoisomers in liver and thigh of chickens. *Br. J. Nutr.* **92,** 295–301.

DellaPenna, D. (2005). A decade of progress in understanding vitamin E synthesis in plants. *J. Plant Physiol.* **162,** 729–737.

Derendorf, H., Lesko, L. J., Chaikin, P., Colburn, W. A., Lee, P., Miller, R., Powell, R., Rhodes, G., Stanski, D., and Venitz (2000). Pharmacokinetic/pharmacodynamic modeling in drug research and development. *J. Clin. Pharmacol.* **40,** 1399–1418.

Drotleff, A. M., and Ternes, W. (2001). Determination of RS, E/Z-tocotrienols by HPLC. *J. Chromatogr. A* **909,** 215–223.

Evans, H. M., and Bishop, K. S. (1923). Existence of a hitherto unknown dietary factor essential for reproduction. *J. Am. Med. Assoc.* **81,** 889–892.

Fechner, H., Schlame, M., Guthmann, F., Stevens, P. A., and Rüstow, B. (1998). α- and δ-tocopherol induce expression of hepatic α-tocopherol-tranfer protein mRNA. *Biochem. J.* **331,** 577–581.

Harris, P. L., and Ludwig, M. I. (1949). Relative vitamin E potency of natural and of synthetic alpha-tocopherol. *J. Biol. Chem.* **179,** 1111–1115.

Hidiroglou, N., McDowell, L. R., Papas, A. M., Antapli, M., and Wilkinson, N. S. (1992). Bioavailability of vitamin E compounds in lambs. *J. Anim. Sci.* **70,** 2556–2561.

Hosomi, A., Arita, M., Sato, Y., Kiyose, C., Ueda, T., Igarashi, O., Arai, H., and Inoue, K. (1997). Affinity for α-tocopherol transfer protein as a determinant of the biological activities of vitamin E analogs. *FEBS Lett.* **409,** 105–108.

Hosomi, A., Goto, K., Kondo, H., Iwatsubo, T., Yokota, T., Ogawa, M., Arita, M., Aoki, J., Arai, H., and Inoue, K. (1998). Localization of α-tocopherol transfer protein in rat brain. *Neurosci. Lett.* **256,** 159–162.

Ingold, K. U., Burton, G. W., Foster, D. O., Hughes, L., Lindsay, D. A., and Webb, A. (1987). Biokinetics of and discrimination between dietary RRR- and SRR-alpha-tocopherols in the male rat. *Lipids* **22,** 163–172.

IUPAC-IUB Joint Commission on Biochemical Nomenclature (1981). "Nomenclature of Tocopherols and Related Compounds" (G.P. Moss, Ed.). http://www.chem.qmul.ac.uk./iupac/misc/toc.html

Jensen, S. K., Engberg, R. M., and Hedemann, M. S. (1999). *All-rac*-alpha-tocopherol acetate is a better vitamin E source than all-*rac*-alpha-tocopherol succinate for broilers. *J. Nutr.* **129,** 1355–1360.

Jensen, S. K., Kristensen, N. B., Lauridsen, C., and Sejersen, K. (2005). Enrichment of cows' milk with natural or synthetic vitamin E. *In* "Proceedings of the 10th Symposium on Vitamins and Additives in Nutrition of Man and Animal." 28–29 September, pp. 78–83. Jena/Thuringia, Germany.

Jensen, S. K., Nørgaard, J. V., and Lauridsen, C. (2006). Bioavailability of α-tocopherol stereoisomers in rats depends on dietary doses of *all-rac*-a-tocopheryl acetate or RRR-a-tocopheryl acetate. *Br. J. Nutr.* **95,** 477–487.

Kim, H.-S., Arai, H., Arita, M., Sato, Y., Ogihara, T., Inoue, K., Mino, M., and Tamai, H. (1998). Effect of α-tocopherol status on α-tocopherol transfer protein expression and its messenger RNA level in rat liver. *Free Radic. Res.* **28,** 87–92.

Kiyose, C., Muramatsu, R., Ueda, T., and Igarashi, O. (1995). Rats after intravenous administration. *Biosci. Biotech. Biochem.* **59,** 791–795.

Kiyose, C., Muramatsu, R., Kameyama, Y., Ueda, T., and Igarashi, O. (1997). Biodiscriminaion of α-tocopherol stereoisomers in humans after oral administration. *Am. J. Clin. Nutr.* **65,** 785–789.

Knarreborg, A., Lauridsen, C., Engberg, R. M., and Jensen, S. K. (2004). Dietary antibiotic growth promoters enhance the bioavailability of α-tocopheryl acetate in broilers through mediations of the lipid absorption processes. *J. Nutr.* **134,** 1487–1492.

Lauridsen, C., and Jensen, S. K. (2005). Influence of supplementation of *all-rac-α*-tocopheryl acetate preweaning and vitamin C postweaning on α-tocopherol and immune responses of piglets. *J. Anim. Sci.* **83,** 1274–1286.

Lauridsen, C., and Jensen, S. K. (2006a). Transfer of vitamin E in milk to the newborn. *In* "The Encyclopedia of Vitamin E" (V. R. Preedy and R. R. Watson, Eds.), Chapter 47, pp. 508–516. CABI International, Wallingford, Oxfordshire, UK.

Lauridsen, C., and Jensen, S. K. (2006b). Forms of vitamin E and α-tocopherol status of weaned piglets. *In* "Proceedings of the 19th International Pig Veterinary Society Congress," 16–19 July, p. 293. Copenhagen, Denmark.

Lauridsen, C., Hedemann, M. S., and Jensen, S. K. (2001a). Hydrolysis of tocopheryl and retinyl esters by porcine caboxyl ester hydrolase is affected by their carboxylate moiety and bile acids. *J. Nutr. Biochem.* **12,** 219–224.

Lauridsen, C., Leonard, S. W., Griffin, D. A., Liebler, D. C., McClure, T. D., and Traber, M. G. (2001b). Quantitative analysis by liquied chromatography-tandem mass spectrometry of deuterium-labeled and unlabeled vitamin E in biological samples. *Anal. Biochem.* **289,** 89–95.

Lauridsen, C., Engel, H., Morrie Craig, A., and Traber, M. G. (2002a). Relative bioactivity of dietary RRR- over *all-rac-α*-tocopheryl acetateles in swine assessed with deuterium labeled vitamin E. *J. Anim. Sci.* **80,** 702–707.

Lauridsen, C., Engel, H., Jensen, S. K., Morrie Craig, A., and Traber, M. G. (2002b). Lactating sows and suckling piglets preferentially incorporate RRR- over *all-rac-α*-tocopherol into milk, plasma and tissues. *J. Nutr.* **132,** 1258–1264.

Lodge, J. K. (2005). Vitamin E bioavailability in humans. *J. Plant Physiol.* **162,** 790–796.

Mahan, D. C., Kim, Y. Y., and Stuart, R. L. (2000). Effect of vitamin E sources (RRR- or *all-rac*-α-tocopheryl acetate) and levels on sow reproductive performance, serum, tissue, and milk α-tocopherol contents over a five-parity period, and the effects on the progeny. *J. Anim. Sci.* **78,** 110–119.

Meglia, G. E., Jensen, S. K., Lauridsen, C., and Waller, K. P. (2006). α-Tocopherol concentration and stereoisomer composition in plasma and milk from dairy cows fed natural or synthetic vitamin E around calving. *J. Dairy Res.* **73,** 227–234.

Moore, A., Burton, G. W., and Ingold, K. U. (1994). Modulation of cholesterol esterase's diastereoselectivity by chiral auxiliaries. The very large effect of a dihydroxy bile acid on the hydrolysis of α-tocoperyl acetates. *J. Am. Chem. Soc.* **116,** 6945–6946.

Müller, D. P. R., Manning, J. A., Mathias, P. M., and Harries, J. T. (1976). Studies of the intestinal hydrolysis of tocopheryl esters. *Int. J. Vitam. Nutr. Res.* **46,** 207–210.

Nelis, H. J., D'Haese, E., and Vermis, K. (2000). Vitamin E. *In* "Modern Chromatographic Analysis of Vitamins" (A. P. Leenheer, W. E. Lambert, and J. F. van Bocxlaer, Eds.), 3rd ed., pp. 143–228. Marcel Dekker, Inc., New York.

Nielsen, H. E., Danielsen, V., Simesen, M. G., Gissel-Nielssen, G., Hjarde, W., Leth, T., and Basse, A. (1979). Selenium and vitamin E deficiency in pigs. 1. Influence on growth and reproduction. *Acta Vet. Scand.* **20,** 276–288.

NRC (1998). "Nutrient Requirements of Swine," 10th ed. National Academy Press, Washington, DC.

Ong, D. E. (1985). Vitamin A binding proteins. *Nutr. Rev.* **43,** 225–232.

Packer, L. (1994). Vitamin E is nature's master antioxidant. *Sci. Am. Sci. Med.* **1,** 54–63.

Scherf, H., Machlin, L. J., Frye, T. M., Krautmann, B. A., and Williams, S. N. (1996). Vitamin E biopotency: Comparison of various 'natural-derived' and chemically synthesized alpha-tocopherols. *Anim. Feed Sci. Tech.* **59,** 115–126.

Traber, M. G. (2000). The bioavailability bugaboo. *Am. J. Clin. Nutr.* **71,** 1029–1030.

Traber, M. G., Sokol, R. J., Burton, G. W., Ingold, K. U., Papas, A. M., Huffaker, J. E., and Kayden, H. J. (1990). Impaired ability of patients with familiar isolated vitamin E deficiency to incorporate α-tocopherol into lipoproteins secreted by the liver. *J. Clin. Investi.* **85,** 397–407.

Ueda, T., Ichikawa, H., and Igarashi, O. (1993). Determination of α-tocopherol-tocopherol stereoisomers in biological speciments using chiral phase igh-performance liquid chromatography. *J. Nutr. Vitaminol.* **39,** 207–219.

Vecchi, M., Walther, W., Glinz, E., Netscher, T., Schmid, R., Lalonde, M., and Wetter, W. (1990). Chromatographische Trennung und quantitative Bestimmung aller acht Stereoisomeren von α-Tocopherol (Separation and quantitation of all eight stereoisomers of α-tocopherol by chromatography). *Helv. Chim. Acta* **73,** 782–789.

VERIS Research summary (1999). A comparison of natural and synthetic vitamin E. http://www.veris-online.org

Weiser, H., and Vecchi, M. (1981). Stereoisomers of α-tocopheryl acetate. II. Characterization of samples by physico-chemical methods and determination of biological acitivities in the rat resorption-gestation tests. *Int. J. Vitam. Nutr. Res.* **51,** 100–113.

Weiser, H., and Vecchi, M. (1982). Stereoisomers of α-tocopheryl acetate. II. Biopotencies of all eight stereoisomers, induvidually or in mixtures, as dertermined by rat resorption-gestation tests. *Int. J. Vitam. Nutr. Res.* **52,** 351–370.

Weiser, H., Riss, G., and Kormann, A. W. (1996). Biodiscrimination of the eight alpha-tocopherol stereosismers result in preferential accumulation of the four 2R forms in tissues and plasma of rats. *J. Nutr.* **126,** 2539–2549.

Yamauchi, J., Iwamoto, T., Kida, S., Massushige, S., Yamada, K., and Esashi, T. (2001). Tocopherol-associated protein is a ligand-dependent transcriptional activator. *Biochem. Biophys. Res. Commun.* **285,** 295–299.

Zimmer, S., Stocker, A., Sarbolouki, M. N., Spycher, S. E., Sassoon, J., and Azzi, A. (2000). A novel human tocopherol-associated protein: Cloning, *in vitro* expression, and characterization. *J. Biol. Chem.* **275,** 25672–25680.

11

Addition Products of α-Tocopherol with Lipid-Derived Free Radicals

Ryo Yamauchi

Department of Applied Life Science, Faculty of Applied Biological Sciences Gifu University, 1-1 Yanagido, Gifu City, Gifu 501-1193, Japan

The addition products of α-tocopherol with lipid-derived free radicals have been reviewed. Free radical scavenging reactions of α-tocopherol take place via the α-tocopheroxyl radical as an intermediate. If a suitable free radical is present, an addition product can be formed from the coupling of the free radical with the α-tocopheroxyl radical. The addition products of α-tocopherol with lipid-peroxyl radicals are 8*a*-(lipid-dioxy)-α-tocopherones, which are hydrolyzed to α-tocopherylquinone. On the other hand, the carbon-centered radicals of lipids prefer to react with the phenoxyl radical of α-tocopherol to form 6-*O*-lipid-α-tocopherol

Vitamins and Hormones, Volume 76

0083-6729/07 $35.00
DOI: 10.1016/S0083-6729(07)76011-9

under anaerobic conditions. The addition products of α-tocopherol with peroxyl radicals (epoxylinoleoyl–peroxyl radicals) produced from cholesteryl ester and phosphatidylcholine were detected in the peroxidized human plasma using a high-sensitive HPLC procedure with postcolumn reduction and electrochemical detection. Thus, the formation of these addition products provides us with much information on the antioxidant function of vitamin E in biological systems.

I. INTRODUCTION

Peroxidation of lipids in biological tissues and fluids is implicated in a variety of damaging pathological events (Sevanian and Hochstein, 1985; Slater, 1984). Lipid peroxidation is a chain reaction that proceeds in three major reactions: initiation, propagation, and termination (Porter *et al.*, 1995). In the initiation reaction, a carbon-centered lipid radical (an alkyl radical, L•) is produced by the abstraction from an unsaturated fatty acid moiety of lipid (LH). The initiation process is very slow and can be catalyzed by heat, light, trace of transition metal ions, and certain enzymes. In the propagation step, the alkyl radical (L•) reacts with molecular oxygen at a very high rate, giving a peroxyl radical (LOO•). The peroxyl radical, a chain-carrying radical, is able to attack another polyunsaturated lipid molecule. Although the initial peroxyl radical is converted to a hydroperoxide, this process produces a new alkyl radical, which is rapidly converted into another peroxyl radical. Thus, the chain reaction does not stop until the chain-carrying peroxyl radical meets and combines with another radical to form nonradical products (termination reaction).

α-Tocopherol (TOH, **1**), the most active form of vitamin E in biological systems, is known to be a chain-breaking antioxidant to inhibit the propagation step by trapping peroxyl radicals (Burton and Ingold, 1981; Mukai *et al.*, 1986; Niki *et al.*, 1984). Figure 1 shows the possible pathways for the reactions of α-tocopherol with lipid-derived free radicals. α-Tocopherol efficiently transfers a hydrogen atom to a peroxyl radical (LOO•) giving a hydroperoxide (LOOH) and an α-tocopheroxyl radical (TO•, **2**). α-Tocopherol also may react with alkyl radical (L•) rapidly (Evans *et al.*, 1992; Franchi *et al.*, 1999). The α-tocopheroxyl radical is very reactive and may undergo radical–radical coupling with other radicals forming adducts (Remorova and Roginskii, 1991). The α-tocopheroxyl radical reacts differently with peroxyl and carbon-centered radicals. Oxygen-centered peroxyl radicals (either diene peroxyl or epoxyene peroxyl, LOO•) tend to add at the 8*a* position of α-tocopheroxyl radical, forming 8*a*-(lipid-dioxy)-α-tocopherones (**3**) (Liebler and Burr, 1992; Liebler *et al.*, 1989, 1990, 1991; Matsuo *et al.*, 1989; Winterle *et al.*, 1984; Yamauchi *et al.*, 1989, 1990, 1994, 1996, 1998b, 2002b).

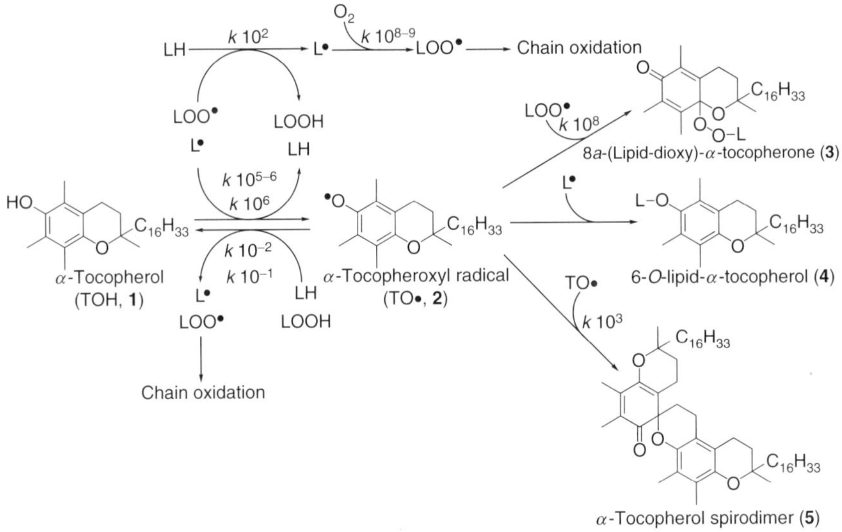

FIGURE 1. Reaction of α-tocopherol during the peroxidation of unsaturated lipids. LH, polyunsaturated lipid; L•, carbon-centered radical of lipid; LOO•, peroxyl radical of lipid; LOOH, lipid hydroperoxide; TO•, α-tocopheroxyl radical; k, rate constant in $M^{-1} s^{-1}$.

On the other hand, carbon-centered radicals (either pentadienyl or epoxyallylic, L•), formed under anaerobic conditions, tend to add to the α-tocopheroxyl oxygen to form 6-*O*-lipid-α-tocopherol adducts (**4**) (Evans *et al.*, 1992; Yamauchi *et al.*, 1993). α-Tocopherol also can undergo a self-coupling reaction to form dimer (**5**), if the concentration in a reaction mixture is high (Burton *et al.*, 1985; Yamauchi *et al.*, 1988). Therefore, the formation of these addition products is consistent with the behavior of α-tocopherol as a chain-breaking antioxidant in biological systems.

This chapter describes the addition products of α-tocopherol with lipid-derived free radicals, methyl linoleate-derived peroxyl and alkyl radicals, peroxyl radicals of phosphatidylcholine (PC), and peroxyl radicals of cholesteryl esters (CE). Detection of the addition products during the peroxidation of human plasma is also described.

II. ADDITION PRODUCTS OF α-TOCOPHEROL WITH METHYL LINOLEATE-DERIVED FREE RADICALS

The primary products of lipid peroxidation are hydroperoxides (LOOH), which can dissociate into free radicals. Lipid hydroperoxides are stable at physiological temperatures, and a major role of transition metals is to catalyze their decomposition. Transition metal ions catalyze homolysis of lipid hydroperoxides that are cleaved to alkylperoxyl radicals (LOO•) by metal ions in the oxidized state such as ferric ion, whereas reduced metal ions, such as ferrous ion, lead to alkoxyl radicals (LO•) (Halliwell and Gutteridge, 1984). The free radicals produced in these processes are believed to stimulate the chain reaction of lipid peroxidation by abstracting further hydrogen from unoxidized lipids. If α-tocopherol is present in such conditions, it can trap these free radicals and terminate lipid peroxidation (Gardner *et al.*, 1972; Kaneko and Matsuo, 1985).

α-Tocopherol was reacted with methyl linoleate (13*S*)-hydroperoxide in the presence of iron-chelate, Fe(III)-acetylacetonate, in a homogeneous solution (Yamauchi *et al.*, 1995). The main products under aerobic conditions were methyl (13*S*)-(8*a*-dioxy-α-tocopherone)-(9*Z*,11*E*)-octadecadienoate (**6**), a mixture of methyl 9-(8*a*-dioxy-α-tocopherone)-(12*S*,13*S*)-epoxy-(10*E*)-octadecenoate and methyl 11-(8*a*-dioxy-α-tocopherone)-(12*S*,13*S*)-epoxy-(9*Z*)-octadecenoate (**7**), and α-tocopherol dimer (**5**). On the other hand, when the reaction proceeded under anaerobic conditions, the products were a mixture of methyl 9- and 13-(α-tocopheroxy)-octadecadienoates (**8**) and a mixture of 9- and 11-(α-tocopheroxy)-(12*S*,13*S*)-epoxyoctadecenoates (**9**), in addition to 8*a*-(lipid-dioxy)-α-tocopherone (**6**).

Figure 2 shows the possible mechanisms of iron-catalyzed reaction of α-tocopherol with methyl linoleate (13*S*)-hydroperoxide (Yamauchi *et al.*, 1995). The first reaction is iron-dependent free radical formation by the hydroperoxide being oxidized to a peroxyl radical or reduced to an alkoxyl radical (Halliwell and Gutteridge, 1984). The 13-peroxyl radical reacts with the 8*a*-carbon-centered radical of α-tocopherol (T•) to form 8*a*-(lipid-dioxy)-α-tocopherone (**6**). The alkoxyl radical, produced by the Fe^{2+}-catalyzed reduction of the hydroperoxide, adds to the adjacent double bond to form the 12,13-epoxycarbon-centered radical. In the presence of oxygen, this carbon-centered radical reacts with molecular oxygen to form epoxyalkyl–peroxyl radicals followed by the addition reaction with T• to form 8*a*-(epoxylipid-dioxy)-α-tocopherones (**7**) (Gardner, 1989). All the carbon-centered radicals produced are expected to react very rapidly with oxygen molecule to form peroxyl radicals. However, the epoxycarbon-centered radical can react with the phenoxyl radical of α-tocopherol (TO•) and forms 6-*O*-epoxylipid-α-tocopherol (**9**) when the oxygen pressure is lowered. In addition, radical elimination of the peroxyl radical may result in the formation of a carbon-centered radical and oxygen which reacts with TO• to form 6-*O*-lipid-α-tocopherol (**8**) (Yamauchi *et al.*, 1995).

Lipid hydroperoxides can be dissociated into free radicals by heat during food processing (Gardner, 1987). α-Tocopherol can suppress further reactions of lipid hydroperoxides by donating a hydrogen atom to the free radicals. It has been reported that α-tocopherol can suppress the thermal decomposition of lipid hydroperoxides and inhibit the formation of volatile and nonvolatile decomposition products (Frankel and Gardner, 1989;

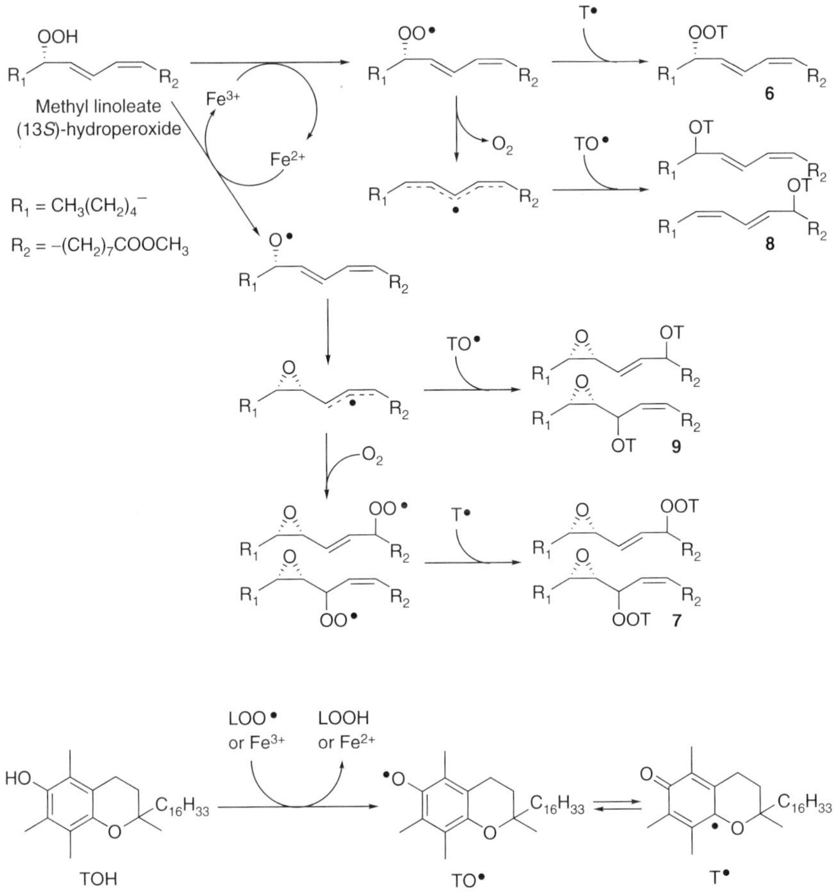

FIGURE 2. Iron-catalyzed decomposition of methyl linoleate (13*S*)-hydroperoxide and reaction with α-tocopherol. TOH, α-tocopherol; TO•, the phenoxyl radical of α-tocopherol; T•, the 8*a*-carbon-centered radical of α-tocopherol.

Hopia *et al.*, 1996). The reaction products of α-tocopherol during the thermal decomposition of methyl linoleate (13*S*)-hydroperoxide were almost the same as those of α-tocopherol in the iron-catalyzed reactions. The main products were 6-*O*-alkyl-α-tocopherols (**8** and **9**) (Yamauchi *et al.*, 1998a). Alkoxyl radicals of linoleate tend to rearrange into epoxyallylic radicals, even in the presence of compounds with a readily abstractable hydrogen, like α-tocopherol (Gardner *et al.*, 1972; Kaneko and Matsuo, 1985). Without oxygen, the resulting epoxyallylic radicals react with α-tocopheroxyl radical to form 6-*O*-epoxyalkyl-α-tocopherol (**9**). Thus, the alkoxyl radical rearrangement is an important pathway in the homolysis of the hydroperoxide by heat. The formation of peroxyl radicals is the other important pathway of the thermal decomposition of lipid hydroperoxides. Since the peroxyl radicals are unstable at high temperature, radical elimination of the peroxyl radical results in the release of oxygen molecule as a leaving group to form carbon-centered radicals (Porter and Wujek, 1984). The carbon-centered radicals then react with the phenoxyl radical of α-tocopherol to form 6-*O*-alkyl-α-tocopherols (**8**).

III. Addition Products of α-Tocopherol with PC-Peroxyl Radicals in Liposomes

Phospholipid liposomal systems are generally accepted to be a suitable model for studying the membrane structure and properties, given that they are surrounded by a lipid bilayer that is structurally similar to the lipidic matrix of a cell membrane (Fiorentini *et al.*, 1994). Phospholipid liposomal systems have been employed to model the peroxylradical scavenging reactions of α-tocopherol (Liebler and Burr, 1992; Liebler *et al.*, 1991; Takahashi *et al.*, 1989; Thomas *et al.*, 1992; Yamauchi *et al.*, 1994, 1996, 1998b). Animal tissue contains phospholipids of molecular species defined by a saturated fatty acid, usually 16 or 18 carbons in length, at the *sn*-1 position of glycerol and an unsaturated fatty acid at the *sn*-2 position (Takamura *et al.*, 1986). Therefore, it is important to elucidate the reaction of α-tocopherol with peroxyl radicals derived from such molecular species of PC in liposomes.

Multilamellar liposomes of 1-palmitoyl-2-linoleoyl-3-*sn*-phosphatidylcholine (PLPC, **10**) containing small amounts of α-tocopherol and a free radical initiator, 2,2′-azobis(2,4-dimethylvaleronitrile) (AMVN), were incubated at 37°C for 15 h (Yamauchi *et al.*, 1998b). The addition products of α-tocopherol with PLPC-peroxyl radicals were obtained and their structures were determined to be a mixture of 1- palmitoyl -2-[9-(8*a*-dioxy-α-tocopherone)-10,12-octadecadienoyl]-3-*sn*-PC (**11a**) and 1-palmitoyl-2-[13-(8*a*-dioxy-α-tocopherone)-9,11-octadecadienoyl]-3-*sn*-PC (**11b**) (TOO-PLPC) (Fig. 3).

The antioxidative efficiency of α-tocopherol against the peroxidation of PLPC and 1-palmitoyl-2-arachidonoyl-3-*sn*-PC (PAPC) in large unilamellar liposomes was assessed by the formation of α-tocopherol products (Yamauchi *et al.*, 1998b). When membrane peroxidation was initiated by a constant flux of peroxyl radicals from a water-soluble free radical initiator, 2,2′-azobis (2-amidinopropane) dihydrochloride (AAPH), α-tocopherol could scavenge both the AAPH-derived and the PC-derived peroxyl radicals to form TOO-PC (**11**), α-tocopherylquinone (**12**), and epoxy-α-tocopherylquinones (**13**). Since PLPC and PAPC exist in larger amounts than α-tocopherol, most of the AAPH-derived peroxyl radicals would attack the PC molecules, and the resulting PC-peroxyl radicals would react with α-tocopherol (Yamauchi *et al.*, 1990, 1994). However, the formation of TOO-PC was only 6–10% of the consumed α-tocopherol due to the restricted mobility of α-tocopherol and

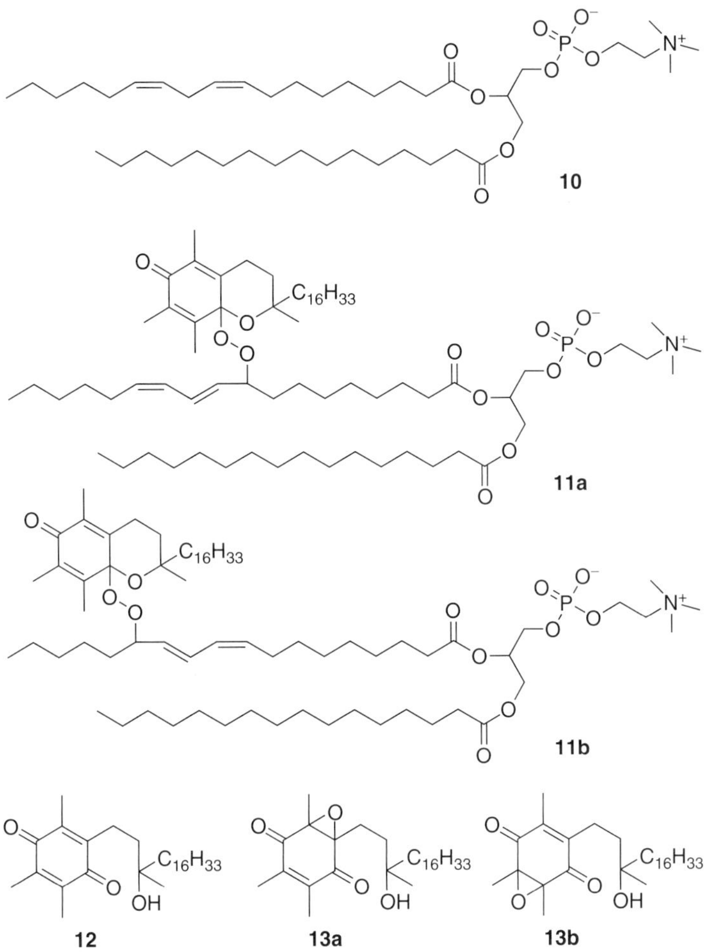

10

11a

11b

12 13a 13b

FIGURE 3. Structures of 1-palmitoyl-2-linoleoyl-3-*sn*-phosphatidylcholine (PLPC, **10**), its addition products with α-tocopherol, 1-palmitoyl-2-[9-(8*a*-dioxy-α-tocophene)-10,12-octadecadienoyl]-3-*sn*-phosphatidylcholine (**11a**) and 1-palmitoyl-2-[13-(8*a*-dioxy-α-tocophene)-9,11-octadecadienoyl]-3-*sn*-phosphatidylcholine (**11b**) (TOO-PLPC), α-tocopherylquinone (**12**), 2,3-epoxy-α-tocopherylquinone (**13a**), and 5,6-epoxy-α-tocopherylquione (**13b**).

PC-peroxyl radicals in the membrane structure. Iron-dependent α-tocopherol oxidation in the PLPC and PAPC liposomes also yielded TOO-PC (**11**), α-tocopherylquinone (**12**), and epoxy α tocopherylquinones (**13**), all of which were formed in the AMVN- and AAPH-initiated oxidation products (Liebler *et al.*, 1991; Yamauchi *et al.*, 1996). Thus, peroxyl radicals from the unsaturated PC molecules can oxidize α-tocopherol in this iron-dependent system.

Transition metal ions have been effective in catalyzing free radical reactions of methyl linoleate hydroperoxides and α-tocopherol in benzene

(Yamauchi *et al.*, 1995). Since phospholipids are known to aggregate in nonpolar solvent such as benzene and form inverted micelles (Porter and Wagner, 1986), methanol has been chosen as the solvent where phospholipids are monomeric states (Yamauchi *et al.*, 2002a). The iron-catalyzed reaction of α-tocopherol with a PC hydroperoxide, 1-palmitoyl-2-[(13*S*)-hydroperoxy-(9*Z*,11*E*)-octadecadienoyl]-3-*sn*-PC (13-PLPCOOH, **14**), in methanol yielded a mixture of 1-palmitoyl-2-[9-(8*a*-dioxy-α-tocopherone)-(12*S*,13*S*)-epoxy-(10*E*)-octadecenoyl]-3-*sn*-PC (**15a**) and 1-palmitoyl-2-[11-(8*a*-dioxy-α-tocopherone)-(12*S*,13*S*)-epoxy-(9*Z*)-octadecenoyl]-3-*sn*-PC (**15b**) (TOO-epoxyPLPC), in which the 12,13-epoxyalkyl-peroxyl radicals derived from 13-PLPCOOH attacked the 8*a*-position of the α-tocopheroxyl radical (Fig. 4). Although the formation of other addition products between

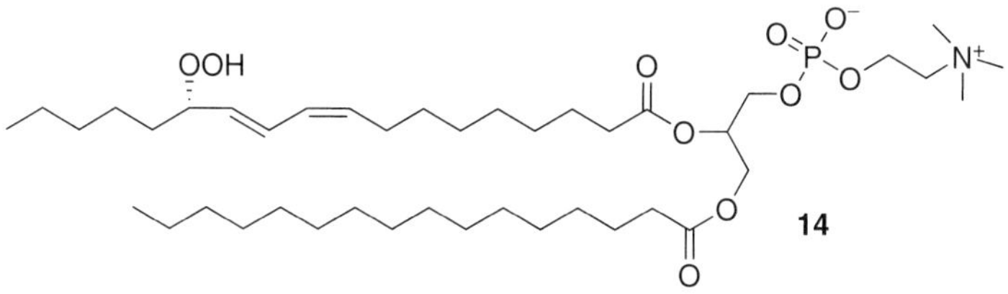

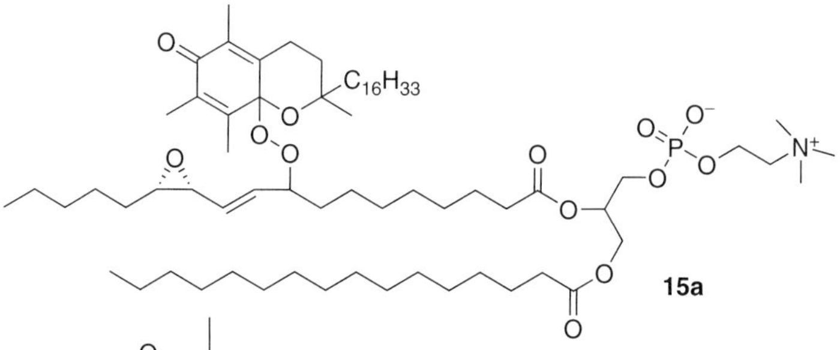

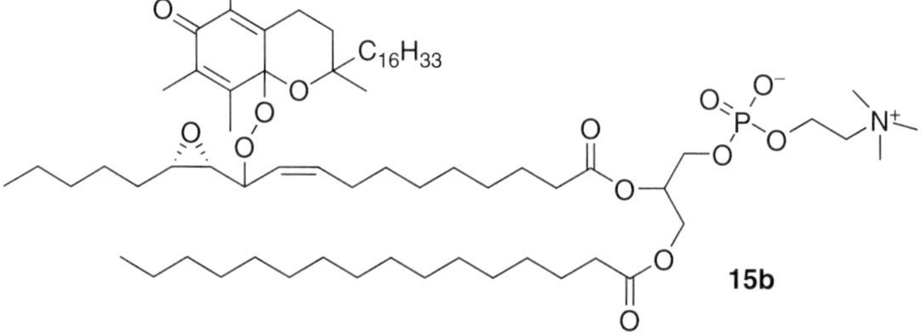

FIGURE 4. Structures of 1-palmitoyl-2-[(13*S*)-hydroperoxy-(9*Z*,11*E*)-octadecadienoyl]-3-*sn*-phosphatidylcholine (13-PLPCOOH, **14**) and its addition products with α-tocopherol, 1-palmitoyl-2-[9-(8*a*-dioxy-α-tocopherone)-(12*S*,13*S*)-epoxy-(10*E*)-octadecenoyl]-3-*sn*-phosphatidylcholine (**15a**) and 1-palmitoyl-2-[11-(8*a*-dioxy-α-tocopherone)-(12*S*,13*S*)-epoxy-(9*Z*)-octadecenoyl]-3-*sn*-phosphatidylcholine (**15b**) (TOO-epoxyPLPC).

α-tocopheroxyl radical and 13-peroxyl radicals of PLPC (TOO-PLPC) was also expected, such products could not be detected in this reaction system.

The ferrous iron-catalyzed reaction of α-tocopherol with 13-PLPCOOH (**14**) was conducted in large unilamellar liposomes (Yamauchi *et al.*, 2002a). When the decomposition of 13-PLPCOOH by ferrous iron was performed in saturated 1,2-dimyristoyl-3-*sn*-PC (DMPC) liposomes, α-tocopherol could react with epoxyPLPC-peroxyl radicals derived from 13-PLPCOOH to form TOO-epoxyPLPC (**15**). This indicates that the formation of alkoxyl radicals by the ferrous iron-catalyzed reductive cleavage of 13-PLPCOOH is the first reaction in saturated DMPC liposomes. On the other hand, the iron-dependent reaction of α-tocopherol with 13-PLPCOOH in unsaturated PLPC liposomal systems yielded two types of 8*a*-(PC-dioxy)-α-tocopherones: TOO-PLPC (**11**) and TOO-epoxyPLPC (**15**). In this reaction, free radicals formed by the iron-catalyzed decomposition of 13-PLPCOOH react primarily with linoleoyl chains of PLPC, which are present in 1000-fold excess over α-tocopherol. The resulting PLPC radicals combine with oxygen to form PLPC-peroxyl radicals, and react with α-tocopheroxyl radical to form TOO-PLPC (**11**) as the product. In addition, the direct reaction between α-tocopherol and the 13-PLPCOOH-derived epoxyPLPC-peroxyl radicals yielded TOO-epoxyPLPC (**15**) (Yamauchi *et al.*, 1998b). It also detected α-tocopherylquinone (**12**) and epoxy-α-tocopherylquinones (**13**) as the reaction products of α-tocopherol in these liposomal systems. α-Tocopherylquinone is formed from the hydrolysis of 8*a*-substituted α-tocopherones (Liebler and Burr, 1992; Liebler *et al.*, 1991; Yamauchi *et al.*, 1996). The formation of epoxy-α-tocopherylquinones contributes to the antioxidant chemistry of α-tocopherol, although the mechanism by which these epoxy-α-tocopherylquinones are formed remains unknown (Liebler and Burr, 1995).

IV. Addition Products of α-Tocopherol with CE-Peroxyl Radicals

Low-density lipoprotein (LDL) is the major carrier of CE in human blood plasma, and the free radical-mediated modification of LDL may play an important role in the development of atherosclerosis (Thomas and Stocker, 2000; Thomas *et al.*, 1995). Peroxidation of LDL lipids affords hydroperoxides mainly from CE, especially cholesteryl linoleate (CL, **16**), which is the predominant lipid component in the LDL and other lipoproteins (Frei *et al.*, 1988). Since α-tocopherol is the most abundant lipid-soluble antioxidant associated with plasma and LDL, it has attracted most attention with respect to antioxidation of plasma lipids (Thomas *et al.*, 1995). Therefore, the reaction products of α-tocopherol with CE hydroperoxides (CEOOH) have been investigated (Yamauchi *et al.*, 2002b).

Racemic CL hydroperoxides (CLOOH), a mixture of cholesteryl 9-hydroperoxy-(10*E*,12*Z*)-octadecadienoate and cholesteryl 13-hydroperoxy-(9*Z*,11*E*)-octadecadienoate, were prepared by the α-tocopherol-controlled autoxidation of CL (Yamamoto *et al.*, 1987). The addition products of α-tocopherol with CL-peroxyl radicals were obtained by reacting α-tocopherol with CLOOH in the presence of Fe(III)-acetylacetonate (Yamauchi *et al.*, 2002b). They were a mixture of cholesteryl 9-(8*a*-dioxy-α-tocopherone)-(10*E*,12*Z*)-octadecadienoate (**17a**) and cholesteryl 13-(8*a*-dioxy-α-tocopherone)-(9*Z*, 11*E*)-octadecadienoate (**17b**) (TOO-CL), and a mixture of cholesteryl 9-(8*a*-dioxy-α-tocopherone)-12,13-epoxy-(10*E*)-octadecenoate (**18a**), cholesteryl 11-(8*a*-dioxy-α-tocopherone)-12,13-epoxy-(9*E*)-octadecenoate (**18b**), cholesteryl 13-(8*a*-dioxy-α-tocopherone)-9,10-epoxy-(11*E*)-octadecenoate (**18c**), and cholesteryl 11-(8*a*-dioxy-α-tocopherone)-9,10-epoxy-(12*E*)-octadecenoate (**18d**) (TOO-CepoxyL) (Fig. 5).

TOO-CL (**17**) is produced from the reaction between an α-tocopheroxyl radical with the CLOOH-derived peroxyl radicals. The alkoxyl radical produced from CLOOH adds to the adjacent double bond to form the epoxycarbon-centered radicals (CepoxyL•) as the same mechanism described in Fig. 2. CepoxyL• then couples with oxygen to form epoxylinoleoyl-peroxyl radicals (CepoxyLOO•). The 8*a*-carbon-centered radical of α-tocopherol reacts rapidly and irreversibly with these peroxyl radicals to form TOO-CepoxyL (**18**). Therefore, the formation of these addition products would provide us much information on the antioxidant function of α-tocopherol in biological systems, especially in blood plasma.

V. DETECTION OF THE ADDITION PRODUCTS OF α-TOCOPHEROL WITH LIPID-PEROXYL RADICALS IN BIOLOGICAL SAMPLES

α-Tocopherol is the principal lipid-soluble biological antioxidant and an important cellular protectant against oxidative damage. α-Tocopherol exerts antioxidant effects primarily by trapping lipid-peroxyl radicals. Direct observation of such reactions in biological systems is generally not feasible, but evidence for tocopherol antioxidant reactions in tissues may be obtained by analyzing the reaction products (Liebler *et al.*, 1996). In most biological samples, the amount of α-tocopherol is very small compared with the lipid contents (Evarts and Bieri, 1974; Kelley *et al.*, 1995), and we need a highly sensitive and specific methodology to analyze the oxidation products of α-tocopherol in biological samples. HPLC techniques have been widely accepted and applied for the analysis of tocopherols and α-tocopherylquinone (Bieri and Tolliver, 1981; Leray *et al.*, 1998; Murphy and Kehrer, 1987; Takeda *et al.*, 1996; Vatassery, 1994). The combination of ultraviolet or fluorescent detection with HPLC separation lacked sufficient sensitivity for

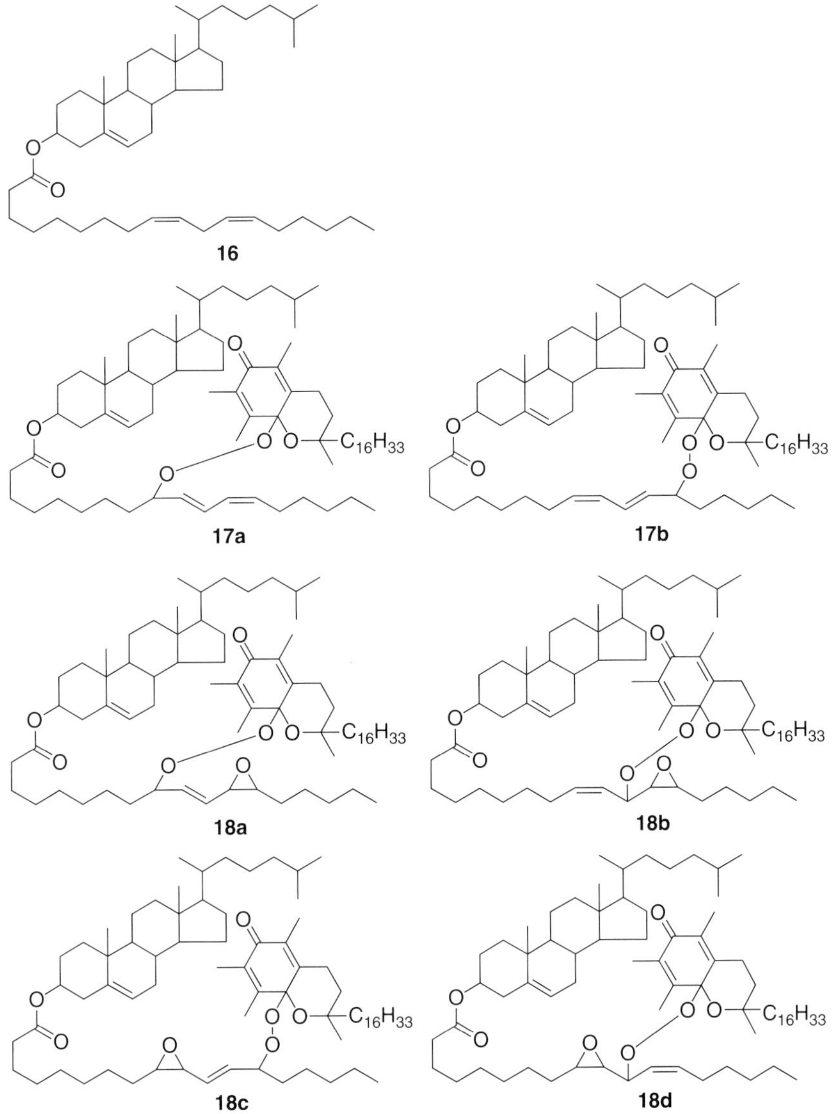

16

17a 17b

18a 18b

18c 18d

FIGURE 5. Structures of cholesteryl linoleate (CL, **16**) and the addition products with α-tocopherol, cholesteryl 9-(8*a*-dioxy-α-tocopherone)-(10*E*,12*Z*)-octadecadienoate (**17a**) and cholesteryl 13-(8*a*-dioxy-α-tocopherone)-(9*Z*,11*E*)-octadecadienoate (**17b**) (TOO-CL), and cholesteryl 9–8*a*-dioxy-α-tocopherone)-12,13-epoxy-(10*E*)-octadecenoate (**18a**), cholesteryl 11-(8*a*-dioxy-α-tocopherone)-12,13-epoxy-(9*E*)-octadecenoate (**18b**), cholesteryl 13-(8*a*-dioxy-α-tocopherone)-9,10-epoxy-(11*E*)-octadecenoate (**18c**), and cholesteryl 11-(8*a*-dioxy-α-tocopherone)-9,10-epoxy-(12*E*)-octadecenoate (**18d**) (TOO-CepoxyL).

biological samples (Bieri and Tolliver, 1981; Vatassery, 1994). On the other hand, electrochemical detection provides the most sensitive measurement of tocopherols and α-tocopherylquinone (Leray *et al.*, 1998; Murphy and Kehrer, 1987; Takeda *et al.*, 1996). Therefore, the HPLC system with a postcolumn reactor and electrochemical detection (HPLC-ECD) has been developed to measure the addition products of α-tocopherol (Yamauchi *et al.*, 2002c,d). After the separation on a reversed-phase column, online platinum-catalyzed reduction allowed the detection of 8*a*-(lipid-dioxy)-α-tocopherones (Fig. 6). The lowest detectable level of each compound was about 0.2 pmol at the signal-to-noise ratio of 3.

This HPLC methodology was applied for the analysis of the oxidative fate of endogenous α-tocopherol in human blood plasma (Yamauchi *et al.*, 2002c). The human plasma was reacted with peroxyl radicals that had been generated in the aqueous phase by the thermal decomposition of AAPH. The addition products of α-tocopherol in the 6-h incubation sample were analyzed by the HPLC-ECD (Figs. 7 and 8). On the analysis of CE products, one major product peak was obtained and identified to be TOO-CepoxyL from its APCI-MS analysis (Fig. 7). The PC fraction also gave one product peak that was identified to be TOO-epoxyPLPC from its ESI-MS analysis (Fig. 8).

The disappearance of endogenous antioxidants in human plasma has been reported in relation to the appearance of various lipid hydroperoxides (Frei *et al.*, 1988). The consumption of endogenous α-tocopherol in human plasma

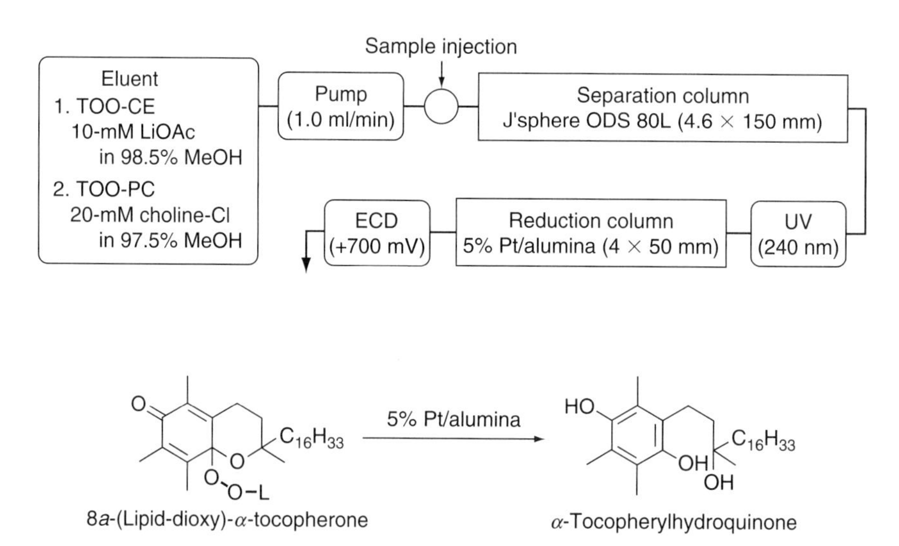

FIGURE 6. HPLC-ECD system for the detection of the addition products of α-tocopherol with lipid-peroxyl radicals.

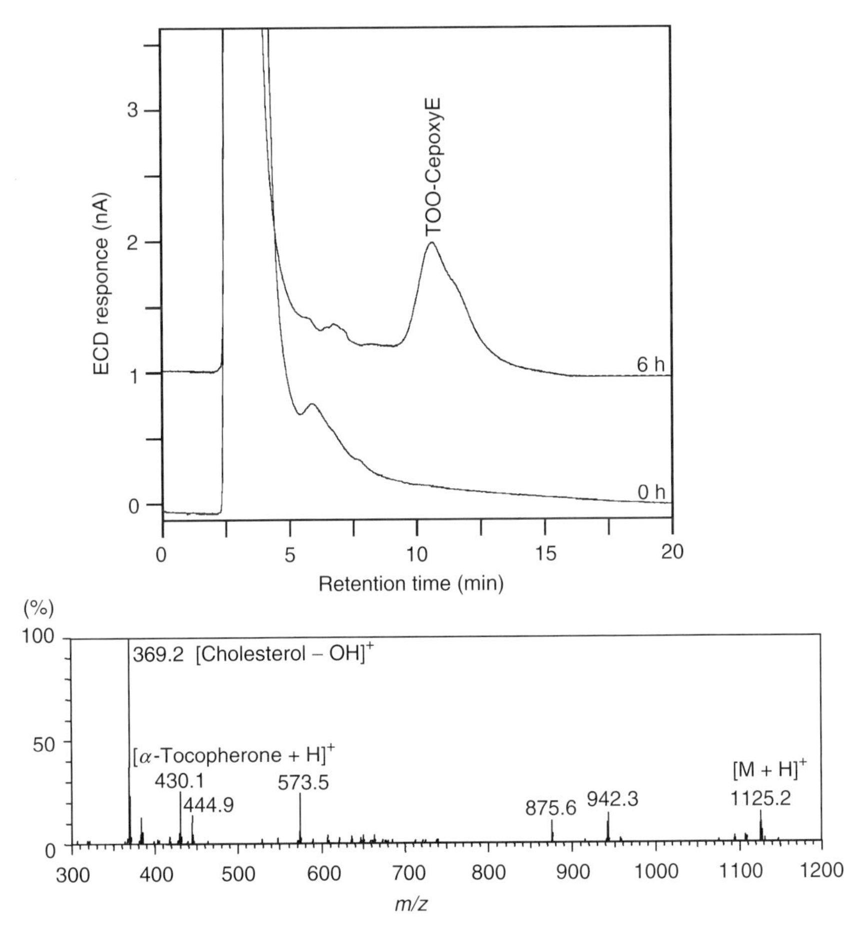

FIGURE 7. HPLC of the addition products of α-tocopherol with cholesterol ester-derived peroxyl radicals from the peroxidized human plasma using postcolumn reduction and electrochemical detection. Human plasma (4.0 ml) was incubated at 37°C for 6 h by the addition of 100-mM AAPH (1.0 ml). The peak corresponding to TOO-CepoxyE was subjected to the positive APCI-MS analysis with methanol/2-propanol/hexane (50:35:15, v/v) as the solvent.

was observed with the formation of lipid hydroperoxides, CEOOH and PCOOH, when the peroxidation was initiated with AAPH (Yamauchi *et al.*, 2002c) or $CuSO_4$ (Yamauchi *et al.*, 2002d) (Table I). The oxidation products, α-tocopherylquinone (**12**), epoxy-α-tocopherylquinones (**13**), TOO-CepoxyE (**18**), and TOO-epoxyPC (**15**), appeared with depletion of α-tocopherol. The similar product distribution has been reported in the AAPH-initiated peroxidation of PLPC liposomes containing α-tocopherol, although the type of 8*a*-substituted tocopherones was different (Yamauchi *et al.*, 1998b).

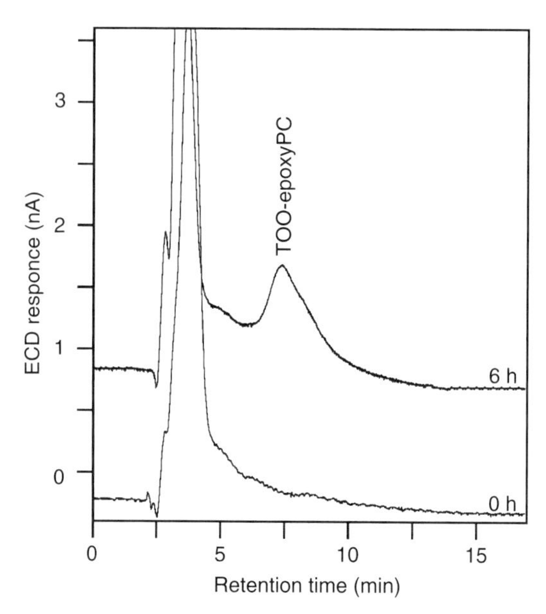

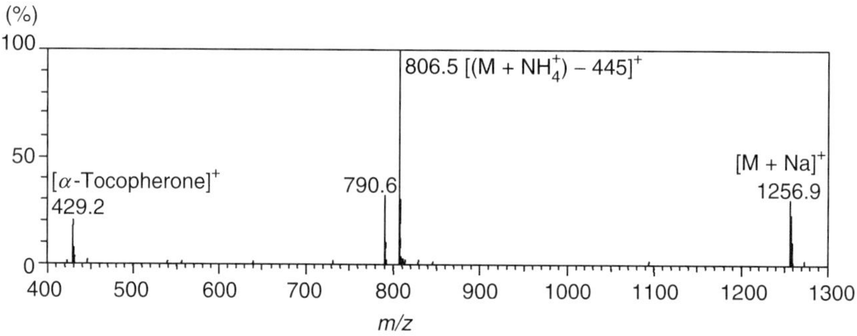

FIGURE 8. HPLC of the addition products of α-tocopherol with PC-derived peroxyl radicals from the peroxidized human plasma using postcolumn reduction and electrochemical detection. Human plasma (4.0 ml) was incubated at 37°C for 6 h by the addition of 100-mM AAPH (1.0 ml). The peak corresponding to TOO-epoxyPC was subjected to the positive ESI-MS analysis with 10-mM ammonium acetate in methanol as the solvent.

The classical action of α-tocopherol is that of a chain-breaking antioxidant that reflects its ability to react rapidly with chain-carrying lipid-peroxyl radicals (Burton and Ingold, 1986; Burton and Traber, 1990; Kamel-Eldin and Appelqvist, 1996; Liebler, 1993). Alternatively, α-tocopherol can directly scavenge the radical that initiates peroxidation (Evans *et al.*, 1992). In both cases, the formed α-tocopheroxyl radical rapidly reacts with another available radical to yield nonradical products. If the α-tocopheroxyl radical traps the chain-propagating peroxyl radicals formed from CL or PLPC directly, the primary products are 8*a*-(CL-dioxy)-α-tocopherones (**17**) or

TABLE I. Concentrations of Lipid Hydroperoxides, Tocopherols, and the Oxidation Products of α-Tocopherol After Incubation of Human Plasma with 20-mM AAPH or 1-mM $CuSO_4$

	Compounds (μM)								
	CEOOH[a]	PCOOH	α-TOH	γ-TOH	TOO-CepoxyE	TOO-epoxyPC	α-TQ	2,3-Epoxy-α-TQ	5,6-Epoxy-α-TQ
20-mM-AAPH[b]									
0 h			22.3 ± 2.1[c]	4.10 ± 0.63					
6 h	34.7 ± 2.2	11.8 ± 1.6	1.2 ± 0.4	2.46 ± 0.10	2.55 ± 0.20	0.48 ± 0.02	1.89 ± 0.12	1.53 ± 0.08	0.30 ± 0.02
1-mM-$CuSO_4$[d]									
0 h			20.7 ± 1.5	4.34 ± 0.67					
6 h	51.0 ± 17.2	7.8 ± 4.2	3.9 ± 1.0	2.10 ± 0.10	0.31 ± 0.08	0.38 ± 0.09	1.15 ± 1.06	1.08 ± 0.50	0.75 ± 0.85

[a] CEOOH—cholesteryl ester hydroperoxide; PCOOH—phosphatidylcholine hydroperoxide; α-TOH—α-tocopherol; γ-TOH—γ-tocopherol; TOO-CepoxyE—the adducts of α-tocopherol with cholesteryl epoxyesters; TOO-epoxyPC—the adducts of α-tocopherol with epoxyphosphatidylcholines; α-TQ, α-tocopherylquinone; 2,3-epoxy-α-TQ—2,3-epoxy-α-tocopherylquinone; 5,6-epoxy-α-TQ—5,6-epoxy-α-tocopherylquinone.

[b] Human plasma (4.0 ml) was incubated at 37°C by the addition of 1.0 ml of 100-mM 2,2′-azobis(2-amidinopropane) dihydrochloride (AAPH).

[c] Each value is expressed as mean ± SD of three different experiments.

[d] Human plasma (2.5 ml) was incubated at 37°C by the addition of 0.5 ml of 5-mM $CuSO_4$.

8*a*-(PLPC-dioxy)-α-tocopherones (**11**). However, such 8*a*-substituted tocopherone products could not be detected in the peroxidized plasma (Yamauchi *et al.*, 1998b, 2002b). Instead, the products detected were the addition products with epoxylipid-peroxyl radicals, TOO-CepoxyL (**18**) and TOO-epoxyPLPC (**15**). The epoxylipid-peroxyl radicals can be produced from the metal-ion-catalyzed decomposition of lipid hydroperoxides such as CEOOH and PCOOH (Gardner, 1989; Wilcox and Marnett, 1993). Thus, the α-tocopherol in human plasma may only react with such secondary-produced peroxyl radicals. Frei *et al.* (1988) demonstrated that α-tocopherol in plasma was considerably less effective in preventing peroxidative damage to lipids induced by aqueous oxidants. Moreover, α-tocopherol can exert prooxidant activity for lipoprotein lipids and the model of tocopherol-mediated peroxidation has been developed to explain such prooxidant activity of α-tocopherol during lipoprotein peroxidation (Bowry and Stocker, 1993; Thomas and Stocker, 2000; see Fig. 1). Therefore, the accumulation of hydroperoxides before the depletion of α-tocopherol and the formation of 8*a*-(epoxylipid-dioxy)-α-tocopherones indicate that α-tocopherol is less effective in preventing lipid peroxidation in human plasma.

It has been reported that 8*a*-substituted α-tocopherones and epoxytocopherones, rather than their hydrolyzed products, might be the principal oxidation products of α-tocopherol in biological systems (Ham and Liebler, 1997; Liebler *et al.*, 1996). However, this was only assumed from the evidence that the additional release of α-tocopherylquinones was observed in the acid-treated samples. Terentis *et al.* (2002) measured the amounts of tocopherol and lipid peroxidation products in homogenate of human aortic tissue exhibiting varying degrees of disease. At all stages, α-tocopherylquinone was the major oxidation product found in atherosclerotic lesions. Although maximal α-tocopherol oxidation was observed at the earliest stage of atherogenesis, their observation that α-tocopherol oxidation exceeded lipid oxidation during early atherogenesis might suggest that α-tocopherol effectively protects arterial lipids from oxidation (Upston *et al.*, 2003).

The reaction between α-tocopherol with carbon-centered radicals might be important under anaerobic conditions. However, the formation of the addition products of α-tocopherol with alkyl radicals in biological samples is still unknown. We need further studies on the detection of α-tocopherol products to elucidate the antioxidant action of α-tocopherol *in vivo* systems.

REFERENCES

Bieri, J. G., and Tolliver, T. J. (1981). On the occurrence of α-tocopherylquinone in rat tissue. *Lipids* **16,** 777–779.

Bowry, V. W., and Stocker, R. (1993). Tocopherol-mediated peroxidation. The prooxidant effect of vitamin E on the radical-initiated oxidation of human low-density lipoprotein. *J. Am. Chem. Soc.* **115,** 6029–6044.

Burton, G. W., and Ingold, K. U. (1981). Autoxidation of biological molecules. 1. The antioxidant activity of vitamin E and related chain-breaking phenolic antioxidants *in vitro*. *J. Am. Chem. Soc.* **103,** 6472–6477.

Burton, G. W., and Ingold, K. U. (1986). Vitamin E: Application of the principles of physical organic chemistry to the exploration of its structure and function. *Acc. Chem. Res.* **19,** 194–201.

Burton, G. W., and Traber, M. G. (1990). Vitamin E: Antioxidant activity, biokinetics, and bioavailability. *Annu. Rev. Nutr.* **10,** 357–382.

Burton, G. W., Doba, T., Gabe, E. J., Hughes, L., Lee, F. L., Prasad, L., and Ingold, K. U. (1985). Autoxidation of biological molecules. 4. Maximizing the antioxidant activity of phenols. *J. Am. Chem. Soc.* **107,** 7053–7065.

Evans, C., Scaiano, J. C., and Ingold, K. U. (1992). Absolute kinetics of hydrogen abstraction from α-tocopherol by several reactive species including an alkyl radical. *J. Am. Chem. Soc.* **114,** 4589–4593.

Evarts, R. P., and Bieri, J. G. (1974). Ratios of polyunsaturated fatty acids to α-tocopherol in tissues of rats fed corn or soybean oils. *Lipids* **9,** 860–864.

Fiorentini, D., Cipollone, M., Galli, M. C., Pugnaloni, A., Biagini, G., and Landi, L. (1994). Characterization of large unilamellar vesicles as models for studies of lipid peroxidation initiated by azocompounds. *Free Radic. Res.* **21,** 329–339.

Franchi, P., Lucarini, M., Pedulli, G. F., Valgimigli, L., and Lunelli, B. (1999). Reactivity of substituted phenols toward alkyl radicals. *J. Am. Chem. Soc.* **121,** 507–514.

Frankel, E. N., and Gardner, G. W. (1989). Effect of α-tocopherol on the volatile thermal decomposition products of methyl linoleate hydroperoxides. *Lipids* **24,** 603–608.

Frei, B., Stocker, R., and Ames, B. N. (1988). Antioxidant defenses and lipid peroxidation in human blood plasma. *Proc. Natl. Acad. Sci. USA* **85,** 9748–9752.

Gardner, H. W. (1987). Reaction of hydroperoxides–products of high molecular weight. *In* "Autoxidation of Unsaturated Lipids" (H. W. S. Chan, Ed.), pp. 51–93. London, Academic Press.

Gardner, H. W. (1989). Oxygen radical chemistry of polyunsaturated fatty acids. *Free Radic. Biol. Med.* **7,** 65–86.

Gardner, H. W., Eskins, K., Grams, G. W., and Inglet, G. E. (1972). Radical addition of linoleic hydroperoxides to α-tocopherol or the analogous hydroxychroman. *Lipids* **7,** 324–334.

Halliwell, B., and Gutteridge, J. M. C. (1984). Oxygen toxicity, oxygen radicals, transition metals and disease. *Biochem. J.* **219,** 1–14.

Ham, A. J. L., and Liebler, D. C. (1997). Antioxidant reactions of vitamin E in the perfused rat liver: Product distribution and effect of dietary vitamin E supplementation. *Arch. Biochem. Biophys.* **339,** 157–164.

Hopia, A., Huang, S.-W., and Frankel, E. N. (1996). Effect of α-tocopherol and Trolox on the decomposition of methyl linoleate hydroperoxides. *Lipids* **31,** 357–365.

Kamel-Eldin, A., and Appelqvist, L.-Å. (1996). The chemistry and antioxidant properties of tocopherols and tocotrienols. *Lipids* **31,** 671–701.

Kaneko, T., and Matsuo, M. (1985). The radical-scavenging reactions of a vitamin E model compound, 2,2,5,7,8-pentamethylchroman-6-ol, with radicals from the Fe(II)-induced decomposition of a linoleic acid hydroperoxide, (9*Z*,11*E*)-13-hydroperoxy-9,11-octadecadienoic acid. *Chem. Pharm. Bull.* **33,** 1899–1905.

Kelley, E. E., Buettner, G. R., and Burns, C. P. (1995). Relative α-tocopherol deficiency in cultured cells: Free radical-mediated lipid peroxidation, lipid oxidizability, and cellular polyunsaturated fatty acid content. *Arch. Biochem. Biophys.* **319,** 102–109.

Leray, C., Andriamampandry, M. D., Freund, M., Gachet, C., and Cazenave, J.-P. (1998). Simultaneous determination of homologues of vitamin E and coenzyme Q and products of α-tocopherol oxidation. *J. Lipid Res.* **39,** 2099–2105.

Liebler, D. C. (1993). The role of metabolism in the antioxidant function of vitamin E. *Crit. Rev. Toxicol.* **23,** 147–169.

Liebler, D. C., and Burr, J. A. (1992). Oxidation of vitamin E during iron-catalyzed lipid peroxidation: Evidence for electron-transfer reactions of the tocopheroxyl radical. *Biochemistry* **31,** 8278–8284.

Liebler, D. C., and Burr, J. A. (1995). Antioxidant stoichiometry and the oxidative fate of vitamin E in peroxyl radical scavenging reactions. *Lipids* **30,** 789–793.

Liebler, D. C., Kaysen, K. L., and Kennedy, T. A. (1989). Redox cycles of vitamin E: Hydrolysis and ascorbic acid dependent reduction of 8*a*-(alkyldioxy)tocopherones. *Biochemistry* **28,** 9772–9777.

Liebler, D. C., Baker, P. F., and Kaysen, K. L. (1990). Oxidation of vitamin E: Evidence for competing autoxidation and peroxyl radical trapping reactions of the tocopheroxyl radical. *J. Am. Chem. Soc.* **112,** 6995–7000.

Liebler, D. C., Kaysen, K. L., and Burr, J. A. (1991). Peroxyl radical trapping and autoxidation reactions of α-tocopherol in lipid bilayers. *Chem. Res. Toxicol.* **4,** 89–93.

Liebler, D. C., Burr, J. A., Philips, L., and Ham, A. J. L. (1996). Gas chromatography-mass spectrometry analysis of vitamin E and its oxidation products. *Anal. Biochem.* **236,** 27–34.

Matsuo, M., Matsumoto, S., Iitaka, Y., and Niki, E. (1989). Radical-scavenging reactions of vitamin E and its model compound, 2,2,5,7,8-pentamethylchroman-6-ol, in a *tert*-butylperoxyl radical-generating system. *J. Am. Chem. Soc.* **111,** 7179–7185.

Mukai, K., Watanabe, Y., Uemoto, Y., and Ishizu, K. (1986). Stopped-flow investigation of antioxidant activity of tocopherols. *Bull. Chem. Soc. Jpn.* **59,** 3113–3116.

Murphy, M. E., and Kehrer, J. P. (1987). Simultaneous measurement of tocopherols and tocopheryl quinones in tissue fractions using high-performance liquid chromatography with redox-cycling electrochemical detection. *J. Chromatogr.* **421,** 71–82.

Niki, E., Saito, T., Kawakami, A., and Kamiya, Y. (1984). Inhibition of oxidation of methyl linoleate in solution by vitamin E and vitamin C. *J. Biol. Chem.* **259,** 4177–4182.

Porter, N. A., and Wujek, D. G. (1984). Autoxidation of polyunsaturated fatty acids, an expanded mechanistic study. *J. Am. Chem. Soc.* **106,** 2626–2629.

Porter, N. A., and Wagner, C. R. (1986). Phospholipid autoxidation. *Adv. Free Radic. Biol. Med.* **2,** 283–323.

Porter, N. A., Caldwell, S. E., and Mills, K. A. (1995). Mechanism of free radical oxidation of unsaturated lipids. *Lipids* **30,** 277–290.

Remorova, A. A., and Roginskii, V. A. (1991). Rate constants for the reaction of α-tocopherol phenoxy radicals with unsaturated fatty acid esters, and the contribution of this reaction to the kinetics of inhibition of lipid oxidation. *Kinet. Catal.* **30,** 726–731.

Sevanian, A., and Hochstein, P. (1985). Mechanisms and consequences of lipid peroxidation in biological systems. *Annu. Rev. Nutr.* **5,** 365–390.

Slater, T. F. (1984). Free-radical mechanisms in tissue injury. *Biochem. J.* **222,** 1–15.

Takahashi, M., Tsuchiya, J., and Niki, E. (1989). Scavenging of radicals by vitamin E in the membranes as studied by spin labeling. *J. Am. Chem. Soc.* **111,** 6350–6353.

Takamura, H., Narita, H., Urade, R., and Kito, M. (1986). Quantitative analysis of polyenoic phospholipids molecular species by high performance liquid chromatography. *Lipids* **21,** 356–361.

Thomas, C. E., McLean, L. R., Parker, R. A., and Ohlweiler, D. F. (1992). Ascorbate and phenolic antioxidant interactions in prevention of liposomal oxidation. *Lipids* **27,** 543–550.

Thomas, S. R., and Stocker, R. (2000). Molecular action of vitamin E in lipoprotein oxidation: Implications for atherosclerosis. *Free Radic. Biol. Med.* **28,** 1795–1805.

Thomas, S. R., Neuzil, J., Mohr, D., and Stocker, R. (1995). Coantioxidants make α-tocopherol an efficient antioxidant for low-density lipoprotein. *Am. J. Clin. Nutr.* **62,** 1357S–1364S.

Takeda, H., Shibuya, T., Yanagawa, K., Kanoh, H., and Takasaki, M. (1996). Simultaneous determination of α-tocopherol and α-tocopherylquinone by high-performance liquid chromatography and coulometric detection in the redox mode. *J. Chromatogr.* **722,** 287–294.

Terentis, A. C., Thomas, S. R., Burr, A. A., Liebler, D. C., and Stocker, R. (2002). Vitamin E oxidation in human atherosclerotic lesions. *Circ. Res.* **90,** 333–339.

Upston, J. M., Kritharides, L., and Stocker, R. (2003). The role of vitamin E in atherosclerosis. *Prog. Lipid Res.* **42,** 405–422.

Vatassery, G. T. (1994). Determination of tocopherols and tocopherolquinone in human red blood cell and platelet samples. *Methods Enzymol.* **234,** 327–331.

Wilcox, A. L., and Marnett, L. J. (1993). Polyunsaturated fatty acid alkoxyl radicals exist as carbon-centered epoxyallylic radicals a key step in hydroperoxide-amplified lipid peroxidation. *Chem. Res. Toxicol.* **6,** 413–416.

Winterle, J., Dulin, D., and Mill, T. (1984). Products and stoichiometry of reaction of vitamin E with alkylperoxyl radicals. *J. Org. Chem.* **49,** 491–495.

Yamamoto, Y., Brodsky, M. H., Baker, J. C., and Ames, B. N. (1987). Detection and characterization of lipid hydroperoxides at picomole levels by high-performance liquid chromatography. *Anal. Biochem.* **160,** 7–13.

Yamauchi, R., Kato, K., and Ueno, Y. (1988). Formation of trimers of α-tocopherol and its model compound, 2,2,5,7,8-tetramethylchroman-6-ol, in autoxidizing methyl linoleate. *Lipids* **23,** 779–783.

Yamauchi, R., Matsui, T., Kato, K., and Ueno, Y. (1989). Reaction of α-tocopherol with 2,2′-azobis(2,4-dimethylvaleronitrile) in benzene. *Agric. Biol. Chem.* **53,** 3257–3262.

Yamauchi, R., Matsui, T., Kato, K., and Ueno, Y. (1990). Reaction products of α-tocopherol with methyl linoleate-peroxyl radicals. *Lipids* **25,** 152–158.

Yamauchi, R., Miyake, N., Kato, K., and Ueno, Y. (1993). Reaction of α-tocopherol with alkyl and alkylperoxyl radicals of methyl linoleate. *Lipids* **28,** 201–206.

Yamauchi, R., Yagi, Y., and Kato, K. (1994). Isolation and characterization of addition products of α-tocopherol with peroxyl radicals of dilinoleoylphosphatidylcholine in liposomes. *Biochim. Biophys. Acta* **1212,** 43–49.

Yamauchi, R., Yamamoto, N., and Kato, K. (1995). Iron-catalyzed reaction products of α-tocopherol with methyl 13(*S*)-hydroperoxy-9(*Z*),11(*E*)-octadecadienoate. *Lipids* **30,** 395–404.

Yamauchi, R., Yagi, Y., and Kato, K. (1996). Oxidation of α-tocopherol during the peroxidation of dilinoleoylphosphatidylcholine in liposomes. *Biosci. Biotechnol. Biochem.* **60,** 616–620.

Yamauchi, R., Goto, K., and Kato, K. (1998a). Reaction of α-tocopherol in heated bulk phase in the presence of methyl linoleate (13*S*)-hydroperoxide or methyl linoleate. *Lipids* **33,** 77–85.

Yamauchi, R., Mizuno, H., and Kato, K. (1998b). Preparation and characterization of 8*a*-(phosphatidylcholine-dioxy)-α-tocopherones and their formation during the peroxidation of phosphatidylcholine in liposomes. *Biosci. Biotechnol. Biochem.* **62,** 1293–1300.

Yamauchi, R., Ozaki, K., Shimoyamada, M., and Kato, K. (2002a). Iron-catalyzed reaction products of α-tocopherol with 1-palmitoyl-2-linoleoyl-3-*sn*-phosphatidylcholine (13*S*)-hydroperoxide. *Chem. Phys. Lipids* **114,** 193–201.

Yamauchi, R., Kamatani, T., Shimoyamada, M., and Kato, K. (2002b). Preparation of the addition products of α-tocopherol with cholesteryl linoleate-peroxyl radicals. *Biosci. Biotechnol. Biochem.* **66,** 670–673.

Yamauchi, R., Noro, H., Shimoyamada, M., and Kato, K. (2002c). Analysis of vitamin E and its oxidation products by HPLC with electrochemical detection. *Lipids* **37,** 515–522.

Yamauchi, R., Ozaki, K., Shimoyamada, M., Nagaoka, S., and Kato, K. (2002d). Oxidation products of α-tocopherol on the copper-initiated peroxidation of human plasma and rat serum. *In* "XI Biennial Meeting of the Society for Free Radical Research International" (C. Pasquier, Ed.), pp. 497–501. Monduzzi Editore, Bologna.

12

VITAMIN E AND APOPTOSIS

PAUL W. SYLVESTER

College of Pharmacy, University of Louisiana at Monroe
Monroe, Louisiana 71209

Vitamin E is a generic term that refers to a family of compounds that is further divided into two subgroups called tocopherols and tocotrienols. All natural forms of tocopherols and tocotrienols are potent antioxidants that regulate peroxidation reactions and controls free radical production within the body. However, it is now firmly established that many of the

Vitamins and Hormones, Volume 76

0083-6729/07 $35.00
DOI: 10.1016/S0083-6729(07)76012-0

biological actions mediated by individual vitamin E isoforms are not dependent on their antioxidant activity. Furthermore, synthetic ether derivatives of vitamin E that no longer possess antioxidant activity also display a wide range of biological activities. One of the most intriguing therapeutic applications for natural vitamin E and vitamin E derivatives currently being investigated is their use as anticancer agents. Specific forms of vitamin E display potent apoptotic activity against a wide range of cancer cell types, while having little or no effect on normal cell function or viability. Experimental studies have also determined that the intracellular mechanisms mediating the apoptotic effects of specific vitamin E compounds display great diversity in different types of caner cells and has been found to restore multidrug resistant tumor cells sensitivity to chemotherapeutic agents. These findings strongly suggest that some natural and synthetic analogues of vitamin E can be used effectively as anticancer therapy either alone or in combination to enhance the therapeutic efficacy and reduce toxicity of other anticancer agents.

I. INTRODUCTION

More than 80 years ago, vitamin E was first recognized as an essential dietary supplement required for normal reproductive function in rats (Evans and Bishop, 1922). Since that time, an extensive volume of research has uncovered numerous other physiological actions for vitamin E. The premise has been established that vitamin E reduces oxidative stress and dietary supplementation may be beneficial for the maintenance of good health and disease prevention, particularly cancer, cardiovascular disease, and neurodegenerative diseases (Anderson *et al.*, 2004b; Packer, 1991; Prasad and Edwards-Prasad, 1992; Sylvester and Theriault, 2003). However, a recent report using meta-analysis of dose–response data obtained from randomized, controlled clinical trials have challenged this premise and has suggested that high-dose (>400 IU/day) vitamin E supplements should be avoided because they have a detrimental impact on human health and significantly increase all-cause mortality (Miller *et al.*, 2005). Although independent investigations have yet confirmed or disproved these conclusions and recommendations, this meta-analysis study has ignited a great deal of controversy and has polarized physicians, scientists, nutritionist, and the general public regarding the use of vitamin E supplements in the prevention and treatment clinical disorders. Additional well-designed and unbiased studies are clearly needed to clarify this issue and definitively determine whether high-dose vitamin E supplements provides significant health benefits or induce detrimental effects on morbidity and mortality (Houston, 2005). Furthermore, although experimental evidence obtained from animal and cell culture studies has clearly demonstrated the potential beneficial effects of vitamin E (Anderson *et al.*, 2004b; Packer, 1991; Prasad and Edwards-Prasad, 1992; Sylvester and Theriault, 2003), clinical studies have produced inconsistent and conflicting results (Bostick *et al.*, 1993; Houston, 2003; Klein, 2006; Schneider, 2005; Sung *et al.*, 2003; Vivekananthan *et al.*, 2003). The present chapter will focus on recent findings related to the intracellular mechanisms mediating vitamin E-induced apoptosis in cancer cells in various experimental model systems. Although the anticancer efficacy of vitamin E observed in animal and *in vitro* studies remains to be confirmed in clinical trials, it is reasonable to speculate that understanding the mechanisms mediating vitamin E-induced apoptosis will provide useful insights for basing effective strategies for use of vitamin E in the prevention and treatment of human cancer.

Therapeutic use of vitamin E either alone or in combination with other anticancer agents has recently become a subject of great interest. Traditional chemotherapeutic agents, while effective is suppressing the growth and progression of many types of cancer, are relatively nonspecific in their mechanism of action and characteristically have toxic effects on both normal and neoplastic tissues. In contrast, tumor cells have been shown to be significantly more sensitive than normal cells to the anticancer effects of specific forms of vitamin E. Cell culture studies have shown that treatment with natural forms and synthetic derivatives of vitamin E significantly inhibits growth and initiates apoptosis in neoplastic cells using treatment doses that have little or no effect on normal cell growth or viability (McIntyre *et al.*, 2000a,b; Neuzil *et al.*, 2001a; Weber *et al.*, 2002). The exact mechanism responsible for mediating this selective cytotoxicity of vitamin E against cancer versus normal cells is presently unknown and may differ in various types of cancer. Investigations have also shown that combined treatment of vitamin E with other chemotherapeutic agents resulted in a significant decrease in viable tumor cell number as compared to either treatment alone, and strongly suggests that vitamin E can be used effectively to enhance therapeutic efficacy and reduce toxicity of other anticancer agents (Anderson *et al.*, 2004a; Guthrie *et al.*, 1997; Kline *et al.*, 2004; Zhang *et al.*, 2004b; Zu and Ip, 2003).

II. VITAMIN E AND VITAMIN E DERIVATIVES

A. NATURAL ISOFORMS

Vitamin E is a general term used to describe a family of eight different naturally occurring compounds that are very closely related in chemical structure (Schneider, 2005). The vitamin E family of compounds is further subdivided into two subgroups call tocopherols and tocotrienols. Both tocopherols and tocotrienols contain a chromanol ring with different substitution patterns of methyl groups at positions 5, 7, and 8 of the head group to result

Tocopherols

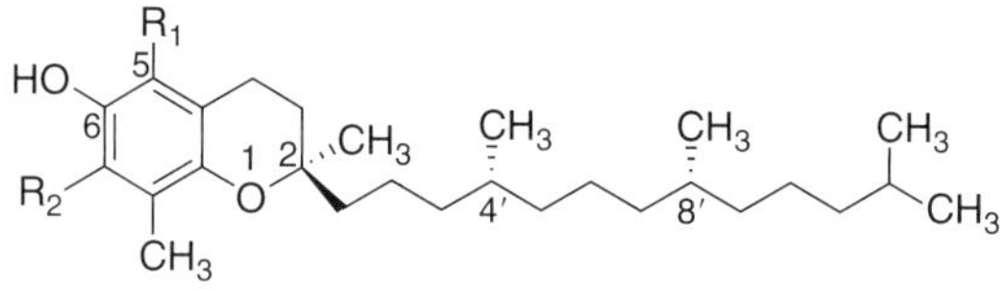

Tocotrienols

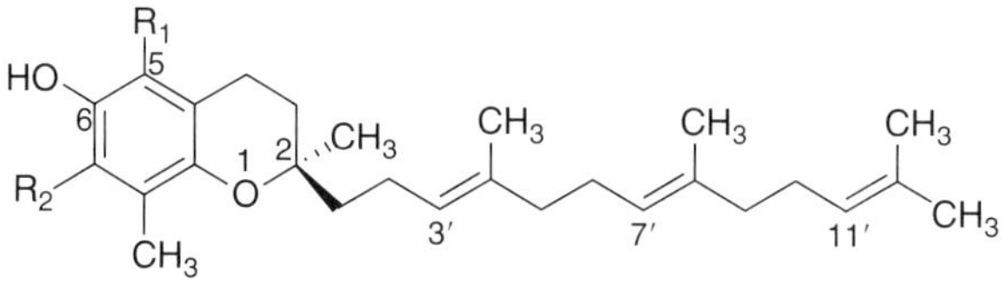

Compound	R_1	R_2	Phytyl chain
α-Tocopherol	CH_3	CH_3	Saturated
β-Tocopherol	CH_3	H_3	Saturated
γ-Tocopherol	H_3	CH_3	Saturated
δ-Tocopherol	H_3	H_3	Saturated
α-Tocotrienol	CH_3	CH_3	Unsaturated
β-Tocotrienol	CH_3	H_3	Unsaturated
γ-Tocotrienol	H_3	CH_3	Unsaturated
δ-Tocotrienol	H_3	H_3	Unsaturated

FIGURE 1. Generalized structure of vitamin E compounds.

in four individual isoforms in each subgroup classified as α-, β-, γ-, and δ-tocopherols and α-, β-, γ-, and δ-tocotrienols (Fig. 1). Vitamin E compounds also contain a 16-carbon phytyl side chain. The difference between tocopherols and tocotrienols is that the phytyl side chain in tocopherols is saturated and contains three chiral centers at carbons 2, 4′, and 8′ (Fig. 1). The naturally occurring tocopherols isoforms have the *R*-configuration at all three positions, whereas synthetic tocopherol is a racemic mixture containing all eight stereoisomers. Since only the *RRR* stereoisomer is biologically active, synthetic α-tocopherol is composed of 7/8 (87.5%) nonbiologically active compound. In contrast, tocotrienols contain three unsaturated double bonds at the 3′, 7′, and 11′ positions on the phytyl side chain (Fig. 1). Another important chemical feature is the phenolic hydroxyl group located on the number 6 carbon of the chromane ring, which functions as the reactive site mediating antioxidant activity for all vitamin E compounds (Fig. 1). However, individual tocopherol and tocotrienol isoforms display a wide range in antioxidant potency, and antioxidant potency shows a poor correlation to the biopotency of many antioxidant-independent actions of these compounds (Schneider, 2005). Evidence suggests that the level of phytyl chain saturation and/or chromane ring methylation is critical in determining the differential biopotencies demonstrated by individual vitamin E isoforms (Serbinova and Packer, 1994; Serbinova *et al.*, 1991; Suzuki *et al.*, 1993; Sylvester and Shah, 2005b; Sylvester and Theriault, 2003). Direct comparisons between the two vitamin E subclasses have shown that tocotrienols are significantly more potent than tocopherols in most biological actions, and display a consistent relationship in most instances corresponding to δ-tocotrienol $\geq$ γ-tocotrienol $>$ α-tocotrienol $>$ δ-tocopherol $>$ γ-, and α-tocopherols (McIntyre *et al.*, 2000a,b). Although the exact reason why tocotrienols are more potent than tocopherols is presently unknown, it is well established that plasma and tissue concentrations of individual tocopherol and tocotrienol isoforms is limited by the specificity and saturability of these particular transfer proteins and transport mechanisms within the body that display significant preference for α-tocopherol (Herrera and Barbas, 2001; Hosomi *et al.*, 1997; Traber and Packer, 1995).

B. SYNTHETIC DERVIVATIVES

Natural and synthetic forms of vitamin E readily undergo oxidation when exposed to air. Therefore, these compounds are commonly modified for commercial use to product uncharged acetate or succinate esters at the site of the hydroxyl group located at position 6 of the chromanol ring. Since the acetate and succinate esters are significantly more stable, they are commonly used as a vitamin E supplement in vitamin pills. However, these acetate and succinate moieties must be removed by esterases in the blood and tissues in order to restore their function as antioxidants (Kline *et al.*, 2003). It is now well established that many of the biological actions of vitamin E and vitamin E derivatives are independent of their antioxidant potency. Although natural α-tocopherol is a potent antioxidant, the majority of studies have shown that it has little or no apoptotic activity on cancer cells (Kline *et al.*, 2004; Neuzil *et al.*, 2002a; Sylvester and Shah, 2005a; Sylvester and Theriault, 2003; Sylvester *et al.*, 2005). In contrast, both acetate and succinate ether derivatives of α-tocopherol display significantly greater anticancer and apoptotic activity than natural or synthetic α tocopherol (Anderson *et al.*, 2004b; Birringer *et al.*, 2003; Kline *et al.*, 2003; Lawson *et al.*, 2003; Neuzil *et al.*, 2001b, 2002a; Shun *et al.*, 2004; Zhang *et al.*, 2004b). Subsequent studies have shown that analogues of α-tocopheryl succinate (α-TS) show differential anticancer potencies (Birringer *et al.*, 2003). A reduction in TS chromanol ring methylation was found to decrease the apoptotic potency. In addition, replacement of the succinyl group with a maleyl group enhanced, while replacement with a glutaryl group decreased apoptotic potency of α-tocopheryl derivatives (Birringer *et al.*, 2003). These studies also clarified the relative importance of the other structural components that make up the α-TS molecule in mediating apoptotic effects. Methylation of the free succinyl carboxyl group on α-TS was found to block apoptotic activity, indicating that the intact succinyl ether moiety was essential for α-TS-induced apoptosis. Trolox is a vitamin E derivative of α-tocopherol that lacks a phytyl tail and although is a potent antioxidant, Trolox does not induce apoptosis or display anticancer activity. The succinylated derivative of Trolox was also found to have no apoptotic activity, indicating that the he phytyl tail is also required for α-TS-induced apoptosis (Birringer *et al.*, 2003). However, phytyl succinate alone does not induce apoptosis (Birringer *et al.*, 2003). Other studies have shown that the nonhydrolyzable ether analogue of α-tocopherol, α-tocopherol ether linked acetic acid (α-TEA), is also a potent inducer of apoptosis in breast cancer cells, while having little or no effect on normal cells (Kline *et al.*, 2004; Lawson *et al.*, 2003). In contrast to α- tocopherol, natural forms of α-, γ, and δ-tocotrienols display potent apoptotic activity (McIntyre *et al.*, 2000a,b; Sylvester and Shah, 2005a; Sylvester *et al.*, 2005), whereas succinate ether derivative of γ-tocotrienol displays more potent apoptotic activity than the parent compound (Birringer *et al.*, 2003). These findings demonstrate that the apoptotic potency of vitamin E and vitamin E derivatives is independent of antioxidant potency and that modifications of different functional groups within the vitamin E molecule significantly modulates apoptotic activity. These finding have stimulated interest and efforts to develop and synthesize additional novel analogues of vitamin E that display greater apoptotic and anticancer activity.

III. VITAMIN E ANTIOXIDANT POTENCY

Various forms of natural vitamin E and vitamin E derivatives have been shown to be potent inducers of apoptosis in cancer cells. However, vitamins E has also been shown to induce protective effects and prevent apoptosis in some experimental model systems. These antiapoptotic effects of vitamin E are primarily associated with its antioxidant activity. All vitamin E compounds containing a free hydroxyl group at position 6 on the chromanol ring possess antioxidant activity. Dietary supplements containing acetate or succinate ether derivatives of α-tocopherol are not antioxidants in their present form but are subsequently reformed to their free alcohol form by esterases in the small intestine (Nakamura *et al.*, 1975). It is well established that α-tocopherol is an important antioxidant that regulates peroxidation reactions and free radical production (Packer, 1991). Uncontrolled or excessive free radical production can ultimately lead to cellular damage, dysfunction, or death (Packer, 1991). Since free radicals can form DNA adducts and lead to gene mutation or dysfunction, the antioxidant potency of vitamin E compounds may play an important initial first step in the prevention of neoplastic transformation of normal cells.

Vitamin E primary function as an antioxidant is to react with fatty acid peroxyl radicals produced from lipid peroxidation (Burton and Ingold, 1981). Direct comparisons between the various tocopherol and tocotrienol isoforms have shown large differences in antioxidant activity. Specifically, α-tocotrienol was found to be between 40 and 60 times more potent in preventing lipid peroxidation in rat liver microsomal membranes and six to seven times better in protecting cytochrome P450 oxidative damage compared to α-tocopherol (Serbinova and Packer, 1994; Serbinova *et al.*, 1991; Suzuki *et al.*, 1993). The α-isoforms of tocopherol and tocotrienol are the most potent lipid soluble antioxidants. This is due to the level of methylation of the chromanol ring, which greatly enhances the reactivity of the hydroxyl group and facilitates the transfer of the hydrogen to a peroxyl radical to form a tocopheroxyl or tocotrienoxyl radical (Burton and Ingold, 1981; Schneider, 2005). Tocopherols and tocotrienols can be restored by reduction of tocopheroxyl and tocotrienoxyl radical by other antioxidants such as vitamin C and ubiquinol (Niki *et al.*, 1984; Shi *et al.*, 1999). Tocopheroxyl and tocotrienoxyl radical can also react with a second peroxyl radical to produce nonradical products, terminating the antioxidant action of these vitamin E compounds and demonstrating that one vitamin E molecule can neutralize two peroxyl radical molecules (Schneider, 2005).

Several factors appear to be responsible for the greater antioxidant biopotency displayed by α-tocotrienol. It was shown that α-tocotrienol displayed a significantly higher recycling efficiency than α-tocopherol (Serbinova and Packer, 1994; Serbinova *et al.*, 1991; Suzuki *et al.*, 1993). In addition, α-tocotrienol was shown to have a more uniform distribution within the microsomal membrane bilayer, and that α-tocotrienol displayed a more efficient interaction with lipid free radicals, as compared to α-tocopherol (Serbinova and Packer, 1994; Serbinova *et al.*, 1991; Suzuki *et al.*, 1993). However, biochemical analysis of the antioxidant potency of vitamin E compounds can be misleading because α-tocopherol is preferentially absorption and transport of by the body. Furthermore, α-tocopherol transfer protein (α-TTP) has been shown to play a critical role in regulating plasma concentrations of vitamin E and experimental data demonstrates that α-tocopherol has a high affinity, whereas tocotrienols have a relative low affinity for this carrier protein in the liver (Hosomi *et al.*, 1997). Therefore, in terms of bioavailability and bioactivity, α-tocopherol is biologically and functionally the most important and most active antioxidant as compared to all other vitamin E isoforms in humans (Kayden and Traber, 1993). Although the other natural forms of tocopherols and tocotrienols are readily absorbed by the small intestine, they do not display the selective and preferential lipoprotein transport in the blood and distribution to target tissues (Kayden and Traber, 1993).

IV. VITAMIN E AS AN ANTICANCER AGENT

A. NATURAL FORMS

Initial studies demonstrated that dietary α-tocopherol supplementation was somewhat effective in suppressing carcinogen-induced cancers in rodents (Haber and Wissler, 1962). Subsequent studies showed that α-tocopherol supplements E reduced carcinogen-induced hamster cheek tumors (Shklar, 1982) and carcinogen-induced stomach tumors in mice colon cancer (Wattenberg, 1972). Dietary supplementation with α-tocopherol was not found to inhibit carcinogen-induced mammary tumors when given alone (Ip, 1982; King and McCay, 1983), but was effective when given in combination with selenium supplements (Horvath and Ip, 1983). Although these studies strongly suggested that dietary supplementation with α-tocopherol may protect against cancer risk in humans, there is little direct epidemiological evidence to support a role for α-tocopherol in preventing human cancer (Hunter *et al.*, 1993). In general, *in vivo* studies investigating the inhibitory effects of α-tocopherol in mammary tumorigenesis have shown for the most part, negative results (Goh *et al.*, 1994; Gould *et al.*, 1991; Nesaretnam *et al.*, 1992; Zurinah *et al.*, 1991). Furthermore, *in vitro* studies have conclusively demonstrated that α-tocopherol has little or no cytotoxic activity against tumor cells even at extremely high concentrations (Schwartz and Shklar, 1992; Sigounas *et al.*, 1997; Sylvester and Shah, 2005b; Sylvester and Theriault, 2003).

In contrast to α-tocopherol, tocotrienols have been shown to be a potent anticancer agent (Kline *et al.*, 2001, 2004; Prasad and Edwards-Prasad, 1982; Sylvester and Shah, 2005b; Sylvester and Theriault, 2003). The anticancer effects of tocotrienols were first discovered in studies investigating the role of high dietary fat intake on carcinogen-induced mammary cancer development and growth in rats (Sylvester *et al.*, 1986). In general, different high fat diets were found to stimulate tumor development regardless of whether the diets were formulated with different animal versus vegetable or saturated versus unsaturated fats. The notable exception to this finding was the observation that high dietary intake of palm oil suppressed carcinogen-induced mammary tumorigenesis in experimental animals (Sylvester *et al.*, 1986). Palm oil naturally contains high levels of tocotrienols (Cottrell, 1991; Packer *et al.*, 2001) and palm oil diets stripped of tocotrienols were found to enhance mammary tumorigenesis in rats (Nesaretnam *et al.*, 1992). Vitamin E content in various types of dietary oils and fats is shown in Table I.

Cell culture experiments using vitamin E extracted from palm oil or what has become to be known as the tocotrienol-rich fraction (TRF) of palm oil, is composed of a mixture of tocotrienols (80%) and α-tocopherol (20%),

TABLE I. Concentration of Vitamin E Isoforms (mg/liter) in Various Dietary Oils and Fats[a]

	Tocopherols					Tocotrienols					
Dietary oil	α	β	γ	δ	Total tocopherol	α	β	γ	δ	Total tocotrienol	Total vitamin E
Palm	152	–	–	–	152	205	–	439	94	738	890
Coconut	5	–	–	6	11	5	1	19	–	25	36
Palm kernel	13	–	–	–	21	21	–	–	–	21	42
Coco butter	11	–	170	17	198	2	–	–	–	2	200
Corn	112	50	620	18	800	–	–	–	–	0	802
Cottonseed	389	–	387	–	776	–	–	–	–	0	776
Peanut	130	–	216	21	367	–	–	–	–	0	367
Olive	51	–	–	–	51	–	–	–	–	0	51
Safflower	387	–	174	240	801	–	–	–	–	0	807
Soybean	101	–	593	264	958	–	–	–	–	0	958
Sunflower	487	–	51	8	546	–	–	–	–	0	546
Lard	12	–	7	–	19	7	7		26		

[a]Selected information obtained from reference Cottrell (1991).

showed that TRF treatment significantly inhibited mammary tumor cell proliferation and induced cell death in a dose-responsive manner (McIntyre *et al.*, 2000b; Nesaretnam *et al.*, 1998; Sylvester *et al.*, 2002; Yu *et al.*, 1999b). Since TRF contains a mixture of α-tocopherol and α-, γ-, and δ-tocotrienols, subsequent studies were conducted to determine if one or all of these vitamin E isoforms were responsible for mediating the antitumor effects of TRF. Direct comparisons between the two vitamin E subclasses showed that tocotrienols were significantly more potent in suppressing growth and inducing cell death than tocopherols (McIntyre *et al.*, 2000b), and at treatment doses that had little effect on normal mammary epithelial cell growth or viability (McIntyre *et al.*, 2000a; Sylvester *et al.*, 2001). Although these studies strongly suggest that tocotrienols are mediating the anticancer effects observed in treatment groups fed high palm oil diets, this hypothesis has been difficult to prove *in vivo*. Investigations have shown that dietary supplementation with isolated tocotrienols or TRF had little or no effect in suppressing mammary tumor growth and development (Gould *et al.*, 1991).

B. SYNTHETIC DERIVATIVES

During the past 30 years, studies have shown that α-TS treatment is more potent than α-tocopherol acetate in suppressing the growth and viability of various neoplastic cells including leukemia (Anderson *et al.*, 2004b; Fariss *et al.*, 1994), breast cancer (Kline *et al.*, 2001), prostate cancer (Zu and Ip, 2003), mesothelioma (Tomasetti *et al.*, 2004), liver cancer (Sakai *et al.*, 2004), bone cancer (Alleva *et al.*, 2005), and others (Schneider, 2005). The succinate ester moiety is required for α-TS-induced apoptosis because the resulting hydrolysis products α-tocopherol and succinic acid are devoid of cytotoxic activity (Djuric *et al.*, 1997; Fariss *et al.*, 1994). α-TEA, the nonhydrolyzable ether-linked acetic acid ester of α-tocopherol, is significantly more potent than the hydrolyzable α-TS in inducing apoptosis (Anderson *et al.*, 2004a,b; Lawson *et al.*, 2003). Since α-TS and α-tocopherol acetate undergo hydrolysis in the small intestine, it has been difficult to obtain high circulating levels of these esters in the blood following oral administration, and suggest that α-TEA may have significant advantages in terms of bioavailability and anticancer efficacy as compared to the hydrolyzable α-tocopherol derivatives.

Taken together, experimental evidence obtained from cell culture studies indicates that α-tocopherol treatment alone is not an effective anticancer agent, whereas tocotrienols and α-tocopherol derivatives appear to be strong candidates for use as cancer chemotherapeutic agents. This suggestion is even more attractive when taking into account the low toxicity displayed by tocotrienols and α-tocopherol derivatives against normal cells. However, *in vivo* studies in animals and humans have produced conflicting results. These findings again point to the need for more extensive research in areas related to tocotrienol and α-tocopherol derivative pharmaceutics and pharmacokinetics in order to optimized formulation, dosage, and delivery.

V. APOPTOSIS

Programed cell death or apoptosis is a normal and essential aspect of organ development and remodeling that is initiated at birth and continues throughout life (Raff, 1998). Apoptosis is also a mechanism by which neoplastic cells can be eliminated from the body (Kerr *et al.*, 1972). Morphological and biochemical characteristics distinguish apoptosis from neurosis in terms of nuclear and cytoplasmic condensation, DNA fragmentation, dilation of the endoplasmic reticulum, and alterations in the cell membrane composition (Heermeier *et al.*, 1996; Hockenbery *et al.*, 1990; Nass *et al.*, 1996; Oltvai and Korsmeyer, 1994). During apoptosis, the plasma membrane forms small blebs and vesicles, loses contact with the extracellular matrix, and ultimately shrinks and disintegrates without releasing its content into the surrounding area. The deteriorating fragments of cellular debris are then quickly scavenged by the neighboring cells and macrophages in the absence of an inflammatory response (Raff, 1998). In contrast, death caused by necrosis is characterized by the cell-undergoing lysis, releasing their contents into the surrounding area, and the initiation of an inflammatory response that serves to initiate the process of tissue repair and wound healing (Raff, 1998).

Initiation of programed cell death involves the activation cysteine proteases known as caspases (Cohen, 1997; Nicholson, 1999). Procaspases are activated by cleavage at a specific site either by themselves or by other caspases (Cohen, 1997; Nicholson, 1999). There are two basic types of caspases. Initiator caspases exclusively cleave and activate other caspases, and effector caspases that cleave other proteins. Initiator procaspases remain in an inactive state though the binding to inhibitory proteins that prevent dimerization and activation. Activation of initiator caspases results from the downregulation or destruction of these caspase inhibitory proteins. Initiator caspases (caspase-2, -8, -9, and -10) activate effector caspases (caspase-3, 6, and 7), which in turn cleave structural and regulatory proteins (DFF45/ICAD, PARP, lamins, cytokeratins, and so on) and are responsible for most features described above (Cohen, 1997; Muzio *et al.*, 1998; Nicholson, 1999; Srinivasula *et al.*, 1998). Different caspases have different roles during the initiation and progression of apoptosis, including the destruction of caspase inhibitors, Bcl-2 antiapoptotic proteins, mitogenic or survival signal transduction pathways, cytoskeleton proteins, and DNA repair and replication enzymes (Ashkenazi and Dixit, 1998; Nicholson, 1999; Salvesen and Dixit, 1999; Thornberry and Lazebnik, 1998).

There are at least three general mechanisms involved in caspases activation and the induction of apoptosis. The first mechanism involves an absence or severe reduction of external mitogenic or survival factor stimulation. It has been established that mitogen-starved normal cells will undergo apoptosis hours after mitogen has been removed from the culture medium (Nass *et al.*, 1996). Various mitogens such as EGF or TGF-α have been shown to induce high cytosolic levels of Bcl-2 and Bcl-xL proteins that inhibit apoptosis by preventing procaspase dimerization and activation (Nass *et al.*, 1996). In the absence of mitogens, the intracellular levels of these antiapoptotic Bcl-2 family member proteins become significantly reduced, which ultimately lead to the induction of apoptosis in these cells. In contrast, many types of tumor cells characteristically overexpress antiapoptotic Bcl-2 family member proteins as a result of paracrine or autocrine growth factor stimulation, a process that inhibits apoptosis and enhances tumor survival (Ashkenazi and Dixit, 1998; Nicholson, 1999; Salvesen and Dixit, 1999; Thornberry and Lazebnik, 1998).

The second mechanism for caspase activation is mediated through the activation of death receptors. Death receptors are members of the tumor necrosis factor (TNF) receptor superfamily and are activated by their corresponding death receptor ligands. Death receptors within TNF superfamily of receptors that includes TNF-R1, Fas, TNF-related apoptosis-inducing ligand (TRAIL) receptors, TRAIL-R1 and TRAIL-R2 (Schulze-Osthoff *et al.*, 1998; Walczak and Krammer, 2000). Specific ligands to these death receptors include TNF-α that activates TNF-R1, FasL that activates Fas, and TRAIL that activates TRAIL-R1 and TRAIL-R2 (Schulze-Osthoff *et al.*, 1998; Walczak and Krammer, 2000). Following activation, death receptors undergo trimerization and recruit adapter proteins such as FADD through homophilic DD interactions, and initiator procaspases such as procaspase-8 to form the death-inducing signaling complex (DISC) at the membrane that promotes caspases-8 activation (Ashkenazi and Dixit, 1998). Within the DISC, procaspase-8 is cleaved to form the active p20 subunit of caspase-8 that subsequently cleave and activate effector caspases such as caspase-3, -6, or -7, and ultimately apoptosis. Alternatively, activation of specific membrane bound receptors by their ligands, such as TGF-β, also induce apoptosis in normal and neoplastic cells even in the presence of high levels of mitogens (Heermeier *et al.*, 1996). Although the growth inhibitory effects of TGF-β on epithelial cells are well documented, the exact role of TGF-β in apoptosis is not presently understood because the TGF-β receptor is not a member of the TNF receptor superfamily of death receptors and activation is not primarily associated with caspase activation.

The third mechanism for caspase activation is mitochondrial stress. Mitochondrial stress-induced caspase activation and apoptosis can be initiated by numerous cellular signals that cause perturbations in the mitochondrial membrane resulting in the release of proapoptotic molecules, such as apoptosis-inducing factor (AIF) and cytochrome *c*, from the intermembrane space into the cytoplasm (Green and Reed, 1998; Susin *et al.*, 1999). Cytochrome *c* then interacts with Apaf-1, dATP/ATP, and procaspase-9 to form a complex called the apoptosome, which leads to the activation of caspase-9, which then activates effector caspase-3 and/or -7, and ultimately leads to apoptosis (Li *et al.*, 1997, 1998; Zou *et al.*, 1997). Cytochrome *c* release from the mitochondria occurs through the permeability transition pore complex (PTPC), a structure that bridges both the inner and outer mitochondrial membranes (Kroemer, 1999; Susin *et al.*, 1999). Cytochrome *c* is required for apoptosome formation, and it is tightly controlled by the Bcl-2 family of proteins that exits to inhibit (Bcl-2, Bcl-xL, Bcl-w, Mcl-1, and A1) or promote (Bax, Bak, Bok, Bik, Hrk, Bim, and Bad) apoptosis through the regulation of PTPC formation (Kroemer, 1999; Oltvai and Korsmeyer, 1994; Susin *et al.*, 1999). Phosphorylation of Bad by Akt, an enzyme association with mitogenesis and cell survival, prevents the translocation and binding of Bad to the mitochondrial membrane and assisting Bax in the formation and opening of the PTPC (Kroemer, 1999; Susin *et al.*, 1999).

VI. Vitamin E Suppression of Apoptosis

Although cancer is a disease that is often associated with defective intracellular mechanisms involved in the induction and progression of apoptosis, many degenerative diseases, such as AIDS, Alzheimer's disease, Parkinson's disease, and aplastic anemia are associated with increased apoptosis. Treatments that prevent the excessive apoptosis that occurs in degenerative disease can provide significant benefit and induce remission or slowing down the progression of these diseases. Vitamin E has been shown to provide protective effects in some types of degenerative pathologies.

Vitamin E plays an important role in maintaining normal neurological health and function is an important (Muller, 1990; Muller and Goss-Sampson, 1990). Since the majority of vitamin E-sensitive neurological disorders are associated with elevations in oxidative damage, it was hypothesized that the neuroprotective effects of α-tocopherol were mediated by its antioxidant activity (Gonzalez-Polo *et al.*, 2003; Roy *et al.*, 2002). However, studies have shown that tocotrienols display significantly greater neuroprotective potency than tocopherols (Sen *et al.*, 2000). In these studies, pregnant animals were fed vitamin E-deficient diets for 2 weeks, resulting in very low levels of vitamin E levels in the fetal brain, but not the mother's brain. Subsequent feeding with TRF, a mixture of α-tocopherol and tocotrienols showed a significantly higher uptake of α-tocotrienol by the fetal brain than by the adult brain (Roy *et al.*, 2002). Furthermore, α-tocotrienol was also found to be significantly more potent in suppressing glutamate-induced neurodegeneration, whereas α-tocopherol was found to have no such neuroprotective effect (Sen *et al.*, 2000). This neuroprotective effect of α-tocotrienol did not result from antioxidant activity, but through the suppression of specific signal transduction targets including c-Src kinase and 12-lipoxygenase (Khanna *et al.*, 2003; Sen *et al.*, 2006). These results strongly suggest that α-tocotrienol may provide nutrition-based protection against glutamate-induced neurotoxicity.

Additional *in vitro* studies using primary neuronal cultures of rat striatum showed that α-, γ-, and δ-tocotrienols, but not α-tocopherol, inhibited oxidative stress-induced apoptosis (Osakada *et al.*, 2004). γ-Tocotrienol also provides significant protection against oxidative stress-induced apoptosis in astrocytes, while α-tocopherol was much less effective (Mazlan *et al.*, 2006). Since uptake of γ-tocotrienol and α-tocopherol by astrocytes was similar, it was hypothesized that the protective effects of γ-tocotrienol might be mediated through both antioxidant-dependent and -independent mechanisms (Mazlan *et al.*, 2006).

In other studies, thioacetamide-induced caspase activation, liver cell apoptosis, and death in rats were found to be significantly attenuated by antioxidants such as vitamin E (Sun *et al.*, 2000). TNF-induced apoptosis of human monocytic U937 cells was also found to be reduced by α-tocopherol, suggesting that α-tocopherol may provide protection from degenerative pathologies associated with the overexpression of the death receptor ligand, TNF (Yano *et al.*, 2000). Likewise, a protective effect of α-tocopherol was also observed in studies examining nicotine-induced lung damage in rats (Demiralay *et al.*, 2006). These studies showed that α-tocopherol significantly reduced nicotine-induced pulmonary epithelial cell apoptosis by reducing associated increases in TNF and myeloperoxidase levels (Demiralay *et al.*, 2006). These findings provide further evidence that α-tocopherol may inhibit death receptor-mediated apoptosis. Treatment with α-tocopherol was also found to be effective in preventing 7-ketocholesterol-induced mitochondrial cytochrome *c* release, caspase activation, and apoptosis in U937 human lymphoma cells, suggesting that antioxidants such as α-tocopherol may also function to prevent apoptosis and enhance cellular survival by preventing mitochondrial dysfunction resulting from oxidative stress (Lizard *et al.*, 2000). Heart failure is another degenerative disease associated with ventricular remodeling and myocardial dysfunction. Studies have shown that following induced myocardial infarction in rabbits, dietary supplementation with α-tocopherol and vitamin C diminished oxidative stress, increased mitochondrial Bcl-2 levels, decreased Bax levels, reduced mitochondrial cytochrome *c* release, attenuated caspase-9 and -3 activation, and significantly reduced myocyte apoptosis (Qin *et al.*, 2006). These data provide additional evidence to support the hypothesis that α-tocopherol may be involved in preventing oxidative stress-induced mitochondrial stress-induced apoptosis in some types of cells.

VII. VITAMIN E-INDUCED APOPTOSIS

A. NATURAL FORMS

The preponderance of evidence clearly shows that natural α-tocopherol displays little or no apoptotic activity (Kimmick *et al.*, 1997; Kline *et al.*, 2004; Neuzil *et al.*, 2002b; Schneider, 2005; Sylvester and Shah, 2005b; Sylvester and Theriault, 2003). Cell culture experiments using α-tocopherol on a series of well-established cancer cell lines that included two erythroleukemia (HEL and OCIM-1) cell lines and a hormone-responsive breast (MCF-7) and prostate cancer (CRL-1740) cell lines showed that tumor cell proliferation was inhibited with only very high doses (mM) of this form of vitamin E (Sigounas *et al.*, 1997). The physiological significance of these studies is unclear. However, a few studies have shown that γ-tocotrienol may have some protective action in the prevention of prostate cancer (Galli *et al.*, 2004; Gysin *et al.*, 2002). Cell culture experiments demonstrated that γ-tocopherol caused a downregulation in cyclin D1 and E levels and induced a blockade cell cycle progression in DU-145 prostate carcinoma cells (Gysin *et al.*, 2002). Similar results were observed in PC-3 prostate cancer cells (Galli *et al.*, 2004). Others have also shown that δ-tocopherol displays apoptotic activity against preneoplastic and neoplastic mammary epithelial cells, but these effects were significantly less potent as compared to similar treatment with tocotrienols (McIntyre *et al.*, 2000b).

Several years ago, studies were conducted to directly compare the apoptotic potency of individual tocopherol and tocotrienol isoforms on normal, preneoplastic, neoplastic, and highly malignant mouse mammary epithelial cells *in vitro* (McIntyre *et al.*, 2000a,b). Results showed that treatment with very high doses of α- or γ-tocopherol had no effect on viability in any of these types of cells. In contrast, δ-tocopherol and α-, γ-, and δ-tocotrienols were found to reduce normal, preneoplastic, neoplastic, and malignant mammary epithelial cell viability in a dose-responsive manner, but the tocotrienols were found to be significantly more potent that δ-tocopherol. These data also showed that the highly malignant cells were the most sensitive, whereas normal cells were the least sensitive to the apoptotic effects of these individual vitamin E isoforms (McIntyre *et al.*, 2000a,b). These findings were of particular interest because tocotrienol-induced apoptosis was observed in preneoplastic and neoplastic mammary epithelial cells using treatment doses that had no effect on normal cells viability.

Subsequent studies conducted to determine the intracellular mechanisms mediating tocotrienol-induced apoptosis in neoplastic mammary epithelial cells showed that γ-tocotrienol induced caspase-8 and -3, but not caspase-9 activation, suggesting that tocotrienol-induced apoptosis is mediated through death receptor activation (Shah *et al.*, 2003). In contrast, α-tocopherol had no effect on caspase activation. Furthermore, γ-tocotrienol-induced caspase activation and apoptosis was blocked when treatment was given in combination with specific caspase-8 and -3, but not caspase-9 inhibitors (Shah *et al.*, 2003). Since caspase-8 processing and activation are associated with death receptor-mediated apoptotic signaling, while caspase-9 activation and processing is associated with mitochondrial stress-mediated apoptotic signaling, these findings strongly suggest that tocotrienol-induced apoptosis is mediated by receptor-induced caspase activation and occurs independently of mitochondrial stress-related apoptotic mechanisms in these cells.

The suggestion that tocotrienol-induced cytotoxicity in mouse neoplastic mammary epithelial cells is unrelated to mitochondrial stress apoptotic signaling was later confirmed in recent publication (Shah and Sylvester, 2005). This study clearly demonstrated that exposure to cytotoxic doses of γ-tocotrienol-induced apoptosis, but did not result in a disruption or loss of mitochondrial membrane potential or the release of cytochrome *c* from the mitochondria into the cytoplasm of neoplastic mammary epithelial cells (Shah and Sylvester, 2005). Furthermore, results also showed that this treatment caused a paradoxical increase in mitochondrial antiapoptotic and corresponding decrease in mitochondrial proapoptotic Bcl-2 family member proteins. The relative intracellular levels of anti- and proapoptotic Bcl-2 proteins determine mitochondrial membrane stability and modulate cytochrome *c* release into the cytoplasm (Antonsson and Martinou, 2000; Hsu *et al.*, 1997; Jaattela, 1999; Putcha *et al.*, 1999). These findings suggested that mitochondrial membrane stability and integrity might actually be enhanced for a limited period of time following exposure to cytotoxic doses of γ-tocotrienol (Shah and Sylvester, 2005) and that tocotrienols may play a role in suppressing mitochondrial stress-mediated apoptosis (Brigelius-Flohe and Traber, 1999; Yoshida *et al.*, 2003). Other investigations have shown γ-tocotrienol-induced apoptosis in human breast cancer cells is associated with a slight disruption in mitochondrial membrane potential and cytochrome *c* release into the cytoplasm, but tocotrienols failed to alter Bcl-2 protein levels and did not induce caspase-9 activation (Takahashi and Loo, 2004), whereas others have reported that tocotrienols altered the ratio of Bax/Bcl-2 to induce mitochondrial disruption, cytochrome *c* release, and caspase-9 activation in human colon carcinoma cells (Agarwal *et al.*, 2004). At present, it is not understood why different intracellular signaling mechanisms appear to be involved in mediating the apoptotic effects of tocotrienols in different types of cancer cells. Additional studies are required to determine the specific intracellular sites of action that these different vitamin E compounds target in order to fully understand the specific mechanisms of action mediating their anticancer and apoptotic effects, as well as to further clarify the potential value of these compounds as chemotherapeutic agents in the prevention and treatment of breast cancer.

Since previous studies showed that tocotrienol-induced apoptosis was mediated through the activation of caspase-8 and -3 in neoplastic mammary epithelial cells and caspase-8 activation is associated with death receptors apoptotic signaling, additional studies were conducted to determine the exact death receptor/ligand involved in tocotrienol-induced apoptosis (Shah and Sylvester, 2004; Shah *et al.*, 2003). Treatment with high doses of TNF-α, FasL, TRAIL, or apoptosis-inducing Fas antibody failed to induce death in +SA mouse neoplastic mammary epithelial cells, indicating that this particular mammary tumor cell line is resistant to death receptor-induced apoptosis (Shah and Sylvester, 2004). Furthermore, treatment with a cytotoxic dose of γ-tocotrienol had no effect on total, membrane or cytosolic levels of Fas, FasL, or FADD, and did not induce translocation of Fas, FasL, or FADD from the cytosolic to the membrane fraction (Shah and Sylvester, 2004). However, it was also shown that tocotrienol treatment also induced a large decrease in phosphatidylinositol 3-kinase (PI3K)/PI3K-dependent kinase (PDK-1)/Akt mitogenic signaling and the subsequent downregulation of FLIP, an endogenous inhibitor of caspase-8 processing and activation (Shah and Sylvester, 2004). Since stimulation of the PI3K/PDK/Akt mitogenic pathway is associated with increased FLIP expression and enhanced cellular proliferation and survival, it was concluded that tocotrienol-induced caspase-8 activation and apoptosis in malignant +SA mammary epithelial cells is associated with suppression in PI3K/PDK-1/Akt mitogenic signaling and subsequent reduction in intracellular FLIP levels. These findings are significant in light of knowledge that resistance to death receptor-induced apoptosis is associated with enhanced tumorigenesis, multidrug resistance, and enhanced survival in a number of tumor cell types cancers (Hughes *et al.*, 1997; Reed, 1999; Yu *et al.*, 1999a). A reduction in the expression of death receptors and/or their ligands, as well as mutations within genes encoding for death receptors, is associated with insensitive or nonfunctional death receptors in various types of tumor cells (Fisher *et al.*, 1995; Hughes *et al.*, 1997; Reed, 1999; Watanabe-Fukunaga *et al.*, 1992). In addition, resistance to apoptotic stimuli can be acquired through upregulation and enhanced activity of various mitogen-dependent signaling pathways, particularly the PI3K/PDK/Akt mitogenic signaling pathway (Downward, 1998; Hermanto *et al.*, 2001; Kandel and Hay, 1999; Neve *et al.*, 2002; Panka *et al.*, 2001; Suhara *et al.*, 2001; Vanhaesebroeck and Waterfield, 1999; Varadhachary *et al.*, 1999, 2001). Studies have also shown that nonfunctional death receptors are associated with cancer cell escape and survival from normal immunosurveillance mechanisms (Hughes *et al.*, 1997; Reed, 1999; Yu *et al.*, 1999a). Furthermore, tocotrienols have been found to significantly inhibit human breast cancer cell growth and viability regardless of estrogen receptor status (Guthrie *et al.*, 1997; Nesaretnam *et al.*, 1995, 1998, 2000). Therefore, tocotrienols have a unique property of inducing apoptosis in a wide variety of tumor cells that possess insensitive or nonfunctional death receptors or display multidrug resistance to traditional chemotherapeutic agents.

Tocotrienol-induced apoptosis has been demonstrated in other forms of cancer. TRF has been shown to selectively induce cell cycle arrest and apoptosis in several types of human prostate cancer cells, while having little effect on normal prostate cell viability (Srivastava and Gupta, 2006). Others have shown that oral administration of tocotrienols can significantly inhibit liver and lung carcinogenesis in mice, as well as significant inhibit human hepatocellular carcinoma HepG2 cell growth and viability *in vitro* (Wada *et al.*, 2005). Tocotrienol treatment was also found to induce apoptosis in the rat hepatoma dRLh-84 cell line, by inducing caspase-8 and -3 activation (Sakai *et al.*, 2004), and induce apoptosis in mouse BNL 1ME A.7R.1 liver cancer cells without adversely affecting normal liver cell viability (Har and Keong, 2005). TRF-induced apoptosis in human colon carcinoma RKO cells was found to be associated with an increase expression of p53 and WAF1/p21 and proapoptotic Bcl-2 family member proteins, and a corresponding increase in mitochondrial cytochrome *c* release, followed by caspase-9 and -3 activation, but independent of mechanisms associated with cell cycle arrest (Agarwal *et al.*, 2004). Taken together, natural tocotrienols are significantly more potent than natural tocopherols in inducing apoptosis in various cancer cell types. However, there does not appear to be a common pathway or mechanism mediating tocotrienol-induced apoptosis in these different types of cancer cells. These findings are very interesting in that they suggest that tocotrienols may be effective in treating a broad range of tumor types regardless of their characteristic defects in normal apoptotic signaling mechanisms.

B. SYNTHETIC DERIVATIVES

Nearly 25 years ago, α-TS was first shown to inhibit cancer cell growth (Prasad and Edwards-Prasad, 1982). Since that time, numerous investigation have confirmed and extended these findings (Kline *et al.*, 2004; Neuzil *et al.*, 2002a,b). Similar to that observed in tocotrienol-induced apoptosis, the molecular mechanism mediating α-TS-induced apoptosis appear to be different in different types of cancers. In human MDA-MB-435 breast cancer cells, α-TS appears to induce mitochondrial stress apoptotic signaling mechanisms (You *et al.*, 2001; Yu *et al.*, 2003). Conversely, others have provided evidence that Fas death receptors mediate α-TS-induced apoptosis in these cells (Turley *et al.*, 1997a; Yu *et al.*, 1999a). In addition α-TS was found to initiate mitochondrial stress-mediated apoptosis in HER2/erbB2-overexpressing human breast cancer cells (Wang *et al.*, 2005). However, in BT-20 estrogen-negative human breast cancer cells, α-TS was found to induce cell cycle blockade by inhibiting cyclin A activity (Turley *et al.*, 1997b), while in MCF-7 human breast cancer cells, α-TS was found to induce apoptosis by lysosomal destabilization (Neuzil *et al.*, 2002b). Succinate derivatives of tocotrienols have also been shown to enhance apoptotic activity, as compared to the parent compound (Birringer *et al.*, 2003).

In prostate cancer cell lines, α-TS was found to cause mitochondrial cytochrome *c* release and apoptosis (Zu and Ip, 2003) and to inhibit prostate cancer cell invasiveness through the downregulation of extracellular matrix metalloproteinase enzymes (Zhang *et al.*, 2004a). However, α-TS treatment was found to induce apoptosis in human LNCaP and PC-3 prostate carcinoma cells, but not normal PrEC human prostate epithelial cells (Israel *et al.*, 2000). Treatment with agonistic anti-Fas antibody was found to trigger apoptosis in PC-3 cells, but not LNCaP and PrEC prostate cells. However, combined treatment with α-TS and agonistic anti-Fas antibodies resulted in an additive apoptotic effect on Fas-sensitive PC-3 cells, a synergistic apoptotic effect on LNCaP cells, but had no effect on normal PrEC cells. These findings strongly suggest that α-TS initiated apoptosis in neoplastic, but not normal human prostate cells, is mediated through Fas death receptor activation (Israel *et al.*, 2000).

In other cancer cell types, α-TS-induced apoptosis in malignant mesothelioma cells was found to be mediated through mitochondrial stress and caspase activation, and overexpression of tocopherol-associated protein-1 further enhanced α-TS cytotoxicity (Neuzil *et al.*, 2006), whereas α-TS was found to induce expression of DR4/DR5 in a p53-dependent manner and reestablishes sensitivity of TRAIL-resistant malignant mesothelioma cells to TRAIL-induced apoptosis (Tomasetti *et al.*, 2004, 2006). Studies also showed that α-TS induced apoptosis in osteosarcoma cells carrying wild-type p53 gene, while cells containing mutant p53 cells were resistant (Alleva *et al.*, 2005, 2006). Furthermore, α-TS treatment was found to sensitize resistant osteosarcoma cells to methotrexate-induced apoptosis (Alleva *et al.*, 2006). α-TS treatment is also effective in reducing cell viability of leukemia cells, apparently by activation of Fas death receptors, and elevations in the expression of p21 and protein kinase C activity (Bang *et al.*, 2001; Fariss *et al.*, 1994).

Additional vitamin E analogues are also effective in inducing apoptosis in cancer cells. α-TEA, a nonhydrolyzable ether-linked acetic acid analogue of α-tocopherol has been found to sensitize Fas resistant A2780S and A2780/cp70R human ovarian cancer cells to Fas-mediated apoptotic signaling (Yu *et al.*, 2006). α-TEA treatment increased intracellular levels of Fas, induced a conformational change in Bax, altered Bid cleavage, and induce the activation of caspases-8, -9, and -3 (Yu *et al.*, 2006). These effects were associated with decreased phosphorylation of Akt and ERK1/2, as well as causing a reduction in the levels of the caspase inhibitory proteins, FLIP and Survivin. Knockdown of Akt and ERK activity was found to further enhanced, while overexpression of Akt blocked α-TEA-induced apoptosis in these cells (Yu *et al.*, 2006). The antitumor actions of α-TEA *in vitro* have also been confirmed *in vivo*. α-TEA was found to be effective in reducing mammary and ovarian tumor burden and metastasis in experimental animals, suggesting that this vitamin E derivative is a strong candidate for use as a anticancer agent in humans (Anderson *et al.*, 2004b; Lawson *et al.*, 2003). Other derivatives of α-tocopherol, such as acetate and glutarate, have been found to be less potent, whereas maleate esters were found to be more potent than α-TS in inducing apoptosis in cancer cells (Birringer *et al.*, 2003). However, little information presently exists to explain the intracellular mechanism of action of these α-tocopherol derivatives. A nonhydrolyzable α-tocotrienol derivative, 6-*O*-carboxypropyl-α-tocotrienol, was found to be effective in reducing viability of A549 human lung adenocarcinoma cells by reducing Ras and RhoA activity, and reducing antiapoptotic Bcl-xL protein levels (Yano *et al.*, 2005). Since lung adenocarcinoma with mutation of ras genes characteristically has a poor prognosis, this tocotrienol derivative may be useful in treatment lung cancer patients.

VIII. CONCLUSION

Clinical and epidemiological studies have not produced consistent results or clearly demonstrated a protective effect of α-tocopherol against the development and progression of cancer (Bostick *et al.*, 1993; Sung *et al.*, 2003; Virtamo *et al.*, 2003). These finding are not surprising in light of the fact that α-tocopherol has not shown significant anticancer or apoptotic activity in the majority of experimental studies (Kline *et al.*, 2004; Neuzil *et al.*, 2002a; Sylvester and Shah, 2005b; Sylvester and Theriault, 2003). However, experimental data obtained in cell culture, and to a lesser extent animal studies, has provided strong evidence to support the suggestion that therapeutic treatment with natural tocotrienols, or synthetic vitamin E derivatives, such as α-TS and α-TEA, may provide significant value as therapeutic agents in the prevention and/or treatment of cancer. However, *in vivo* studies in animals and humans have so far produced conflicting results. It is possible that the discrepancies in the findings obtained from *in vitro* versus *in vivo* studies may result from inefficient or suboptimal delivery of natural tocotrienols and synthetic α-tocopherol derivatives to target tissues within the body. Further pharmaceutics research is clearly needed, particularly in the areas of vitamin E kinetics, formulation, and drug delivery. Important information obtained from these pharmaceutics studies could then be used to conduct meaningful and highly controlled long-term intervention studies in humans. Mechanistic studies must also be continued in order to achieve a more comprehensive understanding of how natural and analogue forms of vitamin E act in different types of tumor cells. The most intriguing aspect of these compounds is their characteristic selectivity to induce apoptosis in cancer, but not normal cells, and their unique ability to resensitize resistant tumor cells to death receptor ligands and/or other traditional cancer chemotherapeutic agents. These findings strongly suggest that tocotrienols and synthetic vitamin E derivatives can be used effectively as anticancer therapy either alone or in combination to enhance the therapeutic efficacy and reduce toxicity of other anticancer agents

REFERENCES

Agarwal, M. K., Agarwal, M. L., Athar, M., and Gupta, S. (2004). Tocotrienol-rich fraction of palm oil activates p53, modulates Bax/Bcl2 ratio and induces apoptosis independent of cell cycle association. *Cell Cycle* **3,** 205–211.

Alleva, R., Benassi, M. S., Tomasetti, M., Gellert, N., Ponticelli, F., Borghi, B., Picci, P., and Neuzil, J. (2005). Alpha-tocopheryl succinate induces cytostasis and apoptosis in osteosarcoma cells: The role of E2F1. *Biochem. Biophys. Res. Commun.* **331,** 1515–1521.

Alleva, R., Benassi, M. S., Pazzaglia, L., Tomasetti, M., Gellert, N., Borghi, B., Neuzil, J., and Picci, P. (2006). Alpha-tocopheryl succinate alters cell cycle distribution sensitising human osteosarcoma cells to methotrexate-induced apoptosis. *Cancer Lett.* **232,** 226–235.

Anderson, K., Lawson, K. A., Simmons-Menchaca, M., Sun, L., Sanders, B. G., and Kline, K. (2004a). Alpha-TEA plus cisplatin reduces human cisplatin-resistant ovarian cancer cell tumor burden and metastasis. *Exp. Biol. Med.* **229,** 1169–1176.

Anderson, K., Simmons-Menchaca, M., Lawson, K. A., Atkinson, J., Sanders, B. G., and Kline, K. (2004b). Differential response of human ovarian cancer cells to induction of apoptosis by vitamin E Succinate and vitamin E analogue, alpha-TEA. *Cancer Res.* **64,** 4263–4269.

Antonsson, B., and Martinou, J. D. (2000). The Bcl-2 protein family. *Exp. Cell Res.* **256,** 50–57.

Ashkenazi, A., and Dixit, V. M. (1998). Death receptors: Signaling and modulation. *Science* **281,** 1305–1308.

Bang, O. S., Park, J. H., and Kang, S. S. (2001). Activation of PKC but not of ERK is required for vitamin E-succinate-induced apoptosis of HL-60 cells. *Biochem. Biophys. Res. Commun.* **288,** 789–797.

Birringer, M., EyTina, J. H., Salvatore, B. A., and Neuzil, J. (2003). Vitamin E analogues as inducers of apoptosis: Structure-function relation. *Br. J. Cancer* **88,** 1948–1955.

Bostick, R. M., Potter, J. D., McKenzie, D. R., Sellers, T. A., Kushi, L. H., Steinmetz, K. A., and Folsom, A. R. (1993). Reduced risk of colon cancer with high intake of vitamin E: The Iowa Women's Health Study. *Cancer Res.* **53,** 4230–4237.

Brigelius-Flohe, R., and Traber, M. G. (1999). Vitamin E: Function and metabolism. *FASEB J.* **13,** 1145–1155.

Burton, G. W., and Ingold, K. U. (1981). Autoxidation of biological molecules. 1. The antioxidant activity of vitamin E and related chain-breaking phenolic antioxidants *in vivo*. *J. Am. Chem. Soc.* **103,** 6472–6477.

Cohen, G. M. (1997). Caspases: The executioners of apoptosis. *Biochem. J.* **326,** 1–16.

Cottrell, R. C. (1991). Introduction: Nutritional aspects of palm oil. *Am. J. Clin. Nutr.* **53,** 989S–1009S.

Demiralay, R., Gursan, N., and Erdem, H. (2006). The effects of erdosteine, N-acetylcysteine, and vitamin E on nicotine-induced apoptosis of pulmonary cells. *Toxicology* **219,** 197–207.

Djuric, Z., Heilbrun, L. K., Lababidi, S., Everett-Bauer, C. K., and Fariss, M. W. (1997). Growth inhibition of MCF-7 and MCF-10A human breast cells by alpha-tocopheryl hemisuccinate, cholesteryl hemisuccinate and their ether analogs. *Cancer Lett.* **111,** 133–139.

Downward, J. (1998). Mechanisms and consequences of activation of protein kinase B/Akt. *Curr. Opin. Cell Biol.* **10,** 262–267.

Evans, H. M., and Bishop, K. S. (1922). On the existence of a hetherto unrecognized dietary factor essential for reproduction. *Science* **56,** 650–651.

Fariss, M. W., Fortuna, M. B., Everett, C. K., Smith, J. D., Trent, D. F., and Djuric, Z. (1994). The selective antiproliferative effects of alpha-tocopheryl hemisuccinate and cholesteryl hemisuccinate on murine leukemia cells result from the action of the intact compounds. *Cancer Res.* **54,** 3346–3351.

Fisher, G. H., Rosenberg, F. J., Straus, S. E., Dale, J. K., Middleton, L. A., Lin, A. Y., Strober, W., Lenardo, M. J., and Puck, J. M. (1995). Dominant interfering Fas gene mutations impair apoptosis in a human autoimmune lymphoproliferative syndrome. *Cell* **81,** 935–946.

Galli, F., Stabile, A. M., Betti, M., Conte, C., Pistilli, A., Rende, M., Floridi, A., and Azzi, A. (2004). The effect of alpha- and gamma-tocopherol and their carboxyethyl hydroxychroman metabolites on prostate cancer cell proliferation. *Arch. Biochem. Biophys.* **423,** 97–102.

Goh, S. H., Hew, N. F., Norhanom, A. W., and Yadav, M. (1994). Inhibition of tumour promotion by various palm-oil tocotrienols. *Int. J. Cancer* **57,** 529–531.

Gonzalez-Polo, R. A., Soler, G., Alvarez, A., Fabregat, I., and Fuentes, J. M. (2003). Vitamin E blocks early events induced by 1-methyl-4-phenylpyridinium (MPP+) in cerebellar granule cells. *J. Neurochem.* **84,** 305–315.

Gould, M. N., Haag, J. D., Kennan, W. S., Tanner, M. A., and Elson, C. E. (1991). A comparison of tocopherol and tocotrienol for the chemoprevention of chemically induced rat mammary tumors. *Am. J. Clin. Nutr.* **53,** 1068S–1070S.
Green, D. R., and Reed, J. C. (1998). Mitochondria and apoptosis. *Science* **281,** 1309–1312.
Guthrie, N., Gapor, A., Chambers, A. F., and Carroll, K. K. (1997). Inhibition of proliferation of estrogen receptor-negative MDA-MB-435 and -positive MCF-7 human breast cancer cells by palm oil tocotrienols and tamoxifen, alone and in combination. *J. Nutr.* **127,** 544S–548S.
Gysin, R., Azzi, A., and Visarius, T. (2002). Gamma-tocopherol inhibits human cancer cell cycle progression and cell proliferation by down-regulation of cyclins. *FASEB J.* **16,** 1952–1954.
Haber, S. L., and Wissler, R. W. (1962). Effect of vitamin E on carcinogenicity of methylcholanthrene. *Proc. Soc. Exp. Biol. Med.* **111,** 774–775.
Har, C. H., and Keong, C. K. (2005). Effects of tocotrienols on cell viability and apoptosis in normal murine liver cells (BNL CL.2) and liver cancer cells (BNL 1ME A.7R.1), *in vitro*. *Asia Pac. J. Clin. Nutr.* **14,** 374–380.
Heermeier, K., Benedict, M., Li, M., Furth, P., Nunez, G., and Hennighausen, L. (1996). Bax and Bcl-xs are induced at the onset of apoptosis in involuting mammary epithelial cells. *Mech. Dev.* **56,** 197–207.
Hermanto, U., Zong, C. S., and Wang, L. H. (2001). ErbB2-overexpressing human mammary carcinoma cells display an increased requirement for the phosphatidylinositol 3-kinase signaling pathway in anchorage-independent growth. *Oncogene* **20,** 7551–7562.
Herrera, E., and Barbas, C. (2001). Vitamin E: Action, metabolism and perspectives. *J. Physiol. Biochem.* **57,** 43–56.
Hockenbery, D., Nunez, G., Milliman, C., Schreiber, R. D., and Korsmeyer, S. J. (1990). Bcl-2 is an inner mitochondrial membrane protein that blocks programmed cell death. *Nature* **348,** 334–336.
Horvath, P. M., and Ip, C. (1983). Synergistic effect of vitamin E and selenium in the chemoprevention of mammary carcinogenesis in rats. *Cancer Res.* **43,** 5335–5341.
Hosomi, A., Arita, M., Sato, Y., Kiyose, C., Ueda, T., Igarashi, O., Arai, H., and Inoue, K. (1997). Affinity for alpha-tocopherol transfer protein as a determinant of the biological activities of vitamin E analogs. *FEBS Lett.* **409,** 105–108.
Houston, M. (2005). Meta-analysis, metaphysics and mythology: Scientific and clinical perspective on the controversies regarding vitamin E for the prevention and treatment of disease in humans. *JANA* **8,** 4–7.
Houston, M. C. (2003). The role of antioxidants in the prevention and treatment of coronary heart disease. *JANA* **6,** 15–21.
Hsu, Y. T., Wolter, K. G., and Youle, R. J. (1997). Cytosol-to-membrane redistribution of Bax and Bcl-X(L) during apoptosis. *Proc. Natl. Acad. Sci. USA* **94,** 3668–3672.
Hughes, S. J., Nambu, Y., Soldes, O. S., Hamstra, D., Rehemtulla, A., Iannettoni, M. D., Orringer, M. B., and Beer, D. G. (1997). Fas/APO-1 (CD95) is not translocated to the cell membrane in esophageal adenocarcinoma. *Cancer Res.* **57,** 5571–5578.
Hunter, D. J., Manson, J. E., Colditz, G. A., Stampfer, M. J., Rosner, B., Hennekens, C. H., Speizer, F. E., and Willett, W. C. (1993). A prospective study of the intake of vitamins C, E, and A and the risk of breast cancer. *N. Engl. J. Med.* **329,** 234–240.
Ip, C. (1982). Dietary vitamin E intake and mammary carcinogenesis in rats. *Carcinogenesis* **3,** 1453–1456.
Israel, K., Yu, W., Sanders, B. G., and Kline, K. (2000). Vitamin E succinate induces apoptosis in human prostate cancer cells: Role for Fas in vitamin E succinate-triggered apoptosis. *Nutr. Cancer* **36,** 90–100.
Jaattela, M. (1999). Escaping cell death: Survival proteins in cancer. *Exp. Cell Res.* **248,** 30–43.
Kandel, E. S., and Hay, N. (1999). The regulation and activities of the multifunctional serine/threonine kinase Akt/PKB. *Exp. Cell Res.* **253,** 210–229.

Kayden, H. J., and Traber, M. G. (1993). Absorption, lipoprotein transport, and regulation of plasma concentrations of vitamin E in humans. *J. Lipid Res.* **34,** 343–358.

Kerr, J. F., Wyllie, A. H., and Currie, A. R. (1972). Apoptosis: A basic biological phenomenon with wide-ranging implications in tissue kinetics. *Br. J. Cancer* **26,** 239–257.

Khanna, S., Roy, S., Ryu, H., Bahadduri, P., Swaan, P. W., Ratan, R. R., and Sen, C. K. (2003). Molecular basis of vitamin E action: Tocotrienol modulates 12-lipoxygenase, a key mediator of glutamate-induced neurodegeneration. *J. Biol. Chem.* **278,** 43508–43515.

Kimmick, G. G., Bell, R. A., and Bostick, R. M. (1997). Vitamin E and breast cancer: A review. *Nutr. Cancer* **27,** 109–117.

King, M. M., and McCay, P. B. (1983). Modulation of tumor incidence and possible mechanisms of inhibition of mammary carcinogenesis by dietary antioxidants. *Cancer Res.* **43,** 2485s–2490s.

Klein, E. A. (2006). Chemoprevention of prostate cancer. *Annu. Rev. Med.* **57,** 49–63.

Kline, K., Yu, W., and Sanders, B. G. (2001). Vitamin E: Mechanisms of action as tumor cell growth inhibitors. *J. Nutr.* **131,** 161S–163S.

Kline, K., Lawson, K. A., Yu, W., and Sanders, B. G. (2003). Vitamin E and breast cancer prevention: Current status and future potential. *J. Mammary Gland Biol. Neoplasia* **8,** 91–102.

Kline, K., Yu, W., and Sanders, B. G. (2004). Vitamin E and breast cancer. *J. Nutr.* **134,** 3458S–3462S.

Kroemer, G. (1999). Mitochondrial control of apoptosis: An overview. *Biochem. Soc. Symp.* **66,** 1–15.

Lawson, K. A., Anderson, K., Menchaca, M., Atkinson, J., Sun, L., Knight, V., Gilbert, B. E., Conti, C., Sanders, B. G., and Kline, K. (2003). Novel vitamin E analogue decreases syngeneic mouse mammary tumor burden and reduces lung metastasis. *Mol. Cancer Ther.* **2,** 437–444.

Li, H., Zhu, H., Xu, C. J., and Yuan, J. (1998). Cleavage of BID by caspase 8 mediates the mitochondrial damage in the Fas pathway of apoptosis. *Cell* **94,** 491–501.

Li, P., Nijhawan, D., Budihardjo, I., Srinivasula, S. M., Ahmad, M., Alnemri, E. S., and Wang, X. (1997). Cytochrome c and d ATP-dependent formation of Apaf-1/caspase-9 complex initiates an apoptotic protease cascade. *Cell* **91,** 479–489.

Lizard, G., Miguet, C., Bessede, G., Monier, S., Gueldry, S., Neel, D., and Gambert, P. (2000). Impairment with various antioxidants of the loss of mitochondrial transmembrane potential and of the cytosolic release of cytochrome c occuring during 7-ketocholesterol-induced apoptosis. *Free Radic. Biol. Med.* **28,** 743–753.

Mazlan, M., Sue Mian, T., Mat Top, G., and Zurinah Wan Ngah, W. (2006). Comparative effects of alpha-tocopherol and gamma-tocotrienol against hydrogen peroxide induced apoptosis on primary-cultured astrocytes. *J. Neurol. Sci.* **243,** 5–12.

McIntyre, B. S., Briski, K. P., Tirmenstein, M. A., Fariss, M. W., Gapor, A., and Sylvester, P. W. (2000a). Antiproliferative and apoptotic effects of tocopherols and tocotrienols on normal mouse mammary epithelial cells. *Lipids* **35,** 171–180.

McIntyre, B. S., Briski, K. P., Gapor, A., and Sylvester, P. W. (2000b). Antiproliferative and apoptotic effects of tocopherols and tocotrienols on preneoplastic and neoplastic mouse mammary epithelial cells. *Proc. Soc. Exp. Biol. Med.* **224,** 292–301.

Miller, E. R., III, Pastor-Barriuso, R., Dalal, D., Riemersma, R. A., Appel, L. J., and Guallar, E. (2005). Meta-analysis: High-dosage vitamin E supplementation may increase all-cause mortality. *Ann. Intern. Med.* **142,** 37–46.

Muller, D. P. (1990). Antioxidant therapy in neurological disorders. *Adv. Exp. Med. Biol.* **264,** 475–484.

Muller, D. P., and Goss-Sampson, M. A. (1990). Neurochemical, neurophysiological, and neuropathological studies in vitamin E deficiency. *Crit. Rev. Neurobiol.* **5,** 239–263.

Muzio, M., Stockwell, B. R., Stennicke, H. R., Salvesen, G. S., and Dixit, V. M. (1998). An induced proximity model for caspase-8 activation. *J. Biol. Chem.* **273,** 2926–2930.

Nakamura, T., Aoyama, Y., Fujita, T., and Katsui, G. (1975). Studies on tocopherol derivatives: V. Intestinal absorption of several d,1–3,4–3H2-alpha-tocopheryl esters in the rat. *Lipids* **10,** 627–633.

Nass, S. J., Li, M., Amundadottir, L. T., Furth, P. A., and Dickson, R. B. (1996). Role for Bcl-xL in the regulation of apoptosis by EGF and TGF beta 1 in c-myc overexpressing mammary epithelial cells. *Biochem. Biophys. Res. Commun.* **227,** 248–256.

Nesaretnam, K., Khor, H. T., Ganeson, J., Chong, Y. H., Sundram, K., and Gapor, A. (1992). The effects of vitamin E tocotrienols from palm oil on chemically induced mammary carcingenesis in female rats. *Nutr. Res.* **12,** 879–892.

Nesaretnam, K., Guthrie, N., Chambers, A. F., and Carroll, K. K. (1995). Effect of tocotrienols on the growth of a human breast cancer cell line in culture. *Lipids* **30,** 1139–1143.

Nesaretnam, K., Stephen, R., Dils, R., and Darbre, P. (1998). Tocotrienols inhibit the growth of human breast cancer cells irrespective of estrogen receptor status. *Lipids* **33,** 461–469.

Nesaretnam, K., Dorasamy, S., and Darbre, P. D. (2000). Tocotrienols inhibit growth of ZR-75–1 breast cancer cells. *Int. J. Food Sci. Nutr.* **51,** S95–S103.

Neuzil, J., Weber, T., Gellert, N., and Weber, C. (2001a). Selective cancer cell killing by alpha-tocopheryl succinate. *Br. J. Cancer* **84,** 87–89.

Neuzil, J., Weber, T., Terman, A., Weber, C., and Brunk, U. T. (2001b). Vitamin E analogues as inducers of apoptosis: Implications for their potential antineoplastic role. *Redox Rep.* **6,** 143–151.

Neuzil, J., Kagedal, K., Andera, L., Weber, C., and Brunk, U. T. (2002a). Vitamin E analogs: A new class of multiple action agents with anti-neoplastic and anti-atherogenic activity. *Apoptosis* **7,** 179–187.

Neuzil, J., Zhao, M., Ostermann, G., Sticha, M., Gellert, N., Weber, C., Eaton, J. W., and Brunk, U. T. (2002b). Alpha-tocopheryl succinate, an agent with in vivo anti-tumour activity, induces apoptosis by causing lysosomal instability. *Biochem. J.* **362,** 709–715.

Neuzil, J., Dong, L. F., Wang, X. F., and Zingg, J. M. (2006). Tocopherol-associated protein-1 accelerates apoptosis induced by alpha-tocopheryl succinate in mesothelioma cells. *Biochem. Biophys. Res. Commun.* **343,** 1113–1117.

Neve, R. M., Holbro, T., and Hynes, N. E. (2002). Distinct roles for phosphoinositide 3-kinase, mitogen-activated protein kinase and p38 MAPK in mediating cell cycle progression of breast cancer cells. *Oncogene* **21,** 4567–4576.

Nicholson, D. W. (1999). Caspase structure, proteolytic substrates, and function during apoptotic cell death. *Cell Death Differ.* **6,** 1028–1042.

Niki, E., Saito, T., Kawakami, A., and Kamiya, Y. (1984). Inhibition of oxidation of methyl linoleate in solution by vitamin E and vitamin C. *J. Biol. Chem.* **259,** 4177–4182.

Oltvai, Z. N., and Korsmeyer, S. J. (1994). Checkpoints of dueling dimers foil death wishes. *Cell* **79,** 189–192.

Osakada, F., Hashino, A., Kume, T., Katsuki, H., Kaneko, S., and Akaike, A. (2004). Alpha-tocotrienol provides the most potent neuroprotection among vitamin E analogs on cultured striatal neurons. *Neuropharmacology* **47,** 904–915.

Packer, L. (1991). Protective role of vitamin E in biological systems. *Am. J. Clin. Nutr.* **53,** 1050S–1055S.

Packer, L., Weber, S. U., and Rimbach, G. (2001). Molecular aspects of α-tocotrienol antioxidant action and cell signaling. *J. Nutr.* **131,** 369S–373S.

Panka, D. J., Mano, T., Suhara, T., Walsh, K., and Mier, J. W. (2001). Phosphatidylinositol 3-kinase/Akt activity regulates c-FLIP expression in tumor cells. *J. Biol. Chem.* **276,** 6893–6896.

Prasad, K. N., and Edwards-Prasad, J. (1982). Effects of tocopherol (vitamin E) acid succinate on morphological alterations and growth inhibition in melanoma cells in culture. *Cancer Res.* **42,** 550–555.

Prasad, K. N., and Edwards-Prasad, J. (1992). Vitamin E and cancer prevention: Recent advances and future potentials. *J. Am. Coll. Nutr.* **11,** 487–500.
Putcha, G. V., Deshmukh, M., and Johnson, E. M., Jr. (1999). BAX translocation is a critical event in neuronal apoptosis: Regulation by neuroprotectants, BCL-2, and caspases. *J. Neurosci.* **19,** 7476–7485.
Qin, F., Yan, C., Patel, R., Liu, W., and Dong, E. (2006). Vitamins C and E attenuate apoptosis, beta-adrenergic receptor desensitization, and sarcoplasmic reticular Ca^{2+} ATPase down-regulation after myocardial infarction. *Free Radic. Biol. Med.* **40,** 1827–1842.
Raff, M. (1998). Cell suicide for beginners. *Nature* **396,** 119–122.
Reed, J. C. (1999). Dysregulation of apoptosis in cancer. *J. Clin. Oncol.* **17,** 2941–2953.
Roy, S., Lado, B. H., Khanna, S., and Sen, C. K. (2002). Vitamin E sensitive genes in the developing rat fetal brain: A high-density oligonucleotide microarray analysis. *FEBS Lett.* **530,** 17–23.
Sakai, M., Okabe, M., Yamasaki, M., Tachibana, H., and Yamada, K. (2004). Induction of apoptosis by tocotrienol in rat hepatoma dRLh-84 cells. *Anticancer Res.* **24,** 1683–1688.
Salvesen, G. S., and Dixit, V. M. (1999). Caspase activation: The induced-proximity model. *Proc. Natl. Acad. Sci. USA* **96,** 10964–10967.
Schneider, C. (2005). Chemistry and biology of vitamin E. *Mol. Nutr. Food Res.* **49,** 7–30.
Schulze-Osthoff, K., Ferrari, D., Los, M., Wesselborg, S., and Peter, M. E. (1998). Apoptosis signaling by death receptors. *Eur. J. Biochem.* **254,** 439–459.
Schwartz, J., and Shklar, G. (1992). The selective cytotoxic effect of carotenoids and alpha-tocopherol on human cancer cell lines *in vitro*. *J. Oral Maxillofac. Surg.* **50,** 367–373.
Sen, C. K., Khanna, S., Roy, S., and Packer, L. (2000). Molecular basis of vitamin E action. Tocotrienol potently inhibits glutamate-induced pp60(c-Src) kinase activation and death of HT4 neuronal cells. *J. Biol. Chem.* **275,** 13049–13055.
Sen, C. K., Khanna, S., and Roy, S. (2006). Tocotrienols: Vitamin E beyond tocopherols. *Life Sci.* **78,** 2088–2098.
Serbinova, E. A., and Packer, L. (1994). Antioxidant properties of alpha-tocopherol and alpha-tocotrienol. *Methods Enzymol.* **234,** 354–366.
Serbinova, E., Kagan, V., Han, D., and Packer, L. (1991). Free radical recycling and intramembrane mobility in the antioxidant properties of alpha-tocopherol and alpha-tocotrienol. *Free Radic. Biol. Med.* **10,** 263–275.
Shah, S., and Sylvester, P. W. (2004). Tocotrienol-induced caspase-8 activation is unrelated to death receptor apoptotic signaling in neoplastic mammary epithelial cells. *Exp. Biol. Med.* **229,** 745–755.
Shah, S., Gapor, A., and Sylvester, P. W. (2003). Role of caspase-8 activation in mediating vitamin E-induced apoptosis in murine mammary cancer cells. *Nutr. Cancer* **45,** 236–246.
Shah, S. J., and Sylvester, P. W. (2005). Tocotrienol-induced cytotoxicity is unrelated to mitochondrial stress apoptotic signaling in neoplastic mammary epithelial cells. *Biochem. Cell Biol.* **83,** 86–95.
Shi, H., Noguchi, N., and Niki, E. (1999). Comparative study on dynamics of antioxidative action of alpha-tocopheryl hydroquinone, ubiquinol, and alpha-tocopherol against lipid peroxidation. *Free Radic. Biol. Med.* **27,** 334–346.
Shklar, G. (1982). Oral mucosal carcinogenesis in hamsters: Inhibition by vitamin E. *J. Natl. Cancer Inst.* **68,** 791–797.
Shun, M. C., Yu, W., Gapor, A., Parsons, R., Atkinson, J., Sanders, B. G., and Kline, K. (2004). Pro-apoptotic mechanisms of action of a novel vitamin E analog (alpha-TEA) and a naturally occurring form of vitamin E (delta-tocotrienol) in MDA-MB-435 human breast cancer cells. *Nutr. Cancer* **48,** 95–105.
Sigounas, G., Anagnostou, A., and Steiner, M. (1997). dl-Alpha-tocopherol induces apoptosis in erythroleukemia, prostate, and breast cancer cells. *Nutr. Cancer* **28,** 30–35.

Srinivasula, S. M., Ahmad, M., Fernandes-Alnemri, T., and Alnemri, E. S. (1998). Autoactivation of procaspase-9 by Apaf-1-mediated oligomerization. *Mol. Cell* **1,** 949–957.

Srivastava, J. K., and Gupta, S. (2006). Tocotrienol-rich fraction of palm oil induces cell cycle arrest and apoptosis selectively in human prostate cancer cells. *Biochem. Biophys. Res. Commun.* **346,** 447–453.

Suhara, T., Mano, T., Oliverira, B. E., and Walsh, K. (2001). Phosphatidylinositol 3-kinase/Akt signaling coontrols endothelial cell sensitivity to Fas-mediated apoptosis via regulation of FLICE-inhibitory protein (FLIP). *Circ. Res.* **89,** 13–19.

Sun, F., Hayami, S., Ogiri, Y., Haruna, S., Tanaka, K., Yamada, Y., Tokumaru, S., and Kojo, S. (2000). Evaluation of oxidative stress based on lipid hydroperoxide, vitamin C and vitamin E during apoptosis and necrosis caused by thioacetamide in rat liver. *Biochim. Biophys. Acta* **1500,** 181–185.

Sung, L., Greenberg, M. L., Koren, G., Tomlinson, G. A., Tong, A., Malkin, D., and Feldman, B. M. (2003). Vitamin E: The evidence for multiple roles in cancer. *Nutr. Cancer* **46,** 1–14.

Susin, S. A., Lorenzo, H. K., Zamzami, N., Marzo, I., Snow, B. E., Brothers, G. M., Mangion, J., Jacotot, E., Costantini, P., Loeffler, M., Larochette, N., Goodlett, D. R., *et al.* (1999). Molecular characterization of mitochondrial apoptosis-inducing factor. *Nature* **397,** 441–446.

Suzuki, Y. J., Tsuchiya, M., Wassall, S. R., Choo, Y. M., Govil, G., Kagan, V. E., and Packer, L. (1993). Structural and dynamic membrane properties of alpha-tocopherol and alpha-tocotrienol: Implication to the molecular mechanism of their antioxidant potency. *Biochemistry* **32,** 10692–10699.

Sylvester, P. W., and Shah, S. (2005a). Intracellular mechanisms mediating tocotrienol-induced apoptosis in neoplastic mammary epithelial cells. *Asia Pac. J. Clin. Nutr.* **14,** 366–373.

Sylvester, P. W., and Shah, S. J. (2005b). Mechanisms mediating the antiproliferative and apoptotic effects of vitamin E in mammary cancer cells. *Front. Biosci.* **10,** 699–709.

Sylvester, P. W., and Theriault, A. (2003). Role of tocotrienols in the prevention of cardiovascular disease and breast cancer. *Curr. Top. Nutraceutical. Res.* **1,** 121–136.

Sylvester, P. W., Russell, M., Ip, M. M., and Ip, C. (1986). Comparative effects of different animal and vegetable fats fed before and during carcinogen administration on mammary tumorigenesis, sexual maturation, and endocrine function in rats. *Cancer Res.* **46,** 757–762.

Sylvester, P. W., McIntyre, B. S., Gapor, A., and Briski, K. P. (2001). Vitamin E inhibition of normal mammary epithelial cell growth is associated with a reduction in protein kinase C (alpha) activation. *Cell Prolif.* **34,** 347–357.

Sylvester, P. W., Nachnani, A., Shah, S., and Briski, K. P. (2002). Role of GTP-binding proteins in reversing the antiproliferative effects of tocotrienols in preneoplastic mammary epithelial cells. *Asia Pac. J. Clin. Nutr.* **11,** S452–S459.

Sylvester, P. W., Shah, S. J., and Samant, G. V. (2005). Intracellular signaling mechanisms mediating the antiproliferative and apoptotic effects of gamma-tocotrienol in neoplastic mammary epithelial cells. *J. Plant Physiol.* **162,** 803–810.

Takahashi, K., and Loo, G. (2004). Disruption of mitochondria during tocotrienol-induced apoptosis in MDA-MB-231 human breast cancer cells. *Biochem. Pharmacol.* **67,** 315–324.

Thornberry, N. A., and Lazebnik, Y. (1998). Caspases: Enemies within. *Science* **281,** 1312–1316.

Tomasetti, M., Rippo, M. R., Alleva, R., Moretti, S., Andera, L., Neuzil, J., and Procopio, A. (2004). Alpha-tocopheryl succinate and TRAIL selectively synergise in induction of apoptosis in human malignant mesothelioma cells. *Br. J. Cancer* **90,** 1644–1653.

Tomasetti, M., Andera, L., Alleva, R., Borghi, B., Neuzil, J., and Procopio, A. (2006). Alpha-tocopheryl succinate induces DR4 and DR5 expression by a p53-dependent route: Implication for sensitisation of resistant cancer cells to TRAIL apoptosis. *FEBS Lett.* **580,** 1925–1931.

Traber, M. G., and Packer, L. (1995). Vitamin E: Beyond antioxidant function. *Am. J. Clin. Nutr.* **62,** 1501S–1509S.

Turley, J. M., Fu, T., Ruscetti, F. W., Mikovits, J. A., Bertolette, D. C., III, and Birchenall-Roberts, M. C. (1997a). Vitamin E succinate induces Fas-mediated apoptosis in estrogen receptor-negative human breast cancer cells. *Cancer Res.* **57,** 881–890.

Turley, J. M., Ruscetti, F. W., Kim, S. J., Fu, T., Gou, F. V., and Birchenall-Roberts, M. C. (1997b). Vitamin E succinate inhibits proliferation of BT-20 human breast cancer cells: Increased binding of cyclin A negatively regulates E2F transactivation activity. *Cancer Res.* **57,** 2668–2675.

Vanhaesebroeck, B., and Waterfield, M. D. (1999). Signaling by distinct classes of phosphoinositide 3-kinases. *Exp. Cell Res.* **253,** 239–254.

Varadhachary, A. S., Peter, M. E., Perdow, S. N., Krammer, P. H., and Salgame, P. (1999). Selective up-regulation of phosphatidylinositol 3′-kinase activity in Th2 cells inhibits caspase-8 cleavage at the death-inducing complex: A mechanism for Th2 resistance from Fas-mediated apoptosis. *J. Immunol.* **163,** 4772–4779.

Varadhachary, A. S., Edidin, M., Hanlon, A. M., Peter, M. E., Krammer, P. H., and Salgame, P. (2001). Phosphatidylinositol 3′-kinase blocks CD95 aggregation and caspase-8 cleavage at the death-inducing signaling complex by modulating lateral diffusion of CD95. *J. Immunol.* **166,** 6564–6569.

Virtamo, J., Pietinen, P., Huttunen, J. K., Korhonen, P., Malila, N., Virtanen, M. J., Albanes, D., Taylor, P. R., and Albert, P. (2003). Incidence of cancer and mortality following alpha-tocopherol and beta-carotene supplementation: A postintervention follow-up. *JAMA* **290,** 476–485.

Vivekananthan, D. P., Penn, M. S., Sapp, S. K., Hsu, A., and Topol, E. J. (2003). Use of antioxidant vitamins for the prevention of cardiovascular disease: Meta-analysis of randomised trials. *Lancet* **361,** 2017–2023.

Wada, S., Satomi, Y., Murakoshi, M., Noguchi, N., Yoshikawa, T., and Nishino, H. (2005). Tumor suppressive effects of tocotrienol *in vivo* and *in vitro*. *Cancer Lett.* **229,** 181–191.

Walczak, H., and Krammer, P. H. (2000). The CD95 (APO-1/Fas) and the TRAIL (APO-2L) apoptosis systems. *Exp. Cell Res.* **256,** 58–66.

Wang, X. F., Witting, P. K., Salvatore, B. A., and Neuzil, J. (2005). Vitamin E analogs trigger apoptosis in HER2/erbB2-overexpressing breast cancer cells by signaling via the mitochondrial pathway. *Biochem. Biophys. Res. Commun.* **326,** 282–289.

Watanabe-Fukunaga, R., Brannan, C. I., Copeland, N. G., Jenkins, N. A., and Nagata, S. (1992). Lymphoproliferation disorder in mice explained by defects in Fas antigen that mediates apoptosis. *Nature* **356,** 314–317.

Wattenberg, L. W. (1972). Inhibition of carcinogenic and toxic effects of polycyclic hydrocarbons by phenolic antioxidants and ethoxyquin. *J. Natl. Cancer Inst.* **48,** 1425–1430.

Weber, T., Lu, M., Andera, L., Lahm, H., Gellert, N., Fariss, M. W., Korinek, V., Sattler, W., Ucker, D. S., Terman, A., Schroder, A., Erl, W., *et al.* (2002). Vitamin E succinate is a potent novel antineoplastic agent with high selectivity and cooperativity with tumor necrosis factor-related apoptosis-inducing ligand (Apo2 ligand) *in vivo*. *Clin. Cancer Res.* **8,** 863–869.

Yano, M., Kishida, E., Iwasaki, M., Kojo, S., and Masuzawa, Y. (2000). Docosahexaenoic acid and vitamin E can reduce human monocytic U937 cell apoptosis induced by tumor necrosis factor. *J. Nutr.* **130,** 1095–1101.

Yano, Y., Satoh, H., Fukumoto, K., Kumadaki, I., Ichikawa, T., Yamada, K., Hagiwara, K., and Yano, T. (2005). Induction of cytotoxicity in human lung adenocarcinoma cells by 6-O-carboxypropyl-alpha-tocotrienol, a redox-silent derivative of alpha-tocotrienol. *Int. J. Cancer* **115,** 839–846.

Yoshida, Y., Niki, E., and Noguchi, N. (2003). Comparative study on the action of tocopherols and tocotrienols as antioxidant: Chemical and physical effects. *Chem. Phys. Lipids* **123,** 63–75.

You, H., Yu, W., Sanders, B. G., and Kline, K. (2001). RRR-alpha-tocopheryl succinate induces MDA-MB-435 and MCF-7 human breast cancer cells to undergo differentiation. *Cell Growth Differ.* **12,** 471–480.

Yu, W., Israel, K., Liao, Q. Y., Aldaz, C. M., Sanders, B. G., and Kline, K. (1999a). Vitamin E succinate (VES) induces Fas sensitivity in human breast cancer cells: Role for Mr 43,000 Fas in VES-triggered apoptosis. *Cancer Res.* **59,** 953–961.

Yu, W., Simmons-Menchaca, M., Gapor, A., Sanders, B. G., and Kline, K. (1999b). Induction of apoptosis in human breast cancer cells by tocopherols and tocotrienols. *Nutr. Cancer* **33,** 26–32.

Yu, W., Sanders, B. G., and Kline, K. (2003). RRR-alpha-tocopheryl succinate-induced apoptosis of human breast cancer cells involves Bax translocation to mitochondria. *Cancer Res.* **63,** 2483–2491.

Yu, W., Shun, M. C., Anderson, K., Chen, H., Sanders, B. G., and Kline, K. (2006). Alpha-TEA inhibits survival and enhances death pathways in cisplatin sensitive and resistant human ovarian cancer cells. *Apoptosis* **11**(10), 1813–1823.

Zhang, M., Altuwaijri, S., and Yeh, S. (2004a). RRR-alpha-tocopheryl succinate inhibits human prostate cancer cell invasiveness. *Oncogene* **23,** 3080–3088.

Zhang, S., Lawson, K. A., Simmons-Menchaca, M., Sun, L., Sanders, B. G., and Kline, K. (2004b). Vitamin E analog alpha-TEA and celecoxib alone and together reduce human MDA-MB-435-FL-GFP breast cancer burden and metastasis in nude mice. *Breast Cancer Res. Treat.* **87,** 111–121.

Zou, H., Henzel, W. J., Liu, X., Lutschg, A., and Wang, X. (1997). Apaf-1, a human protein homologous to C. elegans CED-4, participates in cytochrome c-dependent activation of caspase-3. *Cell* **90,** 405–413.

Zu, K., and Ip, C. (2003). Synergy between selenium and vitamin E in apoptosis induction is associated with activation of distinctive initiator caspases in human prostate cancer cells. *Cancer Res.* **63,** 6988–6995.

Zurinah, W., Ngah, W., Jarien, Z., San, M. M., Marzuki, A., Top, G. D., Shamaan, N. A., and Kadir, K. A. (1991). Effect of tocotrienols on hepatocarcinogenesis induced by 2-acetylaminofluorene in rats. *Am. J. Clin. Nutr.* **53,** 1076S–1081S.

13

Vitamin E During Pre- and Postnatal Periods

Cathy Debier

Institut des Sciences de la Vie, Unité de Biochimie de la Nutrition
Université catholique de Louvain, Croix du Sud 2/8
B-1348 Louvain-la-Neuve, Belgium

Vitamin E is a fat-soluble nutrient that is extremely important during the early stages of life, from the time of conception to the postnatal development of the infant. The mechanisms involved in its placental and mammary uptake appear to be allowed by the presence of lipoprotein

Vitamins and Hormones, Volume 76

0083-6729/07 $35.00
DOI: 10.1016/S0083-6729(07)76013-2

receptors (LDL-receptor, VLDL-receptor, scavenger receptor class B type I) together with lipoprotein lipase at the placental and mammary barriers. In addition, α-tocopherol transfer protein has been described as playing an essential role in the selective transfer of *RRR*-α-tocopherol across the placenta. Lower α-tocopherol concentrations are found in cord blood as compared to maternal circulation. The ingestion of colostrum which contains very high levels of vitamin E is therefore of utmost importance to supply the newborn with an essential defense against oxygen toxicity. Pregnancy is sometimes associated with complications that may lead to a premature delivery of the baby. Preterm infants are usually facing an oxidative stress that is among others related to a deficiency in α-tocopherol, as it accumulates mainly during the third trimester of pregnancy. Despite vitamin E supplementation, preterm infants usually require significantly longer to replenish their serum α-tocopherol levels than full-term infants. The use of vitamin E as a therapeutic agent in preeclampsia, which induces high maternal and fetal morbidity and mortality, has been discussed in numerous papers. This disorder is indeed associated with an important oxidative stress in the placenta and maternal circulation. However, the most recent studies did not show a beneficial effect of vitamin E administration in this pathology.

I. INTRODUCTION

Vitamin E is a naturally occurring fat-soluble nutrient that is involved in several essential biological functions. Its antioxidant role is very well known as it accounts for most of the fat-soluble chain-breaking antioxidant in membranes and blood lipoprotein (Bramley *et al.,* 2000; Chow, 2004; Hacquebard and Carpentier, 2005). It is also essential for normal function of the immune system (Bramley *et al.*, 2000). Vitamin E has been shown to regulate cell proliferation and gene expression (Azzi *et al.*, 2004; Chow, 2004; Gimeno *et al.*, 2004; Hacquebard and Carpentier, 2005; Ricciarelli *et al.*, 2002; Traber, 2005; Zingg and Azzi, 2004).

The absorption as well as transport and the distribution of vitamin E are related to those of lipids (Lodge, 2005). There is indeed no specific carrier protein for vitamin E which circulates with lipids into lipoproteins (chylomicrons, VLDL, LDL, and HDL) (Hacquebard and Carpentier, 2005; Mardones and Rigotti, 2004). In man, HDL and LDL appear to carry the major part of tocopherol (Traber *et al.*, 1992). The most biologically active form of vitamin E is *RRR*-α-tocopherol (Bramley *et al.*, 2000; Chow, 2004; Hacquebard and Carpentier, 2005; Lodge, 2005; Traber and Arai, 1999). Some γ-tocopherol can also be found in the circulation but it is usually 5–10 times lower than the levels of α-tocopherol. The fact that the natural *RRR*-α-tocopherol form has the highest biopotency results from the discrimination that occurs in the liver and is due to a specific protein, the α-tocopherol transfer protein (α-TTP) which has a higher affinity for the *RRR* form. This protein enables the transfer of α-tocopherol across membranes and is also involved in its incorporation into nascent VLDL (Bramley *et al.*, 2000; Chow, 2004; Hacquebard and Carpentier, 2005; Lodge, 2005; Traber and Arai, 1999).

Vitamin E is extremely important during the early stages of life, from the time of conception to the postnatal development of the infant. The exact moment and place that require vitamin E during pregnancy are however not fully clarified, but several key stages have been identified. After fertilization, vitamin E has a positive impact, at least *in vitro*, on the development of early embryos (blastocyst stage) which are susceptible to be damaged by reactive oxygen species (ROS) (Ashworth and Antipatis, 2001; Wang *et al.*, 2002). Moreover, it appears to play an essential role during the critical period of implantation (Jishage *et al.*, 2001; Kaempf-Rotzoll *et al.*, 2002). It is also indispensable for placental maturation (Jishage *et al.*, 2005) and to protect the fetus from an irreparable oxidative stress (Jishage *et al.*, 2001; Kaempf-Rotzoll *et al.*, 2002). Finally, at birth, an adequate supply of vitamin E to the newborn is of utmost importance to protect the organism against oxygen toxicity and to stimulate the development of its immune system (Babinszky *et al.*, 1991; Ostrea *et al.*, 1986). This essential molecule must therefore be provided in adequate amounts throughout pregnancy and lactation.

This chapter focuses on the prenatal and postnatal aspects of the transfer of vitamin E. The dynamics and the mechanisms involved are among others developed. The use of vitamin E in critical situations that may be associated to the process of giving birth are also discussed.

II. Prenatal Transfer of Vitamin E

A. Biochemical Aspects of the Transfer

The placenta plays a key role in fetal nutrition. An *in vitro* study investigating the transfer of vitamin E across the term, normal human placenta, has shown that vitamin E is transferred slowly, at a rate of only 10% of L-glucose (Schenker *et al.*, 1998).

To date, the biochemical mechanisms involved in the placental transfer of this essential nutrient are not fully understood but are most probably related to the metabolism of lipoproteins. The utilization of maternal lipoproteins by human placental cells is allowed by the presence of lipoprotein receptors (LDL-receptor or LDL-R, VLDL-receptor or VLDL-R, scavenger receptor class B type I or SR-BI) together with lipoprotein lipase (LPL) at the placental barrier (Bonet *et al.*, 1992; Hatzopoulos *et al.*, 1998; Herrera, 2002, Herrera *et al.*, 2006;

Winkel *et al.*, 1980; Wittmaack *et al.*, 1995; Wyne and Woollett, 1998). These molecules are involved in the uptake of maternal lipids and may thus also allow the delivery of α-tocopherol to the placental cells.

α-TTP also appears to play an important role in the placental transfer of vitamin E. This protein has been first illustrated in the liver (Section I). However, its presence in the human term placenta has been discovered and seems to play a major role in maintaining adequate α-tocopherol levels in the fetus by facilitating the transfer of the hydrophobic nutrient between maternal and fetal circulation (Kaempf-Rotzoll *et al.*, 2003; Muller-Schmehl *et al.*, 2004). The expression of α-TTP in human term placenta may be linked to the fact that the fetus accumulates α-tocopherol mainly during the last part of pregnancy. It is expressed in the cytosol and nuclei of the trophoblast. Its presence in the endothelium of fetal vessels has however been described only in the experiment of Muller-Schmehl *et al.* (2004). It is important to notice that this protein has never been detected in mice placenta, reflecting differences in the transport mechanisms of α-tocopherol across the human syncytiotrophoblast and murine labyrinthine trophoblast.

According to Muller-Schmehl *et al.* (2004), placental α-TTP would be in charge of the stereoselective transfer of maternal VLDL-bound *RRR*-α-tocopherol which is taken up by the VLDL-R at the syncytiotrophoblast. As for the liver, the presence of α-TTP at the placental level seems to be involved in the discrimination of the transfer according to the structure of the molecule. Several studies have investigated whether the form of supplemental vitamin E (natural versus synthetic) administered during pregnancy affects the human placenta's ability to deliver the vitamin to the fetus (Acuff *et al.*, 1998). As most prenatal vitamin supplements contain *all*-rac-α-tocopheryl acetate as the vitamin E source, it is indeed important to investigate the efficiency of its transfer as compared with the natural *RRR* non-rac form. In the study of Acuff *et al.* (1998), deuterium-labeled isotopes of *RRR* and *all*-rac-α-tocopheryl acetate were given in capsules containing different doses of the two isomers (1:1, by weight) to women 5 days before delivery. At delivery, both maternal and cord plasma as well as lipoproteins presented higher concentrations of the *RRR* form as compared to the *all*-rac form. In addition, the ratio *RRR* to *all*-rac form was greater in cord blood (average 3.4) as compared to maternal blood (average 1.9). This difference means that the placental–fetal unit, the fetal liver, or both discriminate beyond what the maternal organism already does during first-pass metabolism of the stereoisomers of vitamin E. The free vitamin, rather than the acetate, has also been shown to be transferred best (Schenker *et al.*, 1998).

α-TTP seems to play an important role also at earlier stages of pregnancy. It has indeed been identified in the secretory epithelium of the human uterine glands as well as in the secondary yolk sac during early pregnancy, reflecting their importance in supplying vitamin E to the fetus during that period (Jauniaux *et al.*, 2004). α-TTP has also been localized in pregnant mouse uterus (Jishage *et al.*, 2001; Kaempf-Rotzoll *et al.*, 2002). Experiments on α-TTP−/− female mice have shown that the animals presented fertility problems with defective labyrinthine trophoblast, resulting in the death of the embryos at midgestation. The expression of α-TTP gene was observed in the uterus and its levels transiently increased after implantation, suggesting that the placental oxidative stress is protected by vitamin E which is essential for the formation of labyrinthine trophoblasts. The authors concluded that α-tocopherol plays a role in the critical process of implantation.

B. MATERNAL VITAMIN E LEVELS INCREASE DURING PREGNANCY

Vitamin E concentrations in maternal circulation increase during gestation (Al Senaidy, 1996; Gonzalez-Corbella *et al.*, 1998; Herrera *et al.*, 2004; Lachili *et al.*, 1999; Mino and Nagamatu, 1986; Oostenbrug *et al.*, 1998; von Mandach *et al.*, 1994). LDL is the major carrier for maternal α-tocopherol at the end of pregnancy, as reflected by the higher proportion of vitamin E in this kind of lipoprotein as compared to the others (chylomicrons, HDL, VLDL) (Acuff *et al.*, 1998).

The rise of α-tocopherol seems to be associated to a gestational secondary hyperlipidemia which is typical in humans. Indeed, the levels of cholesterol, phospholipids and especially triglycerides of all major lipoprotein classes increase significantly during pregnancy (Al Senaidy, 1996; Herrera, 2002; Mino and Nagamatu, 1986; Okazaki *et al.*, 2004; Oostenbrug *et al.*, 1998). Indeed, at late gestation, there is an enhanced lipolytic activity in maternal adipose tissue in order to transfer long-chain polyunsaturated fatty acids (PUFAs) to the fetus that is at its maximal growth (Herrera *et al.*, 2006). This lipolytic activity contributes to the development of maternal hyperlipidemia (Al Senaidy, 1996; Herrera *et al.*, 2006).

Whether the increase of vitamin E compensates exactly the increase of lipids will determine if the mother may be at risk for a rise in oxidative stress at the end of gestation. There is agreement that the levels of lipid peroxides in blood are generally higher in pregnant women than in nonpregnant women. During gestation, elevations appear by the second trimester and may lessen later in gestation, decreasing further after delivery. Lipid peroxides are also produced in placenta but their pattern of change over the course of pregnancy differs from the one of maternal blood (Little and Gladen, 1999). Some studies have shown that the ratio of plasma α-tocopherol concentration to plasma total lipid content remains unchanged during gestation, while the red blood cell (RBC) α-tocopherol levels (per unit of cells) somewhat decrease during the last trimester. As a result, the α-tocopherol that is available for biological function in the biomembrane may drop because of hyperlipidemia (Mino and Nagamatu, 1986). In another study, a significantly lower vitamin E/total lipid ratio was observed in maternal plasma at late pregnancy (right at delivery), reflecting a relative vitamin E deficiency, despite the fact that α-tocopherol concentrations were higher during that period (Lachili *et al.*, 1999). Antagonist results have however been found in the study of Oostenbrug *et al.* (1998), where the increase in plasma tocopherol levels was still observed even when adjusted for the increase in plasma lipid unsaturation, which brought the authors to the conclusion that the rise in phospholipid-PUFA levels during pregnancy does not compromise vitamin E status.

The dynamics throughout pregnancy appears to be different for γ-tocopherol. In the study of Al Senaidy (1996), the α-tocopherol levels steadily increased, reaching maximum level at late gestation and then gradually decreased after delivery. On the other hand, the optimum level of γ-tocopherol was at midgestation, followed by a progressive decrease until 1 month after delivery. These profile differences may be attributed to their different responses to the changes in maternal lipids during pregnancy.

C. VITAMIN E RESERVES OF FETUSES AND NEWBORNS

Even if vitamin E has been shown to be essential to the fetus during pregnancy, the net placental transfer appears to be low and is not or only slightly influenced by variations in maternal vitamin E intake or supplementation. As a result, lower vitamin E concentrations are found in cord as compared to maternal plasma. This phenomenon has been shown for various mammalian species including humans (Baydas *et al.*, 2002; Gonzalez-Corbella *et al.*, 1998; Kiely *et al.*, 1999; Léger *et al.*, 1998; Mahan and Vallet, 1997; Njeru *et al.*, 1994; Oostenbrug *et al.*, 1998; Pazak and Scholz, 1996; Sanchez-Vera *et al.*, 2005; Schenker *et al.*, 1998; von Mandach *et al.*, 1994).

Several potential mechanisms responsible for this phenomenon have been suggested. The paucity of lipids (triglycerides, total cholesterol, phospholipids) in the circulation of fetuses and neonates and its inefficient transplacental transfer may among others contribute to the low level of vitamin E in neonate plasma (Jauniaux *et al.*, 2004; Kiely *et al.*, 1999; Léger *et al.*, 1998; Pazak and Scholz, 1996). Cholesterol and triglyceride concentrations, which are strongly correlated to the levels of α-tocopherol, are indeed lower in cord as compared to maternal plasma (Kiely *et al.*, 1999). The use of lipid adjusted α-tocopherol is therefore considered to be more reliable in determining α-tocopherol status (Godel, 1989). Another potential explanation for the low levels of vitamin E in cord plasma may result from the fact that hepatic α-TTP may not be well developed in the fetus, leading to low levels of α-tocopherol in VLDL and hence in cord plasma (Kiely *et al.*, 1999). The saturation of placental α-TTP has also been suggested to account for the low effect of maternal supplementation on cord blood vitamin E levels, in case of high α-tocopherol concentrations on the maternal side of the placenta. This hypothesis has however been brought forward in a study investigating the transfer in swines and not in humans (Pinelli-Saavedra and Scaife, 2005). The conventional determination of plasma vitamin E may however not be relevant to assess the protection of cell membranes from lipid peroxidation (Cachia *et al.*, 1995). Accordingly, the measurement of vitamin levels in RBC seems to be a more appropriate indicator of the vitamin E status of the infant as it represents the site of action of the molecule (Kelly *et al.*, 1990). Vitamin E content in maternal RBC has been shown to account for fetal RBC content much better than plasma content does (Cachia *et al.*, 1995; Gonzalez-Corbella *et al.*, 1998; Léger *et al.*, 1998).

The important thing to figure out is whether the low plasma and RBC vitamin E levels of neonates reflect a potential deficiency. In the study of Léger *et al.* (1998), the ratio between the RBC vitamin E and the highly oxidizable carbons of PUFAs was shown to be much lower in neonates as compared to their mothers (1/1300 versus 1/500). This ratio's value may be an indicator for a deficiency in other cell membranes. The same tendency was also observed for plasma in Oostenbrug *et al.* (1998). In the study of Jain *et al.* (1996), vitamin E levels in the newborn RBC was similar to maternal RBC when expressed per unit of nmol/ml packed cells but was significantly lower when expressed per μmol total lipids. Elevated blood levels of vitamin E–quinone suggest an increased oxidative stress in newborns as compared to their mothers. Because vitamin E levels in RBC of newborns are significantly related to vitamin E levels in RBC of their mothers, an increase in vitamin E supplementation to mothers during pregnancy may increase vitamin E levels in the newborn and help impede the effect of extrauterine oxygen toxicity (Jain *et al.*, 1996).

Studies conducted on laboratory animals have allowed to show some aspects of the transfer of α-tocopherol that are impossible to investigate in humans for very obvious reasons. Its efficiency seems to differ among others from one tissue to the other and from one species to the other. In rats, for example, the transfer appears higher in lung and heart tissue as compared to liver for which it is very low, suggesting that there is some preferential incorporation into some tissues as compared to others. On the other hand, there is a sharp rise in liver vitamin E levels following the ingestion of colostrum (Pazak and Scholz, 1996). Other results have however been found for guinea pigs which seem to be characterized by higher placental transfer of α-tocopherol than that reported for other species (Hidiroglou *et al.*, 2001; Kelly *et al.*, 1992). In the study of Kelly *et al.* (1992), it was shown that, at late gestation, the guinea pig fetal liver appears to act as a storage site for α-tocopherol, the majority of which being released immediately following birth. In contrast, lung and brain vitamin E levels are relatively constant over the final period of gestation and during early neonatal life. The ontogeny of α-tocopherol in brain and lung appears to be similar to that of RBC, while plasma α-tocopherol content varies considerably and does not accurately reflect tissue α-tocopherol status. Surprisingly, fetal and maternal lung α-tocopherol concentrations are similar at all time points considered, whereas fetal liver α-tocopherol status is always considerably greater than maternal liver α-tocopherol content. By contrast, in pigs, all tissue levels of vitamin E appear to be low at birth (muscle, liver, lung, brain, heart, kidney) (Hidiroglou *et al.*, 1993).

III. POSTNATAL TRANSFER OF VITAMIN E

A. MAMMARY GLAND UPTAKE

Newborns being characterized by low levels of plasma α-tocopherol as compared to their mothers (von Mandach *et al.*, 1994), a supply through the consumption of colostrum and milk is thus very important, among others because their erythrocyte membrane is particularly susceptible to oxidative damage (Jain, 1989). Vitamin E also plays an important role in the development of the immature immune system of the newborn (Kolb and Seehawer, 1998; Rajaraman *et al.*, 1997).

As for the placental transfer, the mechanisms involved in the uptake of α-tocopherol by the mammary gland are not completely understood. An important part of α-tocopherol is transported into the LDL fraction, together with cholesterol (Acuff *et al.*, 1998). The LDL receptor is a well-known pathway for the uptake of cholesterol by tissues (Goldstein *et al.*, 1985). It also appear to play a role in the delivery of α-tocopherol to extrahepatic tissues (Hacquebard and Carpentier, 2005; Mardones and Rigotti, 2004). After binding, the lipoprotein is internalized into the endocytic compartment where it dissociates from the receptor which can then be recycled to the cell surface (Goldstein *et al.*, 1985). A study conducted by Monks *et al.* (2001) provided evidence that the mammary epithelium of the lactating mouse is also able to take up and internalize LDL from the plasma by a non-LDL receptor-mediated process.

The SR-BI is a cell surface receptor that can bind HDL and LDL and is specifically involved in the selective uptake of lipids from lipoproteins to cells (without internalization of the lipoprotein as opposed to the classical receptor-mediated endocytic pathway) (Fidge, 1999; Hacquebard and Carpentier, 2005; Landschulz *et al.*, 1996; Mardones and Rigotti, 2004; Stangl *et al.*, 1999). This receptor is expressed in high levels in the mammary gland of pregnant rats (Landschulz *et al.*, 1996) and has been suggested as a physiologically relevant cell surface receptor for α-tocopherol uptake (Mardones and Rigotti, 2004). In addition, another member of this family, CD36, has also been described in mammary epithelium (Aoki *et al.*, 1997). This protein has a high affinity for HDL, LDL, and VLDL (Calvo *et al.*, 1998) and has been suggested to be involved in the transport of fatty acids to the mammary cells. CD36 and SR-BI may thus both be involved in the uptake of vitamin E by the mammary gland.

α-Tocopherol can also be taken up by the mammary gland through the LPL pathway as suggested by Martinez *et al.* (2002). They indeed found an increase of mammary LPL activity and α-tocopherol concentrations in 20-day pregnant rats as compared to virgin rats. Mammary α-tocopherol and LPL activity were also higher in lactating rats as compared to litter-removed rats.

Finally, the presence of VLDL-R has been described in mammary epithelial cell lines (Simonsen *et al.*, 1994). In case this receptor is also present in breast epithelium *in vivo*, this mechanism could also contribute to the uptake of α-tocopherol.

As for placenta, the transfer appears selective, with the *RRR* form being preferentially incorporated in milk as compared to *all*-rac-α-tocopherol (Lauridsen *et al.*, 2002). The presence of an α-TTP like mechanism in the mammary gland that would facilitate α-tocopherol secretion into the milk is therefore not excluded (Debier and Larondelle, 2005; Lauridsen *et al.*, 2002).

B. Milk Vitamin E and the Effect of Suckling on Vitamin E Status of Newborns

Human milk and particularly colostrum contains very high concentrations of vitamin E. Even though term infants are born with only one-third of the vitamin E circulating levels of their mothers, they attain serum levels comparable to those of adults within 4–6 days of breast-feeding (Ostrea *et al.*, 1986). The seeming barrier in the fetus to access the antioxidants such as vitamin E thus appears to be rapidly corrected and the substances are replenished postnatally through breast-feeding. The same phenomenon has been observed in the plasma and tissues of other species such as pigs, rats, and seals (Debier *et al.*, 2002; Hidiroglou *et al.*, 1993; Lauridsen *et al.*, 2002; Pazak and Scholz, 1996).

Colostrum contains higher concentrations than mature milk (Chappell *et al.*, 1985; Macias and Schweigert, 2001; Schweigert *et al.*, 2004). Usually, with the progression of lactation, triglycerides and the percentage of medium-chain fatty acids increase, whereas tocopherols, cholesterol, and the percentage of long-chain PUFAs decrease in human milk (Barbas and Herrera, 1998; Boersma *et al.*, 1991). These changes indicate that the diameter of milk fat globules increases as milk matures (Bitman *et al.*, 1986; Boersma *et al.*, 1991; Ruegg and Blanc, 1981). Indeed, colostrum contains higher concentrations of components considered to be derived from the fat-soluble membrane (tocopherol, cholesterol, long-chain PUFAs). Vitamin E content as well as vitamin E/linoleic acid ratio are lower in mature milk as compared to colostrum, evidencing the efficient mechanism of mammary gland vitamin E uptake around parturition (Barbas and Herrera, 1998).

IV. VITAMIN E IN CRITICAL SITUATIONS

A. PRETERM INFANTS: MORE AT RISK FOR A VITAMIN E DEFICIENCY

Neonates are facing an increase in oxidative aggression through the process of childbirth. The fetus is indeed exposed *in utero* to an hypoxic environment and low concentrations of free radicals. The levels then increase dramatically after birth, which may cause an oxidative injury. The potential physiological oxidative stress resulting from the birth process may be particularly serious in preterm infants for whom the transfer of α-tocopherol as well as the maturation of the antioxidant enzymatic systems (superoxide dismutase, glutathione peroxidase, catalase) may not be complete. Indeed, vitamin E content of the fetus increases mainly during the last trimester of pregnancy together with the fetal fat mass (Baydas *et al.*, 2002; Bohles, 1997; Chan *et al.*, 1999).

Premature or low birth weight infants are usually characterized by lower plasma and RBC α-tocopherol levels as compared to term babies (Baydas *et al.*, 2002; Robles *et al.*, 2001). In addition, they require significantly longer to replenish their serum vitamin E levels than normal term infants, probably because of inadequate storage in adipose tissue, reduced ability to absorb vitamin E as well as higher vitamin E requirement (Kelly *et al.*, 1990; Robles *et al.*, 2001; von Mandach *et al.*, 1994). As an example, in a comparative study investigating the vitamin E levels in preterm versus term infants, from birth to 8 weeks of age, term infants were able to replete their vitamin E stores (in plasma and RBCs) within the first week following birth, whereas it was much longer for preterm babies (Kelly *et al.*, 1990). Over the 3 weeks following birth, RBC vitamin E concentrations in the premature infants increased to adult values, while plasma vitamin E concentration did not reach the adult range until 8 weeks postterm. These slow changes in plasma vitamin E status occurred even though the vitamin E intake of these infants was similar to that proving adequate for term infants. Slightly different results have however been obtained in the study of Kaempf and Linderkamp (1998), where they found that, even if preterm infants exhibited a subclinical or biochemical vitamin E deficiency in the first 6 weeks of life in plasma and buccal mucosal cells, the other cells (RBC, platelets, monocytes, and polymorphonuclear leukocytes) showed no such deficiency.

The oxidative state of term versus preterm neonates has been investigated in several studies which revealed higher levels of markers of oxidative stress (hydroperoxides, malondialdehyde MDA) in preterm as compared to term infants (Baydas *et al.*, 2002; Friel *et al.*, 2002; Ochoa *et al.*, 2003; Robles *et al.*, 2001). The high levels of markers of oxidative stress in preterm infants are inversely related to gestational age (Robles *et al.*, 2001). Oxidative injury (plasma MDA) is correlated with the mean daily oxygen exposure, and inversely correlated with plasma vitamin E, suggesting that oxygen radical disease during the first week may be related to antioxidant status (Friel *et al.*, 2002). There are indeed several diseases in very low birth weight infants that have been linked to free radical-mediated pathology. These include pulmonary oxygen injury, intraventricular hemorrhage, retinopathy of prematurity, necrotizing enterocolitis, hemolytic anemia, post-asphyxial central nervous system injury, and acute tubular necrosis (Olivares, 1995; Warner and Wispe, 1992). This deficiency may even be aggravated by the oxygen-based postnatal therapy that is regularly carried out on premature infants. In addition, the use of iron and PUFA supplementation in formula may also increase the risk (Gross and Gabriel, 1985).

Inferring that preterm infants are able to recover "normal" vitamin E levels after a few weeks following birth, assumes the fact that RBC vitamin E concentrations reflect those of other tissues and that the PUFA composition of the preterm infant's RBC is similar to that of adults, which may not necessarily be the case (Kelly *et al.*, 1990). Indeed, if the higher ratio between PUFAs and antioxidants, that is observed in newborns as compared to adults (Oostenbrug *et al.*, 1998) is also true for preterm infants, they would require higher vitamin E levels in their RBC membranes in order to protect the high amounts of PUFAs. This means that they could still be "vitamin E deficient" even though their concentrations have reached adult levels (Kelly *et al.*, 1990).

It is often recommended to supplement preterm infants during their first month of life (Steichen *et al.*, 1987). Nutritional intervention with vitamin E or other antioxidants may improve the antioxidant defenses of premature infants during the first week of their life (Friel *et al.*, 2002). According to the authors, this period may act as a "therapeutic window" where the effects of free radical-mediated damage could be alleviated by increasing vitamin E levels in plasma. However, they noticed that infants receiving the same dose of vitamin E responded with different plasma levels suggesting that infants do not necessarily handle vitamin E in the same manner (Friel *et al.*, 2002). Other authors like Kaempf and Linderkamp (1998) suggest that healthy breast-fed preterm infants do not need vitamin E supplementation.

B. VITAMIN E AND PREECLAMPSIA

Preeclampsia is a disorder that induces the highest maternal and fetal morbidity and mortality of all pregnancy-related complications (Gupta *et al.*, 2005). Pregnancy is always accompanied by an increased intake and utilization of oxygen which generates an increased oxidative burden. In preeclampsia, the maternal-circulating levels of lipid peroxides are increased and the net antioxidant activity is decreased as compared to normal (El-Salahy *et al.*, 2001; Hubel *et al.*, 1996; Walsh, 1994). There is therefore little doubt that oxidative stress is a significant contributor in the pathogenesis of this disorder (Perkins, 2006). The level of lipid peroxidation is positively correlated to the severity of preeclampsia (El-Salahy *et al.*, 2001). Source of circulating levels of lipid peroxides in pregnancy are the placenta (Kim *et al.*, 2006; Raijmakers *et al.*, 2004; Walsh, 1994), as well as maternal leukocytes and endothelium (Raijmakers *et al.*, 2004).

Because of its strong antioxidant power as well as its inhibiting action on the inflammatory response, vitamin E has been suggested to play a role in preventing preeclampsia (Raijmakers *et al.*, 2004). Some studies have indeed shown that vitamin E levels are higher in normotensive pregnant women than in those with mild or severe preeclampsia, suggesting that measurement of vitamin E concentration in plasma may be useful as a prognostic marker of the likely development of preeclampsia (Akyol *et al.*, 2000). However, other studies have shown opposite tendencies (Bowen *et al.*, 2001; Schiff *et al.*, 1996). Although preeclampsia is characterized by increased lipid peroxidation and diminished antioxidant capacity, there is no consensus regarding causality of lipid peroxidation (i.e., vitamin E deficiency) (Gupta *et al.*, 2005).

Several trials investigating the role of antioxidant supplementation in preeclampsia have been carried out. Small studies have shown a reduced incidence of preeclampsia in women at risk who were taking vitamins E and C supplements (Chappell *et al.*, 2002; Holmes and McCance, 2005). A very large study has been conducted by Poston *et al.* (2006) to investigate the potential effect of a supplementation of antioxidants (vitamin E: 400IU and vitamin C: 1000 mg, on a daily basis) from the second trimester of pregnancy until delivery in women identified as being at increased risk of preeclampsia. The results were quite surprising and did not confirm the previous studies. Concomitant supplementation with vitamins C and E did not prevent preeclampsia in women at risk but did increase the rate of babies born with a low birth weight. As such, use of these high-dose antioxidants does not seem to be justified in pregnancy (Poston *et al.*, 2006). These results provide an example of the lack of efficacy of high-dose antioxidants in prevention of disease despite consistent evidence for a state of oxidative stress.

REFERENCES

Acuff, R. V., Dunworth, R. G., Webb, L. W., and Lane, J. R. (1998). Transport of deuterium-labeled tocopherols during pregnancy. *Am. J. Clin. Nutr.* **67,** 459–464.

Akyol, D., Mungan, T., Gorkemli, H., and Nuhoglu, G. (2000). Maternal levels of vitamin E in normal and preeclamptic pregnancy. *Arch. Gynecol. Obstet.* **263,** 151–155.

Al Senaidy, A. M. (1996). Plasma α- and δ-tocopherol have different pattern during normal human pregnancy. *Mol. Cell. Biochem.* **154,** 71–75.

Aoki, N., Ishii, T., Ohira, S., Yamaguchi, Y., Negi, M., Adachi, T., Nakamura, R., and Matsuda, T. (1997). Stage specific expression of milk fat globule membrane glycoproteins in mouse mammary gland: Comparison of MFG-E8, butyrophilin, and CD36 with a major milk protein, β-casein. *Biochim. Biophys. Acta Gen. Subj.* **1334,** 182–190.

Ashworth, C. J., and Antipatis, C. (2001). Micronutrient programming of development throughout gestation. *Reproduction* **122,** 527–535.

Azzi, A., Gysin, R., Kempna, R., Munteanu, A., Villacorta, L., Visarius, T., and Zingg, J. M. (2004). Regulation of gene expression by α-tocopherol. *Biol. Chem.* **385,** 585–591.

Babinszky, L., Langhout, D. J., Verstegen, M. W. A., den Hartog, L. A., Joling, P., and Niewl, M. (1991). Effect of vitamin E and fat source in sows' diets on immune response of suckling and weaned piglets. *J. Anim. Sci.* **69,** 1833–1842.

Barbas, C., and Herrera, E. (1998). Lipid composition and vitamin E content in human colostrum and mature milk. *J. Physiol. Biochem.* **54,** 167–173.

Baydas, G., Karatas, F., Gursu, M. F., Bozkurt, H. A., Ilhan, N., Yasar, A., and Canatan, H. (2002). Antioxidant vitamin levels in term and preterm infants and their relation to maternal vitamin status. *Arch. Med. Res.* **33,** 276–280.

Bitman, J., Freed, L. M., Neville, M. C., Wood, D. L., Hamosh, P., and Hamosh, M. (1986). Lipid composition of prepartum human mammary secretion and postpartum milk. *J. Pediatr Gastroenterol. Nutr.* **5,** 608–615.

Boersma, E. R., Offringa, P. J., Muskiet, F. A., Chase, W. M., and Simmons, I. J. (1991). Vitamin E, lipid fractions, and fatty acid composition of colostrum, transitional milk, and mature milk: An international comparative study. *Am. J. Clin. Nutr.* **53,** 1197–1204.

Bohles, H. (1997). Antioxidative vitamins in prematurely and maturely born infants. *Int. J. Vitam. Nutr. Res.* **67,** 321–328.

Bonet, B., Brunzell, J. D., Gown, A. M., and Knopp, R. H. (1992). Metabolism of very low-density-lipoprotein triglyceride by human placental cells—the role of lipoprotein-lipase. *Metab. Clin. Exp.* **41,** 596–603.

Bowen, R. S., Moodley, J., Dutton, M. F., and Theron, A. J. (2001). Oxidative stress in pre-eclampsia. *Acta Obstet. Gynecol. Scand.* **80,** 719–725.

Bramley, P. M., Elmadfa, I., Kafatos, A., Kelly, F. J., Manios, Y., Roxborough, H. E., Schuch, W., Sheehy, P. J. A., and Wagner, K. H. (2000). Vitamin E. *J. Sci. Food Agric.* **80,** 913–938.

Cachia, O., Leger, C. L., Boulot, P., Vernet, F., Depaulet, A. C., and Descomps, B. (1995). Red-blood-cell vitamin-E concentrations in fetuses are related to but lower than those in mothers during gestation—a possible association with maternal lipoprotein (A) plasma-levels. *Am. J. Obstet. Gynecol.* **173,** 42–51.

Calvo, D., Gomez-Coronado, D., Suarez, Y., Lasuncion, M. A., and Vega, M. A. (1998). Human CD36 is a high affinity receptor for the native lipoproteins HDL, LDL, and VLDL. *J. Lipid Res.* **39,** 777–788.

Chan, D. K. L., Lim, M. S. F., Choo, S. H. T., and Tan, I. K. (1999). Vitamin E status of infants at birth. *J. Perinat. Med.* **27,** 395–398.

Chappell, J. E., Francis, T., and Clandinin, M. T. (1985). Vitamin A and E content of human milk at early stages of lactation. *Early Hum. Dev.* **11,** 157–167.

Chappell, L. C., Seed, P. T., Kelly, F. J., Briley, A., Hunt, B. J., Charnock-Jones, D. S., Mallet, A., and Poston, L. (2002). Vitamin C and E supplementation in women at risk of preeclampsia is associated with changes in indices of oxidative stress and placental function. *Am. J. Obstet. Gynecol.* **187,** 777–784.

Chow, C. K. (2004). Biological functions and metabolic fate of vitamin E revisited. *J. Biomed. Sci.* **11,** 295–302.

Debier, C., and Larondelle, Y. (2005). Vitamins A and E: Metabolism, roles and transfer to offspring. *Br. J. Nutr.* **93,** 153–174.

Debier, C., Pomeroy, P. P., Baret, P. V., Mignolet, E., and Larondelle, Y. (2002). Vitamin E status and the dynamics of its transfer between mother and pup during lactation in grey seals (*Halichoerus grypus*). *Can. J. Zool.* **80,** 727–737.

El-Salahy, E. M., Ahmed, M. I., El-Gharieb, A., and Tawfik, H. (2001). New scope in angiogenesis: Role of vascular endothelial growth factor (VEGF), NO, lipid peroxidation, and vitamin E in the pathophysiology of pre-eclampsia among Egyptian females. *Clin. Biochem.* **34,** 323–329.

Fidge, N. H. (1999). High density lipoprotein receptors, binding proteins, and ligands. *J. Lipid Res.* **40,** 187–201.

Friel, J. K., Widness, J. A., Jiang, T. N., Belkhode, S. L., Rebouche, C. J., and Ziegler, E. E. (2002). Antioxidant status and oxidant stress may be associated with vitamin E intakes in very low birth weight infants during the first month of life. *Nutr. Res.* **22,** 55–64.

Gimeno, A., Zaragoza, R., Vivo-Sese, I., Vina, J. R., and Miralles, V. J. (2004). Retinol, at concentrations greater than the physiological limit, induces oxidative stress and apoptosis in human dermal fibroblasts. *Exp. Dermatol.* **13,** 45–54.

Godel, J. C. (1989). Vitamin E status of northern Canadian newborns: Relation of vitamin E to blood lipids. *Am. J. Clin. Nutr.* **50,** 375–380.

Goldstein, J. L., Brown, M. S., Anderson, R. G., Russel, D. W., and Schneider, W. J. (1985). Receptor-mediated endocytosis: Concepts emerging from the LDL receptor system. *Annu. Rev. Cell Biol.* **1,** 1–39.

Gonzalez-Corbella, M. J., Lopez-Sabater, M. C., Castellote-Bargallo, A. I., Campoy-Folgoso, C., and Rivero-Urgell, M. (1998). Influence of caesarean delivery and maternal factors on fat-soluble vitamins in blood from cord and neonates. *Early Hum. Dev.* **53,** S121–S134.

Gross, S. J., and Gabriel, E. (1985). Vitamin E status in preterm infants fed human milk or infant formula. *J. Pediatr.* **106,** 635–639.

Gupta, S., Agarwal, A., and Sharma, R. K. (2005). The role of placental oxidative stress and lipid peroxidation in preeclampsia. *Obstet. Gynecol. Surv.* **60,** 807–816.

Hacquebard, M., and Carpentier, Y. A. (2005). Vitamin E: Absorption, plasma transport and cell uptake. *Curr. Opin. Clin. Nutr. Metab. Care* **8,** 133–138.

Hatzopoulos, A. K., Rigotti, A., Rosenberg, R. D., and Krieger, M. (1998). Temporal and spatial pattern of expression of the HDL receptor SR-BI during murine embryogenesis. *J. Lipid Res.* **39,** 495–508.

Herrera, E. (2002). Lipid metabolism in pregnancy and its consequences in the fetus and newborn. *Endocrine* **19,** 43–55.

Herrera, E., Ortega, H., Alvino, G., Giovannini, N., Amusquivar, E., and Cetin, I. (2004). Relationship between plasma fatty acid profile and antioxidant vitamins during normal pregnancy. *Eur. J. Clin. Nutr.* **58,** 1231–1238.

Herrera, E., Amusquivar, E., Lopez-Soldado, I., and Ortega, H. (2006). Maternal lipid metabolism and placental lipid transfer. *Horm. Res.* **65,** 59–64.

Hidiroglou, M., Farnworth, E., and Butler, G. (1993). Vitamin E and fat supplementation of sows and the effect on tissue vitamin E concentrations in their progeny. *Reprod. Nutr. Dev.* **33,** 557–565.

Hidiroglou, N., Madere, R., and McDowell, L. (2001). Maternal transfer of Vitamin E to fetal and neonatal guinea pigs utilizing a stable isotopic technique. *Nutr. Res.* **21,** 771–783.

Holmes, V. A., and McCance, D. R. (2005). Could antioxidant supplementation prevent pre-eclampsia? *Proc. Nutr. Soc.* **64,** 491–501.

Hubel, C. A., McLaughlin, M. K., Evans, R. W., Hauth, B. A., Sims, C. J., and Roberts, J. M. (1996). Fasting serum triglycerides, free fatty acids, and malondialdehyde are increased in preeclampsia, are positively correlated, and decrease within 48 hours post partum. *Am. J. Obstet. Gynecol.* **174,** 975–982.

Jain, S. K. (1989). Hyperglycemia can cause membrane lipid peroxidation and osmotic fragility in human red blood cells. *J. Biol. Chem.* **264,** 21340–21345.

Jain, S. K., Wise, R., and Bocchini, J. J. (1996). Vitamin E and vitamin E-quinone levels in red blood cells and plasma of newborn infants and their mothers. *J. Am. Coll. Nutr.* **15,** 44–48.

Jauniaux, E., Cindrova-Davies, T., Johns, J., Dunster, C., Hempstock, J., Kelly, F. J., and Burton, G. J. (2004). Distribution and transfer pathways of antioxidant molecules inside the first trimester human gestational sac. *J. Clin. Endocrinol. Metab.* **89,** 1452–1458.

Jishage, K., Arita, M., Igarashi, K., Iwata, T., Watanabe, M., Ogawa, M., Ueda, O., Kamada, N., Inoue, K., Arai, H., and Suzuki, H. (2001). α-Tocopherol transfer protein is important for the normal development of placental labyrinthine trophoblasts in mice. *J. Biol. Chem.* **276,** 1669–1672.

Jishage, K., Tachibe, T., Ito, T., Shibata, N., Suzuki, S., Mori, T., Hani, T., Arai, H., and Suzuki, H. (2005). Vitamin E is essential for mouse placentation but not for embryonic development itself. *Biol. Reprod.* **73,** 983–987.

Kaempf-Rotzoll, D. E., Igarashi, K., Aoki, J., Jishage, K., Suzuki, H., Tamai, H., Linderkamp, O., and Arai, H. (2002). α-Tocopherol transfer protein is specifically localized at the implantation site of pregnant mouse uterus. *Biol. Reprod.* **67,** 599–604.

Kaempf-Rotzoll, D. E., Horiguchi, M., Hashiguchi, K., Aoki, J., Tamai, H., Linderkamp, O., and Arai, H. (2003). Human placental trophoblast cells express α-tocopherol transfer protein. *Placenta* **24,** 439–444.

Kaempf, D. E., and Linderkamp, O. (1998). Do healthy premature infants fed breast milk need vitamin E supplementation: α- and γ-tocopherol levels in blood components and buccal mucosal cells. *Pediatr. Res.* **44,** 54–59.

Kelly, F. J., Rodgers, W., Handel, J., Smith, S., and Hall, M. A. (1990). Time course of vitamin-E repletion in the premature-infant. *Br. J. Nutr.* **63,** 631–638.

Kelly, F. J., Safavi, M., and Cheeseman, K. H. (1992). Tissue α-tocopherol status during late fetal and early neonatal life of the guinea-pig. *Br. J. Nutr.* **67,** 457–462.

Kiely, M., Cogan, P. F., Kearney, P. J., and Morrissey, P. A. (1999). Concentrations of tocopherols and carotenoids in maternal and cord blood plasma. *Eur. J. Clin. Nutr.* **53,** 711–715.

Kim, Y. H., Kim, C. H., Cho, M. K., Kim, K. M., Lee, S. Y., Ahn, B. W., Yang, S. Y., Kim, S. M., and Song, T. B. (2006). Total peroxyl radical-trapping ability and anti-oxidant vitamins of the umbilical venous plasma and the placenta in pre-eclampsia. *J. Obstet. Gynaecol. Res.* **32,** 32–41.

Kolb, E., and Seehawer, J. (1998). The development of the immune system and vitamin levels in the bovine fetus and neonate: A review including the effect of vitamins on the immune system. *Tierarztl. Umsch.* **53,** 723–730.

Lachili, B., Faure, H., Smail, A., Zama, N., Benlatreche, C., Favier, A., and Roussel, A. M. (1999). Plasma vitamin A, E, and β-carotene levels in adult post-partum Algerian women. *1: Int. J. Vitam. Nutr. Res.* **69,** 239–242.

Landschulz, K. T., Pathak, R. K., Rigotti, A., Krieger, M., and Hobbs, H. H. (1996). Regulation of scavenger receptor, class B, type I, a high density lipoprotein receptor, in liver and steroidogenic tissues of the rat. *J. Clin. Invest.* **98,** 984–995.

Lauridsen, C., Engel, H., Jensen, S. K., Craig, A. M., and Traber, M. G. (2002). Lactating sows and suckling piglets preferentially incorporate RRR- over all-rac-α-tocopherol into milk, plasma and tissues. *J. Nutr.* **132,** 1258–1264.

Léger, C. L., Dumontier, C., Fouret, G., Boulot, P., and Descomps, B. (1998). A short term supplementation of pregnant women before delivery does not improve significantly the vitamin E status of neonates—low efficiency of the vitamin E placental transfer. *Int. J. Vitam. Nutr. Res.* **68,** 293–299.

Little, R. E., and Gladen, B. C. (1999). Levels of lipid peroxides in uncomplicated pregnancy: A review of the literature. *Reprod. Toxicol.* **13,** 347–352.

Lodge, J. K. (2005). Vitamin E bioavailability in humans. *J. Plant Physiol.* **162,** 790–796.

Macias, C., and Schweigert, F. J. (2001). Changes in the concentration of carotenoids, vitamin A, α-tocopherol and total lipids in human milk throughout early lactation. *Ann. Nutr. Metab.* **45,** 82–85.

Mahan, D. C., and Vallet, J. L. (1997). Vitamin and mineral transfer during fetal development and the early postnatal period in pigs. *J. Anim. Sci.* **75,** 2731–2738.

Mardones, P., and Rigotti, A. (2004). Cellular mechanisms of vitamin E uptake: Relevance in α-tocopherol metabolism and potential implications for disease. *J. Nutr. Biochem.* **15,** 252–260.

Martinez, S., Barbas, C., and Herrera, E. (2002). Uptake of α-tocopherol by the mammary gland but not by white adipose tissue is dependent on lipoprotein lipase activity around parturition and during lactation in the rat. *Metab. Clin. Exp.* **51,** 1444–1451.

Mino, M., and Nagamatu, M. (1986). An evaluation of nutritional status of vitamin E in pregnant women with respect to red blood cell tocopherol level. *Int. J. Vitam. Nutr. Res.* **56,** 149–153.

Monks, J., Huey, P. U., Hanson, L., Eckel, R. H., Neville, M. C., and Gavigan, S. (2001). A lipoprotein-containing particle is transferred from the serum across the mammary epithelium into the milk of lactating mice. *J. Lipid Res.* **42,** 686–696.

Muller-Schmehl, K., Beninde, J., Finckh, B., Florian, S., Dudenhausen, J. W., Brigelius-Flohe, R., and Schuelke, M. (2004). Localization of α-tocopherol transfer protein in trophoblast, fetal capillaries' endothelium and amnion epithelium of human term placenta. *Free Radic. Res.* **38,** 413–420.

Njeru, C. A., McDowell, L. R., Wilkinson, N. S., Linda, S. B., and Williams, S. N. (1994). Prepartum and postpartum supplemental Dl-α-tocopheryl acetate effects on placental and mammary vitamin-E transfer in sheep. *J. Anim. Sci.* **72,** 1636–1640.

Ochoa, J. J., Ramirez-Tortosa, M. C., Quiles, J. L., Palomino, N., Robles, R., Mataix, J., and Huertas, J. R. (2003). Oxidative stress in erythrocytes from premature and full-term infants during their first 72h of life. *Free Radic. Res.* **37,** 317–322.

Okazaki, M., Usui, S., Tokunaga, K., Nakajima, Y., Takeichi, S., Nakano, T., and Nakajima, K. (2004). Hypertriglyceridemia in pregnancy does not contribute to the enhanced formation of remnant lipoprotein particles. *Clin. Chim. Acta* **339,** 169–181.

Olivares, M. (1995). Anémies nutritionnelles. *Nutrition du jeune enfant* **2,** 561–575.

Oostenbrug, G. S., Mensink, R. P., Al, M. D. M., van Houwelingen, A. C., and Hornstra, G. (1998). Maternal and neonatal plasma antioxidant levels in normal pregnancy, and the relationship with fatty acid unsaturation. *Br. J. Nutr.* **80,** 67–73.

Ostrea, E. M., Balun, J. E., Winkler, R., and Porter, T. (1986). Influence of breast-feeding on the restoration of the low serum concentration of vitamin E and β-carotene in the newborn infant. *Am. J. Obstet. Gynecol.* **154,** 1014–1017.

Pazak, H. E., and Scholz, R. W. (1996). Effects of maternal vitamin E and selenium status during the perinatal period on age-related changes in tissue concentration of vitamin E in rat pups. *Int. J. Vitam. Nutr. Res.* **66,** 126–133.

Perkins, A. V. (2006). Endogenous anti-oxidants in pregnancy and preeclampsia. *Aust. N. Z. J. Obstet. Gynecol.* **46,** 77–83.

Pinelli-Saavedra, A., and Scaife, J. R. (2005). Pre- and postnatal transfer of vitamins E and C to piglets in sows supplemented with vitamin E and vitamin C. *Livest. Prod. Sci.* **97,** 231–240.

Poston, L., Briley, A. L., Seed, P. T., Kelly, F. J., and Shennan, A. H. (2006). Vitamin C and vitamin E in pregnant women at risk for pre-eclampsia (VIP trial): Randomised placebo-controlled trial. *Lancet* **367,** 1145–1154.

Raijmakers, M. T. M., Dechend, R., and Poston, L. (2004). Oxidative stress and preeclampsia—Rationale for antioxidant clinical trials. *Hypertension* **44,** 374–380.

Rajaraman, V., Nonnecke, B. J., and Horst, R. L. (1997). Effects of replacement of native fat in colostrum and milk with coconut oil on fat-soluble vitamins in serum and immune function in calves. *J. Dairy Sci.* **80,** 2380–2390.

Ricciarelli, R., Zingg, J. M., and Azzi, A. (2002). The 80th anniversary of vitamin E: Beyond its antioxidant properties. *Biol. Chem.* **383,** 457–465.

Robles, R., Palomino, N., and Robles, A. (2001). Oxidative stress in the neonate. *Early Hum. Dev.* **65,** S75–S81.

Ruegg, M., and Blanc, B. (1981). The fat globule size distribution in human milk. *Biochim. Biophys. Acta* **666,** 7–14.

Sanchez-Vera, I., Bonet, B., Viana, M., Quintanar, A., and Lopez-Salva, A. (2005). Increased low-density lipoprotein susceptibility to oxidation in pregnancies and fetal growth restriction. *Obstet. Gynecol.* **106,** 345–351.

Schenker, S., Yang, Y., Perez, A., Acuff, R. V., Papas, A. M., Henderson, G., and Lee, M. P. (1998). Antioxidant transport by the human placenta. *Clin. Nutr.* **17,** 159–167.

Schiff, E., Friedman, S. A., Stampfer, M., Kao, L., Barrett, P. H., and Sibai, B. M. (1996). Dietary consumption and plasma concentrations of vitamin E in pregnancies complicated by preeclampsia. *Am. J. Obstet. Gynecol.* **175,** 1024–1028.

Schweigert, F. J., Bathe, K., Chen, F., Buscher, U., and Dudenhausen, J. W. (2004). Effect of the stage of lactation in humans on carotenoid levels in milk, blood plasma and plasma lipoprotein fractions. *Eur. J. Nutr.* **43,** 39–44.

Simonsen, A. C. W., Heegaard, C. W., Rasmussen, L. K., Ellgaard, L., Kjoller, L., Christensen, A., Etzerodt, M., and Andreasen, P. A. (1994). Very-low-density lipoprotein receptor from mammary-gland and mammary epithelial-cell lines binds and mediates endocytosis of M(R)-40,000 receptor-associated protein. *FEBS Lett.* **354,** 279–283.

Stangl, H., Hyatt, M., and Hobbs, H. (1999). Transport of lipids from high and low density lipoproteins via scavenger receptor-BI. *J. Biol. Chem.* **274,** 32692–32698.

Steichen, J. J., Krug-Wispe, S. K., and Tsang, R. C. (1987). Breastfeeding the low birth weight preterm infant. *Clin. Perinatol.* **14,** 131–171.

Traber, M. G. (2005). Vitamin E regulation. *Curr. Opin. Gastroenterol.* **21,** 223–227.

Traber, M. G., and Arai, H. (1999). Molecular mechanisms of vitamin E transport. *Annu. Rev. Nutr.* **19,** 343–355.

Traber, M. G., Cohn, W., and Muller, D. P. R. (1992). Absorption, transport and delivery to tissues. *In* "Vitamin E in Health and Disease" (L. Packer and J. Fuchs, Eds.), pp. 35–53. Marcel Dekker, New York.

von Mandach, U., Huch, R., and Huch, A. (1994). Maternal and cord serum vitamin-E levels in normal and abnormal pregnancy. *Int. J. Vit. Nutr. Res.* **64,** 26–32.

Walsh, S. W. (1994). Lipid-peroxidation in pregnancy. *Hypertens. Pregnancy* **13,** 1–32.

Wang, X., Falcone, T., Attaran, M. D., Goldberg, J. M., Agarwal, A., and Sharma, R. K. (2002). Vitamin C and Vitamin E supplementation reduce oxidative stress-induced embryo toxicity and improve the blastocyst development rate. *Fertil. Steril.* **78,** 1272–1277.

Warner, B. B., and Wispe, J. R. (1992). Free radical-mediated diseases in pediatrics. *Semin. Perinatol.* **16,** 47–57.

Winkel, C. A., Gilmore, J., MacDonald, P. C., and Simpson, E. R. (1980). Uptake and degradation of lipoproteins by human trophoblastic cells in primary culture. *Endocrinology* **107,** 1892–1898.

Wittmaack, F. M., Gafvels, M. E., Bronner, M., Matsuo, H., McCrae, K. R., Tomaszewski, J. E., Robinson, S. L., Strickland, D. K., and Strauss, J. F. (1995). Localization and regulation of the human very low density lipoprotein/apoprotein-E receptor: Trophoblast expression predicts a role for the receptor in placental lipid transport. *Endocrinology* **136,** 340–348.

Wyne, K. L., and Woollett, L. A. (1998). Transport of maternal LDL and HDL to the fetal membranes and placenta of the Golden Syrian hamster is mediated by receptor-dependent and receptor-independent processes. *J. Lipid Res.* **39,** 518–530.

Zingg, J. M., and Azzi, A. (2004). Non-antioxidant activities of vitamin E. *Curr. Med. Chem.* **11,** 1113–1133.

14

α-TOCOPHEROL: A MULTIFACETED MOLECULE IN PLANTS

SERGI MUNNÉ-BOSCH

Departament de Biologia Vegetal, Facultat de Biologia, Universitat de Barcelona Avinguda Diagonal 645, E-08028 Barcelona, Spain

α-Tocopherol, which belongs to the vitamin E group of compounds, is a lipophilic antioxidant that has a number of functions in plants. Synthesized from homogentisic acid and isopentenyl diphosphate in the chloroplast envelope, α-tocopherol is essential to maintain the integrity of photosynthetic membranes and plays a major role in photo- and antioxidant protection. α-Tocopherol scavenges lipid peroxy radicals, thereby preventing the propagation of lipid peroxidation, and protects lipids and other membrane components by physically quenching and reacting chemically with singlet oxygen. Moreover, given that α-tocopherol increases membrane rigidity,

0083-6729/07 $35.00
DOI: 10.1016/S0083-6729(07)76014-4

its concentration, together with that of the other membrane components, may be regulated to afford adequate fluidity for membrane function. Furthermore, recent studies on tocopherol-deficient plants indicate that α-tocopherol may affect cellular signaling in plants. Evidence thus far indicates that the effects of this compound in plant cellular signaling may be linked to the control of redox homeostasis. α-Tocopherol may influence cellular signaling by controlling the propagation of lipid peroxidation in chloroplasts, therefore modulating the formation of oxylipins such as the phytohormone jasmonic acid.

I. INTRODUCTION

Tocopherols and tocotrienols, collectively known as tocochromanols, are lipid-soluble molecules that belong to the group of vitamin E compounds, and they play an essential role in human nutrition and health. The term "vitamin E" was first introduced by Evans and Bishop (1922) to describe an important dietary factor for animal reproduction. More than 40 years passed before the vitamin E was associated with an antioxidant property (Epstein *et al.*, 1966). Thereafter, and especially during the last 30 years, the roles of this vitamin and other antioxidants have been extensively studied in humans. Reviews of several studies on tocopherols and tocotrienols during this period, which had distinct objectives but which were complementary, have been made. Most of the research into tocopherols and tocotrienols has focused on the fundamental chemistry that explains their antioxidant properties (Kamal-Eldin and Appelqvist, 1996), their specific location, and their role in biological membranes (Wang and Quinn, 2000), and particularly on the benefits of these compounds for human health (Fuchs, 1998; Pryor, 2000). Consequently, many studies have been carried out to obtain tocopherols and tocotrienols from plant extracts, chemically or via microalgal culture (Vandamme, 1992). Moreover, advances in the molecular biology of plants have provided new insights into the manipulation of the synthesis of α-tocopherol to increase the amount of vitamin E in food and thereby prevent nutrient deficiencies (DellaPenna, 1999, 2005; Grusak and DellaPenna, 1999). However, although plants are the sole source of vitamin E, fewer studies have focused on the function of tocopherols and tocotrienols in these organisms. In the 1990s, the function of α-tocopherol in plants was thought to be associated with only its antioxidant activity in the maintenance of membrane integrity (Fryer, 1992). It has been proposed that beside its photo- and antioxidant protective function α-tocopherol could play a role in cellular signaling in plants (Munné-Bosch, 2005; Munné-Bosch and Alegre, 2002; Munné-Bosch and Falk, 2004). Recent evidence obtained in tocopherol-deficient plants demonstrates that α-tocopherol actually plays a role in cellular signaling in plants, thus shedding new light on the multiple functions this compound has in plants. This chapter will focus on the functions of α-tocopherol in plants and will emphasize advances in research over the last few years.

II. OCCURRENCE AND ANTIOXIDANT FUNCTION OF α-TOCOPHEROL IN PLANTS

All vitamin E compounds (tocopherols and tocotrienols) are formed by a chromanol head group and a phytyl side chain. They are amphipatic molecules in which the hydrophobic phytyl tail associates with membrane lipids, and the polar chromanol head groups are exposed to the membrane surface. Tocopherols differ from tocotrienols only in the degree of saturation of their hydrophobic tails. Tocopherols have been found in photosynthetic bacteria, fungi, algae, plants, and animals, despite the inability of the latter to synthesize them (Grusak and DellaPenna, 1999; Lichtenthaler, 1968; Singh *et al.*, 1990). These compounds have been detected in all the photosynthetic organisms examined (Lichtenthaler, 1968), except in the cyanobacterium *Anacystis nidulans* (Omata and Murata, 1984). Tocopherols have been found in seeds and fruits, flowers (e.g., sepals and petals), roots, tubers, cotyledons, hypocotyls, stems, and particularly in leaves of higher plants. In leaves, they predominate in the α-tocopherol form, though in some cases significant amounts of its precursor, γ-tocopherol, have also been found. Plant tissues vary enormously in their total tocopherol content with total concentrations ranging from extremely low levels in potato tuber (<1-μg/g dry weight) to very high levels in leaves and seeds (>1-mg/g dry weight) (Munné-Bosch and Alegre, 2002).

α-Tocopherol is found in the chloroplasts of leaves. α-Tocopherol is synthesized in the chloroplast envelope from homogentisic acid and isopentenyl diphosphate (Arango and Heise, 1998; Soll *et al.*, 1985). Furthermore, α-tocopherol is found in plastoglobuli of the chloroplast stroma, where it is stored (Grumbach, 1983; Lichtenthaler *et al.*, 1981). Plastoglobuli are lipoprotein particles inside chloroplasts. During oxidative stress and senescence, they increase in number and form linkage groups that are attached to each other and remain continuous with the thylakoid membrane by extensions of the half-lipid bilayer (Austin *et al.*, 2006). Finally, α-tocopherol is found in thylakoid membranes, where it exerts its functions (Fryer, 1992; Havaux, 1998). Most of the α-tocopherol synthesized is partitioned between the chloroplastic envelope and the thylakoids, and is stored in plastoglobuli only in some cases, particularly during periods of oxidative stress and senescence. In spinach chloroplasts, one-third of the total α-tocopherol is located in envelope membranes, and the remaining two-thirds in thylakoids (Wise and Naylor, 1987). Within the membranes, tocopherols are thought to be restricted to the lipid matrix of thylakoids and chloroplastic envelope membranes, rather than to protein domains (Havaux, 1998; Havaux *et al.*, 2000).

By using tocopherol-deficient mutants of the cyanobacterium *Synechocystis*, Maeda *et al.* (2005) showed that tocopherol deficiency enhances the sensitivity to linoleic or linolenic acid treatments in combination with high light, consistent with tocopherols playing a crucial role in protecting cells from lipid peroxidation. The tocopherol-deficient mutants were also more susceptible to high-light treatment in the presence of sublethal levels of norflurazon, an inhibitor of carotenoid synthesis, suggesting carotenoids and tocopherols functionally interact in protecting *Synechocystis* from lipid peroxidation and high-light stress.

The principal role of tocopherols as antioxidants is believed to be in the scavenging of lipid peroxy radicals, which are responsible for propagating lipid peroxidation (Liebler, 1993). Ingold *et al.* (1986, 1990) showed that the antioxidant activity of tocopherols as free radical scavengers is associated with the ability to donate its phenolic hydrogen to lipid-free radicals and with specific requirements of the molecule, that is (1) the degree of methylation in the aromatic ring ($\alpha > \beta = \gamma > \delta$), (2) the size of the heterocyclic ring, (3) the stereochemistry at position 2, and (4) the length of the phytyl chain (optimum between 11 and 13 carbons). It is generally agreed that the antioxidant activities of tocopherols against lipid peroxidation *in vivo* are $\alpha > \beta > \gamma > \delta$ (Kamal-Eldin and Appelqvist, 1996).

Lipid peroxidation can be divided into three phases: initiation, propagation, and termination. It begins by generating a radical, generally an alkyl radical from a (poly)unsaturated fatty acid (PUFA) by the action of an initiator, which could be various reactive oxygen species (ROS), lipoxygenase, heat, light, and/or trace metals. The alkyl radical formed during initiation is highly reactive and combines with oxygen to form peroxy radicals. These, in turn, abstract the hydrogen from PUFAs and give rise to lipid hydroperoxides and new alkyl radicals, which propagate the reaction chain (Fig. 1). Alternatively, singlet oxygen ($^1O_2$) can react with PUFAs and give rise to lipid hydroperoxides. Lipid hydroperoxides can be converted, among other products, to (1) alcohols by hydroperoxide glutathione peroxidase or 2-Cys peroxiredoxin, which are localized in chloroplasts (Baier and Dietz, 1999); (2) jasmonic acid by allene oxide synthase and other enzymes found in chloroplasts, cytoplasm, and peroxisomes (Schaller, 2001); or (3) *n*-hexanal, traumatic acid, and other compounds by the action of hydroperoxide lyase and other enzymes (Matsui *et al.*, 1996, 1999). Tocopherols scavenge the lipid peroxy radical before it can abstract hydrogen from the target lipids. The chromanol heads of tocopherols lose a hydrogen atom that is given to the lipid peroxy radical, and tocopheroxyl radicals are then formed (Fig. 1). The ascorbate–glutathione cycle recycles tocopheroxyl radicals to tocopherols (Smirnoff and Wheeler, 2000).

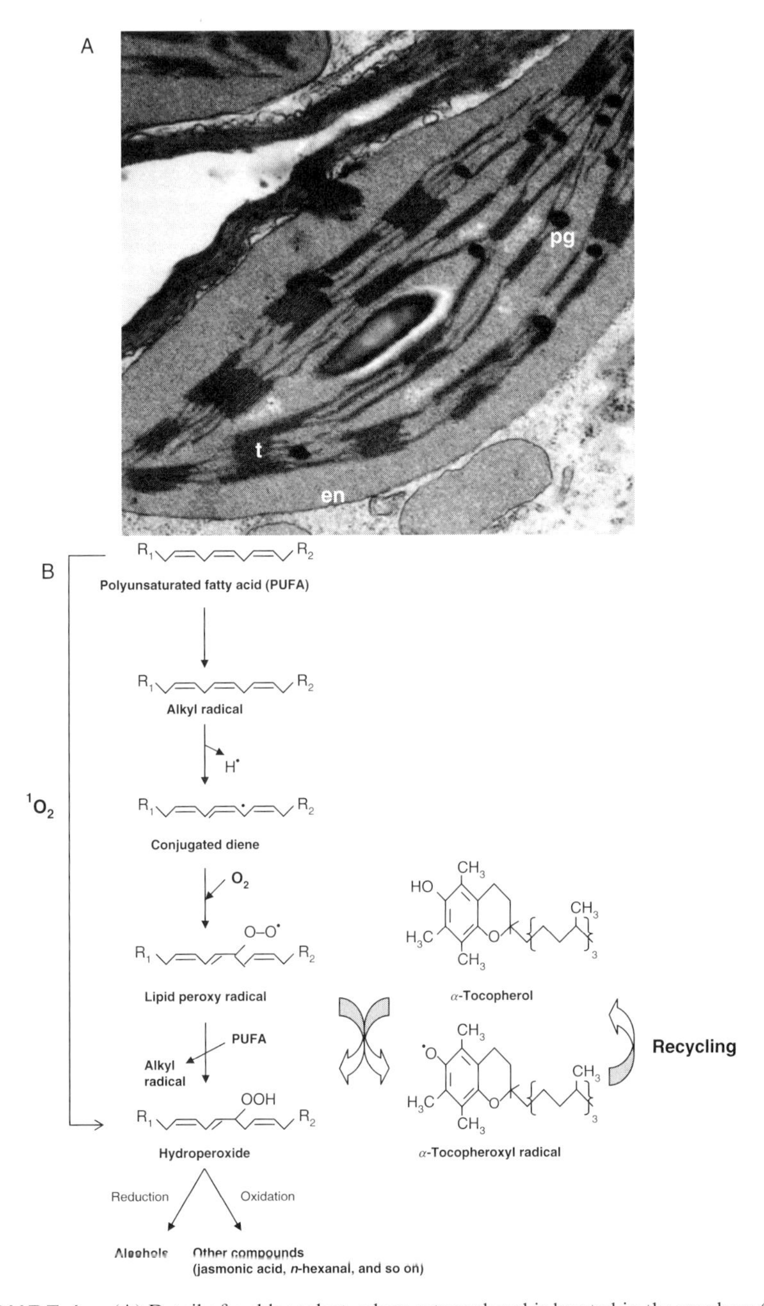

FIGURE 1. (A) Detail of a chloroplast, where α-tocopherol is located in the envelope (en), thylakoids (t), and plastoglobuli (pg). (B) α-Tocopherol prevents the propagation of lipid peroxidation in photosynthetic thylakoid membranes by scavenging lipid peroxy radicals. Photograph is a courtesy from Tana Jubany.

Tocopherols also play a key role as antioxidants because they physically quench or chemically scavenge $^1O_2$. Production of $^1O_2$ in thylakoids occurs because of the interaction of triplet excited reaction center chlorophyll (^{3}P680*) with $^3O_2$. $^1O_2$ production is enhanced when photon energy is in excess of the CO_2 assimilation and the intersystem electron carriers Q_A, Q_B, or plastoquinone of photosystem II occur at the reduced state (Hideg *et al.*, 2000; Melis, 1999). Alternatively, $^1O_2$ has also been proposed as a by-product of membrane lipid peroxidation, arising from disproportionation of peroxy radicals (Cadenas, 1989; Halliwell, 1981). $^1O_2$ can oxidize membrane lipid, protein, amino acids, nucleic acids, nucleotides, pyridine nucleotides, carbohydrates, and thiols (Halliwell and Gutteridge, 1999; Straight and Spikes, 1985). α-Tocopherol can physically quench and therefore deactivate $^1O_2$ in chloroplasts. During quenching, $^1O_2$ is deactivated to $^3O_2$ through a charge transfer mechanism. An electron is lost from the tocopherol and is donated to the electron-deficient $^1O_2$, thereby forming a charge transfer exciplex, which subsequently undergoes intersystem crossing and then dissociates into α-tocopherol and $^3O_2$ (Thomas and Foote, 1978; Yamauchi and Matsushita, 1977). It has been estimated that, before being degraded, one molecule of α-tocopherol can deactivate up to 120 $^1O_2$ molecules by resonance energy transfer (Fahrenholtz *et al.*, 1974). Furthermore, tocopherols also react chemically with $^1O_2$ and are destroyed (Fukuzawa *et al.*, 1997). The reaction occurs through an intermediate hydroperoxydienone, which decomposes to form tocopherol quinone and tocopherol quinone epoxides (Murkovic *et al.*, 1997; Neely *et al.*, 1988) (Fig. 2). α-Tocopherol quinone can be enzymatically converted to α-tocopherol quinol in an NADH- or NADPH-dependent reaction (Kruk and Strzalka, 1995). Experiments using model membranes suggest that α-tocopherol quinol is a potent antioxidant comparable to α-tocopherol (Bindoli *et al.*, 1985; Kruk *et al.*, 1997a; Mukai *et al.*, 1992). However, further experiments on α-tocopherol quinol are required to confirm its antioxidant activity *in vivo*.

III. PHOTOPROTECTIVE FUNCTION OF α-TOCOPHEROL IN PLANTS

The advent of photosynthesis is a central event in the early development of life on Earth. Photosynthesis is all about the collection of solar energy and its conversion into chemical energy. As with other natural light processors such as the human eye, photosynthesis would have not been possible without a safety valve that dissipates excess excitation energy in a harmless way. Photosynthesis in nature thus operates in a constantly shifting balance between efficient capture of solar energy and quick loss of that energy when it is captured in excess. Under optimal growth conditions, plants use most of the absorbed energy for photosynthesis, and photoprotection

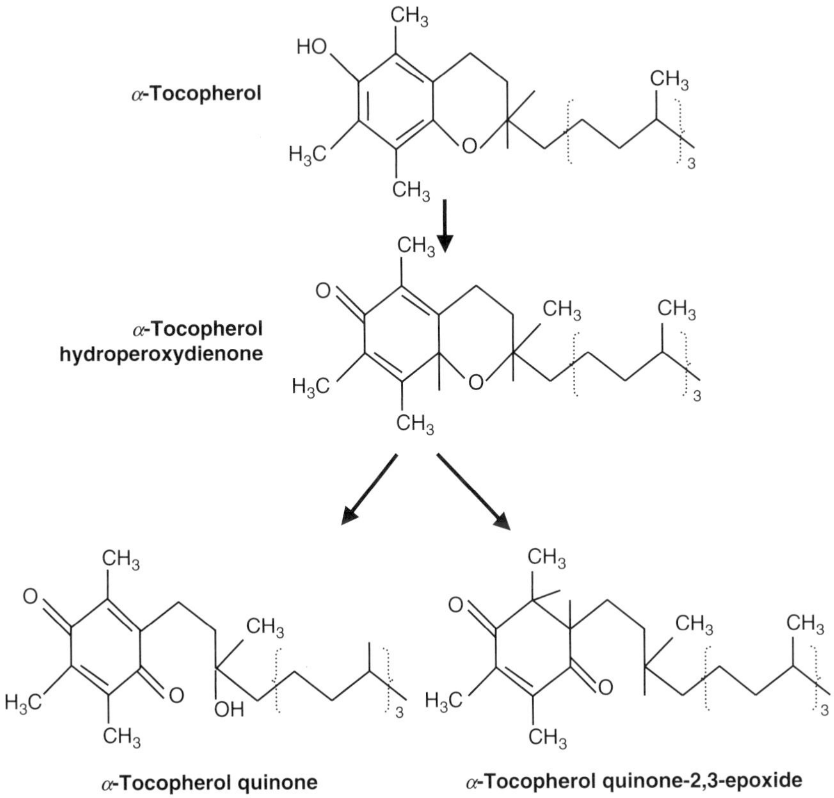

FIGURE 2. Scavenging of singlet oxygen by α-tocopherol and formation of α-tocopherol quinone.

mechanisms are hardly needed. On the other hand, when plants are exposed to adverse environmental conditions (drought, extreme temperatures, nutrient deficit, air pollutants, or pathogen attack), photosynthesis is reduced and photoprotection mechanisms are needed to allow safe dissipation of excess energy.

Plants have evolved several mechanisms to avoid or to get rid of excess energy in photosynthetic membranes. These mechanisms include, among others, alterations in leaf structure (Terashima *et al.*, 2001; Weston *et al.*, 2000), leaf and chloroplast movements (Kasahara *et al.*, 2002; Satter and Galston, 1981), changes in antenna size and pigment composition (Demmig-Adams and Adams, 2002; Havaux, 1998), light attenuation by anthocyanins or other compounds (Steyn *et al.*, 2002), and the regulation of light energy utilization and dissipation (Asada, 1999; Munné-Bosch *et al.*, 2005; Niyogi, 1999). The formation of ROS and its consequent detoxification by antioxidants represents an efficient mechanism of dissipation of excess energy in chloroplasts

(Asada, 2006). Consequently, the detoxification of $^1O_2$ and other ROS by tocopherols afford protection to the photosynthetic apparatus and it serves to dissipate excess energy in chloroplasts.

The formation of α-tocopherol quinone can play an additional role in photoprotection. Using lyophilized, petroleum ether-extracted thylakoid membranes that preserve the structure of thylakoids, Kruk *et al.* (1998) found that α-tocopherol and α-tocopherol quinone interact with photosynthetic electron transport. It was suggested that cyclic electron transport around photosystem II was inhibited by α-tocopherol and stimulated by α-tocopherol quinone (Kruk *et al.*, 1997b). Later Kruk and Strzalka (2001) confirmed that α-tocopherol quinone efficiently oxidizes the reduced cytochrome b559, and thus play a role in cyclic electron flow around photosystem II when the photosynthetic electron transport chain is over-reduced. Therefore, this compound contributes to the dissipation of excess energy in thylakoids, thereby conferring photoprotection onto photosynthetic apparatus.

Furthermore, it has been suggested that the presence of α-tocopherol in thylakoids could reduce the permeability of these membranes to ions. It could therefore affect the maintenance of the light-generated transmembrane proton gradient (Fryer, 1992), which is responsible for ATP synthesis, and the conversion of violaxanthin to zeaxanthin in the xanthophyll cycle, which is involved in the harmless dissipation of excess excitation energy in thylakoids (Demmig-Adams and Adams, 1992; Eskling *et al.*, 1997; Horton *et al.*, 1996). Therefore, an accumulation of α-tocopherol in thylakoids could afford photoprotection by (1) scavenging $^1O_2$ and other ROS, (2) activating cyclic electron flow around photosystem II through the formation of α-tocopherol quinone, and (3) reducing the permeability of thylakoid membranes to protons, which would favor the acidification of the thylakoid lumen in high light and activate violaxanthin de-epoxidase.

By using tocopherol-deficient mutants, Havaux *et al.* (2005) showed that tocopherols protect *Arabidopsis thaliana* plants against photooxidative stress. Leaf disks of two tocopherol mutants, a tocopherol cyclase mutant (*vte1*) and a homogentisate phytyl transferase mutant (*vte2*), were exposed to high-light stress at low temperatures, which resulted in bleaching and lipid photodestruction. However, this was not observed in whole plants exposed to long-term high-light stress, unless the stress conditions were very severe, suggesting compensatory mechanisms for tocopherol deficiency under physiological conditions. These authors identified two such compensatory mechanisms: xanthophyll cycle-dependent energy dissipation by nonphotochemical quenching of photosystem II and synthesis of zeaxanthin. Also, it has been shown that *A. thaliana* mutants lacking xanthophyll cycle-dependent energy dissipation accumulate higher amounts of tocopherols under high light (Golan *et al.*, 2006). It appears therefore that tocopherols are part of an intricate network of photoprotection mechanisms that act in concert to protect photosynthetic membranes from photooxidation. Munné-Bosch and Cela (2006) showed that xanthophyll cycle-dependent excess energy dissipation precedes oxidation of α-tocopherol to its quinone in water-stressed sage plants, thus indicating that the formation of ROS and its consequent detoxification by α-tocopherol become relevant when violaxanthin is completely de-epoxidized to zeaxanthin and xanthophyll cycle-dependent energy dissipation cannot increase further.

IV. α-TOCOPHEROL AND THE STABILITY OF PHOTOSYNTHETIC MEMBRANES

α-Tocopherol strongly interacts with membrane lipids and increases the rigidity of the membrane. Thus, its presence at high amounts in photosynthetic membranes during specific periods of plant development or stress could be detrimental to plant function in terms of membrane stability. However, plants show changes not only in α-tocopherol content but also in the lipid composition of the membrane, and in the concentrations of β-carotene and other lipophilic antioxidants, which also affect membrane fluidity and counteract the rigidifying effects of α-tocoperol (Munné-Bosch and Alegre, 2000; Quartacci *et al.*, 1997). The degree of lipid peroxidation in membranes has been shown to depend not only on the amount of ROS and antioxidants but also on the composition of the membrane (McKersie *et al.*, 1990). Increased fatty acid unsaturation in the membranes, which tends to maintain the liquid crystalline phase, has been correlated with increased stress tolerance in plants (McKersie *et al.*, 1988).

V. ROLE OF α-TOCOPHEROL IN CELLULAR SIGNALING

Recent studies in mutant and transgenic plants have made a significant contribution to furthering our understanding of the role of antioxidants in plants. In particular, it has been shown that antioxidants such as ascorbic acid and glutathione regulate signal transduction and gene expression, particularly in plant responses to stress (Ball *et al.*, 2004; Chen and Gallie, 2004). Furthermore, given that α-tocopherol affects oxidative stress and the extent of lipid peroxidation in chloroplasts, it was proposed that this compound could also affect intracellular signaling in plants (Munné-Bosch, 2005; Munné-Bosch and Alegre, 2002; Munné-Bosch and Falk, 2004). Recent evidence has now emerged indicating that tocopherols participate in cellular signaling in plants, which has several implications for our understanding of the role of tocopherols in plant development and stress tolerance.

The first indication toward identifying a role of tocopherols in cellular signaling came from studies using the *sxd1* mutant of maize (C4 plant). This mutant carries a defect in the *sxd1* gene, which encodes for tocopherol cyclase (SXD1), an enzyme that is essential for the formation of the chromanol ring of tocopherols. An overall growth reduction and source leaf-specific accumulation of anthocyanins and starch characterize this mutant, which is deficient in tocopherols. In addition, minor veins of maturating leaf blades exhibit ultrastructural alterations and callose occlusion of a specific class of plasmodesmata between bundle sheath and vascular parenchyma cells of this mutant, thus leading to a blocking of sucrose transport into the phloem (Botha *et al.*, 2000; Russin *et al.*, 1996).

These results were confirmed in potato plants, which, in contrast to maize, have a C3 type of photosynthetic metabolism. By using an RNAi-silencing approach, Hofius *et al.* (2004) showed that tocopherol deficiency leads to a photoassimilate export-deficient phenotype that is very similar to that observed in maize. These transgenic potato plants show enhanced callose deposition in source leaves, lower photosynthetic capacity, and altered gene expression compared to the wild type. The transcription of the photosynthesis-related *rbcS* and *cab* genes is reduced, while transcripts of the defense-related proteinase inhibitor II (*pin2*) and of proline (*p5cs*), and jasmonic acid (*aoc*) biosyntheses are induced in source leaves. This study provides evidence that the impact of tocopherol deficiency in plasmodesmata function and carbohydrate metabolism is similar in monocot (maize) and dicot (potato) species and cannot be assigned to specific anatomical or biochemical features of C4 metabolism.

A similar phenotype has been observed in *vte1* mutants of *A. thaliana*. This mutant was discovered during a screen for altered tocopherol content, and it has been shown that VTE1 and SXD1 are single-copy orthologues, both encoding an enzyme with tocopherol cyclase activity (Porfirova *et al.*, 2002). Tocopherol deficiency in the *vte1* mutant of *A. thaliana* leads not only to a slightly reduced growth and enhanced susceptibility to photooxidative stress (Porfirova *et al.*, 2002) but also to a photoassimilate export-deficient phenotype characterized by anthocyanin accumulation at low temperatures (Maeda *et al.*, 2006). Similar low-temperature-induced accumulation of anthocyanins has been observed in a regulatory *vte1 A. thaliana* mutant, which has an insertion in the promoter region of the gene-encoding tocopherol cyclase (Munné-Bosch *et al.*, 2007). In these mutants, it is shown additionally that this enhanced accumulation of anthocyanins is triggered by a transient accumulation of jasmonic acid in tocopherol-deficient plants. This study demonstrates that tocopherols may play a role in cellular signaling by altering phytohormone levels in plants. Furthermore, it indicates that tocopherols exert effects on cellular signaling by modulating jasmonic acid levels in plants, rather than directly regulating gene expression. By controlling ROS levels, the extent of lipid peroxidation, and thus hydroperoxide contents in chloroplasts, tocopherols may not only indirectly regulate the amounts of jasmonic acid in leaves but also affect jasmonic acid-dependent gene expression in the nucleus.

Enhanced callose synthesis, and the consequent occlusion of specific plasmodesmata, is the most significant characteristic of the photoassimilate export-deficient phenotype in tocopherol-deficient plants. Callose synthesis is a specific plant response to distinct abiotic and biotic stress factors to control the size exclusion limits of plasmodesmata, and therefore the type and amount of molecules transported from cell to cell. Callose formation is strongly correlated with oxidative damage to membrane lipids and changes in intracellular calcium homeostasis, indicating a mechanistic link between lipid peroxidation and callose synthesis. By controlling the extent of lipid peroxidation, and therefore hydroperoxide content in chloroplasts, toco pherols may indirectly regulate the amounts of jasmonic acid in leaves and affect jasmonic acid-dependent gene expression involved, for instance, in wound stress response. In addition to enhanced callose synthesis, tocopherol-deficient plants show increased accumulation of jasmonic acid-responsive *Pin2* transcripts and upregulation of the jasmonic acid biosynthetic gene *AOC* (Hofius *et al.*, 2004), which supports this contention. Jasmonic acid also regulates the expression of anthocyanin biosynthetic genes (Creelman and Mullet, 1997), which explains the relationship between tocopherol deficiency and anthocyanin accumulation in the source leaves of plants with the photoassimilate export-deficient phenotype.

However, this photoassimilate export-deficient phenotype has not been observed in the *vte1* mutants of *A. thaliana* grown at 22°C under low-light conditions (Sattler *et al.*, 2003), thus indicating that the phenotype observed in tocopherol-deficient plants is strongly dependent on climatic growth conditions. Compelling evidence indicates that the phenotypes described thus far for tocopherol deficiency, characterized by reduced growth, photoassimilate export deficiency, and higher susceptibility to photooxidation, (1) are species-specific, (2) depend on the extent of tocopherol deficiency, (3) depend on growth conditions (light intensity, photoperiod, temperature, and so on) and plant developmental stage, and (4) are more evident as growth conditions become more stressful (Hofius *et al.*, 2004; Munné-Bosch and Falk, 2004; Porfirova *et al.*, 2002).

VI. Have the Functions of Tocopherols Been Evolutionary Conserved?

Among tocopherols, α-tocopherol has been shown to accumulate selectively in the human body and to be the most effective form as an antioxidant and in regulating cell signaling in animals (Brigelius-Flohé *et al.*, 2002). Since α-tocopherol is also the major form in the embryos of seeds and photosynthetic tissues of plants, it appears that plants and animals have converged in selectively using α-tocopherol among all tocopherols.

The antioxidant activity of tocopherols seems to be highly conserved throughout evolution. It has been shown that both plants and animals use tocopherols to reduce the levels of $^1O_2$ and other ROS and to inhibit the propagation of lipid peroxidation within the cell. The aspects related to this antioxidant activity at the membrane level seem similar, although the cells and tissues bearing such membranes are different. The most evident difference is that tocopherols protect photosynthetic membranes in plants, thus it seems that compared to animals, plants will need higher amounts of tocopherols (or additional antioxidants) to cope with potentially higher amount of photogenerated $^1O_2$.

The mechanisms of action of tocopherols in cellular signaling are much less understood in plants than in animals. Specific lipoxygenases, responsible for the initiation of lipid peroxidation, are inhibited by α-tocopherol, at least in part, by efficiently reducing the active site Fe^{3+} of the enzyme to the inactive Fe^{2+} (Cucurou *et al.*, 1991). This study shows that soybean and potato lipoxygenases are inhibited by α-tocopherol, which supports the contention that, in conjunction with the scavenging of lipid peroxyl radicals which propagate lipid peroxidation, α-tocopherol suppresses hydroperoxide formation by inactivation of lipoxygenases. Oxilipins such as jasmonic acid and other lipid peroxidation products activate many genes in plants, especially those involved in plant responses to environmental stress (Weber *et al.*, 2004). Thus, the regulation of lipid peroxidation by α-tocopherol may strongly modulate signal transduction and gene expression not only in animals but also in plants.

Studies on animal models suggest additional mechanisms of action of tocopherols that are independent of their antioxidant functions in the regulation of gene expression (Azzi *et al.*, 1998; Brigelius-Flohé *et al.*, 2002). The inhibition of protein kinase C by tocopherols is one of the best-studied effects of these antioxidants in cellular signaling in animals. Tocopherols may inhibit protein kinase C as a result of the dephosphorylation of the enzyme via activation of protein phosphatase 2A (Clement *et al.*, 1997; Ricciarelli *et al.*, 1998). Alternatively, it has also been proposed that the inhibition of protein kinase C is caused by the activation of diacylglycerol kinase by α-tocopherol, consequently decreasing diacylglycerol and leading to protein kinase C inhibition (Koya *et al.*, 1997). Several studies have examined the effects of α-tocopherol on phospholipase A_2 activity and have shown that a degree of inhibition occurs both *in vitro* and in animal systems when the concentration of α-tocopherol is increased. Grau and Ortiz (1998) suggest that tocopherols inhibit phospholipase A_2 activity by altering the physical properties of the membrane. Chandra *et al.* (2002) have provided the first structural evidence of a specific inhibition of phospholipase A_2 by α-tocopherol. Furthermore, some genes are affected by α-tocopherol at the transcriptional level in animal cells (Azzi *et al.*, 1998; Carlberg, 1999). α- and β-tocopherols show differential effects, which have been attributed to a nonantioxidant mechanism of α-tocopherol in gene regulation. Tocopherol-associated proteins (TAPs) translocate from the cytosol to the nucleus in animal cells, where they may activate gene transcription in an α-tocopherol-dependent manner (Yamaguchi *et al.*, 2001).

Evidence of nonantioxidant functions of tocopherols in higher plants is still limited, although it has been proposed that tocopherols may exert nonantioxidant functions and control photosynthesis and nutrient homeostasis in the cyanobacterium *Synechocystis* sp. PCC 6803 (Sakuragi *et al.*, 2006). Evidence obtained thus far in higher plants, yet limited to the model plant *A. thaliana*, maize, and potato plants, tend to favor a role of tocopherols in regulating the cell redox homeostasis and exert effects on cell signaling by modulating oxylipins levels in plants, rather than directly regulating gene expression. By controlling ROS levels, the extent of lipid peroxidation, and thus hydroperoxide contents in chloroplasts, tocopherols may not only indirectly regulate the amounts of oxylipins, such as jasmonic acid in leaves, but also affect jasmonic acid-dependent gene expression in the nucleus. Jasmonic acid is known as a growth inhibitor and regulates the expression of anthocyanin biosynthetic genes (Creelman and Mullet, 1997), thereby supporting the relationship between tocopherol deficiency, reduced growth, and anthocyanin accumulation described in previous studies (Hofius *et al.*, 2004; Munné-Bosch *et al.*, 2007). Tocopherols therefore influence cellular signaling in plants and may modulate gene expression in the nucleus by affecting lipid peroxidation, and therefore the levels of oxylipins such as jasmonic acid.

VII. Future Perspectives

The recent findings indicating the involvement of tocopherols in cellular signaling open a new field of research to explore the effects of tocopherols in plant development and stress responses. Studies aimed at elucidating the role of tocopherols in plants should not only consider the photo- and antioxidant protective function of these molecules but also their role in the regulation of signal transduction and gene regulation. Although nonantioxidant functions of tocopherols have not been demonstrated thus far in higher plants, the presence of TAPs and tocopherol-specific motifs in the promoters of genes in plants should be investigated. The elucidation of signal transduction pathways and gene expression regulated by tocopherols may probably provide some of the most exciting discoveries awaiting plant biologists in the near future.

References

Arango, Y., and Heise, K. (1998). Localisation of α-tocopherol synthesis in chromoplast envelope membranes of *Capsicum annuum* L. fruits. *J. Exp. Bot.* **49,** 1259–1262.

Asada, K. (1999). The water-water cycle in chloroplasts: Scavenging of active oxygens and dissipation of excess photons. *Annu. Rev. Plant Physiol. Plant Mol. Biol.* **50,** 601–639.

Asada, K. (2006). Production and scavenging of reactive oxygen species in chloroplasts and their functions. *Plant Physiol.* **141,** 391–396.

Austin, J. R., II, Frost, E., Vidi, P., Kessler, F., and Staehelin, L. A. (2006). Plastoglobules are lipoprotein subcompartments of the chloroplast that are permanently coupled to thylakoid membranes and contain biosynthetic enzymes. *Plant Cell* **18,** 21–28.

Azzi, A., Boscoboinik, D., Fazzio, A., Marilley, D., Maroni, P., Ozer, N. K., Spycher, S., and Tasinato, A. (1998). *RRR*-α-Tocopherol regulation of gene transcription in response to the cell oxidant status. *Z. Ernährwiss.* **37,** 21–28.

Baier, M., and Dietz, K. J. (1999). Alkyl hydroperoxides reductases: The way out of the oxidative breakdown of lipids in chloroplasts. *Trends Plant Sci.* **4,** 166–168.

Ball, L., Accotto, G., Bechtold, U., Creissen, G., Funck, D., Jiménez, A., Kular, B., Leyland, N., Mejía-Carranza, J., Reynolds, H., Karpinski, S., and Mullineaux, P. M. (2004). Evidence for a direct link between glutathione biosynthesis and stress defence gene expression in Arabidopsis. *Plant Cell* **16,** 2448–2462.

Bindoli, A., Valente, M., and Cavallini, L. (1985). Inhibition of lipid peroxidation by α-tocopherolquione and α-tocopherolhydroquinone. *Biochem. Int.* **10,** 753–761.

Botha, C. E. J., Cross, R. H. M., van Bel, A. J. E., and Peter, C. I. (2000). Phloem loading in the sucrose-export-defective (SXD-1) mutant maize is limited by callose deposition at plasmodesmata in bundle sheath-vascular parenchyma interface. *Protoplasma* **214,** 65–72.

Brigelius-Flohé, R., Kelly, F. J., Salonen, J. T., Neuzil, J., Zingg, J., and Azzi, A. (2002). The European perspective on vitamin E: Current knowledge and future research. *Am. J. Clin. Nutr.* **76,** 703–716.

Cadenas, E. (1989). Biochemistry of oxygen toxicity. *Annu. Rev. Biochem.* **58,** 79–110.

Carlberg, C. (1999). Lipid soluble vitamins in gene regulation. *Biofactors* **10,** 91–97.

Chandra, V., Jasti, J., Kaur, P., Betzel, C., Srinivasan, A., and Singh, T. P. (2002). First structural evidence of a specific inhibition of phospholipase A_2 by α-tocopherol (vitamin E) and its implications in inflammation: Crystal structure of the complex formed between phospholipase A_2 and α-tocopherol at 1.8 A resolution. *J. Mol. Biol.* **320,** 215–222.

Chen, Z., and Gallie, D. R. (2004). The ascorbic acid redox state controls guard cell signaling and stomatal movement. *Plant Cell* **16,** 1143–1162.

Clement, S., Tasinato, A., Boscoboinik, D., and Azzi, A. (1997). The effect of alpha-tocopherol on the synthesis, phosphorilation and activity of protein kinase C in smooth muscle cells after phorbol 12-myriastate 13-acetate down-regulation. *Eur. J. Biochem.* **246,** 745–749.

Cucurou, C., Battioni, J. P., Daniel, R., and Mansuy, D. (1991). Peroxidase-like activity of lipoxygenases: Different substrate specificity of potato 5-lipoxygenase and soybean 15-lipoxygenase and particular affinity of vitamin E derivatives for the 5-lipoxygenase. *Biochim. Biophys. Acta* **1081,** 99–105.

Creelman, R. A., and Mullet, J. E. (1997). Biosynthesis and action of jasmonates in plants. *Annu. Rev. Plant Physiol. Plant Mol. Biol.* **48,** 355–381.

DellaPenna, D. (1999). Nutritional genomics: Manipulating plant micronutrients to improve human health. *Science* **285,** 375–379.

DellaPenna, D. (2005). Progress in the dissection and manipulation of vitamin E synthesis. *Trends Plant Sci.* **10,** 574–579.

Demmig-Adams, B., and Adams, W. W., III (1992). Photoprotection and other responses of plants to high light stress. *Annu. Rev. Plant Physiol. Plant Mol. Biol.* **43,** 599–626.

Demmig-Adams, B., and Adams, W. W., III (2002). Antioxidants in photosynthesis and human nutrition. *Science* **298,** 2149–2153.

Epstein, S. S., Forsyth, J., Saporoschetz, I. B., and Mantel, N. (1966). An exploratory investigation on the inhibition of selected photosensitizers by agents of varying antioxidant activity. *Radiat. Res.* **28,** 322–335.

Eskling, M., Arvidsson, P. O., and Akerlund, H. E. (1997). The xanthophyll cycle, its regulation and components. *Physiol. Plant.* **100,** 806–816.

Evans, H. M., and Bishop, K. S. (1922). Fetal resorption. *Science* **55,** 650–651.
Fahrenholtz, S. R., Doleiden, F. H., Tozzolo, A. M., and Lamola, A. A. (1974). On the quenching of singlet oxygen by α-tocopherol. *Photochem. Photobiol.* **20,** 505–509.
Fryer, M. J. (1992). The antioxidant effects of thylakoid vitamin E (α-tocopherol). *Plant Cell Environ.* **15,** 381–392.
Fuchs, J. (1998). Potentials and limitations of the natural antioxidants RRR-alpha-tocopherol, L-ascorbic acid and β-carotene in cutaneous photoprotection. *Free Radic. Biol. Med.* **25,** 848–873.
Fukuzawa, K., Matsuura, K., Tokumura, A., Suzuki, A., and Terao, J. (1997). Kinetics and dynamics of singlet oxygen scavenging by α-tocopherol in phospholipid model membranes. *Free Radic. Biol. Med.* **22,** 923–930.
Golan, T., Müller-Moulé, P., and Niyogi, K. K. (2006). Photoprotection mutants of *Arabidopsis thaliana* acclimate to high light by increasing photosynthesis and specific antioxidants. *Plant Cell Environ.* **29,** 879–887.
Grau, A., and Ortiz, A. (1998). Dissimilar protection of tocopherol isomers against membrane hydrolysis by phospholipase A_2. *Chem. Phys. Lipids* **91,** 109–118.
Grumbach, K. H. (1983). Distribution of chlorophylls, carotenoids, and quinones in chloroplasts of higher plants. *Z. Naturforsch.* **38c,** 996–1002.
Grusak, M. A., and DellaPenna, D. (1999). Improving the nutrient composition of plants to enhance human nutrition and health. *Annu. Rev. Plant Physiol. Plant Mol. Biol.* **50,** 133–161.
Halliwell, B. (1981). "Chloroplast Metabolism: The Structure and Function of Chloroplasts in Green Leaf Cells." Clarendon Press, Oxford.
Halliwell, B., and Gutteridge, J. M. C. (1999). "Free Radicals in Biology and Medicine." Oxford University Press (Clarendon), New York.
Havaux, M. (1998). Carotenoids as membrane stabilizers in chloroplasts. *Trends Plant Sci.* **3,** 147–151.
Havaux, M., Bonfils, J., Lütz, C., and Niyogi, K. (2000). Photodamage of the photosynthetic apparatus and its dependence on the leaf development stage in the *npq1 Arabidopsis* mutant deficient in the xanthophyll cycle enzyme violaxanthin de-epoxidase. *Plant Physiol.* **124,** 273–284.
Havaux, M., Eymery, F., Porfirova, S., Rey, P., and Dörmann, P. (2005). Vitamin E protects against photoinhibition and photooxidative stress in *Arabidopsis thaliana*. *Plant Cell* **17,** 3451–3469.
Hideg, E., Barber, J., Heber, U., Asada, K., and Allen, J. F. (2000). Supermolecular structure of photosystem II and location of the PsbS protein—Discussion. *Philos. Trans. R. Soc. Lond. B Biol. Sci.* **355,** 1343–1344.
Hofius, D., Hajirezaei, M., Geiger, M., Tschiersch, H., Melzer, M., and Sonnewald, U. (2004). RNAi-mediated tocopherol deficiency impairs photoassimilate export in transgenic potato plants. *Plant Physiol.* **135,** 1256–1268.
Horton, P., Ruban, A. V., and Walters, R. G. (1996). Regulation of light harvesting in green plants. *Annu. Rev. Plant Physiol. Plant Mol. Biol.* **47,** 655–682.
Ingold, K. U., Burton, G. W., Foster, D. O., Zuker, M., Hughes, L., Lacelle, S., Lusztyk, E., and Slaby, M. (1986). A new vitamin E analogue more active than α-tocopherol in the rat curative myopathy bioassay. *FEBS Lett.* **205,** 117–120.
Ingold, K. U., Burton, G. W., Foster, D. O., and Hughes, L. (1990). Is methyl-branching in α-tocopherol's "tail" important for its *in vivo* activity? Rat curative bioassay measurements of the vitamin E activity of three 2RS-*n*-alkyl-2,5,7,8-tetramethyl-6-hydroxychromans. *Free Radic. Biol. Med.* **9,** 205–210.
Kamal-Eldin, A., and Appelqvist, L. (1996). The chemistry and antioxidant properties of tocopherols and tocotrienols. *Lipids* **31,** 671–701.
Kasahara, M., Kagawa, T., Oikawa, K., Suetsugu, N., Miyao, M., and Wada, M. (2002). Chloroplast avoidance movement reduces photodamage in plants. *Nature* **420,** 829–832.

Koya, D., Lee, I. K., Ishii, H., Kanoh, H., and King, G. L. (1997). Prevention of glomerular dysfunction in diabetic rats by treatment with *d*-alpha-tocopherol. *J. Am. Soc. Nephrol.* **8,** 426–435.

Kruk, J., and Strzalka, K. (1995). Occurrence and function of α-tocopherol quinone in plants. *J. Plant Physiol.* **145,** 405–409.

Kruk, J., and Strzalka, K. (2001). Redox changes of cytochrome b559 in the presence of plastoquinones. *J. Biol. Chem.* **276,** 86–91.

Kruk, J., Jamiola-Rzeminska, M., and Strzalka, K. (1997a). Plastoquinol and α-tocopherol quinol are more active than ubiquinol and α-tocopherol in inhibition of lipid peroxidation. *Chem. Phys. Lipids* **87,** 73–80.

Kruk, J., Burda, K., Radunz, A., Strzalka, K., and Schmid, G. H. (1997b). Antagonistic effects of α-tocopherol and α-tocopherol quinone in the regulation of cyclic electron transport around photosystem II. *Z. Naturforsch.* **52c,** 766–774.

Kruk, J., Burda, K., Schmid, G. H., Radunz, A., and Strzalka, K. (1998). Function of plastoquinones B and C as electron acceptors in photosystem II and fatty acid analysis of plastoquinone B. *Photosynth. Res.* **58,** 203–209.

Lichtenthaler, H. K. (1968). Verbreitung und relative konzentration der lipophilen plastidenchinone in grünen pflanzen. *Planta* **81,** 140–152.

Lichtenthaler, H. K., Prenzel, U., Douce, R., and Joyard, J. (1981). Localisation of prenylquinones in the envelope of spinach chloroplasts. *Biochim. Biophys. Acta* **641,** 99–105.

Liebler, D. C. (1993). The role of metabolism in the antioxidant function of vitamin E. *Crit. Rev. Toxicol.* **23,** 147–169.

Maeda, H., Sakuragi, Y., Bryant, D. A., and DellaPenna, D. (2005). Tocopherols protect *Synechocystis* sp. strain PCC 6803 from lipid peroxidation. *Plant Physiol.* **138,** 1422–1435.

Maeda, H., Song, W., Sage, T. L., and DellaPenna, D. (2006). Tocopherols play a crucial role in low-temperature adaptation and phloem loading in *Arabidopsis. Plant Cell* **18,** 2710–2712.

Matsui, K., Shibutani, M., Hase, T., and Kajiwara, T. (1996). Bell pepper fruit fatty acid hydroperoxide lyase is a cytochrome P450 (CYP74B). *FEBS Lett.* **394,** 21–24.

Matsui, K., Wilkinson, J., Hiatt, B., Knauf, V., and Kajiwara, T. (1999). Molecular cloning and expression of *Arabidopsis* fatty acid hydroperoxide lyase. *Plant Cell Physiol.* **40,** 477–481.

McKersie, B. D., Senaratna, T., Walker, M. A., Kendall, E. J., and Hetherington, P. R. (1988). *In* "Senescence and Aging in Plants" (L. D. Noodén and A. C. Leopold, Eds.), pp. 441–464. Academic Press, New York.

McKersie, B. D., Hoekstra, F. A., and Krieg, L. C. (1990). Differences in the susceptibility of plant membrane lipids to peroxidation. *Biochim. Biophys. Acta* **1030,** 119–126.

Melis, A. (1999). Photosystem II-damage and repair cycle in chloroplasts: What modulates the rate of photodamage *in vivo*? *Trends Plant Sci.* **4,** 130–135.

Mukai, K., Itoh, S., and Morimoto, H. (1992). Stopped-flow kinetic-study of vitamin E regeneration reaction with biological hydroquinones (reduced forms of ubiquinone, vitamin K, and tocopherolquinone) in solution. *J. Biol. Chem.* **267,** 22277–22281.

Munné-Bosch, S. (2005). Linking tocopherols with cellular signaling in plants. *New Phytol.* **166,** 363–366.

Munné-Bosch, S., and Alegre, L. (2000). Changes in carotenoids, tocopherols and diterpenes during drought and recovery, and the biological significance of chlorophyll loss in *Rosmarinus officinalis* plants. *Planta* **210,** 925–931.

Munné-Bosch, S., and Alegre, L. (2002). The function of tocopherols and tocotrienols in plants. *Crit. Rev. Plant Sci.* **21,** 31–57.

Munné-Bosch, S., and Cela, J. (2006). Effects of water deficit on photosystem II photochemistry and photoprotection during acclimation of lyreleaf sage (*Salvia lyrata* L.) plants to high light. *J. Photochem. Photobiol. B Biol.* **85,** 191–197.

Munné-Bosch, S., and Falk, J. (2004). New insights into the function of tocopherols in plants. *Planta* **218,** 323–326.

Munné-Bosch, S., Shikanai, T., and Asada, K. (2005). Enhanced ferredoxin-dependent cyclic electron flow around photosystem I and α-tocopherol quinone accumulation in water-stressed *ndhB*-inactivated mutants. *Planta* **222,** 502–511.

Munné-Bosch, S., Weiler, E. W., Alegre, L., Müller, M., Düchting, P., and Falk, J. (2007). α-Tocopherol may influence cellular signaling by modulating jasmonic acid levels in plants. *Planta.* **225,** 681–691.

Murkovic, M., Wiltschko, D., and Pfannhauser, W. (1997). Formation of alpha-tocopherolquinone and alpha-tocopherolquinone epoxides in plant oil. *Fett* **99,** 165–169.

Neely, W. C., Martin, J. M., and Barker, S. A. (1988). Products and relative reaction rates of the oxidation of tocopherols with singlet molecular oxygen. *Photochem. Photobiol.* **48,** 423–428.

Niyogi, K. K. (1999). Photoprotection revisited: Genetic and molecular approaches. *Annu. Rev. Plant Physiol. Plant Mol. Biol.* **50,** 333–359.

Omata, T., and Murata, N. (1984). Cytochromes and prenylquinones in preparations of cytoplasmic and thylakoid membranes from the cyanobacterium (blue-green alga). *Anacystis nidulans. Biochim. Biophys. Acta* **766,** 395–402.

Porfirova, S., Bergmüller, E., Tropf, S., Lemke, R., and Dörmann, P. (2002). Isolation of an Arabidopsis mutant lacking vitamin E and identification of a cyclase essential for all tocopherol biosynthesis. *Proc. Natl. Acad. Sci. USA* **99,** 12495–12500.

Pryor, W. A. (2000). Vitamin E and heart disease: Basic science to clinical intervention trials. *Free Radic. Biol. Med.* **28,** 141–161.

Quartacci, M. F., Forli, M., Rascio, N., Della Vecchia, F., Boschichio, A., and Navari-Izzo, F. (1997). Desiccation tolerant *Sporolobus stapfianus*: Lipid composition and cellular ultrastructure during dehydration and rehydration. *J. Exp. Bot.* **48,** 1269–1279.

Ricciarelli, R., Tasinato, A., Clement, S., Özer, N. K., Boscoboinik, D., and Azzi, A. (1998). Alpha-tocopherol specifically inactivates cellular protein kinase C alpha by changing its phosphorilation state. *Biochem. J.* **334,** 243–249.

Russin, W. A., Evert, R. E., Vanderveer, P. J., Sharkey, T. D., and Briggs, S. P. (1996). Modification of a specific class of plasmodesmata and loss of sucrose export ability in a *sucrose export defective1* maize mutant. *Plant Cell* **8,** 645–658.

Sakuragi, Y., Maeda, H., DellaPenna, D., and Bryant, D. A. (2006). α-Tocopherol plays a role in photosynthesis and macronutrient homeostasis of the cyanobacterium *Synechocystis* sp. PPC 6803 that is independent of its antioxidant function. *Plant Physiol.* **141,** 508–521.

Satter, R. L., and Galston, A. W. (1981). Mechanisms of control of leaf movements. *Annu. Rev. Plant Physiol.* **32,** 83–110.

Sattler, S. E., Cahoon, E. B., Coughlan, S. J., and DellaPenna, D. (2003). Characterization of tocopherol cyclases from higher plants and cyanobacteria. Evolutionary implications for tocopherol synthesis and function. *Plant Physiol.* **132,** 2184–2195.

Schaller, F. (2001). Enzymes of the biosynthesis of octadecanoid-derived signalling molecules. *J. Exp. Bot.* **52,** 11–23.

Singh, B., Singh, S., and Behl, A. (1990). Lipid temperature relations of six thermophilous fungi. *Acta Bot. Ind.* **18,** 204–208.

Smirnoff, N., and Wheeler, G. L. (2000). Ascorbic acid in plants: Biosynthesis and function. *Crit. Rev. Plant Sci.* **19,** 267–290.

Soll, J., Schultz, G., Joyard, J., Douce, R., and Block, M. A. (1985). Localisation and synthesis of prenylquinones in isolated outer and inner envelope membranes from spinach chloroplasts. *Arch. Biochem. Biophys.* **238,** 290–299.

Steyn, W. J., Wand, S. J. E., Holcroft, D. M., and Jacobs, G. (2002). Anthocyanins in vegetative tissues: A proposed unified function in photoprotection. *New Phytol.* **155,** 349–361.

Straight, R., and Spikes, J. (1985). Photosensitized oxidation of biomolecules. *In* "Singlet O_2" (A. Frimer, Ed.), Vol. 4, pp. 91–143. CRC Press, Boca Raton.

Terashima, I., Miyazawa, S., and Handa, Y. T. (2001). Why are sun leaves thicker than shade leaves? Consideration based on analyses of CO_2 diffusion in the leaf. *J. Plant Res.* **114,** 93–105.

Thomas, M., and Foote, C. S. (1978). Chemistry of oxygen. XXVI: Photooxygenation of phenols. *Photochem. Photobiol.* **27,** 683–693.

Vandamme, E. J. (1992). Production of vitamins, coenzymes and related biochemicals by biotechnological processes. *J. Chem. Technol. Biotechnol.* **53,** 313–327.

Wang, X., and Quinn, P. J. (2000). The location and function of vitamin E in membranes. *Mol. Membr. Biol.* **17,** 143–156.

Weber, H., Chételat, A., Reymond, P., and Farmer, E. E. (2004). Selective and powerful stress gene expression in Arabidopsis in response to malondialdehyde. *Plant J.* **37,** 877–888.

Weston, E., Thorogood, K., Vinti, G., and López-Juez, E. (2000). Light quantity controls leaf-cell and chloroplast development in *Arabidopsis thaliana* wild type and blue-light-perception mutants. *Planta* **211,** 807–815.

Wise, R. R., and Naylor, A. W. (1987). Chilling enhanced photooxidation. Evidence for the role of singlet oxygen and superoxide in the breakdown of pigments and endogenous antioxidants. *Plant Physiol.* **83,** 278–282.

Yamaguchi, J., Iwamoto, T., Kida, S., Masushige, S., Yamada, K., and Esashi, T. (2001). Tocopherol-associated protein is a ligand-dependent transcriptional activator. *Biochem. Biophys. Res. Commun.* **285,** 295–299.

Yamauchi, R., and Matsushita, S. (1977). Quenching effect of tocopherols on the methyl linoleate photooxidation and their oxidation products. *Agric. Biol. Chem.* **41,** 1425–1430.

15

VITAMIN E AND MAST CELLS

JEAN-MARC ZINGG

Institute of Biochemistry and Molecular Medicine
University of Bern, 3012 Bern, Switzerland

Mast cells play an important role in the immune system by interacting with B and T cells and by releasing several mediators involved in activating other cells. Hyperreactivity of mast cells and their uncontrolled accumulation in tissues lead to increased release of inflammatory

Vitamins and Hormones, Volume 76

0083-6729/07 $35.00
DOI: 10.1016/S0083-6729(07)76015-6

mediators contributing to the pathogenesis of several diseases such as rheumatoid arthritis, atherosclerosis, multiple sclerosis, and allergic disorders such as asthma and allergic rhinitis. Interference with mast cell proliferation, survival, degranulation, and migration by synthetic or natural compounds may represent a preventive strategy for the management of these diseases. Natural vitamin E covers a group of eight analogues—the α-, β-, γ-, and δ-tocopherols and the α-, β-, γ-, and δ-tocotrienols, but only α-tocopherol is efficiently retained by the liver and distributed to peripheral tissues. Mast cells preferentially locate in the proximity of tissues that interface with the external environment (the epithelial surface of the skin, the gastrointestinal mucosa, and the respiratory system), what may render them accessible to treatments with inefficiently retained natural vitamin E analogues and synthetic derivatives. In addition to scavenging free radicals, the natural vitamin E analogues differently modulate signal transduction and gene expression in several cell lines; in mast cells, protein kinase C, protein phosphatase 2A, and protein kinase B are affected by vitamin E, leading to the modulation of proliferation, apoptosis, secretion, and migration. In this chapter, the possibility that vitamin E can prevent diseases with mast cells involvement by modulating signal transduction and gene expression is evaluated.

I. INTRODUCTION

A. MAST CELLS

Mast cells play an important role in the immune system by interacting directly with B and T cells and by releasing several mediators (cytokines, chemokines, growth factors, histamine, proteases) involved in the activation of other cells (reviewed in Krishnaswamy *et al.*, 2006; Marone *et al.*, 2000; Mecheri and David, 1999; Schneider *et al.*, 2002). When released in a controlled manner, these mediators initiate the inflammatory response with consequent increased vasodilatation, vascular permeability, and bronchial smooth muscle contraction; they also modulate allergic inflammation, contribute to autoimmunity (Benoist and Mathis, 2002), and participate in the immune response to bacterial and parasitic infection (Feger *et al.*, 2002; Henz *et al.*, 2001). The higher presence of mast cells in tissues that are portals of infections and their ability to phagocytose and to present antigens indicates that they mediate also the host response in innate immunity (Feger *et al.*, 2002; Henz *et al.*, 2001). Hyperreactivity of mast cells and their uncontrolled accumulation in tissues lead to increased release of inflammatory mediators contributing to the pathogenesis of several diseases, such as rheumatoid arthritis (Lee *et al.*, 2002; Woolley, 2003), various forms of mastocytosis

(Valent *et al.*, 2003), atopic dermatitis (Okayama, 2005; Okayama and Kawakami, 2006), interstitial cystitis (Theoharides, 1996), scleroderma (Theoharides, 1996), multiple sclerosis (Zappulla *et al.*, 2002), atherosclerosis (Kelley *et al.*, 2000), Crohn's disease (Schattenfroh *et al.*, 1994), autoimmune diseases (Benoist and Mathis, 2002), and allergic disorders such as asthma and allergic rhinitis (Metcalfe *et al.*, 1997). In the acute situation, the mediators released from mast cells also exert the anaphylactic reaction (Ring *et al.*, 2004). Mast cells induce also angiogenesis by releasing proteases, growth factors, cytokines, and chemokines; enabling the growth, invasion, and metastasis of various cancers; inflammatory processes; and wound repair (Hiromatsu and Toda, 2003). Mastocytosis is associated with increased mast cell proliferation, survival, and migration with consequent increased mast cell number in skin and other tissues (Okayama, 2005; Okayama and Kawakami, 2006; Valent *et al.*, 2003). In some of these diseases, it can be assumed that mast cell signaling and proliferation are deregulated and normalization could play a beneficial role.

B. VITAMIN E

The term vitamin E covers a group of eight lipid-soluble compounds—the α-, β-, γ-, and δ-tocopherols and the α-, β-, γ-, and δ-tocotrienols (Fig. 1). α-Tocopherol is the most biologically active form since it is specifically retained in the body by the liver α-tocopherol transfer protein (α-TTP), thus reaching an average plasma concentration of 23 μM. The plasma concentration of the other tocopherols and tocotrienols is usually below 2 μM because they are not efficiently retained by the liver, metabolized, and predominantly eliminated (Arita *et al.*, 1995). Therefore, most of the studies with vitamin E have been done with α-tocopherol, and stabilized forms of vitamin E are mainly derived from α-tocopherol. However, recent research suggests that each of the tocopherols and tocotrienols can exert analogue-specific cellular effects.

For some of these natural tocopherol and tocotrienol analogues, as well as for synthetic vitamin E derivatives (reviewed in Zingg, 2007), the plasma concentration with biological activity is difficult to be reached, even by means of dietary supplementation; this may be different for mast cells that preferentially locate in the proximity of tissues that interface with the external environment (the epithelial surface of the skin, the gastrointestinal mucosa, and the respiratory system), what may render these cells accessible to the natural vitamin E analogues and synthetic derivatives not reaching sufficiently high plasma concentrations, even by oral supplementation. In view of that topic administrations of some vitamin E analogues may be particularly applicable for mast cells in the skin (Thiele *et al.*, 2005). Aerosolized applications of vitamin E analogues to the respiratory tract may be useful for treatments in asthma and possibly nasal allergies (Hybertson *et al.*, 1995, 2005; Shahar *et al.*, 2004; Suchankova *et al.*, 2006; Zheng *et al.*, 1999).

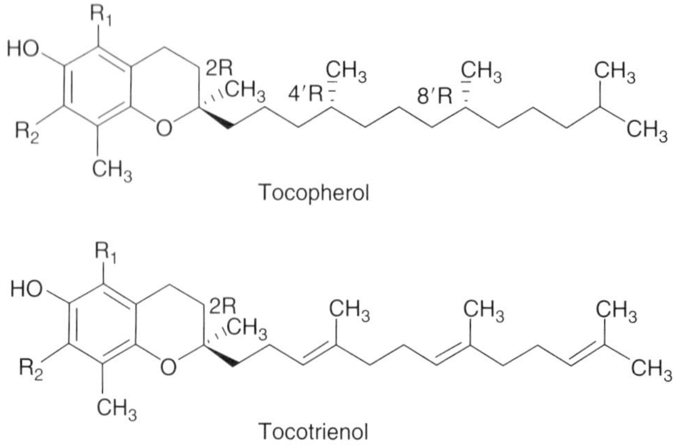

R_1	R_2	
CH_3	CH_3	α
CH_3	H	β
H	CH_3	γ
H	H	δ

Tocopherol

Tocotrienol

FIGURE 1. Natural tocopherols and tocotrienols. Natural vitamin E comprises eight different analogues, the α-, β-, γ-, and δ-tocopherols and the α-, β-, γ-, and δ-tocotrienols. These vitamin E analogues have essentially equal antioxidant potency, but they affect signal transduction and gene expression with different potency as measured by *in vitro* experiments. *In vivo*, the vitamin E analogue with the highest biological activity is α-tocopherol since it is selectively enriched by the liver; the other vitamin E analogues and also excess α-tocopherol are converted to several metabolites and eliminated. Since mast cells preferentially locate in the proximity of tissues that interface with the external environment (the epithelial surface of the skin, the gastrointestinal mucosa, and the respiratory system), they may be accessible for treatments with inefficiently retained natural vitamin E analogues and synthetic derivatives.

Moreover, phenolic compounds are enriched in the gastrointestinal tract (Halliwell *et al.*, 2005), suggesting that the inefficiently absorbed vitamin E analogues could be considered as preventive treatments against intestinal inflammatory and allergic events and colon cancer.

It has also to be considered that vitamin E analogues with low plasma levels can perform cellular functions in tissues at very low concentrations (nanomolar), and the continuous replenishment of tissues and skin may ultimately exert a beneficial effect (Richelle *et al.*, 2006; Theoharides and Bielory, 2004). γ-Tocopherol is enriched in the skin by yet unknown mechanisms, suggesting that it may exert cellular functions relevant for skin, for example the activation of PPARγ expression as shown in keratinocytes and other cells (De Pascale *et al.*, 2006; Thiele *et al.*, 1999; Zingg and Azzi, 2004, 2006). Studies have shown that sufficient amounts of tocotrienols, given as dietary supplements, can reach the brain and perform neuroprotective functions, despite their inefficient uptake (Khanna *et al.*, 2005).

Vitamin E was reported to have beneficial effects in several diseases (reviewed in Azzi *et al.*, 2002a; Brigelius-Flohe *et al.*, 2002; Zingg and Azzi, 2004), including atherosclerosis (Lonn *et al.*, 2005; Munteanu *et al.*, 2004), cancer (Lonn *et al.*, 2005), inflammation (Azzi *et al.*, 2004; Meydani *et al.*, 1997; Singh *et al.*, 2005), Alzheimer's disease (Knopman, 2006), asthma

(Florence, 1995), allergies (Centanni *et al.*, 2001; Montano Velazquez *et al.*, 2006), and wound healing (Musalmah *et al.*, 2002), although in clinical studies these effects of vitamin E were not always evident (Coulter *et al.*, 2006; Hathcock *et al.*, 2005; Lonn *et al.*, 2005; Miller *et al.*, 2005).

At the molecular level, vitamin E has been suggested to play a central role in disease prevention by reducing lipid peroxidation. In addition to such antioxidant effects, vitamin E also exerts nonantioxidant activities suggesting alternative molecular pathways for disease prevention (reviewed in Azzi *et al.*, 2002a; Rimbach *et al.*, 2002; Zingg and Azzi, 2004). Vitamin E modulates signal transduction and gene expression, and these events most likely reflect specific interactions of vitamin E with enzymes, structural proteins, lipids, and transcription factors (Zingg and Azzi, 2004, 2006). The natural vitamin E analogues affect the cellular behavior differentially implying that they are not the result of a general antioxidant action. Over the last decade, vitamin E has been shown to have specific effects on several enzymes such as protein kinase C (PKC), protein kinase B (PKB), protein phosphatase 2A (PP2A), diacylglycerol kinase (DAGK), 5-lipoxygenase (5-LOX), phospholipase A2 (PLA2), and cyclooxygenase 2 (COX-2) (reviewed in Azzi *et al.*, 2002a; Brigelius-Flohe *et al.*, 2002; Ricciarelli *et al.*, 2001; Zingg and Azzi, 2004, 2006). These effects of vitamin E on enzymes involved in signal transduction translate at the cellular level into the modulation of the expression of several genes. Some enzymes, transcription factors, or receptors are modulated by tocopherols at the transcriptional level such as the HMG-CoA reductase, peroxisome proliferators activated receptor gamma (PPARγ), the low density lipoprotein (LDL) receptor, the scavenger receptors CD36 and SR-BI, or the cytochrome P450 enzymes (CYP3A4 and CYP3A5) (reviewed in Azzi *et al.*, 2002a; Zingg and Azzi, 2004).

In the following sections, the main effects of vitamin E on mast cells are summarized; furthermore, the possibility that vitamin E can prevent diseases with mast cells involvement by modulating signal transduction and gene expression is evaluated.

II. CELLULAR EFFECTS OF VITAMIN E IN MAST CELLS

A. INHIBITION OF PROLIFERATION AND SURVIVAL OF MAST CELLS BY VITAMIN E

In human mast cells and several other cell lines, signal transduction through phosphatidylinositol-3-kinase (PI3K) and PKB was described to be modulated by vitamin E (Kempna *et al.*, 2004; Munteanu *et al.*, 2006; Numakawa *et al.*, 2006; Shah and Sylvester, 2005; Sylvester and Shah, 2005). The PI3K/PKB signaling pathway plays an important role in many processes, including cardiovascular physiology, proliferation of cancer cells, cellular migration, apoptosis, survival, and secretion (Hill and Hemmings, 2002; Oudit *et al.*, 2004). PKB or Akt has a wide range of cellular targets, and its increased activity can be found during atherosclerosis and tumorigenesis (Yang *et al.*, 2004).

In the human mastocytoma cell line (HMC-1), vitamin E inhibits the PI3K/PKB signaling pathway with consequent inhibition of proliferation and reduced survival (Fig. 2) (Kempna *et al.*, 2004). The four tocopherols inhibit cell proliferation with different potency ($\delta > \alpha = \gamma > \beta$); δ-tocopherol even leads to apoptosis at higher concentrations (Kempna *et al.*, 2004). However, since δ-tocopherol is not efficiently retained and distributed to peripheral cells by the liver, mast cells may only be exposed to sufficient amounts of δ-tocopherol by using specific applications, for example topic treatments for skin, aerosolized δ-tocopherol for the respiratory tract, or possibly stabilized forms for the gastrointestinal tract.

Neither PKC nor PP2A is involved in the observed effects on cell proliferation, as judged from using inhibitors of PKC and PP2A. Other pathways, such as the Ras-stimulated ERK1/2 (extracellular signal responsive kinase)

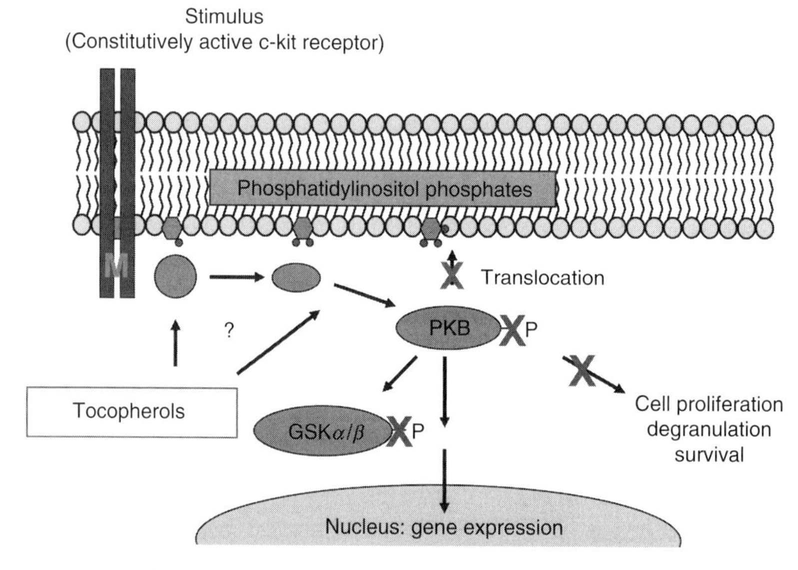

FIGURE 2. The c-kit/PI3K/PKB signal transduction pathway in HMC-1 mastocytoma cells. The c-kit receptor tyrosine kinase is constitutively active as a result of a mutation (M), maintaining the PI3K/PKB signal transduction pathway active, and leading to continuous cell proliferation and increased survival independent of other growth factors. The tocopherols interfere with the c-kit/PI3K/PKB signal transduction cascades by interfering with PKB membrane translocation and PKB (Ser473) phosphorylation (Kempna *et al.*, 2004), ultimately leading to decreased proliferation, induction of apoptosis, and possibly less severe mast cell degranulation. (See Color Insert.)

pathway, are also not affected by tocopherol treatment in these cells. However, the growth inhibition correlates with the reduction of PKB Ser473 phosphorylation by the different tocopherols. The reduction of PKB phosphorylation leads to a decrease of its activity, as judged from a parallel reduction of glycogen synthase kinase 3 alpha/beta (GSK3α/β) phosphorylation. Moreover, the translocation of PKB to the plasma membrane is inhibited by the tocopherols in nonstimulated HMC-1 cells, as well as in cells with an NGFβ-stimulated TrkA receptor known to induce exocytotic signaling (Kempna *et al.*, 2004; Seebeck *et al.*, 2001).

Similar to the results with mastocytoma cells, induction of apoptosis by γ- and δ-tocopherols was shown with prostate cancer cells (Jiang *et al.*, 2004), mouse-activated macrophages (McCormick and Parker, 2004), and mammary epithelial cells (McIntyre *et al.*, 2000). In androgen-sensitive prostate cancer cells (LNCaP), the inhibition of dihydroceramide desaturase is involved in the induction of apoptosis by γ- and δ-tocopherols (Jiang *et al.*, 2004).

B. POSSIBLE MOLECULAR TARGETS FOR VITAMIN E IN MAST CELLS

The molecular targets affected by the tocopherol analogues in HMC-1 mastocytoma cells have yet to be identified. The tocopherols interfere with PKB Ser473 phosphorylation and reduce proliferation of HMC-1 cells, possibly by modulating either c-kit tyrosine kinase directly, PI3K, a kinase phosphorylating PKB (PDK1/2), or a phosphatase dephosphorylating it. In the following paragraphs, the possible molecular targets of vitamin E within the c-kit/PI3K/PKB signaling cascade in HMC-1 mastocytoma cells are analyzed in detail (Fig. 3).

1. Modulation of Tyrosine Kinases and Phosphatases

HMC-1 is a mastocytoma cell line with a gain of function mutation in the c-kit receptor tyrosine kinase [stem cell factor (SCF) receptor] (Furitsu *et al.*, 1993), which leads to constitutive tyrosine kinase activity and activation of PI3K independent of the c-kit ligand, SCF (Fig. 2), with consequent growth factor independent cell proliferation and increased survival (reviewed in Roskoski, 2005). The c-kit receptor tyrosine kinase is present in the majority of hematopoietic cells, playing indispensable functions in their proliferation, differentiation, survival, degranulation, migration, and tissue homing (Galli and Hammel, 1994; Huizinga *et al.*, 1995; Okayama and Kawakami, 2006).

Some evidence for the possible involvement of tyrosine phosphorylation in the observed effects of vitamin E has been described, albeit in other cells; α-tocopherol was shown to inhibit Tyk2 tyrosine kinase activity in oxLDL-stimulated macrophages (Venugopal *et al.*, 2004), and tyrosine phosphorylation of JAK2, STAT1, and STAT3 is decreased by α-tocopherol in

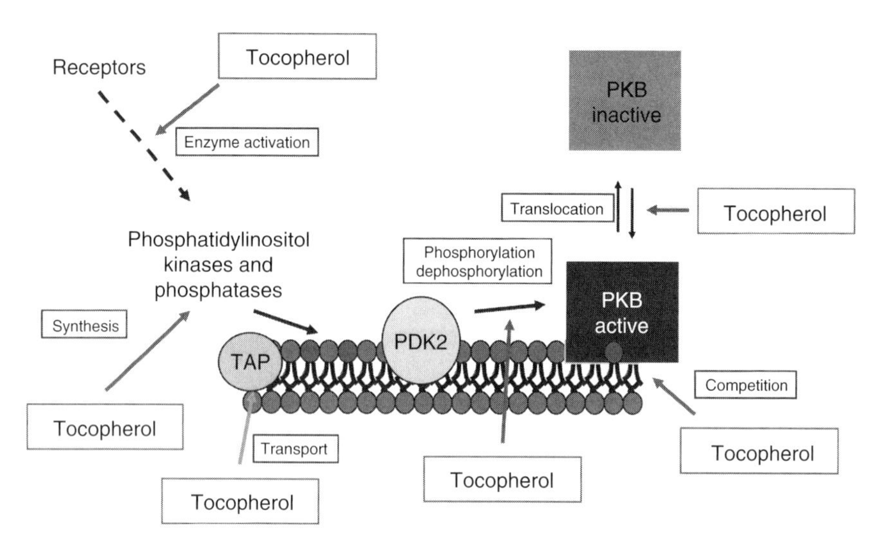

FIGURE 3. Some of the possible molecular targets of vitamin E within the c-kit/PI3K/PKB signal transduction pathway. The tocopherols may interfere with PKB enzyme activation by interfering with translocation to the plasma membrane, by competition with lipid mediator binding to the pleckstrin homology (PH) domain of PKB, by inhibition of PDK2 involved in PKB (Ser473) phosphorylation, by activating a phosphatase against phospho-Ser473 of PKB, by affecting the transport and distribution of lipid mediators (by competition with binding to TAP proteins), by modulating the synthesis of lipid mediators via PI3K or lipid phosphatases, or by earlier steps leading to activation of PI3K (by inhibiting the c-kit receptor tyrosine kinase or by activating a tyrosine phosphatase). (See Color Insert.)

oxLDL-stimulated MRC5 fibroblasts (Maziere *et al.*, 2001). Related to that, in HT4 hippocampal neuronal cells, glutamate-stimulated pp60(c-Src) tyrosine kinase activity is normalized by α-tocotrienol, but not by α-tocopherol (Sen *et al.*, 2000). Tyrosine phosphorylation is also decreased by α-tocopheryl succinate in human neutrophils via activation of a tyrosine phosphatase (Chan *et al.*, 2001). Since classes I and II PI3K are regulated by tyrosine phosphorylation, it can be speculated that inhibition of tyrosine kinase activity by tocopherols may ultimately lead to reduced PKB membrane translocation and phosphorylation (Oudit *et al.*, 2004).

It remains to be shown whether similar mechanisms apply to c-kit tyrosine kinase activity in mastocytoma cells. As for most of the receptors with tyrosine kinase activity, stimulation of c-kit leads to the activation of several signaling pathways, including the Ras/MAPK/ERK1/2 and the PI3K/PKB pathways (Furitsu *et al.*, 1993; Jacobs-Helber *et al.*, 1997). However, since the ERK pathway is not affected in our study with HMC-1 cells (Kempna *et al.*, 2004), it can be assumed that c-kit may not be the direct target of vitamin E, although further tests need to corroborate this.

2. Modulation of PKB Enzyme Activation

The tocopherols could directly interfere with several steps required for PKB enzyme activation in mast cells. Activation of PKB involves a membrane translocation step, followed by phosphorylation of two key regulatory sites, Ser473 and Thr308. The pleckstrin homology domain (PH domain) present in the PKB molecule binds phosphatidylinositol di- and triphosphates (PI34P and PI345P), produced by activated PI3K at the plasma membrane. By the same mechanisms PI3K-dependent kinase 1 (PDK-1), a kinase phosphorylating Thr308 in PKB, becomes active. Phosphorylation of Thr308 leads, however, only to partial activation of PKB. Only after phosphorylation at the second site (Ser473) by a yet unidentified kinase ["PDK-2," such as ATM (Guinea Viniegra *et al.*, 2004), or DNA-dependent protein kinase (Feng *et al.*, 2004), ILK (Delcommenne *et al.*, 1998), PKCα (Partovian and Simons, 2004), PKCβ (Kawakami *et al.*, 2004), PKCε (Lu *et al.*, 2006), or a rictor–mTOR complex (Sarbassov *et al.*, 2005)], the enzyme becomes fully active (Scheid and Woodgett, 2003; Whiteman *et al.*, 2002). Once active, PKB can be inactivated by PP2A or the PTEN lipid phosphatase, which hydrolyzes the products of PI3K (Andjelkovic *et al.*, 1996).

Other possibilities, such as direct binding of tocopherol analogues to PKB preventing its translocation to the plasma membrane as a consequence of competition with lipid mediators, inhibition of proteins involved in membrane translocation of PKB, or activation of lipid phosphatases like PTEN and SHIP1/2, have also to be considered (reviewed in Zingg, 2006). Moreover, it remains to be solved how different tocopherols can inhibit PKB phosphorylation with different efficiencies.

3. Modulation of the Amount and Distribution of Membrane Lipids

Several proteins mediate the uptake of distribution of vitamin E (Fig. 4), and these proteins may play a role in the modulation of signal transduction and gene expression by vitamin E in mast cells. In addition to LDL receptor (LDLR), the uptake of vitamin E is mediated by the scavenger receptor SR-BI in enterocytes as well as in brain capillary endothelial cells at the blood–brain barrier (Goti *et al.*, 2000; Kolleck *et al.*, 1999). Moreover, the plasma phospholipid transfer protein (PL-TP) enhances the vitamin E exchange between lipoproteins and between lipoproteins and cells (Kostner *et al.*, 1995).

Three tocopherol-associated proteins are able to bind vitamin E analogues and various other lipids (Kempna *et al.*, 2003b, 2004; Panagabko *et al.*, 2003). The tocopherol-associated protein 1 [TAP1, SPF, or Sec14-like 2, not to be mixed up with the ATP-binding cassette transporter, antigen peptide transporter protein 1 or 2, which are also designated TAP1 or TAP2, respectively (Neefjes *et al.*, 1993)] is expressed in mouse mast cells (Ikeda *et al.*, 2004). Primary human mast cells and HMC-1 mastocytoma cells are also expressing significant amounts of hTAP (unpublished).

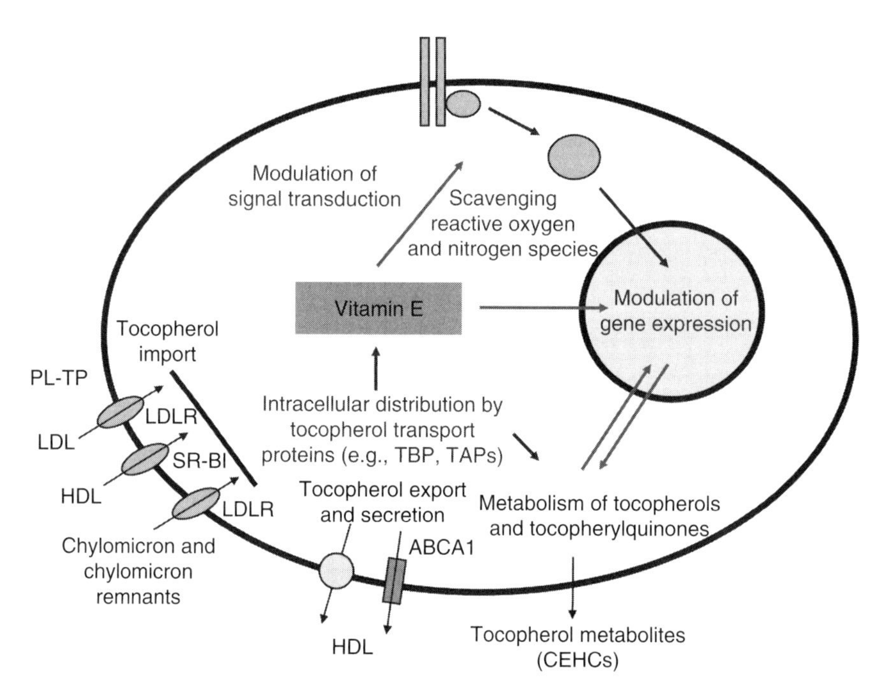

FIGURE 4. Modulation of signal transduction and gene expression by vitamin E. Vitamin E may either specifically modulate signal transduction and gene expression by directly interacting with transcription factors or signal transduction enzymes, or in a more general manner, by reducing damage to enzymes and transcription factors induced by reactive oxygen and nitrogen species. The intracellular concentration and localization of vitamin E may determine the cellular and enzymatic events affected. Several vitamin E-binding proteins (α-TTP, TBP, TAP) are possibly involved in vitamin E uptake and distribution to organelles (mitochondria, nucleus, enzyme complexes) and in modulating the activity of specific enzymes involved in signal transduction (e.g., PI3K). Moreover, proteins involved in the import, export, and metabolism of vitamin E may influence the intracellular concentration of vitamin E and determine whether there is sufficient vitamin E within a cell to affect signal transduction and gene expression (reviewed in Zingg, 2007). (See Color Insert.)

The TAP proteins may be involved in the regulation of cellular tocopherol concentration, tocopherol transport, and tocopherol-mediated signaling and gene expression (Azzi *et al.*, 2002b; Zingg and Azzi, 2006). In line with this, the expression of TAP1 increases the cellular level of α-tocopherol and α-tocopheryl succinate, suggesting that intracellular vitamin E transport is one of the functions of the TAP proteins (Neuzil *et al.*, 2006; Ni *et al.*, 2005; Fig. 4). Moreover, the TAP proteins interact and modulate PI3K, suggesting a direct role of these proteins in modulating signal transduction by vitamin E (Kempna *et al.*, 2003b; Ni *et al.*, 2005). These proteins are also known to bind squalene and phospholipids, such as phosphatidylglycerol, phosphatidylinositol, phosphatidylserine, and phosphatidic acid (Caras *et al.*, 1980), and competition with these lipids by vitamin E may interfere with signal transduction mediated by these lipids (Kempna *et al.*, 2003a,b; Ni *et al.*, 2005). Although there is still limited data about the function of the TAP proteins, the available results suggest that these proteins are involved in uptake of tocopherol and tocopheryl succinate, and may also influence PI3K activity, but a role in mast cell signal transduction requires further experimental proof.

4. Modulation of Other Events in Mast Cells

Vitamin E may also indirectly influence signal transduction in mast cells, for example, by influencing the expression of antibodies or cytokines relevant to mast cell development. Human and mouse models have shown that high vitamin E intake correlates with low IgE concentration and reduced prevalence of allergic reactions (Li-Weber *et al.*, 2002). In peripheral blood T cells, vitamin E inhibits interleukin-4 (IL-4) gene expression by blocking binding of transcription factors to two important IL-4 promoter-binding sites for NF-κB and AP-1 and interferes with promoter activity on T-cell activation (Li-Weber *et al.*, 2002). IL-4 plays a pivotal role in the development of allergic inflammation via induction of IgE isotype switching, increase of IgE receptor expression, promoting Th2-cell differentiation, and stimulating several genes involved in atopic disorders, giving experimental support for a beneficial effect of dietary vitamin E on atopic disorders.

Mast cells express several peroxisome proliferators-activated receptors (PPARβ and PPARγ1/2), which play a role in the negative regulation of mast cell activation (Sugiyama *et al.*, 2000). PPARγ expression is induced by tocopherol analogues (Campbell *et al.*, 2003; Davies *et al.*, 2002; De Pascale *et al.*, 2006; Hsieh *et al.*, 2006), suggesting that mast cell activation may be influenced by vitamin E analogues via modulation of PPARγ.

C. MAST CELL DEGRANULATION AND VITAMIN E

1. Modulation of Signal Transduction Involved in Mast Cell Degranulation by Vitamin E

Other events influenced by signal transduction enzymes such as secretion of inflammatory cytokines and proteases, migration, or modulation of differentiation and survival could also be affected by vitamin E. Several studies have shown that vitamin E inhibits PKC activity via activation of PP2A leading to dephosphorylation of PKC (Egger *et al.*, 2003; Neuzil *et al.*, 2001; Ricciarelli *et al.*, 1998). Interestingly, serine/threonine phosphatases and in particular PP2A play an important role in mast cell degranulation (reviewed in Sim *et al.*, 2006), and it remains to be shown whether vitamin E can affect the degranulation process via activation of PP2A or via inhibition of PKC and PKB activity.

Degranulation and cytokine production are promoted in synergy by SCF via c-kit receptor and by antigen exposure via FcϵRI, leading to activation of phospholipase Cγ, calcium mobilization, activation of several PKC isoforms, mitogen-activated protein kinases (MAPK), and the PI3K/PKB signal transduction pathway (Hundley *et al.*, 2004; Seebeck *et al.*, 2001). PKC is well known to be inhibited by vitamin E via activation of PP2A (Egger *et al.*, 2003; Neuzil *et al.*, 2001; Ricciarelli *et al.*, 1998), and it is open whether mast cell degranulation after stimulation with SCF and antigen exposure is affected by vitamin E.

PKB is activated by enzyme translocation to the plasma membrane, where it recognizes phosphatidylinositol-3,4,5-phosphate and phosphatidylinositol-3,4-phosphate, and subsequent phosphorylation at Thr308 and Ser473; these events are dependent onphosphatidylinositol-3-kinase which generates phosphatidylinositol-3,4,5-phosphate, and the Src homology 2-containing inositol phosphatase (SHIP-2) which hydrolizes it, and both enzymes have been shown to modulate mast cell degranulation (Huber *et al.*, 1998, 2000; Wymann *et al.*, 2003). In addition to that, phosphatidylinositol-3-kinase leads also to activation of phospholipase Cγ1, which hydrolyzes phosphatidylinositol-3,4,5-phosphate and generates inositol phosphates leading to Ca^{2+} mobilization and subsequent degranulation (Djouder *et al.*, 2001). That PKB can modulate secretion was shown in FcϵRI-stimulated mast cells, in which PKB regulates the transcriptional activation of cytokine genes via NF-κB, NF-AT, and AP-1, thus affecting the production and secretion of IL-2 and TNF-α (Kitaura *et al.*, 2000).

2. Modulation of Mast Cell Degranulation by Vitamin E in Cellular and Animal Studies

In several studies, vitamin E was indeed found to modulate degranulation of mast cells induced by specific triggers. In many of these studies, vitamin E reduced mast cell degranulation by scavenging free radicals; since vitamin E also reduces the production of free radicals by NADPH-oxidase (Cachia *et al.*, 1998) or mitochondria (Chow, 2001), activates several drug-metabolizing enzymes (cytochrome P450 CYP3A and CYP4F2) (reviewed by Brigelius-Flohe, 2005), and modulates signal transduction and gene expression (reviewed by Zingg and Azzi, 2004, 2006), alternative modes of action can also be envisioned.

The effect of vitamin E on mediator activity and release was studied in a canine mastocytoma cell line (C2) stimulated by the wasp venom peptide mastoparan as a model for canine atopic dermatitis. Vitamin E reduced histamine and prostaglandin D2 (PGD2) release in both nonstimulated as well as mastoparan-stimulated C2 cells; vitamin E also decreased the activity of chymase, whereas tryptase activity was not influenced (Gueck *et al.*, 2002). Similar effects of vitamin E were described in rat mast cells stimulated with compound 48/80 (C48/80), a mast cell degranulator. In these cells, vitamin E inhibited histamine release by 70% without impairing the uptake of Ca^{2+}. In the absence of the stimulus, vitamin E increased the cell-associated Ca^{2+}; however, it had no effect on spontaneous release of histamine (Ranadive and Lewis, 1982). Taken together, vitamin E decreased the production and release of inflammatory mediators in mast cells, suggesting that vitamin E might have a possible beneficial effect in inflammatory and allergic diseases.

In rats treated once with C48/80, the formation and progression of gastric mucosal lesions was associated with increased lipid peroxide content and xanthine oxidase and myeloperoxidase activities, and decreased vitamin E and hexosamine contents and selenium-dependent glutathione peroxidase activity (Ohta *et al.*, 1997). Vitamin E (α-tocopherol) administered at 0.5 h after C48/80 treatment reduced progressive gastric mucosal lesions dose dependent, similar to superoxide dismutase plus catalase administered at the same time point. In the gastric mucosa of C48/80-untreated rats, orally administered with vitamin E, thiobarbituric acid-reactive substances content decreased with an increase in vitamin E content. These results indicate that orally administered vitamin E prevents acute gastric mucosal lesion progression in C48/80-treated rats possibly by suppressing oxidative stress, neutrophil infiltration, and mucus depletion in the gastric mucosa (Ohta *et al.*, 2006). In these experiments, only α-tocopherol was used, and it is possible that other vitamin E analogues and derivatives with increased activity on signal transduction and gene expression may show even a better performance.

Several drugs, such as paracetamol, phenobarbital, morphine, cocaine, and methadone, induced the release of histamine after metabolic oxidation and the generation of reactive oxygen intermediates. The incubation of isolated rat serosal mast cells with paracetamol evoked the release of histamine only in the presence of (S10) liver microsomes obtained from PCB or phenobarbital-treated rats. The release of histamine was not accompanied by a leakage of lactate dehydrogenase and was blocked by α-tocopherol and by D-mannitol, a hydroxyl free radicals scavenger, suggesting that mast cell exocytosis is activated by the generation of reactive intermediates after metabolic oxidation of paracetamol (Masini *et al.*, 1986). Similar to that polyunsaturated fatty acids (PUFA: arachidonic and linoleic acid) released histamine from isolated purified rat serosal mast cells only in the presence of oxidizing systems such as phenobarbital-induced rat liver microsomes, prostaglandin-H-synthetase (PHS), or soybean lipoxygenase. The secretion of histamine was inhibited by anti-free radical interventions such as D-mannitol, reduced glutathione, and α-tocopherol. Some cyclooxygenase and lipoxygenase inhibitors, cimetidine and carnitine derivatives, were differentially active in the inhibition of mast cell histamine release by activated arachidonic acid. These results suggest that free radical derivatives of PUFA, generated by metabolic activation, triggered mast cell histamine release (Masini *et al.*, 1990). Similar to that the release leukotrienes (LTB4) from rat RBL-2H3 basophilic leukemia cell, induced by treatment with short-chain fatty acids, was inhibited by α-tocopherol (Yamada *et al.*, 1996b).

Incubation of isolated purified rat serosal mast cells with morphine, cocaine, methadone, and oxidative enzymes (PHS; rat liver homogenate fraction S10-mix) led to release of mast cell histamine and the generation of malonedialdehyde (MDA) in the presence of these oxidative enzymes. Histamine release and MDA generation were abated by the free radical scavengers: reduced glutathione, 100-μM GSH, and 100-μM α-tocopherol; and by 100 μM of the spin trapper 5.5-dimethyl-1-pyrroline-*N*-oxide (DMPO), suggesting that morphine, cocaine, and methadone were activated into free radicals which produce membrane lipid perturbation and histamine release (Di Bello *et al.*, 1998). No effect of α-tocopherol was found on histamine secretion induced by the antineoplastic drug adriamycin (Decorti *et al.*, 1987) or by hydrogen peroxide (H_2O_2) (Ogasawara *et al.*, 1986).

Among several vitamin E glycosides, synthesized as potential prodrugs of vitamin E, two glycosides, DL-α-tocopherylglucoside and DL-α-tocopherylmannoside, showed strong inhibitory action on histamine release from mast cells, and DL-α-tocopherylmannoside in addition showed a suppressive action on IgE antibody formation. Thus, certain vitamin E analogues such as DL-α-tocopherylmannoside may have antiallergic and antiinflammatory activity (Satoh *et al.*, 2001).

III. PREVENTIVE EFFECTS OF VITAMIN E ON DISEASES WITH MAST CELL INVOLVEMENT

A. VITAMIN E, MAST CELLS, AND ASTHMA

Asthma is associated with chronic inflammation and hyperreactivity of the airways caused by the recruitment and interplay of several inflammatory cells (eosinophils, neutrophils, macrophages, mast cells, as well as bronchial smooth muscle cells), that secrete a variety of cytokines, chemokines, and proteases (Bochner and Busse, 2005; Sim *et al.*, 2006). The effects of α-tocopherol or probucol supplementation in an asthma mouse model on airway hyperresponsiveness were recently investigated; α-tocopherol supplementation reduced IL-4, IL-5, and diminished inflammatory cells and mucus secretion in lung tissues (Okamoto *et al.*, 2006). No effect was observed on serum IgE, 8-isoprostane, and acrolein levels in bronchoalveolar lavage fluid, suggesting that the observed events are cause by immunmodulation unrelated to antioxidation (Okamoto *et al.*, 2006).

Leukotrienes are involved in the pathogenesis of asthma, and recent data indicate that individuals with asthma may have enhanced basal excretion of urinary leukotriene E4 compared with normal individuals. In individuals with asthma, receptor-mediated activation of neutrophils resulted in the synthesis of leukotrienes from arachidonic acid via the 5-LOX pathway, an enzyme that is inhibited by α-tocopherol and α-tocopheryl acetate. This activation was inhibited by α-tocopherol in a concentration-dependent manner, suggesting possible beneficial effects in asthmatic/allergic patients (Centanni *et al.*, 2001).

Increased dietary vitamin E intake is associated with a reduced incidence of asthma, and combinations of antioxidant supplements, including vitamin E, are effective in reducing ozone-induced bronchoconstriction. However, no benefit of dietary supplementation with vitamin E was found in a parallel group randomized placebo-controlled trial in adults with mild to moderate asthma (Pearson *et al.*, 2004). The inflammatory response to ozone in atopic asthma suggests that soluble mediators of inflammation are released in response to oxidant stress. The impact of antioxidant supplementation on the nasal inflammatory response to ozone exposure in atopic asthmatic children was investigated in a randomized trial using a double-blinded design. The supplementation of vitamins C and E above the minimum dietary requirement in asthmatic children with a low intake of vitamin E provided some protection against the nasal acute inflammatory response to ozone (Sienra-Monge *et al.*, 2004). In a mouse model for nasal allergy, vitamin E supplementation suppressed nasal allergic response, and decreased splenic lymphproliferation and the production of IL-4 and IL-5 and of total serum IgE (Zheng *et al.*, 1999). In a study in patients with seasonal allergic rhinitis, vitamin E supplementation reduced nasal symptom scores (Shahar *et al.*, 2004), whereas in another study with patients with perennial allergic rhinitis, there were no significant effects (Montano Velazquez *et al.*, 2006). Plasma γ-tocopherol levels were negatively associated with allergic sensitization as categorized by IgE levels, whereas α-tocopherol showed no association (Kompauer *et al.*, 2006).

B. VITAMIN E, MAST CELLS, AND SKIN DISEASES

Mast cells accumulate not only around cutaneous malignancies, such as atopic dermatitis, contact dermatitis, but also around skin tumors such as basal cell carcinoma, squamous cell carcinoma, and malignant melanoma. As reviewed by Ch'ng *et al.* (2006), the accumulated mast cells may exert either promoting or inhibitory effects on these tumors. The tumor microenvironment is changed by mast cells by releasing several cytokines, growth factors, proteoglycans, and proteases, that can lead to local, ultraviolet (UV) B-induced immunosuppression, increased angiogenesis at the tumor–host interface, degradation of extracellular matrix facilitating tumor invasion, leading to remodeling and release of sequestered latent stores of angiogenic factors and increased mitogenesis. Mast cells in skin could be influenced either directly by topical application or indirectly by oral consumption of dietary supplements and their possible metabolism to bioactive compounds (Richelle *et al.*, 2006; Theoharides and Bielory, 2004).

In a model for canine atopic dermatitis, vitamin E inhibited histamine, PGD2, and chymase release both in unstimulated and mastoparan-stimulated canine mastocytoma cells (C2), but the signaling pathways involved have not been resolved in detail (Gueck *et al.*, 2002). A correlation between vitamin E intake, IgE levels, and the clinical manifestations of atopy suggests that vitamin E supplemenation could alleviate atopic dermatitis (Tsoureli-Nikita *et al.*, 2002). In allergic model mice using subcutaneously immunization with ovalbumin (OVA), the IgE production was reduced by vitamin E. The combination of dietary vitamin E and β-carotene suppressed IgE production with possible preventive effects on the allergic reaction (Bando *et al.*, 2003). An inverse association between dietary vitamin E intake and the serum IgE concentrations and the frequency of allergen skin sensitization was also found in a random sample of 2633 adults (Fogarty *et al.*, 2000). Vitamin E inhibited IgE production in rat gut lymphocytes (Yamada *et al.*, 1996a), and reduced total serum IgE, nasal responsiveness, and sneezing response in a mouse model of allergy (Zheng *et al.*, 1999).

Occupational exposure to metal working fluids (MWFs) causes allergic and irritant contact dermatitis. Dermal exposure of mice to 5% MWFs for 3 months resulted in accumulation of mast cells and elevation of histamine in the skin. In vitamin E-deficient mice, the exposure of skin with MWFs revealed a 53% increase in mast cell accumulation with a concomitant decrease of total antioxidant reserve in skin of 66% as compared to vitamin E-sufficient mice, suggesting that topical exposure to MWFs increases oxidative stress thus enhancing mast cell accumulation (Shvedova *et al.*, 2002).

Topical application of the γ-tocopherol ester-linked derivative γ-tocopheryl-*N,N*-dimethylglycinate hydrochloride (γ-TDMG) prevents UV-induced inflammation by acting as an antioxidant, suppresses iNOS expression, and directly inhibits COX-2 activity (Yoshida *et al.*, 2006). Although vitamin E esters have been widely and safely used for decades in dermatologic preparations and cosmetics, a few cases of contact dermitis linked to treatment with tocopheryl esters (tocopheryl acetate, tocopheryl nicotinate, tocopheryl linoleate) have been described, possibly involving mast cells (de Groot *et al.*, 1991; Perrenoud *et al.*, 1994).

C. VITAMIN E, MAST CELLS, AND ATHEROSCLEROSIS

Atherosclerosis is a multifactorial process with many characteristics of an inflammatory disease, major cellular participants include monocytes, macrophages, mast cells, endothelial cells, T lymphocytes, platelets, and vascular smooth muscle cells (VSMCs) (Kelley *et al.*, 2000; Ross, 1995). Activation of these cells leads to release of hydrolytic enzymes, cytokines, chemokines, and growth factors that can result in further injury. Elevated blood histamine and isoprostane (8-iso$PGF_{2\alpha}$) levels are associated with coronary artery disease, cardiac events, inflammation, and atherosclerosis, suggesting the involvement of histamine-producing cells (mast cells, monocytes/macrophages, T cells) in these diseases (Clejan *et al.*, 2002; Schneider *et al.*, 2002).

Increased numbers of mast cells were found in atherosclerotic lesions when compared with normal intima and are often associated with macrophages and extracellular lipids (Jeziorska *et al.*, 1997), in particular in fatty streaks and the shoulder regions of atheromas (Kaartinen *et al.*, 1994). Activated mast cells contribute to foam cell and fatty streak formation by stimulating LDL modification and uptake by macrophages (Kovanen, 1996). Moreover, oxidized lipoproteins (oxLDL) activate mast cells and increase inflammatory IL-8 expression (Kelley *et al.*, 2006). In human blood vessels, mast cells have been observed in the intima of carotic arteries at sites of hemodynamic stress, together with monocytes, T lymphocytes, and dendritic cells (Waltner-Romen *et al.*, 1998). It is still unclear, to what degree a higher density of mast cells in plaques is the result of increased recruitment or the consequence of higher proliferation (Frangogiannis and Entman, 2000). Activated mast cells can contribute to foam cells and fatty streak formation by stimulating LDL modification and uptake by macrophages (Kovanen, 1996), by secreting a variety of inflammatory mediators (histamine, cytokines, chemokines, leukotrienes, prostaglandins, platelet activating factor) and enzymes (tryptase, chymase, carboxypeptidase, and cathepsin G) (reviewed in Kelley *et al.*, 2000), possibly leading to weakening and rupture of atherosclerotic plaques (Leskinen *et al.*, 2003).

In this context it is interesting to note that VSMCs of the media were found to express c-kit and SCF suggesting the existence of mast cell–VSMC interaction and of an autocrine loop of c-kit and its ligand on the surface of VSMC (Miyamoto *et al.*, 1997). Injured vessels revealed that c-kit expression within the media and neointima is significantly increased following injury (Hollenbeck *et al.*, 2004) and during in-stent restenosis (Hibbert *et al.*, 2004). It can be assumed that mast cells signaling and proliferation are deregulated during the progression of atherosclerosis, and normalization of mast cell signal transduction by vitamin E could thus play a beneficial role.

IV. Summary

The tocopherols affect the proliferation and survival of HMC-1 mastocytoma cells by influencing signal transduction. All four tocopherols are able to inhibit cell proliferation, albeit with different timing and potency; δ-tocopherol even leads to the induction of apoptosis at later time points. Repression of growth is mediated via inhibition of the c-kit/PI3K/PKB pathway, as judged by analysis of PKB phosphorylation, without the involvement of the PKC and ERK pathways and independent of the radical scavenging activities of the tocopherols. The inhibitory effect can be localized in the events before PKB activation, but after the bifurcation of the c-kit signal transduction to the Ras/MAPK/ERK1/2 and the PI3K/PKB/GSK3α/β pathway. By affecting the activity of PP2A, PKC, and PKB, vitamin E could furthermore modulate secretion, migration, and degranulation of mast cells.

The finding that the tocopherols and certain natural compounds only partially inhibit the PKB signaling pathway could be a key for their daily use as dietary chemopreventive agents against diseases like atherosclerosis and allergies. These compounds may normalize aberrant signaling and gene expression resulting from inherent cellular heterogeneity or environmental cues; these compounds may also interfere with activation of cell signaling resulting from inflammatory processes in the prepathological state which, if not normalized, would turn into overt pathology. Specific and more potent inhibitors would then be required capable of stopping disease progression by completely blocking a deregulated signaling pathway.

The known data about uptake and metabolism of the natural and synthetic vitamin E analogues suggest that the concentrations affecting mast cell signaling are difficult to be reached in human plasma even after supplementation. However, the preferential location of mast cells in the proximity of tissues that interface with the external environment (the epithelial surface of the skin, the gastrointestinal mucosa, and the respiratory system) may render these cells accessible to bioactive compounds such as δ-tocopherol or the tocotrienols, known not to reach high plasma concentrations, even after oral supplementation. Such compounds could be considered for topic application, for example for cosmetics as anti-inflammatory or antiallergic lotions, or for aerosolized applications in the respiratory tract. Moreover, since phenolic compounds have been demonstrated to be enriched in the gastrointestinal tract (Halliwell *et al.*, 2005), they could also be considered as preventive treatment against intestinal inflammatory and allergic events or colon cancer. It furthermore appears to be possible that even a constant and suboptimal uptake of these compounds may lead to sufficient amounts in peripheral tissues able to prevent over time inflammation, allergy, cancer, as well as cardiovascular and neurodegenerative diseases.

REFERENCES

Andjelkovic, M., Jakubowicz, T., Cron, P., Ming, X. F., Han, J. W., and Hemmings, B. A. (1996). Activation and phosphorylation of a pleckstrin homology domain containing protein kinase (RAC-PK/PKB) promoted by serum and protein phosphatase inhibitors. *Proc. Natl. Acad. Sci. USA* **93,** 5699–5704.

Arita, M., Sato, Y., Miyata, A., Tanabe, T., Takahashi, E., Kayden, H. J., Arai, H., and Inoue, K. (1995). Human alpha-tocopherol transfer protein: cDNA cloning, expression and chromosomal localization. *Biochem. J.* **306,** 437–443.

Azzi, A., Gysin, R., Kempna, P., Ricciarelli, R., Villacorta, L., Visarius, T., and Zingg, J. M. (2002a). Regulation of gene and protein expression by vitamin E. *Free Radic. Res.* **36,** 30–35.

Azzi, A., Ricciarelli, R., and Zingg, J. M. (2002b). Non-antioxidant molecular functions of alpha-tocopherol (vitamin E). *FEBS Lett.* **519,** 8–10.

Azzi, A., Gysin, R., Kempna, P., Munteanu, A., Negis, Y., Villacorta, L., Visarius, T., and Zingg, J. M. (2004). Vitamin E mediates cell signaling and regulation of gene expression. *Ann. N Y Acad. Sci.* **1031,** 86–95.

Bando, N., Yamanishi, R., and Terao, J. (2003). Inhibition of immunoglobulin E production in allergic model mice by supplementation with vitamin E and beta-carotene. *Biosci. Biotechnol. Biochem.* **67,** 2176–2182.

Benoist, C., and Mathis, D. (2002). Mast cells in autoimmune disease. *Nature* **420,** 875–878.

Bochner, B. S., and Busse, W. W. (2005). Allergy and asthma. *J. Allergy Clin. Immunol.* **115,** 953–959.

Brigelius-Flohe, R. (2005). Induction of drug metabolizing enzymes by vitamin E. *J. Plant Physiol.* **162,** 797–802.

Brigelius-Flohe, R., Kelly, F. J., Salonen, J. T., Neuzil, J., Zingg, J. M., and Azzi, A. (2002). The European perspective on vitamin E: Current knowledge and future research. *Am. J. Clin. Nutr.* **76,** 703–716.

Cachia, O., Benna, J. E., Pedruzzi, E., Descomps, B., Gougerot-Pocidalo, M. A., and Leger, C. L. (1998). Alpha-tocopherol inhibits the respiratory burst in human monocytes. Attenuation of p47(phox) membrane translocation and phosphorylation. *J. Biol. Chem.* **273,** 32801–32805.

Campbell, S. E., Stone, W. L., Whaley, S. G., Qui, M., and Krishnan, K. (2003). Gamma (γ) tocopherol upregulates peroxisome proliferator activated receptor (PPAR) gamma (γ) expression in SW 480 human colon cancer cell lines. *BMC Cancer* **3**(15), 1–13.

Caras, I. W., Friedlander, E. J., and Bloch, K. (1980). Interactions of supernatant protein factor with components of the microsomal squalene epoxidase system. Binding of supernatant protein factor to anionic phospholipids. *J. Biol. Chem.* **255,** 3575–3580.

Centanni, S., Santus, P., Di Marco, F., Fumagalli, F., Zarini, S., and Sala, A. (2001). The potential role of tocopherol in asthma and allergies: Modification of the leukotriene pathway. *BioDrugs* **15,** 81–86.

Chan, S. S., Monteiro, H. P., Schindler, F., Stern, A., and Junqueira, V. B. (2001). Alpha-tocopherol modulates tyrosine phosphorylation in human neutrophils by inhibition of protein kinase C activity and activation of tyrosine phosphatases. *Free Radic. Res.* **35,** 843–856.

Ch'ng, S., Wallis, R. A., Yuan, L., Davis, P. F., and Tan, S. T. (2006). Mast cells and cutaneous malignancies. *Mod. Pathol.* **19,** 149–159.

Chow, C. K. (2001). Vitamin E regulation of mitochondrial superoxide generation. *Biol. Signals Recept.* **10,** 112–124.

Clejan, S., Japa, S., Clemetson, C., Hasabnis, S. S., David, O., and Talano, J. V. (2002). Blood histamine is associated with coronary artery disease, cardiac events and severity of inflammation and atherosclerosis. *J. Cell. Mol. Med.* **6,** 583–592.

Coulter, I. D., Hardy, M. L., Morton, S. C., Hilton, L. G., Tu, W., Valentine, D., and Shekelle, P. G. (2006). Antioxidants vitamin C and vitamin E for the prevention and treatment of cancer. *J. Gen. Intern. Med.* **21,** 735–744.

Davies, G. F., McFie, P. J., Khandelwal, R. L., and Roesler, W. J. (2002). Unique ability of troglitazone to up-regulate peroxisome proliferator-activated receptor-gamma expression in hepatocytes. *J. Pharmacol. Exp. Ther.* **300,** 72–77.

de Groot, A. C., Berretty, P. J., van Ginkel, C. J., den Hengst, C. W., van Ulsen, J., and Weyland, J. W. (1991). Allergic contact dermatitis from tocopheryl acetate in cosmetic creams. *Contact Dermatitis* **25,** 302–304.

De Pascale, M. C., Bassi, A. M., Patrone, V., Villacorta, L., Azzi, A., and Zingg, J. M. (2006). Increased expression of transglutaminase-1 and PPARgamma after vitamin E treatment in human keratinocytes. *Arch. Biochem. Biophys.* **447,** 97–106.

Decorti, G., Klugmann, F. B., Candussio, L., and Baldini, L. (1987). Adriamycin-induced histamine release from rat peritoneal mast cells: Role of free radicals. *Anticancer Res.* **7,** 497–499.

Delcommenne, M., Tan, C., Gray, V., Rue, L., Woodgett, J., and Dedhar, S. (1998). Phosphoinositide-3-OH kinase-dependent regulation of glycogen synthase kinase 3 and protein kinase B/AKT by the integrin-linked kinase. *Proc. Natl. Acad. Sci. USA* **95,** 11211–11216.

Di Bello, M. G., Masini, E., Ioannides, C., Fomusi Ndisang, J., Raspanti, S., Bani Sacchi, T., and Mannaioni, P. F. (1998). Histamine release from rat mast cells induced by the metabolic activation of drugs of abuse into free radicals. *Inflamm. Res.* **47,** 122–130.

Djouder, N., Schmidt, G., Frings, M., Cavalie, A., Thelen, M., and Aktories, K. (2001). Rac and phosphatidylinositol 3-kinase regulate the protein kinase B in Fc epsilon RI signaling in RBL 2H3 mast cells. *J. Immunol.* **166,** 1627–1634.

Egger, T., Schuligoi, R., Wintersperger, A., Amann, R., Malle, E., and Sattler, W. (2003). Vitamin E (alpha-tocopherol) attenuates cyclo-oxygenase 2 transcription and synthesis in immortalized murine BV-2 microglia. *Biochem. J.* **370,** 459–467.

Feger, F., Varadaradjalou, S., Gao, Z., Abraham, S. N., and Arock, M. (2002). The role of mast cells in host defense and their subversion by bacterial pathogens. *Trends Immunol.* **23,** 151–158.

Feng, J., Park, J., Cron, P., Hess, D., and Hemmings, B. A. (2004). Identification of a PKB/Akt hydrophobic motif Ser-473 kinase as DNA-dependent protein kinase. *J. Biol. Chem.* **279,** 41189–41196.

Florence, T. M. (1995). The role of free radicals in disease. *Aust. NZJ. Ophthalmol.* **23,** 3–7.

Fogarty, A., Lewis, S., Weiss, S., and Britton, J. (2000). Dietary vitamin E, IgE concentrations, and atopy. *Lancet* **356,** 1573–1574.

Frangogiannis, N. G., and Entman, M. L. (2000). Mast cells in myocaridal ischemia and reperfusion. *In* "Mast Cells and Basophils" (G. Marone, L. M. Lichtenstein, and S. J. Galli, Eds.), pp. 507–522. Academic Press, Harcourt Publishers, London.

Furitsu, T., Tsujimura, T., Tono, T., Ikeda, H., Kitayama, H., Koshimizu, U., Sugahara, H., Butterfield, J. H., Ashman, L. K., Kanayama, Y., Matsuzawa, Y., Kitamura, Y., *et al.* (1993). Identification of mutations in the coding sequence of the proto-oncogene c-kit in a human mast cell leukemia cell line causing ligand-independent activation of c-kit product. *J. Clin. Invest.* **92,** 1736–1744.

Galli, S. J., and Hammel, I. (1994). Mast cell and basophil development. *Curr. Opin. Hematol.* **1,** 33–39.

Goti, D., Hammer, A., Galla, H. J., Malle, E., and Sattler, W. (2000). Uptake of lipoprotein-associated alpha-tocopherol by primary porcine brain capillary endothelial cells. *J. Neurochem.* **74,** 1374–1383.

Gueck, T., Aschenbach, J. R., and Fuhrmann, H. (2002). Influence of vitamin E on mast cell mediator release. *Vet. Dermatol.* **13,** 301–305.

Guinea Viniegra, J., Martinez, N., Modirassari, P., Hernandez Losa, J., Parada Cobo, C., Sanchez-Arevalo Lobo, V. J., Avceves-Luquero, C. I., Alavrez-Vallina, L., Ramon, Y. C. S., Rojas, J. M., and Sanchez-Prieto, R. (2004). Full activation of PKB/Akt in response to insulin or ionizing radiation is mediated through ATM. *J. Biol. Chem.* **280,** 429–436.

Halliwell, B., Rafter, J., and Jenner, A. (2005). Health promotion by flavonoids, tocopherols, tocotrienols, and other phenols: Direct or indirect effects? Antioxidant or not? *Am. J. Clin. Nutr.* **81,** 268S–276S.

Hathcock, J. N., Azzi, A., Blumberg, J., Bray, T., Dickinson, A., Frei, B., Jialal, I., Johnston, C. S., Kelly, F. J., Kraemer, K., Packer, L., Parthasarathy, S., *et al.* (2005). Vitamins E and C are safe across a broad range of intakes. *Am. J. Clin. Nutr.* **81,** 736–745.

Henz, B. M., Maurer, M., Lippert, U., Worm, M., and Babina, M. (2001). Mast cells as initiators of immunity and host defense. *Exp. Dermatol.* **10,** 1–10.

Hibbert, B., Chen, Y. X., and O'Brien, E. R. (2004). c-Kit-immunopositive vascular progenitor cells populate human coronary in-stent restenosis but not primary atherosclerotic lesions. *Am. J. Physiol. Heart Circ. Physiol.* **287,** H518–H524.

Hill, M. M., and Hemmings, B. A. (2002). Inhibition of protein kinase B/Akt. Implications for cancer therapy. *Pharmacol. Ther.* **93,** 243–251.

Hiromatsu, Y., and Toda, S. (2003). Mast cells and angiogenesis. *Microsc. Res. Tech.* **60,** 64–69.

Hollenbeck, S. T., Sakakibara, K., Faries, P. L., Workhu, B., Liu, B., and Kent, K. C. (2004). Stem cell factor and c-kit are expressed by and may affect vascular SMCs through an autocrine pathway. *J. Surg. Res.* **120,** 288–294.

Hsieh, C. C., Huang, C. J., and Lin, B. F. (2006). Low and high levels of alpha-tocopherol exert opposite effects on IL-2 possibly through the modulation of PPAR-gamma, IkappaBalpha, and apoptotic pathway in activated splenocytes. *Nutrition* **22,** 433–440.

Huber, M., Helgason, C. D., Damen, J. E., Liu, L., Humphries, R. K., and Krystal, G. (1998). The src homology 2-containing inositol phosphatase (SHIP) is the gatekeeper of mast cell degranulation. *Proc. Natl. Acad. Sci. USA* **95,** 11330–11335.

Huber, M., Hughes, M. R., and Krystal, G. (2000). Thapsigargin-induced degranulation of mast cells is dependent on transient activation of phosphatidylinositol-3 kinase. *J. Immunol.* **165,** 124–133.

Huizinga, J. D., Thuneberg, L., Kluppel, M., Malysz, J., Mikkelsen, H. B., and Bernstein, A. (1995). W/kit gene required for interstitial cells of Cajal and for intestinal pacemaker activity. *Nature* **373,** 347–349.

Hundley, T. R., Gilfillan, A. M., Tkaczyk, C., Andrade, M. V., Metcalfe, D. D., and Beaven, M. A. (2004). Kit and FcepsilonRI mediate unique and convergent signals for release of inflammatory mediators from human mast cells. *Blood* **104,** 2410–2417.

Hybertson, B. M., Leff, J. A., Beehler, C. J., Barry, P. C., and Repine, J. E. (1995). Effect of vitamin E deficiency and supercritical fluid aerosolized vitamin E supplementation on interleukin-1-induced oxidative lung injury in rats. *Free Radic. Biol. Med.* **18,** 537–542.

Hybertson, B. M., Chung, J. H., Fini, M. A., Lee, Y. M., Allard, J. D., Hansen, B. N., Cho, O. J., Shibao, G. N., and Repine, J. E. (2005). Aerosol-administered alpha-tocopherol attenuates lung inflammation in rats given lipopolysaccharide intratracheally. *Exp. Lung Res.* **31,** 283–294.

Ikeda, T., Murakami, M., and Funaba, M. (2004). Expression of tocopherol-associated protein in mast cells. *Clin. Diagn. Lab. Immunol.* **11,** 1189–1191.

Jacobs-Helber, S. M., Penta, K., Sun, Z., Lawson, A., and Sawyer, S. T. (1997). Distinct signaling from stem cell factor and erythropoietin in HCD57 cells. *J. Biol. Chem.* **272,** 6850–6853.

Jeziorska, M., McCollum, C., and Woolley, D. E. (1997). Mast cell distribution, activation, and phenotype in atherosclerotic lesions of human carotid arteries. *J. Pathol.* **182,** 115–122.

Jiang, Q., Wong, J., Fyrst, H., Saba, J. D., and Ames, B. N. (2004). γ-Tocopherol or combinations of vitamin E forms induce cell death in human prostate cancer cells by interrupting sphingolipid synthesis. *Proc. Natl. Acad. Sci. USA* **101,** 17825–17830.

Kaartinen, M., Penttila, A., and Kovanen, P. T. (1994). Accumulation of activated mast cells in the shoulder region of human coronary atheroma, the predilection site of atheromatous rupture. *Circulation* **90,** 1669–1678.

Kawakami, Y., Nishimoto, H., Kitaura, J., Maeda-Yamamoto, M., Kato, R. M., Littman, D. R., Rawlings, D. J., and Kawakami, T. (2004). Protein kinase C betaII regulates Akt phosphorylation on Ser-473 in a cell type- and stimulus-specific fashion. *J. Biol. Chem.* **279,** 47720–47725.

Kelley, J. L., Chi, D. S., Abou-Auda, W., Smith, J. K., and Krishnaswamy, G. (2000). The molecular role of mast cells in atherosclerotic cardiovascular disease. *Mol. Med. Today* **6,** 304–308.

Kelley, J., Hemontolor, G., Younis, W., Li, C., Krishnaswamy, G., and Chi, D. S. (2006). Mast cell activation by lipoproteins. *Methods Mol. Biol.* **315,** 341–348.

Kempna, P., Cipollone, R., Villacorta, L., Ricciarelli, R., and Zingg, J. M. (2003a). Isoelectric point mobility shift assay for rapid screening of charged and uncharged ligands bound to proteins. *IUBMB Life* **55,** 103–107.

Kempna, P., Zingg, J. M., Ricciarelli, R., Hierl, M., Saxena, S., and Azzi, A. (2003b). Cloning of novel human SEC14p-like proteins: Cellular localization, ligand binding and functional properties. *Free Radic. Biol. Med.* **34,** 1458–1472.

Kempna, P., Reiter, E., Arock, M., Azzi, A., and Zingg, J. M. (2004). Inhibition of HMC-1 mast cell proliferation by vitamin E: Involvement of the protein kinase B pathway. *J. Biol. Chem.* **279,** 50700–50709.

Khanna, S., Patel, V., Rink, C., Roy, S., and Sen, C. K. (2005). Delivery of orally supplemented alpha-tocotrienol to vital organs of rats and tocopherol-transport protein deficient mice. *Free Radic. Biol. Med.* **39,** 1310–1319.

Kitaura, J., Asai, K., Maeda-Yamamoto, M., Kawakami, Y., Kikkawa, U., and Kawakami, T. (2000). Akt-dependent cytokine production in mast cells. *J. Exp. Med.* **192,** 729–740.

Knopman, D. S. (2006). Current treatment of mild cognitive impairment and Alzheimer's disease. *Curr. Neurol. Neurosci. Rep.* **6,** 365–371.

Kolleck, I., Schlame, M., Fechner, H., Looman, A. C., Wissel, H., and Rustow, B. (1999). HDL is the major source of vitamin E for type II pneumocytes. *Free Radic. Biol. Med.* **27,** 882–890.

Kompauer, I., Heinrich, J., Wolfram, G., and Linseisen, J. (2006). Association of carotenoids, tocopherols and vitamin C in plasma with allergic rhinitis and allergic sensitisation in adults. *Public Health Nutr.* **9,** 472–479.

Kostner, G. M., Oettl, K., Jauhiainen, M., Ehnholm, C., Esterbauer, H., and Dieplinger, H. (1995). Human plasma phospholipid transfer protein accelerates exchange/transfer of alpha-tocopherol between lipoproteins and cells. *Biochem. J.* **305,** 659–667.

Kovanen, P. T. (1996). Mast cells in human fatty streaks and atheromas: Implications for intimal lipid accumulation. *Curr. Opin. Lipidol.* **7,** 281–286.

Krishnaswamy, G., Ajitawi, O., and Chi, D. S. (2006). The human mast cell: An overview. *Methods Mol. Biol.* **315,** 13–34.

Lee, D. M., Friend, D. S., Gurish, M. F., Benoist, C., Mathis, D., and Brenner, M. B. (2002). Mast cells: A cellular link between autoantibodies and inflammatory arthritis. *Science* **297,** 1689–1692.

Leskinen, M. J., Kovanen, P. T., and Lindstedt, K. A. (2003). Regulation of smooth muscle cell growth, function and death *in vitro* by activated mast cells–a potential mechanism for the weakening and rupture of atherosclerotic plaques. *Biochem. Pharmacol.* **66,** 1493–1498.

Li-Weber, M., Giaisi, M., Treiber, M. K., and Krammer, P. H. (2002). Vitamin E inhibits IL-4 gene expression in peripheral blood T cells. *Eur. J. Immunol.* **32,** 2401–2408.

Lonn, E., Bosch, J., Yusuf, S., Sheridan, P., Pogue, J., Arnold, J. M., Ross, C., Arnold, A., Sleight, P., Probstfield, J., and Dagenais, G. R. (2005). Effects of long-term vitamin E supplementation on cardiovascular events and cancer: A randomized controlled trial. *JAMA* **293,** 1338–1347.

Lu, D., Huang, J., and Basu, A. (2006). Protein kinase C-epsilon activates protein kinase B/AKT via DNA-PK to protect against TNF-induced cell death. *J. Biol. Chem.* **281,** 22799–22807.

Marone, G., Lichtenstein, L. M., and Galli, S. J. (2000). "Mast Cells and Basophils." Academic Press, Hartcourt Publisher, London.

Masini, E., Lodovici, M., Fantozzi, R., Brunelleschi, S., Conti, A., and Mannaioni, P. F. (1986). Histamine release by free radicals: Paracetamol-induced histamine release from rat peritoneal mast cells after in vitro activation by monooxygenase. *Agents Actions* **18,** 85–88.

Masini, E., Palmerani, B., Gambassi, F., Pistelli, A., Giannella, E., Occupati, B., Ciuffi, M., Sacchi, T. B., and Mannaioni, P. F. (1990). Histamine release from rat mast cells induced by metabolic activation of polyunsaturated fatty acids into free radicals. *Biochem. Pharmacol.* **39,** 879–889.

Maziere, C., Conte, M. A., and Maziere, J. C. (2001). Activation of JAK2 by the oxidative stress generated with oxidized low-density lipoprotein. *Free Radic. Biol. Med.* **31,** 1334–1340.

McCormick, C. C., and Parker, R. S. (2004). The cytotoxicity of vitamin E is both vitamer- and cell-specific and involves a selectable trait. *J. Nutr.* **134,** 3335–3342.

McIntyre, B. S., Briski, K. P., Gapor, A., and Sylvester, P. W. (2000). Antiproliferative and apoptotic effects of tocopherols and tocotrienols on preneoplastic and neoplastic mouse mammary epithelial cells. *Proc. Soc. Exp. Biol. Med.* **224,** 292–301.

Mecheri, S., and David, B. (1999). Functional and ultrastructural insights into the immunobiology of mouse and human mast cells. *Allergy Clin. Immunol. Int.* **11,** 226–233.

Metcalfe, D. D., Baram, D., and Mekori, Y. A. (1997). Mast cells. *Physiol. Rev.* **77,** 1033–1079.

Meydani, S. N., Meydani, M., Blumberg, J. B., Leka, L. S., Siber, G., Loszewski, R., Thompson, C., Pedrosa, M. C., Diamond, R. D., and Stollar, B. D. (1997). Vitamin E supplementation and *in vivo* immune response in healthy elderly subjects. A randomized controlled trial. *JAMA* **277,** 1380–1386.

Miller, E. R., III, Pastor-Barriuso, R., Dalal, D., Riemersma, R. A., Appel, L. J., and Guallar, E. (2005). Meta-analysis: High-dosage vitamin E supplementation may increase all-cause mortality. *Ann. Intern. Med.* **142,** 37–46.

Miyamoto, T., Sasaguri, Y., Sasaguri, T., Azakami, S., Yasukawa, H., Kato, S., Arima, N., Sugama, K., and Morimatsu, M. (1997). Expression of stem cell factor in human aortic endothelial and smooth muscle cells. *Atherosclerosis* **129,** 207–213.

Montano Velazquez, B. B., Jauregui-Renaud, K., Banuelos Arias Adel, C., Ayala, J. C., Martinez, M. D., Campillo Navarrete, R., Rosalia, I. S., Salazar Mdel, R., Serrano, H. A., Mondragon, A. O., and Perez, R. L. (2006). Vitamin E effects on nasal symptoms and serum specific IgE levels in patients with perennial allergic rhinitis. *Ann. Allergy Asthma Immunol.* **96,** 45–50.

Munteanu, A., Zingg, J. M., and Azzi, A. (2004). Anti-atherosclerotic effects of vitamin E—myth or reality? *J. Cell. Mol. Med.* **8,** 59–76.

Munteanu, A., Taddei, M., Tamburini, I., Bergamini, E., Azzi, A., and Zingg, J. M. (2006). Antagonistic effects of oxidized low density lipoprotein and alpha-tocopherol on CD36 scavenger receptor expression in monocytes: Involvement of protein kinase B and peroxisome proliferator-activated receptor-gamma. *J. Biol. Chem.* **281,** 6489–6497.

Musalmah, M., Fairuz, A. H., Gapor, M. T., and Wan Ngah, W. Z. (2002). Effect of vitamin E on plasma malondialdehyde, antioxidant enzyme levels and the rates of wound closures during wound healing in normal and diabetic rats. *Asia Pac. J. Clin. Nutr.* **11**(Suppl. 7), S448–S451.

Neefjes, J. J., Momburg, F., and Hammerling, G. J. (1993). Selective and ATP-dependent translocation of peptides by the MHC-encoded transporter. *Science* **261,** 769–771.

Neuzil, J., Weber, T., Schroder, A., Lu, M., Ostermann, G., Gellert, N., Mayne, G. C., Olejnicka, B., Negre-Salvayre, A., Sticha, M., Coffey, R. J., and Weber, C. (2001). Induction of cancer cell apoptosis by alpha-tocopheryl succinate: Molecular pathways and structural requirements. *FASEB J.* **15,** 403–415.

Neuzil, J., Dong, L. F., Wang, X. F., and Zingg, J. M. (2006). Tocopherol-associated protein-1 accelerates apoptosis induced by alpha-tocopheryl succinate in mesothelioma cells. *Biochem. Biophys. Res. Commun.* **343,** 1113–1117.

Ni, J., Wen, X., Yao, J., Chang, H. C., Yin, Y., Zhang, M., Xie, S., Chen, M., Simons, B., Chang, P., di Sant'agnese, A., Messing, E. M., *et al.* (2005). Tocopherol-associated protein suppresses prostate cancer cell growth by inhibition of the phosphoinositide 3-kinase pathway. *Cancer Res.* **65,** 9807–9816.

Numakawa, Y., Numakawa, T., Matsumoto, T., Yagasaki, Y., Kumamaru, E., Kunugi, H., Taguchi, T., and Niki, E. (2006). Vitamin E protected cultured cortical neurons from oxidative stress-induced cell death through the activation of mitogen-activated protein kinase and phosphatidylinositol 3-kinase. *J. Neurochem.* **97,** 1191–1202.

Ogasawara, H., Fujitani, T., Drzewiecki, G., and Middleton, E., Jr. (1986). The role of hydrogen peroxide in basophil histamine release and the effect of selected flavonoids. *J. Allergy Clin. Immunol.* **78,** 321–328.

Ohta, Y., Kobayashi, T., Nishida, K., and Ishiguro, I. (1997). Relationship between changes of active oxygen metabolism and blood flow and formation, progression, and recovery of lesions is gastric mucosa of rats with a single treatment of compound 48/80, a mast cell degranulator. *Dig. Dis. Sci.* **42,** 1221–1232.

Ohta, Y., Kobayashi, T., Imai, Y., Inui, K., Yoshino, J., and Nakazawa, S. (2006). Effect of oral vitamin E administration on acute gastric mucosal lesion progression in rats treated with compound 48/80, a mast cell degranulator. *Biol. Pharm. Bull.* **29,** 675–683.

Okamoto, N., Murata, T., Tamai, H., Tanaka, H., and Nagai, H. (2006). Effects of alpha tocopherol and probucol supplements on allergen-induced airway inflammation and hyper-responsiveness in a mouse model of allergic asthma. *Int. Arch. Allergy Immunol.* **141,** 172–180.

Okayama, Y. (2005). Oxidative stress in allergic and inflammatory skin diseases. *Curr. Drug Targets Inflamm. Allergy* **4,** 517–519.

Okayama, Y., and Kawakami, T. (2006). Development, migration, and survival of mast cells. *Immunol. Res.* **34,** 97–115.

Oudit, G. Y., Sun, H., Kerfant, B. G., Crackower, M. A., Penninger, J. M., and Backx, P. H. (2004). The role of phosphoinositide-3 kinase and PTEN in cardiovascular physiology and disease. *J. Mol. Cell. Cardiol.* **37,** 449–471.

Panagabko, C., Morley, S., Hernandez, M., Cassolato, P., Gordon, H., Parsons, R., Manor, D., and Atkinson, J. (2003). Ligand specificity in the CRAL-TRIO protein family. *Biochemistry* **42,** 6467–6474.

Partovian, C., and Simons, M. (2004). Regulation of protein kinase B/Akt activity and Ser473 phosphorylation by protein kinase Calpha in endothelial cells. *Cell. Signal.* **16,** 951–957.

Pearson, P. J., Lewis, S. A., Britton, J., and Fogarty, A. (2004). Vitamin E supplements in asthma: A parallel group randomised placebo controlled trial. *Thorax* **59,** 652–656.

Perrenoud, D., Homberger, H. P., Auderset, P. C., Emmenegger, R., Frenk, E., Saurat, J. H., and Hauser, C. (1994). An epidemic outbreak of papular and follicular contact dermatitis to tocopheryl linoleate in cosmetics. Swiss Contact Dermatitis Research Group. *Dermatology* **189,** 225–233.

Ranadive, N. S., and Lewis, R. (1982). Differential effects of antioxidants and indomethacin on compound 48/80 induced histamine release and Ca2+ uptake in rat mast cells. *Immunol. Lett.* **5,** 145–150.

Ricciarelli, R., Tasinato, A., Clement, S., Ozer, N. K., Boscoboinik, D., and Azzi, A. (1998). Alpha-tocopherol specifically inactivates cellular protein kinase C alpha by changing its phosphorylation state. *Biochem. J.* **334,** 243–249.

Ricciarelli, R., Zingg, J. M., and Azzi, A. (2001). Vitamin E: Protective role of a Janus molecule. *FASEB J.* **15,** 2314–2325.

Richelle, M., Sabatier, M., Steiling, H., and Williamson, G. (2006). Skin bioavailability of dietary vitamin E, carotenoids, polyphenols, vitamin C, zinc and selenium. *Br. J. Nutr.* **96,** 227–238.

Rimbach, G., Minihane, A. M., Majewicz, J., Fischer, A., Pallauf, J., Virgli, F., and Weinberg, P. D. (2002). Regulation of cell signalling by vitamin E. *Proc. Nutr. Soc.* **61,** 415–425.

Ring, J., Brockow, K., and Behrendt, H. (2004). History and classification of anaphylaxis. *Novartis. Found. Symp.* **257,** 6–16; discussion 16–24, 45–50, 276–285.

Roskoski, R., Jr. (2005). Signaling by kit protein-tyrosine kinase—the stem cell factor receptor. *Biochem. Biophys. Res. Commun.* **337,** 1–13.

Ross, R. (1995). Cell biology of atherosclerosis. *Annu. Rev. Physiol.* **57,** 791–804.

Sarbassov, D. D., Guertin, D. A., Ali, S. M., and Sabatini, D. M. (2005). Phosphorylation and regulation of Akt/PKB by the rictor-mTOR complex. *Science* **307,** 1098–1101.

Satoh, T., Miyataka, H., Yamamoto, K., and Hirano, T. (2001). Synthesis and physiological activity of novel tocopheryl glycosides. *Chem. Pharm. Bull. (Tokyo)* **49,** 948–953.

Schattenfroh, S., Bartels, M., and Nagel, E. (1994). Early morphological changes in Crohn's disease. Transmission electron-microscopic findings and their interpretation: An overview. *Acta Anat. (Basel)* **149,** 237–246.

Scheid, M. P., and Woodgett, J. R. (2003). Unravelling the activation mechanisms of protein kinase B/Akt. *FEBS Lett.* **546,** 108–112.

Schneider, E., Rolli-Derkinderen, M., Arock, M., and Dy, M. (2002). Trends in histamine research: New functions during immune responses and hematopoiesis. *Trends Immunol.* **23,** 255–263.

Seebeck, J., Westenberger, K., Elgeti, T., Ziegler, A., and Schutze, S. (2001). The exocytotic signaling pathway induced by nerve growth factor in the presence of lyso-phosphatidylserine in rat peritoneal mast cells involves a type D phospholipase. *Regul. Pept.* **102,** 93–99.

Sen, C. K., Khanna, S., Roy, S., and Packer, L. (2000). Molecular basis of vitamin E action. Tocotrienol potently inhibits glutamate-induced pp60(c-Src) kinase activation and death of HT4 neuronal cells. *J. Biol. Chem.* **275,** 13049–13055.

Shah, S. J., and Sylvester, P. W. (2005). Gamma-tocotrienol inhibits neoplastic mammary epithelial cell proliferation by decreasing Akt and nuclear factor kappaB activity. *Exp. Biol. Med. (Maywood)* **230,** 235–241.

Shahar, E., Hassoun, G., and Pollack, S. (2004). Effect of vitamin E supplementation on the regular treatment of seasonal allergic rhinitis. *Ann. Allergy Asthma Immunol.* **92,** 654–658.

Shvedova, A. A., Kisin, E., Murray, A., Smith, C., Castranova, V., and Kommineni, C. (2002). Enhanced oxidative stress in the skin of vitamin E deficient mice exposed to semisynthetic metal working fluids. *Toxicology* **176,** 135–143.

Sienra-Monge, J. J., Ramirez-Aguilar, M., Moreno-Macias, H., Reyes-Ruiz, N. I., Del Rio-Navarro, B. E., Ruiz-Navarro, M. X., Hatch, G., Crissman, K., Slade, R., Devlin, R. B., and Romieu, I. (2004). Antioxidant supplementation and nasal inflammatory responses among young asthmatics exposed to high levels of ozone. *Clin. Exp. Immunol.* **138,** 317–322.

Sim, A. T., Ludowyke, R. I., and Verrills, N. M. (2006). Mast cell function: Regulation of degranulation by serine/threonine phosphatases. *Pharmacol Ther.* **112**(2), 425–439.

Singh, U., Devaraj, S., and Jialal, I. (2005). Vitamin E, oxidative stress, and inflammation. *Annu. Rev. Nutr.* **25,** 151–174.

Suchankova, J., Voprsalova, M., Kottova, M., Semecky, V., and Visnovsky, P. (2006). Effects of oral alpha-tocopherol on lung response in rat model of allergic asthma. *Respirology* **11,** 414–421.

Sugiyama, H., Nonaka, T., Kishimoto, T., Komoriya, K., Tsuji, K., and Nakahata, T. (2000). Peroxisome proliferator-activated receptors are expressed in human cultured mast cells: A possible role of these receptors in negative regulation of mast cell activation. *Eur. J. Immunol.* **30,** 3363–3370.

Sylvester, P. W., and Shah, S. J. (2005). Mechanisms mediating the antiproliferative and apoptotic effects of vitamin E in mammary cancer cells. *Front Biosci.* **10,** 699–709.

Theoharides, T. C. (1996). The mast cell: A neuroimmunoendocrine master player. *Int. J. Tissue React.* **18,** 1–21.

Theoharides, T. C., and Bielory, L. (2004). Mast cells and mast cell mediators as targets of dietary supplements. *Ann. Allergy Asthma Immunol.* **93,** S24–S34.

Thiele, J. J., Weber, S. U., and Packer, L. (1999). Sebaceous gland secretion is a major physiologic route of vitamin E delivery to skin. *J. Invest. Dermatol.* **113,** 1006–1010.

Thiele, J. J., Hsieh, S. N., and Ekanayake-Mudiyanselage, S. (2005) Vitamin E: Critical review of its current use in cosmetic and clinical dermatology. *Dermatol. Surg.* **31,** 805–813; discussion 813.

Tsoureli-Nikita, E., Hercogova, J., Lotti, T., and Menchini, G. (2002). Evaluation of dietary intake of vitamin E in the treatment of atopic dermatitis: A study of the clinical course and evaluation of the immunoglobulin E serum levels. *Int. J. Dermatol.* **41,** 146–150.

Valent, P., Akin, C., Sperr, W. R., Horny, H. P., Arock, M., Lechner, K., Bennett, J. M., and Metcalfe, D. D. (2003). Diagnosis and treatment of systemic mastocytosis: State of the art. *Br. J. Haematol.* **122,** 695–717.

Venugopal, S. K., Devaraj, S., and Jialal, I. (2004). RRR-alpha-tocopherol decreases the expression of the major scavenger receptor, CD36, in human macrophages via inhibition of tyrosine kinase (Tyk2). *Atherosclerosis* **175,** 213–220.

Waltner-Romen, M., Falkensammer, G., Rabl, W., and Wick, G. (1998). A previously unrecognized site of local accumulation of mononuclear cells. The vascular-associated lymphoid tissue. *J. Histochem. Cytochem.* **46,** 1347–1350.

Whiteman, E. L., Cho, H., and Birnbaum, M. J. (2002). Role of Akt/protein kinase B in metabolism. *Trends Endocrinol. Metab.* **13,** 444–451.

Woolley, D. E. (2003). The mast cell in inflammatory arthritis. *N. Engl. J. Med.* **348,** 1709–1711.

Wymann, M. P., Bjorklof, K., Calvez, R., Finan, P., Thomast, M., Trifilieff, A., Barbier, M., Altruda, F., Hirsch, E., and Laffargue, M. (2003). Phosphoinositide 3-kinase gamma: A key modulator in inflammation and allergy. *Biochem. Soc. Trans.* **31,** 275–280.

Yamada, K., Hung, P., Yoshimura, K., Taniguchi, S., Lim, B. O., and Sugano, M. (1996a). Effect of unsaturated fatty acids and antioxidants on immunoglobulin production by mesenteric lymph node lymphocytes of Sprague-Dawley rats. *J. Biochem. (Tokyo)* **120,** 138–144.

Yamada, K., Mori, M., Matsuo, N., Shoji, K., Ueyama, T., and Sugano, M. (1996b). Effects of fatty acids on accumulation and secretion of histamine in RBL-2H3 cells and leukotriene release from peritoneal exudate cells isolated from Wistar rats. *J. Nutr. Sci. Vitaminol. (Tokyo)* **42,** 301–311.

Yang, Z. Z., Tschopp, O., Baudry, A., Dummler, B., Hynx, D., and Hemmings, B. A. (2004). Physiological functions of protein kinase B/Akt. *Biochem. Soc. Trans.* **32,** 350–354.

Yoshida, E., Watanabe, T., Takata, J., Yamazaki, A., Karube, Y., and Kobayashi, S. (2006). Topical application of a novel, hydrophilic gamma-tocopherol derivative reduces photo-inflammation in mice skin. *J. Invest. Dermatol.* **126**(7), 1633–1640.

Zappulla, J. P., Arock, M., Mars, L. T., and Liblau, R. S. (2002). Mast cells: New targets for multiple sclerosis therapy? *J. Neuroimmunol.* **131,** 5–20.

Zheng, K., Adjei, A. A., Shinjo, M., Shinjo, S., Todoriki, H., and Ariizumi, M. (1999). Effect of dietary vitamin E supplementation on murine nasal allergy. *Am. J. Med. Sci.* **318,** 49–54.

Zingg, J. M. (2006). Modulation of signal transduction by vitamin E. *In* "XIII Congress of the Society for Free Radical Research International," pp. 47–60. Medimond International Proceedings, Davos.

Zingg, J. M. (2007). Molecular and cellular activities of vitamin E analogoues. *In* "Mini Reviews in Medicinal Chemistry," in press.

Zingg, J. M., and Azzi, A. (2004). Non-antioxidant activities of vitamin E. *Curr. Med. Chem.* **11,** 1113–1133.

Zingg, J. M., and Azzi, A. (2006). Modulation of cellular signalling and gene expression by vitamin E. *In* "New Topics in Vitamin E Research" (O. H. Bendrick, Ed.). NOVA Publishers, New York.

16

TOCOTRIENOLS IN CARDIOPROTECTION

SAMARJIT DAS,* KALANITHI NESARETNAM,† AND DIPAK K. DAS*

*Cardiovascular Research Center, University of Connecticut School of Medicine Farmington, Connecticut 06030

†Malaysian Palm Oil Board, Kuala Lumpur, Malaysia

Tocotrienols, a group of Vitamin E stereoisomers, offer many health benefits including their ability to lower cholesterol levels, and provide anticancer and tumor-suppressive activities. Several recent studies determined the cardioprotective abilities of tocotrienols, although the number

Vitamins and Hormones, *Volume 76*

0083-6729/07 $35.00
DOI: 10.1016/S0083-6729(07)76016-8

is only 1% compared to the study with tocopherols. Both in acute perfusion experiments and in chronic models, tocotrienols attenuate myocardial ischemia-reperfusion injury, artherosclerosis, and reduced ventricular arrythmias. Apart from the antioxidative role of tocotrienols, it appears that tocotrienols mediated cardioprotection is also achieved through the preconditioning-like effect, the best yet devised method of cardioprotection. Hence, tocotrienols likely fulfills the definition of a pharmacological preconditioning agent and give a tremendous opportunity to place tocotrienols as an important therapeutic option in cardiovascular system.

I. INTRODUCTION

Tocotrienols, a group of vitamin E stereoisomers, offer many health benefits including their ability to lower cholesterol levels, and provide anticancer and tumor-suppressive activities. A diet rich in tocotrienols, especially dietary tocotrienols from a tocotrienol-rich fraction (TRF) of palm oil, reduced the concentration of plasma cholesterol and apolipoprotein B, platelet factor 4, and thromboxane B_2, indicating its ability to protect against platelet aggregation and endothelial dysfunction (Qureshi *et al.*, 1991a,b). Red palm oil is one of the richest sources of carotenoids; together with vitamin E, tocotrienols, and ascorbic acid present in this oil, it represents a powerful network of antioxidants, which can protect various tissue and cells from oxidative damage (Edem, 2002; Hendrich *et al.*, 1994; Krinsky, 1992; Packer, 1992). For rat hearts, α-tocotrienol were more proficient in the protection against oxidative stress induced by ischemia-reperfusion than α-tocopherol (Serbinova *et al.*, 1992). Tocotrienols are found to be more effective in central nervous system protection compared to α-tocopherol (Sen *et al.*, 2004). In another study, TRF is found to inhibit the glutamate-induced pp60c-src kinase activation in HT4 neuronal cells (Sen *et al.*, 2000). A recent study has indicated that TRF was able to reduce myocardial infarct size and improve postischemic ventricular dysfunction, and reduce the incidence of ventricular arrhythmias (Das *et al.*, 2005b). TRF was also shown to stabilize 20S and 26S proteasome activities and reduce the ischemia-reperfusion-induced increase in c-Src phosphorylation (Das *et al.*, 2005b).

The growing interest in tocotrienols among all other vitamin E isoforms is the purpose of this chapter.

II. A BRIEF HISTORY OF VITAMIN

Hippocrates (460–377 BC), the father of medicine said, "Let food be thy medicine and medicine be thy food." In the eighteenth century, it was found that the intake of citrus fruits can reduce the development of scurvy. In 1905, a British clinician, William Fletcher, who was working with the disease Beriberi, discovered that taking unpolished rice prevented Beriberi and taking polished rice did not. On the basis of this finding, he concluded that if some special factors were removed from the foods, there are high chances to have diseases. The very next year, Dr. Fletcher's hypothesis became stronger when another British biochemist, Sir Frederick Gowland Hopkins, found that foods contained necessary "accessory factors" in addition to proteins, carbohydrates, fats, minerals, and water.

In 1911, Polish chemist Casimir Funk discovered that the anti-beriberi substance in unpolished rice was an amine, so Dr. Funk named the special amine as "vitamine" for "vita amine" after "vita" means life and "amine" which he found in the unpolished rice, a nitrogen-containing substance. It was later discovered that many vitamins do not contain nitrogen, and, therefore, not all vitamins are amine. Because of its widespread use, Funk's term continued to be applied, but the final letter "e" was dropped.

In 1912, Hopkins and Funk further advanced the vitamin hypothesis of deficiency, a theory that postulates that the absence of sufficient amounts of a particular vitamin in a system may lead to certain diseases. During the early 1900s, through experiments in which animals were deprived of certain types of foods, scientists succeeded in isolating and identifying the various vitamins recognized today.

III. Vitamin E, Now and Then

In 1922 at Berkeley University in California, a physician scientist, Dr. Herbert M. Evans and his assistant Katherine S. Bishop, discovered a fat-soluble alcohol that functioned as an antioxidant, which they named "Factor X" (Papas, 1999). Evans and Bishop were feeding rats a semipurified diet when they noticed that the female rats were unable to produce offspring because the pups died in the womb. They then fed the female rats lettuce and wheat germ, and observed that healthy offspring were produced. During their research, Evans and Bishop discovered that "Factor X" was contained in the lipid extract of the lettuce and concluded that this "Factor X" was fat-soluble (Papas, 1999). In 1924, Dr. Bennett Sure renamed "Factor X" as Vitamin E. The first component identified was α-tocopherol. It was named as such from the Greek *tokos* (offspring) and *pheros* (to bear) and the *ol* ending was added to indicate the alcoholic properties of the molecules. For over more than 30 years, it was well believed that vitamin E existed in only one forms, α-tocopherol. As a result vitamin E named as tocopherol. It is the most abundant form of vitamin E found in blood and body tissue. But in 1956, scientist J. Green discovered the eight isoform of vitamin E, four tocopherol isomers (α, β, γ, and δ) and four tocotrienol isomers (α, β, γ, and δ), split into two different categories: tocopherols and tocotrienols which are corresponding stereoisomers. Tocopherols and tocotrienols are very similar, except for the fact that tocopherols have a saturated phytyl tail, and tocotrienols have an isoprenoid tail with three unsaturated points. In addition, on the chromanol nucleus, the various isoforms differ in their methyl substitutions (Fig. 1). Tocotrienols are initially named as ζ, ε, or η-tocopherols. The δ-form has one methyl group, the γ- and β-forms have two methyl groups, and α-form contains three methyl groups on its chromanol head. Tocopherols

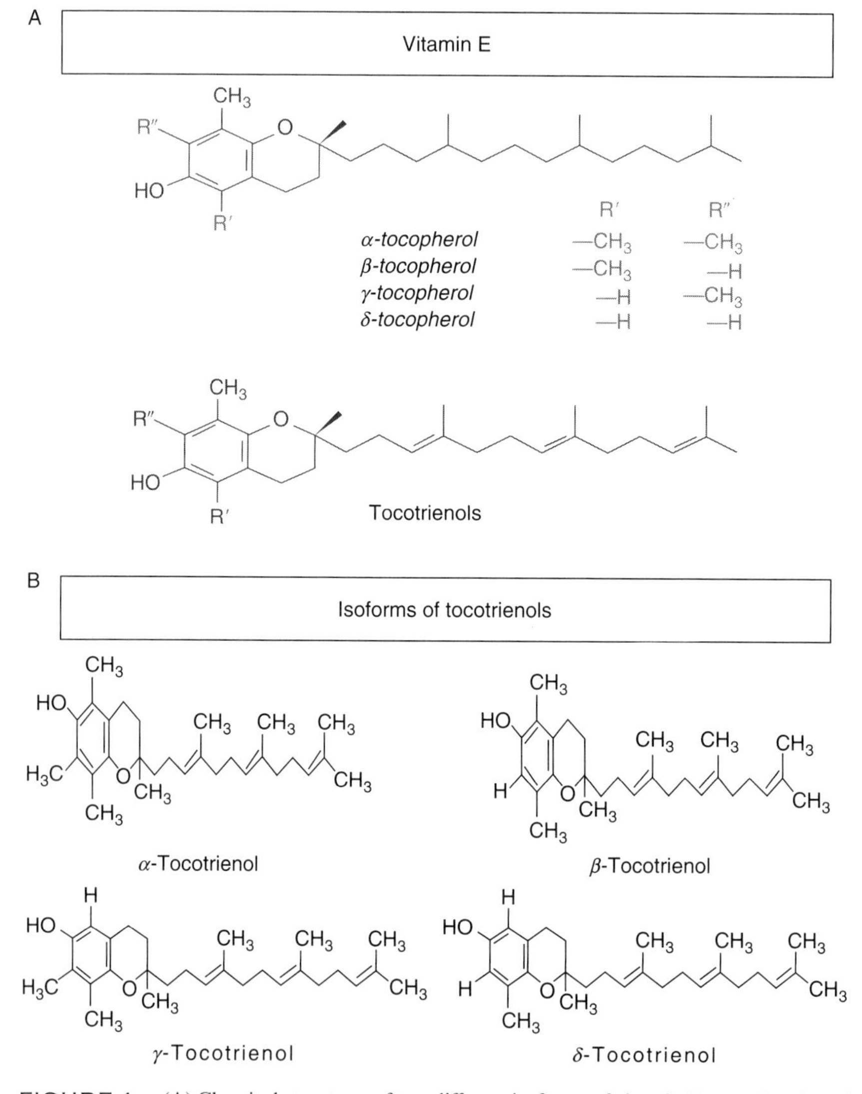

FIGURE 1. (A) Chemical structures of two different isoforms of vitamin E, tocopherols and tocotrienols. (B) Four different isoforms of tocotrienols, α, β, γ, and δ, different by their methyl group position in their respective ring structure.

and tocotrienols share a common chromanol head and a side chain at the C2 position (Theriault *et al.*, 1999). Very recently, two new isomers of tocotrinols have been found and are present in TRF of rice bran oil, desmethyl (D-P_{21}-T3) and didesmethyl (D-P_{25}-T3) tocotrienols (Das *et al.*, 2005a).

The therapeutic application of vitamin E was first shown by Kamimura (1977). The inhibitory effect of the unsaturated fatty acid by α-tocopherol was well established by Tappel (1953, 1954, 1955). This observation was repeated in humans with the same result by Horwitt *et al.*, in the very next year when Tappel identified the fact that the deficiency of α-tocopherol may lead to the high levels of oxidative lipid damage (Horwitt *et al.*, 1956). Antioxidative effect of vitamin E can be due to the equal contribution of phenolic head as well as the phytyl tail was explained in both *in vitro* and *in vivo* studies (Burton and Ingold, 1989). The important discoveries of various aspects of α-tocopherols are listed in Table I. Since α-tocopherol is the most abundant vitamin E in the body, its activity as an antioxidant and its role in protection from oxidative stress have been studied more extensively than other forms of vitamin E. Studies showed that α-tocopherols are protective against atherosclerosis. A study (Devaraj and Jialal, 2005) of α-tocopherol's effect on important proinflammatory cytokine, like tumor necrosis factor-α (TNF-α), which is released from human monocytes, demonstrated that α-tocopherols inhibited the release of TNF-α via inhibition of 5-lipoxygenase. Inhibition of 5-lipoxygenase also significantly reduced TNF mRNA and NF-κB-binding activity. Other study (Meydani, 2004) showed how α-tocopherol inhibits the activation of endothelial cells stimulated by high levels of low-density lipoprotein (LDL) cholesterol and proinflammatory cytokines. This inhibition is associated with the suppression of chemokines, the expression of cell surface adhesion molecules and the adhesion of leukocytes to endothelial cells, all of which contribute to the development of lesions in the arterial wall. While the benefits of tocopherols have been studied for years, health benefits of the other four forms of vitamin E are only recently being explored. Just like cholesterol, tocopherols also influence the biophysical membrane characteristics like fluidity (Sen *et al.*, 2006).

But for the last few years researchers have been focusing more towards tocotrienols compared to tocopherols because of the fact that tocotrienols have a more potent antioxidative property than α-tocopherols (Serbinova and Packer, 1994; Serbinova *et al.*, 1991). Still there is not enough research going on with tocotrienols as compared to the extensive work done on tocopherols.

IV. TOCOTRIENOLS VERSUS TOCOPHEROLS

There are at least eight isoforms that are commonly found to have vitamin E's activity: α-, β-, γ-, and δ-tocopherols and α-, β-, γ- and δ-tocotrienols (Fig. 1). Tocotrienols differ from tocopherols by having a farnesyl (isoprenoid) structure compare to saturated phytyl side chain. Yet, the focus on tocopherols is much higher than that of tocotrienols. Out of the studies done on tocopherols

TABLE I. Historical Background of Vitamin E

1922	Food factor <X> discovered by H. M. Evans and L. S. Bishop as a substance essential for rat pregnancy.
	Food factor <X> found in yeast and lettuce by H. A. Martill.
1923	Food factor <X> found in alfalfa, wheat, oats, and butter by H. M. Evans *et al.*
1924	Food factor <X> named vitamin E by B. Sure.
1936	α-Tocopherol extracted from wheat germ oil by H. M. Evans *et al.*
1938	Chemical structure of vitamin E determined by E. Fenholz.
	DL-α-Tocopherol synthesized by P. Karrer.
1950	Research on application of vitamin E in treating frostbite started by M. Kamimura.
1956	<Free radical theory of the aging process> proposed by D. Harmann.
	Eight homologues of vitamin E (tocopherols and tocotrienols) discovered by J. Green.
1961	Vitamin E (DL-α-tocopherol) admitted to the Japanese Pharmacopoeia.
1962	Antioxidant activity in the body suggested by A. L. Tappel.
1968	Recommended dietary allowance (RDA) of vitamin E set at 30 IU (20-mg α-TE) in the United States.
1972	Recommended dietary allowance (RDA) of vitamin E revised to 10 IU (7-mg α-TE) in the United States.
1988	The approval standards for vitamin products revised and the daily intake of vitamin E as an OTC product set at 300 mg/day in Japan.
1991	Vitamin E shown in MONICA Study to reduce risk of coronary disorders.
	α-Tocopherol transport protein (α-TTP), which selectively transports α-tocopherol, isolated from the liver.
1993	Familial vitamin E deficiency reported by C. Ben Hamida *et al.*
1994	Vitamin E intake reported to reduce mortality from coronary heart disorders by M. C. Bellizzi *et al.*—European PARADOX.
1996	Vitamin E shown in CHAOS Study to reduce risk of myocardial infarction.
1997	Vitamin E reported by M. Sano *et al.* to delay progression of Alzheimer's disease.
	Vitamin E reported by S. N. Meydani *et al.* to activate immunological competence in the elderly.
	Vitamin E reported by A. Herday *et al.* to improve liver function in patients with hepatitis C.
1999	Vitamins C and E reported by L. C. Chappell *et al.* to relieve preeclampsia.
	Recommended dietary allowances of vitamin E set at 10 mg for males and 8 mg for females in Japan by the sixth revision of nutrition requirements.
2000	Vitamins E reported by M. Boaz *et al.* to reduce risk of cardiovascular disease in hemodialysis patients.
2004	Vitamin E reported by S. N. Meydani *et al.* to lower the incidence of common colds in elderly nursing home residents.

only 1% have been done on tocotrienols (Sen *et al.*, 2006). But for the last few years, there has been a growing interest among researchers on tocotrienols as compared to tocopherols.

The abundance of α-tocopherol in the living cells compares to other isoforms and of course the maximum half-life period of the same isoform may be the major cause of its research importance among the various disciplines of clinical research. But it is well established that the antioxidative power of tocotrienols is 1600 times more than that of α-tocopherol (Serbinova and Packer, 1994). There is evidence that tocotrienols are more potent when compared with tocopherols. The reason of this increased efficacy is the unsaturation in the aliphatic tail which facilitates easier penetration into the tissue (Suzuki *et al.*, 1993) and also because of unsaturation in the aliphatic tail tocotrienols are a more potent antioxidant than tocopherols. These important findings may be attracting many other researchers to consider tocotrienols as a better therapeutic agent than tocopherols. It has already been proved by various research groups that tocotrienols possess neuroprotective, anticancer and also cholesterol-lowering properties as compared to its other isoform, tocopherols.

It is only tocotrienols, which at nanomolar concentration protect the neuronal cells from glutamate-induced cell death (Khanna *et al.*, 2003; Roy *et al.*, 2002; Sen *et al.*, 2000). In a very interesting study, Sen *et al.* showed that tocotrienols, but not tocopherols, inhibit the activation of pp60 (c-Src), which is a key regulator of glutamate-induced neuronal cell death (Sen *et al.*, 2000). In another study, it was found that tocotrienols, and not tocopherols, protect the neurons from glutamate-induced 12-lipoxygenase (12-Lox) activation (Khanna *et al.*, 2003). This 12-Lox takes very important part in signal transduction pathway to kill the neurons. The molecular level of target for the neuroprotective effect of tocotrienols, mainly α-tocotrienol, is cytosol, but not at the nucleus (Khanna *et al.*, 2003; Sen *et al.*, 2000). It is now a well-established fact that it is tocotrienols, mainly α-tocotrienol, which possess a potent neuroprotection at very low concentration, but not any other tocopherol (Khanna *et al.*, 2005).

The anticarcinogenic property of tocotrienols has been established. Many studies have shown tocotrienols provide better protection against cancer than tocopherols do. In mice, tocotrienols were compared with α-tocopherols, and interestingly it was found that i.p. administration of α- and γ-tocotrienols, and not α-tocopherols, showed a slight life-prolonging effect in mice from transplanted tumors (Komiyama *et al.*, 1989). Similar observation was found by Gould *et al.* (1991) in rats for chemoprevention of chemically induced mammary tumors. Also, in human study, it was shown that tocotrienols significantly suppress growth of breast cancer cells in culture, whereas tocopherols fail to show similar action under identical conditions (Nesaretnam *et al.*, 1995). In another study, it was shown that the anticarcinogenic property of tocotrienols may be a better option than tamoxifen, from breast cancer prevention (Guthrie *et al.*, 1997). γ- and δ-tocotrienols are considered as the most effective isoform among the all eight isoforms of vitamin E, for there physiological role in modulating normal mammary gland growth, function, and remodeling (McIntyre *et al.*, 2000). Relative to tocopherols, tocotrienols are a more potent suppressor of EGF-dependent normal mammary epithelial cell growth, the mechanism of which is, by the deactivation of PKC-α (Sylvester *et al.*, 2002).

The hypocholesterolemic effect of tocotrienols is also found to be more potent than that of tocopherols. Due to the presence of three double bonds in the isoprenoid chain, tocotrienols can lower cholesterol level much more effectively compared to tocopherols (Qureshi *et al.*, 1986). Tocotrienols significantly reduced the concentration of plasma cholesterol and apolipoprotein B, platelet factor 4, and thromboxane B_2, indicating its ability to protect against platelet aggregation and endothelial dysfunction (Qureshi *et al.*, 1991a,b). It was found that tocotrienols and not tocopherols suppress the HMG-CoA reductase, which directly inhibits the biosynthesis of cholesterol (Parker *et al.*, 1993; Pearce *et al.*, 1992, 1994). Later on, the significant hypocholesterolemic effect of tocotrienols was compared with tocopherols in humans (Qureshi *et al.*, 1995, 2001b, 2002), chicken (Qureshi and Peterson, 2001), hypercholesterolemic rat (Iqbal *et al.*, 2003), swine (Qureshi *et al.*, 2001a), and as well as hamster's plasma (Raederstorff *et al.*, 2002). In conclusion, researchers have shown the lowering the cholesterol, tocotrienols are the better option than tocopherols.

V. SOURCES OF TOCOTRIENOLS

Tocotrienols are mainly found in the seed endosperm of almost all the monocots such as wheat, rice, barley oat, rye, and sour cherry. In some dicots, endosperm also contain tocotrienols. Some Apiaceae species and also in some Solanaceae species such as tobacco (Sen *et al.*, 2006). Tocotrienols cannot be present in the endosperms, but always present as a mixture of tocopherol–tocotrienol. That is why researchers normally use tocotrienols rich factor (TRF), the ratio between tocotrienols to tocopherols. The TRF of rice bran oil, 90:10, is the maximum so far identified. In this particular oil, apart from the normal four isoforms of tocotrienols, there are two new isoforms also found as well, desmo and didesmo-tocotrienols. Crude palm oil extract from the fruit of *Elaeis guineensis* also contains higher concentration of TRF, almost 80:20. Normally, the major components of palm-derived TRF extract contain mainly 36% γ-tocotrienol, 26–30% α-tocotrienol, and 20–22% α-tocopherol, and 12% δ-tocotrienol (Kamat *et al.*, 1997).

VI. TOCOTRIENOLS IN FREE RADICAL SCAVENGING AND ANTIOXIDANT ACTIVITY

TRF has excellent free radical scavenging capacity (Kamat *et al.*, 1997). Numerous studies (Ikeda *et al.*, 2003; Kamat *et al.*, 1997) show that TRF is a potent inhibitor of lipid peroxidation and protein peroxidation in rat microsomes and mitochondria. At low concentrations of 5 μM, TRF, mainly

δ-tocotrienol and to a lesser extent α- and δ-tocotrienols, significantly inhibited oxidative damage to both lipids and proteins in rat brain mitochondria. Studies of the effect of γ-tocotrienols on endothelial nitric oxide synthase (eNOS) activity in spontaneously hypertensive rats have reported that on treatment with antioxidant γ-tocotrienol increased the nitric oxide (NO) activity and concomitantly reduced the blood pressure and enhanced total antioxidant status in plasma and blood vessels (Ikeda *et al.*, 2003). In general, TRF has significantly higher antioxidant ability as compared to tocopherols. This can be explained by the structural difference between the saturated side chain of tocopherols and the unsaturated side chain of tocotrienols. The molecular mobility of polyenoic lipids in the membrane bilayer (composed mainly of unsaturated fatty acid) is much higher than that of saturated lipids, and hence tocotrienols are more mobile and less restricted in their interaction with lipid radicals in membranes than tocopherols. This is further supported by the higher effectiveness of tocotrienols in processes that may involve oxidative stress such as in red blood cells where tocotrienols have more potency against oxidative hemolysis than α-tocopherols (Kamat *et al.*, 1997). In an *in vitro* study, the potent free radical scavenging property of α-tocopherol was found 1600 times more compare to free radical scavenging property of α-tocotrienols (Serbinova and Packer, 1994). In another study, it was found the potent antioxidative property of γ-tocotrienols significantly protect the spontaneously hypertensive rats (Newaz *et al.*, 2003).

VII. TOCOTRIENOLS AND CARDIOPROTECTION

Since cardiovascular disease contributes in a major way to the morbidity and mortality, it is becoming a strain on the economy of many countries worldwide. Various factors have been identified as possible causes of different cardiac diseases such as heart failure and ischemic heart disease. As discussed earlier, tocotrienols are very poorly studied compare to tocopherols. Due to this reason there is very few evidence of cardioprotective effect of tocotrienols, whereas the cardioprotective effect of tocopherols is immense.

VIII. ATHEROSCLEROSIS

Atherosclerosis is the process by which the deposition of cholesterol plaques on the wall of blood vessels and make those vessels narrow and ultimately getting block by those fatty deposit. Atherosclerosis finally leads to ischemia in the heart muscle and can cause damage to the heart muscle. The complete blockage of the arteries leads to myocardial infarction (MI).

According to the World Health Organization, the major cause of death in the world as a whole by the year 2020 will be acute coronary occlusion (Murray and Lopez, 1997).

As mentioned earlier, tocotrienols differ from tocopherols only in three double bonds in the isoprenoid chain which appear to be essential for the inhibition of cholesterogenesis by higher cell penetration and followed by better interaction with the deposited plaques (Qureshi *et al.*, 1986). In some clinical trials with hypercholesterolemics patients, tocotrienols significantly reduced the serum cholesterols (Qureshi *et al.*, 1991b). In another similar kind of clinical trial, tocotrienols lowered both the serum cholesterols, total cholesterol (TC), more interestingly the LDL cholesterols (Tan *et al.*, 1991). In the late 1980s, it was found that the one of the major cause of lipid oxidation was the oxidation of LDL. Therefore, the observation by Tan *et al.* (1991) draws many researchers attention toward tocotrienols as a better antilipid oxidative agent. Later on, a diet rich in tocotrienols, especially dietary tocotrienols from a TRF of palm oil, reduced the concentration of plasma cholesterol and apolipoprotein B, platelet factor 4, and thromboxane B_2, indicating its ability to protect against platelet aggregation and endothelial dysfunction (Qureshi *et al.*, 1991a,b). In mammalian cells, 3-hydroxy-3-methylglutaryl coenzyme A (HMG-CoA) reductase, enzyme was found to regulate the cholesterol production. Tocotrienols, mainly γ-isoform or the tocotrienols mixture, significantly suppress the secretion of HMG-CoA reductase, which ultimately lowers the production of cholesterols in the cells (Parker *et al.*, 1993; Pearce *et al.*, 1992, 1994). Another possible mechanism of protection from lipid peroxidation by tocotrienols was found by isoprenoid-mediated suppression of mevalonate synthesis depletes tumor tissues of two intermediate products, farnesyl pyrophosphate and geranylgeranyl pyrophosphate, which are incorporated posttranslationally into growth control associated proteins (Elson and Qureshi, 1995). From the above observations, researchers are also started to compare tocotrienols with any statin group of medicine. In one of the study, Qureshi *et al.* showed that in chicken, when tocotrienols were applied with lovastatin or when lovastatin was compared to tocotrienol action, there was no difference in terms of cholesterol-lowering power (Qureshi and Peterson, 2001). Very interestingly, it was found apart from these two mechanism of tocotrienols, tocotrienols are also protecting from hypercholesterolemic phase by activating the conversion of LDL to HDL through the interphase VLDL–VDL and finally HDL (Qureshi *et al.*, 1995, 2001a). In hypercholesterolemic phase, it was also observed either γ-tocotrienol or the tocotrienols mixture increases the number of HDL, which then interact with LDL to reduce the concentration of LDL in the plasma (Qureshi *et al.*, 1995), HDL may also go by phagocytosis to lower the LDL concentration. In another clinical trial, 100 mg/day of TRF derived from rice bran oil effectively lowers the serum cholesterol in hypercholesterolemic patients (Qureshi *et al.*, 2002). The same study showed that α-tocopherol induces the HMG-CoA reductase and that is why in higher doses of TRF, the opposite effect is observed to some extent compare to 100 mg/day of TRF (Qureshi *et al.*, 2002). This may be due the fact that tocotrienols were found to go on conversion into tocopherols *in vivo* (Qureshi *et al.*, 2001). This study clearly showed it is only tocotrienols which are responsible for the lowering of serum cholesterol, but not with tocopherols. Tocopherols may increase the cholesterol level by inducing HMG-CoA reductase (Qureshi *et al.*, 2002).

IX. Tocotrienols in Ischemic Heart Disease

Ischemia is a stage when there is no blood flow in a cell; as blood is the only carrier of air or oxygen, cells become subject to a lot of stress due to lack of oxygen. When this kind of situation arises in the heart, the disease is known as ischemic heart disease. Apart from atherosclerotic plaque deposition, oxidative stress is also considered as one of the major causes of ischemic heart disease.

The excellent free radical scavenging property of tocotrienols attenuates the oxidative stress better compared to tocopherols. That is why, recently researchers are considering tocotrienols as a better therapeutic option from ischemic heart disease compared to tocopherols.

γ-Tocotrienols are found to act as a myocardial preconditioning agent by activating the eNOS expression (Ikeda *et al.*, 2003). eNOS is considered one of the major cause of intracellular NO generator. This NO then goes on vasodialation and protects the heart from ischemic phase. Due to the eNOS-regulating property, γ-tocotrienol is now considered as an important pharmacological preconditioning agent. In a very recent study, it was shown for the first time that beneficial effects of tocotrienol derived from palm oil are due to its ability to reduce c-Src activation, which is linked with the stabilization of proteasomes, mainly 20S and 26S (Das *et al.*, 2005b). Tocotrienols have extremely short half-lives; after oral ingestion, they are not recognized by α-tocotrienol transport protein, which also accounts for their low bioavailability. For this reason, TRF was used in an acute experiment to determine its immediate effects on the ischemic-reperfused myocardium. The results indicate that tocotrienol readily blocks the ischemia-reperfusion-mediated increase in Src kinase activation and proteasome inactivation, thereby providing cardioprotection (Das *et al.*, 2005b). After this observation, in the continuing study by the same group, but this time with gavaging of the TRF derived from palm oil for 15 days protects the heart from ischemia-reperfusion injury was observed (Das *et al.*, 2005a). In this chronic experiment, it was also observed that the key mechanism may be the inhibition of Src activation by TRF. Myocardial ischemia/reperfusion caused an induction of the expression of c-Src protein (Hattori *et al.*, 2001) inhibition of c-Src with PPI reduces the extent of cellular injury. The ability of TRF to block the increased phosphorylation of c-Src appears to play a crucial role in its ability to protect the heart from ischemia-reperfusion injury.

X. Summary and Conclusion

It should be clear from the above discussion that tocotrienols as TRF, provide cardioprotection not only by its cholesterol lowering property or by its reducing oxidative stress, but also through their ability to performing redox signaling by potentiating an antideath signal through the reduction of proapoptotic factors, at least cSrc was identified, thereby leading to the decrease in cardiomyocytes apoptosis. Out of a minimum of four different isoforms of tocotrienols, α- and γ-tocotrienols are considered as the effective isoforms especially, which possess the cardioprotective abilities. Both α- and γ-isoforms are found to possess antiatherosclerotic properties not only by reducing the LDL cholesterols but also by increasing the number of HDL cholesterols and also simultaneous induction of HMG-CoA reductase activity. Apart from antiatherosclerotic property, TRF was found to be protective both acutely and chronically, from ischemia-reperfusion-mediated cardiac dysfunction by inhibiting the phosphorylation of c-Src expression significantly with both 20S and 26S proteasome stabilization.

Acknowledgments

This study was supported by NIH HL 34360, HL 22559, HL 33889, and HL 56803.

References

Burton, G. W., and Ingold, K. U. (1989). Vitamin E as an *in vitro* and *in vivo* antioxidant. *Ann. NY Acad. Sci.* **570,** 7–22.

Das, S., Nesaretam, K., and Das, D. K. (2005a). Cardioprotective abilities of palm oil derived tocotrienol rich factor. Proc. Malaysian Palm Oil Board. 12–19.

Das, S., Powell, S. R., Wang, P., Divald, A., Nasaretnam, K., Tosaki, A., Cordis, G. A., Maulik, N., and Das, D. K. (2005b). Cardioprotection with palm tocotrienol: Andioxidant activity of tocotrienol is linked with its ability to stabilize proeasomes. *Am. J. Physiol. Heart Circ. Physiol.* **289,** H361–H367.

Devaraj, S., and Jialal, I. (2005). Alpha-tocopherol decreases tumor necrosis factor-alpha mRNA and protein from activated human monocytes by inhibition of 5-lipoxygenase. *Free Radic. Biol. Med.* **38,** 1212–1220.

Edem, D. O. (2002). Palm oil: Biochemical, physiological, nutritional, hematological, and toxicological aspects: A review. *Plant Foods Hum. Nutr.* **57,** 319–341.

Elson, C. E., and Qureshi, A. A. (1995). Coupling the cholesterol- and tumor-suppressive actions of palm oil to the impact of its minor constituents on 3-hydroxy-3-methylglutaryl coenzyme A reductase activity. *Prostaglandins Leukot. Essent. Fatty Acids* **52,** 205–207.

Gould, M. N., Haag, J. D., Kennan, W. S., Tanner, M. A., and Elson, C. E. (1991). A comparison of tocopherol and tocotrienol for the chemoprevention of chemically induced rat mammary tumors. *Am. J. Clin. Nutr.* **53**(4 Suppl.), 1068S–1070S.

Guthrie, N., Gapor, A., Chambers, A. F., and Carroll, K. K. (1997). Inhibition of proliferation of estrogen receptornegative MDA-MB-435 and-positive MCF-7 human breast cancer cells by palm oil tocotrienols and tamoxifen, alone and in combination. *J. Nutr.* **127,** 544S–548S.

Hattori, R., Otani, H., Uchiyama, T., Imamura, H., Cui, J., Maulik, N., Cordis, G. A., Zhu, L., and Das, D. K. (2001). Src tyrosine kinase is the trigger but not the mediator of ischemic preconditioning. *Am. J. Physiol. Heart Circ. Physiol.* **281,** H1066–H1074.

Hendrich, S., Lee, K., Xu, X., Wang, H., and Murphy, P. A. (1994). Defining food components as new nutrients. *J. Nutr.* **124,** 1789S–1792S.

Horwitt, M. K., Harvey, C. C., Duncan, G. D., and Wilson, W. C. (1956). Effects of limited tocopherol intake in man with relationships to erythrocyte hemolysis and lipid oxidations. *Am. J. Clin. Nutr.* **4,** 408–419.

Ikeda, S., Tohyama, T., Yoshimura, H., Hamamura, K., Abe, K., and Yamashita, K. (2003). Dietary alpha-tocopherol decreases alpha-tocotrienol but not gamma-tocotrienol concentration in rats. *J. Nutr.* **133,** 428–434.

Iqbal, J., Minhajuddin, M., and Beg, Z. H. (2003). Suppression of 7,12-dimethylbenz[alpha] anthracene-induced carcinogenesis and hypercholesterolaemia in rats by tocotrienol-rich fraction isolated from rice bran oil. *Eur. J. Cancer Prev.* **12,** 447–453.

Kamat, J. P., Sarma, H. D., Devasagayam, T. P., Nesaretnam, K., and Basiron, Y. (1997). Tocotrienols from palm oil as effective inhibitors of protein oxidation and lipid peroxidation in rat liver microsomes. *Mol. Cell. Biochem.* **170,** 131–137.

Kamimura, M. (1977). Physiology and clinical use of vitamin E (author's transl.). *Hokkaido Igaku Zasshi* **52,** 185–188.

Khanna, S., Roy, S., Ryu, H., Bahadduri, P., Swaan, P. W., Ratan, R. R., and Sen, C. K. (2003). Molecular basis of vitamin E action: Tocotrienol modulates 12-lipoxygenase, a key mediator of glutamate-induced neurodegeneration. *J. Biol. Chem.* **278,** 43508–43515.

Khanna, S., Roy, S., Slivka, A., Craft, T. K., Chaki, S., Rink, C., Notestine, M. A., DeVries, A. C., Parinandi, N. L., and Sen, C. K. (2005). Neuroprotective properties of the natural vitamin E alpha-tocotrienol. *Stroke* **36,** 2258–2264.

Komiyama, K., Iizuka, K., Yamaoka, M., Watanabe, H., Tsuchiya, N., and Umezawa, I. (1989). Studies on the biological activity of tocotrienols. *Chem. Pharm. Bull. (Tokyo)* **37,** 1369–1371.

Krinsky, N. I. (1992). Mechanism of action of biological antioxidants. *Proc. Soc. Exp. Biol. Med.* **200,** 248–254.

McIntyre, B. S., Briski, K. P., Tirmenstein, M. A., Fariss, M. W., Gapor, A., and Sylvester, P. W. (2000). Antiproliferative and apoptotic effects of tocopherols and tocotrienols on normal mouse mammary epithelial cells. *Lipids* **35,** 171–180.

Meydani, M. (2004). Vitamin E modulation of cardiovascular disease. *Ann. NY Acad. Sci.* **1031,** 271–279.

Murray, C. J., and Lopez, A. D. (1997). Alternate projections of mortality and disability by cause 1990–2020: Global burden of disease study. *Lancet* **349,** 1498–1504.

Nesaretnam, K., Guthrie, N., Chambers, A. F., and Carroll, K. K. (1995). Effect of tocotrienols on the growth of a human breast cancer cell line in culture. *Lipids* **30,** 1139–1143.

Newaz, M. A., Yousefipour, Z., Nawal, N., and Adeeb, N. (2003). Nitric oxide synthase activity in blood vessels of spontaneously hypertensive rats: Antioxidant protection by gamma-tocotrienol. *J. Physiol. Pharmacol.* **54,** 319–327.

Packer, L. (1992). Interactions among antioxidants in health and disease. Vitamin E and the redox cycle. *Proc. Soc. Exp. Biol. Med.* **200,** 271–276.

Papas, A. (1999). "The Vitamin E Factor." Harper Collins Publishers, Inc., New York.

Parker, R. A., Pearce, B. C., Clark, R. W., Gordon, D. A., and Wright, J. J. (1993). Tocotrienols regulate cholesterol production in mammalian cells by post-transcriptional suppression of 3-hydroxy-3-methylglutarylcoenzyme A reductase. *J. Biol. Chem.* **268,** 11230–11238.

Pearce, B. C., Parker, R. A., Deason, M. E., Qureshi, A. A., and Wright, J. J. (1992). Hypocholesterolemic activity of synthetic and natural tocotrienols. *J. Med. Chem.* **35,** 3595–3606.

Pearce, B. C., Parker, R. A., Deason, M. E., Dischino, D. D., Gillespie, E., Qureshi, A. A., Volk, K., and Wright, J. J. (1994). Inhibitors of cholesterol biosynthesis. 2. Hypocholesterolemic and antioxidant activities of benzopyran and tetrahydronaphthalene analogues of the tocotrienols. *J. Med. Chem.* **37,** 526–541.

Qureshi, A. A., and Peterson, D. M. (2001). The combined effects of novel tocotrienols and lovastatin on lipid metabolism in chickens. *Atherosclerosis* **156,** 39–47.

Qureshi, A. A., Burger, W. C., Peterson, D. M., and Elson, C. E. (1986). The structure of an inhibitor of cholesterol biosynthesis isolated from barley. *J. Biol. Chem.* **261,** 10544–10550.

Qureshi, A. A., Qureshi, N., Hasler-Rapacz, J. O., Weber, F. E., Chaudhary, V., Crenshaww, T. D., Gapor, A., Ong, A. S., Chong, Y. H., Peterson, D., and Rapacz, J. (1991a). Dietary tocotrienols reduce concentrations of plasma cholesterol, apolipoprotein B, thromboxane B_2, and platelet factor 4 in pigs with inherited hyperlipidemias. *Am. J. Clin. Nutr.* **53**(4 Suppl.), 1042S–1046S.

Qureshi, A. A., Qureshi, N., Wright, J. J., Shen, Z., Kramer, G., Gapor, A., Chong, Y. H., DeWitt, G., Ong, A., and Peterson, D. M. (1991b). Lowering of serum cholesterol in hypercholesterolemic humans by tocotrienols (palmvitee). *Am. J. Clin. Nutr.* **53,** 1021S–1026S.

Qureshi, A. A., Bradlow, B. A., Brace, L., Manganello, J., Peterson, D. M., Pearce, B. C., Wright, J. J., Gapor, A., and Elson, C. E. (1995). Response of hypercholesterolemic subjects to administration of tocotrienols. *Lipids* **30,** 1171–1177.

Qureshi, A. A., Peterson, D. M., Hasler-Rapacz, J. O., and Rapacz, J. (2001a). Novel tocotrienols of rice bran suppress cholesterogenesis in hereditary hypercholesterolemic swine. *J. Nutr.* **131,** 223–230.

Qureshi, A. A., Sami, S. A., Salser, W. A., and Khan, F. A. (2001b). Synergistic effect of tocotrienol-rich fraction (TRF(25)) of rice bran and lovastatin on lipid parameters in hypercholesterolemic humans. *J. Nutr. Biochem.* **12,** 318–329.

Qureshi, A. A., Sami, S. A., Salser, W. A., and Khan, F. A. (2002). Dose-dependent suppression of serum cholesterol by tocotrienol-rich fraction (TRF25) of rice bran in hypercholesterolemic humans. *Atherosclerosis* **161,** 199–207.

Raederstorff, D., Elste, V., Aebischer, C., and Weber, P. (2002). Effect of either gamma-tocotrienol or a tocotrienol mixture on the plasma lipid profile in hamsters. *Ann. Nutr. Metab.* **46,** 17–23.

Roy, S., Lado, B. H., Khanna, S., and Sen, C. K. (2002). Vitamin E sensitive genes in the developing rat fetal brain: A high-density oligonucleotide microarray analysis. *FEBS Lett.* **530,** 17–23.

Sen, C. K., Khanna, S., Roy, S., and Packer, L. (2000). Molecular basis of vitamin E action. Tocotrienol potently inhibits glutamate-induced pp60(c-Src) kinase activation and death of HT4 neuronal cells. *J. Biol. Chem.* **275,** 13049–13055.

Sen, C. K., Khanna, S., and Roy, S. (2004). Tocotrienol: The natural vitamin E to defend the nervous system? *Ann. NY Acad. Sci.* **1031,** 127–142.

Sen, C. K., Khanna, S., and Roy, S. (2006). Tocotrienols: Vitamin E beyond tocopherols. *Life Sci.* **78,** 2088–2098.

Serbinova, E. A., and Packer, L. (1994). Antioxidant properties of alpha-tocopherol and alpha-tocotrienol. *Methods Enzymol.* **234,** 354–366.

Serbinova, E., Kagan, V., Han, D., and Packer, L. (1991). Free radical recycling and intramembrane mobility in the antioxidant properties of alpha-tocopherol and alpha-tocotrienol. *Free Radic. Biol. Med.* **10,** 263–275.

Serbinova, E., Khwaja, S., Catudioc, J., Ericson, J., Torres, Z., Gapor, A., Kagan, V., and Packer, L. (1992). Palm oil vitamin E protects against ischemia/reperfusion injury in the isolated perfused Langendorff heart. *Nutr. Res.* **12,** S203–S215.

Suzuki, Y. J., Tsuchiya, M., Wassall, S. R., Choo, Y. M., Govil, G., Kagan, V. E., and Packer, L. (1993). Structural and dynamic membrane properties of alpha-tocopherol and alpha-tocotrienol: Implication to the molecular mechanism of their antioxidant potency. *Biochemistry* **32,** 10692–10699.

Sylvester, P. W., Nachnani, A., Shah, S., and Briski, K. P. (2002). Role of GTP-binding proteins in reversing the antiproliferative effects of tocotrienols in preneoplastic mammary epithelial cells. *Asia Pac. J. Clin. Nutr.* **11**(Suppl. 7), S452–S459.

Tan, D. T., Khor, H. T., Low, W. H., Ali, A., and Gapor, A. (1991). Effect of a palm-oil-vitamin E concentrate on the serum and lipoprotein lipids in humans. *Am. J. Clin. Nutr.* **53**(4 Suppl.), 1027S–1030S.

Tappel, A. L. (1953). The inhibition of hematin-catalyzed oxidations by alpha-tocopherol. *Arch. Biochem. Biophys.* **47,** 223–225.

Tappel, A. L. (1954). Studies of the mechanism of vitamin E action. II. Inhibition of unsaturated fatty acid oxidation catalyzed by hematin compounds. *Arch. Biochem. Biophys.* **50,** 473–485.

Tappel, A. L. (1955). Studies of the mechanism of vitamin E action. III. *In vitro* copolymerization of oxidized fats with protein. *Arch. Biochem.* **54,** 266–280.

Theriault, A., Chao, J. T., Wang, Q., Gapor, A., and Adeli, K. (1999). Tocotrienol: A review of its therapeutic potential. *Clin. Biochem.* **32,** 309–319.

17

VITAMIN E AND CANCER

KIMBERLY KLINE,* KARLA A. LAWSON,† WEIPING YU,‡ AND BOB G. SANDERS‡

*Division of Nutrition, University of Texas at Austin, Austin, Texas 78712
†Cancer Prevention Fellowship Program, National Cancer Institute National Institutes of Health, Bethesda, Maryland 20892
‡School of Biological Sciences, University of Texas at Austin Austin, Texas 78712

Vitamins and Hormones, Volume 76

0083-6729/07 $35.00
DOI: 10.1016/S0083-6729(07)76017-X

Perhaps not surprisingly, vitamin E which has been touted to be potentially beneficial for a variety of disorders, including cancer, heart disease, and even Alzheimer's disorder, based on its function as an antioxidant has failed to withstand the scrutiny of recent, double-blinded, placebo-controlled clinical trials, including failure to provide science-based support for vitamin E as a potent anticancer agent. Although less studied, vitamin E forms other than *RRR*-α-tocopherol or synthetic *all-rac*-α-tocopherol show promise as anticancer agents in preclinical studies. This chapter will (1) review basic information about natural and synthetic vitamin E compounds as well as vitamin E analogues, (2) summarize the current status of human intervention trials, (3) review data from preclinical cell culture and animal model studies of vitamin E compounds and novel vitamin E-based analogues in regards to future potential for cancer treatment, and (4) summarize some of the insights that have been gained into the anticancer mechanisms of action of vitamin E-based compounds which are providing interesting insights into their potent proapoptotic effects, which include restoration of apoptotic signaling pathways and blockage of prosurvival signaling events.

I. BASIC INFORMATION ABOUT VITAMIN E

A. HOW MANY VITAMIN Es ARE THERE?

One thing that continues to hamper the understanding of vitamin E by laypeople, scientists, and clinicians is that vitamin E comes in multiple naturally occurring forms called tocopherols and tocotrienols which have different molecular structures, different levels of bioavailability, different biological activities, and exhibit different mechanisms of action, especially regarding anticancer actions (Fig. 1). Furthermore, unlike other vitamins, the synthetic form of vitamin E is not equivalent in chemical structure or biological activity to the form of vitamin E recognized as the most biologically important, namely *RRR*-α-tocopherol (Fig. 1). The synthetic form,

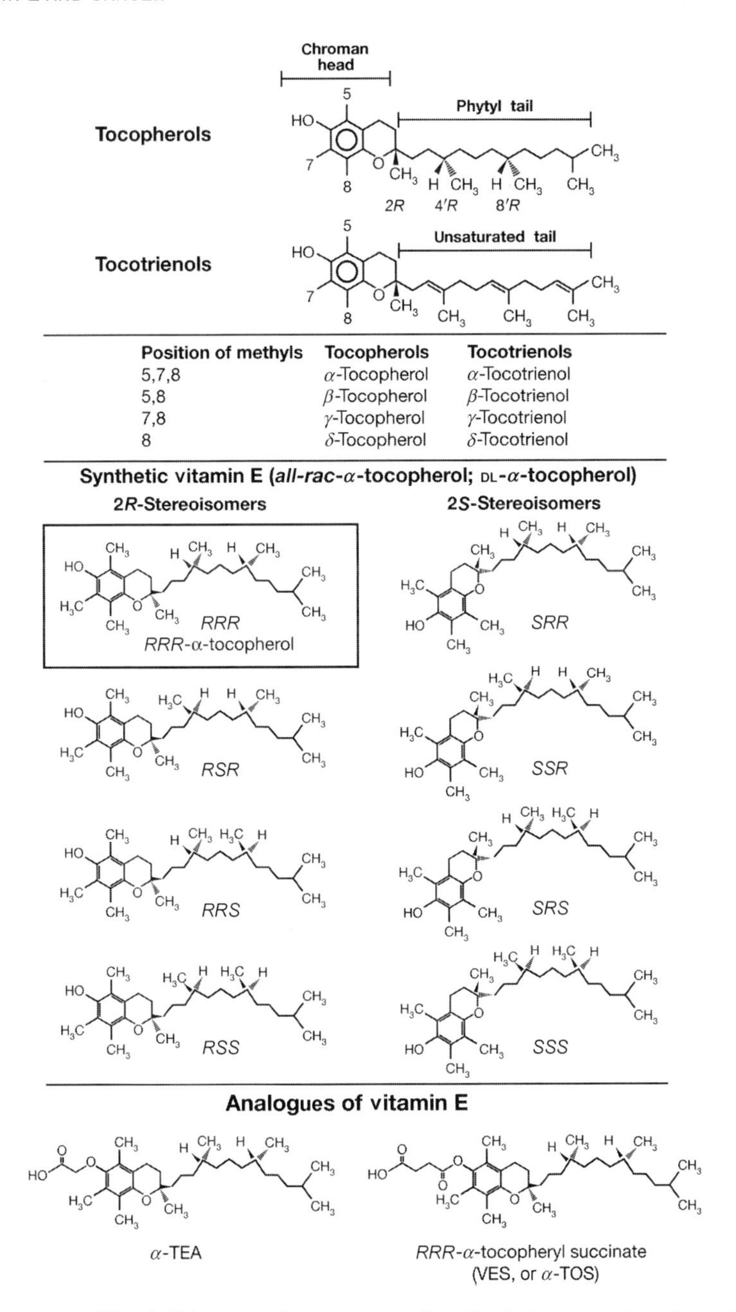

FIGURE 1. Vitamin E is a general term used to refer collectively to a number of structurally and functionally different small lipids. Chemical structures of plant derived, naturally occurring vitamin E compounds (tocopherols and tocotrienols), commercially prepared synthetic vitamin E (*all-rac*-α-tocopherol/DL-α-tocopherol), and analogues of vitamin E [α-TEA/2,5,7,8-tetramethyl-2*R*-(4′*R*,8′*R*,12′-trimethyltridecyl)chroman-6-yloxy acetic acid and *RRR*-α-tocopheryl succinate/

properly referred to as *all-rac*-α-tocopherol (for *all-racemic*) but also frequently called DL-α-tocopherol, is a mixture of eight stereoisomers, only one of which is *RRR*-α-tocopherol (Fig. 1). Acetate and succinate esters of both *RRR*-α-tocopherol and *all-rac*-α-tocopherol are used as vitamin E sources in commercial supplements because they are more stable in the presence of air (oxygen). Of these two esters, *RRR*-α-tocopheryl succinate [referred to as vitamin E succinate (VES) or α-TOS] has been demonstrated to have anticancer activity (reviewed by Kline *et al.*, 2004; Prasad *et al.*, 2003; Wang *et al.*, 2006). The protective acetate or succinate moieties are expected to be removed by general esterase activity during digestion, releasing the free tocopherol for absorption (Institute of Medicine, 2000). Additionally, a number of novel analogues of *RRR*-α-tocopherol have been designed, synthesized, and tested for anticancer properties (Arya *et al.*, 1998; Lawson *et al.*, 2003; Shiau *et al.*, 2006; Thompson and Wilding, 2003; Wang *et al.*, 2006). One analogue called α-tocopherol acetic acid analogue (α-TEA; Fig. 1) has been characterized using cell culture and preclinical animal models and exhibits potent anticancer properties (Anderson *et al.*, 2004a,b; Best, 2005; Jia, 2006; Kline *et al.*, 2003, 2004; Lawson *et al.*, 2003, 2004a,b; Shun, 2005; Shun *et al.*, 2004; Snyder, 2005; Wang, 2006; Yu *et al.*, 2006; Zhang *et al.*, 2004b).

B. WHY IT IS SO IMPORTANT THAT THE FORM OF VITAMIN E BE IDENTIFIED PROPERLY?

Given the fact that there are all these different vitamin E compounds and that they are not equivalent makes it imperative that the form being used in scientific studies be accurately identified. For example, if a scientific paper only states that α-tocopherol was used, what really was used... *RRR*-α-tocopherol or synthetic *all-rac*-α-tocopherol? To illustrate the magnitude of this problem further, let us consider some recent human intervention trials (see Section II for more details on these intervention trials). Several human intervention trials investigating the effect of supplemental vitamin E on breast cancer risk have shown no beneficial effect. The acetate derivative of natural source of vitamin E, namely, *RRR*-α-tocopheryl acetate was used in these studies. Would the outcome have been different if the acetate derivative of synthetic vitamin E (*all-rac*-α-tocopheryl acetate) had been used? On the flip side of this question, would the ATBC Trial (α-Tocopherol, β-Carotene

---

vitamin E succinate (VES)/α-TOS]. There are eight naturally occurring forms of vitamin E which differ in number and position of methyl groups on the chroman head (α-, β-, γ-, δ-) and presence or absence of double bonds in the phytyl tail (saturated = tocopherols; unsaturated = tocotrienols). Synthetic vitamin E is a mixture of approximately equal amounts of eight stereoisomers, only one of which is identical to *RRR*-α-tocopherol, which is recognized as having the highest biopotency. Two analogues of *RRR*-α-tocopherol which possess anticancer activities are: α-TEA which is an ether-linked acetic acid analogue and *RRR*-α-tocopheryl succinate which is an ester-linked succinic acid analogue and referred to in the literature as VES or α-TOS.

Cancer Prevention Study) finding of 32% fewer cases of prostate cancer and 41% fewer deaths from prostate cancer been observed if they had supplemented with *RRR*-α-tocopheryl acetate rather than the acetate derivative of synthetic vitamin E (*all-rac*-α-tocopheryl acetate)? This points out some of the major challenges present in the vitamin E field.

C. WHY IS *RRR*-α-TOCOPHEROL THE MOST BIOAVAILABLE FORM OF VITAMIN E?

Although it appears that all "natural source" (namely, α-, β-, γ-, δ-tocopherols and tocotrienols produced by plants) and synthetic vitamin E stereoisomers are absorbed by intestinal enterocytes and secreted into chylomicrons more or less equally for exogenous fat transport; *RRR*-α-tocopherol is preferentially transported in the serum during post-liver endogenous fat transport due to the selective actions of the liver α-tocopherol transfer protein (α-TPP), making *RRR*-α-tocopherol the dominant form in the serum and tissues (Traber and Arai, 1999). For purposes of establishing the human requirement for vitamin E, the Food and Nutrition Board Panel limited the definition of vitamin E activity to *RRR*-α-tocopherol and the other three synthetic 2*R*-stereoisomer forms (*RSR*-, *RRS*-, and *RSS*-) of *all-rac*-α-tocopherol (Institute of Medicine, 2000). They set the recommended dietary allowance (RDA) for vitamin E at 15 mg/day for both males and females 14 years of age and older. Also in recognition that high levels of supplemental vitamin E may have the adverse effect of increased tendency for hemorrhage, a tolerable upper intake level of 1000 mg was set. Defining vitamin E activity as only *RRR*-α-tocopherol and including only the 2*R*-stereoisomer forms of synthetic vitamin E represents a change that has been questioned by some investigators who believe other forms of vitamin E, especially γ-tocopherol, the major form of vitamin E in the US diet, may have unique benefits that *RRR*-α-tocopherol and synthetic vitamin E do not have (Campbell *et al.*, 2003a; Hensley *et al.*, 2004; Jiang *et al.*, 2001; Wagner *et al.*, 2004).

In addition to the selection of *RRR*-α-tocopherol by the liver α-TTP, studies have characterized vitamin E metabolism and shown that they are metabolized via degradation of the side chain by an initial ω-oxidation followed by β-oxidation (Landes *et al.*, 2003; Zhou *et al.*, 2004). Increased degradation of vitamin E forms other than *RRR*-α-tocopherol by liver P450 enzymes also contributes to the predominance of plasma and tissue *RRR*-α-tocopherol over other vitamin E forms (Brigelius-Flohe, 2005; Zhou *et al.*, 2004).

The reason why *RRR*-α-tocopherol is selected over all other tocopherol and tocotrienol forms found in the diet is not understood. The preference for *RRR*-α-tocopherol does not appear to be on the basis of antioxidant properties since the antioxidant properties of γ-tocopherol, for example, are reported to be equal or better than *RRR*-α-tocopherol (Huang and Appel, 2003). Although vitamin E compounds have historically and sometimes dogmatically been pigeonholed as antioxidants, Azzi and coworkers have consistently presented the case that *RRR*-α-tocopherol has important nonantioxidant activities (Ricciarelli *et al.*, 2001, 2002; Zingg and Azzi, 2004). Whether these nonantioxidant functions provide the basis for the biological preference for *RRR*-α-tocopherol is unknown. Clearly from a vitamin E and cancer perspective, studies addressing selective tissue availability of vitamin E forms other than *RRR*-α-tocopherol that exhibit marked anticancer activities are badly needed.

D. CHALLENGES FOR *IN VIVO* TESTING OF VITAMIN E FORMS OTHER THAN *RRR*-α-TOCOPHEROL

Since *RRR*-α-tocopherol does not exhibit significant anticancer properties, its preferred selection over other vitamin E forms possesses major challenges for coming up with strategies that will permit *in vivo* bioavailability of vitamin E forms that have potential to be effective chemopreventive or chemotherapeutic agents. One strategy for increasing the bioavailability of γ-tocopherol is the inhibition of P450 xenobiotic metabolizing enzymes with sesame seeds or sesamin, the lignan found in sesame seeds. Studies show that dietary sesame seeds as well as sesamin lead to elevated levels of tissue and plasma γ-tocopherol due to the inhibition of P450 CYP 3A metabolism of γ-tocopherol (Ikeda *et al.*, 2002; Parker *et al.*, 2000). We are unaware of any vitamin E cancer studies conducted where P450 enzymes have been inhibited.

Other strategies include special formulations of vitamin E compounds in liposomes or nanoparticles and use of different routes of administration, for example aerosol delivery (Lawson *et al.*, 2003; Wang, 2006). Why liposome/nanoparticle formulations and aerosol delivery? Liposomes and nanoparticles serve as controlled release carriers as well as biocompatible solubilizing vehicles for small lipid, water-insoluble agents such as α-TEA or VES. Favorable pharmacokinetic characteristics of liposomes/nanoparticles include long systemic circulation time, enhanced tumor permeability, accumulation and retention, improved therapeutic efficacy with reduced therapeutic dosage, reduced toxicity, and controlled delivery of combinations of anticancer agents (Lee, 2006). Likewise aerosol delivery offers several advantages. The epithelial surface area of the adult human lung is large, equivalent to the surface area of a standard single court tennis (100 m^2), and it is rich in blood and lymph flow, thereby, providing a rich environment for aerosol delivery of liposome or nanoparticle formulated anticancer agents such as α-TEA (Patton, 1997). Furthermore, aerosol pulmonary delivery enhances drug bioavailability, enhances rapid drug action and distribution to tissues, avoids first-pass hepatic effects, provides enhanced anticancer efficacy with lower dosage, and has been shown to be a cost-effective method for drug delivery for home treatment. Further research into strategies to increase the bioavailability of vitamin E-based compounds showing promising antitumor efficacy in preclinical studies is needed.

II. INTERVENTION TRIALS

A. WHAT HAVE WE LEARNED ABOUT VITAMIN E AND CANCER FROM HUMAN INTERVENTION TRIALS?

A number of randomized clinical trials have been conducted to investigate the effect of supplemental vitamin E on cancer development. In 2000, the Dietary Reference Intakes for antioxidant vitamins and minerals was published, summarizing the outcome for one large randomized trial, the ATBC Cancer Prevention Study (The ATBC Cancer Prevention Study Group, 1994), two small trials regarding supplemental vitamin E either mammary dysplasia or benign breast disease, and a handful of trials looking at the effect of supplemental vitamin E on colon polyp formation (Institute of Medicine, 2000). For the most part, these trials showed no effect of supplemental vitamin E on cancer development (Institute of Medicine, 2000) other than findings in the ATBC trial of a statistically significant 32% lower prostate cancer incidence and a 41% reduction in prostate cancer deaths observed in male smokers, given 50 mg/day synthetic DL-α-tocopherol acetate (terminology as cited in report) compared to those not taking vitamin E (Heinonen *et al.*, 1998). The report from the Institute of Medicine (2000) concluded that data regarding supplemental vitamin E and prevention of cancer was weak and evidence from long-term randomized clinical trials was lacking. Since this report, two large randomized control trials have reported findings of vitamin E supplementation on cancer incidence, the Heart Outcomes Prevention Evaluation (HOPE) trial and the Women's Health Study (WHS) trial (HOPE Study Investigators, 2005; Lee *et al.*, 2005).

The HOPE trial and an extended follow-up with a limited number of original participants, the HOPE-TOO trial, investigated the effect of *RRR*-α-tocopheryl acetate (400 IU/day) on both heart disease and cancer outcomes (HOPE Study Investigators, 2005). More than 9000 male and female participants in the original trial and 4000 participants (~74% males and 26% females for both trials) who agreed to continued follow-up in the HOPE-TOO trial were at least 55 years of age at enrollment and had history of ischemic heart disease, stroke, peripheral artery disease, or diabetes mellitus and one additional cardiovascular disease risk factor. Participants were followed for an average of 4.5 years in the HOPE trial and an average of 7.0 years of follow-up was available for those who agreed to continued intervention in the HOPE-TOO trial. Both the HOPE and HOPE-TOO trials found no significant association between daily vitamin E supplementation and cancer incidence. At the conclusion of the HOPE trial, 11.6% ($n = 552$) of participants given vitamin E were diagnosed with cancer compared to 12.3% ($n = 586$) of those taking placebo. The relative risk (RR) of any cancer diagnosis for those taking vitamin E was 6% lower than those taking placebo, which was not statistically significant (RR = 0.94, 95% confidence interval = 0.84–1.06, $p = 0.30$). With additional follow-up for the HOPE-TOO participants, risk was 4% decreased when compared to placebo (RR = 0.96, 95% confidence interval = 0.84–1.09, $p = 0.50$). Findings here failed to substantiate those seen in the ATBC trial for prostate cancer, but did show a small decrease in the number of lung cancers diagnosed in those taking vitamin E supplementation. The decrease in lung cancer seen here failed to meet the requirements for statistical significance based on the presence of multiple comparisons. Study investigators concluded that long-term vitamin E supplementation did not prevent cancer in this population.

The WHS, a randomized, double-blinded, placebo-controlled trial of vitamin E and aspirin, followed more than 19,000 women randomized to receive 600 IU of supplemental vitamin E (*RRR*-α-tocopherol acetate) every other day, for more than 10 years (Lee *et al.*, 2005). The trial population was unique in that those being studied were healthy at baseline, with no history of coronary heart disease, stroke, cancer, or other chronic diseases. After an average of 10.1 years of follow-up, cancer incidence was not lower for those taking vitamin E supplements when compared to placebo controls. In the supplemental vitamin E group, there were 1437 cases of cancer among 19,937 women, compared to 1428 of 19,939 women taking placebo (RR = 1.01, 95% confidence interval = 0.94–1.08). There was no statistically significant difference in the number of breast, lung, or colon cancers diagnosed for those women taking supplements when compared to participants taking placebo.

Additional randomized trials have reported results regarding cancer incidence with long-term vitamin E supplementation as part of a combination of antioxidant vitamins. The Women's Angiographic Vitamin and Estrogen (WAVE) trial tested 800-IU vitamin E (form not designated) daily in combination with vitamin C and showed no protective effect for the combination of vitamins E and C on cancer incidence (Waters *et al.*, 2002). The SU.VI.MAX trial of antioxidant supplements tested 30-mg vitamin E (form not designated) daily in combination with ascorbic acid, β-carotene, selenium, and zinc and showed a significantly lower cancer incidence in men taking the antioxidant mixture, but not in women (Hercberg *et al.*, 2004). Little knowledge on the effect of individual vitamin E supplementation on cancer prevention can be gleaned from these trials, as a variety of antioxidants were administered simultaneously.

The Selenium and Vitamin E Cancer Prevention Trial (SELECT), a randomized, double-blinded, clinical trial, is currently under way to test the hypothesis that either selenium (200-μg/day L-selenomethionine) or vitamin E (400-IU/day all-*rac*-α-tocopheryl acetate) singly or in combination will be effective at reducing prostate cancer incidence and mortality when administered over a 12-year period (Lippman *et al.*, 2005).

In summary, data from randomized, double-blinded, placebo-controlled trials do not support a beneficial effect of long-term supplementation with *RRR*-α-tocopherol or synthetic *all-rac*-α-tocopheryl acetate for the prevention of a variety of cancer types, with the possible exception of synthetic vitamin E for prostate cancer. Secondary analysis of the ATBC trial points to a reduction in prostate cancer with synthetic vitamin E supplementation in heavy smokers, though results from the HOPE trial with *RRR*-α-tocopheryl acetate do not support these findings.

B. CONCLUSIONS

The failure to find convincing and consistent clinical evidence supportive of chemopreventive actions for vitamin E may reflect the fact that the wrong form of vitamin E or insufficient dosages have been tested on the wrong target populations. Alternatively, these clinical findings may be another example of preclinical basic science studies and initial clinical trails indicating a potential role which does not translate into successful clinical outcomes like the retinoic acid paradox in cancer prevention (Freemantle *et al.*, 2006). Answers will require a more complete understanding of the basic biology of the various vitamin E compounds: tissue selective bioavailability, retention, and intracellular localization, biochemical and molecular mechanisms of action, validated biomarkers for each vitamin E form to verify compliance and anticancer activity, and identification and removal of competing antagonistic actions by other dietary factors.

III. PRECLINICAL STUDIES

A. LACK OF EVIDENCE FOR ANTICANCER EFFECTS BY *RRR*-α-TOCOPHEROL OR *All-rac*-α-TOCOPHEROL

Preclinical studies have shown limited or no evidence for a significant anticancer effect by *RRR*-α-tocopherol, except for studies in human prostate and colon cancer cells in culture where *RRR*-α-tocopherol inhibited cancer cell growth or induced apoptosis (Gysin *et al.*, 2002; Miyoshi *et al.*, 2005). Indeed, some studies show that *RRR*-α-tocopherol blocks other forms of vitamin E anticancer effects when cancer cells are cotreated in cell culture (Kline *et al.*, 2004; Weber *et al.*, 2003). Furthermore, preclinical animal studies do not provide consistent, compelling evidence for a significant anticancer role for either *RRR*-α-tocopherol or synthetic *all-rac*-α–tocopherol for breast cancer, and there are only a few studies showing a beneficial effect for colon and prostate (Campbell *et al.*, 2003a; Kelloff and Boone, 1994; Kline *et al.*, 2003; Siler *et al.*, 2004; Venkateswaran *et al.*, 2004).

B. EVIDENCE FOR ANTICANCER EFFECTS OF γ-TOCOPHEROL

γ-Tocopherol, the most common form of vitamin E in the US diet and the second most common form in human tissues, has drawn attention for its antitumor activity due to several unique features, including detoxification of reactive nitrogen species and anti-inflammatory and anticancer activities (Brigelius-Flohe and Traber, 1999; Campbell *et al.*, 2003a,b; Cooney *et al.*, 1993; Hensley *et al.*, 2004; Jiang *et al.*, 2000, 2001; Stone *et al.*, 2004; Wagner *et al.*, 2004). γ-Tocopherol and its metabolic product carboxyethyl hydroxychroman (CEHC) have been reported to be effective inhibitors of human prostate cancer cell proliferation (Galli *et al.*, 2004). γ-Tocopherol, but not *RRR*-α-tocopherol, induces apoptosis in several human colon cancer cells containing different genetic alterations without damage to normal colon cells (Campbell *et al.*, 2006). γ-Tocopherol was shown to be more effective than *RRR*-α-tocopherol in inhibiting proliferation and inducing apoptosis of human prostate and colon cancer cells in culture (Gysin *et al.*, 2002; Jiang *et al.*, 2004). Data from our laboratory show that γ-tocopherol induces human breast cancer cells to undergo apoptosis in cell culture and show that γ-tocopherol induces apoptosis by activation of TRAIL/DR4/5 proapoptotic signaling and suppression of FLIP and survivin antiapoptotic signaling (unpublished data).

In summary, γ-tocopherol shows promise for treatment of breast, prostate, and colon cancer. Preclinical animal studies are needed to show *in vivo* efficacy and to monitor tumor tissue levels of γ-tocopherol versus *RRR*-α-tocopherol. A major challenge will be to determine if there is tissue selective uptake, retention, and unique subcellular localization of γ-tocopherol, and to overcome the preferential uptake and retention of *RRR*-α-tocopherol over γ-tocopherol.

C. EVIDENCE FOR ANTICANCER EFFECTS OF VITAMIN E METABOLITES

Metabolism of vitamin E compounds results in production of carboxyethylhydroxy chroman (CEHC) derivatives that have antioxidant properties as well as ability to inhibit proinflammatory processes (Hensley *et al.*, 2004; Jiang *et al.*, 2000). CEHC derivatives circulate and are excreted in the urine. As expected, urine contains more γ-CEHC than α-CEHC supporting data showing γ-tocopherol to be metabolized in preference to *RRR*-α-tocopherol (reviewed by Hensley *et al.*, 2004). To date, little information is available on the efficacy of vitamin E metabolites as anticancer agents. Studies are definitely needed to determine if vitamin E metabolites play a role in cancer.

D. TOCOTRIENOLS AS POTENTIAL ANTICANCER AGENTS

Tocotrienols are of interest, in part, because of their strong antioxidant activity, but also for their potential anticancer properties. The literature on tocotrienols and cancer is limited. Reviews of the tocotrienols and their role in pathogenesis conclude that tocotrienols possess neuroprotective, anticancer, and cholesterol-lowering properties that are often not exhibited by tocopherols (Guthrie and Carroll, 1999; Papas, 2002; Sen *et al.*, 2006). Of the vitamin E forms, the rate of metabolism is highest for the tocotrienols. All four tocotrienols specifically bind to and activate steroid and xenobiotic receptor (SXR)/pregnane X receptor (PXR) and show tissue-specific induction of SXR target genes, especially CYP3A4 (Zhou *et al.*, 2004). As strong activators of metabolic enzymes, tocotrienols in combination with chemotherapeutics have the potential to decrease drug efficacy (Zhou *et al.*, 2004). In our hands, low levels of tocotrienols induce human breast cancer cells in culture to undergo cell death by apoptosis (Yu *et al.*, 1999b). In rank order, the most effective vitamin E forms for inhibition of colony formation of human breast cancer cells are α-TEA; VES; δ-, γ-, and α-tocotrienols; and δ- and γ-tocopherols. *RRR*-α-tocopherol was noneffective (Kline *et al.*, 2003).

In summary, based on data from cell culture studies, tocotrienols show promise for anticancer activity but supportive animal studies are needed.

E. VITAMIN E ANALOGUES AS POTENTIAL ANTICANCER AGENTS

The synthesis and testing of novel vitamin E compounds for anticancer efficacy is an ongoing endeavor by several laboratories (Anderson *et al.*, 2004a,b; Arya *et al.*, 1998; Birringer *et al.*, 2003; Lawson *et al.*, 2003; Neuzil *et al.*, 2001b, 2004; Shiau *et al.*, 2006; Thompson and Wilding, 2003; Tomic-Vatic, 2005; Wang *et al.*, 2006). To date, two vitamin E analogues, VES and α-TEA, have been studied in greatest detail and have shown potent anticancer properties in preclinical animal models. VES administered intraperitoneally has been demonstrated to inhibit human breast, colon, and mesothelioma cancer and mouse lung, melanoma, and mammary cancer in animal models (reviewed by Wang *et al.*, 2006).

In an effort to develop a clinically useful vitamin E-based chemotherapeutic agent and to administer it in a clinically relevant manner, a nonhydrolyzable ether analogue of *RRR*-α-tocopherol, namely, α-TEA has been produced. The parent compound for VES and α-TEA is *RRR*-α-tocopherol. α-TEA differs from VES in that it has an acetic acid moiety linked to the phenolic oxygen at carbon 6 of the chroman head by an ether linkage, whereas VES has a succinic acid moiety linked by an ester linkage at this site (Lawson *et al.*, 2003). The basic structure of VES has the potential of compromising its anticancer efficacy *in vivo* in that the ester linkage can be hydrolyzed by cellular esterases, yielding

RRR-α-tocopherol and succinic acid, neither of which exhibit anticancer properties (Anderson *et al.*, 2004a). VES is ineffective as an anticancer agent in human ovarian cancer cells in which cellular esterases hydrolyze the ester linkage (Anderson *et al.*, 2004a). Furthermore, VES has been shown to lose its anticancer effectiveness when delivered orally, presumably due to intestinal esterases (Lawson *et al.*, 2004a).

As α-TEA is a lipid that is insoluble in water, aerosol delivery of liposomal preparations was chosen as a potentially effective, clinically relevant method of delivery. α-TEA administered by aerosol (depositing 36–72 μg of α-TEA into the respiratory tract for 17–25 days) or gavage (orally administering 5–6 mg/day) either separately or in combination with other chemotherapeutics has been demonstrated to inhibit tumor growth and metastases in human breast, ovarian, and prostate cancer in xenografts; mammary cancer in a syngeneic mouse model; and UV-induced skin cancer in a mouse model (Anderson *et al.*, 2004a,b; Best, 2005; Jia, 2006; Kline *et al.*, 2003, 2004; Lawson *et al.*, 2003, 2004a,b; Snyder, 2005; Wang, 2006; Zhang *et al.*, 2004b). α-TEA liposomal formulations delivered by aerosol (72 μg deposited into the respiratory tract for 25 days) or gavage (5 mg/day for 25 days) did not show liver, kidney, or bone marrow toxicity (unpublished data).

In summary, cell culture and preclinical animal studies show vitamin E analogues, VES and α-TEA, to exhibit potent anticancer efficacy. Further development and clinical testing are needed.

IV. ANTICANCER MECHANISMS OF ACTION OF VITAMIN E-BASED COMPOUNDS

The pleiotropic anticancer actions of vitamin E compounds for human cancer cells are summarized in Table I. An important qualifier for the information listed in Table I is that in a majority of the cases the essential or contributory nature of the cell component/biochemical event in vitamin E compound-mediated anticancer effects has not been rigorously examined. Until appropriate overexpression and depletion (knockout/knockdown) studies are conducted to demonstrate physiological relevance, their role must be considered correlative. Furthermore, demonstration of some of these potential mechanisms of action needs to be validated in more than one cell type to demonstrate the universality of the effect. Finally, these potential mechanisms need to be validated using *in vivo* models to increase their potential translational significance. Vitamin E compounds have been demonstrated to be potent proapoptotic agents triggering apoptosis via both death receptor-mediated and mitochondrial-mediated pathways (Table I). Major roles for restoration of death receptor signaling as well as cross talk to mitochondrial-dependent events in epithelial-based cancers like human breast, prostate, and ovarian cancer cells have been made; while roles for

TABLE I. Mechanisms of Anticancer Actions of Vitamin E Compounds[a]

Cell component	I/D	Vitamin E compound	Human cancer cell type	References
Induction of Apoptosis				
Cell Components Associated with Death Receptor-Mediated Pathways of Apoptosis				
TGF-β type II receptor	I	VES, α-TEA, δ-T3	Breast	Charpentier *et al.*, 1993, 1996; Shun *et al.*, 2004; Yu *et al.*, 1997
TGF-β (secretion and activation)	I	VES	Breast, gastric	Charpentier *et al.*, 1993; Wu *et al.*, 2001
Fas (CD95) cell surface expression	I	VES, α-TEA	Breast, gastric, ovarian, prostate, leukemia	Bang *et al.*, 2001; Israel *et al.*, 2000; Jia, 2006; Turley *et al.*, 1997b; Wu *et al.*, 2002; Yu *et al.*, 1999a, 2006
Fas ligand expression	I	VES, α-TEA	Breast, prostate	Israel *et al.*, 2000; Jia, 2006; Turley *et al.*, 1997b
Fas ligand secretion	I	VES	Prostate	Salih *et al.*, 2001
TRAIL DR4 and DR5	I	VES	Mesothelioma	Tomasetti *et al.*, 2006
Daxx associated with Fas	I	α-TEA	Prostate	Jia, 2006
FADD associated with Fas	I	VES, α-TEA	Gastric, prostate	Jia, 2006; Wu *et al.*, 2002
Caspase-8 activation	Y	VES, α-TEA, γ-T	Breast, colon, gastric, ovarian, prostate	Campbell *et al.*, 2006; Jia, 2006; Wu *et al.*, 2002; Yu *et al.*, 2006
Caspase-4	I	VES	Prostate	Malafa *et al.*, 2006
Cell Components Associated with Mitochondrial-Mediated Pathways of Apoptosis				
ROS generation/accumulation	Y	VES	Leukemia, neuroblastoma	Alleva *et al.*, 2001; Swettenham *et al.*, 2005; Weber *et al.*, 2003
Ask 1	I	VES	Prostate	Zu *et al.*, 2005
GADD45β	I	VES	Prostate	Zu *et al.*, 2005

(*Continues*)

TABLE I. (*Continued*)

Cell component	I/D	Vitamin E compound	Human cancer cell type	References
Phospho-Sek1		VES	Prostate	Zu *et al.*, 2005
Phospho-JNK	I	VES, α-TEA, δ-T3	Breast, gastric, ovarian, prostate	Shun *et al.*, 2004; Wu *et al.*, 2004; Yu *et al.*, 1998, 2001, 2006; Zhao *et al.*, 2006; Zu *et al.*, 2005
Bax conformational change	Y	VES, α-TEA	Breast, ovarian, prostate	Jia, 2006; Yu *et al.*, 2003, 2006
Bax translocation to mitochondria	Y	VES	Leukemia, breast, prostate	Weber *et al.*, 2003; Yu *et al.*, 2003; Zu *et al.*, 2005
Bid truncation (activation)	Y	α-TEA	Ovarian, prostate	Jia, 2006; Yu *et al.*, 2006
Bak translocation to mitochondria	Y	VES	Breast	Yu *et al.*, 2003
Blockage of Bak binding to Bcl-2/Bcl-xL	Y	VES	Prostate	Shiau *et al.*, 2006
NOXA	I	α-TEA	Breast	Wang, 2006
Phospho-Bim EL in mitochondria	Y	VES	Prostate	Zu *et al.*, 2005
Phospho-Bcl-2(Ser70)	Y	VES	Prostate	Zu *et al.*, 2005
Increased mitochondrial membrane permeabilization	Y	VES, γ-T3	Breast	Takahashi and Loo, 2004; Yu *et al.*, 2003
Loss of mitochondrial inner membrane potential	Y	VES	Leukemia, neuroblastoma	Alleva *et al.*, 2001; Neuzil *et al.*, 1999; Swettenham *et al.*, 2005
Cytochrome *c* release	Y	VES, α-TEA, γ-T	Breast, ovarian, prostate, leukemia	Alleva *et al.*, 2001; Jiang *et al.*, 2004; Shiau *et al.*, 2006; Takahashi and Loo, 2004; Tomasetti *et al.*, 2004; Weber *et al.*, 2003; Yu *et al.*, 2003; Zu *et al.*, 2005
Caspase-9 activation	Y	VES, γ-T, α-TEA, TRF	Breast, prostate, leukemia, colon, ovarian, neuroblastoma	Agarwal *et al.*, 2004; Jia, 2006; Jiang *et al.*, 2004; Swettenham *et al.*, 2005; Weber *et al.*, 2003; Yu *et al.*, 2003, 2006

Lysosomal Instability in Apoptosis				
Destabilization of lysosomal membranes	Y	VES	Leukemia	Neuzil *et al.*, 1999, 2002
Execution Phase Mediators of Apoptosis				
Caspase-3 protein	I	α-T	Prostate	Miyoshi *et al.*, 2005
Caspase-3 activation	Y	VES, α-TEA, γ-T, TRF	Breast, colon, leukemia, ovarian, prostate	Agarwal *et al.*, 2004; Akazawa *et al.*, 2002; Alleva *et al.*, 2001; Bang *et al.*, 2001; Campbell *et al.*, 2006; Jia, 2006; Jiang *et al.*, 2004; Miyoshi *et al.*, 2005; Neuzil *et al.*, 1999, 2002; Weber *et al.*, 2003; Yu *et al.*, 2003, 2006
Caspase-7 activation	Y	γ-T	Colon	Campbell *et al.*, 2006
Cathepsin D needed	Y	VES	Leukemia	Neuzil *et al.*, 2002
Transcription Factor Mediators of Apoptosis				
c-Jun or phospho-c-Jun	I	VES, α-TEA, δ-T3	Breast, gastric, prostate, ovarian	Jia, 2006; Shun *et al.*, 2004; Wu *et al.*, 2001; Yu *et al.*, 1998, 2001, 2006 Zhao *et al.*, 1997, 2002
c-Fos	D	VES	Breast	Zhao *et al.*, 1997
AP-1 consensus sequence binding or transactivation	I	VES	Breast	Yu *et al.*, 1998; Zhao *et al.*, 1997
ATF-2	I	VES	Breast	Yu *et al.*, 2001
E2F1	D	VES	Osteosarcoma	Alleva *et al.*, 2005
p73	I	α-TEA	Breast	Wang, 2006
Phospho-FOXO1	D	α-TEA	Prostate	Jia, 2006
p53	I	VES, TRF	Colon, mesothelioma	Agarwal *et al.*, 2004; Tomasetti *et al.*, 2006

(*Continues*)

TABLE I. (*Continued*)

Cell component	I/D	Vitamin E compound	Human cancer cell type	References
Other Mediators of Apoptosis				
Sphingomyelin metabolism accumulation of ceramide	I	VES	Leukemia	Weber *et al.*, 2003
Sphingolipid intermediates	Y	γ-T	Prostate	Jiang *et al.*, 2004
Protein kinase Cα	D/I	VES, γ-T3, δ-T3	Breast, leukemia, colon	Bang *et al.*, 2001; Guthrie and Carroll, 1999; Neuzil *et al.*, 2001a
Protein phosphatase 2A activity	I	VES	Leukemia, colon	Neuzil *et al.*, 2001a
Phospho-ERK1/2	I	VES	Breast, gastric	Yu *et al.*, 2001; Zhao *et al.*, 2006
Blockage of Survival				
EGFR-2 (Her-2/neu)	D	TE, VES, α-TEA	Breast, prostate	Akazawa *et al.*, 2002; Jia, 2006; You *et al.*, 2001
K-Ras	D	VES	Breast, colon	Donapaty *et al.*, 2006
Phospho-Akt	D	TE, α-TEA	Breast, ovarian, prostate	Akazawa *et al.*, 2002; Jia, 2006; Shun, 2005; Wu *et al.*, 2004; Yu *et al.*, 2006
Survivin protein	D	VES, α-TEA	Ovarian, prostate	Jia, 2006; Shun, 2005; Yu *et al.*, 2006
c-FLIP	D	α-TEA	Ovarian, prostate	Jia, 2006; Shun, 2005; Yu *et al.*, 2006
Activation of NF-κB by TRAIL or TNF	D	VES	Leukemia	Dalen and Neuzil, 2003
Blockage of Cellular Proliferation: Inhibition of DNA Synthesis/Induction of Cell Cycle Arrest				
Estrogen receptor expression	D	All-*rac*-T	Breast	Chamras *et al.*, 2005
Androgen receptor expression	D	VES	Prostate	Zhang *et al.*, 2002
PPARγ	I	γ-T	Colon	Campbell *et al.*, 2003b, 2006
COX-2 activity	D	γ-T	Lung	Jiang *et al.*, 2000

PGE2 synthesis	D	γ-T	Lung	Jiang *et al.*, 2000
FGF-2	D	VES	Mesothelioma	Stapelberg *et al.*, 2005
FGF-R2	D	VES	Mesothelioma	Stapelberg *et al.*, 2004
IGFBP	NC	Tocotrienols	Breast	Nesaretnam *et al.*, 1998
Oxidative stress	I	VES	Mesothelioma	Stapelberg *et al.*, 2005
erg1	D	VES	Mesothelioma	Stapelberg *et al.*, 2005
E2F1	D	VES	Breast, mesothelioma	Donapaty *et al.*, 2006; Stapelberg *et al.*, 2004; Turley *et al.*, 1997a
c-Myc	D	VES	Breast	Donapaty *et al.*, 2006
DNA synthesis arrest	Y	α-T, β-T, γ-T, VES	Breast, prostate, colon	Gysin *et al.*, 2002; Ni *et al.*, 2003; Turley *et al.*, 1997a; Yu *et al.*, 2002
Cyclin D1	D	α-T, γ-T, VES	Prostate, breast	Azzi *et al.*, 2004; Donapaty *et al.*, 2006; Gysin *et al.*, 2002; Ni *et al.*, 2003
Cyclin E	D	α-T, γ-T, VES	Prostate	Gysin *et al.*, 2002; Ni *et al.*, 2003
cdk2 and cdk4	D	VES	Prostate	Ni *et al.*, 2003
cdk2/cyclin A complex	D	VES	Breast	Turley *et al.*, 1997a
Cdki: p21 (Waf1/Cip1)	I	VES	Breast, colon	Agarwal *et al.*, 2004; Turley *et al.*, 1997a; Yu *et al.*, 2002
Cdki: p27	I	VES	Prostate	Venkateswaran *et al.*, 2002
Phospho-Rb	D	VES	Prostate, leukemia	Bang *et al.*, 2001; Ni *et al.*, 2003
Induction of Differentiation				
ERK1/2	I	VES	Breast, leukemia	Lee *et al.*, 2002; You *et al.*, 2002
p21 (Wafl/Cip1)	I	VES	Breast, leukemia	Lee *et al.*, 2002; You *et al.*, 2002
c-Jun	I	VES	Breast	You *et al.*, 2002

(*Continues*)

TABLE I. (*Continued*)

Cell component	I/D	Vitamin E compound	Human cancer cell type	References
Blockage of Metastasis				
MMP-9	D	VES	Prostate	Zhang *et al.*, 2004a
VEGF release	D	VES	Breast, glioma	Schindler and Mentlein, 2006
VEGF mRNA	D	VES	Breast	Malafa and Neitzel, 2000
Sensitization of Tumor Cells to Killing by Other Endogenous Death Mediators or Drugs				
Fas	Y	VES, α-TEA, α-T	Breast, ovarian, prostate	Jia, 2006; Miyoshi *et al.*, 2005; Yu *et al.*, 1999a, 2006
TRAIL	Y	VES	Colon, leukemia, mesothelioma	Dalen and Neuzil, 2003; Tomasetti *et al.*, 2004; Weber *et al.*, 2002
Cisplatin	Y	α-TEA	Ovarian	Anderson *et al.*, 2004b
5-Fluorouracil	Y	VES	Colon	Chinery *et al.*, 1997
Cox-2 inhibitor	Y	α-TEA	Breast	Zhang *et al.*, 2004b
Etoposide	Y	α-T	Prostate	Miyoshi *et al.*, 2005
Adriamycin	Y	TE	Breast	Nishikawa *et al.*, 2003

[a] I—increase; D—decrease; Y—yes; NC—no change; α-T—*RRR*-α-tocopherol; β-T—*RRR*-β-tocopherol; γ-T—*RRR*-γ-tocopherol; α-T3—α-tocotrienol; γ-T3—γ-tocotrienol; δ-T3—δ-tocotrienol; α-TEA—*RRR*-α-tocopherol acetic acid analogue; TE—α-tocopheryloxybutyric acid; TRF—tocotrienol-rich fraction of palm oil; VES—vitamin E succinate or *RRR*-α-tocopheryl succinate.

sphingomyelin metabolism, and lysosomal and mitochondrial destablization have been made for human cancer cells derived from the hematopoietic–lymphoid lineage (Table I).

Downregulation of survival factors also contributes to their prodeath abilities. Furthermore, blockage of cellular proliferation via inhibition of DNA synthesis by multiple factors, induction of differentiation, and blockage of metastasis also contribute to their anticancer effects. Of special note is the ability of certain vitamin E compounds to sensitize tumor cells to killing by clinically relevant drugs or endogenous mediators of cell death (Table I).

In summary, the study of vitamin E compounds is contributing to our general understanding of anticancer mechanisms of dysregulated signaling in cancer cells, as well as showing the great promise they have as future chemopreventive and therapeutic agents.

V. What About Vitamin E Supplementation and Cancer Survivorship?

We have no science-based answers to this important question. Thanks to advances in detection and treatment, the number of cancer survivors in the United States has more than tripled over the past 30 years, with ~10 million survivors (Rowland *et al.*, 2004). Data show that 64% of adults whose cancers are diagnosed today can expect to be living 5 years; breast cancer survivors make up the largest group of cancer survivors (22%), followed by prostate cancer (17%), and colorectal cancer (11%). Sixty-one percent of cancer survivors are aged 65 and older; and an estimated one of every six people over age 65 is a cancer survivor. Seventy-nine percent of childhood cancer survivors will be living 5 years after diagnosis and nearly 75% will be living 10 years following diagnosis. There is a growing need to promote health and ensure the well-being of cancer survivors. Given these increases, the lack of research addressing type of vitamin E, amount, and how lifestyle factors, including diet and physical activity, can be individually tailored to reduce the risk of cancer recurrence is a critical concern.

VI. Conclusions

Vitamin E is a generic term used to describe a number of chemically and functionally different compounds. Vitamin E supplementation remains in the may or may not reduce risks of cancer category. More basic information about different vitamin E forms including analogues regarding effective routes of their administration, optimal dosages, as well as better, more in-depth understanding of anticancer mechanisms of action is needed.

One interesting property exhibited by certain vitamin E compounds, namely γ-tocopherol, δ-tocopherol, the tocotrienols, the *RRR*-α-tocopherol derivative, VES, and the novel vitamin E analogue α-TEA, is the ability to selectively induce cancer cells to undergo apoptosis.

Further studies to better understand the diverse functions of the diverse vitamin E forms are needed. Proper formulation of these anticancer vitamin E compounds to ensure adequate bioavailability may yet lead to clinical success in cancer prevention and treatment; however, based on what we know so far, it is hard to provide good evidence-based advice. Preclinical investigations continue to provide proof that certain vitamin E forms are effective as monotherapy and as combination therapies. Also these studies show that vitamin E compounds are useful tools for better understanding dysregulated signaling pathways in cancer cells and are effective proof-of-principle examples of how single agents can effectively restore dysregulated prodeath signaling in human breast, prostate, and ovarian cancer cells.

It is important to increase our understanding of the molecular mechanisms that contribute to the anticancer effects of vitamin E compounds. Additionally, we need to better understand the tissue distribution of the different vitamin E compounds and the factors that control this distribution so that they can be manipulated to promote the uptake and retention of forms with the highest anticancer activity. Better understanding of synergy and antagonism among the various vitamin E forms and their metabolites is also needed.

A common approach for improved cancer treatment is the use of combinations of agents which produce additive or synergistic combinatorial effects. In theory, this should permit the use of lower drug doses and help minimize toxic off-target effects. Alternatively, drugs can be used to sensitize cancer cells to killing by endogenous-based death mediators such as TGF-β, Fas, or TRAIL. In this regard, α-TEA and VES are remarkable in that both "sensitize" cancer cells that are resistant or low responders to these endogenous death mediators.

ACKNOWLEDGMENTS

This work is supported by Public Health Service Grant CA59739 (to K.K. and B.G.S.), the Foundation for Research (to K.K. and B.G.S.), and American Institute for Cancer Research Grant (to W.Y.).

REFERENCES

Agarwal, M. K., Agarwal, M. L., Athar, M., and Gupta, S. (2004). Tocotrienol-rich fraction of palm oil activates p53, modulates Bax/Bcl2 ratio and induced apoptosis independent of cell cycle association. *Cell Cycle* **3,** 205–211.

Akazawa, A., Nishikawa, K., Suzuki, K., Asano, R., Kumadaki, I., Satoh, H., Hagiwara, K., Shin, S. J., and Yano, T. (2002). Induction of apoptosis in a human breast cancer cell overexpressing ErbB-2 receptor by α-tocopheryloxybutyric acid. *Jpn. J. Pharmacol.* **89,** 417–421.

Alleva, R., Tomasetti, M., Andera, L., Gellert, N., Borghi, B., Weber, C., Murphy, M. P., and Neuzil, J. (2001). Coenzyme Q blocks biochemical but not receptor-mediated apoptosis by increasing mitochondrial antioxidant protection. *FEBS Lett.* **503,** 46–50.

Alleva, R., Benassi, M. S., Tomasetti, M., Gellert, N., Ponticelli, F., Borghi, B., Picci, P., and Neuzil, J. (2005). α-Tocopheryl succinate induces cytostasis and apoptosis in osteosarcoma cells: The role of E2F1. *Biochem. Biophys. Res. Commun.* **331,** 1515–1521.

Anderson, K., Simmons-Menchaca, M., Lawson, K. A., Atkinson, J., Sanders, B. G., and Kline, K. (2004a). Differential response of human ovarian cancer cells to induction of apoptosis by vitamin E succinate and vitamin E analogue, α-TEA. *Cancer Res.* **64,** 4263–4269.

Anderson, K., Lawson, K. A., Simmons-Menchaca, M., Sun, L.-Z., Sanders, B. G., and Kline, K. (2004b). α-TEA plus cisplatin reduces human cisplatin-resistant ovarian cancer cell tumor burden and metastasis. *Exp. Biol. Med.* **229,** 1169–1176.

Arya, P., Alibhai, N., Quin, H., and Burton, G. W. (1998). Design and synthesis of analogs of vitamin E: Antiproliferative activity against human breast adenocarcinoma cells. *Bioorg. Med. Chem. Lett.* **8,** 2433–2438.

ATBC Cancer Prevention Study Group (1994). The effect of vitamin E and beta carotene on the incidence of lung cancer and other cancers in male smokers. *N. Engl. J. Med.* **330,** 1029–1035.

Azzi, A., Gysin, R., Kempna, P., Munteanu, A., Villacorta, L., Visarius, T., and Zingg, J.-M. (2004). Regulation of gene expression by α-tocopherol. *Biol. Chem.* **385,** 585–591.

Bang, O.-K., Park, J.-H., and Kang, S.-S. (2001). Activation of PKC but not of ERK is required for vitamin E-succinate-induced apoptosis of HL-60 cells. *Biochem. Biophys. Res. Commun.* **288,** 789–797.

Best, S. R. (2005). α-TEA & Celecoxib in Skin Cancer Prevention & Therapy. Master of Arts Thesis. University of Texas at Austin.

Birringer, M., EyTina, J. H., Salvatore, B. A., and Neuzil, J. (2003). Vitamin E analogues as inducers of apoptosis: Structure-function relation. *Br. J. Cancer* **88,** 1948–1955.

Brigelius-Flohe, R. (2005). Induction of drug metabolizing enzymes by vitamin E. *J. Plant Physiol.* **162,** 797–802.

Brigelius-Flohe, R., and Traber, M. G. (1999). Vitamin E: Function and metabolism. *FASEB J.* **13,** 1145–1155.

Campbell, S. E., Stone, W., Whaley, S., and Krishnan, K. (2003a). Development of gamma (γ)-tocopherol as a colorectal cancer chemopreventive agent. *Crit. Rev. Oncol. Hematol.* **47,** 249–259.

Campbell, S. E., Stone, W. L., Whaley, S. G., Qui, M., and Krishnan, K. (2003b). γ-Tocopherol upregulates peroxisome proliferators activated receptor (PPAR) gamma expression in SW480 human colon cancer cell lines. *BMC Cancer* **3,** 25.

Campbell, S. E., Stone, W. L., Lee, S., Whaley, S., Yang, H., Qui, M., Goforth, P., Sherman, D., HcHaffie, D., and Krishnan, K. (2006). Comparative effects of RRR-alpha- and RRR-gamma-tocopherol on proliferation and apoptosis in human colon cancer cell lines. *BMC Cancer* **6,** 13.

Chamras, H., Barsky, S. H., Ardashian, A., Navasartian, D., Heber, D., and Glaspy, J. A. (2005). Novel interactions of vitamin E and estrogen in breast cancer. *Nutr. Cancer* **52,** 43–48.

Charpentier, A., Groves, S., Simmons-Menchaca, M., Turley, J., Zhao, B., Sanders, B. G., and Kline, K. (1993). RRR-α-tocopheryl succinate inhibits proliferation and enhances secretion of transforming growth factor-β (TGF-β) by human breast cancer cells. *Nutr. Cancer* **19,** 225–239.

Charpentier, A., Simmons-Menchaca, M., Yu, W., Zhao, B., Qian, M., Heim, K., Sanders, B. G., and Kline, K. (1996). RRR-α-tocopheryl succinate enhances TGF-β1, -β2, and -β3 and TGF-βR-II expression by human MDA-MB-435 breast cancer cells. *Nutr. Cancer* **26,** 237–250.

Chinery, R., Brockman, J. A., Peeler, M. O., Shyr, Y., Beauchamp, R. D., and Coffey, R. J. (1997). Antioxidants enhance the cytotoxicity of chemotherapeutic agents in colorectal cancer: A p53-independent induction of p21WAF1/CIP1 via c/EBPbeta. *Nat. Med.* **3,** 1233–1241.

Cooney, R. V., Franke, A. A., Harwood, P. J., Hatch-Pigott, V., Custer, L. J., and Mordan, L. J. (1993). Gamma-tocopherol detoxification of nitrogen dioxide: Superiority to alpha-tocopherol. *Proc. Natl. Acad. Sci. USA* **90,** 1771–1775.

Dalen, H., and Neuzil, J. (2003). α-Tocopheryl succinate sensitizes a T lymphoma cell line to TRAIL-induced apoptosis by suppressing NF-κB activation. *Br. J. Cancer* **88,** 153–158.

Donapaty, S., Louis, S., Horvath, E., Kun, J., Sebti, S. M., and Malafa, M. (2006). RRR-α-tocopherol succinate down-regulates oncogenic Ras signaling. *Mol. Cancer Ther.* **5,** 309–316.

Freemantle, S. J., Dragnev, K. H., and Dmitrovsky, E. (2006). The retinoic acid paradox in cancer chemoprevention. *J. Natl. Cancer Inst.* **98,** 426–427.

Galli, F., Stabile, A. M., Betti, M., Conte, C., Pistilli, A., Rende, M., Floridi, A., and Azzi, A. (2004). The effect of alpha- and gamma-tocopherol and their carboxyethyl hydroxychroman metabolites on prostate cancer cell proliferation. *Arch. Biochem. Biophys.* **423,** 97–102.

Guthrie, N., and Carroll, K. K. (1999). Tocotrienols and cancer. *In* "Biological Oxidants and Antioxidants: Molecular Mechanisms and Health Effects" (L. Packer and A. S. H. Ong, Eds.), pp. 257–264. AOCS Press, Champaign.

Gysin, R., Azzi, A., and Visarius, T. (2002). γ-Tocopherol inhibits human cancer cell cycle progression and cell proliferation by down-regulation of cyclins. *FASEB J.* **16,** 1952–1954.

Heinonen, O. P., Albanes, D., Virtamo, J., Taylor, P. R., Huttunen, J. K., Hartman, A. M., Haapakoski, J., Malila, N., Rautalahti, M., Ripatti, S., Maenpaa, H., Teerenhovi, L., *et al.* (1998). Prostate cancer and supplementation with α-tocopherol and β-carotene: Incidence and mortality in a controlled trial. *J. Natl Cancer Inst.* **90,** 440–446.

Hensley, K., Benaksas, E. J., Bolli, R., Comp, P., Grammas, P., Hamdeydari, L., Mou, S., Pye, Q. N., Stoddard, M. F., Wallis, G., Williamson, K. S., West, M., *et al.* (2004). New perspectives on vitamin E: Gamma-tocopherol and carboxyethylhydroxychroman metabolites in biology and medicine. *Free Radic. Biol. Med.* **136,** 1–15.

Hercberg, S., Galan, P., Preziosi, P., Bertrais, S., Mennen, L., Malvy, D., Roussel, A.-M., Favier, A., and Briancon, S. (2004). The SU.VI.MAX stduy: A randomized, placebo-controlled trial of the health effects of antioxidant vitamins and minerals. *Arch. Intern. Med.* **164,** 2335–2342.

Hope Study Investigators (2005). Effect of long-term vitamin E supplementation on cardiovascular events and cancer. *JAMA* **293,** 1338–1347.

Huang, H.-Y., and Appel, L. J. (2003). Supplementation of diets with α-tocopherol reduces serum concentrations of γ- and δ-tocopherol in humans. *J. Nutr.* **133,** 3137–3140.

Ikeda, S., Tohyama, T., and Yamashita, K. (2002). Dietary sesame seed and its lignans inhibit 2,7,8-trimethyl-2(2′-carboxyethyl)-6-hydroxychroman excretion into urine of rats fed γ-tocopherol. *J. Nutr.* **132,** 961–966.

Institute of Medicine, Food and Nutrition Board, Panel on dietary antioxidants and related compounds (2000). Dietary Reference Intakes for Vitamin C, Vitamin E, Selenium, and Carotenoids, pp. 1–486. National Academy Press, Washington, DC.

Israel, K., Yu, W., Sanders, B. G., and Kline, K. (2000). Vitamin E succinate induces apoptosis in human prostate cancer cells: Role for Fas in vitamin E succinate-triggered apoptosis. *Nutr. Cancer* **36,** 90–100.

Jia, L. (2006). Proliferation Suppression Activities of α-TEA, a Derivative of RRR-α-Tocopherol, Inividually or in Combination with Selenium, in Human Prostate Cancer Cells. Ph.D. Dissertation. University of Texas at Austin.

Jiang, Q., Elson-Schwab, I., Courtemanche, C., and Ames, B. N. (2000). γ-Tocopherol and its major metabolite, in contrast to α-tocopherol, inhibit cyclooxygenae activity in macrophages and epithelial cells. *Proc. Natl. Acad. Sci. USA* **97,** 11494–11499.

Jiang, Q., Christen, S., Shigenaga, M. K., and Ames, B. N. (2001). Gamma-tocopherol, the major form of vitamin E in the US diet, deserves more attention. *Am. J. Clin. Nutr.* **74,** 714–722.

Jiang, Q., Wong, J., Fyrst, H., Saba, J. D., and Ames, B. N. (2004). γ-Tocopherol or combinations of vitamin E forms induce cell death in human prostate cancer cells by interrupting sphingolipid synthesis. *Proc. Natl. Acad. Sci. USA* **101,** 17825–17830.

Kelloff, G. J., and Boone, C. W. (1994). Cancer chemopreventive agents: Drug development status and future prospects. *J. Cell Biochem. Suppl.* **20,** 282–294.

Kline, K., Lawson, K. A., Yu, W., and Sanders, B. G. (2003). Vitamin E and breast cancer prevention: Current status and future potential. *J. Mammary Gland Biol. Neoplasia* **8,** 91–102.

Kline, K., Yu, W., and Sanders, B. G. (2004). Vitamin E and breast cancer. *J. Nutr.* **134,** 3458S–3462S.

Landes, N., Pfluger, P., Kluth, D., Birringer, M., Ruhl, R., Bol, G.-F., Glatt, H., and Brigelius-Flohe, R. (2003). Vitamin E activates gene expression via the pregnane X receptor. *Biochem. Pharmacol.* **65,** 269–273.

Lawson, K. A., Anderson, K., Menchaca, M., Atkinson, J., Sun, L., Knight, V., Gilbert, B. E., Conti, C., Sanders, B. G., and Kline, K. (2003). Novel vitamin E analogue decreases syngeneic mouse mammary tumor burden and reduces lung metastasis. *Mol. Cancer Ther.* **2,** 437–444.

Lawson, K. A., Anderson, K., Simmons-Menchaca, M., Atkinson, J., Sun, L.-Z., Sanders, B. G., and Kline, K. (2004a). Comparison of vitamin E derivatives α-TEA and VES in reduction of mouse mammary tumor burden and metastasis. *Exp. Biol. Med.* **229,** 954–963.

Lawson, K. A., Anderson, K., Snyder, R. M., Simmons-Menchaca, M., Atkinson, J., Sun, L.-Z., Bandyopadhyay, A., Knight, V., Gilbert, B. E., Sanders, B. G., and Kline, K. (2004b). Novel vitamin E analogue and 9-nitro-camptothecin administered as liposome aerosols decrease syngeneic mouse mammary tumor burden and inhibit metastasis. *Cancer Chemother. Pharmacol.* **54,** 421–431.

Lee, I.-M., Cook, N. R., Gaziano, J. M., Gordon, D., Ridker, P. M., Manson, J. E., Hennekens, C. H., and Buring, J. E. (2005). Vitamin E in the primary prevention of cardiovascular disease and cancer. The Women's Health Study: A randomized controlled trial. *JAMA* **294,** 83–88.

Lee, J.-K., Jung, J. C., Chun, J.-S., Kang, S.-S., and Bang, O.-S. (2002). Expression of p21Waf1 is dependent on the activation of ERK during vitamin E-succinate-induced monocytic differentiation. *Mol. Cells* **13,** 125–129.

Lee, R. J. (2006). Liposomal delivery as a mechanism to enhance synergism between anticancer drugs. *Mol. Cancer Ther.* **5,** 1639–1640.

Lippman, S. M., Goodman, P. J., Klein, E. A., Parnes, H. L., Thompson, I. M., Jr., Kristal, A. R., Santella, R. M., Probstfield, J. L., Moinpour, C. M., Albanes, D., Taylor, P. R., Minasian, L. M., *et al.* (2005). Designing the Selenium and Vitamin E Cancer Prevention Trial (SELECT). *J. Natl. Cancer Inst.* **97,** 94–102.

Malafa, M. P., and Neitzel, L. T. (2000). Vitamin E succinate promotes breast cancer tumor dormancy. *J. Surg. Res.* **93,** 163–170.

Malafa, M. P., Fokum, F. D., Andoh, J., Neitzel, L. T., Bandyopadhyay, S., Zhan, R., Iizumi, M., Furuta, E., Horvath, E., and Watabe, K. (2006). Vitamin E succinate suppresses prostate tumor growth by inducing apoptosis. *Int. J. Cancer* **118,** 2441–2447.

Miyoshi, N., Naniwa, K., Kumagai, T., Uchida, K., Osawa, T., and Nakamura, Y. (2005). Alpha-tocopherol-mediated caspase-3 up-regulation enhances susceptibility to apoptotic stimuli. *Biochem. Biophys. Res. Commun.* **334,** 466–473.

Nesaretnam, K., Stephen, R., Dils, R., and Darbre, P. (1998). Tocotrienols inhibit the growth of human breast cancer cells irrespective of estrogen receptor status. *Lipids* **33,** 461–469.

Neuzil, J., Svensson, I., Weber, T., Weber, C., and Brunk, U. T. (1999). α-Tocopheryl succinate-induced apoptosis in Jurkat T cells involves caspase-3 activation, and both lysosomal and mitochondrial destabilisation. *FEBS Lett.* **445,** 295–300.

Neuzil, J., Weber, T., Schroder, A., Lu, M., Ostermann, G., Gellert, N., Mayne, G. C., Olejnicka, B., Negre-Salvayre, A., Sticha, M., Coffey, R. J., and Weber, C. (2001a). Induction of cancer cell apoptosis by α-tocopheryl succinate: Molecular pathways and structural requirements. *FASEB J.* **15,** 403–415.

Neuzil, J., Weber, T., Terman, A., Weber, C., and Brunk, U. T. (2001b). Vitamin E analogues as inducers of apoptosis: Implications for their potential antineoplastic role. *Redox Rep.* **6,** 143–151.

Neuzil, J., Zhao, M., Ostermann, G., Sticha, M., Gellert, N., Weber, C., Eaton, J. W., and Brunk, U. T. (2002). α-Tocopheryl succinate, an agent with *in vivo* anti-tumour activity, induces apoptosis by causing lysosomal instability. *Biochem. J.* **32,** 709–715.

Neuzil, J., Tomasetti, M., Mellick, A. S., Alleva, R., Salvatore, B. A., Birringer, M., and Fariss, M. W. (2004). Vitamin E analogues: A new class of inducers of apoptosis with selective anti-cancer effects. *Curr. Cancer Drug Targets* **4,** 355–372.

Ni, J., Chen, M., Zhang, Y., Li, R., Huang, J., and Yeh, S. (2003). Vitamin E succinate inhibits human prostate cancer cell growth via modulating cell cycle regulatory machinery. *Biochem. Biophys. Res. Commun.* **300,** 357–363.

Nishikawa, K., Satoh, H., Hirai, A., Suzuzki, K., Asano, R., Kumadaki, I., Hagiwara, K., and Yano, T. (2003). Alpha-tocopheryloxybutyric acid enhances necrotic cell death in breast cancer cells treated with chemotherapy agent. *Cancer Lett.* **201,** 51–56.

Parker, R. S., Sontag, T. J., and Swanson, J. E. (2000). Cytochrome P4503A-dependent metabolism of tocopherols and inhibition by sesamin. *Biochem. Biophys. Res. Commun.* **277,** 531–534.

Papas, A. M. (2002). Beyond α-tocopherol: The role of the other tocopherols and tocotrienols. *In* "Phytochemicals in Nutrition and Health" (M. S. Meskin, Ed.), pp. 61–77. CRC Press, Boco Raton.

Patton, J. S. (1997). Deep-lung delivery of therapeutic proteins. *Chemtech* **27,** 34–38.

Prasad, K. N., Kuman, B., Yan, X. D., Hanson, A. J., and Cole, W. C. (2003). Alpha-tocopheryl succinate, the most effective form of vitamin E for adjuvant cancer treatment: A review. *J. Am. Coll. Nutr.* **22,** 108–117.

Ricciarelli, R., Zingg, J.-M., and Azzi, A. (2001). Vitamin E 80th anniversary: A double life, not only fighting radicals. *IUBMB Life* **2,** 71–76.

Ricciarelli, R., Zingg, J.-M., and Azzi, A. (2002). The 80th anniversary of vitamin E: Beyond its antioxidant properties. *Biol. Chem.* **383,** 457–465.

Rowland, J., Mariotto, A., Aziz, N., Tesauro, G., Feuer, E. J., Blackman, D., Thompson, P., and Pollack, L. A. (2004). Cancer survivorship—U.S. 1971–2001. *Morb. Mortal. Wkly. Rep.* **53,** 526–529.

Salih, H. R., Starling, G. C., Knauff, M., Llewellyn, M.-B., Davis, P. M., Pitts, W. J., Aruffo, A., and Kiener, P. A. (2001). Retinoic acid and vitamin E modulate expression and release of CD178 in carcinoma cells: Consequences for induction of apoptosis in CD95-sensitive cells. *Exp. Cell Res.* **270,** 248–258.

Sen, C. K., Khasnna, S., and Roy, S. (2006). Tocotrienols: Vitamin E beyond tocopherols. *Life Sci.* **78,** 2088–2098.

Schindler, R., and Mentlein, R. (2006). Flavonoids and vitamin E reduce the release of the angiogenic peptide vascular endothelial growth factor from human tumor cells. *J. Nutr.* **136,** 1477–1482.

Shiau, C.-W., Huang, J.-W., Wang, D.-S., Weng, J.-R., Yang, C.-C., Lin, C.-H., Li, C., and Chen, C.-S. (2006). α-Tocopheryl succinate induces apoptosis in prostate cancer cells in part through inhibition of Bcl-xL/Bcl-2 function. *J. Biol. Chem.* **281,** 11819–11825.

Shun, M.-C. (2005). Investigation of the Molecular Mechanisms of Apoptosis Induced by a Novel Vitamin E Derivative (α-TEA) in Human Breast and Ovarian Cancer Using Cell Culture. Ph.D. Dissertation. University of Texas at Austin.

Shun, M.-C., Yu, W., Gapor, A., Parsons, R., Atkinson, J., Sanders, B. G., and Kline, K. (2004). Pro-apoptotic mechanisms of action of a novel vitamin E analog (α-TEA) and a naturally occurring form of vitamin E (δ-tocotrienol) in MDA-MB-435 human breast cancer cells. *Nutr. Cancer* **48,** 95–105.

Siler, U., Barella, L., Spitzer, V., Schnorr, J., Lein, M., Goralczyk, R., and Wertz, K. (2004). Lycopene and vitamin E interfere with autocrine/paracrine loops in the Dunning prostate cancer model. *FASEB J.* **18,** 1019–1021.

Snyder, R. (2005). Ph.D. Dissertation. University of Texas at Austin.

Stapelberg, M., Tomasetti, M., Alleva, R., Gellert, N., Procopio, A., and Neuzil, J. (2004). α-Tocopheryl succinate inhibits prolideration of mesothelioma cells by selective down-regulation of fibroblast growth factor receptors. *Biochem. Biophys. Res. Commun.* **318,** 636–641.

Stapelberg, M., Gellert, N., Swettenham, E., Tomasetti, M., Witting, P. K., Procopio, A., and Neuzil, J. (2005). α-Tocopheryl succinate inhibits malignant mesothelioma by distrupting the fibroblast growth factor autocrine loop. *J. Biol. Chem.* **280,** 25369–25376.

Stone, W. L., Krishnan, K., Campbell, S. E., Qui, M., Whaley, S. G., and Yang, H. (2004). Tocopherols and the treatment of colon cancer. *Ann. NY Acad. Sci.* **1031,** 223–233.

Swettenham, E., Witting, P. K., Salvatore, B. A., and Neuzil, J. (2005). α-Tocopheryl succinate selectively induces apoptosis in neuroblastoma cells; potential therapy of malignancies of the nervous system. *J. Neurochem.* **94,** 144–1456.

Takahashi, K., and Loo, G. (2004). Disruption of mitochondria during tocotrienol-induced apoptosis in MDA-MB-231 human breast cancer cells. *Biochem. Pharmacol.* **67,** 315–324.

Thompson, T. A., and Wilding, G. (2003). Androgen antagonist activity by the antioxidant moiety of vitamin E, 2,2,5,7,8-pentamethyl-6-chromanol in human prostate cancer cells. *Mol. Cancer Ther.* **2,** 797–803.

Tomasetti, M., Rippo, M. R., Alleva, R., Moretti, S., Andera, L., Neuzil, J., and Procopio, A. (2004). α-Tocopheryl succinate and TRAIL selectively synergise in induction of apoptosis in human malignant mesothelioma cells. *Br. J. Cancer* **90,** 1644–1653.

Tomasetti, M., Andera, L., Alleva, R., Borghi, B., Nuezil, J., and Procopio, A. (2006). α-Tocopheryl succinate induces DR4 and DR5 expression by a p53-dependent route: Implication for sensitization of resistant cancer cells to TRAIL apoptosis. *FEBS Lett.* **580,** 1925–1931.

Tomic-Vatic, A., EyTina, J., Chapman, J., Mahdavian, E., Neuzil, J., and Salvatore, B. A. (2005). Vitamin E amides, a new class of vitamin E analogues with enhanced proapoptotic activity. *Int. J. Cancer* **117,** 188–193.

Traber, M. G., and Arai, H. (1999). Molecular mechanisms of vitamin E transport. *Annu. Rev. Nutr.* **19,** 343–355.

Turley, J. M., Ruscetti, F. W., Kim, S.-J., Fu, T., Gou, F. V., and Birchenall-Roberts (1997a). Vitamin E succinate inhibits proliferation of BT-20 human breast cancer cells: Increased binding of cyclin A negatively regulates E2F transactivation activity. *Cancer Res.* **57,** 2668–2675.

Turley, J. M., Fu, T., Ruscetti, F. W., Mikovits, J. A., Bertolette, D. C., and Birchenall-Roberts, M. C. (1997b). Vitamin E succinate induces Fas-mediated apoptosis in estrogen receptor-negative human breast cancer cells. *Cancer Res.* **57,** 881–890.

Venkateswaran, V., Fleshner, N. E., and Klotz, L. H. (2002). Modulation of cell proliferation and cell cycle regulators by vitamin E in human prostate carcinoma cell lines. *J. Urol.* **168,** 1578–1582.

Venkateswaran, V., Fleshner, N. E., Sugar, L. M., and Klotz, L. H. (2004). Antioxidants block prostate cancer in *Lady* transgenic mice. *Cancer Res.* **54,** 5891–5896.

Wagner, K. H., Kamal-Eldin, A., and Elmadfa, I. (2004). Gamma-tocopherol: An underestimated vitamin? *Ann. Nutr. Metab.* **48,** 169–188.

Wang, P. (2006). Studies of the Antitumor Activity of α-TEA in Human Breast Cancer Cells. Ph.D. Dissertation. University of Texas at Austin.

Wang, X.-F., Dong, L., Zhao, Y., Tomasetti, M., Wu, K., and Neuzil, J. (2006). Vitamin E analogues as anticancer agents: Lessons from studies with α-tocopheryl succinate. *Mol. Nutr. Food Res.* **50,** 675–685.

Waters, D. D., Alderman, E. L., Hsia, J., Howard, B. V., Cobb, F. R., Rogers, W. J., Ouyang, P., Thompson, P., Tardif, J. C., Higginson, L., Bittner, V., Steffes, M., *et al.* (2002). Effects of hormone replacement therapy and antioxidant vitamin supplements on coronary atherosclerosis in postmenopausal women: A randomized controlled trial. *JAMA* **288,** 2432–2440.

Weber, T., Lu, M., Andera, L., Lahm, H., Gellert, N., Fariss, M. W., Korinek, V., Sattler, W., Ucker, D. S., Terman, A., Schroder, A., Erl, W., *et al.* (2002). Vitamin E succinate is a potent novel antineoplastic agent with high selectivity and cooperativity with tumor necrosis factor-related apoptosis-inducing ligand (Apo2 ligand) *in vivo*. *Clin. Cancer Res.* **8,** 863–869.

Weber, T., Dalen, H., Andera, L., Negre-Salvayre, A., Auge, N., Sticha, M., Lloret, A., Terman, A., Witting, P. K., Higuchi, M., Plasilova, M., Zivny, J., *et al.* (2003). Mitochondria play a central role in apoptosis induced by α-tocopheryl succinate, an agent with antineoplastic activity: Comparison with receptor-mediated pro-apoptotic signaling. *Biochemistry* **42,** 4277–4291.

Wu, K., Liu, B. H., Zhao, D. Y., and Zhao, Y. (2001). Effects of vitamin E succinate on expression of TGF-β1, c-Jun and JNK1 in human gastric cancer SGC-7901 cells. *World J. Gastroenterol.* **7,** 83–87.

Wu, K., Li, Y., Zhao, Y., Shan, Y.-J., Xia, W., Yu, W.-P., and Zhao, L. (2002). Roles of Fas signaling pathway in vitamin E succinate-induced apoptosis in human gastric cancer SGC-7901 cells. *World J. Gastroenterol.* **8,** 982–986.

Wu, K., Zhao, Y., Li, G. C., and Yu, W. P. (2004). c-Jun N-terminal kinase is required for vitamin E succinate-induced apoptosis in human gastric cancer cells. *World J. Gastroenterol.* **10,** 1110–1114.

You, H., Yu, W., Sanders, B. G., and Kline, K. (2001). RRR-α-tocopheryl succinate induces MDA-MB-435 and MCF-7 human breast cancer cells to undergo differentiation. *Cell Growth Differ.* **12,** 471–480.

You, H., Yu, W., Munoz-Medellin, D., Brown, P. H., Sanders, B. G., and Kline, K (2002). Role of extracellular signal-regulated kinase pathway in RRR-α-tocopheryl succinate-induced differentiation of human MDA-MB-435 breast cancer cells. *Mol. Carcinog.* **33,** 228–236.

Yu, W., Heim, K., Qian, M., Simmons-Menchaca, M., Sanders, B. G., and Kline, K. (1997). Evidence for role of transforming growth factor-β in RRR-α-tocopheryl succinate-induced apoptosis of human MDA-MB-435 breast cancer cells. *Nutr. Cancer* **27,** 267–278.

Yu, W., Simmons-Menchaca, M., You, H., Brown, P., Birrer, M. J., Sanders, B. G., and Kline, K. (1998). RRR-α-tocopheryl succinate induction of prolonged activation of c-jun amino-terminal kinase and c-jun during induction of apoptosis in human MDA-MB-435 breast cancer cells. *Mol. Carcinog.* **22,** 247–257.

Yu, W., Israel, K., Liao, Q. Y., Aldaz, C. M., Sanders, B. G., and Kline, K. (1999a). Vitamin E succinate (VES) induces Fas sensitivity in human breast cancer cells: Role for Mr 43,000 Fas in VES-triggered apoptosis. *Cancer Res.* **59,** 953–961.

Yu, W., Simmons-Menchaca, M., Gapor, A., Sanders, B. G., and Kline, K. (1999b). Induction of apoptosis in human breast cancer cells by tocopherols and tocotrienols. *Nutr. Cancer* **33,** 26–32.

Yu, W., Liao, Q. Y., Hantash, F. M., Sanders, B. G., and Kline, K. (2001). Activation of extracellular signal-regulated kinase and c-Jun-NH2-terminal kinase but not p38 mitogen-activated protein kinases is required for RRR-α-tocopheryl succinate-induced apoptosis of human breast cancer cells. *Cancer Res.* **61,** 6569–6576.

Yu, W., Sanders, B. G., and Kline, K. (2002). RRR-α-tocopheryl succinate induction of DNA synthesis arrest of human MDA-MB-435 cells involves TGF-β-independent activation of p21Waf1/Cip1. *Nutr. Cancer* **43,** 227–236.

Yu, W., Sanders, B. G., and Kline, K. (2003). RRR-α-tocopheryl succinate-induced apoptosis of human breast cancer cells involves Bax translocation to mitochondria. *Cancer Res.* **63,** 2483–2491.

Yu, W., Shun, M.-C., Anderson, K., Chen, H., Sanders, B. G., and Kline, K. (2006). α-TEA inhibits survival and enhances death pathways in cisplatin sensitive and resistant human ovarian cancer cells. *Apoptosis* **11**(10), 1813–1823.

Zhang, Y., Ni, J., Messing, E. M., Chang, E., Yang, C.-R., and Yeh, S. (2002). Vitamin E succinate inhibits the function of androgen receptor and the expression of prostate-specific antigen in prostate cancer cells. *Proc. Natl. Acad. Sci.* **99,** 7408–7413.

Zhang, M., Altuwaijri, S., and Yeh, S. (2004a). RRR-α-tocopheryl succinate inhibits human prostate cancer cell invasiveness. *Oncogene* **23,** 3080–3088.

Zhang, S., Lawson, K. A., Simmons-Menchaca, M., Sun, L.-Z., Sanders, B. G., and Kline, K. (2004b). Vitamin E analog α-TEA and celecoxib alone and together reduce human MDA-MB-435-FL-GFP breast cancer burden and metastasis in nude mice. *Breast Cancer Res. Treat.* **87,** 111–121.

Zhao, B., Yu, W., Qian, M., Simmons-Menchaca, M., Brown, P., Birrer, M. J., Sanders, B. G., and Kline, K. (1997). Involvement of activator protein (AP-1) in induction of apoptosis by vitamin E succinate in human breast cancer cells. *Mol. Carcinog.* **19,** 180–190.

Zhao, Y., Wu, K., Xia, W., Shan, Y.-J., Wu, L.-J., and Yu, W.-P. (2002). The effects of vitamin E succinate on the expression of c-jun gene and protein in human gastric cancer SGC-7901 cells. *World J. Gastroenterol.* **8,** 782–786.

Zhao, Y., Zhao, X., Yang, B., Neuzil, J., and Wu, K. (2006). Alpha-tocopheryl succinate-induced apoptosis in human gastric cancer cells is modulated by ERK1/2 and c-Jun N-terminal kinase in a biphasic manner. *Cancer Lett.* **247**(2), 345–352.

Zhou, C., Tabb, M. M., Sadatrafiei, A., Grun, F., and Blumberg, B. (2004). Tocotrienols activate the steroid and xenobiotic receptor, SXR, and selectively regulate expression of its target genes. *Drug Metab. Dispos.* **32,** 1075–1082.

Zingg, J.-M., and Azzi, A. (2004). Non-antioxidant activities of vitamin E. *Curr. Med. Chem.* **11,** 1113–1133.

Zu, K., Hawthorn, L., and Ip, C. (2005). Up-regulation of *c-Jun*-NH2-kinase pathway contributes to the induction of mitochondria-mediated apoptosis by α-tocopheryl succinate in human prostate cancer cells. *Mol. Cancer Ther.* **4,** 43–50.

18

Vitamin E Analogues and Immune Response in Cancer Treatment

Marco Tomasetti* and Jiri Neuzil[†,‡]

*Department of Molecular Pathology and Innovative Therapies
Polytechnic University of Marche, Ancona, Italy
[†]Apoptosis Research Group, School of Medical Science
Griffith University, Southport, Qld, Australia
[‡]Molecular Therapy Group, Institute of Molecular Genetics
Czech Academy of Sciences, Prague, Czech Republic

Vitamins and Hormones, Volume 76

0083-6729/07 $35.00
DOI: 10.1016/S0083-6729(07)76018-1

Chemotherapeutic drugs induce both proliferation arrest and apoptosis; however, some cancer cells escape drug toxicity and become resistant. The suppression of the immune system by chemotherapeutic agents and radiation promotes the development and propagation of various malignancies via "mimicry-induced" autoimmunity, and maintain a cytokine milieu that favors proliferation by inhibiting apoptosis. A novel, efficient approach is based on a synergistic effect of different anticancer agents with different modes of action. Recently, a redox-silent analogue of vitamin E, α-tocopheryl succinate (α-TOS), has come into focus due to its anticancer properties. α-TOS behaves in a very different way than its redox-active counterpart, α-tocopherol, since it promotes cell death. It exerts pleiotrophic responses in malignant cells leading to cell cycle arrest, differentiation, and apoptosis. Apart from its role in killing cancer cells via apoptosis, α-TOS affects expression of genes involved in cell proliferation and cell death in a "subapoptotic" manner. For example, it modulates the cell cycle machinery, resulting in cell cycle arrest. The ability of α-TOS to induce a prolonged *S* phase contributes to sensitization of cancer cells to drugs destabilizing DNA during replication. A cooperative antitumor effect was observed also when α-TOS was combined with immunological agents. α-TOS and TRAIL synergize to kill cancer cells either by upregulating TRAIL death receptors or by amplifying the mitochondrial apoptotic pathway without being toxic to normal cells. α-TOS and TRAIL in combination with dendritic cells induce INF-γ production by CD4+ and CD8+ T lymphocytes, resulting in a significant tumor growth inhibition or in complete tumor regression. These findings are indicative of a novel strategy for cancer treatment that involves enhanced immune system surveillance.

I. INTRODUCTION

Despite concerted research and recent advancement in molecular medicine, cancer remains a significant challenge. The main strategy to treat cancer, besides surgery, has been radiotherapy and chemotherapy. These approaches operate primarily by injuring proliferating cancer cells at the level of DNA replication or cell division and inducing apoptotic cell death. Although such treatments can result in tumor stasis or regression, they are rarely curative and often are hampered by the existence or development of resistant tumor cells. In addition, radiotherapy and chemotherapy generally do not distinguish between malignant and nonmalignant types of proliferating cell, thereby causing unwanted toxicity to normal tissues, including bone marrow, gut, kidney, and the heart. In recent years, efforts to improve cancer therapy have focused on developing selective approaches based on biological mechanisms that help overcome tumor resistance while minimizing toxic side effects.

The concept of immune therapy of tumors with recombinant immune-stimulating cytokines has initially raised great hopes, but has not yet translated into clinical application. Associated cytotoxic side effects limit the efficacy and prevent clinical use of cancer immune therapy. The multifunctional cytokines of the tumor necrosis factor (TNF) family posses a strong potential as anti-tumor therapeutic agents, acting in different ways either directly on the cancer cells inducing apoptosis, by affecting tissues surrounding the tumor, in particular the vasculature stroma, or by promoting the induction of tumor-directed immune responses.

The clinical use of TNF ligands is hampered due to lack of tumor selectivity. Although TNF is presently one of few cytokines that has made it into the clinic, its use as a cancer therapeutic is restricted due to its toxicity at higher doses. In addition, cytokines can protect both normal and cancer cells against apoptosis induced by various cytotoxic agents. Under some conditions, the various immune response-polarizing cytokines that tumor cells secrete (Linker-Israeli, 1992; Moqattash and Lutton, 1998) inhibit chemotherapy- or radiation-induced apoptosis (Sachs and Lotem, 1993). The persistence of infectious agents and chronic inflammation in cancer patients promotes NF-κB activation and production of inflammatory cytokines, thereby suppressing apoptosis of malignant cells (Greten *et al.*, 2004; Pikarsky *et al.*, 2004).

A member of the TNF ligand family, the TNF-related apoptosis-inducing ligand (TRAIL), has drawn considerable attention as a potential effective antitumor therapeutic agent (Ashkenazi *et al.*, 1999; Pitti *et al.*, 1996). Although TRAIL is a potent anticancer agent in preclinical models, it has been shown that some tumor cells are resistant to the ligand. This resistance can be inherent or acquired during exposure to TRAIL. On the basis of evidence that cytotoxic agents can restore immune sensitivity to otherwise immune-resistant cancers (LeBlanc *et al.*, 2002; Wang and El-Deiry, 2003), there is a great potential for combinations of immunological agents with chemotherapy or radiotherapy.

A redox-silent analogue of vitamin E, α-tocopheryl succinate (α-TOS), has come into focus due to its anticancer properties. The provitamin α-TOS behaves different to its redox-active counterpart, α-tocopherol. Thus, rather than protecting, α-TOS promotes cell death by exerting pleiotrophic responses in malignant cells that lead to cell cycle arrest, differentiation, and apoptosis

(Birringer *et al.*, 2003; Neuzil *et al.*, 2001a,b; Ni *et al.*, 2003; Turley *et al.*, 1997; You *et al.*, 2002). Apart from its role in killing cancer cells via apoptosis, α-TOS affects gene expression involved in cell proliferation and cell death at concentrations below those that trigger apoptotic cell death. For example, it modulates the cell cycle machinery, resulting in cell cycle blockage (Alleva *et al.*, 2005). The ability of α-TOS to induce a prolonged *S* phase contributes to sensitization of cancer cells to drugs destabilizing DNA during replication, as shown for methotrexate in osteosarcoma cells (Alleva *et al.*, 2006). A cooperative antitumor effect was observed also when α-TOS was combined with immunological agents (Tomasetti *et al.*, 2004a). This chapter presents results from *in vitro* and *in vivo* studies that support the use of the vitamin E analogue as an adjuvant in therapy to improve the efficacy of various approaches by increasing tumor responses and decreasing the associated toxicity.

II. VITAMIN E ANALOGUES AS ANTICANCER AGENTS

A. VITAMIN E ANALOGUES: THEIR STRUCTURE AND BIOLOGICAL ACTIVITY

Vitamin E is a lipid-soluble micronutrient consumed on regular bases and its intake can be increased by food supplementation with no secondary deleterious effects. It is a generic term used to refer to a group of different naturally occurring compounds known as tocopherols (α-TOH, β-TOH, γ-TOH, δ-TOH) and tocotrienols (α-T3H, β-T3H, γ-T3H, δ-T3H). Structural features, consisting of Domains I, II, and III, play essential roles in activities of vitamin E and its analogues (Neuzil *et al.*, 2004; Fig. 1). Domain I, also referred to as the functional domain, makes vitamin E an antioxidant due to the redox-active hydroxyl group. Vitamin E exhibits its antioxidant properties by scavenging free radicals that can attack DNA and cause mutations. A benefit from cancer prevention has been suggested in several studies (Kline *et al.*, 2003; Knekt, 1991; Prasad and Edwards-Prasad, 1992). Although a few reviews have focused on vitamin E and cancer prevention, the role of vitamin E in cancer treatment has not received adequate attention. This may be due to the fact that the most widely used forms of vitamin E exhibited only marginal, if any, anticancer activity on tumor cells in culture or *in vivo* (Rautalahti *et al.*, 1999). The situation is completely different when modifications are made in the functional domain. In the case of α-tocopheryl succinate (α-TOS), the hydroxyl group within Domain I is esterified with a succinyl moiety that makes the analogue redox-silent and endows it with strong apoptogenic activity.

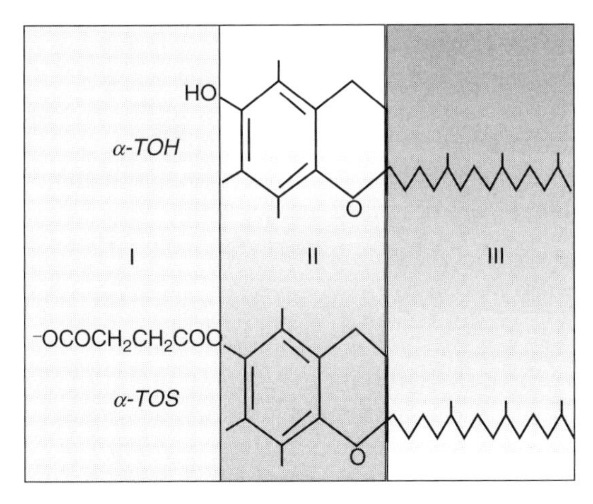

FIGURE 1. Scheme of major domains in α-tocopherol and α-tocopheryl succinate. Both α-TOH and α-TOS comprise three major domains. Domain I (functional domain) provides the analogues with their major biological activity. In case of α-TOH, it is the hydroxyl group that gives it its redox activity, while in α-TOS, the succinyl moiety provides the agent with strong apoptogenic efficacy. Domain II (signaling domain) is involved in modulation of signaling pathways, such as the protein phosphatase 2A/protein kinase C pathway. Domain III (hydrophobic domain) is responsible for docking the agents in circulating lipoproteins and in biological membranes.

A large number of *in vitro* and *in vivo* data reveal that α-TOS displays proapoptotic activity to malignant cells. Typical morphological and biochemical alterations, characterized by chromatin condensation, chromatin crescent formation and/or margination, DNA fragmentation, and apoptotic body formation, occur when apoptosis is triggered by α-TOS in a variety of types of tumor cells. In fact, α-TOS has shown high levels of apoptosis in at least 50 types of cancer cells tested thus far (human, murine, and avian) and tissue type (breast, prostate, lung, stomach, ovary, lymphoma, colon, and even mesothelium) (Anderson *et al.*, 2004; Israel *et al.*, 2000; Kline *et al.*, 2004; Neuzil, 2003; Neuzil *et al.*, 2001a, 2004; Stapelberg *et al.*, 2005; Tomasetti *et al.*, 2004a,b; Weber *et al.*, 2002; Wu *et al.*, 2002; You *et al.*, 2001; Yu *et al.*, 2003). Diverse types of malignant cells show different susceptibility to α-TOS. The apoptotic rate induced by exposure to α-TOS at 50 μM for 12 h varied from 30 to 60% in different malignant cells. About 50% of apoptosis was induced by α-TOS treatment at 20 μg/ml (38 μM) for 48 h in MDA-MB-435 human breast cancer cells (Yu *et al.*, 2003). Exposure to α-TOS at 40 μM for 24 and 48 h triggered 50 and 90% of human mesothelioma cells to undergo apoptosis, respectively, but α-TOS at the same dosage showed virtually no apoptotic effect on nonmalignant mesothelial cells (Tomasetti *et al.*, 2004a). α-TOS is not harmful toward normal cells and tissues with apoptotic rates less than 5% (Neuzil *et al.*, 2001c). In summary, α-TOS is a potent apoptosis inducer highly selective for malignant cells.

However, the exact mechanisms by which α-TOS triggers apoptosis are still unclear. Two pathways, associated with the caspase cascade, have been intensively investigated, viz. the intrinsic or mitochondria-mediated mechanism and the extrinsic or death receptor-mediated route. Some aspects of these signaling pathways, in relation to apoptosis induction in cancer cells by vitamin E analogues, are described below.

B. INITIATION OF APOPTOTIC PATHWAY BY MITOCHONDRIA DESTABILIZATION

Mitochondria play an important role in programed cell death (Green and Kroemer, 2004), and the role of mitochondria has also been demonstrated for apoptosis induced by vitamin E analogues (Neuzil *et al.*, 2004). Thus, vitamin E analogues belong to the class of "mitocans," that is anticancer agents that initiate cell death by targeting mitochondria (Neuzil *et al.*, 2006, 2007).

The first event observed on exposure of cells to α-TOS is the activation of sphingomyelinase (SMase), an enzyme that converts sphingomyelin to the lipid second messenger ceramide, a strong inducer of apoptosis (Ogretmen and Hannun, 2004). Jurkat cells treated with α-TOS resulted in activation of SMase within 15–30 min in a caspase-independent manner, possibly, as a direct target of the vitamin E analogue (Weber *et al.*, 2003). It is also plausible that the SMase activation was a response to changes in membrane fluidity on incorporation of the lipophilic α-TOS (cf. Dimanche-Boitrel *et al.*, 2005). There is evidence that treatment of cells with α-TOS causes generation of reactive oxygen species (ROS) (Kogure *et al.*, 2001; Stapelberg *et al.*, 2005; Wang *et al.*, 2005; Weber *et al.*, 2003).

The major form of ROS generated by cells in response to α-TOS appears to be superoxide. The addition of superoxide dismutase removed the radicals and also inhibited apoptosis in cells exposed to α-TOS (Kogure *et al.*, 2001; Wang *et al.*, 2005). Experiments, in which mitochondrially targeted coenzyme Q (Kelso *et al.*, 2001) suppressed radical formation and inhibited α-TOS-induced apoptosis (Alleva *et al.*, 2001; Wang *et al.*, 2005; Weber *et al.*, 2003), demonstrated that mitochondria are the site of superoxide generation as well as the target of ROS. It is not clear at present, whether the initiation of apoptotic pathways leading to mitochondria-dependent events is a direct response to the challenge of α-TOS or whether this is mediated via ceramide formation. However, in both cases, this results in destabilization of the mitochondrial membrane.

We have recently studied the possibility that superoxide generated as a response of cancer cells to α-TOS is due to an effect of the provitamin on the mitochondrial electron redox chain, whose interruption would result in superoxide generation. Combining biochemical approaches and molecular modeling and using complex II-deficient cells, we identified the complex II ubiquinone-binding sites as the target of α-TOS and, possibly, also other vitamin E analogues that induce apoptosis (Dong *et al.*, submitted for publication). We thus suggested a hypothesis, according to which oxidative stress due to interference with the mitochondrial redox chain will lead to superoxide generation (Neuzil *et al.*, 2006). These ROS will then, in the form of hydrogen peroxide, diffuse into the cytoplasm, where they will be presumably (by the action of redox-active iron) converted into more reactive radical species that will catalyze dimerization of the proapoptotic proteins, such as Bax, that will then move to the mitochondrial outer membrane to form a megachannel (D'Alessio *et al.*, 2005). The mitochondria-derived superoxide will also activate the latent oxidase activity of cytochrome *c* that results in hydroperoxidation of cardiolipin, a phospholipids that holds cytochrome *c* in the proximity of the intermembrane face of the mitochondrial inner membrane (Kagan *et al.*, 2005). As a result, cytochrome *c* can translocate into the cytosol via the Bax channel and the downstream mitochondrial pathway will then be initiated.

Mitochondrial destabilization as a response to α-TOS comprises mobilization of apoptotic mediators, which include cytochrome *c*, the apoptosis-inducing factor (AIF) and Smac/Diablo (reviewed in Neuzil *et al.*, 2004). Cytochrome *c*, on cytosolic translocation, forms a ternary complex with Apaf-1 and procaspase-9, leading to autoactivation of the initiator caspase-9 with ensuing activation of the effector caspase-3, -6, or -7. At this stage, the cell enters the "point of no return," that is the irreversible phase of the apoptotic pathway. Smac/Diablo is an important agonist of the caspase-dependent apoptotic signaling, since it antagonizes the caspase inhibitory members of the inhibitors of apoptosis proteins (IAPs) family. Thus, cytosolic translocation of Smac/Diablo may promote inhibition of the survival pathway in apoptosis induced by α-TOS, which could maximize the apoptogenic potential in resistant cells. Another mitochondrial protein amplifying apoptosis in cells exposed to α-TOS is AIF (Weber *et al.*, 2003) that translocates directly into the nuclei, thereby bypassing the caspase activation cascade. Once in the nucleus, AIF triggers cleavage of chromatin in a caspase-independent manner (Cande *et al.*, 2002). Thus, AIF can induce cell death in cells where mutations in the caspase-dependent signaling render the cancer cells resistant to apoptosis. The mitochondrial pro- and antiapoptotic proteins, including Bax, Bcl-2, Mcl-1, and Bcl-xL are important factors related to mitochondrial apoptotic signaling pathway, as long as formation of a megachannel across the mitochondrial outer membrane is not compromized (Cory *et al.*, 2003). Generation of the mitochondrial permeability transition pore has been suggested in cells exposed to α-TOS (Yamamoto *et al.*, 2000). It is likely that this is modulated by a cross-talk between the mitochondrial pro- and antiapoptotic proteins (Weber *et al.*, 2003; Yamamoto *et al.*, 2000). Overexpression of Bax sensitized cells to α-TOS-induced apoptosis, whereas overexpression of Bcl-2 or Bcl-xL protected them from the vitamin E analogue (Weber *et al.*, 2003;

Yu *et al.*, 2003). Similarly, downregulation of Bcl-2 by antisense oligodoxynucleotide treatment sensitized cells to α-TOS-induced apoptosis (Neuzil *et al.*, 2001a,c; Weber *et al.*, 2003).

Shiau *et al.* (2006) published an intriguing report, according to which α-TOS and similar compounds act as the so-called "BH3 mimetics" in that they bind to the BH3 domains on Bcl-2 and Bcl-xL. This, in fact, prevents binding to the antiapoptotic proteins of Bax or Bak that are then free to form channels in mitochondria. Thus, α-TOS acts as an inducer of apoptosis and as a BH3 mimetic, which endows it with a unique activity thus far not found in any other anticancer agent.

A compelling evidence for mitochondria as major transmitters of apoptotic signaling by vitamin E analogues follows from experiments in which mtDNA-deficient (ρ^0) cells were found to be resistant to α-TOS when compared to their wild-type and revertant counterparts (Wang *et al.*, 2005; Weber *et al.*, 2003). Cancer cells lacking mtDNA, resistant to apoptosis (Dey and Moraes, 2000), failed to translocate cytochrome *c* when challenged with α-TOS, unlike the apoptosis-sensitive parental and revertant cells (Weber *et al.*, 2003). Thus, mitochondria are indisputably the major intracellular organelles that relay the initial apoptotic signals downstream to the stage at which the cell enters the apoptosis commitment stage. Most probably, other organelles may be involved in the apoptosis process induced by vitamin E analogues such as lysosomes as documented earlier (Neuzil *et al.*, 1999, 2002). Notwithstanding, mitochondria are obligatory for transmission of the early apoptogenic events in cells challenged with α-TOS, probably amplified by mediators released from organelles like lysosomes or the endoplasmatic reticulum (Fig. 2).

C. DEREGULATION OF SIGNALING PATHWAYS BY VITAMIN E ANALOGUES

Of the signaling pathways modulated by vitamin E analogues, some are implicated in resistance of cancers to established treatments. One of them, encountered in breast cancer, stems from overexpression of erbB2, a receptor tyrosine kinase proto-oncogene. ErbB2 is a member of the epithelial growth factor receptor superfamily and a product of the *c-neu* gene (Roskoski, 2004). This transmembrane protein is overexpressed in $\geq$30% of primary breast cancers. The major complication associated with erbB2 overexpression is linked to spontaneous activation of Akt (protein kinase B) via the phosphatidylinositol 3-kinase pathway (Vivanco and Sawyers, 2002; Zhou and Hung, 2003). Akt is a serine/threonine kinase that promotes cellular survival. Once activated, Akt exerts antiapoptotic effects through phosphorylation of proteins such as Bad or caspase-9. Moreover, Akt causes activation of the transcriptional factor nuclear factor-κB (NF-κB) (Kane *et al.*, 1999) that controls expression of prosurvival genes, including the IAP family members (LaCasse *et al.*, 1998). In most nontransformed cells, NF-κB complexes are largely cytoplasmic. Following activation, the inhibitory IκB proteins become

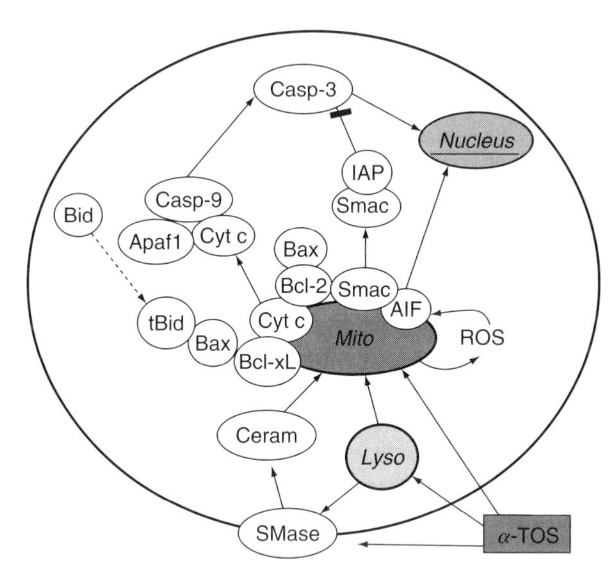

FIGURE 2. Possible pathways in apoptosis induction by α-TOS. (1) Upstream apoptosis signaling from mitochondria: α-TOS translocates to the cell, activates SMase, possibly causing the destabilization of lysosomes, giving rise to the formation of the lipid second message ceramide, leading to the destabilization of the mitochondrial membrane. α-TOS directly and/or via ceramide formation destabilizes mitochondrial membranes, and the ROS generation may amplify this process. (2) Downstream apoptosis signaling from mitochondria: Mitochondrial membrane destabilization, likely promoted by leakage by lysosomal proteases, leads to cytosolic relocalization of proapoptotic factors (such as cytochrome *c*, Smac/Diablo or AIF) that can be regulated by Bcl-2 family proteins (including Bcl-2, Bcl-xL, or Mcl-1, which can be compromized by other Bcl-2-related protein Bax, probably mobilized to mitochondria after cleavage of Bid to its proapoptotic form). Cytochrome *c*, Apaf-1 and pro-caspase-9 form a ternary complex, leading to the activation of the initiator caspase-9 that, in turn, results in the activation of the effector caspases. Smac/Diablo may amplify this process by suppressing the caspase inhibitory activity of IAP family proteins, while the IAP family members are supposed to transmit the mitochondrial destabilization to nuclear apoptotic events.

phosphorylated/ubiquitinated and subsequently degraded, which allows NF-κB translocation to the nucleus, where it promotes transcription of target genes, especially those involved in immune activation, cell proliferation, and cell survival.

We showed that vitamin E analogues induced apoptosis at comparable levels in both erbB2-low and -high mouse and human breast cancer cells, regardless of their erbB2 status. One plausible mechanism is that these agents induce relocalization of Smac/Diablo from mitochondria to the cytosol (Wang *et al.*, 2005), where Smac/Diablo binds to IAPs so that caspase-3 is "free" to execute its apoptosis function (Du *et al.*, 2000). Moreover, we have showed that attachment of a short peptide to α-TOS targets the vitamin E analogue specifically to cancer cells with high erbB2 expression (Wang *et al.*, 2007). In another report, it has been shown that α-tocopheryloxybutyric acid, a compound analogous to α-TOS, induced apoptosis in the erbB2-overexpressing human breast cancer cells MDA-MB-453 by simultaneously inhibiting activation of erbB2 (its phosphorylation) and activating the p38 MAP kinase (Akazawa *et al.*, 2002). Several other papers showed modulation of the MAP kinase pathway by vitamin E analogues as a way by which the agents induce apoptosis. Interestingly, Kline's group reported that extracellular signal-regulated kinases (ERKs) and the c-Jun NH_2-terminal kinase (JNK), but not the p38 MAP kinase, were involved in α-TOS-induced apoptosis in the human breast cancer MDA-MB-435 cells, and this activated the transcription factors c-Jun and/or ATF-2 (Yu *et al.*, 2001; Zhao *et al.*, 2006). It is possible that this pathway is targeted by α-TOS in the erbB2-low MDA-MB-435 cells, while the erbB2-high MDA-MB-453 cells activate their apoptotic machinery by concerted deregulation of the erbB2/Akt and p38 pathways when challenged with vitamin E analogues. Activation of JNK by α-TOS has also been shown for gastric cancer cells (Wu *et al.*, 2004), and it may amplify the mitochondrial apoptosis signaling pathway, as shown for prostate cancer cells (Zu *et al.*, 2005).

One of the intriguing targets of vitamin E analogues is the prosurvival transcription factor NF-κB. Inhibition of activation of NF-κB by α-TOS was first documented in the context of cardiovascular diseases (Erl *et al.*, 1997). One possibility is that the vitamin E analogue triggers apoptosis, resulting in activation of caspase-3 that cleaves the obligatory NF-κB subunit p65, rendering it inactive (Levkau *et al.*, 1999). We have shown that α-TOS initiates a "subapoptotic" phenotype in endothelial cells, under which the cells activate their effector caspases but do not enter the commitment phase (Neuzil *et al.*, 2001d), probably because this requires efficient activation of cyclin-dependent kinases (Harvey *et al.*, 2000). Regardless of the precise mechanism, inhibition of NF-κB activation by vitamin E analogues is anti-survival, that is proapoptotic, and can also be implicated in adjuvant cancer therapy, such as shown for the T-lymphoma Jurkat cells, whose treatment with α-TOS sensitized them to TRAIL-dependent killing (Dalen and Neuzil, 2003).

Several reports also implicated inhibition of the cell cycle progression as a means by which vitamin E analogues may induce apoptosis or inhibit proliferation of cancer cells and/or sensitize them to other anticancer drugs. Ni *et al.* (2003) showed that α-TOS inhibits proliferation of prostate cancer cells by downregulating expression of several critical cyclins and the cognate cyclin-dependent kinases (CDK), resulting in hypophosphorylation of the Rb protein and the G1/S arrest. Cell cycle arrest and apoptosis were also induced by α-TOS in osteosarcoma cells via activation of p53 and reduced expression of the transcription factor E2F1, critical for the G1/S transition (Alleva *et al.*, 2005). Exposure of osteosarcoma cells to α-TOS also promoted a prolonged arrest at the S/G2 border, sensitizing the cells to methotrexate-induced apoptosis (Alleva *et al.*, 2006). These finding can be reconciled with an earlier report in which α-TOS suppressed proliferation of breast cancer cells by inhibiting the E2F1-dependent transactivation via increased binding of cyclin A (Turley *et al.*, 1997). Apoptosis induction and inhibition of proliferation by α-TOS have also been shown for malignant mesothelioma cells (Tomasetti *et al.*, 2004a), the latter paradigm being due to selective disruption of the FGF-FGFR autocrine signaling loop, most likely affected by modulation in the E2F1 and egr-1 transactivation activity (Stapelberg *et al.*, 2004, 2005).

Thus, there is ample evidence that vitamin E analogues affect a variety of signaling pathways that either directly trigger apoptosis or, more likely, amplify/complement other apoptotic signaling pathways, in particular those involving mitochondrial destabilization.

III. VITAMIN E ANALOGUES AS ADJUVANTS IN CANCER CHEMOTHERAPY

The potential use of vitamin E analogues as anticancer drugs and adjuvants has been attracting attention for some time. The ability of α-TOS to modulate expression of proteins involved in cell proliferation and cell death, forcing cancer cells to undergo transient growth arrest, differentiation, or senescence (irreversible cell growth arrest) can restore susceptibility to drug resistance. Therefore, α-TOS could be used as an adjuvant, in combination with chemotherapeutic drugs to potentiate the effect of cancer therapy. The vitamin E analogue has been shown to enhance growth-inhibitory effects in cancer cells of several chemotherapeutic agents. For instance, α-TOS augments the effect of adriamycin on human prostate carcinoma (Ripoll *et al.*, 1986) and murine leukemia cells (Fariss *et al.*, 1994), and the effect of *cis*-platin, tamoxifen and decarbazine on human melanoma (Prasad *et al.*, 1994) and human parotid acinar carcinoma cells (Prasad and Kumar, 1996). A S/G2 arrest was observed in osteosarcoma cells sensitized by α-TOS to methotrexate, an anticancer agent whose action is cell cycle dependent (Alleva *et al.*, 2006). Therefore, the ability of α-TOS to modulate cell proliferation and cell death could be utilized to improve the effect of standard cancer therapy which includes immune therapy.

IV. IMMUNOLOGICAL INDUCERS OF APOPTOSIS: MECHANISMS AND CLINICAL APPLICATION IN CANCER

A. DEATH RECEPTOR SIGNALING PATHWAY

The ligands of the TNF family and the corresponding receptors fulfill a variety of immune-regulatory functions (Locksley *et al.*, 2001). The receptors comprise a group of structurally related type I transmembrane proteins characterized by cystein-rich modules in their extracellular domains. A subgroup of these receptors, including TNF receptors (TNFR1 and TNFR2), CD95 (Fas), TNF-related apoptosis-inducing ligand (TRAIL) receptors (DR4 and DR5), and the ectodermal dysplasia receptor (EDAR) share a conserved protein–protein interaction domain in their cytoplasmic tail, which is necessary to convey the apoptotic signal (Thorburn, 2004). The ligands for these receptors belong to a family of related cytokines, including TNF-α, lymphotoxin (LT-α), the Fas ligand (FasL), Apo-3 ligand (Apo-3L), and TRAIL. They act in an autocrine or paracrine manner and, on binding, trigger oligomerization of their respective receptors, an event required to initiate the apoptotic program. The apoptotic signal involves the cytosolic death domains (DDs) that interact with DD-containing proteins such as the Fas-associated death domain (FADD), which comprises the death effector domain (DED) that can interact with similar domains on pro-caspase-8 and -10. Activated initiator caspases then activate the effector caspase-3, -6, and -7 (Ashkenazi, 2002). Active caspase-8 can cleave the proapoptotic Bcl-2 family member Bid, generating the truncated Bid (tBid). tBid then stimulates the intrinsic apoptotic pathway by inducing conformational changes in the proapoptotic Bcl-2 family proteins Bax and Bak, which allows the release of apoptogenic proteins from mitochondria (Cory *et al.*, 2003). Death receptor-induced caspase-8 and tBid-mediated stimulation of the intrinsic pathway can contribute in a widely varying degree to death receptor-induced apoptosis. Two phenotypes of cells have been described. In type I cells, caspase-8 is robustly activated at the death-inducing signaling domain (DISC, a complex composed of the death receptor, the adaptor protein and procaspase-8) and is by itself sufficient to activate effector caspases, allowing the execution of apoptosis (Peter and Krammer, 2003). In type II cells, the death receptor-induced caspase-8 activation alone is not sufficient to activate effector caspases directly to an extent allowing robust apoptosis induction. Under these conditions, the generated tBid transmits the initial death receptor activation to the intrinsic apoptotic pathway.

While CD95 and TRAIL death receptors predominantly act as inducers of apoptosis, TNFR1 is also associated with nonapoptotic events. CD95 and TRAIL death receptors directly interact with FADD, whereas TNFR1 indirectly associates with FADD via the TNF receptor-associated death domain (TRADD). TRADD also mediates recruitment of TNF receptor-associated factor-2 (TRAF2) which interferes with TNFR1-induced apoptosis by two mechanisms. First, TRAF2 and the related TRAF1 protein associate with c-IAP-1 and c-IAP-2 to form a complex that inhibits caspase-8 activation in the context of TNFR1 signaling (Wajant *et al.*, 2003). Second, TRAF2 is implicated in TNFR1-induced activation of the antiapoptotic NF-κB pathway, possibly due to transient association with the silencer of death domain (SODD) protein (Jiang *et al.*, 1999). NF-κB proteins are involved in the transcriptional activation of a number of inflammatory-related genes in response to cytokines (TNF, IL-1), certain forms of physical stress, UV radiation, and ROS (Baud and Karin, 2001). NF-κB also controls expression of many antiapoptotic factors, such as members of the IAP family and a variety of genes involved in the immune system activation, cell proliferation, apoptosis, and cell survival.

Apoptotic signaling is tightly regulated by the balance of pro- and anti-apoptotic proteins and any shift in this balance may cause an onset of pathogenesis of various diseases, including the neoplastic disease. However, deregulated proliferation and other carcinogenetic events do not immediately result in tumor development, as the underlying signals also sensitize for apoptosis induction and allow recognition of aberrantly expressed gene products by the immune system. Therefore, tumor formation is dependent on acquisition of an antiapoptotic state of the transformed cell and/or a "build-up" of immunosuppressive barriers to boost the tumor surveillance mechanisms (Smyth *et al.*, 2002).

B. CD95 ACTIVATION

Like other members of the TNF ligand family, CD95L is mainly expressed as a trimeric type II transmembrane protein, but also occurs as a soluble trimeric protein. While transmembrane CD95L strongly activates CD95, its soluble counterpart does not activate the receptor, but rather acts its competitive inhibitor (Schneider *et al.*, 1998). For practical purposes, soluble CD95L trimers can be transformed into active ligands by antibody-mediated cross-linking or its genetically engineered, enforced expression in the form of a hexamer (Holler *et al.*, 2003; Schneider *et al.*, 1998). Thus, soluble CD95L contains all structural information required for CD95 activation. Although the artificial immobilization of soluble CD95L on the cell surface of tumor cells could restrict CD95 activation to the tumor area, the use of CD95 in tumor therapy was discontinued. Acute hepatotoxicity and rapid death due to liver failure was observed in animals treated with CD95 agonistic antibody (Timmer *et al.*, 2002).

C. TRAIL: A PROMISING CANCER THERAPEUTIC

TRAIL is the most recently identified death ligand. The soluble recombinant TRAIL is probably the best candidate among all death ligands for systemic application in cancer therapy for several reasons. There are few agents that are cancer cell-specific in terms of efficacy and cell death induction. TRAIL kills many transformed cells but is nontoxic to most normal cells (Ashkenazi and Dixit, 1998; French and Tschopp, 1999; Wiley *et al.*, 1995). Notably, administration of soluble recombinant TRAIL in experimental animals, including mice and primates, induced significant tumor regression without systemic toxicity (Ashkenazi *et al.*, 1999; Walczak *et al.*, 1999).

At least five receptors for TRAIL have been identified in humans and two of them, DR4 (TRAIL-R1) and DR5 (TRAIL-R2), are capable of transducing the apoptotic signal (Ashkenazi, 2002; Degli-Esposti, 1999). The other three receptors (TRAIL-R3, TRAIL-R4, and the soluble receptor osteoprotegerin, TRAIL-R5) lack death domains but may serve as decoy receptors to modulate TRAIL-mediated cell death. TRAIL-induced apoptosis involves signaling via DR4 and DR5 to the FADD- and caspase-8-dependent pathway (Bodmer *et al.*, 2000). TRAIL signaling involves two caspase cascades (Suliman *et al.*, 2001). After initial activation of caspase-8 by the TRAIL-associated DISC, divergence of signals occurs in two directions: (1) direct activation of caspase-3 without involvement of mitochondria and (2) formation of the apoptosome, which leads to activation of caspase-9 (described above). These two pathways converge at the level of caspase-3 (Li *et al.*, 1997). The decline of mitochondrial membrane potential ($\Delta\psi_m$) in cells exposed to TRAIL can be blocked by a caspase-8 inhibitor, but not by an inhibitor of caspase-9 (Kim *et al.*, 2001; Suliman *et al.*, 2001). Thus, caspase-8 links the apoptotic signal from the activated TRAIL DRs to mitochondria. The mechanisms controlling the release of mitochondrial proteins in TRAIL-induced apoptosis are currently under intensive investigation.

Bax is required for TRAIL-induced apoptosis of some cancer cells by allowing release of second mitochondria-derived activator of caspases (Smac/Diablo) and antagonizing the IAP family members (Deng *et al.*, 2002). Bax gene ablation can lead to resistance to the death ligand (Burns and El-Deiry, 2001; LeBlanc *et al.*, 2002), and reintroduction of Bax into Bax-deficient cells restores TRAIL sensitivity (Deng *et al.*, 2002). Although one or both TRAIL DRs are expressed in most tumor cell lines and primary tumor samples, same cancer cells are insensitive to the cytotoxic effect of TRAIL.

D. FACTORS INFLUENCING TRAIL SENSITIVITY

The pathways operational in TRAIL-induced apoptosis and their regulation are complex. There is also a great variation in the response to the death ligand. The reasons may include higher expression of decoy receptors in normal cells and the absence or lower expression of these receptors in transformed cells (Ashkenazi and Dixit, 1998, 1999). However, the expression level of DR4/DR5 and DcR1/DcR2 does not always correlate with TRAIL sensitivity (Kim *et al.*, 2000). In general, the sensitivity to TRAIL may be regulated both at the intracellular and receptor level (Srivastava, 2000).

A biphasic curve of dose–response to TRAIL has been observed in several types of malignant cells, indicating either the presence of two different populations (resistant and sensitive) or development of resistance during

TRAIL treatment (Jin *et al.*, 2004; Tomasetti *et al.*, 2004a). Tumor cells can acquire resistance to apoptosis through interference with either intrinsic or extrinsic apoptotic signaling pathways. Mutation of the proapoptotic Bax confers resistance to TRAIL-induced apoptosis in HCT116 cells (LeBlanc *et al.*, 2002). Overexpression of FLIP suppresses DR-induced apoptosis in malignant mesothelioma cells (Rippo *et al.*, 2004). Tumor cells may avoid TRAIL-mediated killing through downregulation of DRs (extrinsic resistance). It was observed that low expression of DRs on the cell surface is responsible for cellular resistance to TRAIL-induced cytotoxicity in human colon cancer cells (Jin *et al.*, 2004). DR5 mutations have been found in head and neck cancer, breast and lung carcinomas, and Hodgkin's lumphoma (El-Deiry, 2001). DR5 has been implicated in the cellular response to DNA-damaging radiation and chemotherapy as a target of p53 (Wu *et al.*, 1997, 2000). Anticancer drugs have been shown to sensitize TRAIL-receptor negative cells to TRAIL-mediated apoptosis by inducing mobilization of death receptors to the plasma membrane (Arizono *et al.*, 2003). Several reports have demonstrated that chemotherapeutic agents augment TRAIL-mediated cytotoxicity both in TRAIL-sensitive and -resistant cancer cells (LeBlanc *et al.*, 2002; Wang and El-Deiry, 2003).

E. EFFECT OF VITAMIN E ANALOGUES ON THE REGULATION OF DEATH RECEPTORS: RELEVANCE FOR CANCER THERAPY

Key elements on the basis of the apoptosis machinery, including signaling through DRs or mitochondria, have been described to play a critical role in tumor surveillance as well as in cancer therapy. However, both immune surveillance and cancer therapies are hampered by the various types of resistance that are developing in human cancer. Since many forms of cancer resistance involve the DR machinery, intensive efforts need to be undertaken to design new therapeutic DR-targeted strategies that overcome this critical problem (Debatin and Krammer, 2004). DRs are downregulated or inactivated in many tumors. For instance, the expression of Fas is reduced in many cancers such as colon and hepatocellular carcinomas (Moller *et al.*, 1994; Volkmann *et al.*, 2001). In addition, deletions and mutations of DR4 have been detected in head and neck and in lung cancer (Fisher *et al.*, 2001). Expression of DR4 and/or DR5 is more regulated than that of Fas. In addition to chemotherapeutic drugs, ionizing radiation can also induce TRAIL DRs (Kim *et al.*, 2000; Wu *et al.*, 1997). Thus, upregulation of DR4 and DR5 by anticancer agents or radiation may further enhance the efficacy of TRAIL.

A synergistic and cooperative effect of α-TOS and TRAIL was observed in mesothelioma cells. Importantly, the effect was selective for cancer cells while not affecting nonmalignant mesothelial cells (Tomasetti *et al.*, 2004b). Mesothelioma is a fatal type of neoplasia with poor therapeutic prognosis, largely due to resistance to apoptosis, and impaired apoptosis renders mesothelioma cells resistant to TRAIL. Subapoptotic doses of α-TOS significantly decrease the relatively high IC_{50} values for TRAIL in malignant mesothelioma cells by the factor of 10–100. The observation that α-TOS and TRAIL synergize in $p53^{wt}$ but not in the $p53^{null}$ mesothelioma cells suggests a role of p53 in transactivation of the proapoptotic genes involved in drug synergism (Tomasetti *et al.*, 2006). The p53 protein is a key component in the cellular "emergency-response" mechanism (Sionov and Haupt, 1999). The most studied function of p53 is its role as a transcription factor that can activate transcription of an ever-increasing number of target genes (Vousden and Lu, 2002).

Although α-TOS has the propensity to induce apoptosis in a p53-independent manner (Weber *et al.*, 2002), it induces expression and activation of p53 at low concentrations, which then triggers expression of DR4 and DR5. Such expression of death receptors is not induced by α-TOS in the $p53^{null}$ mesothelioma cells. Studies using siRNA directed at p53 revealed that the p53 protein contributes significantly to the expression of TRAIL's DRs. Thus, p53-dependent upregulation of DR4 or DR5 may be a basis for sensitization of mesothelioma cells to TRAIL. In addition, it was observed that α-TOS-induced expression and activation of the p53 tumor suppressor protein is enhanced in the presence of a water-soluble compound with a high-reducing potential, such as *N*-acetyl-L-cysteine, which changes the cell's redox state. Regulation of activity of many transcription factors by redox modulators was previously described (Sun and Oberley, 1996). A novel mode of action of α-TOS has been thus described: reduction of the redox-sensitive amino acid residues on the p53 protein leads to an increase in the efficiency of TRAIL's death receptor expression, sensitizing mesothelioma cells to the immunological apoptogens like TRAIL. Elevated expression of the TRAIL's death receptors resulted in a synergistic and cooperative α-TOS/TRAIL effect, which was observed only in mesothelioma cells and in selected TRAIL-resistant mesothelioma cells with functional p53.

Apoptosis induced by DRs can be modulated at several levels. Intracellular antiapoptotic molecules can block the apoptotic signaling pathway or divert them toward alternative responses. Such molecules include the cellular FLICE-like inhibitory protein (c-FLIP), which competes with caspase-8 for binding to FADD (Irmler *et al.*, 1997), or XIAP, c-IAP-1 and c-IAP-2 that directly inhibit caspase activity. A role of FLIP in inhibiting TRAIL-induced cell death has been previously observed in mesothelioma cells (Rippo *et al.*, 2004). Upregulation of TRAIL DRs by α-TOS may contribute to a shift in the anti- and proapoptotic signals in favor of the latter, triggering apoptotic signals, which may then be amplified by the intrinsic pathway.

Kinetic analysis of TRAIL-induced signaling revealed a transient activation of caspase-8, which resulted in induction, albeit low, of apoptosis. Caspase-8 activation was less pronounced in the presence of TRAIL plus α-TOS. Under this setting, activation of the mitochondria-dependent apoptotic pathway, involving Bid cleavage, cytochrome *c* cytosolic mobilization, and finally,

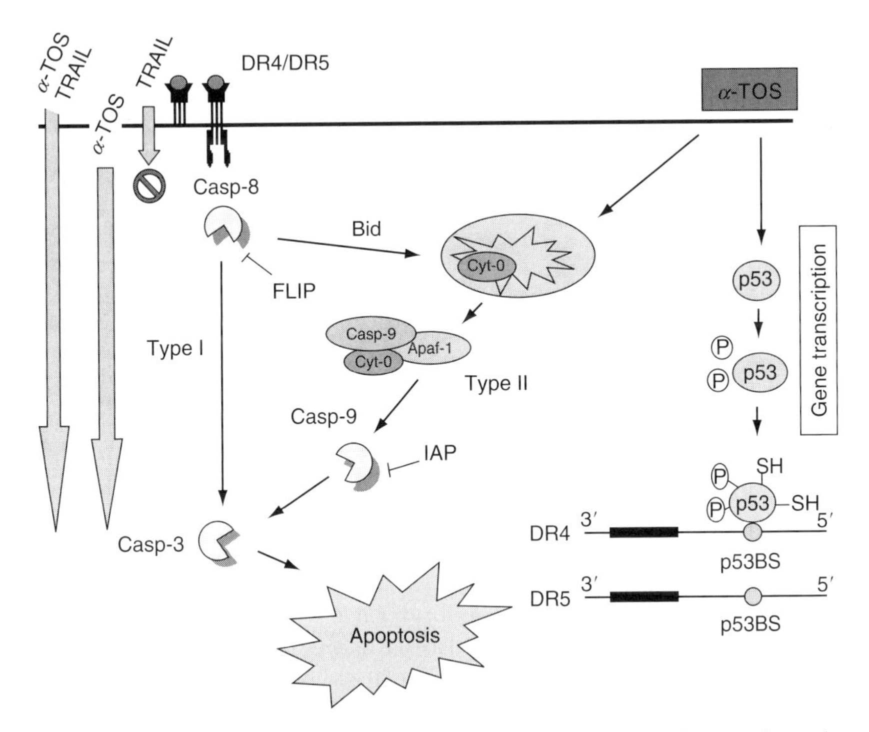

FIGURE 3. Synergistic effects of α-TOS and TRAIL. TRAIL-induced apoptosis requires mitochondria. The activation of caspase-8 by TRAIL is necessary but may not be sufficient to induce apoptosis. Cross-talk between the decoy receptor (DcR) and the mitochondrial pathway is mediated by caspase-8 cleavage of Bid. tBid activates proapoptotic members Bax and Bak to release cytochrome *c* from mitochondria. Bcl-2 and Bcl-xL inhibit α-TOS- and/or TRAIL-induced apoptosis by blocking cytochrome *c* release. In mesothelioma cells, α-TOS directly induces expression and activation of the p53 protein. Phosphorylated p53 accumulates in the nucleus and triggers transcription of DR4 and DR5 via an intronic sequence-specific p53-binding site, which is regulated by a redox mechanism. Increased expression of DRs on the cell surface amplifies the apoptotic signals, leading to activation of caspase-9. Caspase-9 then activates the downstream caspases.

caspase-9 activation, was observed. Bid cleavage may lead to mitochondrial translocation of Bax, as shown for α-TOS in other cancer models (Weber *et al.*, 2003; Yu *et al.*, 2003). Preincubation of mesothelioma cells with the caspase-8 inhibitor Z-IETD-FMK suppressed the synergistic effect of TRAIL and α-TOS, while treatment with the pan-caspase inhibitor Z-VAD-FMK completely inhibited apoptosis initiated by the two inducers both alone and in combination. These results clearly document that caspase-8 activation is essential for TRAIL-dependent apoptosis, which is amplified by activation of caspase-9 in the presence of subapoptotic levels of α-TOS. Thus, there is a cross-talk between α-TOS and TRAIL in potentiation of cell death in TRAIL-resistant mesothelioma cells, linking receptor- and mitochondria-associated events (Fig. 3).

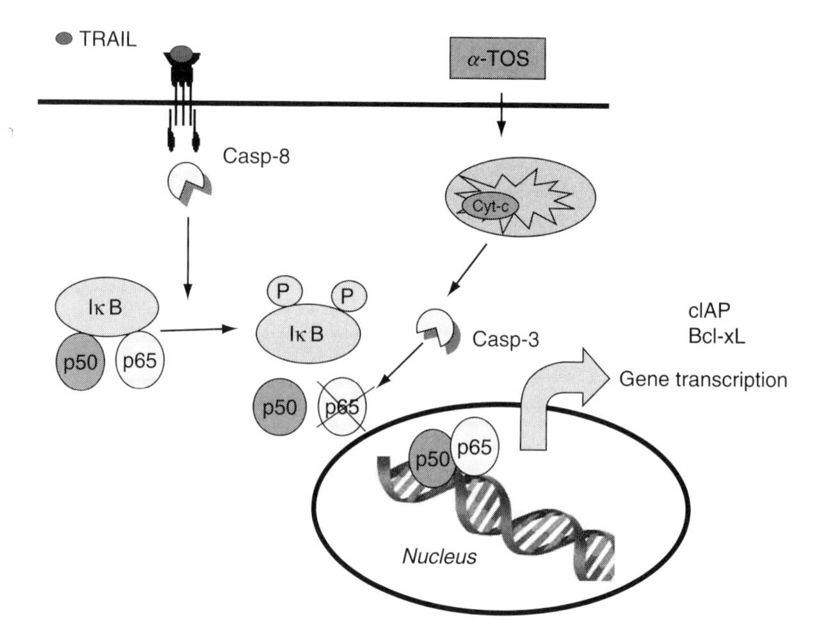

FIGURE 4. Inhibition of NF-κB by vitamin E analogue in TRAIL-induced apoptosis. Binding of TRAIL to DR4 and DR5 results in activation of NF-κB. The p50 and p65 subunits of NF-κB are maintained in the cytoplasm by binding to the IκB inhibitory protein. In response to cellular stimulation, IκB proteins are phosphorylated, ubiquitinated, and degraded. Removal of IκB from NF-κB allows the transcription factor to translocate into the nucleus where it transactivates numerous apoptosis-related genes such as the IAP proteins and Bcl-x_L. α-TOS inhibits activation of NF-κB by disruption of the p65 subunit, whereby amplifying susceptibility of cancer cells to immunological stimuli. (See Color Insert.)

A cooperative proapoptotic effect of α-TOS with immunological apoptogens has been observed in breast cancer (Yu *et al.*, 1999) and colon cancer in an animal model (Weber *et al.*, 2002). The former report showed that α-TOS converted Fas-resistant cells to Fas-sensitive ones via mobilization of the Fas receptor from the cytosol to the plasma membrane. Moreover, α-TOS was reported to enhance sensitivity of Jurkat cells to apoptosis induction by TRAIL, while this effect was abolished in the presence of α-tocopherol (Dalen and Neuzil, 2003). In this report, a transient NF-κB activation occurred when the cells were exposed to TRAIL. As mentioned, NF-κB controls expression of prosurvival genes, including FLIP (Kreuz *et al.*, 2001) and IAPs (Degli-Esposti *et al.*, 1997). α-TOS, by inhibiting TRAIL-induced transient NF-κB activation, may inhibit expression of prosurvival proteins that confer resistance to TRAIL (Fig. 4).

There is thus evidence that α-TOS potentiates immunological inducers of apoptosis both in cell culture and in animal models. Since α-TOS and TRAIL are relatively nontoxic to normal cells, their combination could represent an exiting partnership of high therapeutic relevance.

V. TARGETING IMMUNE SURVEILLANCE

A. IMMUNE SURVEILLANCE AGAINST TUMOR DEVELOPMENT

Immune surveillance in tumorigenesis is mediated by both innate and adaptive components of cellular immunity (Smyth *et al.*, 2001). The adaptive components consist of CD8+ cytotoxic T cells (CTLs) that recognize tumor antigens presented by MHC class I molecules on tumor cells. While natural killer (NK) cells have been implicated in innate immunity against tumors, especially the MHC class I-deficient variants (Karre *et al.*, 1986). CTL and NK cells both act in an integrated manner: tumor cells destroyed by NK cells serve as a source for tumor antigens processed by dendritic cells and NK cell-derived cytokines, in particular interferon-γ (INF-γ), contribute to T cell-activating microenvironment (Smyth *et al.*, 2002). Activated NK cells or activated CTLs can destroy malignant cells by two mechanisms: first, by release of perforin- and granzyme-containing granules and second, by the membrane-bound variants of CD95L and TRAIL. INF-γ produced by activated NK and T cells further enhances the action of CD95L and TRAIL by sensitizing to DR-induced apoptosis and by upregulation of TRAIL. The importance of NK cell-expressed death ligands, especially TRAIL, for immune surveillance of cancer was confirmed in animal models showing that NK cell depletion and neutralization of TRAIL and CD95L enhanced liver metastasis of transplanted L929, CB27.4, and Reca cells (Takeda *et al.*, 2001).

Besides its antimetastatic action, recent studies revealed a role of TRAIL in host surveillance against primary tumor development. Formation of primary sarcomas induced by the carcinogen methylcholanthrene was accelerated in the presence of TRAIL-neutralizing antibodies, and depletion of NK cells did not interfere with the tumor-promoting effect of TRAIL neutralization, suggesting that an NK cell-independent source of TRAIL was responsible for this tumor-suppressive effect (Takeda *et al.*, 2002). The tumor-suppressive effects of TRAIL and CD95 are probably not only dependent on apoptosis induction in sensitive cells, but could also involve local immune responses induced by nonapoptotic CD95/TRAIL receptor signaling in tumor or tumor-associated cells. A key player in the regulation of apoptosis sensitivity is the tumor suppressor p53. It is a transcription factor induced by cellular stress and DNA damage, which controls the expression of target genes implicated in cell cycle arrest and apoptosis induction (Fridman and Lowe, 2003). The p53 apoptotic target genes can be divided into two groups, the first group encoding proteins that act via receptor-mediated signaling, the second group encoding those that regulate apoptotic effector proteins (Bartke *et al.*, 2001; Wu *et al.*, 1997). In addition, p53 has been reported to directly affect apoptosis in a transcription-independent manner by interfering with antiapoptotic and activating the function proapoptotic members of the Bcl-2 family, respectively (Fridman and Lowe, 2003). The various p53-mediated changes in the cellular program may by themselves suffice for intrinsic apoptosis induction or should at least sensitize tumor cells for death ligand-induced apoptosis. The apoptosis-inducing properties of various anticancer drugs acting via DNA damage have been traced back in part to p53-mediated upregulation of CD95 and CD95L (Altucci *et al.*, 2001; Muller *et al.*, 1998). Cooperation of death ligands expressed by NK cells and apoptosis sensitization by p53 has been also suggested due to the observation that efficient apoptosis induction in cancer cells via the extrinsic pathway often requires cotreatment with anticancer drugs (Evdokiou *et al.*, 2002; Frese *et al.*, 2002; Nataska *et al.*, 2004).

B. VITAMIN E ANALOGUES AS ADJUVANTS IN TUMOR VACCINATION

CTL responses are an important effector arm of the antitumor immune response. Hence, vaccination with antigens recognized by tumor-specific CTLs may represent an effective strategy for cancer immunotherapy. The ability of dendritic cells (DCs) to efficiently prime native and memory T lymphocytes in an MHC-restricted fashion has been exploited in the design of cell-based vaccines for cancer immunotherapy (Banchereau and Steinman, 1998; Boczkowski *et al.*, 2000). For these applications, DCs have been pulsed with defined peptides (Burdin and Moingeon, 2001), tumor lysates (Chang *et al.*, 2002), apoptotic tumor cells (Gregoire *et al.*, 2003), tumor RNA (Boczkowski *et al.*, 2000), and cocultured (Coveney *et al.*, 1997) or fused (Orentas *et al.*, 2001) with intact tumor cells. Whereas antigen-pulsed DCs have been shown to be able of suppressing tumor growth or conferring resistance to secondary tumor challenge, they have been less effective in treating established tumors (Coveney *et al.*, 1997; Fields *et al.*, 1998). Efforts to improve the effectiveness of DC-based vaccines in treating established diseases have included the transfer of cytokine (Ju *et al.*, 2001; Liu *et al.*, 2002) or chemokine (Kirk *et al.*, 2001) genes to stimulate T_{H1}-specific responses or promote recruitment of T lymphocytes and DCs (Ju *et al.*, 2001; Liu *et al.*, 2002). Thus, the use of adjuvants to improve the effectiveness of DC-based vaccines is a viable strategy for treating established primary or metastatic tumors.

Preestablished tumors treated with the vitamin E analogue, α-TOS, in combination with nonantigen-pulsed, immature DCs *in vivo* resulted in a significant tumor growth inhibition or in complete tumor regression in some cases (Ramanathapuram *et al.*, 2004). The superior effect of the combination therapy was correlated with increased INF-γ production by CD4+ and CD8+ T lymphocytes compared with controls, suggesting a polarization toward a T_{H1}-cell-mediated immune response. The authors observed that the antitumor activity of immature DCs plus α-TOS was comparable to that of mature tumor-pulsed lysate plus α-TOS, and coincubation of soluble components of α-TOS-treated tumor cells with immature DCs determines upregulation of the costimulatory molecules, CD80 and CD86, on DCs, suggesting that α-TOS may induce DC maturation *in vivo*. The mature DCs are able to migrate to secondary lymphoid organs where they initiate antitumor T-cell responses (Banchereau and Steinman, 1998). The use of α-TOS plus DC-based vaccines represents a novel chemoimmunotherapeutic approach that could be translated to the clinic for treating cancer patients.

VI. CONCLUSIONS

Chemotherapy, immunotherapy, and radiotherapy have been used for several decades as major cancer treatment modalities. However, the lack of cytotoxic drug selectivity and the ensuing toxic side-effects, when high doses of drugs are administrated, often limit the use of established therapies. Drugs at low doses are often not effective, and this approach often results in the patient relapse and development of drug resistance. Tumor resistance to conventional therapies presents a major problem in cancer management. Failure to eradicate advanced resistant tumors has led to exploration of novel approaches. One option is based on the combination of two or more anticancer agents that, each at a low dose, could improve their effects. We propose that a redox-silent analogue of vitamin E, α-TOS, is a feasible adjuvant of standard cancer therapy. When administered at higher doses, the vitamin E analogue is a potent inducer of apoptosis in cancer cells *in vitro*. Conversely, subapoptotic doses of the vitamin E analogue induce cell cycle arrest, differentiation or DNA synthesis blockage. Alteration of cell cycle distribution, and modulation of regulatory proteins involved in cell proliferation and cell death have been shown to enhance apoptotic effects of chemotherapy and immunotherapy.

Unlike many established chemotherapeutic agents, α-TOS is selective for malignant cells. The vitamin E analogue is carried within the bloodstream by circulating lipoproteins and is eventually hydrolyzed to α-TOH via the hepatic system. The redox-active form of vitamin E is then released into circulation, where it boosts antioxidant defenses. It is plausible that α-TOS plays a dual role, inducing apoptosis in malignant cells and protecting normal cells from free radical insult. In the "worst case scenario," the apoptogenic α-TOS is converted (detoxified) into a neutral bystander rather than into toxic secondary metabolites, as is often the case for other antitumor agents. Therefore, α-TOS may present a highly intriguing candidate for cancer therapy either in its own right and/or in combination with other therapeutic modalities.

REFERENCES

Akazawa, A., Nishikawa, K., Suzuki, K., Asano, R., Kumadaki, I., Satoh, H., Hagiwara, K., Shin, S. J., and Yano, T. (2002). Induction of apoptosis in a human breast cancer cell overexpressing ErbB-2 receptor by α-tocopheryloxybutyric acid. *Jpn. J. Pharmacol.* **89**, 417–421.

Alleva, R., Tomasetti, M., Andera, L., Gellert, N., Borghi, B., Weber, C., Murphy, M. P., and Neuzil, J. (2001). Coenzyme Q blocks biochemical but not receptor-mediated apoptosis by increasing mitochondrial antioxidant protection. *FEBS Lett.* **503,** 46–50.

Alleva, R., Benassi, M. S., Tomasetti, M., Gellert, N., Ponticelli, F., Borghi, B., Picci, P., and Neuzil, J. (2005). Alpha-tocopheryl succinate induces cytostasis and apoptosis in osteosarcoma cells: The role of E2F1. *Biochem. Biophys. Res. Commun.* **331,** 1515–1521.

Alleva, R., Benassi, M. S., Tomasetti, M., Gellert, N., Pazzaglia, L., Borghi, B., Picci, P., and Neuzil, J. (2006). α-Tocopheryl succinate alters cell cycle distribution sensitizing human osteosarcoma cells to methotrexate-induced apoptosis. *Cancer Lett.* **232,** 226–235.

Altucci, L., Rossin, A., Raffelsberger, W., Reitmair, A., Chomienne, C., and Gronemeyer, H. (2001). Retinoic acid-induced apoptosis in leukaemia cells is mediated paracrine action of tumor-selective death ligand TRAIL. *Nat. Med.* **7,** 680–686.

Anderson, K., Simmons-Menchaca, M., Lawson, K. A., Atkinson, J., Sanders, B. G., and Kline, K. (2004). Differential response of human ovarian cancer cells to induction of apoptosis by vitamin E succinate and vitamin E analogue, α-TEA. *Cancer Res.* **64,** 4263–4269.

Arizono, Y., Yoshikawa, H., Naganuma, H., Hamada, Y., Nakajima, Y., and Tasaka, K. (2003). A mechanism of resistance to TRAIL/Apo2L-induced apoptosis of newly established glioma cell line and sensitization to TRAIL by genotoxic agent. *Br. J. Cancer* **88,** 298–306.

Ashkenazi, A. (2002). Targeting death and decoy receptors of the tumour-necrosis factor superfamily. *Nat. Rev. Cancer* **2,** 420–430.

Ashkenazi, A., and Dixit, V. M. (1998). Death receptors: Signaling and modulation. *Science* **281,** 1305–1308.

Ashkenazi, A., and Dixit, V. M. (1999). Apoptosis control by death and decoy receptors. *Curr. Opin. Cell Biol.* **11,** 255–260.

Ashkenazi, A., Pai, R. C., Frong, S., Leung, S., Lawrence, D. A., Marsters, S. A., Blackie, C., Chang, L., McMurtrey, A. E., Hebert, A., DeForge, L., Koumenis, I. L., *et al.* (1999). Safety and antitumor activity of recombinant soluble Apo2 ligand. *J. Clin. Invest.* **104,** 155–162.

Banchereau, J., and Steinman, R. M. (1998). Dendritic cells and the control of immunity. *Nature* **392,** 245–252.

Bartke, T., Siegmund, D., Peters, N., Reichwain, M., Henkler, F., Scheurich, P., and Wajant, H. (2001). p53 Upregulates cFLIP, inhibits transcription of NF-kappaB-regulated genes and induces caspase-8-independent cell death in DLD-1 cells. *Oncogene* **20,** 571–580.

Baud, V., and Karin, M. (2001). Signal transduction by tumor necrosis factor and its relatives. *Trends Cell Biol.* **11,** 372–377.

Birringer, M., EyTina, J. H., Salvatore, B. A., and Neuzil, J. (2003). Vitamin E analogues as inducers of apoptosis: Structure-function relation. *Br. J. Cancer* **88,** 1948–1955.

Boczkowski, D., Nair, S. K., Nam, J. H., Lyerly, H. K., and Gilboa, E. (2000). Induction of tumor immunity and cytotoxic T lymphocyte responses using dendritic cells transfected with messenger RNA amplified from tumor cells. *Cancer Res.* **60,** 1028–1034.

Bodmer, J. L., Holler, N., Reynard, S., Vinciguerra, P., Schneider, P., Juo, P., Blenis, J., and Tschopp, J. (2000). TRAIL receptor-2 signals apoptosis through FADD and caspase-8. *Nat. Cell Biol.* **2,** 241–243.

Burdin, N., and Moingeon, P. (2001). Cancer vaccines based on dendritic cells loaded with tumor-associated antigens. *Cell Biol. Toxicol.* **17,** 67–75.

Burns, T. F., and El-Deiry, W. S. (2001). Identification of inhibitors of TRAIL-induced death (ITIDs) in the TRAIL-sensitive colon carcinoma cell line SW480 using a genetic approach. *J. Biol. Chem.* **276,** 37879–37886.

Cande, C., Cecconi, F., Dessen, P., and Kroemer, G. (2002). Apoptosis-inducing factor (AIF): Key to the conserved caspase-independent pathway of cell death? *J. Cell Sci.* **115,** 4727–4734.

Chang, A. E., Redman, B. G., Whitfield, J. R., Nickoloff, B. J., Braun, T. M., Lee, P. P., Geiger, J. D., and Mule, J. J. (2002). A phase I trial of tumor lysate-pulsed dendritic cells in the treatment of advanced cancer. *Clin. Cancer Res.* **8,** 1021–1032.

Cory, S., Huang, D. C., and Adams, J. M. (2003). The Bcl-2 family: Roles in cell survival and oncogenesis. *Oncogene* **22,** 8590–8607.

Coveney, E., Wheatley, G. H., III, and Lyerly, H. K. (1997). Active immunization using dendritic cells mixed with tumor cells inhibits the growth of primary breast cancer. *Surgery* **122,** 228–234.

Dalen, H., and Neuzil, J. (2003). α-Tocopheryl succinate sensitises T lymphoma cells to TRAIL killing by suppressing NF-κB activation. *Br. J. Cancer* **88,** 153–158.

D'Alessio, M., De Nicola, M., Coppola, S., Gualandi, G., Pugliese, L., Cerella, C., Cristofanon, S., Civitareale, P., Ciriolo, M. R., Bergamaschi, A., Magrini, A., and Ghibelli, L. (2005). Oxidative Bax dimerization promotes its translocation to mitochondria independently of apoptosis. *FASEB J.* **19,** 1504–1506.

Debatin, K. M., and Krammer, P. H. (2004). Death receptors in chemotherapy and cancer. *Oncogene* **12,** 2950–2966.

Degli-Esposti, M. (1999). To die or not to die-the quest of the TRAIL receptors. *J. Leukoc. Biol.* **65,** 535–542.

Degli-Esposti, M. A., Dougall, W. C., Smolak, P. J., Waugh, J. Y., Smith, C. A., and Goodwin, R. G. (1997). The novel receptor TRAIL-R4 induces NF-kappaB and protects against TRAIL-mediated apoptosis, yet retains and incomplete death domain. *Immunity* **7,** 813–820.

Deng, Y., Lin, Y., and Wu, X. (2002). TRAIL-induced apoptosis requires Bax-dependent mitochondrial release of Smac/DIABLO. *Genes Dev.* **16,** 33–45.

Dey, R., and Moraes, C. T. (2000). Lack of oxidative phosphorylation and low mitochondrial membrane potential decrease susceptibility to apoptosis and do not modulate the protective effect of Bcl-x_L in osteosarcoma cells. *J. Biol. Chem.* **275,** 7087–7094.

Dimanche-Boitrel, M. T., Meurette, O., Rebillard, A., and Lacour, S. (2005). Role of early plasma membrane events in chemotherapy-induced cell death. *Drug Resist Updat.* **8,** 5–14.

Du, C., Fang, M., Li, Y., Li, L., and Wang, X. (2000). Smac, a mitochondrial protein that promotes cytochrome c-dependent caspase activation by eliminating IAP inhibition. *Cell* **102,** 33–42.

El-Deiry, W. S. (2001). Insights into cancer therapeutic design based on p53 and TRAIL receptor signaling. *Cell Death Differ.* **8,** 1066–1075.

Erl, W., Weber, C., Wardemann, C., and Weber, P. C. (1997). α-Tocopheryl succinate inhibits monocytic cell adhesion to endothelial cells by suppressing NF-κB mobilization. *Am. J. Physiol.* **273,** H634–H640.

Evdokiou, A., Bouralexis, S., Atkins, G. J., Chai, F., Hay, S., Clayer, M., and Findlay, D. M. (2002). Chemotherapeutic agents sensitize osteogenic sarcoma cells, but not normal human bone cells, to Apo2L/TRAIL-induced apoptosis. *Int. J. Cancer* **99,** 491–504.

Fariss, M. W., Fortuna, M. B., Everett, C. K., Smith, J. D., Trent, D. F., and Gjuric, Z. (1994). The selective antiproliferative effects of alpha-tocopheryl hemisuccinate and cholesteryl hemisuccinate on murine leukemia cells result from the action of the intact compounds. *Cancer Res.* **54,** 3346–3351.

Fields, R. C., Shimizu, K., and Mule, J. J. (1998). Murine dendritic cells pulsed with whole tumor lysates mediate potent antitumor immune responses *in vitro* and *in vivo*. *Proc. Natl. Acad. Sci. USA* **95,** 9482–9487.

Fisher, M. J., Birmani, A. K., Wu, L., Aplec, R., Harper, J. C., Powell, S. M., Rebbeck, T. R., Sidransky, D., Gazdar, A. F., and El-Deiry, W. S. (2001). Nucleotide substitution in the ectodomain of trail receptor DR4 is associated with lung cancer and head and neck cancer. *Clin. Cancer Res.* **7,** 1688–1697.

French, L. E., and Tschopp, J. (1999). The TRAIL to selective tumor death. *Nat. Med.* **5,** 146–147.

Frese, S., Brunner, T., Gugger, M., Uduehi, A., and Schmid, R. A. (2002). Enhancement of Apo2L/TRAIL (tumor necrosis factor-related apoptosis-inducing ligand)-induced apoptosis in non-small cell lung cancer cell lines by chemotherapeutic agents without correlation to the expression level of cellular protease caspase-8 inhibitory protein. *J. Thorac. Cardiovasc. Surg.* **123,** 168–174.

Fridman, J. S., and Lowe, S. W. (2003). Control of apoptosis by p53. *Oncogene* **22,** 9030–9040.

Green, D. R., and Kroemer, G. (2004). The pathophysiology of mitochondrial cell death. *Science* **305,** 626–629.

Gregoire, M., Ligeza-Poisson, C., Juge-Morineau, N., and Spisek, R. (2003). Anti-cancer therapy using dendritic cells and apoptotic tumour cells: Pre-clinical data in human mesothelioma and acute myeloid leukaemia. *Vaccine* **21,** 791–794.

Greten, F. R., Eckmann, L., Greten, T. F., Park, J. M., Li, Z. W., Egan, L. J., Kagnoff, M. F., and Karin, M. (2004). IKKbeta links inflammation and tumorigenesis in a mouse model of colitis-associated cancer. *Cell* **118,** 285–296.

Harvey, K. J., Lukovic, D., and Ucker, D. S. (2000). Caspase-dependent Cdk activity is a requisite effector of apoptotic death events. *J. Cell Biol.* **10,** 59–72.

Holler, N., Tardivel, A., Kovacsovics-Bankowski, M., Hertig, S.,Gaide, O., Martinon, F., Tinel, A., Deperthes, D., Calderara, S., Schulthess, T., Engel, J., Schneider, P., *et al.* (2003). Two adjacent trimeric Fas ligands are required for Fas signaling and formation of a death-inducing signaling complex. *Mol. Cell. Biol.* **23,** 1428–1440.

Irmler, M., Thome, M., Hahne, M., Schneider, P., Hofmann, K., Steiner, V., Bodmer, J. L., Schroter, M., Burns, K., Mattmann, C., Rimoldi, D., French, L. E., *et al.* (1997). Inhibition of death receptor signals by cellular FLIP. *Nature* **388,** 190–195.

Israel, K., Yu, W., Sanders, B. G., and Kline, K. (2000). Vitamin E succinate induces apoptosis in human prostate cancer cells: Role for Fas in vitamin E succinate-triggered apoptosis. *Nutr. Cancer* **36,** 90–100.

Jiang, Y., Woronicz, J. D., Liu, W., and Goeddel, D. V. (1999). Prevention of constitutive TNF receptor 1 signaling by silencer of death domains. *Science* **283,** 543–546.

Jin, Z., McDonald, E. R., III, Dicker, D. T., and El-Deiry, W. S. (2004). Deficient TRAIL death receptor transport to the cell surface in human colon cancer cells selected for resistance to TRAIL-induced apoptosis. *J. Biol. Chem.* **279,** 35829–35839.

Ju, D. W., Tao, Q., Lou, G., Bai, M., He, L., Yang, Y., and Cao, X. (2001). Interleukin 18 transfection enhances antitumor immunity induced by dendritic cell-tumor cell conjugates. *Cancer Res.* **61,** 3735–3740.

Kagan, V. E., Tyurin, V. A., Jiang, J., Tyurina, Y. Y., Ritov, V. B., Amoscato, A. A., Osipov, A. N., Belikova, N. A., Kapralov, A. A., Kini, V., Vlasova, I. I., Zhao, Q., *et al.* (2005). Cytochrome c acts as a cardiolipin oxygenase required for release of proapoptotic factors. *Nat. Chem. Biol.* **1,** 223–232.

Kane, L. P., Shapiro, V. S., Stokoe, D., and Weiss, A. (1999). Induction of NF-κB by the Akt/PKB kinase. *Curr. Biol.* **9,** 601–604.

Karre, K., Ljunggren, H. G., Piontek, G., and Kiessling, R. (1986). Selective rejection of H-2-deficient lymphoma variants suggests alternative immune defence strategy. *Nature* **319,** 675–678.

Kelso, G. F., Porteous, C. M., Coulter, C. V., Hughes, G., Porteous, W. K., Ledgerwood, E. C., Smith, R. A. J., and Murphy, M. P. (2001). Selective targeting of a redox-active ubiquinone to mitochondria within cells. *J. Biol. Chem.* **276,** 4588–4596.

Kim, K., Fisher, M. J., Xu, S. Q., and el-Deiry, W. S. (2000). Molecular determinants of response to TRAIL in killing of normal and cancer cells. *Clin. Cancer Res.* **6,** 335–346.

Kim, E. J., Suliman, A., Lam, A., and Srivastava, R. K. (2001). Failure of bcl-2 to block mitochondria dysfunction during TRAIL-induced apoptosis. *Int. J. Oncol.* **18,** 187–194.

Kirk, C. J., Hartigan-O'Connor, D., and Mule, J. J. (2001). The dynamics of the T-cell antitumor response: Chemokine-secreting dendritic cells can prime tumor-reactive T cells extranodally. *Cancer Res.* **61,** 8794–8802.

Kline, K., Lawson, K. A., Yu, W., and Sanders, B. G. (2003). Vitamin E and breast cancer prevention: Current status and future potential. *J. Mammary Gland Biol. Neoplasia* **8,** 91–102.

Kline, K., Yu, W., and Sanders, B. G. (2004). Vitamin E and breast cancer. *J. Nutr.* **134,** 3458S–3462S.

Knekt, P. (1991). Role of vitamin E in the prophylaxis of cancer. *Ann. Med.* **23,** 3–12.

Kogure, K., Morita, M., Nakashima, S., Hama, S., Tokumura, A., and Fukuzawa, K. (2001). Superoxide is responsible for apoptosis in rat vascular smooth muscle cells induced by α-tocopheryl hemisuccinate. *Biochim. Biophys. Acta* **1528,** 25–30.

Kreuz, S., Siegmund, D., Scheurich, P., and Wajant, H. (2001). NF-κB inducers upregulate cFLIP, a cycloheximide-sensitive inhibitor of death receptor signaling. *Mol. Cell. Biol.* **21,** 3964–3973.

LaCasse, E. C., Baird, S., Korneluk, R. G., and MacKenzie, A. E. (1998). The inhibitors of apoptosis (IAPs) and their emerging role in cancer. *Oncogene* **17,** 3247–3259.

LeBlanc, H., Lawrence, D., Varfolomeev, E., Totpal, K., Morlan, J., Schow, P., Fong, S., Schwall, R., Sinicrop, D., and Ashkenazi, A. (2002). Tumor-cell resistance to death receptor-induced apoptosis through mutational inactivation of the pro-apoptotic Bcl-2 homolog Bax. *Nat. Med.* **8,** 274–281.

Levkau, B., Scatena, M., Giachelli, C. M., Ross, R., and Raines, E. W. (1999). Apoptosis overrides survival signals through a caspase-mediated dominant-negative NF-κB loop. *Nat. Cell Biol.* **1,** 227–233.

Li, P., Nijhawan, D., Budihardjo, I., Srinivasula, S. M., Ahmad, M., Alnemri, E. S., and Wang, X. (1997). Cytochrome c and ATP-dependent formation of Apaf-1/caspase-9 complex initiates an apoptotic protease cascade. *Cell* **91,** 479–489.

Linker-Israeli, M. (1992). Cytokine abnormalities in human lupus. *Clin. Immunol. Immunopathol.* **63,** 10–12.

Liu, Y., Zhang, W., Chan, T., Saxena, A., and Xiang, J. (2002). Engineered fusion hybrid vaccine of IL-4 gene-modified myeloma and relative mature dendritic cells enhanced antitumor immunity. *Leuk. Res.* **26,** 757–763.

Locksley, R. M., Killeen, N., and Lenardo, M. J. (2001). The TNF and TNF receptor superfamilies: Integrating mammalian biology. *Cell* **104,** 87–501.

Moqattash, S., and Lutton, J. D. (1998). Leukaemia cells and the cytokine network. *Proc. Soc. Exp. Biol. Med.* **219,** 8–27.

Moller, P., Koretz, K., Leithauser, F., Bruderlein, S., Henne, C., Quentmeier, A., and Krammer, P. H. (1994). Expression of APO-1 (CD95), a member of the NGF/TNF receptor superfamily, in normal and neoplastic colon epithelium. *Int. J. Cancer* **1,** 371–377.

Muller, M., Wilder, S., Bannasch, D., Israeli, D., Lehlback, K., Li-Weber, M., Friedman, S. L., Galle, P. R., Stremmel, W., Oren, M., and Krammer, P. H. (1998). p53 activates the CD95 (APO-1/Fas) gene in response to DNA damage by anticancer drugs. *J. Exp. Med.* **188,** 2033–2045.

Nataska, S., Yoshima, T., Horinaka, M., Shiraishi, T., Wakada, M., and Sakai, T. (2004). Histone deacetylase inhibitors upregulate death receptor 5/TRAIL-R2 and sensitize apoptosis induced by TRAIL/APO2-L in human malignant tumor cells. *Oncogene* **19,** 6261–6271.

Neuzil, J. (2003). Vitamin E succinate and cancer treatment: A vitamin E prototype for selective antitumour activity. *Br. J. Cancer* **89,** 1822–1826.

Neuzil, J., Svensson, I., Weber, T., Weber, C., and Brunk, U. T. (1999). α-Tocopheryl succinate-induced apoptosis in Jurkat T cells involves caspase-3 activation, and both lysosomal and mitochondrial destabilization. *FEBS Lett.* **445,** 295–300.

Neuzil, J., Weber, T., Schroder, A., Lu, M., Ostermann, G., Gellert, N., Mayne, G. C., Olejnicka, B., Negre-Salvayre, A., Sticha, M., Coffey, R. J., and Weber, C. (2001a). Induction of cancer cell apoptosis by alpha-tocopheryl succinate: Molecular pathways and structural requirements. *FASEB J.* **15,** 403–415.

Neuzil, J., Weber, T., Teman, A., Weber, C., and Brunk, U. T. (2001b). Vitamin E analogues as inducers of apoptosis: Implications for their potential antineoplastic role. *Redox Rep.* **6,** 143–151.

Neuzil, J., Weber, T., Gellert, N., and Weber, C. (2001c). Selective cancer killing by alpha-tocopheryl succinate. *Br. J. Cancer* **84,** 87–89.

Neuzil, J., Schröder, A., von Hundelshausen, P., Zernecke, A., Weber, T., Gellert, N., and Weber, C. (2001d). Inhibition of inflammatory endothelial responses by a pathway involving caspase activation and p65 cleavage. *Biochemistry* **40,** 4686–4692.

Neuzil, J., Zhao, M., Ostermann, G., Sticha, M., Gellert, N., Weber, C., Eaton, J. W., and Brunk, U. T. (2002). α-Tocopheryl succinate, an agent with *in vivo* anti-tumour activity, induces apoptosis by causing lysosomal instability. *Biochem. J.* **362,** 709–715.

Neuzil, J., Tomasetti, M., Mellick, A. S., Alleva, R., Salvatore, B. A., Birringer, M., and Fariss, M. W. (2004). Vitamin E analogues: A new class of inducers of apoptosis with selective anti-cancer effects. *Curr. Cancer Drug Targets* **4,** 355–372.

Neuzil, J., Tomasetti, M., Zhao, Y., Dong, L. F., Birringer, M., Wang, X. F., Low, P., Wu, K., Salvatore, B. A., and Ralph, S. J. (2007). Vitamin E analogs, a novel group of "mitocans," as anticancer agents: The importance of being redox-silent. *Mol. Pharmacol.* **71,** 1185–1199.

Neuzil, J., Wang, X. F., Dong, L. F., Low, P., and Ralph, S. J. (2006). Molecular mechanism of 'mitocan'-induced apoptosis in cancer cells epitomizes the multiple roles of reactive oxygen species and Bcl-2 family proteins. *FEBS Lett.* **580,** 5125–5129.

Ni, J., Chen, M., Zhang, Y., Li, R., Huang, J., and Yeh, S. (2003). Vitamin E succinate inhibits human prostate cancer cell growth via modulation cell cycle regulatory machinery. *Biochem. Biophys. Res. Commun.* **300,** 357–363.

Ogretmen, B., and Hannun, Y. A. (2004). Biologically active sphingolipids in cancer pathogenesis and treatment. *Nat. Rev. Cancer* **4,** 604–616.

Orentas, R. J., Schauer, D., Bin, Q., and Johnson, B. D. (2001). Electrofusion of a weakly immunogenic neuroblastoma with dendritic cells produces a tumor vaccine. *Cell. Immunol.* **213,** 4–13.

Peter, M. E., and Krammer, P. H. (2003). The CD95 (APO-1/Fas) DISC and beyond. *Cell Death Differ.* **10,** 26–35.

Pikarsky, E., Porat, R. M., Stein, I., Abramovitch, R., Amit, S., Kasem, S., Gutkovich-Pyest, E., Urieli-Shoval, S., Galun, E., and Ben-Neriah, Y. (2004). NF-kappaB functions as a tumour promoter in inflammation-associated cancer. *Nature* **43,** 461–466.

Pitti, R. M., Marsters, S. A., Ruppert, S., Donahue, C. J., Moore, A., and Ashkenazi, A. (1996). Induction of apoptosis by Apo-2 ligand, a new member of the tumor necrosis factor cytokine family. *J. Biol. Chem.* **271,** 12687–12690.

Prasad, K. N., and Kumar, R. (1996). Effect of individual and multiple antioxidant vitamins on growth and morphology of human nontumorigenic and tumorigenic parotid acinar cells in culture. *Nutr. Cancer* **26,** 11–19.

Prasad, K. N., and Edwards-Prasad, J. (1992). Vitamin E and cancer prevention: Recent advances and future potentials. *J. Am. Coll. Nutr.* **11,** 487–500.

Prasad, K. N., Hernandez, C., Edwards-Prasad, J., Nelson, J., Borus, T., and Robinson, W. A. (1994). Modification of the effect of tamoxifen, cis-platin, DTIC, and interferon-alpha 2b on human melanoma cells in culture by a mixture of vitamins. *Nutr. Cancer* **22,** 233–245.

Ramanathapuram, L. V., Kobie, J. J., Bearss, D., Payne, C. M., Trevor, K. T., and Akporiaye, T. (2004). α-Tocopheryl succinate sensitize established tumors to vaccination with normal dendritic cells. *Cancer Immunol Immunother.* **53,** 580–588.

Rautalahti, M. T., Virtano, J. R., Taylor, P. R., Heinonem, O. P., Albanes, D., Haukka, J. K., Edwards, B. K., Karkkainen, P. A., Stolzenberg-Solomon, R. Z., and Huttunen, J. (1999). The effect of supplementation with a-tocopherol and b-carotene on the incidence and mortality of carcinoma of the pancreas in a randomized, controlled study. *Cancer* **86,** 37–42.

Ripoll, E. A., Rama, B. N., and Webber, M. M. (1986). Vitamin E enhances the chemotherapeutic effets of adriamycin on human prostatic carcinoma cells *in vitro*. *J. Urol.* **136,** 529–531.

Rippo, M. R., Moretti, S., Vescovi, S., Tomasetti, M., Orecchia, S., Amici, G., Catalano, A., and Procopio, A. (2004). FLIP overexpression inhibits death receptor-induced apoptosis in malignant mesothelial cells. *Oncogene* **26,** 7753–7760.

Roskoski, R. (2004). The ErbB/HER receptor protein-tyrosine kinases and cancer. *Biochem. Biophys. Res. Commun.* **319,** 1–11.

Sachs, L., and Lotem, J. (1993). Control of programmed cell death in normal and leukemic cells: New implications for therapy. *Blood* **82,** 15–21.

Shiau, C. W., Huang, J. W., Wang, D. S., Weng, J. R., Yang, C. C., Lin, C. H., Li, C., and Chen, C. S. (2006). Tocopheryl succinate induces apoptosis in prostate cancer cells in part through inhibition of Bcl-x_L/Bcl-2 function. *J. Biol. Chem.* **281,** 11819–11825.

Schneider, P., Holler, N., Bodmer, J. L., Hahne, M., Frei, K., Fontana, A., and Tschopp, J. (1998). Conversion of membrane-bound Fas (CD95) ligand to its soluble form is associated with downregulation of its proapoptotic activity and loss of liver toxicity. *J. Exp. Med.* **187,** 1205–1213.

Sionov, R. V., and Haupt, Y. (1999). The cellular response to p53: The decision between life and death. *Oncogene* **18,** 6145–6157.

Smyth, M. J., Gogfrey, D. I., and Trapani, J. A. (2001). A fresh look at tumor immunosurveillance and immunotherapy. *Nat. Immunol.* **2,** 293–299.

Smyth, M. J., Hayakawa, Y., Takeda, K., and Yagita, H. (2002). New aspects of natural killer-cell surveillance and therapy of cancer. *Nat. Rev. Cancer* **2,** 850–861.

Srivastava, R. K. (2000). Intracellular mechanisms of TRAIL and its role in cancer therapy. *Mol. Cell Biol. Res. Commun.* **4,** 67–75.

Stapelberg, M., Tomasetti, M., Gellert, N., Alleva, R., Procopio, A., and Neuzil, J. (2004). α-Tocopheryl succinate inhibits proliferation of mesothelioma cells by differential down-regulation of fibroblast growth factor receptors. *Biochem. Biophys. Res. Commun.* **318,** 636–641.

Stapelberg, M., Gellert, N., Swetternham, E., Tomasetti, M., Witting, P. K., Procopio, A., and Neuzil, J. (2005). Alpha-tocopheryl succinate inhibits malignant mesothelioma by disrupting the fibroblast growth factor autocrine loop: Mechanism and the role of oxidative stress. *J. Biol. Chem.* **280,** 25369–25376.

Sun, Y., and Oberley, L. W. (1996). Redox regulation of transcriptional activators. *Free Radic. Biol. Med.* **21,** 335–348.

Suliman, A., Lam, A., Datta, R., and Srivastava, R. K. (2001). Intracellular mechanisms of TRAIL: Apoptosis through mitocondrial-dependent and -independent pathways. *Oncogene* **20,** 2122–2133.

Takeda, K., Hayakawa, Y., Smyth, M. J., Kayagaki, N., Yamaguchi, N., Kakuta, S., Iwakura, Y., Yagita, H., and Okumura, K. (2001). Involvement of tumor necrosis factor-related apoptosis-inducing ligand in surveillance of tumor metastasis by liver natural killer cells. *Nat. Med.* **7,** 94–100.

Takeda, K., Smyth, M. J., Cretney, Hayakawa, Y., Kayagaki, N., Yagita, H., and Okumura, K. (2002). Critical role for tumor necrosis factor-related apoptosis-inducing ligand in immune surveillance against tumor development. *J. Exp. Med.* **195,** 161–169.

Thorburn, A. (2004). Death receptor-inducing cell killing. *Cell Signal.* **16,** 139–144.

Timmer, T., de Vries, E. G., and de Jong, S. (2002). Fas receptor-mediated apoptosis: A clinical application? *J. Pathol.* **196,** 125–134.

Tomasetti, M., Rippo, M. R., Alleva, R., Moretti, S., Andera, L., Neuzil, J., and Procopio, A. (2004a). Alpha-tocopheryl succinate and TRAIL selectively synergise in induction of apoptosis in human malignant mesothelioma cells. *Br. J. Cancer* **90,** 1644–1653.

Tomasetti, M., Gellert, N., Procopio, A., and Neuzil, J. (2004b). A vitamin E analogue suppresses malignant mesothelioma in a preclinical model: A future drug against a fatal neoplastic disease? *Int. J. Cancer* **109,** 641–642.

Tomasetti, M., Ladislav, A., Alleva, R., Borghi, B., Neuzil, J., and Procopio, A. (2006). α-Tocopheryl succinate induces DR4 and DR5 expression by a p53-dependent route: Implication for sensitisation of resistant cancer cells to TRAIL apoptosis. *FEBS Lett.* **580,** 1925–1931.

Turley, J. M., Ruscetti, F. W., Kim, S. J., Fu, T., Gou, F. V., and Birchenall-Roberts, M. C. (1997). Vitamin E succinate inhibits proliferation of BT-20 human breast cancer cells: Increased binding of cyclin A negatively regulates E2F transactivation activity. *Cancer Res.* **57,** 2668–2675.

Volkmann, M., Schiff, J. H., Hajjar, Y., Otto, G., Stilgenbauer, F., Fiehn, W., Galle, P. R., and Hofmann, W. J. (2001). Loss of CD95 expression is linked to most but not all p53 mutants in European hepatocellular carcinoma. *J. Mol. Med.* **79,** 594–600.

Vousden, K. H., and Lu, X. (2002). Live or let die: The cell's response to p53. *Nat. Rev. Cancer* **2,** 594–604.

Vivanco, I., and Sawyers, C. L. (2002). The phosphatidylinositol 3-kinase AKT pathway in human cancer. *Nat. Rev. Cancer* **2,** 489–501.

Wajant, H., Pfizenmaier, K., and Scheurich, P. (2003). Tomor necrosis factor signaling. *Cell Death Differ.* **10,** 45–65.

Walczak, H., Miller, R. E., Ariail, K., Gliniak, B., Griffith, T. S., Kubin, M., Chin, W., Jones, J., Woodward, A., Le, T., Smithm, C., Smolak, P., *et al.* (1999). Tumoricidal activity of tumour necrosis factor-related apoptosis-inducing ligand *in vivo. Nat. Med.* **5,** 157–163.

Wang, S., and El-Deiry, W. S. (2003). Requirement of p53 targets in chemosensitization of colonic carcinoma to death ligand therapy. *Proc. Natl. Acad. Sci. USA* **100,** 15095–15100.

Wang, X. F., Witting, P. K., Salvatore, B. A., and Neuzil, J. (2005). Vitamin E analogs trigger apoptosis in HER2/erbB2-overexpressing breast cancer cells by signaling via the mitochondria pathway. *Biochem. Biophys. Res. Commun.* **326,** 282–289.

Wang, X. F., Birringer, M., Dong, L. F., Veprek, P., Low, P., Swettenham, E., Stanic, M., Yuan, L. H., Zobalova, R., Wu, K., Ralph, S. J., Ledvina, M., *et al.* (2007). A peptide conjugate of vitamin E succinate targets breast cancer cells with high erbB2 expression. *Cancer Res.* **67,** 3337–3344.

Weber, T., Lu, M., Andera, L., Lahm, H., Gellert, N., Fariss, M. W., Korinek, V., Sattler, W., Ucker, D. S., Terman, A., Schroder, A., Erl, W., *et al.* (2002). Vitamin E succinate is a potent novel anti-neoplastic agent with high tumor selectivity and cooperativity with tumor necrosis factor-related apoptosis-inducing ligand (Apo2 ligand) *in vivo. Clin. Cancer Res.* **8,** 863–869.

Weber, T., Dalen, H., Andera, L., Nègre-Savayre, A., Augè, N., Sticha, M., Loret, A., Terman, A., Witting, P., Higuchi, M., Plasilova, M., Zivny, J., *et al.* (2003). Mitochondria play a central role in apoptosis induced by α-tocopheryl succinate, an agent with anti-neoplastic activity: Comparison with receptor-mediated pro-apoptotic signalling. *Biochemistry* **42,** 4277–4291.

Wiley, S. R., Schooley, K., Smolak, P. J., Din, W. S., Huang, C. P., Nicholl, J. K., Sutherland, G. R., Smith, T. D., Rauch, C., Smith, C. A., and Goodwin, R. G. (1995). Identification and characterization of a new number of the TNF family that induces apoptosis. *Immunity* **3,** 673–682.

Wu, G. S., Burns, T. F., McDonald, E. R., III, Jiang, W., Meng, R., Krantz, I. D., Kao, G., Gan, D. D., Zhou, J. Y., Muschel, R., Hamilton, S. R., Spinner, N. B., *et al.* (1997). KILLER/DR5 is a DNA damage-inducible p53-regulated death receptor gene. *Nat. Genet.* **17,** 141–143.

Wu, G. S., Kim, K., and El-Deiry, W. S. (2000). KILLER/DR5, a novel DNA-damage inducible death receptor gene, links the p53-tumor suppressor to caspase activation and apoptotic death. *Adv. Exp. Med. Biol.* **465,** 143–151.

Wu, K., Zhao, Y., Liu, B. H., Li, Y., Liu, F., Guo, J., and Yu, W. P. (2002). RRR-α-tocopheryl succinate inhibits human gastric cancer SGC-7901 cell growth by inducing apoptosis and DNA synthesis arrest. *World J. Gastroenterol.* **8,** 26–30.

Wu, K., Zhao, Y., Li, G. C., and Yu, W. P. (2004). c-Jun N-terminal kinase is required for vitamin E succinate-induced apoptosis in human gastric cancer cells. *World J. Gastroenterol.* **10,** 1110–1114.

Yamamoto, S., Tamai, H., Ishisaka, R., Kanno, T., Arita, K., Kobuchi, H., and Utsumi, K. (2000). Mechanism of α-tocopheryl succinate-induced apoptosis of promyelocytic leukaemia cells. *Free Radic. Res.* **33,** 407–418.

You, H., Yu, W., Sanders, B. G., and Kline, K. (2001). RRR-α-Tocopheryl succinate induces MDA-MB-435 and MCF-7 human breast cancer cells to undergo differentiation. *Cell Growth Differ.* **12,** 471–480.

You, H., You, W., Munoz-Medellin, D., Brown, P. H., Sanders, B. G., and Kline, K. (2002). Role of extracellular signal-regulated kinase pathway in RRR−α-tocopheryl succinate-induced differentiation of human MDA-MB-435 breast cancer cells. *Mol. Carcinog.* **33,** 228–236.

Yu, W., Israel, K., Liao, Q. Y., Aldaz, C. M., Sanders, B. G., and Kline, K. (1999). Vitamin E succinate (VES) induces fas sensitivity in human breast cancer cells: Role for Mr 43,000 Fas in VES-triggered apoptosis. *Cancer Res.* **59,** 953–961.

Yu, W., Liao, Q. Y., Hantash, F. M., Sanders, B. G., and Kline, K. (2001). Activation of extracellular signal-regulated kinase and c-jun-NH_2-terminal kinase but not p38 mitogen-activated protein kinases is required for RRR-α-tocopheryl succinate-induced apoptosis of human breast cancer cells. *Cancer Res.* **61,** 6569–6576.

Yu, W., Sanders, B. G., and Kline, K. (2003). α-Tocopheryl succinate-induced apoptosis of human breast cancer cells involves Bax translocation to mitochondria. *Cancer Res.* **63,** 2483–2491.

Zhao, Y., Zhao, X., Yang, B., Neuzil, J., and Wu, K. (2006). α-Tocopheryl succinate-induced apoptosis in human gastric cancer cells is modulated by ERK1/2 and c-Jun N-terminal kinase in a biphasic manner. *Cancer Lett.* **247**(2), 345–352.

Zhou, B. P., and Hung, M. C. (2003). Dysregulation of cellular signaling by HER2/neu in breast cancer. *Semin. Oncol.* **30,** 38–48.

Zu, K., Hawthorn, L., and Ip, C. (2005). Up-regulation of c-Jun-NH_2-kinase pathway contributes to the induction of mitochondria-mediated apoptosis by α-tocopheryl succinate in human prostate cancer cells. *Mol. Cancer Ther.* **4,** 43–50.

19

The Roles of α-Vitamin E and Its Analogues in Prostate Cancer

Jing Ni*,† and Shuyuan Yeh*,†

*Department of Urology, University of Rochester
Rochester, New York 14642

†Department of Pathology, University of Rochester
Rochester, New York 14642

Vitamins and Hormones, Volume 76

0083-6729/07 $35.00
DOI: 10.1016/S0083-6729(07)76019-3

Prostate cancer is the second most commonly diagnosed cancer and the third leading fatal cancer in American men. Comprehensive studies from human epidemiological studies, animal tumor models, and cellular molecular levels suggested that α-vitamin E and its derivatives possess remarkable chemopreventive and chemotherapeutic against prostate cancer. This chapter details the facts of α-vitamin E and its nonantioxidant functions in prostate cancer, focuses on the biological mechanisms for the α-vitamin E and its ester analogue, α-vitamin E succinate (VES), in prevention and therapy of prostate cancer, and raises specific questions that remain for intensive investigation in the future.

I. INTRODUCTION

Vitamin E was first identified as an important nutrient for rat reproductive function in 1922 (Evans and Bishop, 1922). Although its role in human reproductive function is "not" so critical as that in rat, deficiency of vitamin E in human is associated with several human diseases, including ataxia with vitamin E deficiency (AVED), a neurodegeneration disorder (Ouahchi *et al.*, 1995). Moreover, several studies demonstrated that supplementation of α-vitamin E, the major biological form of vitamin E, could reduce risks of infertility, neurological disorders, inflammation, cardiovascular diseases, diabetes, and certain types of cancers in humans (Traber and Sies, 1996; Tucker and Townsend, 2005). In 1994, the unexpected finding from the Alpha-Tocopherol, Beta-Carotene Cancer Prevention study (ATBC trial)[1] indicated that daily supplement of α-vitamin E could reduce the incidence and mortality of prostate cancer in smoking men (The Alpha-Tocopherol,

[1]Abbreviations: ABCA1, ATP-binding cassette transporter A1; AR, androgen receptor; ATBC, Alpha-Tocopherol, Beta-Carotene Cancer Prevention study; HOPE-TOO, Heart Outcomes Prevention Evaluation-The Ongoing Outcome; HPFS, Health Professionals Follow-up Study; IGFBP-3, insulin-like growth factor-binding protein-3; PLCO, Prostate, Lung, Colorectal, and Ovarian cancer-screening trial; PSA, prostate-specific antigen; SELECT, Selenium and Vitamin E Cancer Prevention Trial; SR-BI, scavenger receptor class B type I, TAP, α-tocopherol-associated protein; TBP, α-tocopherol-binding protein; α-vitamin E, α-tocopherol TTP, α-tocopherol transfer protein; VEA, α-vitamin E acetate, α-tocopheryl acetate; VES, α-vitamin E succinate, α-tocopheryl succinate; VDR, vitamin D receptor.

Beta Carotene Cancer Prevention Study Group, 1994). The following epidemiological studies, while controversial, have been supporting the protective roles of α-vitamin E in prostate cancer (Chan *et al.*, 1999; Heart Protection Study Collaborative Group, 2002; Lonn *et al.*, 2005; Kirsh *et al.*, 2006). Furthermore, accumulating *in vitro* and *in vivo* evidences indicated that α-vitamin E and its derivatives may function through nonantioxidant mechanisms to exhibit their antitumor roles in prostate cancer. α-Vitamin E analogues, especially its ester derivative, VES, have been shown to modulate multiple signal transduction pathways to control cell growth, cell cycle, and apoptosis in prostate cancer cells. Therefore, VES may have strong chemopreventive as well as chemotherapeutic effects on prostate cancer. Here, we focus on reviewing the past and most recent studies of α-vitamin E and VES in prostate cancer.

II. Family Members, Source, and Proper Supplemental Dose of Vitamin E

Vitamin E refers to a family of tocopherols and tocotrienols. Each subfamily is composed of α-, β-, γ-, and δ-isoforms. Vitamin E cannot be synthesized by humans and must be obtained from the diet with an abundant source found in vegetable oil, nuts, and egg yolks. The dominant form of vitamin E in the diet is γ-tocopherol (γ-vitamin E). For example, high levels of γ-vitamin E are found in soybean oil (10.8 mg/1 tablespoon) and corn oil (8.2 mg/1 tablespoon), while high levels of α-tocopherol (α-vitamin E) are found in wheat germ oil (20.3 mg/1 tablespoon), sunflower oil (5.6 mg/1 tablespoon), and almonds (7.3 mg/1 ounce) (United States Department of Agriculture (USDA) food composition database).

Although γ-vitamin E is the predominant isoform of vitamin E from food sources, α-vitamin E is the major biological isoform in the human body. In human serum, ~90% of vitamin E is α-vitamin E and ~10% is γ-vitamin E (Brigelius-Flohe and Traber, 1999). The concentration of α-vitamin E in adult human serum is ~19–29 μM (Ford and Sowell, 1999).

Vitamin E, including α-vitamin E, from the natural diet is only in *RRR* configuration, whereas α-vitamin E in most supplements is in synthetic form, containing eight racemic forms. The biological activity of "natural" α-vitamin E is around twofold as high as "synthetic" α-vitamin E. This notion is concluded from the results of classical rat fetal gestation–resorption assay, which determines the ability of different isoforms or analogues of vitamin E to maintain live fetuses in pregnant rats (Leth and Sondergaard, 1977). Vitamin E is easily oxidized and loses its antioxidant activity. To increase the stability of α-vitamin E for supplements, a number of ester analogues have been synthesized, including α-vitamin E acetate (VEA), α-vitamin E succinate (VES), and α-vitamin E nicotinate (tocopheryl nicotinate, TN).

These vitamin E analogues cannot be oxidized because the hydroxyl group, which contributes to the redox activity, has been protected with an ester bond (Prasad and Edwards-Prasad, 1992). Yet, when they go through the gastrointestinal tract, they can be converted to α-vitamin E by esterase *in vivo*. To date, synthetic VEA has been used in several clinical trials, including the current Selenium and Vitamin E Cancer Prevention Trial (SELECT) for prostate cancer in the United States (Klein *et al.*, 2001; Salonen *et al.*, 2000; The Alpha-Tocopherol, Beta Carotene Cancer Prevention Study Group, 1994; Yusuf *et al.*, 2000).

In 2000, the US Recommended Dietary Allowances (RDA) for vitamin E was set at 15-mg α-tocopherol equivalent/day for adult. On the basis of the rat fetal gestation–resorption assay, 1-IU vitamin E is defined as 1 mg (= 2.13 μmole) of synthetic all-*rac*-α-tocopheryl acetate, which is equivalent to 0.45-mg (= 1.05 μmole) *RRR*-α-tocopherol (Panel on Dietary Antioxidants and Related Compounds, 2000). It is reported that 200–400 IU/day of α-vitamin E intake is nontoxic for humans, and this dose is commonly found in the α-vitamin E supplements on the market. In the case of vitamin E deficiency or some antioxidant deficiency, one can consume α-vitamin E at 800–1000 IU/day or even higher (Hathcock *et al.*, 2005). Gabsi *et al.* reported that α-vitamin E supplementation (800 mg/day) in AVED patients can stabilize the neurological signs and lead to mild improvement of cerebellar ataxia, especially in early stages of the disease (Gabsi *et al.*, 2001). Unexpectedly, the results from meta-analysis of 19 clinical trials from 1966 to 2004 suggested that high doses of α-vitamin E (more than 400 IU/day) will cause unexpected death (Miller *et al.*, 2005). However, this conclusion has been questioned by many researchers, particularly about the analysis methods and unhealthy participants in the clinical trials (Blatt and Pryor, 2005; DeZee *et al.*, 2005; Hemila, 2005; Krishnan *et al.*, 2005; Lim *et al.*, 2005; Marras *et al.*, 2005; Meydani *et al.*, 2005). Overall it is generally agreed that proper daily dose of vitamin E is still beneficial to individual's health.

III. GENERAL PHYSIOLOGICAL FUNCTION OF VITAMIN E

Vitamin E in human appears to act as an antioxidant to reduce free radicals. It is the best fat-soluble antioxidant *in vivo* to intercept free radicals and prevent lipid destruction. It protects fatty acids, vitamin A, and nucleic acids from peroxidants and maintains cell membrane integrity (Dieber-Rotheneder *et al.*, 1991; Esterbauer *et al.*, 1992; Herrera and Barbas, 2001). However, vitamin E also exhibits nonantioxidant functions via modulating signaling pathways and gene expression. α-Vitamin E is known to down-regulate PKC and NF-κB signaling (Azzi *et al.*, 2002; Chatelain *et al.*, 1993; Morante *et al.*, 2005; Tasinato *et al.*, 1995) and regulate various gene expressions such as TTP, cytochrome P450-3A (CYP3A), CD36, scavenger receptor class B type I (SR-BI), and collagen α1 (Chojkier *et al.*, 1998; Plymate *et al.*, 2003; Przyklenk and Whittaker, 2000; Ricciarelli *et al.*, 2000; Shaw and Huang, 1998; Teupser *et al.*, 1999; for review, see reference Azzi *et al.*, 2004). Several distinct biological functions of different vitamin E isoforms may be due to their unique nonantioxidant characters (Chatelain *et al.*, 1993; Tasinato *et al.*, 1995).

Among those α-vitamin E ester analogues that are available on the market, VES is the most effective analogue in terms of antitumor activity (Neuzil *et al.*, 2001; Weber *et al.*, 2002; Zhang *et al.*, 2002, 2004; Zu and Ip, 2003). *In vivo* and *in vitro* studies have shown that VES inhibits growth and induces apoptosis only in carcinoma cells or transformed cells, but not in normal cells (Donapaty *et al.*, 2006; Neuzil *et al.*, 2001; Weber *et al.*, 2002; Zhang *et al.*, 2002). Further mechanistic studies indicated that VES induces apoptosis through targeting multiple molecules/signaling pathways, including transforming growth factor-β (TGF-β), Fas (CD95/APO-1), the c-Jun N-terminal kinase (JNK), mitogen-activated protein kinase (MAPK), and Bcl-2 family in various types of carcinoma cells (Shiau *et al.*, 2006; Yu *et al.*, 1997, 1998, 1999; Zhang *et al.*, 2002; Zu *et al.*, 2005). In addition to pro-apoptotic function, the antitumor activity of VES occurs through blocking cell cycle progression (Ni *et al.*, 2003; Turley *et al.*, 1997; Venkateswaran *et al.*, 2002), inducing differentiation (You *et al.*, 2001, 2002), inhibiting invasion (Zhang *et al.*, 2004), and suppressing angiogenesis (Malafa and Neitzel, 2000; Malafa *et al.*, 2002; Schindler and Mentlein, 2006) *in vitro* and/or *in vivo*. Those studies suggested that VES might be applied as a therapeutic agent for cancer treatment.

IV. VITAMIN E ABSORPTION AND TRANSPORT

All vitamin E isoforms from the diet have an equal uptake efficacy in the intestine, and then complex with lipoproteins to form chylomicrons, which are transported to the liver through the lymphatic system. However, in the liver, only α-vitamin E is released from the hepatocytes into circulation and then supplied to the peripheral tissues. In contrast, other isoforms of vitamin E cannot be released from the liver and are metabolized and excreted into the bile and urine (Brigelius-Flohe and Traber, 1999). Therefore, γ-vitamin E is the most dominant form of vitamin E in the food sources, however, 90% of the vitamin E in human serum and tissues is the α-vitamin E (Brigelius-Flohe and Traber, 1999). This selective effect in liver is mainly due to the presence of α-tocopherol transfer protein (TTP) which has a preferentially binding ability to α-vitamin E (Hosomi *et al.*, 1997; Panagabko *et al.*, 2003).

To date, intracellular transport of vitamin E has not been clearly understood. Manor group proposed that in the liver all of vitamin E isoforms are first taken up via endocytosis by SB-RI, then are transported into vesicle organelles, accumulating in lysosomes. In the lysosomes, the lipoproteins are degraded and vitamin E is exposed to TTP. TTP selectively binds and facilitates α-vitamin E release from the hepatocyte via transport vesicles with the assistance of ATP-binding cassette transporter A1 (ABCA1) (Qian *et al.*, 2005). This hypothesis is based on the observation that treatment with antibodies against SR-BI could significantly reduce cellular α-vitamin E amounts, and ABC transporter inhibitor, glyburide, could abrogate TTP-dependent secretion of vitamin E. In addition, they demonstrated that induction of TTP expression increases the rate of α-vitamin E secretion to the media and TTP is located in the lysosomes of hepatocytes (Qian *et al.*, 2005). However, in another study, Miyazono *et al.* showed that TTP generally locates in the cytosol. This notion is concluded by the observation that TTP can translocate to lysosome/endosome after treatment with chloroquine and relocate into the cytosol after washout of chloroquine. Therefore, they proposed that α-vitamin E might bind to TTP in the cytosol instead of the lysosome (Horiguchi *et al.*, 2003). Nevertheless, TTP plays a critical role in the binding, transporting, and secreting of α-vitamin E in hepatocytes.

Furthermore, we found that tocopherol-associated protein (TAP) can facilitate vitamin E uptake in prostate cancer cells (Ni *et al.*, 2005). In addition, TTP, SR-BI, and ABCA1 express differentially in various prostate cancer cell lines, presumably to control the cellular bioavailability of vitamin E in prostate/prostate cancer cells as well (Ni *et al.*, 2007). Therefore, the transport of α-vitamin E may utilize both vitamin E uptake genes (e.g., TAP and SR-BI) and efflux genes (e.g., TTP and ABCA1) in prostate/prostate cancer cells. The roles of those vitamin E binding proteins will be further discussed in the following section.

V. α-VITAMIN E-BINDING PROTEINS

α-Vitamin E is a fat-soluble nutrient; however, its *in vivo* transport into cells is not only a passive process. Accumulating evidence suggested that α-vitamin E transport needs the help of vitamin E-binding proteins and other lipid transfer proteins (Ouahchi *et al.*, 1995; Traber and Sies, 1996). In addition to TTP, there are several other vitamin E-binding and transport proteins, including TAP and α-tocopherol-binding protein (TBP). As we found that differential expression of different vitamin E transport protein may affect cellular bioavailability of vitamin E and its consequent antiproliferative activity, the following sections will focus on discussing the function of those vitamin E-binding/transporter proteins toward vitamin E absorption and transportation.

A. α-TOCOPHEROL TRANSFER PROTEIN

TTP locates at chromosome 8q13, encoding 278 amino acids. TTP has a high expression in the liver and low expression in the brain, intestine, and other organs (Hosomi *et al.*, 1998). TTP contains a CRAL-TRIO (cellular retinal and TRIO guanine exchange factor) domain, which specifically binds small lipids. TTP has at least threefold higher binding affinity for α-vitamin E than others through this CRAL-TRIO domain (Hosomi *et al.*, 1997; Panagabko *et al.*, 2003). The dissociation constant of α-vitamin E to TTP is around 25 nM. Therefore, TTP can specifically bind α-vitamin E and transports it out of the liver cells into circulation (Arita *et al.*, 1997). Functional loss of TTP caused by TTP point mutations in human can result in the extremely low amount of α-vitamin E in serum, and eventual development of the neurological disorder, AVED (Ouahchi *et al.*, 1995; Yokota *et al.*, 2000). Consistently, TTP knockout mice have reduced α-vitamin E levels in serum and peripheral tissues, and develop AVED-like syndromes (Leonard *et al.*, 2002; Yokota *et al.*, 2001).

TTP also expresses in the prostate, although not as high as in the liver. Our recent study confirmed and extended the previous reports that TTP has a more preferentially affinity for α-vitamin E than γ-vitamin E or α-vitamin E analogue, VES. This notion is supported with our finding that TTP expression levels are negatively correlated with cellular α-vitamin E amount in prostate cancer cells. Overexpression or knockdown of TTP inversely regulates the cellular levels of α-vitamin E, but not γ-vitamin E or VES, in prostate cancer cells (Ni, the Prostate in press).

B. α-TOCOPHEROL-ASSOCIATED PROTEIN

TAP, also named as Sec14-like 2 (Sec14l2) or supernatant protein factor (SPF), is located at chromosome 22q12, containing 12 exons and encoding 403 amino acids. TAP has a lower expression in numerous tissues with a higher expression in the liver, brain, and prostate (Zimmer *et al.*, 2000). TAP also contains a CRAL-TRIO domain (Stocker *et al.*, 2002). However, unlike TTP, which preferentially binds α-vitamin E, TAP has a broad affinity for small lipids, including α-vitamin E, γ-vitamin E, and phosphatidylinositol (Kempna *et al.*, 2003). This notion is further supported by our observation that higher TAP expression can facilitate α-, γ-, δ-vitamin E and VES accumulation, and promote their antiproliferative activity in prostate cancer cells (Ni *et al.*, 2005). The dissociation constant of α-vitamin E to TAP is ~0.46 μM (Zimmer *et al.*, 2000), lower than physiological α-vitamin E concentrations (19–29 μM) in human serum (Ford and Sowell, 1999), suggesting that TAP can bind α-vitamin E and mediate vitamin E function *in vivo*. TAP is also involved in cholesterol synthesis, stimulating the activity of squalene monooxygenase and oxidosqualene cyclase (Shibata *et al.*, 2001). However, results from the *in vitro* binding assay indicated that TAP has a weak affinity for many compounds that are involved in cholesterol synthesis (Panagabko *et al.*, 2003). TAP can be cocrystallized with *RRR*-α-tocopheryl quinone (α-TQ), the oxidation product of α-vitamin E *in vivo*. Accordingly, Stocker *et al.* proposed that TAP/(SPF)'s role in cholesterol synthesis is an indirect effect: TAP/(SPF) mediates the transfer of α-TQ to low-density lipoprotein (LDL), while α-TQ can be oxidized to α-TQH_2. Thus α-TQ/α-TQH_2 protects the LDL from oxidation, resulting in reducing cholesterol uptake, eventually inducing cholesterol synthesis (Stocker, 2004; Stocker and Baumann, 2003).

Interestingly, in addition to its function to control cellular bioavailability of vitamin E, TAP can function like a tumor suppressor to control the growth of prostate cancer cells independent of its vitamin E-binding property (Ni *et al.*, 2005).

C. α-TOCOPHEROL-BINDING PROTEIN

TBP, a 14.2-kDa cytosolic protein, was initially isolated from rat liver (Dutta-Roy *et al.*, 1993a). It is reported that TBP expresses in liver, heart, and human placenta (Dutta-Roy *et al.*, 1993a,b), and prefers to bind α-vitamin E *in vitro*. There has been no naturally occurring TBP mutations identified in human. Overall, the function of TBP remains to be elucidated.

D. OTHER VITAMIN E TRANSPORT PROTEINS

In addition to TTP, TAP, and TBP, many lipid transfer proteins, including SR-BI and ABCA1, also have a loose binding affinity for vitamin E.

SR-BI is the high-density lipoprotein (HDL) receptor, which mainly mediates cholestryl ester uptake and can facilitate vitamin E uptake into the liver, brain, and intestine (Goti *et al.*, 2001; Reboul *et al.*, 2006). SR-BI knockout mice have high levels of α-vitamin E in serum with significantly low vitamin E content in several organs, including ovary, testis, lung, and brain (Mardones *et al.*, 2002).

ABCA1 was reported to mediate cellular secretion of α-vitamin E out of fibroblasts and macrophages (Oram *et al.*, 2001). In addition, the secretion of α-vitamin E from the liver into serum is TTP- and ABCA1-dependent (Qian *et al.*, 2005). This is supported by the observation that the glyburide, the inhibitor of ATP-binding cassette transporter, completely abolished

TTP- mediated vitamin E release from hepatocytes (Qian *et al.*, 2005). Consistently, ABCA1$^{-/-}$ mice have undetectable levels of vitamin E in plasma (Orso *et al.*, 2000).

Our recent studies indicated that TTP, TAP, SR-BI, and ABCA1 can express in prostate cancer cells, suggesting that they may be involved in vitamin E transport in normal prostate and prostate cancer.

VI. Vitamin E and Diseases

Vitamin E is essential for maintaining numerous functions in mammals. In general, vitamin E (α-vitamin E) in adult human serum is in the range of 19–29 μM (Ford and Sowell, 1999). Low levels or deficiency of vitamin E will cause or increase the risk for various diseases, including infertility, neurological disorders, inflammation, cardiovascular diseases, diabetes, and certain types of cancers in humans (Traber and Sies, 1996; Tucker and Townsend, 2005). In humans, genetic abnormality generally underlies vitamin E deficiency. It has been reported that mutation of TTP will cause AVED, a neurodegeneration disorder (Ouahchi *et al.*, 1995). Vitamin E deficiency also happens as a consequence of fat malabsorption syndromes (Traber and Sies, 1996). For example, the patients with abetalipoproteinemia, which is due to genetic absence of apolipoprotein B (apoB) and/or apoB-containing lipidproteins, have an undetectable amount of vitamin E in their plasma (Gregg and Wetterau, 1994; Kayden, 1972).

VII. α-Vitamin E Function in Prostate Cancer: Clinical Studies

To address the function of vitamin E in prostate cancer, we first address the outcome of several clinical studies (Table I). In 1994, a research group in Finland reported that daily 50-mg α-vitamin E acetate supplementation decreased prostate cancer incidence by 32%, and mortality by 42% in smoking men in their large clinical trial of ATBC study (Heinonen *et al.*, 1998; The Alpha-Tocopherol, Beta Carotene Cancer Prevention Study Group, 1994). Prostate cancer is the only cancer that can be suppressed by α-vitamin E supported by this clinical trial. In the postintervention follow-up study, they found that the benefits of α-vitamin E on prostate cancer prevention rapidly disappeared after discontinuation of the supplement (Virtamo *et al.*, 2003). Moreover, their results show that low levels of α-vitamin E in serum increased prostate cancer mortality (Eichholzer *et al.*, 1996). In agreement with this notion, results from the β-carotene and retinol efficacy trial showed that the lower levels of α-vitamin E in serum are significantly associated with higher incidence of prostate cancer (Goodman *et al.*, 2003). Consistently, studies

TABLE I. Summary of Clinical Trials for the Effect of α-Vitamin E Supplementation in PCa

Trial (study)	Subjects' description	Dosage and type of vitamin E	Follow-up	Risk ratio (95% CI)	Clinical outcome	References
ATBC trial, $n = 29133$	Male smokers, 50–69 years	50-mg/day *rac*-α-vitamin E acetate ($\pm$VEA)	5–8 years	0.66 (0.52–0.86)	α-Vitamin E supplementation reduced the incidence and mortality of PCa in smoking men.	Study Group (1994), Albanes *et al.*, 1995
					α-Vtiamin E supplementation was associated with high level of α-vitamin E and γ-vitamin E in serum and low risk of PCa.	Weinstein *et al.*, 2005
					α-Vitamin E supplementation is associated with low level of serum hormone (androgen) in old men and a reduced risk of PCa.	Hartman *et al.*, 2001
Postrial, $n = 25563$			6 years	0.88 (0.76–1.03)	α-Vitamin E-preventive role on PCa is transient, diminishes fairly rapidly following cessation of supplementation.	Virtamo *et al.*, 2003

HPFS, $n = 47780$	Healthy men, 40–75 years	0.1–15.0 IU/day	9 years	Never smokers: 0.91, current smokers/quit with past 10 years: 0.78, quit >10 year age: 2.43	α-Vitamin E may not prevent prostate cancer. However, α-vitamin E supplementation is inversely associated with the risk of metastatic or fatal prostate cancer in smoking men.	Chan *et al.*, 1999
		15.1–99.9 IU/day		Never smokers: 1.29, current smokers/quit with past 10 years: 0.51 (0.21–1.23)?, quit >10 year age: 1.33		
		≥100 IU/day		Never smokers: 1.42, current smokers/quit with past 10 years: 0.44 (0.18–1.07)?, quit >10 year age: 1.49 (metastic/fatal cases)		
PLCO trial, $n = 29361$	55–74 years	>0–30 IU/day	8 years	Never smokers: 1.34, current smokers/quit with past 10 years: 0.67, quit >10 year age: 0.63	α-Vitamin E supplementation did no prevent prostate cancer. However, it reduced the risk of prostate cancer in smoking men.	Kirsh *et al.*, 2006
		>30–400 IU/day		Never smokers: 1.16, current smokers/quit with past 10 years: 0.72, quit >10 year age: 1.03		

(*Continues*)

TABLE I. (*Continued*)

Trial (study)	Subjects' description	Dosage and type of vitamin E	Follow-up	Risk ratio (95% CI)	Clinical outcome	References
		>400 IU/day		Never smokers: 1.29, current smokers/quit with past 10 years: 0.29, quit > 10 y age: 0.95 (all advanced cases)		
Heart Protection Study, $n = 15434$	Known CDV or diabetes, 40–80 years	600 mg/day synthetic vitamin E	5 years		α-Vitamin E supplementation reduced 9% nonsignificant incidence of PCa and may not prevent PCa.	Heart Protection Study Collaborative Group, 2002
HOPE trial, $n = 9541$	Known CVD or diabetes, mean age 66 years	400-IU/day *RRR*-α-tocopheryl acetate (+VEA)	4.5 years	0.98 (0.76–1.26)		
Extension (HOPE-TOO trial), $n = 7030$			4.5 + 2.6 years	0.90 (0.68–1.19)	Long-term supplementation of α-vitamin E may not prevent prostate cancer in patients with CVD or diabetes.	Lonn *et al.*, 2005

from Switzerland on 17-year follow-up of the prospective basal study showed that the levels of α-vitamin E, but not vitamin C, retinol, or carotene, were significantly low in prostate cancer patients (Eichholzer *et al.*, 1999). Furthermore, during the follow-up in the ATBC study, they also found that long-term α-vitamin E supplementation decreases serum androgen (testosterone) concentrations from 573 to 539 ng/dl, suggesting that it could be one of the factors associated to the low incidence and mortality of prostate cancer in ATBC studies (Hartman *et al.*, 2001). However, whether the slightly lower concentration of androgen is a contributory factor to lower the incidence and mortality remains to be elucidated. It is worth mentioning that androgen ablation therapy did not really lower the mortality of prostate cancer.

However, recent Heart Outcomes Prevention Evaluation-The Ongoing Outcome (HOPE-TOO), Prostate, Lung, Colorectal, and Ovarian cancer-screening trial (PLCO), and Health Professionals Follow-up Study (HPFS) observed no effects of either dietary or supplemental α-vitamin E on prostate cancer risk. But in the subanalyses of PLCO and HPLS trial, supplementation of α-vitamin E is associated with the low incidence of prostate cancer either at advance stage or at metastic/fatal stage in smoking men and men who recently quit smoking (Chan *et al.*, 1999; Kirsh *et al.*, 2006; Lonn *et al.*, 2005), which are consistent with the results from the ATBC trial. Taken together, those clinical trials suggested that supplement of α-vitamin E might have a beneficial effect against the prostate cancer especially in smoking men. The plausible explanation is that α-vitamin E may prevent prostate cancer through its antioxidant activity against oxidative stress stimulated by smoking. However, such a hypothesis cannot explain why α-vitamin E has specific preventive activity against prostate cancer, but not other cancers in smoking men, suggesting that α-vitamin E might regulate some specific signaling in prostate cancer.

To further elucidate whether α-vitamin E and another nutrient, selenium, alone or in combination, can prevent prostate cancer, in 2001, the National Cancer Institute (NCI) began a large clinical study, SELECT. More than 32,000 men have or will participate in this trial throughout its 14 years time frame (Klein *et al.*, 2001). The trial does not use smoking status as criteria. Another large clinical trial, the Physician's Health Study II, is also ongoing, expected to be finished in 2007 (Christen *et al.*, 2000). Both trials will provide further insights into the role of α-vitamin E on prostate cancer.

In the meantime, these clinical trials may have limitations to reveal the real roles of α-vitamin E on prostate cancer. New clinical trials should take special consideration in (1) what is the dose, (2) which racemic form of α-vitamin E (natural α-vitamin E versus synthetic α-vitamin E) is used, and (3) the targeted populations. More importantly, new clinical trials should go beyond the usage of α-vitamin E. For example, in cultured prostate cancer cells, VES strongly suppresses cell growth, in contrast, α-vitamin E or VEA, which is currently major form used in clinical trials, only slightly inhibits cancer cell growth.

New α-vitamin E/VES analogues that can suppress prostate cancer cell growth *in vitro* and animal models without toxicity might be considered for inclusion in the new clinical trials in the future.

VIII. α-VITAMIN E FUNCTION IN PROSTATE CANCER: ANIMAL STUDIES

Daily oral supplementation of α-vitamin E and its analogues have been applied in preclinical animal studies. For example, Nakamura *et al.* (1991) reported that a diet containing an antioxidant mix, including 0.8% catechol, 0.8% resorcinol, 0.8% hydroquinone, 2-ppm selenium, 2% γ-orysanol, and 1% α-vitamin E, could reduce the incidence and lesions of DMBA-initiated rat prostate carcinogenesis. Venkateswaran *et al.* showed that administration of a mix of three compounds (VES, selenium, and lycopene) in the diet dramatically inhibits prostate cancer development in Lady transgenic mice, which spontaneously develop prostate tumor (Venkateswaran *et al.*, 2004a). However, these two studies did not show whether oral intake of α-vitamin E alone or VES alone has any antitumor activity *in vivo*. Another report showed that daily oral lycopene combined with α-vitamin E (5 mg/kg body weight each) suppressed the orthotopic growth of PC-346C human prostate cancer in nude mice (Limpens *et al.*, 2006), but α-vitamin E alone (5 or 50 mg/kg daily) did not have any impact on the tumor burden in this study. Collectively, those studies indicated that α-vitamin E can be combined with other nutrients (antioxidants) to prevent prostate cancer.

Considering VES is one of the most efficient α-vitamin E analogues in terms of antitumor activity *in vitro*, and has been studied *in vivo* by intraperitoneal (ip) administration in breast cancer and colon cancer studies (Weber *et al.*, 2002), we have investigated whether VES by ip administration has any effect in prostate cancer growth *in vivo*. The transgenic adenocarcinoma of the mouse prostate (TRAMP) model, which mimics the progression of human prostate cancer and has been widely used in chemopreventive and chemotherapeutic studies (Greenberg *et al.*, 1995; Gupta *et al.*, 2001; Huss *et al.*, 2004; Mentor-Marcel *et al.*, 2001), has been applied in our study. Our results showed that ip administration of VES (50 or 100 mg/kg twice per week) could significantly inhibit the tumor burden and metastasis in TRAMP mice. This chemopreventive and therapeutic effects are accompanied by cell proliferation reduction as well as apoptosis induction. In addition, VES can increase insulin-like growth factor-binding protein-3 (IGFBP-3) levels, which are negatively associated with prostate cancer progression (Chan *et al.*, 2002); however, VES does not change IGF-1 levels in TRAMP mice. The extended investigation demonstrated that modulation of IGFBP-3 might represent one of the mechanisms for chemopreventive and chemotherapeutic activity of VES (Yin *et al.*, 2007). Importantly, VES by ip administration of 50 and 100 mg/kg does not show chronic toxicity in mice. This conclusion is obtained from the observation that the body weight, general appearance, and histopathologic examination of liver, kidney, intestine, spinal cord, prostate, testis, and jejunum did not significantly differ between the VES-treated and vehicle control-treated mice. Consistent with the antitumor effect of VES in TRAMP mice, our results indicate that ip administration of VES (100 mg/kg twice per week) could reduce the xenograft tumor growth in nude mice (Yin *et al.*, 2007). Together, VES or analogues with similar functional structures could be a potential chemopreventive and chemotherapeuitc agent for prostate cancer.

IX. α-Vitamin E in Prostate Cancer: Molecular Mechanism Studies in Cancer Cells

Consistent with epidemiology and animal studies, *in vitro* cell culture system studies found that α-vitamin E or its derivative, VES, could inhibit the growth of prostate cancer cells, but not normal prostate cells (Israel *et al.*, 2000; Zhang *et al.*, 2002). Several mechanisms have been proposed by different research groups to account for α-vitamin E/VES antiprostate tumor activity.

A. Cellular Bioavailability of α-Vitamin E and VES

The cellular bioavailability of drugs is an important contributor for the drug efficacy. We have examined α-vitamin E and VES concentrations after treatment in different prostate cancer cells and found that cells' retention ability of α-vitamin E or VES is associated with different transport gene expression and growth inhibition efficacy in prostate cancer cells. For example, prostate cancer DU145 cells are less sensitive to 20-μM α-vitamin E or VES treatment. Consistently, the cellular amount of α-vitamin E and VES in DU145 cells is much lower compared to that in other prostate cancer cells after α-vitamin E/VES treatment (Ni *et al.*, 2007). Therefore, the cell's uptake efficacy might be one of the factors contributing to the growth inhibitory efficacy of α-vitamin E and VES.

B. Cell Cycle Arrest and DNA Synthesis Arrest

It was reported that VES inhibited DNA synthesis in prostate cancer LNCaP, PC-3, and DU145 cells (Israel *et al.*, 1995). In agreement with this notion, we and others showed that VES could block cell cycle progression in prostate cancer cells (Ni *et al.*, 2003; Venkateswaran *et al.*, 2002).

Venkateswaran *et al.* showed that VES could reduce the cells in S phase in both LNCaP and PC-3 cells. This S phase reduction was accompanied by G1/S phase arrest in LNCaP cells and G2/M phase arrest in PC-3 cells. Additional analysis revealed that VES significantly upregulates p27 expression, leading to the inactivation of cyclin E/cdk2, which contributes to the G1 arrest in prostate cancer cells (Venkateswaran *et al.*, 2002). Consistently, we also found that VES blocked the cell cycle at G1/S phase, and extensive examinations revealed that VES could regulate the expression of cyclin D1, D3, cdk2, cdk4, but not cdk6, resulting in disrupting Rb-E2F pathway in prostate cancer LNCaP cells. Interestingly, our results indicated that reduced cyclin D and cdk4 expression are earlier responsiveness for VES treatment (Ni *et al.*, 2003). Together, VES could alter cell cycle progression in prostate cancer cells.

C. APOPTOSIS

In 2000, Gunawardena *et al.* showed that 3 days of racemic α-vitamin E 5-μg/ml treatment can stimulate apoptosis in actively dividing prostate cancer LNCaP cells, but not in confluent quiescent cells. They assessed nucleosome fragmentation by cell death detection ELISA as apoptosis index (Gunawardena *et al.*, 2000). In addition to α-vitamin E, VES has been widely characterized as an apoptosis induction agent through different pathways involvement. In generally, the apoptosis can be induced through mitochondria-mediated, lysosomes-mediated, endoplasmic reticulum stress/ cytokine signaling pathway-mediated, and extracellular death pathway-mediated manners. Zhu *et al.* proposed that VES induces prostate cancer cells apoptosis mainly through the mitochondrial pathway. This conclusion is based on the changed profiles of caspases family after VES treatment in PC-3 cells. They showed that all three executioner caspases, caspase-3, -6, -2, and the initiator caspases, caspase-8, -9, -10, but not caspase-1 or -12, were activated by VES in prostate cancer PC-3 cells (Zu and Ip, 2003). The extended study by Malafa *et al.* (2006) suggested that caspase-4 mRNA expression was also induced after VES treatment, and caspase-4 inhibitor rescued VES-induced apoptosis in PC-3 cells. Furthermore, by the microarray, Zu *et al.* identified other molecules modulated by VES. They revealed that Ask1, GADD45β, and Sek1, three key components of stress-activated mitogen-activated protein kinase MAPK pathway, are new targets of VES. After further analysis, they proposed that upregulated phosphorylation of Sek1 could be responsible for the JNK activation, and the activated JNK then phosphorylates its target genes, Bcl-2 and Bim, resulting in the mitochondrial translocation of Bax and Bim, consequently, inducing mitochondrial-mediated apoptosis (Zu *et al.*, 2005). In agreement with their finding, Shiau *et al.* reported that VES induced apoptosis in prostate cancer cells partly through inhibiting the function of Bcl-2 family. The mechanistic studies revealed that the interaction between Bcl-xL and Bcl-2 was blocked by VES. Overexpression of Bcl-xL significantly rescued VES-induced apoptosis in prostate cancer LNCaP cells. Knockdown of Bcl-xL could sensitize PC-3 cells to VES-induced apoptosis. Interestingly, they used the computer model to analyze the structure of VES and Bcl-xl and found that VES could bind to Bcl-xL. Accordingly, they developed new VES derivatives with truncated side chain aiming to stabilize the binding and found that the new compounds have more proapoptosis activity compared to VES (Shiau *et al.*, 2006). Those interesting findings might need the preclinical animal model to further validate its efficacy *in vivo*.

VES could also regulate the Fas pathway in prostate cancer cells. Studies from Israel *et al.* (Israel *et al.*, 2000) showed that VES treatment triggered Fas translocation from cytosol to membrane as well as induced Fas ligand expression in prostate cancer LNCaP and PC-3 cells. In contrast, VES has no impact on normal prostate epithelial cells.

D. SIGNAL PATHWAY

Androgens and androgen receptor (AR) play important roles in the initiation and progression of prostate cancer. VES has been reported to inhibit the expression of AR and PSA in prostate cancer LNCaP cells. Interestingly, VES only reduced the expression of AR, but not retinol X receptor α (RXRα), peroxisome proliferator-activated receptor α (PPARα), or other steroid hormone receptors, suggesting this downregulation of AR/PSA signaling is specifically regulated by VES. Further characterizations showed that VES could regulate AR at both trancriptional and translational levels. In addition to downregulation of AR, VES upregulated VDR expression (Zhang *et al.*, 2002), suggesting that VES treatment could sensitize prostate cancer cells to low dose of vitamin D-induced antitumor activity. Currently, vitamin D has been applied for phase II clinical trial to treat prostate cancer (Trump *et al.*, 2006). The advantage of the combination treatment of VES and vitamin D is to reduce the side effects, including hypercalciumia, caused by high doses of vitamin D (Yin and Yeh, unpublished data).

The high expression levels of IGFBP-3 are inversely associated with prostate cancer progression (Chan *et al.*, 2002). Targeting on IGFBP-3 has been used as a strategy for prostate cancer therapy (Liu *et al.*, 2005). We also demonstrated that IGFBP-3 is another target gene of VES in prostate cancer cells. IGFBP-3 mRNA and protein levels can be repressed by VES treatment. Importantly, the downregulation of IGFBP-3 mRNA level is the direct VES effect since the protein inhibitor, cycloheximide, did not rescue VES-mediated reduction of IGFBP-3 mRNA levels. Strikingly, this reduction can be confirmed in preclinical animal models TRAMP and xenograft tumor mouse model (Yin *et al.*, 2007).

E. INVASION, METASTASIS, AND ANGIOGENESIS

The primary tumor needs to degrade the extracellular matrix in order to metastasize to distal site. This step involves many matrix metaloproteinases and other proteases. The antiprostate tumor activity of VES might also come from its ability to inhibit the prostate cancer invasion and metastasis. Studies from our laboratory showed that VES could inhibit the invasion of prostate cancer cells. This effect is associated with the reduction of the activity of MMP-9, but not MMP-2, or tissue inhibitors of MMP (TIMPs) (Zhang *et al.*, 2004). Further characterization showed that VES could also inhibit cathepsins B and D activity in prostate cancer PC-3 cells (JN, unpublished data). However, how VES regulates the activities of those molecules is not well known. Angiogenesis is essential for the growth of many primary tumors and their subsequent metastases by providing oxygen and removing waste and is a target of some prostate cancer therapies. Currently, there is no report showing whether VES has antiangiogenesis effects in prostate cancer, but our *in vivo* animal studies indicate that VES might exhibit antiangiogenesis activity in prostate cancer (Yin *et al.*, 2007). Further detailed investigation of these directions will help us get insights into functional mechanisms of VES and its analogue in prostate cancer.

X. SUMMARY AND PERSPECTIVES

In this chapter, we summarized the current knowledge of α-vitamin E and its derivative, VES, in prostate cancer. VES could modulate multiple pathways to inhibit prostate cancer cell growth. However, detailed mechanisms by which VES regulates those pathways remain largely unknown. Unlike vitamins A and D, no vitamin E receptor(s) have been identified. Yet, it was reported that TAP is an α-vitamin E-dependent transcriptional activator (Yamauchi *et al.*, 2001). Furthermore, our data of TAP IHC staining in clinical human prostate sample showed that TAP could locate in the nucleus (Ni *et al.*, 2005). It is possible that α-vitamin E binds to TAP to turn on vitamin E-regulated genes expression. Therefore, to characterize whether TAP is a vitamin E-dependent transcriptional factor, or to identify α-vitamin E receptor, if it exists, is an interesting direction. Alternatively, identification and characterization of α-vitamin E-binding proteins will help obtain insights of α-vitamin E's physiological roles.

It is also of great interest to systematically identify α-vitamin E/VES target genes, such as via microarray methods. This could help us better understand the molecular basis of α-vitamin E/VES function and to design new strategies to combine VES with other drugs. For example, several studies indicated that combination of VES and selenium may have synergistic or additive antitumor effects (Ni *et al.*, 2003; Venkateswaran *et al.*, 2004b; Zu and Ip, 2003).

Those studies suggested that combination therapy might have a benefit for treating prostate cancer, since they target on different molecules and turn on multiple pathways to inhibit prostate cancer cell growth. Currently, the SELECT trial is ongoing in United States and will be concluded in year 2013. The proposed study here may provide the molecular basis for the clinical SELECT trial.

The bioavailability of α-vitamin E is important for its efficacy and biological effects. The retention ability of α-vitamin E/VES is associated with its growth inhibitory efficacy in prostate cancer cells. In addition, our unpublished data indicated that α-vitamin E/VES could accumulate in mouse prostate after ip administration. However, there is no report ever focusing on the detection of α-vitamin E levels in human prostate tissues. It is important for us to examine whether α-vitamin E could accumulate in human prostate and whether the better efficacy of α-vitamin E to treat prostate cancer is associated with higher α-vitamin E accumulation in human prostate, in addition to detecting serum vitamin E.

On the basis of the structural and functional analyses of VES, several VES analogues have been created (Birringer *et al.*, 2003; Shiau *et al.*, 2006; Tomic-Vatic *et al.*, 2005). Those new analogues have higher proapoptotic efficacy compared to VES *in vitro*. Although those have not yet been shown effectively in preclinical prostate cancer animal models, those new compounds will presumably lose their antitumor activity by oral consumption due to the presence of ester bond in structure as VES dose. It is thus inconvenient for them, as well as for VES, to be applied for chemopreventive purpose. Therefore, it is of great interest to develop new α-vitamin E analogues that could inhibit the tumor burden without overt toxicity through oral intake. If the results are promising, those new analogues might be applied as potential chemopreventive as well as chemotherapeutic agents for prostate cancer in the future.

Recent human studies of α-vitamin E yield conflicting data and suggest several factors should be taken into consideration, including target population and the usage of α-vitamin E such as the dose, the form (natural form or racemic form), and stability of α-vitamin E. The metabolism of α-vitamin E affects the bioavailability of vitamin E and its subsequent efficacy. New α-vitamin E analogues with longer half-life should be considered in the future clinical trials. On the other hand, it has been reported that γ-vitamin E scavenges the reactive nitrogen oxide species better than α-vitamin E (Cooney *et al.*, 1993). Tocotrienols have stronger antioxidant function than α-vitamin E (Suzuki *et al.*, 1993). Therefore, γ-vitamin E and tocotrienol may help maintain the intact structure of α-vitamin E to allow it to exhibit its other functions. On another hand, γ-vitamin E and tocotrienols have distinct nonantioxidant functions other than α-vitamin E. How to combine those different forms of vitamin E to prevent and treat prostate cancer might be another direction for future studies.

REFERENCES

Albanes, D., Heinonen, O. P., Huttunen, J. K., Taylor, P. R., Virtamo, J., Edwards, B. K., Haapakoski, J., Rautalahti, M., Hartman, A. M., Palmgren, J., and Greenwald, P. (1995). Effects of alpha-tocopherol and beta-carotene supplements on cancer incidence in the Alpha-Tocopherol Beta-Carotene Cancer Prevention Study. *Am. J. Clin. Nutr.* **62,** 1427S–1430S.

Arita, M., Nomura, K., Arai, H., and Inoue, K. (1997). Alpha-tocopherol transfer protein stimulates the secretion of alpha-tocopherol from a cultured liver cell line through a brefeldin A-insensitive pathway. *Proc. Natl. Acad. Sci. USA* **94,** 12437–12441.

Azzi, A., Ricciarelli, R., and Zingg, J. M. (2002). Non-antioxidant molecular functions of alpha-tocopherol (vitamin E). *FEBS Lett.* **519,** 8–10.

Azzi, A., Gysin, R., Kempna, P., Munteanu, A., Villacorta, L., Visarius, T., and Zingg, J. M. (2004). Regulation of gene expression by alpha-tocopherol. *Biol. Chem.* **385,** 585–591.

Birringer, M., EyTina, J. H., Salvatore, B. A., and Neuzil, J. (2003). Vitamin E analogues as inducers of apoptosis: Structure-function relation. *Br. J. Cancer* **88,** 1948–1955.

Blatt, D. H., and Pryor, W. A. (2005). High-dosage vitamin E supplementation and all-cause mortality. *Ann. Intern. Med.* **143,** 150–151; author reply 156–158.

Brigelius-Flohe, R., and Traber, M. G. (1999). Vitamin E: Function and metabolism. *FASEB J.* **13,** 1145–1155.

Chan, J. M., Stampfer, M. J., Ma, J., Rimm, E. B., Willett, W. C., and Giovannucci, E. L. (1999). Supplemental vitamin E intake and prostate cancer risk in a large cohort of men in the United States. *Cancer Epidemiol. Biomarkers Prev.* **8,** 893–899.

Chan, J. M., Stampfer, M. J., Ma, J., Gann, P., Gaziano, J. M., Pollak, M., and Giovannucci, E. (2002). Insulin-like growth factor-I (IGF-I) and IGF binding protein-3 as predictors of advanced-stage prostate cancer. *J. Natl. Cancer Inst.* **94,** 1099–1106.

Chatelain, E., Boscoboinik, D. O., Bartoli, G. M., Kagan, V. E., Gey, F. K., Packer, L., and Azzi, A. (1993). Inhibition of smooth muscle cell proliferation and protein kinase C activity by tocopherols and tocotrienols. *Biochim. Biophys. Acta* **1176,** 83–89.

Chojkier, M., Houglum, K., Lee, K. S., and Buck, M. (1998). Long- and short-term D-alpha-tocopherol supplementation inhibits liver collagen alpha1(I) gene expression. *Am. J. Physiol.* **275,** G1480–G1485.

Christen, W. G., Gaziano, J. M., and Hennekens, C. H. (2000). Design of Physicians' Health Study II—a randomized trial of beta-carotene, vitamins E and C, and multivitamins, in prevention of cancer, cardiovascular disease, and eye disease, and review of results of completed trials. *Ann. Epidemiol.* **10,** 125–134.

Cooney, R. V., Franke, A. A., Harwood, P. J., Hatch-Pigott, V., Custer, L. J., and Mordan, L. J. (1993). Gamma-tocopherol detoxification of nitrogen dioxide: Superiority to alpha-tocopherol. *Proc. Natl. Acad. Sci. USA* **90,** 1771–1775.

DeZee, K. J., Shimeall, W., Douglas, K., and Jackson, J. L. (2005). High-dosage vitamin E supplementation and all-cause mortality. *Ann. Intern. Med.* **143,** 153–154; author reply 156–158.

Dieber-Rotheneder, M., Puhl, H., Waeg, G., Striegl, G., and Esterbauer, H. (1991). Effect of oral supplementation with D-alpha-tocopherol on the vitamin E content of human low density lipoproteins and resistance to oxidation. *J. Lipid Res.* **32,** 1325–1332.

Donapaty, S., Louis, S., Horvath, E., Kun, J., Sebti, S. M., and Malafa, M. P. (2006). RRR-alpha-tocopherol succinate down-regulates oncogenic Ras signaling. *Mol. Cancer Ther.* **5,** 309–316.

Dutta-Roy, A. K., Leishman, D. J., Gordon, M. J., Campbell, F. M., and Duthie, G. G. (1993a). Identification of a low molecular mass (14.2 kDa) alpha-tocopherol-binding protein in the cytosol of rat liver and heart. *Biochem. Biophys. Res. Commun.* **196,** 1108–1112.

Dutta-Roy, A. K., Gordon, M. J., Leishman, D. J., Paterson, B. J., Duthie, G. G., and James, W. P. (1993b). Purification and partial characterisation of an alpha-tocopherol-binding protein from rabbit heart cytosol. *Mol. Cell. Biochem.* **123,** 139–144.

Eichholzer, M., Stahelin, H. B., Gey, K. F., Ludin, E., and Bernasconi, F. (1996). Prediction of male cancer mortality by plasma levels of interacting vitamins: 17-Year follow-up of the prospective basel study. *Int. J. Cancer* **66,** 145–150.

Eichholzer, M., Stahelin, H. B., Ludin, E., and Bernasconi, F. (1999). Smoking, plasma vitamins C, E, retinol, and carotene, and fatal prostate cancer: Seventeen-year follow-up of the prospective basel study. *Prostate* **38,** 189–198.

Esterbauer, H., Gebicki, J., Puhl, H., and Jurgens, G. (1992). The role of lipid peroxidation and antioxidants in oxidative modification of LDL. *Free Radic. Biol. Med.* **13,** 341–390.

Evans, H., and Bishop, K. (1922). On the existence of a hitherto unrecognized dietary factor essential for reproduction. *Science* **56,** 650–651.

Ford, E. S., and Sowell, A. (1999). Serum alpha-tocopherol status in the United States population: Findings from the third national health and nutrition examination survey. *Am. J. Epidemiol.* **150,** 290–300.

Gabsi, S., Gouider-Khouja, N., Belal, S., Fki, M., Kefi, M., Turki, I., Ben Hamida, M., Kayden, H., Mebazaa, R., and Hentati, F. (2001). Effect of vitamin E supplementation in patients with ataxia with vitamin E deficiency. *Eur. J. Neurol.* **8,** 477–481.

Goodman, G. E., Schaffer, S., Omenn, G. S., Chen, C., and King, I. (2003). The association between lung and prostate cancer risk, and serum micronutrients: Results and lessons learned from beta-carotene and retinol efficacy trial. *Cancer Epidemiol. Biomarkers Prev.* **12,** 518–526.

Goti, D., Hrzenjak, A., Levak-Frank, S., Frank, S., van der Westhuyzen, D. R., Malle, E., and Sattler, W. (2001). Scavenger receptor class B, type I is expressed in porcine brain capillary endothelial cells and contributes to selective uptake of HDL-associated vitamin E. *J. Neurochem.* **76,** 498–508.

Greenberg, N. M., DeMayo, F., Finegold, M. J., Medina, D., Tilley, W. D., Aspinall, J. O., Cunha, G. R., Donjacour, A. A., Matusik, R. J., and Rosen, J. M. (1995). Prostate cancer in a transgenic mouse. *Proc. Natl. Acad. Sci. USA* **92,** 3439–3443.

Gregg, R. E., and Wetterau, J. R. (1994). The molecular basis of abetalipoproteinemia. *Curr. Opin. Lipidol.* **5,** 81–86.

Gunawardena, K., Murray, D. K., and Meikle, A. W. (2000). Vitamin E and other antioxidants inhibit human prostate cancer cells through apoptosis. *Prostate* **44,** 287–295.

Gupta, S., Hastak, K., Ahmad, N., Lewin, J. S., and Mukhtar, H. (2001). Inhibition of prostate carcinogenesis in TRAMP mice by oral infusion of green tea polyphenols. *Proc. Natl. Acad. Sci. USA* **98,** 10350–10355.

Hartman, T. J., Dorgan, J. F., Woodson, K., Virtamo, J., Tangrea, J. A., Heinonen, O. P., Taylor, P. R., Barrett, M. J., and Albanes, D. (2001). Effects of long-term alpha-tocopherol supplementation on serum hormones in older men. *Prostate* **46,** 33–38.

Hathcock, J. N., Azzi, A., Blumberg, J., Bray, T., Dickinson, A., Frei, B., Jialal, I., Johnston, C. S., Kelly, F. J., Kraemer, K., Packer, L., Parthasarathy, S., *et al.* (2005). Vitamins E and C are safe across a broad range of intakes. *Am. J. Clin. Nutr.* **81,** 736–745.

Heart Protection Study Collaborative Group (2002). MRC/BHF Heart Protection Study of antioxidant vitamin supplementation in 20,536 high-risk individuals: A randomised placebo-controlled trial. *Lancet* **360,** 23–33.

Heinonen, O. P., Albanes, D., Virtamo, J., Taylor, P. R., Huttunen, J. K., Hartman, A. M., Haapakoski, J., Malila, N., Rautalahti, M., Ripatti, S., Maenpaa, H., Teerenhovi, L., *et al.* (1998). Prostate cancer and supplementation with alpha-tocopherol and beta-carotene: Incidence and mortality in a controlled trial. *J. Natl. Cancer Inst.* **90,** 440–446.

Hemila, H. (2005). High-dosage vitamin E supplementation and all-cause mortality. *Ann. Intern. Med.* **143,** 151–152; author reply 156–158.

Herrera, E., and Barbas, C. (2001). Vitamin E: Action, metabolism and perspectives. *J. Physiol. Biochem.* **57,** 43–56.

Horiguchi, M., Arita, M., Kaempf-Rotzoll, D. E., Tsujimoto, M., Inoue, K., and Arai, H. (2003). pH-dependent translocation of alpha-tocopherol transfer protein (alpha-TTP) between hepatic cytosol and late endosomes. *Genes Cells* **8,** 789–800.

Hosomi, A., Arita, M., Sato, Y., Kiyose, C., Ueda, T., Igarashi, O., Arai, H., and Inoue, K. (1997). Affinity for alpha-tocopherol transfer protein as a determinant of the biological activities of vitamin E analogs. *FEBS Lett.* **409,** 105–108.

Hosomi, A., Goto, K., Kondo, H., Iwatsubo, T., Yokota, T., Ogawa, M., Arita, M., Aoki, J., Arai, H., and Inoue, K. (1998). Localization of alpha-tocopherol transfer protein in rat brain. *Neurosci. Lett.* **256,** 159–162.

Huss, W. J., Lai, L., Barrios, R. J., Hirschi, K. K., and Greenberg, N. M. (2004). Retinoic acid slows progression and promotes apoptosis of spontaneous prostate cancer. *Prostate* **61,** 142–152.

Israel, K., Sanders, B. G., and Kline, K. (1995). RRR-alpha-tocopheryl succinate inhibits the proliferation of human prostatic tumor cells with defective cell cycle/differentiation pathways. *Nutr. Cancer* **24,** 161–169.

Israel, K., Yu, W., Sanders, B. G., and Kline, K. (2000). Vitamin E succinate induces apoptosis in human prostate cancer cells: Role for Fas in vitamin E succinate-triggered apoptosis. *Nutr. Cancer* **36,** 90–100.

Kayden, H. J. (1972). Abetalipoproteinemia. *Annu. Rev. Med.* **23,** 285–296.

Kempna, P., Zingg, J. M., Ricciarelli, R., Hierl, M., Saxena, S., and Azzi, A. (2003). Cloning of novel human SEC14p-like proteins: Ligand binding and functional properties. *Free Radic. Biol. Med.* **34,** 1458–1472.

Kirsh, V. A., Hayes, R. B., Mayne, S. T., Chatterjee, N., Subar, A. F., Dixon, L. B., Albanes, D., Andriole, G. L., Urban, D. A., and Peters, U. (2006). Supplemental and dietary vitamin E, beta-carotene, and vitamin C intakes and prostate cancer risk. *J. Natl. Cancer Inst.* **98,** 245–254.

Klein, E. A., Thompson, I. M., Lippman, S. M., Goodman, P. J., Albanes, D., Taylor, P. R., and Coltman, C. (2001). Select: The next prostate cancer prevention trial. Selenum and Vitamin E Cancer Prevention Trial. *J. Urol.* **166,** 1311–1315.

Krishnan, K., Campbell, S., and Stone, W. L. (2005). High-dosage vitamin E supplementation and all-cause mortality. *Ann. Intern. Med.* **143,** 151; author reply 156–158.

Leth, T., and Sondergaard, H. (1977). Biological activity of vitamin E compounds and natural materials by the resorption-gestation test, and chemical determination of the vitamin E activity in foods and feeds. *J. Nutr.* **107,** 2236–2243.

Leonard, S. W., Terasawa, Y., Farese, R. V., Jr., and Traber, M. G. (2002). Incorporation of deuterated RRR- or all-rac-alpha-tocopherol in plasma and tissues of alpha-tocopherol transfer protein–null mice. *Am. J. Clin. Nutr.* **75,** 555–560.

Lim, W. S., Liscic, R., Xiong, C., and Morris, J. C. (2005). High-dosage vitamin E supplementation and all-cause mortality. *Ann. Intern. Med.* **143,** 152; author reply 156–158.

Limpens, J., Schroder, F. H., de Ridder, C. M., Bolder, C. A., Wildhagen, M. F., Obermuller-Jevic, U. C., Kramer, K., and van Weerden, W. M. (2006). Combined lycopene and vitamin E treatment suppresses the growth of PC-346C human prostate cancer cells in nude mice. *J. Nutr.* **136,** 1287–1293.

Liu, B., Lee, K. W., Li, H., Ma, L., Lin, G. L., Chandraratna, R. A., and Cohen, P. (2005). Combination therapy of insulin-like growth factor binding protein-3 and retinoid X receptor ligands synergize on prostate cancer cell apoptosis *in vitro* and *in vivo*. *Clin. Cancer Res.* **11,** 4851–4856.

Lonn, E., Bosch, J., Yusuf, S., Sheridan, P., Pogue, J., Arnold, J. M., Ross, C., Arnold, A., Sleight, P., Probstfield, J., and Dagenais, G. R. (2005). Effects of long-term vitamin E supplementation on cardiovascular events and cancer: A randomized controlled trial. *JAMA* **293,** 1338–1347.

Malafa, M. P., and Neitzel, L. T. (2000). Vitamin E succinate promotes breast cancer tumor dormancy. *J. Surg. Res.* **93,** 163–170.

Malafa, M. P., Fokum, F. D., Smith, L., and Louis, A. (2002). Inhibition of angiogenesis and promotion of melanoma dormancy by vitamin E succinate. *Ann. Surg. Oncol.* **9,** 1023–1032.

Malafa, M. P., Fokum, F. D., Andoh, J., Neitzel, L. T., Bandyopadhyay, S., Zhan, R., Iiizumi, M., Furuta, E., Horvath, E., and Watabe, K. (2006). Vitamin E succinate suppresses prostate tumor growth by inducing apoptosis. *Int. J. Cancer* **118,** 2441–2447.

Mardones, P., Strobel, P., Miranda, S., Leighton, F., Quinones, V., Amigo, L., Rozowski, J., Krieger, M., and Rigotti, A. (2002). Alpha-tocopherol metabolism is abnormal in scavenger receptor class B type I (SR-BI)-deficient mice. *J. Nutr.* **132,** 443–449.

Marras, C., Lang, A. E., Oakes, D., McDermott, M. P., Kieburtz, K., Shoulson, I., Tanner, C. M., and Fahn, S. (2005). High-dosage vitamin E supplementation and all-cause mortality. *Ann. Intern. Med.* **143,** 152–153; author reply 156–158.

Mentor-Marcel, R., Lamartiniere, C. A., Eltoum, I. E., Greenberg, N. M., and Elgavish, A. (2001). Genistein in the diet reduces the incidence of poorly differentiated prostatic adenocarcinoma in transgenic mice (TRAMP). *Cancer Res.* **61,** 6777–6782.

Meydani, S. N., Lau, J., Dallal, G. E., and Meydani, M. (2005). High-dosage vitamin E supplementation and all-cause mortality. *Ann. Intern. Med.* **143,** 153; author reply 156–158.

Miller, E. R., III, Pastor-Barriuso, R., Dalal, D., Riemersma, R. A., Appel, L. J., and Guallar, E. (2005). Meta-analysis: High-dosage vitamin E supplementation may increase all-cause mortality. *Ann. Intern. Med.* **142,** 37–46.

Morante, M., Sandoval, J., Gomez-Cabrera, M. C., Rodriguez, J. L., Pallardo, F. V., Vina, J. R., Torres, L., and Barber, T. (2005). Vitamin E deficiency induces liver nuclear factor-kappaB DNA-binding activity and changes in related genes. *Free Radic. Res.* **39,** 1127–1138.

Nakamura, A., Shirai, T., Takahashi, S., Ogawa, K., Hirose, M., and Ito, N. (1991). Lack of modification by naturally occurring antioxidants of 3,2′-dimethyl-4-aminobiphenyl-initiated rat prostate carcinogenesis. *Cancer Lett.* **58,** 241–246.

Neuzil, J., Weber, T., Gellert, N., and Weber, C. (2001). Selective cancer cell killing by alpha-tocopheryl succinate. *Br. J. Cancer* **84,** 87–89.

Ni, J., Chen, M., Zhang, Y., Li, R., Huang, J., and Yeh, S. (2003). Vitamin E succinate inhibits human prostate cancer cell growth via modulating cell cycle regulatory machinery. *Biochem. Biophys. Res. Commun.* **300,** 357–363.

Ni, J., Wen, X., Yao, J., Chang, H. C., Yin, Y., Zhang, M., Xie, S., Chen, M., Simons, B., Chang, P., di Sant'Agnese, A., Messing, E. M., *et al.* (2005). Tocopherol-associated protein suppresses prostate cancer cell growth by inhibition of the phosphoinositide 3-kinase pathway. *Cancer Res.* **65,** 9807–9816.

Ni, J., Pang, S., and Yeh, S. (2007). Differential retention of α-vitamin E is correlated with its transporter gene expression and growth inhibitory efficacy in prostate cancer cells. *The Prostate* **67,** 463–471.

Oram, J. F., Vaughan, A. M., and Stocker, R. (2001). ATP-binding cassette transporter A1 mediates cellular secretion of alpha-tocopherol. *J. Biol. Chem.* **276,** 39898–39902.

Orso, E., Broccardo, C., Kaminski, W. E., Bottcher, A., Liebisch, G., Drobnik, W., Gotz, A., Chambenoit, O., Diederich, W., Langmann, T., Spruss, T., Luciani, M. F., *et al.* (2000). Transport of lipids from golgi to plasma membrane is defective in tangier disease patients and Abc1-deficient mice. *Nat. Genet.* **24,** 192–196.

Ouahchi, K., Arita, M., Kayden, H., Hentati, F., Ben Hamida, M., Sokol, R., Arai, H., Inoue, K., Mandel, J. L., and Koenig, M. (1995). Ataxia with isolated vitamin E deficiency is caused by mutations in the alpha-tocopherol transfer protein. *Nat. Genet.* **9,** 141–145.

Panagabko, C., Morley, S., Hernandez, M., Cassolato, P., Gordon, H., Parsons, R., Manor, D., and Atkinson, J. (2003). Ligand specificity in the CRAL-TRIO protein family. *Biochemistry* **42,** 6467–6474.

Panel on Dietary Antioxidants and Related Compounds, Subcommittees on Upper Reference Levels of Nutrients and Interpretation and Uses of DRIs, Standing Committee on the Scientific Evaluation of Dietary Reference Intakes, Food and Nutrition Board, Institute of Medicine. (2000). Dietary reference intakes for vitamin C, vitamin E, selenium, and carotenoids. National Academic Press, Washington, DC.

Plymate, S. R., Haugk, K. H., Sprenger, C. C., Nelson, P. S., Tennant, M. K., Zhang, Y., Oberley, L. W., Zhong, W., Drivdahl, R., and Oberley, T. D. (2003). Increased manganese superoxide dismutase (SOD-2) is part of the mechanism for prostate tumor suppression by Mac25/insulin-like growth factor binding-protein-related protein-1. *Oncogene* **22,** 1024–1034.

Prasad, K. N., and Edwards-Prasad, J. (1992). Vitamin E and cancer prevention: Recent advances and future potentials. *J. Am. Coll. Nutr.* **11,** 487–500.

Przyklenk, K., and Whittaker, P. (2000). Brief antecedent ischemia enhances recombinant tissue plasminogen activator-induced coronary thrombolysis by adenosine-mediated mechanism. *Circulation* **102,** 88–95.

Qian, J., Morley, S., Wilson, K., Nava, P., Atkinson, J., and Manor, D. (2005). Intracellular trafficking of vitamin E in hepatocytes: The role of tocopherol transfer protein. *J. Lipid Res.* **46,** 2072–2082.

Reboul, E., Klein, A., Bietrix, F., Gleize, B., Malezet-Desmoulins, C., Schneider, M., Margotat, A., Lagrost, L., Collet, X., and Borel, P. (2006). Scavenger receptor class B type I (SR-BI) is involved in vitamin E transport across the enterocyte. *J. Biol. Chem.* **281,** 4739–4745.

Ricciarelli, R., Zingg, J. M., and Azzi, A. (2000). Vitamin E reduces the uptake of oxidized LDL by inhibiting CD36 scavenger receptor expression in cultured aortic smooth muscle cells. *Circulation* **102,** 82–87.

Salonen, J. T., Nyyssonen, K., Salonen, R., Lakka, H. M., Kaikkonen, J., Porkkala-Sarataho, E., Voutilainen, S., Lakka, T. A., Rissanen, T., Leskinen, L., Tuomainen, T. P., Valkonen, T. P., *et al.* (2000). Antioxidant supplementation in atherosclerosis prevention (ASAP) study: A randomized trial of the effect of vitamins E and C on 3-year progression of carotid atherosclerosis. *J. Intern. Med.* **248,** 377–386.

Schindler, R., and Mentlein, R. (2006). Flavonoids and vitamin E reduce the release of the angiogenic peptide vascular endothelial growth factor from human tumor cells. *J. Nutr.* **136,** 1477–1482.

Shaw, H. M., and Huang, C. (1998). Liver alpha-tocopherol transfer protein and its mRNA are differentially altered by dietary vitamin E deficiency and protein insufficiency in rats. *J. Nutr.* **128,** 2348–2354.

Shiau, C. W., Huang, J. W., Wang, D. S., Weng, J. R., Yang, C. C., Lin, C. H., Li, C., and Chen, C. S. (2006). Alpha-tocopheryl succinate induces apoptosis in prostate cancer cells in part through inhibition of Bcl-xL/Bcl-2 function. *J. Biol. Chem.* **281,** 11819–11825.

Shibata, N., Arita, M., Misaki, Y., Dohmae, N., Takio, K., Ono, T., Inoue, K., and Arai, H. (2001). Supernatant protein factor, which stimulates the conversion of squalene to lanosterol, is a cytosolic squalene transfer protein and enhances cholesterol biosynthesis. *Proc. Natl. Acad. Sci. USA* **98,** 2244–2249.

Stocker, A. (2004). Molecular mechanisms of vitamin E transport. *Ann. NY Acad. Sci.* **1031,** 44–59.

Stocker, A., and Baumann, U. (2003). Supernatant protein factor in complex with RRR-alpha-tocopherylquinone: A link between oxidized vitamin E and cholesterol biosynthesis. *J. Mol. Biol.* **332,** 759–765.

Stocker, A., Tomizaki, T., Schulze-Briese, C., and Baumann, U. (2002). Crystal structure of the human supernatant protein factor. *Structure (Camb.)* **10,** 1533–1540.

Suzuki, Y. J., Tsuchiya, M., Wassall, S. R., Choo, Y. M., Govil, G., Kagan, V. E., and Packer, L. (1993). Structural and dynamic membrane properties of alpha-tocopherol and alpha-tocotrienol: Implication to the molecular mechanism of their antioxidant potency. *Biochemistry* **32,** 10692–10699.

Tasinato, A., Boscoboinik, D., Bartoli, G. M., Maroni, P., and Azzi, A. (1995). d-Alpha-tocopherol inhibition of vascular smooth muscle cell proliferation occurs at physiological concentrations, correlates with protein kinase C inhibition, and is independent of its antioxidant properties. *Proc. Natl. Acad. Sci. USA* **92,** 12190–12194.

Teupser, D., Thiery, J., and Seidel, D. (1999). Alpha-tocopherol down-regulates scavenger receptor activity in macrophages. *Atherosclerosis* **144,** 109–115.

The Alpha-Tocopherol Beta Carotene Cancer Prevention Study Group (1994). The effect of vitamin E and beta carotene on the incidence of lung cancer and other cancers in male smokers. *N. Engl. J. Med.* **330,** 1029–1035.

Tomic-Vatic, A., Eytina, J., Chapman, J., Mahdavian, E., Neuzil, J., and Salvatore, B. A. (2005). Vitamin E amides, a new class of vitamin E analogues with enhanced proapoptotic activity. *Int. J. Cancer* **117,** 188–193.

Traber, M. G., and Sies, H. (1996). Vitamin E in humans: Demand and delivery. *Annu. Rev. Nutr.* **16,** 321–347.

Trump, D. L., Potter, D. M., Muindi, J., Brufsky, A., and Johnson, C. S. (2006). Phase II trial of high-dose, intermittent calcitriol (1,25-dihydroxyvitamin D3) and dexamethasone in androgen-independent prostate cancer. *Cancer* **106,** 2136–2142.

Tucker, J. M., and Townsend, D. M. (2005). Alpha-tocopherol: Roles in prevention and therapy of human disease. *Biomed. Pharmacother.* **59,** 380–387.

Turley, J. M., Ruscetti, F. W., Kim, S. J., Fu, T., Gou, F. V., and Birchenall-Roberts, M. C. (1997). Vitamin E succinate inhibits proliferation of BT-20 human breast cancer cells: Increased binding of cyclin A negatively regulates E2F transactivation activity. *Cancer Res.* **57,** 2668–2675.

United States Department of Agriculture (USDA) food composition database. http://www.nal.usda.gov/fnic/foodcomp/search/

Venkateswaran, V., Fleshner, N. E., and Klotz, L. H. (2002). Modulation of cell proliferation and cell cycle regulators by vitamin E in human prostate carcinoma cell lines. *J. Urol.* **168,** 1578–1582.

Venkateswaran, V., Fleshner, N. E., Sugar, L. M., and Klotz, L. H. (2004a). Antioxidants block prostate cancer in lady transgenic mice. *Cancer Res.* **64,** 5891–5896.

Venkateswaran, V., Fleshner, N. E., and Klotz, L. H. (2004b). Synergistic effect of vitamin E and selenium in human prostate cancer cell lines. *Prostate Cancer Prostatic Dis.* **7,** 54–56.

Virtamo, J., Pietinen, P., Huttunen, J. K., Korhonen, P., Malila, N., Virtanen, M. J., Albanes, D., Taylor, P. R., and Albert, P. (2003). Incidence of cancer and mortality following alpha-tocopherol and beta-carotene supplementation: A postintervention follow-up. *JAMA* **290,** 476–485.

Weber, T., Lu, M., Andera, L., Lahm, H., Gellert, N., Fariss, M. W., Korinek, V., Sattler, W., Ucker, D. S., Terman, A., Schroder, A., Erl, W., *et al.* (2002). Vitamin E succinate is a potent novel antineoplastic agent with high selectivity and cooperativity with tumor necrosis factor-related apoptosis-inducing ligand (Apo2 ligand) *in vivo*. *Clin. Cancer Res.* **8,** 863–869.

Weinstein, S. J., Wright, M. E., Pietinen, P., King, I., Tan, C., Taylor, P. R., Virtamo, J., and Albanes, D. (2005). Serum alpha-tocopherol and gamma-tocopherol in relation to prostate cancer risk in a prospective study. *J. Natl. Cancer Inst.* **97,** 396–399.

Yamauchi, J., Iwamoto, T., Kida, S., Masushige, S., Yamada, K., and Esashi, T. (2001). Tocopherol-associated protein is a ligand-dependent transcriptional activator. *Biochem. Biophys. Res. Commun.* **285,** 295–299.

Yin, Y., Ni, J., Guo, Y., DiMaggio, M., Chen, M., and Yeh, S. (2007). The therapeutic and preventive effect of RRR-alpha-vitamin E succinate on prostate cancer *via* induction of insulin like growth factor binding protein 3. *Clin. Cancer Res.* **13,** 2271–2280.

Yokota, T., Uchihara, T., Kumagai, J., Shiojiri, T., Pang, J. J., Arita, M., Arai, H., Hayashi, M., Kiyosawa, M., Okeda, R., and Mizusawa, H. (2000). Postmortem study of ataxia with retinitis pigmentosa by mutation of the alpha-tocopherol transfer protein gene. *J. Neurol. Neurosurg. Psychiatry* **68,** 521–525.

Yokota, T., Igarashi, K., Uchihara, T., Jishage, K., Tomita, H., Inaba, A., Li, Y., Arita, M., Suzuki, H., Mizusawa, H., and Arai, H. (2001). Delayed-onset ataxia in mice lacking alpha-tocopherol transfer protein: Model for neuronal degeneration caused by chronic oxidative stress. *Proc. Natl. Acad. Sci. USA* **98,** 15185–15190.

You, H., Yu, W., Sanders, B. G., and Kline, K. (2001). RRR-alpha-tocopheryl succinate induces MDA-MB-435 and MCF-7 human breast cancer cells to undergo differentiation. *Cell Growth Differ.* **12,** 471–480.

You, H., Yu, W., Munoz-Medellin, D., Brown, P. H., Sanders, B. G., and Kline, K. (2002). Role of extracellular signal-regulated kinase pathway in RRR-alpha-tocopheryl succinate-induced differentiation of human MDA-MB-435 breast cancer cells. *Mol. Carcinog.* **33,** 228–236.

Yu, W., Heim, K., Qian, M., Simmons-Menchaca, M., Sanders, B. G., and Kline, K. (1997). Evidence for role of transforming growth factor-beta in RRR-alpha-tocopheryl succinate-induced apoptosis of human MDA-MB-435 breast cancer cells. *Nutr. Cancer* **27,** 267–278.

Yu, W., Simmons-Menchaca, M., You, H., Brown, P., Birrer, M. J., Sanders, B. G., and Kline, K. (1998). RRR-alpha-tocopheryl succinate induction of prolonged activation of c-jun amino-terminal kinase and c-jun during induction of apoptosis in human MDA-MB-435 breast cancer cells. *Mol. Carcinog.* **22,** 247–257.

Yu, W., Israel, K., Liao, Q. Y., Aldaz, C. M., Sanders, B. G., and Kline, K. (1999). Vitamin E succinate (VES) induces Fas sensitivity in human breast cancer cells: Role for Mr 43,000 Fas in VES-triggered apoptosis. *Cancer Res.* **59,** 953–961.

Yusuf, S., Dagenais, G., Pogue, J., Bosch, J., and Sleight, P. (2000). Vitamin E supplementation and cardiovascular events in high-risk patients. The Heart Outcomes Prevention Evaluation Study Investigators. *N. Engl. J. Med.* **342,** 154–160.

Zhang, M., Altuwaijri, S., and Yeh, S. (2004). RRR-alpha-tocopheryl succinate inhibits human prostate cancer cell invasiveness. *Oncogene* **23,** 3080–3088.

Zhang, Y., Ni, J., Messing, E. M., Chang, E., Yang, C. R., and Yeh, S. (2002). Vitamin E succinate inhibits the function of androgen receptor and the expression of prostate-specific antigen in prostate cancer cells. *Proc. Natl. Acad. Sci. USA* **99,** 7408–7413.

Zimmer, S., Stocker, A., Sarbolouki, M. N., Spycher, S. E., Sassoon, J., and Azzi, A. (2000). A novel human tocopherol-associated protein: Cloning, *in vitro* expression, and characterization. *J. Biol. Chem.* **275,** 25672–25680.

Zu, K., and Ip, C. (2003). Synergy between selenium and vitamin E in apoptosis induction is associated with activation of distinctive initiator caspases in human prostate cancer cells. *Cancer Res.* **63,** 6988–6995.

Zu, K., Hawthorn, L., and Ip, C. (2005). Up-regulation of c-Jun-NH2-kinase pathway contributes to the induction of mitochondria-mediated apoptosis by alpha-tocopheryl succinate in human prostate cancer cells. *Mol. Cancer Ther.* **4,** 43–50.

20

Vitamin E: Inflammation and Atherosclerosis

U. Singh and S. Devaraj

Department of Pathology
Laboratory for Atherosclerosis and Metabolic Research
UC Davis Medical Center, Sacramento, California 95817

Vitamins and Hormones, Volume 76

0083-6729/07 $35.00
DOI: 10.1016/S0083-6729(07)76020-X

Cardiovascular disease (CVD) is the leading cause of morbidity and mortality in the western world with its incidence increasing lately in developing countries. Several lines of evidence support a role for inflammation in atherogenesis. Hence, dietary micronutrients having antiinflammatory properties may have a potential beneficial effect with regard to CVD. Vitamin E is a potent antioxidant with antiinflammatory properties. It comprises eight different isoforms: four tocopherols (T) (α, β, γ, and δ) and four tocotrienols (T3) (α, β, γ, and δ). A wealth of data is available for the preventive efficacy of α-T. α-T supplementation in human subjects and animal models has been shown to be antioxidant and antiinflammatory in terms of decreasing C-reactive protein (CRP) and release of proinflammatory cytokines, the chemokine IL-8 and PAI-1 levels especially at high doses. γ-T is effective in decreasing reactive nitrogen species and also appears to have antiinflammatory properties; however, there are scanty data examining pure γ-T preparations. Furthermore, tocotrienols (α and γ) also have implications for prevention of CVD; however, there are conflicting and insufficient data in the literature with regards to their potency. In this chapter, we have gathered recent emerging data on α-T specifically and also have given a composite view of γ-T and tocotrienols especially with regards to their effect on inflammation as it relates to CVD.

I. INTRODUCTION

Atherosclerosis is a chronic inflammatory disease of the arterial wall characterized by progressive accumulation of lipids, cells [macrophages, T lymphocytes, and smooth muscle cells (SMC)], and extracellular matrix (Ross, 1999). Monocytes–macrophages are pivotal cells in atherosclerosis and participate in all stages of atherosclerosis from initiation of the fatty streak to plaque rupture (Torsney *et al.*, 2007; Viles-Gonzalez *et al.*, 2006). CRP, the prototypic marker of inflammation, in addition to being a risk marker, appears to be an active participant in atherosclerosis (Verma *et al.*, 2006). Epidemiological studies suggest an association between increased intake of vitamin E and reduced morbidity and mortality from coronary artery disease (Gey *et al.*, 1991; Kushi *et al.*, 1996; Stampfer *et al.*, 1993). We have previously published review articles (Kaul *et al.*, 2001; Singh and Jialal, 2004; Singh *et al.*, 2005) on the role of α-tocopherol (α-T) as an antiinflammatory and antioxidant agent in CVD. In this chapter, we will discuss in detail the effects of γ-T and tocotrienols (T3) on inflammation and their preventive efficacy for CVD.

II. Inflammation and Atherosclerosis

Much evidence support a pivotal role for inflammation in all phases of atherosclerosis from the initiation of the fatty streak to the culmination in acute coronary syndromes (plaque rupture) (Libby, 2002; Tziakas *et al.*, 2006). Major cellular participants in atherosclerosis include monocytes, macrophages, activated vascular endothelium, T lymphocytes, platelets, and SMC.

Monocytes and macrophages are critical cells present at all stages of atherogenesis and, when stimulated, can produce biologically active mediators that have a profound influence on the progression of atherosclerosis. Monocytes and macrophages secrete several proinflammatory, proatherogenic cytokines, such as IL-1β, TNF-α, and IL-6, which have been shown to be present in the atherosclerotic lesions and are known to augment monocyte–endothelial adhesion. Supportive evidence for the central role played by IL-1 in the development of atherosclerosis has been documented by Kirii *et al.* (2003) demonstrating decreased severity of atherosclerosis in ApoE-knockout (ApoE KO) mice-deficient for IL-1β. These reports put forth evidence that IL-1 signaling promotes inflammation in the vascular wall. TNF-α has been shown to promote monocyte adhesion to endothelium and contribute to the necrotic core by promoting apoptosis of macrophages and SMC (Jialal *et al.*, 2004). Activated macrophages also release matrix metalloproteinases that cause a rent in the endothelium and tissue factor that promotes thrombus formation.

Atherosclerosis is associated with impaired endothelial dysfunction, and these changes induce adhesion and transendothelial migration of monocytes (Jialal *et al.*, 2004). Both IL-1β and TNF-α stimulate expression of adhesion molecules such as vascular cell adhesion molecule-1 (VCAM-1), intercellular cell adhesion molecule-1 (ICAM-1), and E-selectin. Chemotaxis and entry of monocytes into the subendothelial space is promoted by monocyte chemoattractant protein-1 (MCP-1), interleukin-8 (IL-8), and fractalkine. Several studies have shown a strong association between levels of soluble CAMs (which are shed from activated cells such as endothelial cells) and coronary as well as carotid atherosclerosis as reviewed previously by us (Jialal *et al.*, 2004).

Several large population studies have indicated that biomarkers of inflammation predict an increased risk for CVD (Jialal *et al.*, 2004; Libby, 2002; Tziakas *et al.*, 2006). The prototypic marker of inflammation is CRP, a member of the pentraxin family (Jialal and Devaraj, 2001; Jialal *et al.*, 2004). Numerous prospective studies from populations throughout the world have shown that elevated levels of CRP confer a greater risk of CVD as reviewed previously (Jialal *et al.*, 2004). Plasma CRP is considered to be a sensitive marker of systemic inflammation, and chronically high levels predict increased risk of future cardiovascular events (CVE) (Jialal *et al.*, 2004). Inflammation (as manifested by an increase in CRP) is not only increased in CVD but also in diseases with increased cardiovascular risk, for example end-stage renal disease (ESRD) (Himmelfarb and McMonagle, 2001; Himmelfarb *et al.*, 2003), metabolic syndrome (MetS) (Ridker, 2003), and diabetes (Pickup, 2004). Hence, needless to say dietary micronutrients, especially with antiinflammatory, for example, AT may have potential beneficial effects with regard to cardiovascular disease (CVD).

III. VITAMIN E

A. CHEMICAL FORM AND ABSORPTION

The term vitamin E is a generic description for all tocopherols and tocotrienols derivatives. Both tocophenols and tocotrienols have four isomers differing by the number and position of methyl groups on the chroman ring designated as: trimethyl (α-), dimethyl (β- or γ-), and monomethyl (δ-) tocopherol/tocotrienols forms (Traber and Arai, 1999). Tocopherols contain a saturated phytyl chain, while tocotrienols have a similar chain but with three double bonds at positions 3′, 7′, and 11′. Commercially available vitamin E consists of either a mixture of naturally occurring tocopherols and tocotrienols; *RRR*-α-T (formerly called D-α-T); synthetic α-T (all-*rac*-α-T, formerly called DL-α-T) which consists of the eight possible stereoisomers in equal amounts or their esters such as α-tocopheryl acetate (TA), tocopheryl succinate (TS), or tocopheryl nicotinate (TN). The ester form prevents the oxidation of vitamin E and prolongs its shelf life. Except for individuals with malabsorption syndromes, these esters are readily hydrolyzed in the gut and are absorbed in the unesterified form (Traber and Arai, 1999). The natural vitamin E sources are vegetable oils: safflower seeds oil containing almost exclusively α-T (59.3 mg/1 g of oil), soy oil is rich in γ-, δ-, and α-T (62.4, 20.4, and 11.0 mg/1 g of oil), and palm oil contains tocotrienols (17.2 mg/1 g of oil) in addition to α-T (18.3 mg/g of oil) (Munteanu *et al.*, 2004).

The bioavailability of the different forms of vitamin E is highly differential. For example, although the amount of γ-T in the diet is higher than that of α-T, the plasma γ-T concentration is only 10% of that of α-T which is the most abundant form in plasma. Once ingested with diet, all forms of vitamin E are taken up by intestinal cells and released into the circulation in chylomicrons. The vitamins reach the liver via chylomicron remnants. In the liver, a specific protein, α-tocopherol transfer protein (α-TTP), selectively targets *RRR*-α-T for incorporation into VLDL. Other forms are much less well retained and are excreted via the bile, the urine [as carboxyethyl hydroxychromans (CEHCs)], or unknown routes. Relative affinities of tocopherol analogues for α-TTP calculated from the degree of competition for the α-form are as follows: α-T 100%, β-T 38%, γ-T 9%, and δ-T 2% (Azzi *et al.*, 2002). The significance of α-TTP is well evident from one of the recent reports showing increased basal oxidative stress as well as inflammatory status in α-TTP null mice (Ttpa$^{-/-}$) (Schock *et al.*, 2004). Terasawa *et al.* (2000) also reported that in ApoE KO mice, vitamin E deficiency caused by disruption of the α-TTP gene (*Ttpa*$^{-/-}$) increased the severity of atherosclerotic lesions in the proximal aorta.

Importantly, the antiinflammatory activity is demonstrated by α-T and γ-T as well as tocotrienols. Overall, α-T is the principal and most potent lipid-soluble antioxidant in plasma and LDL. α-T is present in LDL particle in quantities (between five and nine molecules/LDL particle) that can be easily modified by dietary intake or oral supplementation (Jessup *et al.*, 2004). Several lines of evidence support a relationship between low α-T levels and the development of atherosclerosis as reviewed previously by us (Kaul *et al.*, 2001; Singh and Jialal, 2004; Singh *et al.*, 2005).

IV. ANIMAL STUDIES

Animal studies by and large generate useful, and often otherwise unattainable, information on the content of arterial lipids, antioxidants, and lipid oxidation *in vivo*. The data with regards to the role of vitamin E in experimental atherogenesis in animals has also been extensively reviewed earlier by us (Kaul *et al.*, 2001; Singh and Jialal, 2004) as well as other investigators (Upston *et al.*, 2003). The summarized data in this aspect is presented in Table I. However, more recent and well-important data with regards to α-T which was available in literature after 2003/2004 is being discussed. For example, Suarna C *et al.* (2006) have reported findings from their pioneering work carried out in ApoE KO and ApoE/TtPa KO mice fed a high-fat diet

TABLE I. Summarized Results of α-T and Atherosclerosis in Animal Studies

Outcome	Species
↓ 35% in atherosclerotic lesions	Macaques
↓ Restenosis after angioplasty	Rabbits
Less aortic atheroslerosis	Chickens
↓ Aortic lesions even after formation of fatty streaks	Mice (ApoE KO)
↓ MCP-1	Mice (ApoE KO)

(HFD) without or with α-T (0.2%), a synthetic vitamin E analogue –BO-653 (0.2%) or α-T plus BO-653 (α-T + BO-653; 0.2%) for 6 months. Dietary α-T supplements restored circulating and aortic levels of α-T and decreased atherosclerosis in the aortic root of ApoE/Ttpa KO mice to a level comparable to that seen in ApoE KO mice without any effect in ApoE KO mice. However, dietary supplements with a BO-653, either alone or in combination with α-T, decreased atherosclerosis in ApoE KO and in ApoE/Ttpa KO mice. These results suggest that vitamin E supplements ameliorate experimental atherosclerosis in conditions of preexisting severe vitamin E deficiency implicating that vitamin E at best has a modest effect on experimental atherosclerosis in hyperlipidemic mice, and only in situations of severe vitamin E deficiency and independent of lipid oxidation in the vessel wall. These findings have been suggested to be extrapolated to human atherosclerosis to help explain the failure of vitamin E supplements to ameliorate CVD outcome, because in contrast to the ApoE/Ttpa KO mouse model employed here, human atherosclerotic lesions are not deficient in α-T.

A. OTHER FORMS OF α-T AND THEIR SIGNIFICANCE

As outlined previously, α-T also exists in the previously undetected natural form, tocopheryl phosphate (TP). Significantly, TP contains a chroman OH group esterified by phosphoric acid, making the molecule an ideal candidate for a number of cellular functions such as oxidant-protected intracellular transport, enzymatic regulation of its concentration, and cell signaling (Gianello *et al.*, 2005). TP is resistant to the alkaline conditions encountered in detection of α-T which normally include a hydrolysis step as part of the extraction process procedures and is consequently not included with free α-T in typical analyses. Furthermore, alkaline hydrolysis converts TP to a salt form, rendering it relatively insoluble to the organic solvents commonly used for α-T extractions. Therefore, TP is not detected by standard assays for α-T.

TP may represent the storage form or an absorption form of the α-T (Ogru *et al.*, 2003). The possibility remains that TP is a signaling molecule, this being compatible with its very low amounts found in tissues, on the same order of magnitude as known signaling molecules such as inositol phosphate. Should TP operate in a signaling context, hydrolysis of the phosphate group by a phosphatase may represent a mechanism for regulating the levels of the "active" signal. Alkaline phosphatase is a possible candidate as the dephosphorylating agent, as it has been shown to hydrolyze the phosphate group from TP *in vitro* (Topi and Alessandrini, 1953). The existence of a phosphatase(s) and kinase(s), which may be involved in interconverting α-T and TP, need to be explored. Additionally, TS has been also shown to have cell properties far stronger than those of α-T. It has been suggested that TS may substitute for TP at the level of a receptor causing a permanent activation of cellular signals.

B. α-T SUPPLEMENTATION IN HUMANS

Largely, several groups have shown that α-T has antiinflammatory effects both *in vivo* and *in vitro* as summarized in Table II. In addition, recent literature survey revealed various studies in relation to α-T for its effect on inflammation and atherogenesis as discussed herewith. Postprandial oxidative stress is of crucial importance in the pathogenesis of endothelial damage. Owing to the existing relationship between endothelial damage, glycemic control, disorders of lipid metabolism, and coagulative hemostatic disorders, a single-blind, controlled clinical study was conducted (Neri *et al.,* 2005) in patients with early stage, untreated T2DM; subjects with IGT; and healthy controls. Before supplementation, all three groups had significantly increased levels of oxidants, vWF, and VCAM-1 and significantly decreased levels of antioxidants and NO after consumption of a moderate-fat meal. These measures were then reassessed after 15 days of standard antioxidant treatment consisting of a thiol-containing antioxidant (*N*-acetylcysteine 600 g/day),

TABLE II. Summarized Results of α-T and Atherosclerosis in Human Supplementation Studies

Outcome	Dose
↓ CRP (prebreakfast/presupper)	α-T 800 IU *RRR*-α-T
↓ PAI-1 (prebreakfast only)	
↓ Inflammatory status (IL-1, TNF, and Il-1 RA in whole blood)	600 IU all-*rac* for 4 weeks
↓ CRP, no change in IL-6 or sCAMs	400 IU/day *RRR*-α-T for 6 weeks in smokers with ACS
↓ CRP	*RRR*-α-T 800 IU/d for 6 weeks in diabetics
↓ IL-1β, MO-EC adhesion	1200 IU/day α-T (normal and T2DM with and (without macrovascular disease)
↓ CRP	*RRR*-α-T in T2DM
↓ PAI-1 and P-selectin	Versus normal for 4 weeks in diabetics
↓ Platelet adhesion	α-T 200 IU/day
↓ P-selectin (40%)	α-T 600 mg/day in hypercholesterolemic subjects
↓ Cytokines (IL-1, TNF) Chemokines (IL-8)	*RRR*-α-T 600 IU/day for 6 weeks
No change in inflammatory cytokines and oxidative stress	400 IU *RRR*-α-T
No change in MO-EC adhesion sICAM, VCAM with increased lag phase of oxidation	all-*rac* α-T (800 IU/day for 12 weeks)

a bound antioxidant (vitamin E 300 g/day), and an aqueous phase antioxidant (vitamin C 250 mg/day). After 15 days of antioxidant treatment, significant improvements in these measures were seen in all groups which supported the antiinflammatory, in addition to antioxidative potential of different supplements.

Gasparetto *et al.* (2005) analyzed sVCAM-1, reactive oxygen metabolite (ROM) level, total antioxidant status (TAS), and telediastolic left ventricular volume (TLVV) in patients with myocardial infarction undergoing reperfusion therapy and treated with antioxidant vitamins (α-T) or placebo (P) before and for 1 month after reperfusion. After reperfusion, serum VCAM-1, ROM, and TLVV were significantly higher in patients treated with placebo than in those treated with antioxidant vitamins, while TAS was significantly higher in patients treated with antioxidant supplementation. Forty-eight hours after reperfusion, α-T treatment significantly decreased sVCAM-1, ROMs, TAS; TLVV was decreased significantly at 1 week as well as 1 month after reperfusion implying that vitamin treatment improves the antioxidant system and reduces oxidative stress, inflammatory process, and left ventricular remodeling.

It is well known that hemodialysis patients, as opposed to nondialyzed CRF patients have increased oxidative stress in terms of markedly higher levels of Cu/Zn-SOD (Akiyama *et al.*, 2005). Thus, antioxidant effects of oral vitamin E supplementation (600-mg/day VE-PO) were compared with vitamin E coating of a dialyzer (VE-BMD) in terms of their reversal effect, if any, on Cu/Zn-SOD in 31 hemodialysis patients (Akiyama *et al.*, 2005) divided into two groups. VE-PO and VE-BMD showed almost comparable and significantly decreased mRNA as well as content of Cu/Zn-SOD, which reached the level of nondialyzed CRF patients, thus suggesting the positive outcome of vitamin E by both means in CRF patients undergoing dialysis.

Furthermore, chronic exercise or intensive racing is known to result in oxidative stress. Thus, McAnulty *et al.* (2005) examined the effect of 2 months of α-T (800 IU/day) [E] or placebo (P), in 38 triathletes on plasma Hcy concentrations, antioxidant potential, and oxidative stress. Plasma α-T was 75% higher ($p < 0.001$) in E versus P prerace and this group difference was maintained throughout the race. There were no significant time, group, or interaction effects on plasma Hcy concentrations between E and P. Plasma F (2)-isoprostanes increased 181% versus 97% during the race in E versus P, and lipid hydroperoxides were also significantly elevated at 1.5-h postrace in E versus P. Plasma antioxidant potential was significantly higher 1.5-h postrace in E versus P. This study indicates that prolonged large doses of α-T supplementation did not affect plasma Hcy concentrations and exhibited prooxidant characteristics in highly trained athletes during exhaustive exercise. Another study (Mastaloudis *et al.*, 2006) investigated, if 6 weeks of supplementation with vitamins E and C could alleviate exercise-induced lipid peroxidation and inflammation in runners during a 50-km ultramarathon in 22 subjects randomly assigned to one of two groups: (1) placebos (PL) or (2) antioxidants (AO: 1000-mg vitamin C and 300-mg *RRR*-α-T). With supplementation, plasma α-T and AA increased in the AO but not the PL group. Although F2-IsoP levels were similar between groups at baseline, but increased during the run only in the PL group. In PL women, F2-IsoPs were significantly elevated postrace, but returned to prerace concentrations by 2-h postrace. In PL men, F2-IsoP concentrations were higher postrace, 2-h postrace, and 1–4, and 6 days postrace. However, markers of inflammation were increased dramatically in response to the run regardless of treatment group implicating that supplementation prevented endurance exercise-induced lipid peroxidation but had no effect on inflammatory markers.

Furthermore, an attempt was made to clarify whether supplementation of vitamin E can alter the low-density lipoprotein (LDL) oxidation properties and thereby affect endothelial cell function and prostacyclin production in smokers compared to nonsmokers on diets rich in fish in a pilot study (Seppo *et al.*, 2005). The lag phase increased significantly after the supplementation of vitamin E both in smokers and nonsmokers. Native LDL dose dependently tended to reduce the viability of endothelial cells *in vitro* more markedly when isolated from smokers than from nonsmokers. Vitamin E supplementation had no beneficial effect on the cytotoxicity of oxidized LDLs in endothelial cell culture. On the other hand, simultaneous administration of Trolox, the water-soluble analogue of vitamin E, attenuated the LDL cytotoxicity on endothelial cells. The vitamin E supplementation to LDL donors attenuated the increase in prostacyclin production both in smokers and nonsmokers.

Wu *et al.* (2005a) explored the effect of *in vitro* vitamin E at physiologically relevant concentrations (10–60 μM) on the production of the vasodilator PGI2 and PGE2 by ECs as well as its underlying mechanism. Vitamin E dose dependently (10–40 μM) increased the production of both prostanoids by ECs. This was associated with a dose-dependent (10–40 μM) upregulation of cytosolic phospholipase A(2). In contrast, vitamin E dose dependently (10–60 μM) inhibited COX activity but did not affect the expression of either COX-1 or COX-2, indicating that the effect of vitamin E on COX activity was posttranslational. Thus, vitamin E had opposing effects on the two key enzymes in prostanoid biosynthesis; at the concentrations used in this study, this resulted in a net increase in the production of vasodilator prostanoids. The vitamin E-induced increase in PGI2 and PGE2 production may contribute to its suggested beneficial effect in preserving endothelial function.

Another study carried out by van Dam *et al.* (2003) investigated mechanisms via which α-T can directly affect endothelial activation as induced by H(2)O(2) and TNF in a combination of *in vivo* and *in vitro* studies. The study was carried out in 20 healthy volunteers treated with increasing doses of α-T up to 800 IU/ml for 12 weeks, involving the measurement of plasma levels of sCAMs and CRP. α-T attenuated $H_2O_2^{-1}$, but not TNF-induced increases in adhesion molecule expression. In healthy persons, α-T significantly decreased plasma levels of sE-selectin, sVCAM, and sICAM with no change in CRP. It was concluded that α-T specifically inhibits lipid peroxidation-induced endothelial activation *in vitro*. The observed vitamin E-induced decrease in sCAMs in control subjects suggests that lipid peroxidation does also take place in healthy individuals. Although vitamin E supplementation may be especially effective in specific groups of patients exposed to increased oxidative stress, this study suggested that vitamin E supplementation can be of benefit in healthy individuals as well.

Tahir *et al.* (2005) determined the impact of the combination of vitamins C and E or vitamin C only on serum levels of CAMs and CRP in patients with chronic degenerative aortic stenosis (AS), with or without concomitant CAD. One hundred patients with asymptomatic or mildly symptomatic moderate AS were randomized in 2:2:1 format in an open-label trial. Forty-one patients received vitamin E (400 IU) and vitamin C (1000 mg) daily, 39 patients received vitamin C (1000 mg) only, and 20 patients were followed as controls for 6 months. In the vitamins E and C group, there was a significant reduction in serum ICAM-1 with a return to baseline 6 months after cessation of therapy. In the vitamin C group only, there was a reduction in serum P selectin ($p = 0.033$). All the inflammatory markers were unchanged in control group over 6 months of follow-up leading to conclusion that vitamins E and C supplementation had modest antiinflammatory effect in chronic degenerative AS.

Furthermore, Accinni *et al.* (2006) also assessed the effects and the advantages of a combined dietary supplementation involving vitamin E with PUFA n-3, niacin, and γ-oryzanol on lipid profile, inflammatory status, and oxidative balance. Fifty-seven dyslipidemic volunteers were randomly assigned to receive: placebo (A); PUFA n-3 and vitamin E (B); the same as B plus γ-oryzanol and niacin (C) for 4 months. All dyslipidemic subjects showed, at baseline, oxidative stress and, after 4 months, all biochemical markers (lipid profile, ROS, total antioxidant capacity (TAC), vitamin E, IL-1, TNF, and TXB2) improved significantly in groups treated with dietary supplementation. These authors concluded that the strategy of combining different compounds, which protect each other and act together at different levels of the lipid chain production, improves lipid profile, inflammatory, and oxidative status, which might be because of reduction of the dose of each compound below the threshold of its side effects.

Another study was carried out by Boshtam *et al.* (2005) as a triple-blind, placebo-controlled clinical trial to determine the effect of the vitamin E on fasting blood sugar (FBS), serum insulin, and glycated hemoglobin (GHb) in T2DM patients. A total of 100 patients, with no complications, aged 20–60 years old were divided into two groups; treatment (vitamin E tablets [200 IU/day]) and placebo. No effect of vitamin E supplementation in these patients was seen on any of the parameters tested at the end of 27 weeks of treatment reaching to conclusion that a daily vitamin E supplement of 200 IU does not affect insulin, GHb, or FBS in T2DM patients.

It is worthwhile here to mention that among different lipoproteins, HDL, by and large, represents an antiinflammatory molecule. Furthermore, apolipoprotein A1 (Apo-A1) is the major apolipoprotein of HDL. In this context, Aldred *et al.* (2006) examined the effects of α-T supplementation on proapolipoprotein A1 expression by proteomics to improve the understanding of the physiological roles of α-T. A double-blind, randomized, parallel design supplementation trial was carried out in healthy subjects ($n = 32$; 11 males and 21 females) who consumed α-T supplements (134 or 268 mg/day) or placebo capsules for up to 28 days. Using semiquantitative proteomics, the authors report that proapolipoprotein A1 (identified by MS and Western blotting) was altered at least twofold. This was further confirmed using ELISA as revealed by a significant increase in plasma Apo-A1 concentration following α-T supplementation in terms of time as well as dose effect ($p < 0.01$ after 28 days supplementation with 268 mg α-T/day). However, whether α-T supplementation finally result in increased HDL is a question that needs to be addressed in future trials.

Vitamin E is known to inhibit platelet aggregation *in vitro* (Calzada *et al.*, 1997; Steiner, 1983). To further explore this finding, a short-term moderate dosage [800 IU of DL-tocopherol acetate (TA) for 14 days] of synthetic DL-α-TA supplementation trial was carried out (Dereska *et al.*, 2006) to see its effect on platelet aggregation, coagulation profile, and simulated bleeding time in healthy individuals. This study revealed that vitamin E supplementation with moderate dosage did not significantly prolong bleeding or platelet aggregation *in vivo*. Thus, the affect of vitamin E on platelet aggregation *in vitro* does not appear to be reproducible *in vivo* leading to a very important conclusion that perioperative discontinuation of vitamin E may not be necessary.

C. MOLECULAR AND CELLULAR EFFECTS OF α-T

Advances have been made in understanding the molecular effects of α-T beyond that of preventing LDL oxidation. The understanding of various regulatory, nonoxidative response to α-T by crucial cells involved in the pathogenesis of atherosclerosis is very important. As discussed above, such responses include inhibition of SMC proliferation, preservation of endothelial function, inhibition of MO-EC adhesion, inhibition of monocyte ROS and cytokine release, and inhibition of platelet adhesion and aggregation (Calzada *et al.*, 1997; Kaul *et al.*, 2001; Singh and Jialal, 2004; Singh *et al.*, 2005; Steiner, 1983). These cellular responses to α-T are associated with transcriptional and posttranscriptional events. Activation of diacylglycerol kinase and protein phosphatase 2A (PP2A), and the inhibition of protein kinase C (PKC), cyclooxygenase, lipoxygenase, tyrosine kinase

TABLE III. α-T and Cell Culture Studies

Cell type	Outcome
ECs	↓ E-selectin, ↓ ICAM/VCAM Inhibition of Ox-LDL mediated ICAM in HUVEC
ECs and MO	↓ LDL-induced MO adhesion to EC
	↓ ICAM
ECs	Inhibition of Ox-LDL mediated ICAM in HUVEC
MO/macrophages	↓ CD11b and VLA4 via inhibition of NF-κB activity, ↓ tissue factor
Platelets	↓ Platelet aggregation *in vitro* and *in vivo*, delays intraarterial thrombogenesis
Smooth muscle cells	↓ Proliferation

(Tyk2)-phsphorylation, and cytokine release by α-T are all examples of post-transcriptional regulation which have been discussed previously (Kaul *et al.*, 2001). Please see Table III for various reported studies in cell culture models.

V. INTERVENTION STUDIES

Inflammation is recognized as an overwhelming burden to the healthcare status of our population and the underlying basis of a significant number of diseases. The elderly generally bear the burden of morbidity and mortality, which may be reflective of elevated markers of inflammation resulting from decades of lifestyle choices. Lower cancer rates are associated with diets high in fiber, fruits, vegetables, and tea. CVD, metabolic syndrome, hypertension, diabetes, and hyperlipidemia may be ameliorated by treating the underlying cause: inflammation caused by visceral adipose tissue. Although there is much more to understand, we have enough information presently to make the necessary changes in our lifestyles to significantly affect the inflammatory process and potentially live longer, healthier lives, with fewer burdens to an overburdened and failing medical system.

While α-T has several beneficial effects on oxidation and on different cells that participate in atherogenesis, the results of randomized clinical trials have been equivocal and have been reviewed in detail previously (Jialal and Devaraj, 2003; Kaul *et al.*, 2001). However, prospective-controlled clinical trials have failed to demonstrate a benefit of antioxidant vitamin supplementation in primary or secondary prevention of CVD. Furthermore, Hatzigeorgiou *et al.* (2006) have also reported a negative finding from their study done in 865 consecutive patients, 39–45 years of age, without known coronary artery disease and presented for a periodic physical examination. Antioxidant intake was assessed with the Block Dietary Questionnaire, and coronary atherosclerosis was identified by measuring coronary artery calcification using electron beam computed tomography. The mean age for the study participants was 42 with 83% male, and the prevalence of coronary artery calcification was 20%. Vitamin supplements were used by 56% of the participants, and the mean daily intake (dietary plus supplemental) of vitamins A, C, and E were 1683, 371, and 97 mg, respectively. No significant correlation was found between coronary artery calcification score and individual vitamin or total antioxidant vitamin intake, even after adjusting for traditional cardiac risk factors. The highest quartile of vitamin E was positively associated with calcification. The authors have concluded that high doses of vitamin E may confer an increased risk of calcified atherosclerosis. Thus, based on totality of all evidences, the use of vitamin E (α-T) supplements as a preventive or therapeutic intervention remains controversial and cannot be recommended.

VI. OTHER FORMS OF VITAMIN E

A. γ-TOCOPHEROL

γ-T is the most abundant form of vitamin E in the US diet; however, it has received little attention since the discovery of vitamin E in 1922 and is not included in the current dietary intake recommendations. This is mainly due to the lower bioavailability and bioactivity of γ-T than α-T (Traber and Arai, 1999). γ-T is the most prevalent form of vitamin E in plant seeds and in products derived from them (Wagner *et al.*, 2004). Vegetable oils such as corn, soybean, and sesame, and nuts such as walnuts, pecans, and peanuts are rich sources of γ-T. Because of the widespread use of these plant products, GT represents 70% of the vitamin E consumed in the typical US diet (Devaraj and Traber, 2003). Recent evidence suggests that GT has properties that may be important to human health and that are not shared by AT.

B. ABSORPTION AND AVAILABILITY

Studies using deuterium-labeled tocopherols have led to the understanding of the absorption and metabolism of γ-T. Both α-T and γ-T are absorbed equally well by the gut, but the TTP responsible for packaging tocopherols into VLDL has a higher affinity for α-T over γ-T (Hensley *et al.*, 2004). For this reason, α-T supplementation decreases circulating γ-T (Hensley *et al.*, 2004). γ-T appears to be degraded largely to the hydrophilic γ-CEHC (Himmelfarb and McMonagle, 2001) by a cytochrome P450-dependent process (Jiang *et al.*, 2000) and is then primarily excreted in urine (Hasselwander and Young, 1998). Plasma γ-CEHC concentrations are reported to be 50–100 nM in humans (Traber and Arai, 1999). In human urine, γ-CEHC exists predominantly as a glucuronide conjugate with concentrations ranging from 4 to 33 μM (Himmelfarb and McMonagle, 2001), which increased to >100 μM after supplementation with γ-T (Eisengart *et al.*, 1956). Finally, in addition to the urinary excretion of γ-T as γ-CEHC, biliary excretion may be an alternative route for eliminating excess γ-T (Jiang *et al.*, 2001). Excess γ-T secreted into feces during supplementation may play a role in eliminating fecal mutagens and thus reduce colon cancer. Thus, the biological disposition and retention of γ-T appear to be regulated by a metabolism that is quite different from that of α-T.

C. γ-T AND ANTIINFLAMMATORY EFFECTS

γ-T has been shown to inhibit SMC proliferation by inhibiting PKC activity, while β-T had no effect, indicating that this effect is independent of antioxidant activity. Jiang *et al.* (2000) found that both γ-T and γ-CEHC possess antiinflammatory activity. Both inhibited PGE2 synthesis via inhibition of COX-2 activity in LPS-stimulated macrophages, while α-T slightly decreased PGE2 and in this study, had no effect on COX activity. In a subsequent study of Carrageenan induced inflammation in rats, γ-T supplementation (33 or 100 mg/kg) but not α-T (33 mg/kg) decreased PGE2 synthesis at the site of inflammation in addition to decreasing eight-isoprostanes, TNF, and total nitrates/nitrites (Jiang and Ames, 2003).

D. γ-T AND OTHER BENEFICIAL EFFECTS

γ-T is known to have antioxidant activity mainly due to its ability to donate phenolic hydrogens (electrons) to lipid radicals. Because of its lack of one of the electron-donating methyl groups on the chromanol ring, γ-T is somewhat less potent in donating electrons than is α-T and is, thus, a slightly less powerful antioxidant (Jiang *et al.*, 2001). However, the unsubstituted C5-position of γ-T appears to make it better able to trap lipophilic electrophiles such as reactive nitrogen species (RNS). In pioneering studies, Cooney *et al.* (1993) found that γ-T is superior to α-T in detoxifying nitrogen dioxide. They showed that γ-T reduces nitrogen dioxide to the less harmful nitric oxide or traps nitrogen dioxide to form 5-nitro-γ-T (5-N-γ-T), analogous to the nitration of tyrosine (Goss *et al.*, 1999). Hensley *et al.* (2000) reported an HPLC method for measuring 5-N-γ-T in which they reported an increase in 5-N-γ-T (both unadjusted and adjusted for γ-T) in rat astrocytes stimulated with bacterial lipopolysaccharide. Jiang and Ames (2003) showed that supplementation of rats with 90 mg γ-T/kg diet significantly inhibited protein nitration as evidenced by decreased levels of 3-nitro-tyrosine in plasma, liver, and kidney.

The precise location of α-T and γ-T in the lipid environment may be partially responsible for their different reactivities (Chatelain *et al.*, 1993). The lack of a methyl group makes γ-T relatively less hydrophobic, which may affect its location and interaction with lipids and aqueous-phase components. A superior effect of γ-T was also observed when lipid peroxidation was initiated by peroxynitrite in brain homogenates (Williamson *et al.*, 2002). γ-T is predominantly located in the biomembrane of brain homogenates, which have a lipid arrangement similar to that of the liposome model. Differences in the liposomal and LDL particle lipid environments and the arrangement of tocopherol species with in these particles could play an important role in the protective effects observed, the understanding of which requires further investigation. The major pathway for the metabolism of γ-T is the cytochrome P450-dependent oxidation to its water-soluble metabolite γ-CEHC, which is excreted in urine.

Furthermore, Wu *et al.* (2005b) worked on the hypothesis that increased levels of nitrated γ-T (5-NO_2-γ-T) are present in smokers and individuals with conditions associated with elevated nitrative stress. Thus, the monitoring of 5-NO_2-γ-T and its possible metabolite(s) represents a useful marker of RNS generation *in vivo*. While γ-CEHC was abundant in urine from healthy volunteers, as well as patients with CHD and T2DM, 5-NO_2-γ-CEHC was undetectable (limit of detection of 5 nM) by proton NMR and MS. To understand this observation, the uptake and metabolism of γ-T and 5-NO_2-γ-T by HepG2 cells was examined. γ-T was readily incorporated into cells and metabolized to γ-CEHC over a period of 48 h. In contrast, 5-NO_2-γ-T was poorly incorporated into HepG2 cells and not metabolized to 5-NO_2-γ-CEHC over the same time period. Thus, it was concluded that nitration of γ-T prevents its incorporation into liver cells and therefore its metabolism to the water-soluble metabolite. Whether 5-NO_2-γ-T could be metabolized via other pathways *in vivo* requires further investigation.

E. γ-TOCOPHEROLS AND CARDIOVASCULAR DISEASE

Several animal studies provide some evidence that γ-T might be beneficial against CVD. Saldeen *et al.* (1999) investigated the effects of α-T and γ-T supplementation on platelet aggregation and thrombosis in Sprague–Dawley rats. They found that γ-T supplementation (100 mg/kg/day) led to a greater decrease in platelet aggregation and delay of arterial thrombogenesis than did α-T supplementation. γ-T supplementation also resulted in stronger inhibition of superoxide generation and lipid peroxidation. Subsequently, they reported that γ-T was significantly more potent than was α-T in enhancing SOD activity in plasma and arterial tissue and in increasing the arterial protein expression of both manganese SOD and Cu/Zn SOD (Liu *et al.*, 2003). Also, γ-T supplementation was associated with increase in eNOS.

However, the relevance of these studies is unclear since most of the studies discussed used either mixed tocopherol (MT) preparations and γ-T-enriched supplements rather than purified γ-T alone.

Although much less is known about γ-T than about α-T, much evidence suggests that γ-T may be important in the defense against CVD. Plasma γ-T concentrations are inversely associated with increased morbidity and mortality due to CVD (Albert *et al.*, 2002). Various investigators (Kontush *et al.*, 1999; Ohrvall *et al.*, 1996) reported that serum concentrations of γ-T, but not of α-T, were lower in CVD patients than in healthy control subjects. In a concomitant cross-sectional study of Swedish and Lithuanian middle-aged men, Kristenson *et al.* (1997) found that plasma γ-T concentrations were twice as high in the Swedish men, who had a 25% lower incidence of CVD-related mortality. In a 7-year follow-up study of 34,486 postmenopausal women, Kushi *et al.* (1996) concluded that the intake of dietary vitamin E (mainly γ-T), but not of supplemental vitamin E (mainly α-T), was significantly inversely associated with increased risk of death by CVD. These investigators (Yochum *et al.*, 2000) further showed that dietary vitamin E was associated with a reduced incidence of death from stroke in postmenopausal women. Regular consumption of nuts, which are an excellent source of γ-T among other ingredients, lowers the risk of MI and death from IHD.

In a recent case-control and follow-up study of antioxidant response to MI, Ruiz Rejón *et al.* (2002) studied 106 MI patients and 104 control subjects. It was found that plasma γ-T was significantly lower by 21% in MI patients relative to matched control patients, while α-T was not different between cases and controls. Contrastingly α-T, β-carotene, lycopene, cryptoxanthine, and lutein were not statistically associated with risk of MI. The weakness of Ruiz Rejón *et al.* (2002) report is that the study design did not allow γ-T measurement in patients before MI, so it is unclear whether low γ-T levels in MI patients reflected a predisposition toward, or a result of cardiac stress.

Another study of γ-T in MI has been published (El-Sohemy *et al.*, 2002) that investigated 475 MI survivors in Costa Rica. The results of this study were less straightforward than those of the Ruiz Rejón *et al* (2002). Costa Rican subjects in the highest quintile of dietary γ-T had a lower risk of MI compared with those in the lowest quintile ($p < 0.02$ for the trend) but the trend was no longer statistically substantial in multivariate analysis. A weak association was found for adipose tissue γ-T in univariate and multivariate models, while a substantial inverse association was found for dietary α-T intake and MI. Interestingly, the inverse association between total vitamin E intake and MI was strengthened when chronic supplement users were excluded. Part of the discrepancy between these two studies might arise because the Costa Rican study relied, in part, on food-frequency questionnaires for evaluation of tocopherol intake and such surrogate indices are often poorly correlated with plasma tocopherol values. Blood tocopherol levels were not reported in this study. Clearly more epidemiological data are needed to assess the value (or lack of value) for γ-T in reducing the risk of MI.

As evident from literature that smokers provide a common at-risk group whose study might shed light on protective factors for CVD. As the major lipophilic antioxidant, requirements for vitamin E may be higher in smokers due to increased utilization. In an observational study, vitamin E status was compared in smokers and nonsmokers using a holistic approach by measuring plasma, erythrocyte, lymphocyte, and platelet α-T and γ-T, as well as the specific urinary vitamin E metabolites α- and γ-CEHC. Fifteen smokers (average age 27 years, smoking time 7.5 years) and nonsmokers of comparable age, gender, and body mass index (BMI) were recruited (Jeanes *et al.*, 2004). No significant differences were found between plasma and erythrocyte α-T and γ-T in smokers and nonsmokers. However, smokers had significantly lower α-T and γ-T levels in their lymphocytes, as well as significantly lower α-T levels in platelets; β-T levels were similar. Interestingly, smokers also had significantly higher excretion of the urinary γ-T metabolite, γ-CEHC compared to nonsmokers, while their α-CEHC (metabolite of α-T) levels were similar. There was no significant difference between plasma ascorbate, urate, and F2-isoprostane levels. Therefore, in this population of cigarette smokers (mean age 27 years, mean smoking duration 7.5 years), alterations to vitamin E status can be observed even without the more characteristic changes to ascorbate and F2-isoprostanes. While several high-profile studies have shown α-T intake somewhat protective against CAD in smokers, no similar studies have been undertaken using γ-T as an independent variable. Further, Morton *et al.* (2002) measured γ-T and 5-NO_2-γ-T in sera and carotid plaques of patients with CVD using LC-MS. Plasma γ-T was 12% reduced in CAD, while 5-NO_2-γ-T was elevated threefold.

F. γ-T SUPPLEMENTATION IN HUMANS

Despite the promises of γ-T as an effective antioxidant and antiinflammatory agent *in vitro*, with regards to γ-T supplementation in humans, the literature is scanty. In a clinical trial, Himmelfarb *et al.* (2003) enrolled 15 uremic patients undergoing dialysis. Five patients were supplemented with *RRR*-α-T (300 mg/day) and 10 received mixture of tocopherols (60% *RRR*-γ-T, 28% *RRR*-δ-T, and 18% *RRR*-α-T) for a duration of 14 days. A potentially important observation in this study is that the administration of the γ-T-enriched preparation, but not the α-T preparation, significantly reduced CRP concentrations in hemodialysis patients. However, owing to the limitation of this study because of small sample size with respect to inflammatory biomarkers, further studies with larger sample sizes will be required to more definitively address these important end points. Furthermore, Liu *et al.* (2003) supplemented healthy subjects with either placebo, all-*rac* α-T (100 mg/day), or MTs (comprising 100 mg γ-T, 20 mg δ-T, and 20 mg α-T) for 8 weeks. Mixed tocopherols but not α-T supplementation decreased ADP-induced platelet aggregation. Both α-T and MTs supplementation resulted in reduced PKC and increased SOD and NO release. These two studies point to an important role for either γ-T or combined α-T + γ-T supplementation on biomarkers of oxidative stress and inflammation and CVD; however this needs to be carefully studied.

Furthermore, another study carried out by Clarke *et al.* (2006) demonstrate that supplementation with MTs increases serum and blood cell GT but does not alter biomarkers of platelet activation in subjects with T2DM. These authors compared the effects of supplementation with α-T (500 mg) and a γ-T-rich compound (500 mg, containing 60% GT) on serum and cellular tocopherol concentrations, urinary tocopherol metabolite excretion, and *in vivo* platelet activation in subjects with T2DM. Fifty-eight subjects were randomly assigned to receive either 500 mg α-T/day, 500 mg MT/day, or matching placebo. Serum, erythrocyte, and platelet tocopherol and urinary metabolite concentrations were measured at baseline and after the 6-week intervention. Soluble CD40 ligand, urinary 11-dehydro-TXB2, serum TXB2, sP-selectin, and vWF were measured as biomarkers of *in vivo* platelet activation. Serum α-T increased with both tocopherol treatments. Serum and cellular γ-T increased fourfold in the MT group, whereas red blood cell γ-T decreased significantly after α-T supplementation. Neither treatment had any significant effect on markers of platelet activation. However, supplementation with α-T decreased red blood cell γ-T, whereas MT increased both serum α-T and serum and cellular γ-T.

Thus, it is clear that while γ-T shows great promise as an antioxidant and antiinflammatory agent, controlled intervention studies in humans are required to clearly establish the benefits of γ-T supplementation. Furthermore, potential synergistic effects between γ-T and α-T and other antioxidants should also be explored. These efforts should help to clarify the role of γ-T in CVD prevention and human health.

VII. Tocotrienols

Tocotrienols are naturally occurring farnesylated unsaturated analogues of α-, β-, γ-, and δ-T. As discussed earlier, tocotrienols differ from the corresponding tocopherols only in their aliphatic tail. Tocopherols have a phytyl side chain attached to their chromanol nucleus, whereas the tail of tocotrienols is unsaturated and forms an isoprenoid chain. In contrast to corn, wheat, and soybean which contain mainly tocopherols; barley, oats, palm, and commercial rice barns contain >70% tocotrienols (known as tocotrienol-rich fraction, TRF), which consists of α-, β-, γ-, and δ-T3. Although α-T and GT are considered to be the more biologically active forms of vitamin E, yet some of the reports published herewith suggests that α-T3 too possess antioxidant activity and may prevent CVD (Kamal-Eldin and Appelqvist, 1996; Servinova *et al.*, 1991; Yoshida *et al.*, 2003). Hypocholesterolemic effects of tocotrienols have also been reported (Pearce *et al.*, 1992; Qureshi *et al.*, 1991a,b, 1995, 2000, 2001a,b,c, 2002). Tocotrienols possess powerful neuroprotective, anticancer, and cholesterol-lowering properties that are often not exhibited by tocopherols. Current developments in vitamin E research clearly indicate that members of the vitamin E family are not redundant with respect to their biological functions. α-T3, γ-T, and δ-T3 have emerged as vitamin E molecules with functions in health and disease that are clearly distinct from that of AT.

VIII. Hypocholesterolemic Effect

The interest in tocotrienol as a hypocholesterolemic compound began when Qureshi *et al.* (1986) attributed the major cholesterol-lowering action of barley to be at the level of cholesterol synthesis. By sequential extraction of barley, α-T3 was identified as the chemical constituent responsible for inhibiting the rate-limiting enzyme of the cholesterol biosynthetic pathway, HMG-CoA reductase (HMGR). In elucidating the molecular mechanism for this suppression, the effect was ascribed to the side chain's unique ability to increase cellular farnesol, a mevalonate-derived product, which signals the proteolytic degradation of HMGR (Qureshi *et al.*, 1993). These authors further isolated two different novel tocotrienols (desmethyl [D-P21-T3] and didesmethyl [D-P25-T3]) from heated rice bran and reported the inhibition of atherosclerotic lesion formation in ApoE$^{-/-}$ mice models as compared to α-T (Qureshi *et al.*, 2001c). Also, another group (Black *et al.*, 2000) supplemented atherogenic diet with 0.5% α-T alone or 0.5% and 1.5% palm tocotrienols (palm E; 33% α-T, 16.1% α-T3, 2.3% β-T3, 32.2% γ-T3, 16.1% δ-T3) in ApoE$^{+/-}$ female mice. These authors reported protection of palm tocotrienols supplementation against diet-induced atherosclerosis both at 0.5 and 1.5% palm tocotrienols diet as apposed to 0.5-g α-T alone. On the basis of this it appears that tocotrienol content of the palm E supplement had an important independent effect. Further, 0.5% palm E-mediated reduction in atheroma formation in mice whose plasma cholesterol concentration and lipoprotein pattern had changed little, if at all, suggesting that attenuation of lesion formation by these diets must have occurred by mechanisms other than cholesterol control alone.

Yun *et al.* (2006) evaluated the safety and efficacy of large supplements of TRF and its constituents in 5-week-old female chickens. Diets supplemented with 50, 100, 250, 500, 1000, or 2000 ppm of tocotrienol-rich-fraction (TRF), α-T, AT3, or GT3 were fed to chickens for 4 weeks. Supplemental TRF produced a dose–response (50–2000 ppm) lowering of serum total and LDL cholesterol levels of 22% and 52%, respectively, compared with the control group. α-T did not affect total or LDL cholesterol levels. Supplemental α-T3 within the 50–500 ppm range produced a dose–response lowering of total (17%) and LDL (33%) cholesterol levels. The more potent γ- and δ-isomers yielded dose–response (50–2000 ppm) reductions of serum total (32%) and LDL (66%) cholesterol levels. Furthermore, serum triglyceride levels were significantly lower in sera of pullets receiving the higher supplements. The safe dose of various tocotrienols for human consumption might be 200–1000 mg/day based on this study.

High-cholesterol diets alter myocardial and vascular NO-cGMP signaling and have been implicated in ischemic/reperfusion injury. In this regard, Esterhuyse *et al.* (2005) investigated the effects of dietary red palm oil (RPO) containing fatty acids, carotonoids, tocopherols, and tocotrienols on myocardial ischemic tolerance and NO-cGMP pathway function in the Wistar rats fed a standard rat chow $\pm$ RPO, or a standard rat chow + cholesterol $\pm$ RPO diet. Myocardial mechanical function and NO-cGMP signaling pathway intermediates were determined before, during, and after 25 min ischemia. RPO-supplementation improved aortic output recovery and increased myocardial ischemic cGMP concentrations. Simulated ischemia (hypoxia) increased cardiomyocyte nitric oxide levels in the two RPO supplemented groups, but not in control nonsupplemented groups. RPO supplementation also increased hypoxic nitric oxide levels in the control diet fed, but not the cholesterol fed rats. These data suggest that dietary RPO may improve myocardial ischemic tolerance by increasing bioavailability of NO and improving NO-cGMP signaling in the heart.

The goal in type 2 diabetics is to reduce LDL-C levels $\leq$100 mg/dl. On the basis of this aspect, the therapeutic impacts of tocotrienols on serum and lipoprotein lipid levels in type 2 diabetic patients were investigated (Baliarsingh *et al.*, 2005). On the basis of known TRF-mediated decrease on elevated blood glucose and glycated hemoglobin A(1C) (HbA(1C)) in diabetic rats, a randomized, double-blind, placebo-controlled study involving 19 T2DM subjects with hyperlipidemia was conducted. After 60 days of TRF treatment, subjects showed an average decline of 23, 30, and 42% in serum total lipids, TC, and LDL-C, respectively. In the present investigation tocotrienols mediated a reduction of LDL-C from an average of 179 to 104 mg/dl. However, hypoglycemic effect of TRF was not observed in these patients because they were glycemically stable and their glucose and HbA(1) levels were close to normal values. Thus, this study made an important hallmark that daily intake of dietary TRF by T2DM subjects can be useful in the prevention and treatment of hyperlipidemia and atherogenesis.

Taking into consideration that the migration of circulating monocytes into the subendothelial space occurs through the expressing of some adhesion molecules on endothelial cells, Naito *et al.* (2005) investigated whether oxysterols (25-OH-cholesterol) can enhance the MO-EC exposed to 25-OHC via increasing expression of VCAM. 25-OHC enhanced surface expression

VCAM-1 mRNA and stimulated MO-EC adhesion in a dose-dependent fashion. The combination treatment with anti-VCAM-1 and anti-CD11b monoclonal antibodies significantly reduced the monocyte adherence to 25-hydroxycholesterol-stimulated ECs. Compared to α-T, tocotrienols displayed a more profound inhibitory effect on adhesion molecule expression and monocytic cell adherence, although δ-T3 exerted a most profound inhibitory action on monocytic cell adherence when compared to α-T and α-, β-, and γ-T3. Tocotrienols accumulated in HAECs to levels ~25–95-fold greater than that of α-T. Hence these results indicated that 25-OHC can enhance the interaction between monocytes and HAECs and that tocotrienols had a profound inhibitory effect on monocytic cell adherence to HAECs relative to α-T via inhibiting the VCAM-1 expression. These superior inhibitory effects of tocotrienols may be dependent on their intracellular accumulation.

Overall, it appears that one or more of the tocotrienols reduced lesion formation in atherosclerosis-prone $ApoE^{+/-}$ mice by at least two mechanisms. One may have been an antioxidant effect with no alteration in hepatic or serum cholesterol or in serum lipoproteins and the second is largely independent of antioxidant action through some other unrelated mechanism.

Further, the ratio of the tocopherols and tocotrienols has been suggested to play an important role in determining the hypocholesterolemic properties of tocotrienols (Qureshi *et al.*, 2000). The presence of more than 20% α-T in TRF from palm oil results in an attenuation of the hypocholesterolemic effect of tocotrienols. These findings have been confirmed by Khor and Ng (2000). Hence it has been reported that TRF may desirably be prepared from any natural source with minimum concentration of tocopherols in the mixture. It has been very convincingly shown by several investigators (Khor *et al.*, 1995; Watkins *et al.*, 1993) that lower doses (5 or 50 mg/kg) of γ-T3 are much more effective in inhibiting the activity of HMG-CoA reductase and lowering the serum total cholesterol, and LDL-cholesterol levels than their respective higher doses (50 or 100 mg/kg) in guinea pigs and rats. The reason for low inhibitory effect on the activity of HMG-CoA reductase by γ-T3 at higher doses is suggested to be due to the bioconversion of γ-T3 to α-T in the body. However, in hypercholesterolemic humans placed on AHA Step 1 diet, it has been reported (Qureshi *et al.*, 2002) that there occurs a dose-dependent suppression of serum cholesterol by TRF of rice bran. A dose of 100 mg/day of TRF produced maximum decreases of 20, 25, 14% ($p < 0.05$), and 12%, respectively, in serum total cholesterol, LDL-cholesterol, apolipoprotein B (Apo-B), and triglycerides compared with the baseline values, suggesting that a dose of 100-mg/day TRF25 plus AHA Step-1 diet may be the optimal dose for controlling the risk of coronary heart disease in hypercholesterolemic human subjects.

It is worthwhile to mention that apart from positive aspect of tocotrienols published from some investigators, various others have reported negative effect. A double-blind, placebo-controlled trial (O'Byrne *et al.*, 2000) was carried out by our group. Subjects were randomly assigned to receive placebo ($n = 13$), α-($n = 13$), γ-($n = 12$), or δ-($n = 13$) tocotrienyl acetate supplements (250 mg/day). All subjects followed a low-fat diet for 4 weeks, then took supplements with dinner for the following 8 weeks while still continuing diet restrictions. Following supplementation in the respective groups plasma concentrations were: α-T3 0.98 ± 0.80 μmol/liter, γ-T3 0.54 ± 0.45 μmol/liter, and δ-T3 0.09 ± 0.07 μmol/liter. α-T3 increased *in vitro* LDL oxidative resistance (+22%, $p < 0.001$) and decreased its rate of oxidation ($p < 0.01$). Neither serum LDL cholesterol nor Apo-B was significantly decreased by tocotrienyl acetate supplements. Our results are supported by another reported investigation that supplementation of different tocotrienols (200 mg/day for 28 days) failed to improve CV risk factors in patients with hypercholesterolemia (Mustad *et al.*, 2002). In an earlier investigation (Mensink *et al.*, 1999), a randomized, double-blind, placebo-controlled trial was conducted in 20 men with mildly elevated serum lipid concentrations which were given a mixture of 140-mg tocotrienols and 80-mg α-T versus 80-mg α-T alone. There was no favorable effect of a tocotrienol-rich palm oil concentrate on serum lipids and lipoproteins. Further, Tan *et al.* (1991) reported that a daily supplement of only one capsule of tocotrienols containing 18-mg tocopherols and 42-mg tocotrienols effectively lowered serum LDL-cholesterol concentrations in both normal and hypercholesterolemic subjects. However, no control group was involved in that study, which makes it impossible to separate possible time or placebo effects from treatment effects. Furthermore, Wahlqvist *et al.* (1992) also failed to demonstrate any hypocholesterolemic effect of tocotrienols. In their study, 44 hypercholesterolemic subjects received either placebo capsules or an increasing dose of tocotrienols. After 16 weeks, serum lipid measures were not significantly different between the two experimental groups. However, Qureshi *et al.* (1995) reported an LDL cholesterol-lowering effect of γ-T3 (in particular) in hypercholesterolemic subjects. From additional studies with chickens (Qureshi *et al.*, 2001a), it was concluded that α-T suppressed the inhibiting effects of γ-T3 on HMG-CoA reductase activity. It was therefore postulated that effective tocotrienol preparations should contain <15–20% (by weight) α-T and 45% γ- plus δ-T3. Probably the variable composition of the tocotrienol supplements explained why some studies claimed effects, whereas other studies did not as discussed above. However, Qureshi *et al.*'s hypothesis is not in line with the results of Atroshi *et al.* (1992), who reported no effect on serum lipoproteins when using capsules containing only 15% γ-tocopherol. Therefore, the suggestion of Qureshi *et al.* (1995) awaits confirmation from studies with different population groups. In this respect, it seems suffice to state that the TRF from rice bran oil might be a more promising hypolipidemic agent because it contains a high amount of γ-T3 and a low amount of α-T (Yoshida *et al.*, 2003). In addition, rice bran oil may contain two other tocotrienol components that have been reported to improve the serum lipoprotein profile (Qureshi *et al.*, 2001b). Thus on the whole, in the context of conflicting and limited literature available herewith for efficacy of tocotrienols as a hypocholesterolemic agent, more controlled trials warrants future investigation in order to reach a firm conclusion for this form of vitamin E.

IX. ANTIINFLAMMATORY EFFECTS

Theriault *et al.* (2002) have shown that α-T3 is effective in reducing MO-EC adhesion. On the basis of this, same group of investigators (Chao *et al.*, 2002) explored efficacy of δ-T3 versus α-T3 on monocyte–endothelial cell adhesion and endothelial cell adhesion molecules using HUVEC cell line as the model system. Relative to α-T3, δ-T3 displayed 1.5-fold more profound inhibitory effect on monocytic cell adherence using a 15-μM concentration within 24 h. This inhibitory action was reversed by coincubation with farnesol and geranylgeraniol, suggesting a role for prenylated proteins in the regulation of monocyte adhesion. Furthermore, another study is published by same investigators (Theriault *et al.*, 1999) delineating the mechanism of action of γ-T3 on hepatic modulation of Apo-B production using cultured HepG2 cells as the model system. Unlike tocopherol, γ-T3 has also been shown to reduce plasma Apo-B levels in hypercholesterolemic subjects. These authors report that γ-T3 stimulates Apo-B degradation possibly as the result of decreased Apo-B translocation into the endoplasmic reticulum lumen. It is speculated that the lack of cholesterol availability reduces the number of secreted Apo-B-containing lipoprotein particles by limiting translocation of Apo-B into the endoplasmic reticulum lumen.

Further, TRF25 and D-P25-T3 have been shown to lower arachidonic acid in various tissues of hereditary hypercholesterolemic swine (Qureshi *et al.*, 1991a). There was reported to be an overall reduction in prostaglandins and leukotrienes, both of which are synthesized from arachidonic acid, and thus a possible reduction in IL-1. Furthermore, pretreatment with novel tocotrienols has been reported to reduce the induction of TNF in response to bacterial LPS in mice (Qureshi *et al.*, 1993). The inhibition of TNF by novel tocotrienols thus has been suggested to affect atherosclerosis (Qureshi *et al.*, 1993).

X. ANTIOXIDANT EFFECT

Furthermore, while there are numerous reports on the antioxidant properties of tocopherols, fewer studies are available for tocotrienols. Again, the reports available are conflicting. For example, it was reported that α-T3 possessed 40- to 60-fold higher antioxidant activity than α-T against ferrous iron/ascorbate- and ferrous iron/NADPH-induced lipid peroxidation in rat liver microsomes (Servinova *et al.*, 1991) followed by another report by Suzuki *et al.* (1993) stating that α-T3 exhibited greater peroxyl radical scavenging potency than α-T in liposomal membranes. Sen *et al.* (2000) reported that α- and γ-T3 were more effective than α-T in preventing glutamate-induced neuronal cell death by regulating signal transduction processes. Kamat and Devasagayam (1995) observed similar results in rat brain mitochondria and noted a stronger effect with α-T3. Furthermore, a clinical trial carried out in the past by Tomeo *et al.* (1995) underscored the antioxidative effect of palm tocotrienols in two subgroups (C versus treated) of 25 patients each with evidence of hyperlipidemia and carotid stenosis. Twenty-eight percent of the tocotrienols supplemented patients revealed positive outcome in terms of carotid artery regression as well as reduction in serum TBARS with no change in serum total cholesterol, LDL-C, and TG content. As described previously, we have also shown that α-T3 supplementation in humans significantly decreased LDL oxidation in patients with hypercholesterolemia. In an elaborate study by Yoshida *et al.* (2003), a number of mechanisms were shown to contribute to the higher antioxidant activity of α-T3 compared to α-T, including: (1) a more uniform distribution in the membrane lipid bilayer, (2) a more efficient interaction of the chromanol ring with lipid radicals, and (3) a higher recycling efficiency from chromanoxyl radicals. Apart from its beneficial effect in CVD, Khanna *et al.* (2006) report the positive outcome of tocotrienols as a potent neuroprotective. On a concentration basis, this finding represents the most potent of all biological functions exhibited by any natural vitamin E molecule. Thus, on the whole, we conclude that tocotrienols because of its unique structure comparable to α-T might possess better antioxidant activity than α-T. However, this needs to be confirmed in human subjects.

XI. MECHANISM OF ACTION AND FUTURE DIRECTION

Although the exact mechanism of inhibition of atherosclerotic lesions by tocotrienols has not been elucidated, there is evidence suggesting that tocotrienols affect several distinct steps in the pathways leading to formation of complex atherosclerotic lesions. The effect of tocotrienols in contrast to that of α-T on the activity of PKC has not been reported. However as discussed earlier, various investigators have reported that tocotrienols too have antioxidant activity and are effective in protecting against some free radical-related diseases.

XII. CONCLUSION

Various different forms of vitamin E exhibit antiinflammatory and inhibit several biological events involved in atherogenesis. Although the studies carried out with cell culture and animal models prove promising antiatherosclerotic effects of α-T, the results of clinical trials are equivocal possibly because of inadequate subjects selection (by gender, vitamin E status, and so on), the dose, timing of intake and chemical form of tocopherol, and so on. Future trials are highly warranted in this context in subjects with increased oxidative stress (such as diabetics or patients on hemodialysis) to unravel the mystery of different forms of vitamin E and reach to a final conclusion per se which has a mixed outcome in present scenario.

REFERENCES

Accinni, R., Rosina, M., Bamonti, F., Della Noce, C., Tonini, A., Bernacchi, F., Campolo, J., Caruso, R., Novembrino, C., Ghersi, L., Lonati, S., and Grossi, S. (2006). Effects of combined dietary supplementation on oxidative and inflammatory status in dyslipidemic subjects. *Nutr. Metab. Cardiovasc. Dis.* **16**(2), 121–127.

Akiyama, S., Inagaki, M., Tsuji, M., Gotoh, H., Gotoh, T., Washio, K., Gotoh, Y., and Oguchi, K. (2005). Comparison of effect of vitamin E-coated dialyzer and oral vitamin E on hemodialysis-induced Cu/Zn-superoxide dismutase. *Am. J. Nephrol.* **25**(5), 500–506.

Albert, C. M., Gaziano, J. M., Willett, W. C., and Manson, J. E. (2002). Nut consumption and decreased risk of sudden cardiac death in the physicians' health study. *Arch. Intern. Med.* **162,** 1382–1387.

Aldred, S., Sozzi, T., Mudway, I., Grant, M. M., Neubert, H., Kelly, F. J., and Griffiths, H. R. (2006). Alpha tocopherol supplementation elevates plasma apolipoprotein A1 isoforms in normal healthy subjects. *Proteomics* **6**(5), 1695–1703.

Atroshi, F., Antila, E., and Sankari, S. (1992). Palm oil vitamin E effects in hypercholesterolemia. *In* "Lipid-Soluble Antioxidants: Biochemistry and Clinical Applications" (A. S. H. Ong and L. Packer, Eds.), pp. 575–580. Birkhäuser Verlag, Basel, Switzerland.

Azzi, A., Ricciarelli, R., and Zingg, J. M. (2002). Non-antioxidant molecular functions of alpha-tocopherol (vitamin E). *FEBS Lett.* **519,** 8–10.

Baliarsingh, S., Beg, Z. H., and Ahmad, J. (2005). The therapeutic impacts of tocotrienols in type 2 diabetic patients with hyperlipidemia. *Atherosclerosis* **182,** 367–374.

Black, T. M., Wang, P., Maeda, N., and Coleman, R. A. (2000). Palm tocotrienols protect ApoE+/− mice from diet-induced atheroma formation. *J. Nutr.* **130,** 2420–2426.

Boshtam, M., Rafiei, M., Golshadi, I. D., Ani, M., Shirani, Z., and Rostamshirazi, M. (2005). Long term effects of oral vitamin E supplement in type II diabetic patients. *Int. J. Vitam. Nutr. Res.* **75,** 341–346.

Calzada, C., Bruckdorfer, K., and Rice Evans, C. (1997). The influence of antioxidant nutrients on platelet function in healthy volunteers. *Atherosclerosis* **128,** 97–105.

Chao, J. T., Gapor, A., and Theriault, A. (2002). Inhibitory effect of delta-tocotrienol, a HMG CoA reductase inhibitor, on monocyte-endothelial cell adhesion. *J. Nutr. Sci. Vitaminol.* **48,** 332–337.

Chatelain, E., Boscoboinik, D. O., Bartoli, G. M., Kagan, V. E., Gey, F. K., Packer, L., and Azzi, A. (1993). Inhibition of smooth muscle cell proliferation and protein kinase C activity by tocopherols and tocotrienols. *Biochim. Biophys. Acta* **1176,** 83–89.

Clarke, M. W., Ward, N. C., Wu, J. H., Hodgson, J. M., Puddey, I. B., and Croft, K. D. (2006). Supplementation with mixed tocopherols increases serum and blood cell gamma-tocopherol but does not alter biomarkers of platelet activation in subjects with type 2 diabetes. *Am. J. Clin. Nutr.* **83,** 95–102.

Cooney, R. V., Franke, A. A., Harwood, P. J., Hatch-Pigott, V., Custer, L. J., and Mordan, L. J. (1993). Gamma-tocopherol detoxification of nitrogen dioxide: Superiority to alpha-tocopherol. *Proc. Natl. Acad. Sci. USA* **90,** 1771–1775.

Dereska, N. H., McLemore, E. C., Tessier, D. J., Bash, D. S., and Brophy, C. M. (2006). Short-term, moderate dosage vitamin E supplementation may have no effect on platelet aggregation, coagulation profile, and bleeding time in healthy individuals. *J. Surg. Res.* **132**(1), 121–129.

Devaraj, S., and Traber, M. G. (2003). Gamma-tocopherol, the new vitamin E? *Am. J. Clin. Nutr.* **77,** 530–531.

Eisengart, A., Milhorat, A. T., Simon, E. J., and Sundheim, L. (1956). The metabolism of vitamin E. II. Purification and characterization of urinary metabolites of alpha-tocopherol. *J. Biol. Chem.* **221,** 807–817.

El-Sohemy, A., Baylin, A., Spiegelman, D., Ascherio, A., and Campos, H. (2002). Dietary and adipose tissue gamma-tocopherol and risk of myocardial infarction. *Epidemiology* **13,** 216–223.

Esterhuyse, A. J., Toit, E. D., and Rooyen, J. V. (2005). Dietary red palm oil supplementation protects against the consequences of global ischemia in the isolated perfused rat heart. *Asia Pac. J. Clin. Nutr.* **14,** 340–347.

Gasparetto, C., Malinverno, A., Culacciati, D., Gritti, D., Prosperini, P. G., Specchia, G., and Ricevuti, G. (2005). Antioxidant vitamins reduce oxidative stress and ventricular remodeling in patients with acute myocardial infarction. *Int. J. Immunopathol. Pharmacol.* **18,** 487–496.

Gey, K., Puska, P., Jordan, P., and Moser, U. K. (1991). Inverse correlation between plasma vitamin E and mortality from ischemic heart disease in cross-cultural epidemiology. *Am. J. Clin. Nutr.* **53**(Suppl.), 326S–334S.

Gianello, R., Libinaki, R., Azzi, A., Gavin, P. D., Negis, Y., Zingg, J. M., Holt, P., Keah, H. H., Griffey, A., Smallridge, A., West, S. M., and Ogru, E. (2005). Alpha-tocopheryl phosphate: A novel, natural form of vitamin E. *Free Radic. Biol. Med.* **39**(7), 970–976.

Goss, S. P., Hogg, N., and Kalyanaraman, B. (1999). The effect of alpha-tocopherol on the nitration of gamma-tocopherol by peroxynitrite. *Arch. Biochem. Biophys.* **363,** 333–340.

Hasselwander, O., and Young, I. S. (1998). Oxidative stress in chronic renal failure. *Free Radic. Res.* **29,** 1–11.

Hatzigeorgiou, C., Taylor, A. J., Feuerstein, I. M., Bautista, L., and O'Malley, P. G. (2006). Antioxidant vitamin intake and subclinical coronary atherosclerosis. *Prev. Cardiol.* **9,** 75–81.

Hensley, K., Williamson, K. S., and Floyd, R. A. (2000). Measurement of 3-nitrotyrosine and 5-nitro-gamma-tocopherol by high-performance liquid chromatography with electrochemical detection. *Free Radic. Biol. Med.* **28,** 520–528.

Hensley, K., Benaksas, E. J., Bolli, R., Comp, P., Grammas, P., Hamdheydari, L., Mou, S., Pye, Q. N., Stoddard, M. F., Wallis, G., Williamson, K. S., West, M., *et al.* (2004). New perspectives on vitamin E: Gamma-tocopherol and carboxyelthylhydroxychroman metabolites in biology and medicine. *Free Radic. Biol. Med.* **36**(1), 1–15. Review.

Himmelfarb, J., and McMonagle, E. (2001). Manifestations of oxidant stress in uremia. *Blood Purif.* **19,** 200–205.

Himmelfarb, J., Kane, J., McMonagle, E., Zaltas, E., Bobzin, S., Boddupalli, S., Phinney, S., and Miller, G. (2003). Alpha and gamma tocopherol metabolism in healthy subjects and patients with end-stage renal disease. *Kidney Int.* **64**(3), 978–991.

Jeanes, Y. M., Hall, W. L., Proteggente, A. R., and Lodge, J. K. (2004). Cigarette smokers have decreased lymphocyte and platelet alpha-tocopherol levels and increased excretion of the gamma-tocopherol metabolite gamma-carboxyethyl-hydroxychroman (gamma-CEHC). *Free Radic. Res.* **38,** 861–868.

Jessup, W., Kritharides, L., and Stocker, R. (2004). Lipid oxidation in atherogenesis: An overview. *Biochem. Soc. Trans.* **32,** 134–138.

Jialal, I., and Devaraj, S. (2001). Inflammation and atherosclerosis: The value of the high-sensitivity C-reactive protein assay as a risk marker. *Am. J. Clin. Pathol.* **116**(Suppl.), S108–S115.

Jialal, I., and Devaraj, S. (2003). Antioxidants and atherosclerosis: Don't throw out the baby with the bath water. *Circulation* **107,** 926–928.

Jialal, I., Devaraj, S., and Venugopal, S. K. (2004). C-reactive protein: Risk marker or mediator in atherothrombosis? *Hypertension* **44,** 1–6.

Jiang, Q., and Ames, B. N. (2003). Gamma-tocopherol, but not alpha-tocopherol, decreases proinflammatory eicosanoids and inflammation damage in rats. *FASEB J.* **17,** 816–822.

Jiang, Q., Elson-Schwab, I., Courtemanch, C., and Ames, B. N. (2000). Gamma-tocopherol and its major metabolite, in contrast to alpha-tocopherol, inhibit cyclooxygenase activity in macrophages and epithelial cells. *Proc. Natl. Acad. Sci. USA* **97**(21), 11495–11499.

Jiang, Q., Christen, S., Shigenaga, M. K., and Ames, B. N. (2001). Gamma-tocopherol, the major form of vitamin E in the US diet, deserves more attention. *Am. J. Clin. Nutr.* **74,** 714–722.

Kamal-Eldin, A., and Appelqvist, L. A. (1996). The chemistry and antioxidant properties of tocopherols and tocotrienols. *Lipids* **31,** 671–701.

Kamat, J. P., and Devasagayam, T. P. (1995). Tocotrienols from palm oil as potent inhibitors of lipid peroxidation and protein oxidation in rat brain mitochondria. *Neurosci. Lett.* **195,** 179–182.

Kaul, N., Devaraj, S., and Jialal, I. (2001). Alpha-tocopherol and atherosclerosis. *Exp. Biol. Med. (Maywood)* **226,** 5–12.

Khanna, S., Roy, S., Parinandi, N. L., Maurer, M., and Sen, C. K. (2006). Characterization of the potent neuroprotective properties of the natural vitamin E alpha-tocotrienol. *J. Neurochem.* **98,** 1474–1486.

Khor, H. T., and Ng, T. T. (2000). Effects of administration of alpha-tocopherol and tocotrienols on serum lipids and liver HMG CoA reductase activity. *Int. J. Food Sci. Nutr.* **51,** S3–S11.

Khor, H. T., Chieng, D. Y., and Ong, K. K. (1995). Tocotrienols inhibit HMG-CoA reductase activity in the guinea pig. *Nutr. Res.* **15,** 537–544.

Kirii, H., Niwa, T., Yamada, Y., Wada, H., Saito, K., Iwakura, Y., Asano, M., Moriwaki, H., and Seishima, M. (2003). Lack of interleukin-1β decreases the severity of atherosclerosis in apoE-deficient mice. *Arterioscler. Thromb. Vasc. Biol.* **23,** 656–660.

Kontush, A., Spranger, T., Reich, A., Baum, K., and Beisiegel, U. (1999). Lipophilic antioxidants in blood plasma as markers of atherosclerosis: The role of alpha-carotene and gamma-tocopherol. *Atherosclerosis* **144,** 117–122.

Kristenson, M., Zieden, B., Kucinskiene, Z., Elinder, L. S., Bergdahl, B., Elwing, B., Abaravicius, A., Razinkovienë, L., Calkauskas, H., and Olsson, A. G. (1997). Antioxidant state and mortality from coronary heart disease in Lithuanian and Swedish men: Concomitant cross sectional study of men aged 50. *Br. Med. J.* **314,** 629–633.

Kushi, L. H., Folsom, A. R., Prineas, R. J., Mink, P. J., Wu, Y., and Bostick, R. M. (1996). Dietary antioxidant vitamins and death from coronary heart disease in postmenopausal women. *New Engl. J. Med.* **334,** 1156–1162.

Libby, P. (2002). Inflammation in atherosclerosis. *Nature* **420,** 868–874.

Liu, M., Wallmon, A., Olsson-Mortlock, C., Wallin, R., and Saldeen, T. (2003). Mixed tocopherols inhibit platelet aggregation in humans: Potential mechanisms. *Am. J. Clin. Nutr.* **77,** 700–706.

Mastaloudis, A., Traber, M. G., Carstensen, K., and Widrick, J. J. (2006). Antioxidants did not prevent muscle damage in response to an ultramarathon run. *Med. Sci. Sports Exerc.* **38,** 72–80.

McAnulty, S. R., McAnulty, L. S., Nieman, D. C., Morrow, J. D., Shooter, L. A., Holmes, S., Heward, C., and Henson, D. A. (2005). Effect of alpha-tocopherol supplementation on plasma homocysteine and oxidative stress in highly trained athletes before and after exhaustive exercise. *J. Nutr. Biochem.* **16**(9), 530–537.

Mensink, R. P., van Houwelingen, A. C., Kromhout, D., and Hornstra, G. (1999). A vitamin E concentrate rich in tocotrienols had no effect on serum lipids, lipoproteins, or platelet function in men with mildly elevated serum lipid concentrations. *Am. J. Clin. Nutr.* **69,** 213–219.

Morton, L. W., Ward, N. C., Croft, K. D., and Puddey, I. B. (2002). Evidence for the nitration of gamma-tocopherol *in vivo*: 5-Nitro-gamma-tocopherol is elevated in the plasma of subjects with coronary heart disease. *Biochem. J.* **364**(Pt. 3), 625–628.

Munteanu, A., Zingg, J. M., and Azzi, A. (2004). Anti-atherosclerotic effects of vitamin E—myth or reality? *J. Cell. Mol. Med.* **8,** 59–76.

Mustad, V. A., Smith, C. A., Ruey, P. P., Edens, N. K., and DeMichele, S. J. (2002). Supplementation with 3 compositionally different tocotrienol supplements does not improve cardiovascular disease risk factors in men and women with hypercholesterolemia. *Am. J. Clin. Nutr.* **76,** 1237–1243.

Naito, Y., Shimozawa, M., Kuroda, M., Nakabe, N., Manabe, H., Katada, K., Kokura, S., Ichikawa, H., Yoshida, N., Noguchi, N., and Yoshikawa, T. (2005). Tocotrienols reduce 25-hydroxycholesterol-induced monocyte-endothelial cell interaction by inhibiting the surface expression of adhesion molecules. *Atherosclerosis* **180,** 19–25.

Neri, S., Signorelli, S. S., Torrisi, B., Pulvirenti, D., Mauceri, B., Abate, G., Ignaccolo, L., Bordonaro, F., Cilio, D., Calvagno, S., and Leotta, C. (2005). Effects of antioxidant supplementation on postprandial oxidative stress and endothelial dysfunction: A single-blind, 15-day clinical trial in patients with untreated type 2 diabetes, subjects with impaired glucose tolerance, and healthy controls. *Clin. Ther.* **27**(11), 1764–1773.

O'Byrne, D., Grundy, S., Packer, L., Devaraj, S., Baldenius, K., Hoppe, P. P., Kraemer, K., Jialal, I., and Traber, M. G. (2000). Studies of LDL oxidation following alpha-, gamma-, or delta-tocotrienyl acetate supplementation of hypercholesterolemic humans. *Free Radic. Biol. Med.* **29,** 834–845.

Ogru, E., Gianello, R., Libinaki, R., Smallridge, A., Bak, R., Geytenbeek, S., and Kannar, D. West, S. (2003). "Vitamin E Phosphate: An Endogenous form of Vitamin E." *Medimond. Med. Pub.* Englewood, New Jersey.

Ohrvall, M., Sundlof, G., and Vessby, B. (1996). Gamma, but not alpha, tocopherol levels in serum are reduced in coronary heart disease patients. *J. Intern. Med.* **239,** 111–117.

Pearce, B. C., Parker, R. A., Deason, M. E., Qureshi, A. A., and Wright, J. J. (1992). Hypocholesterolemic activity of synthetic and natural tocotrienols. *J. Med. Chem.* **35,** 3595–3606.

Pickup, J. C. (2004). Inflammation and activated innate immunity in the pathogenesis of type 2 diabetes. *Diabetes Care* **27,** 813–823.

Qureshi, A. A., Burger, W. C., Peterson, D. M., and Elson, C. E. (1986). The structure of an inhibitor of cholesterol biosynthesis isolated from barley. *J. Biol. Chem.* **261,** 10544–10550.

Qureshi, A. A., Qureshi, N., Hasler-Rapacz, J. O., Weber, F. E., Chaudhary, V., Crenshaw, T. D., Gapor, A., Ong, A. S., Chong, Y. H., and Peterson, D. (1991a). Dietary tocotrienols reduce concentrations of plasma cholesterol, apolipoprotein B, thromboxane B2, and platelet factor 4 in pigs with inherited hyperlipidemias. *Am. J. Clin. Nutr.* **53,** 1042S–1046S.

Qureshi, A. A., Qureshi, N., Wright, J. J., Shen, Z., Kramer, G., Gapor, A., Chong, Y. H., DeWitt, G., Ong, A., and Peterson, D. M. (1991b). Lowering of serum cholesterol in hypercholesterolemic humans by tocotrienols (palmvitee). *Am. J. Clin. Nutr.* **53,** 1021S–1026S.

Qureshi, A. A., Bradlow, B. A., Brace, L., Manganello, J., Peterson, D. M., Pearce, B. C., Wright, J. J. K., Gapor, A., and Elson, C. E. (1995). Response of hypercholesterolemic subjects to administration of tocotrienols. *Lipids* **30,** 1171–1177.

Qureshi, A. A., Mo, H., Packer, L., and Peterson, D. M. (2000). Isolation and identification of novel tocotrienols from rice bran with hypocholesterolemic, antioxidant, and antitumor properties. *J. Agric. Food Chem.* **48,** 3130–3140.

Qureshi, A. A., Peterson, D. M., Hasler-Rapacz, J. O., and Rapacz, J. (2001a). Novel tocotrienols of rice bran suppress cholesterogenesis in hereditary hypercholesterolemic swine. *J. Nutr.* **131,** 223–230.

Qureshi, A. A., Salser, W. A., Parmar, R., and Emeson, E. E. (2001b). Novel tocotrienols of rice bran inhibit atherosclerotic lesions in C57BL/6 ApoE-deficient mice. *J. Nutr.* **131,** 2606–2618.

Qureshi, A. A., Sami, S. A., Salser, W. A., and Khan, F. A. (2001c). Synergistic effect of tocotrienol-rich fraction (TRF(25)) of rice bran and lovastatin on lipid parameters in hypercholesterolemic humans. *J. Nutr. Biochem.* **12,** 318–329.

Qureshi, A. A., Sami, S. A., Salser, W. A., and Khan, F. A. (2002). Dose-dependent suppression of serum cholesterol by tocotrienol-rich fraction (TRF25) of rice bran in hypercholesterolemic humans. *Atherosclerosis* **161,** 199–207.

Qureshi, N., Hofman, J., and Qureshi, A. A. (1993). Inhibition of LPS induced tumor necrosis factor synthesis and hypocholesterolermic effect of novel tocotrienols. *In* "Proceedings of the PORIM International Palm Oil Congress." September 20–25, N16.

Ridker, P. M. (2003). Clinical application of C-reactive protein for cardiovascular disease detection and prevention. *Circulation* **168,** 363–369.

Ross, R. (1999). Atherosclerosis: An inflammatory disease. *N. Engl. J. Med.* **340,** 115–126.

Ruiz Rejón, F., Martin-Pena, G., Granado, F., Ruiz-Galiana, J., Blanco, I., and Olmedilla, B. (2002). Plasma status of retinol, alpha- and gamma-tocopherols, and main carotenoids to first myocardial infarction: Case control and follow-up study. *Nutrition* **18,** 26–31.

Saldeen, T., Li, D., and Mehta, J. L. (1999). Differential effects of alpha-and gamma-tocopherol on low-density lipoprotein oxidation, superoxide activity, platelet aggregation and arterial thrombogenesis. *J. Am. Coll. Cardiol.* **34,** 1208–1215.

Schock, B. C., Van der Vliet, A., Corbacho, A. M., Leonard, S. W., Finkelstein, E., Valacchi, G., Obermueller-Jevic, U., Cross, C. E., and Traber, M. G. (2004). Enhanced inflammatory responses in alpha-tocopherol transfer protein null mice. *Arch. Biochem. Biophys.* **423,** 162–169.

Sen, C. K., Khanna, S., Roy, S., and Packer, L. (2000). Molecular basis of vitamin E action. Tocotrienol potently inhibits glutamate-induced pp60(c-Src) kinase activation and death of HT4 neuronal cells. *J. Biol. Chem.* **275,** 13049–13055.

Seppo, L., Lahteenmaki, T., Tikkanen, M. J., Vanhanen, H., Korpela, R., and Vapaatalo, H. (2005). Effects of vitamin E on the toxicity of oxidized LDL on endothelial cells *in vitro* in smokers vs nonsmokers on diets rich in fish. *Eur. J. Clin. Nutr.* **59**(11), 1282–1290.

Servinova, E., Kagan, V., Han, D., and Packer, L. (1991). Free radical recycling and intramembrane mobility in the antioxidant properties of alpha-tocopherol and alpha-tocotrienol. *Free Radic. Biol. Med.* **10,** 263–275.

Singh, U., and Jialal, I. (2004). Anti-inflammatory effects of alpha-tocopherol. *Ann. NY Acad. Sci.* **1031,** 1–9.

Singh, U., Devaraj, S., and Jialal, I. (2005). Vitamin E, oxidative stress, and inflammation. *Annu. Rev. Nutr.* **25,** 151–174. Review.

Stampfer, M. J., Hennekens, C. H., Manson, J. E., Colditz, G. A., Rosner, B., *et al.* (1993). A prospective study of vitamin E supplementation and risk of coronary disease in women. *N. Engl. J. Med.* **328,** 1444–1449.

Steiner, M. (1983). Effect of alpha tocopherol administration on platelet function in man. *Thromb. Haemost.* **49,** 73–77.

Suarna, C., Wu, B. J., Choy, K., Mori, T., Croft, K., Cynshi, O., and Stocker, R. (2006). Protective effect of vitamin E supplements on experimental atherosclerosis is modest and depends on preexisting vitamin E deficiency. *Free Radic. Biol. Med.* **41,** 722–730.

Suzuki, Y. J., Tsuchiya, M., Wassall, S. R., Choo, Y. M., Govil, G., Kagan, V. E., and Packer, L. (1993). Structural and dynamic membrane properties of α-tocopherol and α-tocotrienol: Implication to the molecular mechanism of their antioxidant potency. *Biochemistry* **32,** 10692–10699.

Tahir, M., Foley, B., Pate, G., Crean, P., Moore, D., McCarroll, N., and Walsh, M. (2005). Impact of vitamin E and C supplementation on serum adhesion molecules in chronic degenerative aortic stenosis: A randomized controlled trial. *Am. Heart J.* **150,** 302–306.

Tan, D. T., Khor, H. T., Low, W. H., Ali, A., and Gapor, A. (1991). Effect of a palm-oil-vitamin E concentrate on the serum and lipoprotein lipids in humans. *Am. J. Clin. Nutr.* **53,** 1027S–1030S.

Terasawa, Y., Ladha, Z., Leonard, S. W., Morrow, J. D., Newland, D., Senan, D., Packer, L., Traber, M. G., and Farese, R. V., Jr. (2000). Increased atherosclerosis in hyperlipidemic mice deficient in alpha-tocopherol transfer protein and vitamin E. *Proc. Natl. Acad. Sci. USA* **97,** 13830–13834.

Theriault, A., Wang, Q., Gapor, A., and Adeli, K. (1999). Effects of gamma-tocotrienol on ApoB synthesis, degradation, and secretion in HepG2 cells. *Arterioscler. Thromb. Vasc. Biol.* **19**(3), 704–712.

Theriault, A., Chao, J. T., Gapor, A., Chao, J. T., and Gapor, A. (2002). Tocotrienol is the most effective vitamin E for reducing endothelial expression of adhesion molecules and adhesion to monocytes. *Atherosclerosis* **160,** 21–30.

Tomeo, A. C., Geller, M., Watkins, T. R., Gapor, A., and Bierenbaum, M. L. (1995). Antioxidant effects of tocotrienols in patients with hyperlipidemia and carotid stenosis. *Lipids* **30,** 1179–1183.

Topi, G. C., and Alessandrini, A. (1953). Enzymic hydrolysis of the phosphoric acid radical of alpha-tocopherol phosphate. *Acta Vitaminol.* **7,** 8–11.

Torsney, E., Mandal, K., Halliday, A., Jahangiri, M., and Xu, Q. (2007). Characterisation of progenitor cells in human atherosclerotic vessels. *Atherosclerosis* **191,** 25–64.

Traber, M. G., and Arai, H. (1999). Molecular mechanisms of vitamin E transport. *Annu. Rev. Nutr.* **19,** 343–355. Review.

Tziakas, D. N., Chalikias, G. K., Kaski, J. C., Kekes, A., Hatzinikolaou, E. I., Stakos, D. A., Tentes, I. K., Kortsaris, A. X., and Hatseras, D. I. (2006). Inflammatory and anti-inflammatory variable clusters and risk prediction in acute coronary syndrome patients: A factor analysis approach. *Atherosclerosis* [Epub ahead of print].

Upston, J. M., Kritharides, L., and Stocker, R. (2003). The role of vitamin E in atherosclerosis. *Prog. Lipid Res.* **42,** 405–422. Review.

van Dam, B., van Hinsbergh, V. W., Stehouwer, C. D., Versteilen, A., Dekker, H., Buytenhek, R., Princen, H., and Schalkwijk, C. (2003). Vitamin E inhibits lipid peroxidation-induced adhesion molecule expression in endothelial cells and decreases soluble cell adhesion molecules in healthy subjects. *Cardiovasc. Res.* **57**(2), 563–571.

Verma, S., Devaraj, S., and Jialal, I. (2006). Is C-reactive protein an innocent bystander or proatherogenic culprit? C-reactive protein promotes atherothrombosis. *Circulation* **113,** 2135–2150. Review.

Viles-Gonzalez, J. F., Fuster, V., and Badimon, J. J. (2006). Links between inflammation and thrombogenicity in atherosclerosis. *Curr. Mol. Med.* **6,** 489–499. Review.

Wagner, K. H., Kamal-Eldin, A., and Elmadfa, I. (2004). Gamma-tocopherol—an underestimated vitamin? *Ann. Nutr. Metab.* **48,** 169–188.

Wahlqvist, M. L., Krivokuca-Bogetic, Z., Lo, C. S., Hage, B., and Smith, R.Lukito, W. (1992). Differential serum responses of tocopherols and tocotrienols during vitamin supplementation in hypercholesterolemic individuals without change in coronary risk factors. *Nutr. Res.* **12** (Suppl.), S181–S201.

Watkins, T., Lenz, P., Gapor, A., Struck, M., Tomeo, A., and Bierenbaum, M. (1993). γ-Tocotrienol as a hypocholesterolemic and antioxidant agent in rats fed atherogenic diets. *Lipids* **28,** 1113–1118.

Williamson, K. S., Gabbita, S. P., Mou, S., West, M., Pye, Q. N., Markesbery, W. R., Cooney, R. V., Grammas, P., Reimann-Philipp, U., Floyd, R. A., and Hensley, K. (2002). The nitration product 5-nitro-gamma-tocopherol is increased in the Alzheimer brain. *Nitric Oxide* **6,** 221–227.

Wu, D., Liu, L., Meydani, M., and Meydani, S. N. (2005a). Vitamin E increases production of vasodilator prostanoids in human aortic endothelial cells through opposing effects on cyclo-oxygenase-2 and phospholipase A2. *J. Nutr.* **135,** 1847–1853.

Wu, J. H., Hodgson, J. M., Ward, N. C., Clarke, M. W., Puddey, I. B., and Croft, K. D. (2005b). Nitration of gamma-tocopherol prevents its oxidative metabolism by HepG2 cells. *Free Radic. Biol. Med.* **39,** 483–494.

Yochum, L. A., Folsom, A. R., and Kushi, L. H. (2000). Intake of antioxidant vitamins and risk of death from stroke in postmenopausal women. *Am. J. Clin. Nutr.* **72,** 476–478.

Yoshida, Y., Niki, E., and Noguchi, N. (2003). Comparative study on the action of tocopherols and tocotrienols as antioxidant: Chemical and physical effects. *Chem. Phys. Lipids* **123,** 63–75.

Yun, S. G., Thomas, A. M., Gapor, A., Tan, B., Qureshi, N., and Qureshi, A. A. (2006). Dose-response impact of various tocotrienols on serum lipid parameters in 5-week-old female chickens. *Lipids* **41,** 453–461.

21

Vitamin E in Chronic Liver Diseases and Liver Fibrosis

Antonio Di Sario, Cinzia Candelaresi, Alessia Omenetti, and Antonio Benedetti

Department of Gastroenterology, Università Politecnica delle Marche, Polo Didattico III, Piano, Via Tronto 10, 60020 Torrette, Ancona, Italy

Liver fibrosis may be considered as a dynamic and integrated cellular response to chronic liver injury. The activation of hepatic stellate cells and the consequent deposition of large amounts of extracellular matrix

Vitamins and Hormones, Volume 76

0083-6729/07 $35.00
DOI: 10.1016/S0083-6729(07)76021-1

play a major role in the fibrogenic process, but it has been shown that other cellular components of the liver are also involved. Although the pathogenesis of liver damage usually depends on the underlying disease, oxidative damage of biologically relevant molecules might represent a common link between different forms of chronic liver injury and hepatic fibrosis. In fact, oxidative stress-related molecules may act as mediators able to modulate all the events involved in the progression of liver fibrosis. In addition, chronic liver diseases are often associated with decreased antioxidant defenses. Although vitamin E levels have been shown to be decreased in chronic liver diseases of different etiology, the role of vitamin E supplementation in these clinical conditions is still controversial. In fact, the increased serum levels of α-tocopherol following vitamin E supplementation not always result in a protective effect on liver damage. In addition, clinical trials have usually been performed in small cohorts of patients, thus making definitive conclusions impossible. At present, treatment with vitamin E or other antioxidant compounds could be proposed for nonalcoholic fatty liver disease (NAFLD), the most frequent hepatic lesion in western countries which can progress to nonalcoholic steatohepatitis and cirrhosis due to the production of large amounts of oxidative stress products. However, although some studies have shown encouraging results, multicentric and long-term clinical trials are needed.

I. FIBROSIS IN CHRONIC LIVER DISEASES

Liver fibrosis is a dynamic and highly integrated cellular response to chronic liver damage and is characterized, from a histological point of view, by progressive accumulation of extracellular matrix (ECM) proteins which distort the hepatic architecture by forming regenerating nodules (Bataller and Brenner, 2005; Fig. 1).

Progressive accumulation of fibrotic tissue in the liver is the consequence of reiterated liver tissue damage due to viral infections (*Hepatitis B virus*—HBV and *Hepatitis C virus*—HCV), toxic compounds, drugs, metabolic and autoimmune diseases. However, although millions of patients worldwide are affected by chronic liver diseases, only a minority develops significant fibrosis and cirrhosis, and this is true especially for HCV-related liver cirrhosis (Pinzani *et al.*, 2005).

Studies performed in the last years have clearly demonstrated that the excessive accumulation of ECM in fibrotic liver diseases is a dynamic process mainly regulated by hepatic stellate cells (HSC), which undergo, because of a chronic liver injury, a process of activation developing a myofibroblast-like

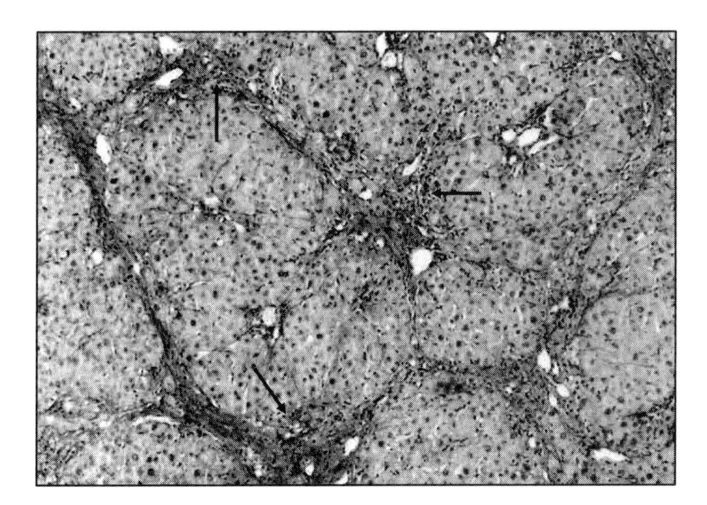

FIGURE 1. Histological picture of hepatic cirrhosis in humans: as evident, the lobular architecture is completely sovverted by the presence of multiple nodules surrounded by fibrotic tissue; an infiltration of inflammatory cells is also evident (arrows). (See Color Insert.)

phenotype associated with increased proliferation and collagen synthesis (Friedman, 2000). Similar morphological and functional modifications can be observed when HSC are cultured on uncoated plastic dishes (Friedman *et al.*, 1989). Using this experimental model, it has been shown that the process of HSC activation results from the complex interplay of different factors such as cytokines (Friedman, 1999), growth factors (Pinzani *et al.*, 1989; Svegliati-Baroni *et al.*, 1999b), oxidative stress (Lee *et al.*, 1995; Svegliati-Baroni *et al.*, 1999a), paracrine stimuli from injured hepatocytes (Svegliati-Baroni *et al.*, 1998), and modification of Na^+/H^+ exchange activity (Benedetti *et al.*, 2001; Di Sario *et al.*, 1997, 1999, 2001, 2003; Fig. 2).

Although HSC activation may represent the final step leading to liver fibrosis, pathogenesis of liver damage usually depends on the underlying disease. In alcoholic liver damage, alcohol increases levels of certain cytokines (i.e., TNF-α) that induce inflammatory infiltration in the liver and stimulate apoptosis of hepatocytes. In addition, acetaldehyde, the major alcohol metabolism product, increases the production of reactive oxygen species (ROS) which stimulate the fibrogenic behavior of HSC (Maher *et al.*, 1994). As regards to HCV-induced liver fibrosis, both hepatocyte damage as well as direct HSC activation by HCV proteins seem to be involved. In chronic cholestatic liver diseases, HSC are mainly activated by fibrogenic mediators secreted by biliary cells. Finally, hepatocyte damage and HSC activation by ROS and proinflammatory cytokines have been shown to induce the appearance of fibrosis in patients with nonalcoholic steatohepatitis (NASH) (Wanless and Shiota, 2004).

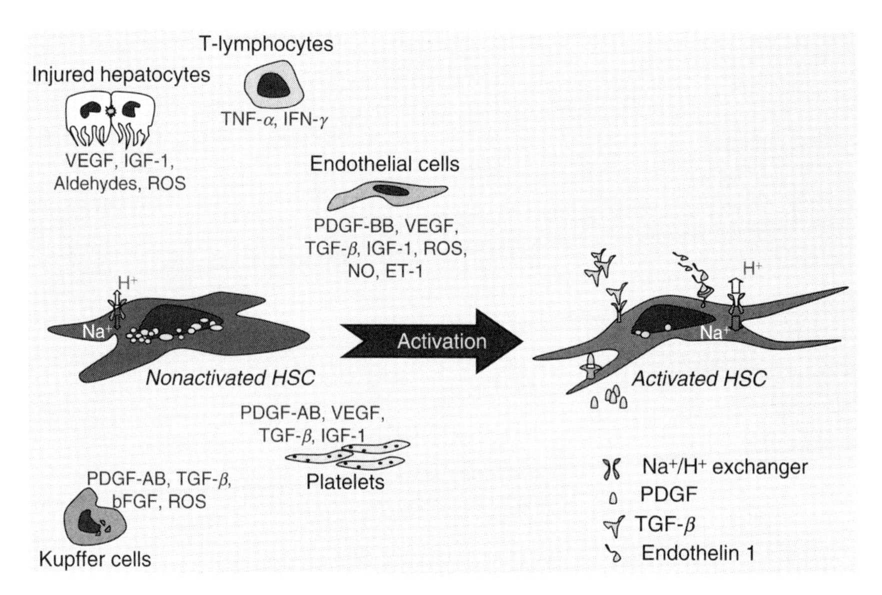

FIGURE 2. Main factors involved in the process of HSC activation during fibrogenesis. As evident, virtually all liver cells are involved in this process through the release of cytokines, growth factors, and oxidative stress products. In addition, the increased activity of the Na^+/H^+ exchanger has been shown to play an important role. VEGF, vascular endothelial growth factor; IGF-1, insulin-like growth factor 1; ROS, reactive oxygen species; TNF-α, tumor necrosis factor α; PDGF, platelet-derived growth factor; TGF-β, transforming growth factor β; NO, nitric oxide; ET-1, endothelin 1; bFGF, basic fibroblast growth factor. (See Color Insert.)

From a clinical point of view, the natural history of liver fibrosis can be influenced by both environmental and genetic factors, the latter explaining the large spectrum of responses to the same etiologic agent found in patients with chronic liver disease (Bataller *et al.*, 2003); however, further studies are needed to clarify the role of genetic variants in liver fibrosis.

Although several noninvasive methods for the assessment of liver fibrosis have been proposed, liver biopsy is still considered the gold-standard method, since it provides information about the necroinflammatory grade and the stage of fibrosis (Di Sario *et al.*, 2004).

II. OXIDATIVE STRESS, CHRONIC LIVER DISEASE, AND LIVER FIBROSIS

Several *in vitro* and *in vivo* observations suggest that oxidative damage of biologically relevant molecules could represent a common link between different forms of chronic liver injury and hepatic fibrosis. In fact, lipid peroxidation has been associated with liver fibrosis caused by iron overload, alcohol,

CCl_4, and HCV infection. In addition, oxidative stress can be considered to be one of the major causes of liver damage and may be involved in hepatic fibrogenesis by the stimulation of collagen gene expression (Bendia *et al.*, 2005). Tissue damage usually occurs when oxidative stress-related molecules, generated within the cell or in the extracellular environment, exceed antioxidant defenses, and many experimental evidences clearly indicate that both hepatic and plasma antioxidant defenses are significantly decreased in chronic liver diseases.

A. OXIDATIVE STRESS IN ALCOHOL-INDUCED LIVER DAMAGE

It is well known that alcohol affects the liver (Fig. 3) mainly through a direct toxicity because of its predominant metabolism in the liver associated with oxidation–reduction (redox) changes and oxidative stress. Changes of the redox status are mediated by alcohol dehydrogenase (ADH), which produces acetaldehyde, whereas oxidative stress is mainly generated by the activity of the microsomal ethanol-oxidizing system (MEOS) and its enzyme cytochrome P450 2E1 (CYP450 2E1), which releases free radicals.

Acetaldehyde produced by the oxidation of alcohol is able to inhibit the repair of alkylated nucleoproteins, to decrease the activity of several enzymes and to damage mitochondria; in addition, it promotes cell death by depleting the concentration of reduced glutathione, by inducing lipid peroxidation, and by increasing the toxic effects of free radicals. Finally, acetaldehyde has

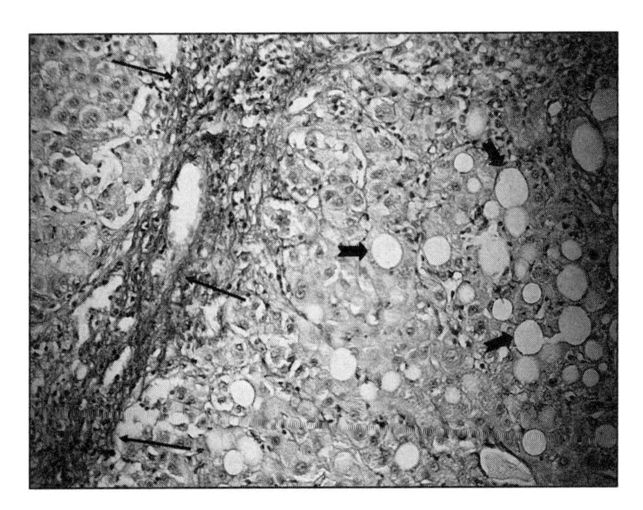

FIGURE 3. Histological appearance of the liver in a patient with alcoholic hepatitis: a massive steatosis is evident (short arrows), together with accumulation of fibrotic tissue (long arrows); no nodules are evident. (See Color Insert.)

been shown to directly stimulate proliferation of HSC and to increase collagen synthesis (Svegliati-Baroni *et al.*, 2001). The oxidative stress caused by CYP450 2E1 induction and mitochondrial injury results in lipid peroxidation and cell membrane damage (Lieber, 2004). Moreover, oxidative stress promotes inflammation and increased production of proinflammatory cytokines, such as TNF-α, which are also able to stimulate collagen synthesis by HSC (Kato *et al.*, 2003).

1. Vitamin E in Alcohol-Induced Liver Damage

Among the dietary antioxidants, vitamin E is one of the most important. α-Tocopherol, the most biologically active form of vitamin E, is the major lipid-soluble antioxidant in serum and has the ability to inhibit generation of singlet oxygen, free radicals, lipid radicals, and lipid hydroperoxides. Several studies have clearly demonstrated that serum levels of vitamin E are significantly reduced in patients with alcoholic liver disease (Bjorneboe Gunn-Elin *et al.*, 1988; Bonjour, 1981; Leo *et al.*, 1993) and that vitamin E levels correlate negatively with production of oxidative stress products and positively with the extent of liver damage (Masalkar and Abhang, 2005). Therefore, maintenance of normal concentrations of vitamin E seems to be essential to prevent lipid peroxidation induced by alcohol consumption. However, both experimental (Sadrzadeh *et al.*, 1995) and clinical studies (de la Maza *et al.*, 1995) have shown that although vitamin E supplementation is able to increase α-tocopherol serum levels, this does not result in a protective effect on alcohol-induced liver damage.

B. OXIDATIVE STRESS IN IRON-INDUCED LIVER DAMAGE

Iron is a vital micronutrient which plays an important role in regulating different cell functions such as DNA synthesis, cell respiration, and transport of oxygen and electrons. However, due to both a primary defect of iron metabolism (i.e., hemochromatosis) and acquired causes (iron-loading anemias, viral and alcoholic liver diseases), iron accumulates in parenchymal organs and affects cell viability, leading to organ fibrosis.

The toxicity of iron is principally due to increased lipid peroxidation in cellular organelles, and especially in mitochondria, which represent not only the primary target of oxidative injury but also a system for ROS generation (Cadenas *et al.*, 1977). It has been shown that lipid peroxidation of mitochondrial membrane may alter the fluidity state of the membrane itself or lead to swelling and lysis of the mitochondria (McKnight and Hunter, 1966).

The occurrence of liver fibrosis and cirrhosis is common in humans with iron overload (Fig. 4), and the availability of animal models has offered the opportunity to study many aspects of the pathogenesis of liver damage. More in detail, it has been shown that enteral iron overload is able to increase

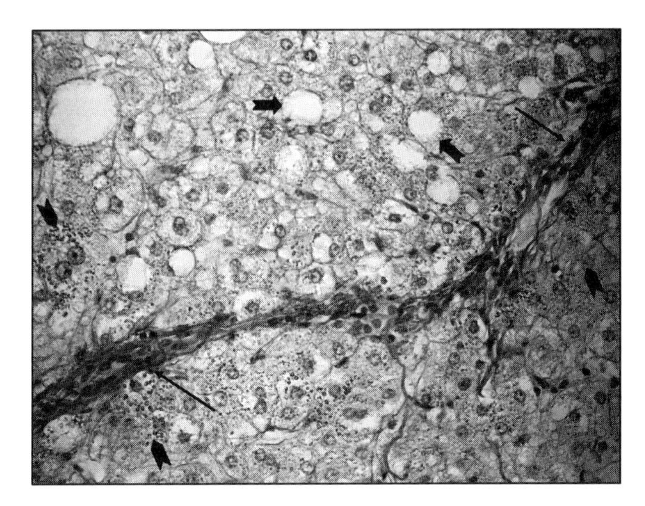

FIGURE 4. Hepatic iron accumulation in a young patient with alteration of liver enzymes: diffuse siderosis (brown staining; arrowheads), macrovescicular steatosis (short arrows), and deposition of fibrotic tissue (long arrows) are evident. (See Color Insert.)

products of lipid peroxidation which in turn activate collagen type I gene expression by HSC acting through both direct (Pietrangelo *et al.*, 1994) and paracrine (Gualdi *et al.*, 1994) mechanisms. Therefore, excessive iron is able to trigger the molecular and cellular events leading to fibrogenesis in the liver, although the full potential of iron as pathogenetic factor in human liver diseases is when acting on a preinjured organ or together with other hepatotoxins (Pietrangelo, 2003).

1. Vitamin E in Iron-Induced Liver Damage

It has been shown that patients with hereditary hemochromatosis have increased levels of iron and lipid peroxidation products and decreased levels of vitamin E (von Herbay *et al.*, 1994; Young *et al.*, 1994). However, the potential therapeutic role of vitamin E supplementation in this clinical condition principally derives from experimental *in vitro* and *in vivo* studies. Sharma *et al.* (1990) have shown that α-tocopherol is able to protect hepatocytes isolated from rats with chronic dietary iron overload against lipid peroxidation, loss of viability, and ultrastructural damage. Whittaker *et al.* (1996) demonstrated that dietary supplementation with α-tocopherol significantly increases liver concentrations of vitamin E and significantly decreases the amount of lipid peroxidation products in the liver.

As regards to iron-induced liver fibrosis, by using two different animal models it has been shown that iron-loading results in a significant decrease in hepatic and plasma vitamin E levels that are overcome by vitamin E supplementation. In addition, administration of vitamin E is able to significantly reduce the number of proliferating HSC and the amount of collagen deposition, thus preventing the occurrence of hepatic cirrhosis (Brown *et al.*, 1997; Pietrangelo *et al.*, 1995). Similar results were obtained by Parola *et al.* (1992), who demonstrated that in the chronic rat model of CCl_4 intoxication vitamin E dietary supplementation is able to almost completely prevent the marked upregulation of hepatic $TGF_{\beta 1}$ and collagen type I genes.

C. OXIDATIVE STRESS AND VITAMIN E IN AUTOIMMUNE HEPATITIS

Autoimmune hepatitis is a chronic liver disease of unknown etiology characterized by circulating autoantibodies, hyperglobulinemia, and interface hepatitis, which is usually responsive to immunosuppressive therapy. The mechanisms of progression to fibrosis and cirrhosis are not completely clear but oxidative stress seems to play a role.

In a paper, Pemberton *et al.* (2004) have shown that antioxidant status is significantly compromised in patients with autoimmune hepatitis, with several components of the antioxidant defense mechanism (i.e., vitamin E) being significantly decreased, and that oxidative stress is not just a feature of late stage decompensated liver disease but a significant feature of early autoimmune hepatitis. In addition, in these patients a clear association between markers of oxidative stress and histologically assessed liver inflammation and fibrosis is evident. Therefore, trials of antioxidant therapy should be indicated, especially in patients who fail to respond to conventional drugs, in order to evaluate whether such treatment could prevent or slow progression to cirrhosis.

D. OXIDATIVE STRESS AND VITAMIN E IN CHOLESTATIC LIVER DISEASES

While with minor impact in comparison to alcohol abuse and hepatotropic viruses, also a severe and prolonged obstruction of either intrahepatic or extrahepatic biliary tree can damage hepatic tissue up to the development of cirrhosis. Among the possible causes of biliary obstruction are atresia of bile ducts, neoplasia, gallstones (secondary biliary cirrhosis) but more often an autoimmune extensive alteration of the biliary tree (primary biliary cirrhosis—PBC) (Fig. 5).

As regards to the pathogenesis of cholestatic liver damage, the hydrophobic bile acids that accumulate are considered the main cause of hepatotoxicity (Balistreri *et al.*, 1996). Among the potential mechanisms of bile acid cytotoxicity, production of ROS may play a role (Sokol *et al.*, 1991), as suggested by the observation of increased lipid peroxide levels in the blood of children with chronic cholestasis (Lubrano *et al.*, 1989) and by the significant peroxidation of membrane lipids in isolated hepatocytes and liver mitochondria following exposure to hydrophobic bile acids (Sokol *et al.*, 1995).

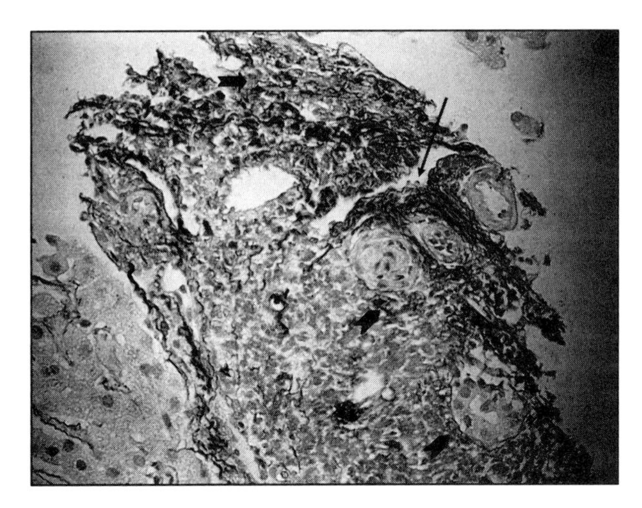

FIGURE 5. Primary biliary cirrhosis (stages I–II) in a female patient with cholestasis: the histological picture shows portal inflammation with lymphoid aggregates around bile ducts (arrowheads), necrotic parenchymal foci in periportal areas (short arrow), and fibrotic septa (long arrow). (See Color Insert.)

Studies on the role of ROS and consequent redox imbalance in the pathogenesis of cholestatic liver diseases have been usually performed using the model system of bile duct ligation (BDL) in the rat, which allows to reproduce in a relatively short period all the events occurring in the human cholestatic syndrome (Fig. 6). A significant increase of lipid peroxidation products in the liver is already evident after 3–7 days from BDL (Parola *et al.*, 1996), and appears to be favored not only by a net increase of lipid peroxides and relative decomposition products but also by a reduced hepatic disposition of lipid peroxidation aldehydes. Although it is still unclear whether oxidative stress is involved in the initiation of the disease process in cholestatic liver diseases, it certainly represents an important factor in disease progression likely through an increased production of fibrotic tissue by activated HSC.

Although antioxidant status is compromised in patients with PBC, information regarding the effect of vitamin E is scanty. Jeffrey *et al.* (1987) found that vitamin E is decreased in 44% of patients with PBC and that this reduction parallels the degree of cholestasis. On the contrary, Kaplan *et al.* (1988) and Aboutwerat *et al.* (2003) demonstrated that patients with primary biliary cirrhosis do not show evidence of significant vitamin E deficiency, leading to the hypothesis that such patients are able to maintain normal vitamin E levels by using alternative synthetic pathways (i.e., vitamin C).

As regards to the potential therapeutic effects of vitamin E in cholestatic liver diseases, only a few information derived from experimental studies are at present available. Muriel and Moreno (2004) have investigated the effect of

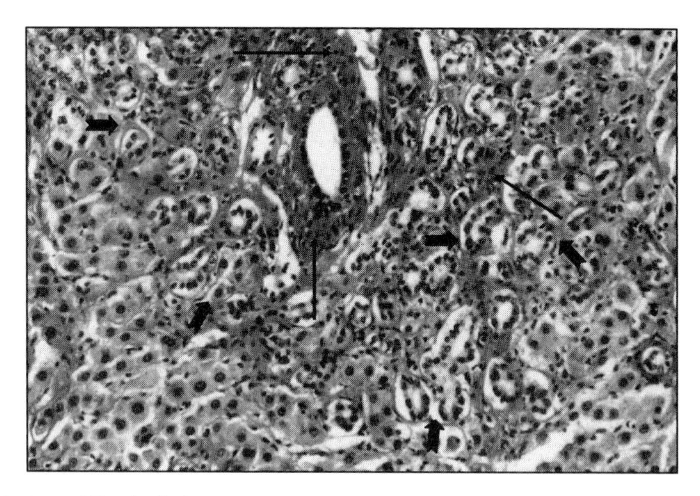

FIGURE 6. Histological appearance of rat liver after 4-week BDL as seen after immunostaining for Sirius red. As evident, the hepatic lobular architecture is massively modified by proliferating ductules (short arrows) accompanied by fibrotic tissue (long arrows). (See Color Insert.)

vitamin E and other antioxidant compounds on experimental cholestasis induced by BDL in the rat. Results of this study clearly show that vitamin E alone is not able to modify serum levels of liver enzymes or lipid peroxidation products or collagen deposition, thus suggesting that a mixture of antioxidants should be needed to counteract the oxidant stress produced by cholestasis.

On the basis of these previous findings, our group (Di Sario *et al.*, 2005) have studied, by using the same experimental model as above, the effect of a new drug in which vitamin E is complexed with silybin and phosphatidylcholine. The results of our study demonstrate that the administration of this complex is able to significantly reduce both HSC proliferation (Fig. 7A and B) and activation, as well as collagen deposition (Fig. 7C and D) and synthesis, thus confirming the hypothesis that an antioxidant mixture rather than a single antioxidant compound should be administered in chronic cholestatic liver diseases.

E. OXIDATIVE STRESS IN HCV-RELATED LIVER DISEASE

The HCV is a single-stranded positive sense RNA virus which has been estimated to infect about 170 million people worldwide (Lauer and Walker, 2001). The natural history of the disease is extremely variable, since several factors, such as the age at infection, gender, alcohol consumption, and coinfection with HBV or HIV, may influence the rate of disease progression (Grob *et al.*, 2000; Fig. 8A and B). The main pathogenetic mechanisms for liver injury and fibrosis in HCV hepatitis include immunologic liver damage,

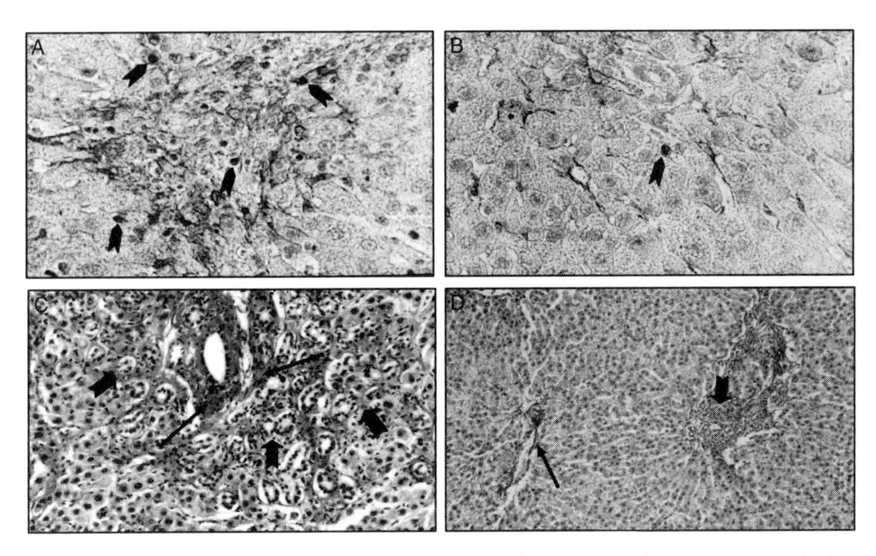

FIGURE 7. Number of proliferating and activated HSC in rat liver (arrowheads) after 4-week BDL (panel A) and after BDL plus treatment with the sylibin–phosphatidylcholine–vitamin E complex (panel B). As evident, the number of proliferating and activated cells, evaluated by means of a double immunohistochemical staining for α-smooth muscle actin and PCNA, is significantly reduced by treatment with the pharmacological complex. In rats treated with BDL (panel C), the lobular architecture is massively modified by proliferating ductules (short arrows) accompanied by fibrotic tissue (long arrows), whereas only a small amount of fibrotic tissue (long arrow) and a reduced number of bile ducts (short arrow) are evident in rats treated with BDL plus the sylibin–phosphatidylcholine–vitamin E complex (panel D). (See Color Insert.)

cytotoxicity mediated by many viral products as well as induction of oxidative stress. This latter aspect is suggested by increased levels of lipid peroxidation products in serum and liver of infected patients with respect to healthy controls (De Maria *et al.*, 1996; Farinati *et al.*, 1995; Paradis *et al.*, 1997); moreover, these levels are positively correlated with the degree of liver inflammation (Yadav *et al.*, 2002). ROS produced by activated macrophages and reactive aldehydes arising from lipid peroxidation can directly activate HSC (Svegliati-Baroni *et al.*, 1998), thus leading to hepatic fibrosis and cirrhosis. In addition, HCV infection is also associated with liver iron accumulation, which in turn generates lipid radicals (Pietrangelo, 1998). On the other hand, oxidative stress may affect mechanisms of immune response, thus making cells more susceptible to apoptosis and impairing the antiviral immune response (Buttke and Sandstorm, 1994).

1. Vitamin E in HCV-Related Liver Disease

Serum levels of vitamin E are significantly decreased in patients with chronic HCV hepatitis and inversely correlated with the degree of fibrosis (Jain *et al.*, 2002; Yadav *et al.*, 2002). Liver levels of vitamin E also are

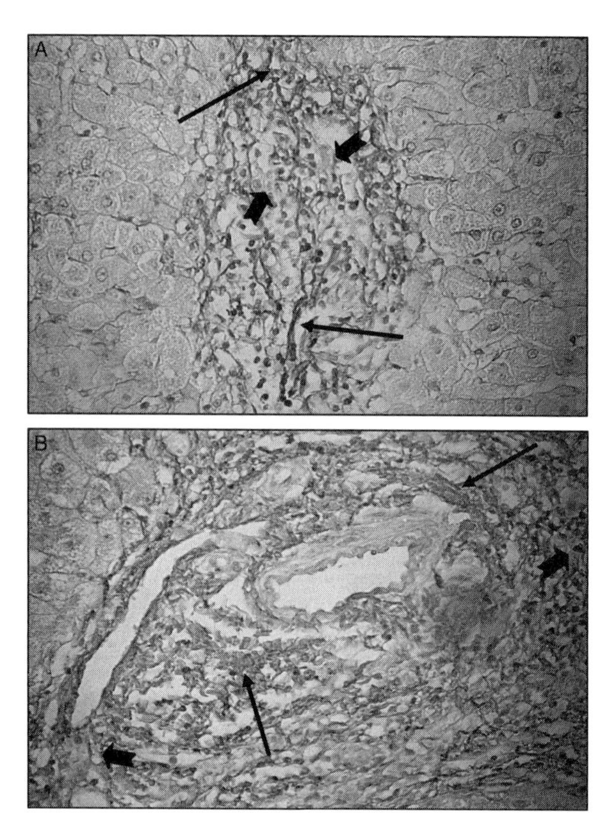

FIGURE 8. Histological pictures of HCV-related liver disease. In panel A, a moderate hepatitis is evident with inflammatory infiltrate in the portal space (short arrows) and evolution toward fibrosis (long arrows); on the contrary, panel B shows a picture of severe hepatitis, with marked enlargement of the portal space, severe inflammation (short arrows), and fibrosis (long arrows). (See Color Insert.)

decreased in HCV-infected patients, with even lower values in patients with established cirrhosis (Jain *et al.*, 2002). This finding suggests that severe liver disease is the result of depletion of antioxidant defense, although this cause–effect relationship is speculative and may be a reflection of increased tissue damage and fibrosis. Therefore, further studies are needed to better clarify this aspect.

Although the recommended treatment for HCV infection is a combination of pegylated interferon and ribavirin, which results in a sustained virological response in ~50–60% of patients (Baker, 2003), treatment with antioxidants may reduce hepatic inflammation and fibrosis, thus slowing or even preventing progression to cirrhosis. Although antioxidants are relatively cheap and much less toxic than the current antiviral therapies, only a few studies are currently available and their results are often inconclusive.

In one clinical trial which compared vitamin E treatment with placebo, a significant reduction in alanine aminotransferase (ALT) levels was observed, but this reduction did not occur in all patients and was not complete (von Herbay *et al.*, 1997); in addition, no virological effects were seen.

In a pilot study performed in HCV patients refractory to interferon therapy (Houglum *et al.*, 1997), vitamin E treatment was able to reduce the degree of oxidative stress and to prevent the fibrogenesis cascade (HSC activation and collagen type I mRNA expression) but did not significantly affect serum ALT levels, HCV titers, or histological degree of hepatocellular inflammation or fibrosis. These effects on fibrogenesis may be explained by the observation, in an animal model of liver damage, that vitamin E is able to inhibit activation and to induce apoptosis of HSC (Shen *et al.*, 2005).

However, in another study of HCV patients unresponsive to conventional therapy, retreatment with interferon plus oral supplementation with vitamin E improved neither biochemical nor virological response (Ideo *et al.*, 1999). In a pilot study using interferon–antioxidant combined therapy in interferon-naive patients, Look *et al.* (1999) concluded that there was a trend to a more favorable outcome with vitamin E supplementation, although no beneficial effect of therapy was registered.

Finally, a recent published paper (Murakami *et al.*, 2006) showed that oral vitamin E supplementation is able to prevent the decrease of eicosapentaenoic acid in phospholipids of mononuclear cells during combination therapy with interferon plus ribavirin in patients with HCV-related chronic liver disease.

It could therefore be concluded that optimal therapy of chronic HCV hepatitis in the future might require treatment with interferon/ribavirin to eliminate the virus together with antioxidant compounds (i.e., vitamin E) in order to reduce oxidative stress and to slow progression to cirrhosis.

F. OXIDATIVE STRESS AND VITAMIN E IN HBV-RELATED LIVER DISEASE

Approximately 350 million people worldwide are chronically infected by HBV, although the clinical course of the disease is extremely variable (Lee, 1997) (Fig. 9). Since HBV is believed not to have a direct cytophatic effect, both viral persistence and liver damage may be related to the weak and poorly effective host's immune response.

In several studies, increased oxidative stress has been suggested to be responsible for the hepatocellular damage caused by chronic HBV infection (Chrobot *et al.*, 2000; Demirdag *et al.*, 2003). Bolukbas *et al.* (2005) have found that total peroxide level, a parameter of oxidative stress, is significantly higher in patients with chronic HBV hepatitis with respect to inactive HBsAg carriers and controls, and positively correlated with ALT levels, thus suggesting that oxidative stress plays a critical role in hepatic injury and is associated with the severity of the disease.

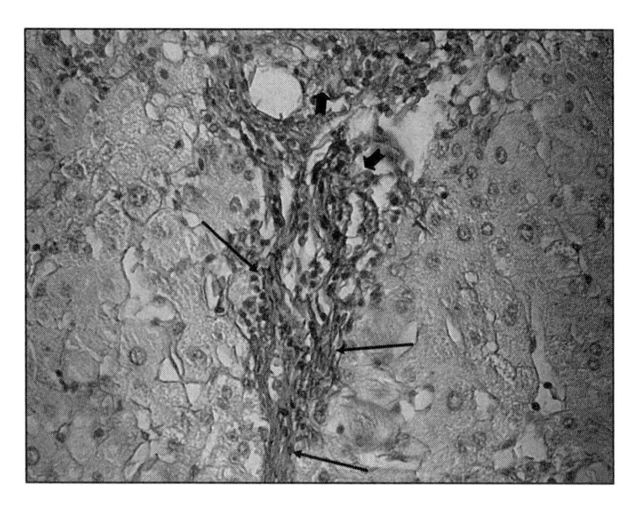

FIGURE 9. Histological picture of HBV-related liver disease: periportal necroinflammation (short arrows) and fibrotic septa (long arrows) are evident. (See Color Insert.)

Since vitamin E has been shown to enhance the cell-mediated immunity (Gogu and Bloomberg, 1993; Meydani, 1995), some clinical trials have tried to investigate the effect of this vitamin in patients with chronic HBV hepatitis.

In a first preliminary study, Andreone *et al.* (1998) evaluated the effect of a 3-month vitamin E administration in 22 patients with active HBV hepatitis who did not respond to a previous treatment with interferon, and found a complete biochemical and virological response in 5 patients, thus suggesting that vitamin E is a safe and useful treatment for chronic HBV hepatitis.

In a further study, the same group evaluated vitamin E supplementation in 32 patients with chronic HBV hepatitis (Andreone *et al.*, 2001). At the end of study period, they found ALT normalization in 47% and HBV-DNA negativization in 53% of patients; a complete response was observed in 47% of patients.

Therefore, these results indicate that vitamin E is of potential interest in the treatment of chronic HBV hepatitis; due to its good tolerability and low price, it might represent an alternative or additional treatment to other therapies, although further studies in a larger number of patients are needed.

G. OXIDATIVE STRESS AND VITAMIN E IN LIVER CIRRHOSIS AND HEPATOCELLULAR CARCINOMA

As already stated, cirrhosis represents the natural evolution of chronic liver diseases. Serum vitamin E levels are significantly reduced in patients with liver cirrhosis (Ferre *et al.*, 2002; Look *et al.*, 1999), and this decrease is related to the degree of liver impairment and is associated with an increased plasma susceptibility to oxidation (Ferre *et al.*, 2002).

Only a few information are at present available about vitamin E in hepatocellular carcinoma, since its serum levels have been shown to be normal (Pan *et al.*, 1993; Yamamoto *et al.*, 1998) or decreased (Rocchi *et al.*, 1997) in these patients. Interestingly, Rocchi *et al.* (1997) have found that vitamin E content is significantly reduced in malignant liver nodules with respect to surrounding cirrhotic tissue, whereas in metastatic liver nodules from digestive neoplasms the vitamin E content is almost twice that of healthy surrounding areas, thus suggesting the need for an accurate evaluation when employing liposoluble vitamins in the chemopreventive treatment of different malignant diseases.

However, due to the fact that vitamin E has been shown to confer protection from lipid peroxidation in hepatocellular carcinoma both *in vitro* (Fantappie *et al.*, 2004) and *in vivo* (Gorozhanskaia *et al.*, 1995), some clinical trials have been performed.

In a first study, the cumulative tumor-free survival and cumulative survival rate tended to be higher in cirrhotic patients treated with vitamin E with respect to patients who were not treated with this vitamin, although this difference was not statistically significant (Takagi *et al.*, 2003).

In a more recent study (Clerici *et al.*, 2004), the administration of vitamin E in combination with tamoxifen was able to increase the survival rate and to improve the clinical outcome in patients with inoperable hepatocellular carcinoma.

H. OXIDATIVE STRESS IN NONALCOHOLIC FATTY LIVER DISEASE

Nonalcoholic fatty liver disease (NAFLD; Fig. 10A) is emerging as the most common chronic liver condition in western countries (Reid, 2001). Although this condition is usually benign, about 20–30% of patients show a progression of the hepatic damage to NASH (Fig. 10B), cirrhosis, and hepatocellular carcinoma. The occurrence of NAFLD is more frequent among people with diabetes and obesity (Bellentani *et al.*, 2000); these later, together with the metabolic syndrome, are also risk factors for NASH and advanced fibrosis on liver biopsy (Angulo *et al.*, 1999).

Although the pathogenesis of NAFLD is not completely understood, alteration of local and systemic factors (insulin resistance) that control the balance between the influx or synthesis of hepatic lipids and their export or oxidation may play an important role. The pathogenesis of insulin resistance involves many genetic polymorphism that influence insulin secretion as well as environmental factors which promote obesity (Choudhury and Sanyal, 2004).

Among the various proinflammatory and profibrotic factors involved in the pathogenesis of NAFLD, adipocytokines (TNF-α, leptin, adiponectin), free fatty acids, and bacterial endotoxins have been shown to play an important

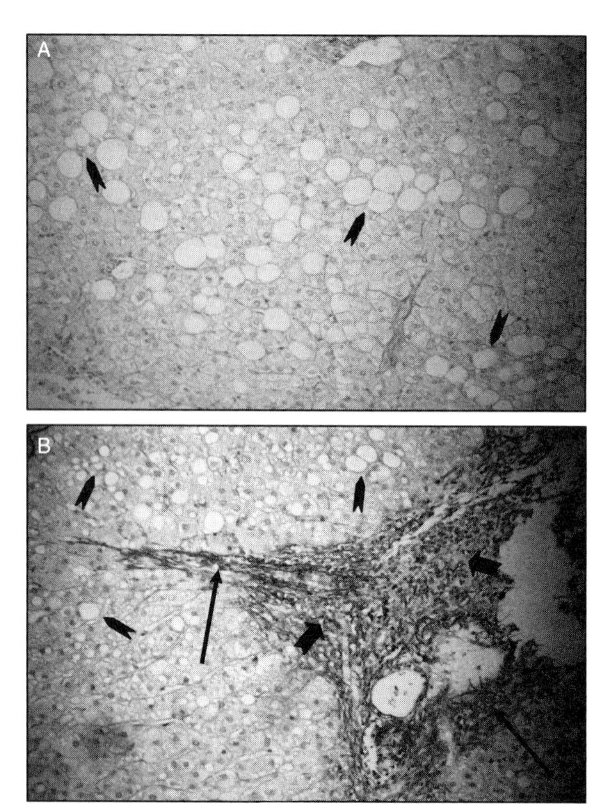

FIGURE 10. (Panel A) Nonalcoholic fatty liver disease in a young woman with increased serum levels of transaminases: a diffuse macrovescicular steatosis is evident (short arrows), whereas fibrosis is not present. (Panel B) NASH in a young obese man with diabetes: diffuse steatosis (arrowheads) with portal space inflammation (short arrows) and deposition of fibrotic tissue (long arrows) are evident. (See Color Insert.)

role (Tilg and Diehl, 2000). All these factors might be directly hepatotoxic or might generate oxygen radicals with subsequent lipid peroxidation, cytokine induction, and liver damage (Day and James, 1998).

By using an experimental model of NASH, Weltman *et al.* (1996) have shown that steatohepatitis is strongly associated with oxidative stress. This association in humans is supported by the detection of lipid peroxidation products in plasma and liver biopsies from patients with NAFLD (Chalasani *et al.*, 2004; Seki *et al.*, 2002). In addition, levels of lipid peroxidation-related antibodies are significantly higher in NAFLD patients than in controls (Albano *et al.*, 2005).

Although the mechanisms responsible for oxidative stress generation in NAFLD are at present not completely clear, induction of CYP2E1 (Weltman *et al.*, 1996), mitochondrial dysfunction (Pessayre *et al.*, 2002), and hepatic iron accumulation may play a role. In addition, a reduced daily dietary intake of vitamin E has been documented in NASH patients, which might contribute in predisposing the fatty livers to oxidative damage (Musso *et al.*, 2003).

The generation of oxidative stress-related molecules has been suggested to be responsible for the progression from steatosis to NASH and fibrosis, since oxidative stress exerts profibrogenic actions on hepatic HSC (Bendia *et al.*, 2005; Svegliati-Baroni *et al.*, 1998, 1999a).

1. Vitamin E in NAFLD

Since a reduced daily dietary intake of vitamin E (Musso *et al.*, 2003) as well as reduced serum levels of this vitamin (Bahcecioglu *et al.*, 2005) have been documented in patients with NASH, several clinical trials have been performed in order to demonstrate a beneficial effect of vitamin E supplementation in this clinical condition.

In a pilot study performed in children with NASH (Lavine, 2000), vitamin E administration was associated with a normalization of serum levels of liver enzymes.

Studies performed in adults showed a beneficial effect of vitamin E supplementation in patients with NAFLD or NASH, although its effect on liver histology and enzymes were quite different. Hasegawa *et al.* (2001) found that 1-year vitamin E treatment was able to improve both liver enzymes and histological findings such as steatosis, inflammation, and fibrosis. Harrison *et al.* (2003) treated 45 patients with vitamin E plus vitamin C for 6 months and found that this treatment was effective in improving fibrosis score but not necroinflammatory activity or serum ALT levels. Vitamin E was also able to increase the effect of ursodeoxycholic acid on liver enzymes in patients with histologically confirmed NAFLD (Madan *et al.*, 2005). Finally, an Italian group has performed a randomized controlled trial in which vitamin E was compared to metformin in a cohort of patients with NAFLD (Bugianesi *et al.*, 2005). The results of this study show only a modest improvement in liver parameters induced by vitamin E administration, thus underlying the need for larger, multicentric, and longer-term clinical trials.

References

Aboutwerat, A., Pemberton, P. W., Smith, A., Burrows, P. C., McMahon, R. F. T., Jain, S. K. J., and Wanes, T. W. (2003). Oxidant stress is a significant feature of primary biliary cirrhosis. *Biochim. Biophys. Acta* **1637,** 142–150.

Albano, E., Mottaran, E., Vidali, M., Reale, E., Saksena, S., Occhino, G., Burt, A. D., and Day, C. P. (2005). Immune response toward lipid peroxidation products as a predictor of the progression of non-alcoholic fatty liver disease (NAFLD) to advanced fibrosis. *Gut* **54,** 987–993.

Andreone, P., Gramenzi, A., and Bernardi, M. (1998). Vitamin E for chronic hepatitis B. *Ann. Intern. Med.* **128,** 156–157.

Andreone, P., Fiorino, S., Cursaro, C., Gramenzi, A., Margotti, M., Di Giammarino, L., Biselli, M., Miniero, R., Gasbarrini, G., and Bernardi, M. (2001). Vitamin E as treatment for chronic hepatitis B: Results of a randomized controlled pilot trial. *Antiviral Res.* **49,** 75–81.

Angulo, P., Keach, J. C., Batts, K. P., and Lindor, K. D. (1999). Independent predictors of liver fibrosis in patients with nonalcoholic steatohepatitis. *Hepatology* **30,** 1356–1362.

Bahcecioglu, I. H., Yalniz, M., Ilhau, N., Ataseven, H., and Ozercan, I. H. (2005). Levels of serum vitamin A, alpha-tocopherol and malondialdehyde in patients with non-alcoholic steatohepatitis: Relationship with histopathologic severity. *Int. J. Clin. Pract.* **59,** 318–323.

Baker, D. E. (2003). Pegylated interferon plus ribavirin for the treatment of chronic hepatitis C. *Rev. Gastroenterol. Disord.* **3,** 93–109.

Balistreri, W. F., Grand, R., Hoofnagle, J. H., Suchy, F. J., Ryckman, F. C., Perlmutter, D. H., and Sokol, R. J. (1996). Biliary atresia: Current concepts and research directions. *Hepatology* **23,** 1682–1692.

Bataller, R., North, K. E., and Brenner, D. A. (2003). Genetic polymorphism and the progression of liver fibrosis: A critical appraisal. *Hepatology* **37,** 493–503.

Bataller, R., and Brenner, D. A. (2005). Liver fibrosis. *J. Clin. Invest.* **115,** 209–218.

Bellentani, S., Saccoccio, G., Masutti, F., Croce, L. S., Brandi, G., Sasso, F., Cristanini, G., and Tiribelli, C. (2000). Prevalence of and risk factors for hepatic steatosis in Northern Italy. *Ann. Intern. Med.* **18,** 112–117.

Bendia, E., Benedetti, A., Svegliati-Baroni, G., Candelaresi, C., Macarri, G., Trozzi, L., and Di Sario, A. (2005). Effect of cyanidin 3-O-β-glucopyranoside on hepatic stellate cell proliferation and collagen synthesis induced by oxidative stress. *Dig. Liver. Dis.* **37,** 342–348.

Benedetti, A., Di Sario, A., Casini, A., Ridolfi, F., Bendia, E., Pigini, P., Tonnini, C., D'Ambrosio, L., Feliciangeli, G., Macarri, G., and Svegliati-Baroni, G. (2001). Inhibition of the Na^+/H^+ exchanger reduces rat hepatic stellate cell activity and liver fibrosis: An *in vitro* and *in vivo* study. *Gastroenterology* **120,** 545–556.

Bjorneboe Gunn-Elin, A., Johnsen, J., Bjorneboe, A., Skylv, N., and Morland, J. (1988). Diminished serum concentration of vitamin E in alcoholics. *Ann. Nutr. Metab.* **32,** 56–61.

Bolukbas, C., Bolukbas, F. F., Horoz, M., Aslan, M., Celik, H., and Erel, O. (2005). Increased oxidative stress associated with the severity of the liver disease in various forms of hepatitis B virus infection. *BMC Infect. Dis.* **5,** 95–101.

Bonjour, J. P. (1981). Vitamins and alcoholism. *Int. J. Vitam. Nutr. Res.* **51,** 307–318.

Brown, K. E., Poulos, J. E., Li, L., Soweid, A. M., Ramm, G. A., O'Neill, R., Britton, R. S., and Bacon, B. R. (1997). Effect of vitamin E supplementation on hepatic fibrogenesis in chronic dietary iron overload. *Am. J. Physiol.* **272,** G116–G123.

Bugianesi, E., Gentilcore, E., Manini, R., Natale, S., Vanni, E., Villanova, N., David, E., Pizzetto, M., and Marchesini, G. (2005). A randomized controlled trial of metformin versus vitamin E or prescriptive diet in nonalcoholic fatty liver disease. *Am. J. Gastroenterol.* **100,** 1082–1090.

Buttke, T. M., and Sandstorm, P. A. (1994). Oxidative stress as a member of apoptosis. *Immunol. Today* **15,** 7–10.

Cadenas, B., Boveris, A., Ragan, C. I., and Stoppani, A. O. (1977). Production of superoxide radicals and hydrogen peroxide by NADH-ubiquinone reductase and ubiquinol-cytochrome c reductase from beef-heart mitochondria. *Arch. Biochem. Biophys.* **180,** 248–257.

Chalasani, N., Deeg, M. A., and Crabb, D. W. (2004). Systemic lipid peroxidation and its metabolic and dietary correlates in patients with non-alcoholic steatohepatitis. *Am. J. Gastroenterol.* **99,** 1497–1502.

Choudhury, J., and Sanyal, A. J. (2004). Insulin resistance and the pathogenesis of non-alcoholic fatty liver disease. *Clin. Liver Dis.* **8,** 575–594.

Chrobot, A. M., Szaflarska-Szczepanik, A., and Drewa, G. (2000). Antioxidant defense in children with chronic viral hepatitis B and C. *Med. Sci. Monit.* **6,** 713–718.

Clerici, G., Castellani, D., Russo, G., Fiorucci, S., Sabatino, G., Giuliano, V., Gentili, G., Morelli, O., Raffo, P., Baldoni, M., Morelli, A., and Toma, S. (2004). Treatment with all-trans retinoic acid plus tamoxifen and vitamin E in advanced hepatocellular carcinoma. *Anticancer Res.* **24,** 1255–1260.

Day, C. P., and James, O. F. (1998). Steatohepatitis: A tale of two "hits"? *Gastroenterology* **114,** 842–845.

de la Maza, M. P., Peterman, M., Bunout, D., and Hirsch, S. (1995). Effects of long-term vitamin E supplementation in alcoholic cirrhotics. *J. Am. Coll. Nutr.* **14,** 192–196.

De Maria, N., Colantoni, A., Fagiuoli, S., Liu, G. J., Rogers, B. K., Farinati, F., Van Thiel, D. H., and Floyd, R. A. (1996). Association between reactive oxygen species and disease activity in chronic hepatitis C. *Free Radic. Biol. Med.* **21,** 291–295.

Demirdag, K., Yilmas, S., Ozdarendeli, A., Ozden, M., Kalkan, A., and Kilic, S. S. (2003). Levels of plasma malondialdehyde and eritrocyte antioxidant enzyme activities in patients with chronic hepatitis B. *Hepatogastroenterology* **50,** 766–770.

Di Sario, A., Svegliati-Baroni, G., Bendia, E., D' Ambrosio, A., Ridolfi, F., Marileo, J. R., Jezequel, A. M., and Benedetti, A. (1997). Characterization of ion transport mechanisms regulating intracellular pH in hepatic stellate cells. *Am. J. Physiol.* **273,** G39–G48.

Di Sario, A., Bendia, E., Svegliati-Baroni, G., Ridolfi, F., Bolognini, L., Feliciangeli, G., Jezequel, A. M., Orlandi, F., and Benedetti, A. (1999). Intracellular pathways mediating Na^+/H^+ activation by platelet-derived growth factor in rat hepatic stellate cells. *Gastroenterology* **116,** 1155–1166.

Di Sario, A., Svegliati-Baroni, G., Bendia, E., Ridolfi, F., Saccomanno, S., Ugili, L., Trozzi, L., Marzioni, M., Jezequel, A. M., Macarri, G., and Benedetti, A. (2001). Intracellular pH regulation and Na^+/H^+ exchange activity in human hepatic stellate cells: Effect of platelet-derived growth factor, insulin-like growth factor 1 and insulin. *J. Hepatol.* **34,** 378–385.

Di Sario, A., Bendia, E., Taffetani, S., Marzioni, M., Candelaresi, C., Pigini, P., Schindler, U., Kleemann, H. W., Trozzi, L., Macarri, G., and Benedetti, A. (2003). Selective Na^+/H^+ exchange inhibition by cariporide reduces liver fibrosis in the rat. *Hepatology* **37,** 256–266.

Di Sario, A., Feliciangeli, G., Bendia, E., and Benedetti, A. (2004). Diagnosis of liver fibrosis. *Eur. Rev. Med. Pharmacol. Sci.* **8,** 11–18.

Di Sario, A., Bendia, E., Taffetani, S., Omenetti, A., Candelaresi, C., Marzioni, M., De Minicis, S., and Benedetti, A. (2005). Hepatoprotective and antifibrotic effect of a new silybin-phosphatidylcholine-vitamin E complex in rats. *Dig. Liver Dis.* **37,** 869–876.

Fantappie, O., Lodovici, M., Fabrizio, P., Marchetta, S., Fabbroni, V., Solazzo, M., Lasagna, N., Pantaleo, P., and Mozzanti, R. (2004). Vitamin E protects DNA from oxidative damage in human hepatocellular carcinoma cell lines. *Free Radic. Res.* **38,** 751–759.

Farinati, F., Cardin, R., De Maria, N., Della Libera, G., Marafin, C., Lecis, E., Burra, P., Floreani, A., Cecchetto, A., and Naccarato, R. (1995). Iron storage, lipid peroxidation and glutathione turnover in chronic anti-HCV positive hepatitis. *J. Hepatol.* **22,** 449–456.

Ferre, N., Camps, J., Prats, E., Girona, J., Gomez, F., Heras, M., Simo, J. M., Ribalta, J., and Joven, J. (2002). Impaired vitamin E status in patients with parenchymal liver cirrhosis: Relationship with lipoprotein compositional alterations, nutritional factors, and oxidative susceptibility of plasma. *Metabolism* **51,** 609–615.

Friedman, S. L. (1999). Cytokines and fibrogenesis. *Semin. Liver Dis.* **19,** 129–140.

Friedman, S. L. (2000). Molecular regulation of hepatic fibrosis: An integrated cellular response to tissue injury. *J. Biol. Chem.* **275,** 2247–2250.

Friedman, S. L., Roll, J. F., Boyles, J., Arenson, D. M., and Bissell, D. M. (1989). Maintenance of differentiated phenotype of cultured rat hepatic lipocytes by basement membrane matrix. *J. Biol. Chem.* **264,** 10756–10762.

Gogu, S. R., and Bloomberg, J. B. (1993). Vitamin E increases interleukin2 dependent cellular growth and glycoprotein glycosylation in murine cytotoxic T-cell line. *Biochem. Biophys. Res. Commun.* **193,** 872–877.

Gorozhanskaia, E. G., Patiutko, I. I., and Sgaidak, I. V. (1995). The role of alpha-tocopherol and retinol in correcting disorders of lipid peroxidation in patients with malignant liver neoplasms. *Vopr. Onkol.* **41,** 47–51.
Grob, P. J., Negro, F., and Renner, E. L. (2000). Hepatitis C virus infection. *Schweiz. Rundsch. Med. Prax.* **89,** 1587–1604.
Gualdi, R., Casalgrandi, G., Montosi, G., Ventura, E., and Pietrangelo, A. (1994). Excess iron into hepatocytes is required for activation of collagen type I gene durino experimental siderosis. *Gastroenterology* **107,** 1118–1124.
Harrison, S. A., Torgerson, S., Hayashi, P., Ward, J., and Schenker, S. (2003). Vitamin E and vitamin C treatment improves fibrosis in patients with non-alcoholic steatohepatitis. *Am. J. Gastroenterol.* **98,** 2485–2490.
Hasegawa, T., Yoneda, M., Nakamura, K., Makino, I., and Terano, A. (2001). Plasma transforming growth factor-β1 level and efficacy of α-tocopherol in patients with non-alcoholic steatohepatitis: A pilot study. *Aliment. Pharmacol. Ther.* **15,** 1667–1672.
Houglum, K., Venkateramani, A., Lyche, K., and Chojkier, M. (1997). *Gastroenterology* **113,** 1069–1073.
Ideo, G., Bellobuono, A., Tempini, S., Mondazzi, L., Airoldi, A., Benetti, G., Bissoli, F., Cestari, C., Colombo, E., Del Poggio, P., Fracassetti, O., Lazzaroni, S., *et al.* (1999). Antioxidant drugs combined with alpha-interferon in chronic hepatitis C not responsive to alpha-interferon alone: A randomized, multicentre study. *Eur. J. Gastroenterol. Hepatol.* **11,** 1203–1207.
Jain, S. K., Pemberton, P. W., Smith, A., McMahon, R. F. T., Burrows, P. C., Aboutwerat, A., and Warnes, T. W. (2002). Oxidative stress in chronic hepatitis C: Not just a feature of late stage disease. *J. Hepatol.* **36,** 805–811.
Jeffrey, G. P., Muller, D. P., Burroughs, A. K., Mattheus, S., Kemp, C., Epstein, O., Metcalfe, T. A., Southam, E., Tazir-Melboucy, M., and Thomas, P. K. (1987). Vitamin E deficiency and its clinical significance in adults with primary biliary cirrhosis and other forms of chronic liver disease. *J. Hepatol.* **4,** 307–317.
Kaplan, M. M., Elta, G. H., Furie, B., Sadowski, J. A., and Russel, R. M. (1988). Fat-soluble vitamin nutriture in primary biliary cirrhosis. *Gastroenterology* **95,** 787–792.
Kato, J., Sato, Y., Inui, N., Nakano, Y., Takimoto, R., Takada, K., Kobune, M., Kuroiva, G., Miyake, S., Kohgo, Y., and Niitsu, Y. (2003). Ethanol induces transforming growth factor-α expression in hepatocytes, leading to stimulation of collagen synthesis by hepatic stellate cells. *Alcohol Clin. Exp. Res.* **27,** S58–S63.
Lauer, G. M., and Walker, B. D. (2001). Hepatitis C virus infection. *N. Engl. J. Med.* **345,** 41–52.
Lavine, J. E. (2000). Vitamin E treatment of non-alcoholic steatohepatitis in children: A pilot study. *J. Pediatr.* **136,** 711–713.
Lee, K. S., Buck, M., Houglum, K., and Chojkier, M. (1995). Activation of hepatic stellate cells by TGFβ and collagen type I is mediated by oxidative stress through c-myb expression. *J. Clin. Invest.* **96,** 2461–2468.
Lee, W. M. (1997). Hepatitis B virus infection. *N. Engl. J. Med.* **337,** 1733–1745.
Leo, M. A., Rosman, A. S., and Lieber, C. S. (1993). Differential depletion of carotenoids and tocopherol in liver disease. *Hepatology* **17,** 977–986.
Lieber, C. S. (2004). Alcoholic fatty liver: Its pathogenesis and mechanism of progression to inflammation and fibrosis. *Alcohol* **34,** 9–19.
Look, M. P., Gerard, A., Rao, G. S., Sudhop, T., Fischer, H. P., Sauerbruch, T., and Spengler, U. (1999). Interferon/antioxidant combination therapy for chronic hepatitis C: A controlled pilot trial. *Antiviral Res.* **43,** 113–122.
Lubrano, R., Frediani, T., Citti, G., Cardi, E., Mannarino, O., Elli, M., and Cozzi, F. (1989). Erythrocyte membrane lipid peroxidation before and after vitamin E supplementation in children with cholestasis. *J. Pediatr.* **115,** 380–384.

Madan, K., Batra, Y., Gupta, D. S., Chander, B., Anand Rajan, K. D., Singh, R., Panda, S. K., and Acharya, S. K. (2005). Vitamin E-based therapy is effective in ameliorating non-alcoholic fatty liver disease. *Indian J. Gastroenterol.* **24,** 251–255.

Maher, J. J., Zia, S., and Tzagarakis, C. (1994). Acetaldehyde-induced stimulation of collagen synthesis and gene expression is dependent on conditions of cell culture: Studies with rat lipocytes and fibroblasts. *Alcohol. Clin. Exp. Res.* **18,** 403–409.

Masalkar, P. D., and Abhang, S. A. (2005). Oxidative stress and antioxidant status in patients with alcoholic liver disease. *Clin. Chim. Acta* **355,** 61–65.

McKnight, R. C., and Hunter, F. E. (1966). Mitochondrial membrane ghosts produced by lipid peroxidation induced by ferrous ion. II. Composition and enzymatic activity. *J. Biol. Chem.* **241,** 2757–2765.

Meydani, M. (1995). Vitamin E. *Lancet* **345,** 170–175.

Murakami, Y., Nagai, A., Kawakami, T., Hino, K., Kitase, A., Hara, Y. I., Okuda, M., Okita, K., and Okita, M. (2006). Vitamin E and C supplementation prevents decrease of eicosapentaenoic acid in mononuclear cells in chronic hepatitis C patients during combination therapy of interferon α-2b and ribavirin. *Nutrition* **22,** 114–122.

Muriel, P., and Moreno, M. G. (2004). Effects of silymarin and vitamin E and C on liver damage induced by prolonged biliary obstruction in the rat. *Basic Clin. Pharmacol. Toxicol.* **94,** 99–104.

Musso, G., Gambino, R., De Micheli, F., Cassader, M., Rizzetto, M., Durazzo, M., Faga, E., Silli, B., and Pagano, G. (2003). Dietary habits and their relations to insulin resistance and postprandial lipemia in nonalcoholic steatohepatitis. *Hepatology* **37,** 909–916.

Pan, W. H., Wang, C. Y., Huang, S. M., Yeh, S. Y., Lin, W. G., Lin, D. I., and Liaw, Y. F. (1993). Vitamin A, vitamin E or beta-carotene status and hepatitis B-related hepatocellular carcinoma. *Ann. Epidemiol.* **3,** 217–224.

Paradis, V., Mathurin, P., Kollinger, M., Imbert-Bismut, F., Charlotte, F., Piton, A., Opolon, P., Holstege, A., Poynard, T., and Bedossa, P. (1997). *In situ* detection of lipid peroxidation in chronic hepatitis C: Correlation with pathological features. *J. Clin. Pathol.* **50,** 401–406.

Parola, M., Muraca, R., Dianzani, I., Barrera, G., Leonarduzzi, G., Bendinelli, P., Piccoletti, R., and Poli, G. (1992). Vitamin E dietary supplementation inhibits transforming growth factor β1 gene expression in the rat liver. *FEBS Lett.* **308,** 267–270.

Parola, M., Leonarduzzi, G., Robino, G., Albano, E., Poli, G., and Dianzani, M. U. (1996). On the role of lipid peroxidation in the pathogenesis of liver damage induced by long standing cholestasis. *Free Radic. Biol. Med.* **20,** 351–359.

Pemberton, P. W., Aboutwerat, A., Smith, A., Burrows, P. C., McMahon, R. F. T., and Warnes, T. W. (2004). Oxidant stress in type I autoimmune hepatitis: The link between necroinflammation and fibrogenesis? *Biochim. Biophys. Acta* **1689,** 182–189.

Pessayre, D., Mansouri, A., and Fromenty, B. (2002). Nonalcoholic steatosis and steatohepatitis: V. Mitochondrial dysfunction in steatohepatitis. *Am. J. Physiol.* **282,** G193–G199.

Pietrangelo, A. (1998). Iron, oxidative stress and liver fibrogenesis. *J. Hepatol.* **28,** S8–S13.

Pietrangelo, A. (2003). Iron-induced oxidant stress in alcoholic liver fibrogenesis. *Alcohol* **30,** 121–129.

Pietrangelo, A., Gualdi, R., Casalgrandi, G., Geerts, A., De Bleser, P., Montosi, G., and Ventura, E. (1994). Enhanced hepatic collagen type I mRNA expression into fat-storing cells in a rodent model of hemochromatosis. *Hepatology* **19,** 714–721.

Pietrangelo, A., Gualdi, R., Casalgrandi, G., Montosi, G., and Ventura, E. (1995). Molecular and cellular aspects of iron-induced hepatic cirrhosis in rodents. *J. Clin. Invest.* **95,** 1824–1831.

Pinzani, M., Gesualdo, L., Sabbah, G. M., and Abboud, H. E. (1989). Effects of platelet-derived growth factor and other polypeptide mitogens on DNA synthesis and growth of cultured rat liver fat-storing cells. *J. Clin. Invest.* **84,** 1786–1793.

Pinzani, M., Rombouts, K., and Colagrande, S. (2005). Fibrosis in chronic liver diseases: Diagnosis and management. *J. Hepatol.* **42,** S22–S36.

Reid, A. E. (2001). Nonalcoholic steatohepatitis. *Gastroenterology* **121,** 711–723.

Rocchi, E., Seium, Y., Camellini, L., Casalgrandi, G., Borghi, A., D'Alimonte, P., and Cioni, G. (1997). Hepatic tocopherol content in primary hepatocellular carcinoma and liver metastases. *Hepatology* **26,** 67–72.

Sadrzadeh, S. M., Maydani, M., Khettry, U., and Nanji, A. A. (1995). High-dose vitamin E supplementation has no effect on ethanol-induced pathological liver injury. *J. Pharmacol. Exp. Ther.* **273,** 455–460.

Seki, S., Kitada, T., Yamada, T., Sakaguchi, H., Nakatani, K., and Wakasa, K. (2002). *In situ* detection of lipid peroxidation and oxidative DNA damage in non-alcoholic fatty liver diseases. *J. Hepatol.* **37,** 56–62.

Sharma, B. K., Bacon, B. R., Britton, R. S., Park, C. H., Magiera, C. J., O'Neill, R., Dalton, N., and Speroff, T. (1990). Prevention of hepatocyte injury and lipid peroxidaiton by iron chelators and alpha-tocopherol in isolated iron-loaded rat hepatocytes. *Hepatology* **12,** 31–39.

Shen, X. H., Cheng, W. F., Li, X. H., Sun, J. Q., Li, F., Ma, L., and Xie, L. M. (2005). Effects of dietary supplementation with vitamin E and selenium on rat hepatic stellate cell apoptosis. *World J. Gastroenterol.* **11,** 4957–4961.

Sokol, R. J., Devereaux, M., and Khandwala, R. A. (1991). Effect of dietary lipid and vitamin E on mitochondrial lipid peroxidation and hepatic injury in the bile duct-ligated rats. *J. Lipid Res.* **32,** 1349–1357.

Sokol, R. J., Devereaux, M., Khandwala, R. A., and O'Brien, K. (1995). Evidence for involvement of oxygen free radicals in bile acid toxicity to isolated rat hepatocytes. *Hepatology* **17,** 869–881.

Svegliati-Baroni, G., D'Ambrosio, L., Ferretti, G., Casini, A., Di Sario, A., Salzano, R., Ridolfi, F., Saccomanno, S., Jezequel, A. M., and Benedetti, A. (1998). Fibrogenic effect of oxidative stress on rat hepatic stellate cells. *Hepatology* **27,** 720–726.

Svegliati-Baroni, G., Di Sario, A., Casini, A., Ferretti, G., D'Ambrosio, L., Ridolfi, F., Bolognini, L., Salzano, R., Orlandi, F., and Benedetti, A. (1999a). The Na^+/H^+ exchanger modulates the fibrogenic effect of oxidative stress in hepatic stellate cells. *J. Hepatol.* **30,** 868–875.

Svegliati-Baroni, G., Ridolfi, F., Di Sario, A., Casini, A., Marucci, L., Gaggiotti, G., Orlandoni, P., Macarri, G., Perego, L., Benedetti, A., and Folli, F. (1999b). Insulin and IGF-1 stimulate proliferation and type I collagen accumulation by human hepatic stellate cells: Differential effects on signal transduction pathways. *Hepatology* **29,** 1743–1751.

Svegliati-Baroni, G., Ridolfi, F., Di Sario, A., Saccomanno, S., Bendia, E., Benedetti, A., and Greenwel, P. (2001). Intracellular signaling pathways involved in acetaldehyde-induced collagen and fibronectin gene expression in human hepatic stellate cells. *Hepatology* **33,** 1130–1140.

Takagi, H., Kakizaki, S., Sohara, N., Sato, K., Tsukioka, G., Tago, Y., Konaka, K., Kabeya, K., Kaneko, M., Takeyama, H., Hashimoto, Y., Yamada, T., *et al.* (2003). Pilot clinical trial of the use of alpha-tocopherol for the prevention of hepatocellular carcinoma in patients with liver cirrhosis. *Int. J. Vitam. Nutr. Res.* **73,** 411–415.

Tilg, H., and Diehl, A. M. (2000). Cytokines in alcoholic and non-alcoholic steatohepatitis. *N. Engl. J. Med.* **343,** 1467–1476.

von Herbay, A., de Groot, H., Hegi, U., Stremmel, W., Strohmeyer, G., and Sies, H. (1994). Low vitamin E content in plasma of patients with alcoholic liver disease, hemochromatosis and Wilson's disease. *J. Hepatol.* **20,** 41–46.

von Herbay, A., Stahl, W., Niederau, C., and Sies, H. (1997). Vitamin E improves the aminotransferase status of patients suffering from viral hepatitis C: A randomised, double blind, placebo-controlled study. *Free Radic. Res.* **27,** 599–605.

Wanless, I. R., and Shiota, K. (2004). The pathogenesis of non-alcoholic steatohepatitis and other fatty liver diseases: A four-step model including the role of lipid release and hepatic venular obstruction in the progression to cirrhosis. *Semin. Liver Dis.* **24,** 99–106.
Weltman, M. D., Farrell, G. C., and Liddle, C. (1996). Increased hepatocyte CYP2E1 expression in a rat model of hepatic steatosis with inflammation. *Gastroenterology* **111,** 1645–1653.
Whittaker, P., Wamer, W. G., Chanderbhan, R. F., and Dunkel, V. C. (1996). Effects of alpha-tocopherol and beta-carotene on hepatic lipid peroxidation and blood lipids in rats with dietary iron overload. *Nutr. Cancer* **25,** 119–128.
Yadav, D., Hertan, H. I., Schweitzer, P., Norkus, E. P., and Pitchumoni, C. S. (2002). Serum and liver micronutrient antioxidants and serum oxidative stress in patients with chronic hepatitis C. *Am. J. Gastroenterol.* **97,** 2634–2639.
Yamamoto, Y., Yamashita, S., Fujisawa, A., Kokura, S., and Yoshikawa, T. (1998). Oxidative stress in patients with hepatitis, cirrhosis, and hepatoma evaluated by plasma antioxidants. *Biochem. Biophys. Res. Commun.* **247,** 166–170.
Young, I. S., Trouton, T. G., Torney, J. J., McMaster, D., Callender, M. E., and Trimble, E. R. (1994). Antioxidant status and lipid peroxidation in hereditary hemochromatosis. *Free Radic. Biol. Med.* **16,** 393–397.

Index

Page numbers followed by f and t indicate figures and tables, respectively.

D

E

F

Q

R

U

V

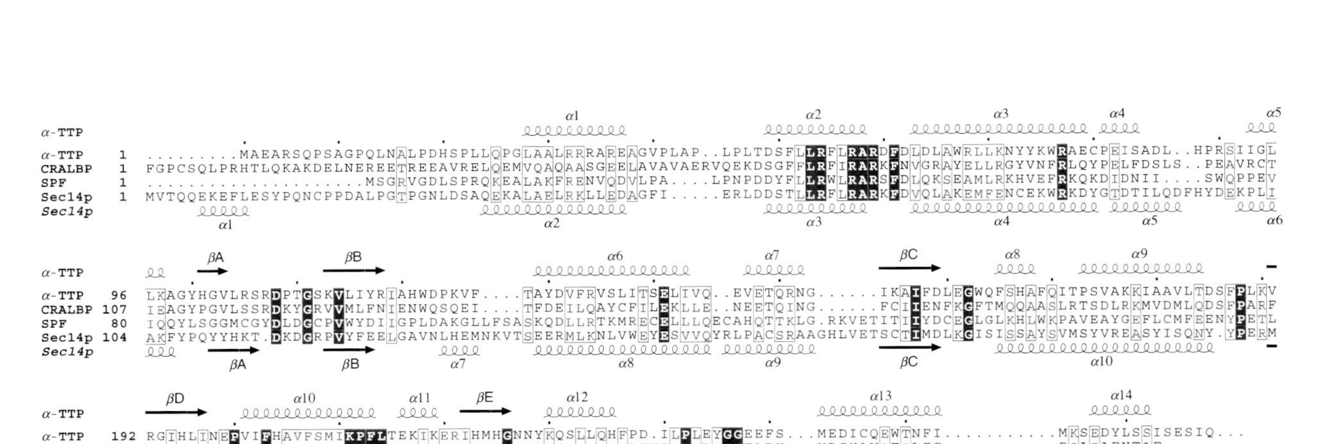

MIN, FIGURE 2. Secondary structure of α-TTP. Sequence alignment of four CRAL-TRIO proteins. The sequences of human α-TTP and yeast Sec14p were aligned using structural information, after which the sequences of human CRALBP and SPF over the relevant region were then also aligned. Secondary structure assignments are indicated above for α-TTP and below for Sec14p. White lettering boxed with a red background indicates residues that are identical in all four proteins, whereas red lettering indicates similar residues. Generated by ESPript (Gouet *et al.*, 1999). With permission from Min (2003) © 2003 *Proceedings of the National Academy of Sciences USA*.

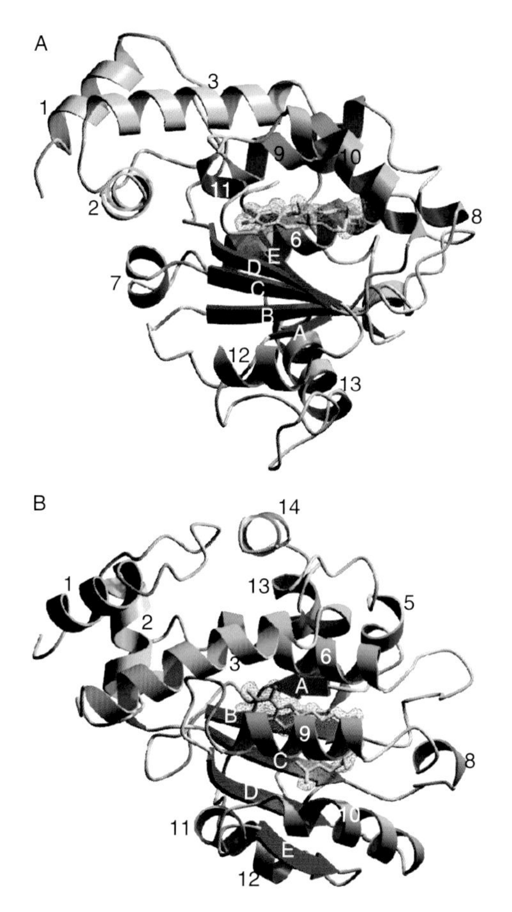

MIN, FIGURE 3. Structure of α-TTP. (A and B) A ribbon diagram of α-TTP (PDB:1R5L) with α-T colored in yellow. The N-terminal domain helices are indicated in green, and the C-terminal domain helices in blue and strands in red. The C-terminal domain forms a binding pocket for α-T in the form of a cage, with a floor formed by the β sheet and a ceiling formed by α-helices. Helices are indicated by numerals and strands by letters, in the same order as in Fig. 2. The $2f_o$-f_c electron density contoured at 1σ is drawn as a blue mesh over α-T. The two views are related by a 90° rotation along the horizontal axis. Ribbon diagram generated with POVScript and rendered with POVRay (Fenn *et al.*, 2003). With permission from Min (2003) © 2003 *Proceedings of the National Academy of Sciences USA*.

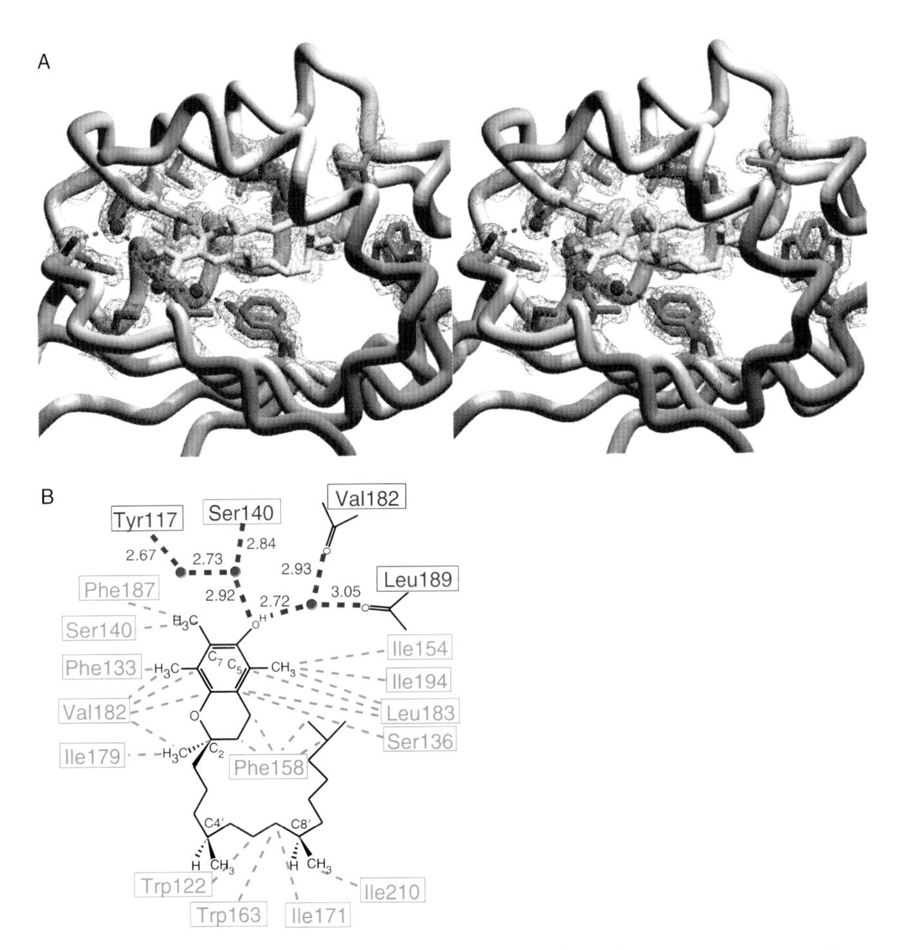

MIN, FIGURE 4. Ligand-binding pocket of α-TTP. (A) Residues that form van der Waals contacts with α-T or form hydrogen-bonding networks with ordered water molecules in the ligand-binding pocket are depicted in this stereo view from PDB:1R5L. The ligand is depicted in yellow in a stick representation with a $2f_o$-f_c map contoured at 1σ drawn in yellow. Side chains are also depicted in stick representation but in green, except for Leu183 in cyan, with a $2f_o$-f_c map contoured at 1σ drawn in blue. Three well-ordered water molecules located in the binding pocket are represented as red spheres. Dashed red lines indicate hydrogen bonds. The protein backbone of the C-terminal domain is shown in light gray, excluding residues 216–220 and 249–275 for clarity. Figure generated in POVScript and rendered in POVRay. (B) A schematic representation of the ligand-binding pocket. (2*R*, 4′*R*, 8′*R*)-α-T are shown with the C2, C4′, and C8′ stereocenters labeled, respectively. The C5- and C7-positions on the chroman ring, where there are differences in methylation with β-, γ-, and δ-T, are also labeled. Side chains are indicated by three-letter codes and the position of the residue in the sequence. The carbonyl oxygen atoms of Val182 and Leu189 are drawn to illustrate the hydrogen bond interactions with one of the water molecules. Three well-ordered water molecules located in the binding pocket are depicted as blue circles. Dashed red lines indicate hydrogen bonds, and the lengths of the bonds are indicated. Dashed green lines indicate van der Waals interactions. With permission from Min (2003) © 2003 *Proceedings of the National Academy of Sciences USA.*

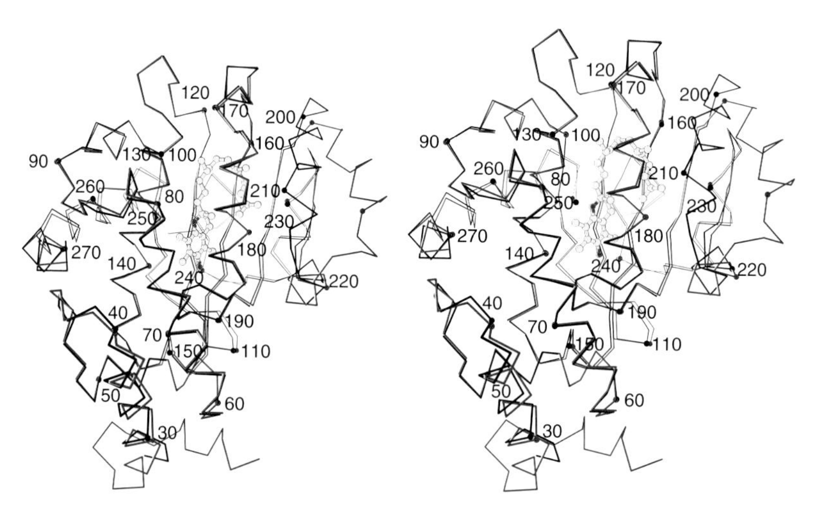

MIN, FIGURE 5. Comparison of two conformations of α-TTP. A stereo diagram of the Cα trace of α-TTP (PDB:1R5L) is drawn in black, with α-T shown in ball-and-stick representation in yellow. Small spheres and a corresponding numeral indicate positions of every 10th Cα atom. The *apo* form of α-TTP (PDB:1OIZ:chain A) is superimposed and colored in red. The diagrams were generated using BOBScript (Esnouf, 1999).

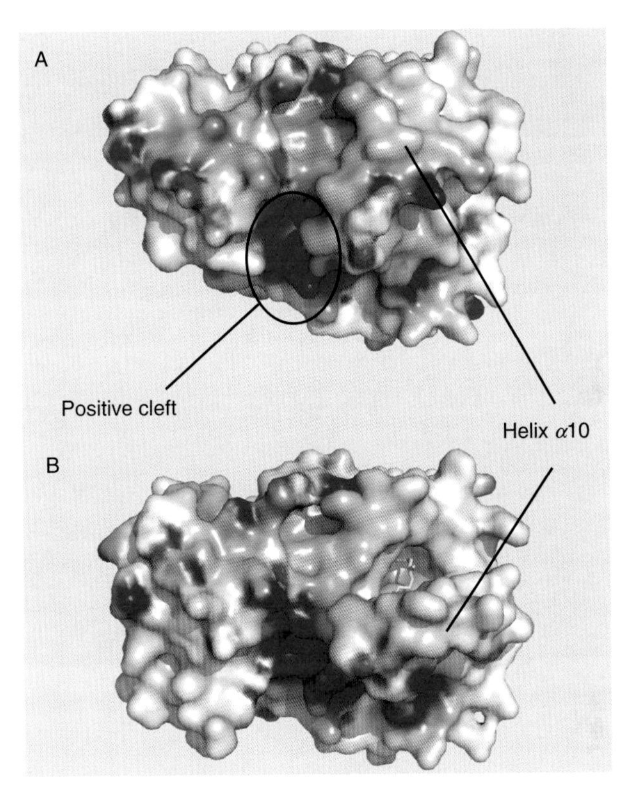

MIN, FIGURE 6. A positively charged cleft located near the proposed conformationally sensitive region of α-TTP. (A) Potential surface map of ligand-bound α-TTP (PDB:1R5L) is shown, colored from red (−) to dark blue (+) (>+15 kT). A strongly positively charged cleft is apparent near helix α10 that changes position in the *apo* form of α-TTP. (B) Potential surface map of the *apo* form of α-TTP (PDB:1OIZ:chain A) is shown, calculated in the same fashion as for A. Here the open view of the ligand-binding pocket is apparent from the movement of helixα10. A stick representation of α-T in yellow is shown in the position it occupies in the ligand-bound structure. Molecular surface and surface potential map calculated with GRASP and rendered with PyMOL (DeLano, 2002; Nicholls *et al.*, 1991).

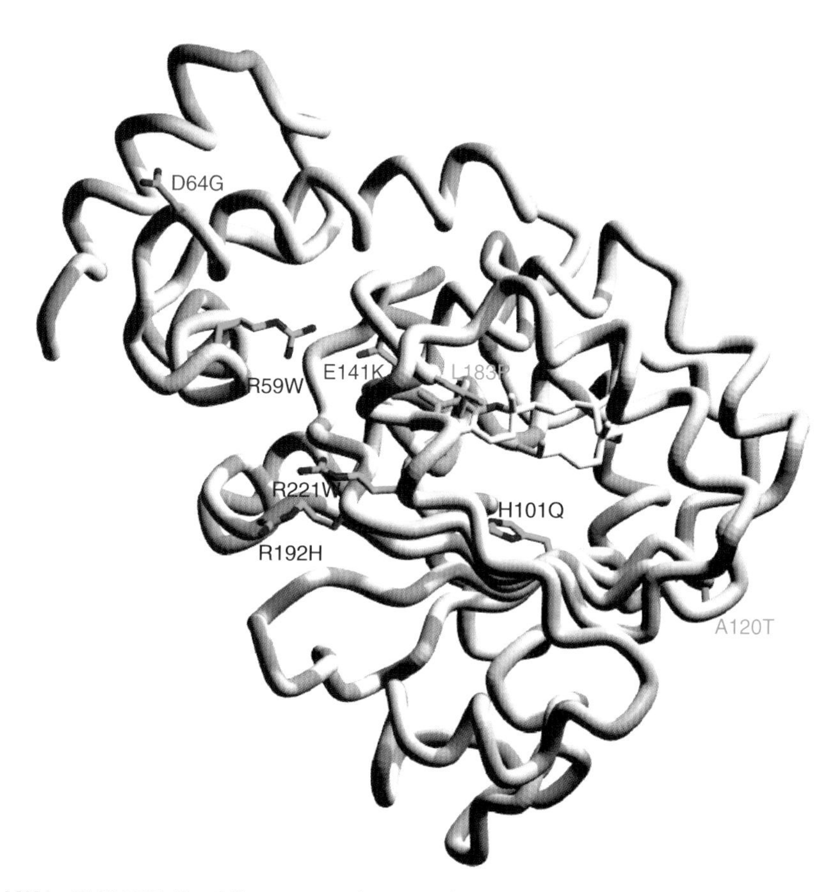

MIN, FIGURE 7. Missense mutations associated with AVED. A coil representation of α-TTP (PDB:1R5L) is presented, with α-T indicated in yellow. The side chains of residues that are mutated in AVED are shown as stick models with the carbon atoms colored green and numbered. Table I contains a list of all known mutations associated with AVED. Drawing generated in POVScript and rendered with POVRay. With permission from Min (2003) © 2003 *Proceedings of the National Academy of Sciences USA*.

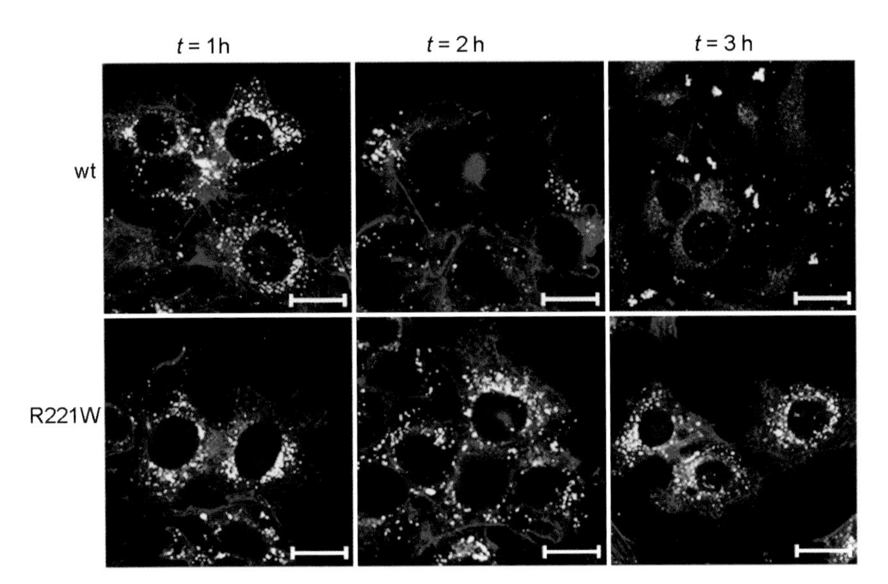

MANOR AND MORLEY, FIGURE 4. TTP-mediated intracellular transport of NBD-TOH. Cells stably expressing the indicated TTP variant were incubated with NBD-TOH and washed. At the indicated times, cells were fixed and imaged under a confocal fluorescence microscope. NBD-TOH fluorescence is shown in green, whereas the actin cytoskeleton is shown in red (Texas Red-conjugated phalloidin stain). Scale bar = 12 μm. See Qian *et al.* (2006) for details.

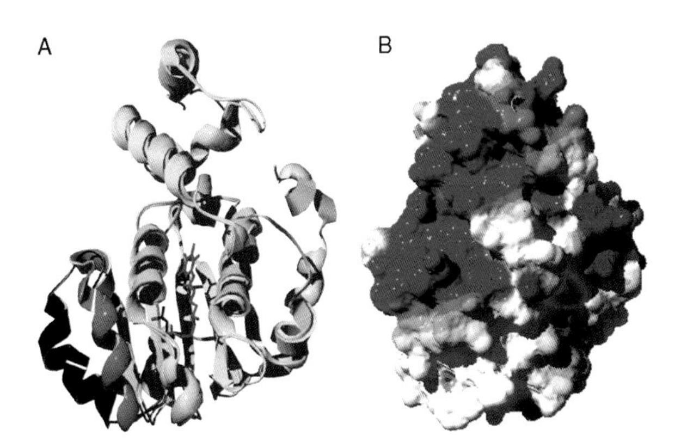

MANOR AND MORLEY, FIGURE 5. Three-dimensional structure and calculated electrostatic surface potential of TTP. (A) Superposition of the apo and holo conformations of TTP. The ligand (*RRR*-α-TOH) is shown in red. The amphipathic lid is colored purple in the holo conformation and black in the apo conformation. (B) Calculated electrostatic surface potential of TTP. Basic residues are shown in blue and acidic residues in red. Drawing and calculations were done using Swiss-PDB software and the 1OIZ and 1OIP coordinate files from the Protein Data Bank (Meier *et al.*, 2003).

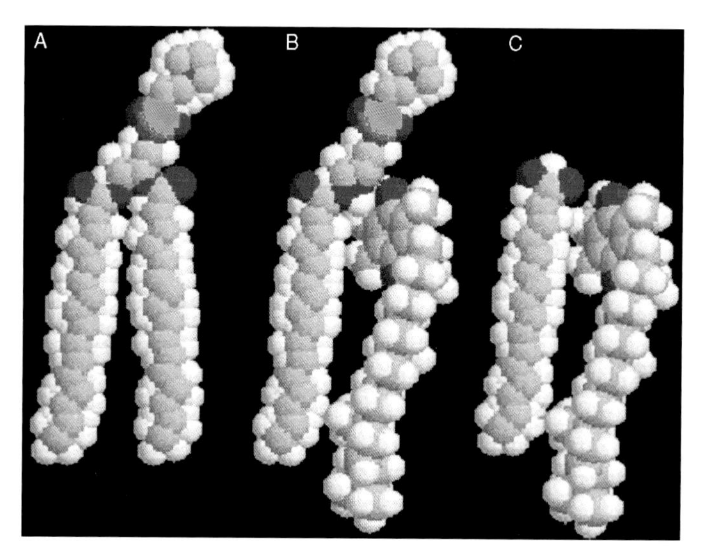

QUINN, FIGURE 7. Space-filling models of (A) phosphatidylcholine, (B) lysophosphatidylcholine–vitamin E complex, and (C) palmitic acid–vitamin E complex.

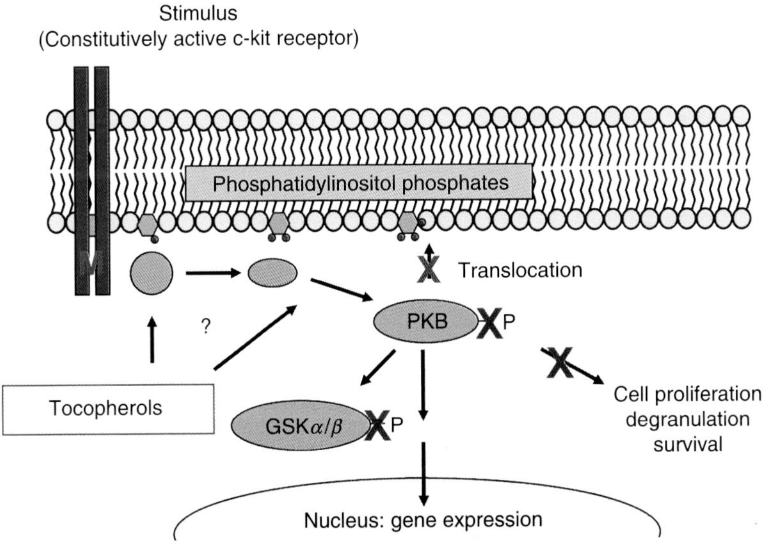

ZINGG, FIGURE 2. The c-kit/PI3K/PKB signal transduction pathway in HMC-1 mastocytoma cells. The c-kit receptor tyrosine kinase is constitutively active as a result of a mutation (M), maintaining the PI3K/PKB signal transduction pathway active, and leading to continuous cell proliferation and increased survival independent of other growth factors. The tocopherols interfere with the c-kit/PI3K/PKB signal transduction cascades by interfering with PKB membrane translocation and PKB (Ser473) phosphorylation (Kempna *et al.*, 2004), ultimately leading to decreased proliferation, induction of apoptosis, and possibly less severe mast cell degranulation.

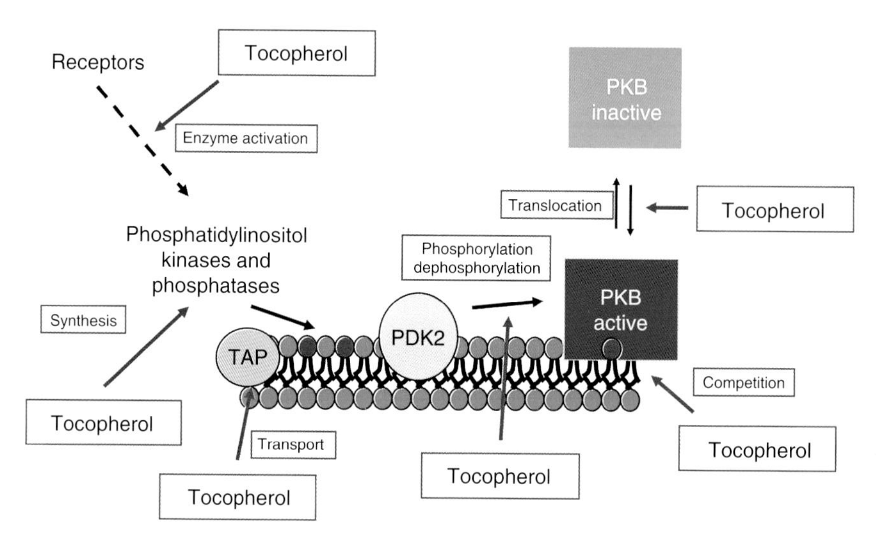

ZINGG, FIGURE 3. Some of the possible molecular targets of vitamin E within the c-kit/PI3K/PKB signal transduction pathway. The tocopherols may interfere with PKB enzyme activation by interfering with translocation to the plasma membrane, by competition with lipid mediator binding to the pleckstrin homology (PH) domain of PKB, by inhibition of PDK2 involved in PKB (Ser473) phosphorylation, by activating a phosphatase against phospho-Ser473 of PKB, by affecting the transport and distribution of lipid mediators (by competition with binding to TAP proteins), by modulating the synthesis of lipid mediators via PI3K or lipid phosphatases, or by earlier steps leading to activation of PI3K (by inhibiting the c-kit receptor tyrosine kinase or by activating a tyrosine phosphatase).

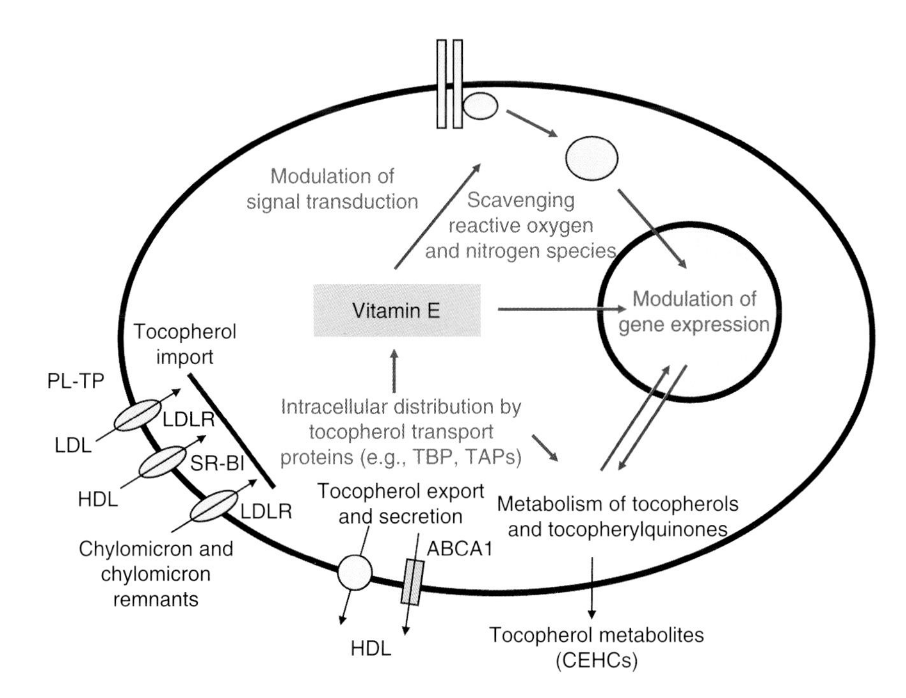

ZINGG, FIGURE 4. Modulation of signal transduction and gene expression by vitamin E. Vitamin E may either specifically modulate signal transduction and gene expression by directly interacting with transcription factors or signal transduction enzymes, or in a more general manner, by reducing damage to enzymes and transcription factors induced by reactive oxygen and nitrogen species. The intracellular concentration and localization of vitamin E may determine the cellular and enzymatic events affected. Several vitamin E-binding proteins (α-TTP, TBP, TAP) are possibly involved in vitamin E uptake and distribution to organelles (mitochondria, nucleus, enzyme complexes) and in modulating the activity of specific enzymes involved in signal transduction (e.g., PI3K). Moreover, proteins involved in the import, export, and metabolism of vitamin E may influence the intracellular concentration of vitamin E and determine whether there is sufficient vitamin E within a cell to affect signal transduction and gene expression (reviewed in Zingg, 2007).

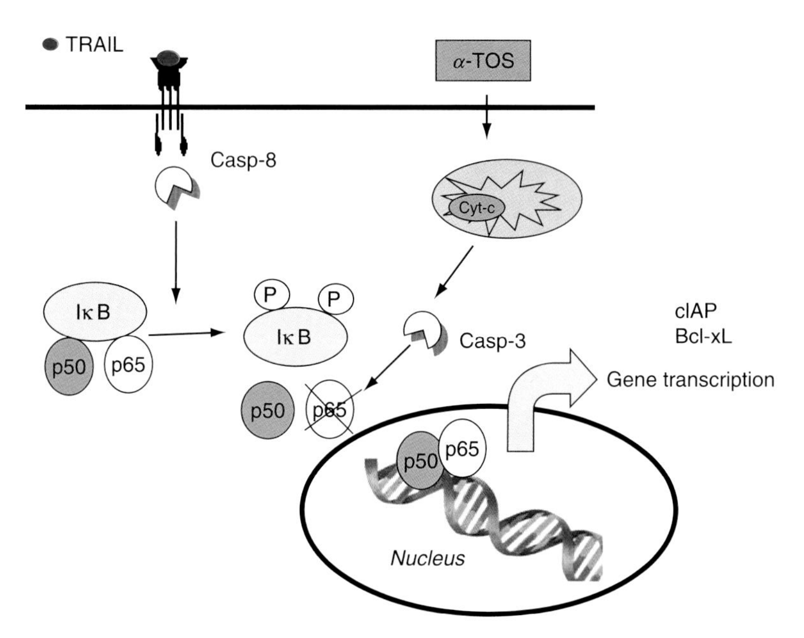

TOMASETTI AND NEUZIL, FIGURE 4. Inhibition of NF-κB by vitamin E analogue in TRAIL-induced apoptosis. Binding of TRAIL to DR4 and DR5 results in activation of NF-κB. The p50 and p65 subunits of NF-κB are maintained in the cytoplasm by binding to the IκB inhibitory protein. In response to cellular stimulation, IκB proteins are phosphorylated, ubiquitinated, and degraded. Removal of IκB from NF-κB allows the transcription factor to translocate into the nucleus where it trans-activates numerous apoptosis-related genes such as the IAP proteins and Bcl-x_L. α-TOS inhibits activation of NF-κB by disruption of the p65 subunit, whereby amplifying susceptibility of cancer cells to immunological stimuli.

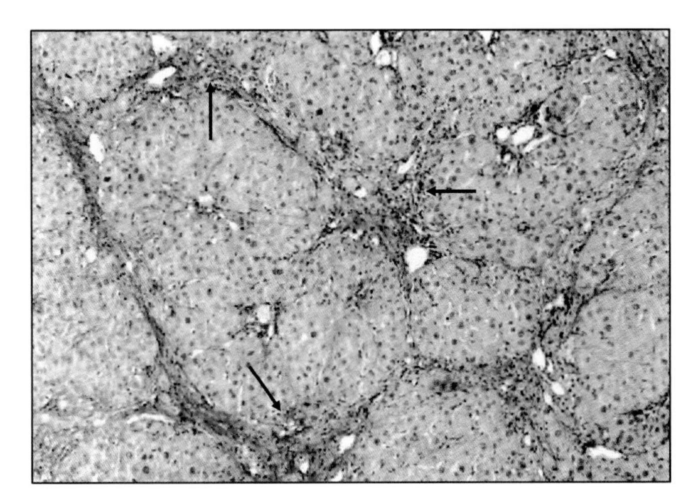

DI SARIO ET AL., FIGURE 1. Histological picture of hepatic cirrhosis in humans: as evident, the lobular architecture is completely sovverted by the presence of multiple nodules surrounded by fibrotic tissue; an infiltration of inflammatory cells is also evident (arrows).

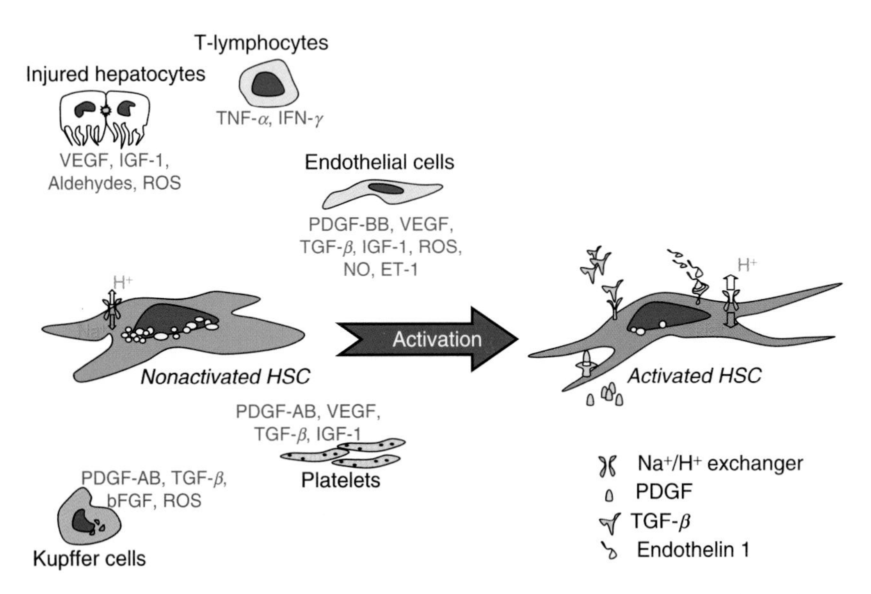

DI SARIO ET AL., FIGURE 2. Main factors involved in the process of HSC activation during fibrogenesis. As evident, virtually all liver cells are involved in this process through the release of cytokines, growth factors, and oxidative stress products. In addition, the increased activity of the Na^+/H^+ exchanger has been shown to play an important role. VEGF, vascular endothelial growth factor; IGF-1, insulin-like growth factor 1; ROS, reactive oxygen species; TNF-α, tumor necrosis factor α; PDGF, platelet-derived growth factor; TGF-β, transforming growth factor β; NO, nitric oxide; ET-1, endothelin 1; bFGF, basic fibroblast growth factor.

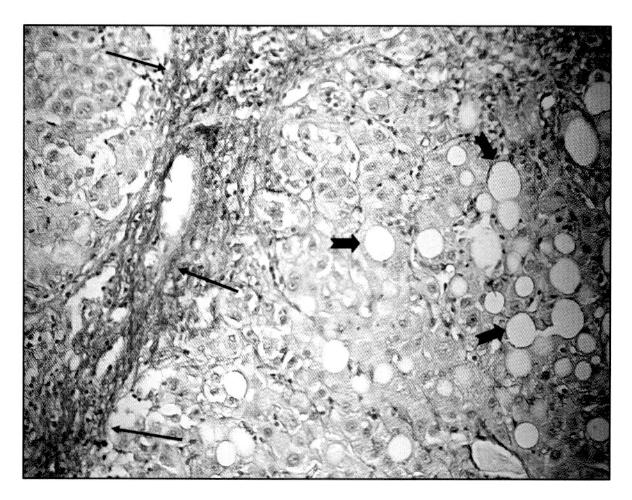

DI SARIO ET AL., FIGURE 3. Histological appearance of the liver in a patient with alcoholic hepatitis: a massive steatosis is evident (short arrows), together with accumulation of fibrotic tissue (long arrows); no nodules are evident.

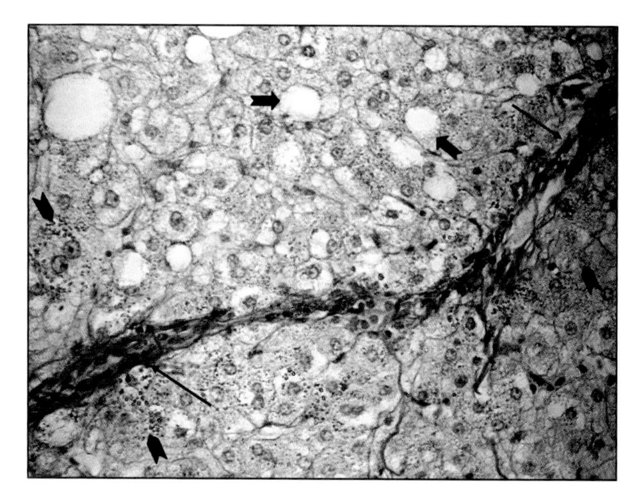

DI SARIO ET AL., FIGURE 4. Hepatic iron accumulation in a young patient with alteration of liver enzymes: diffuse siderosis (brown staining; arrowheads), macrovescicular steatosis (short arrows), and deposition of fibrotic tissue (long arrows) are evident.

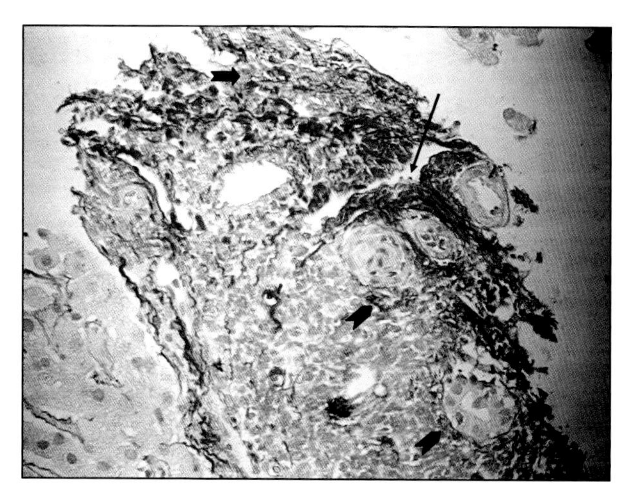

DI SARIO ET AL., FIGURE 5. Primary biliary cirrhosis (stages I–II) in a female patient with cholestasis: the histological picture shows portal inflammation with lymphoid aggregates around bile ducts (arrowheads), necrotic parenchymal foci in periportal areas (short arrow), and fibrotic septa (long arrow).

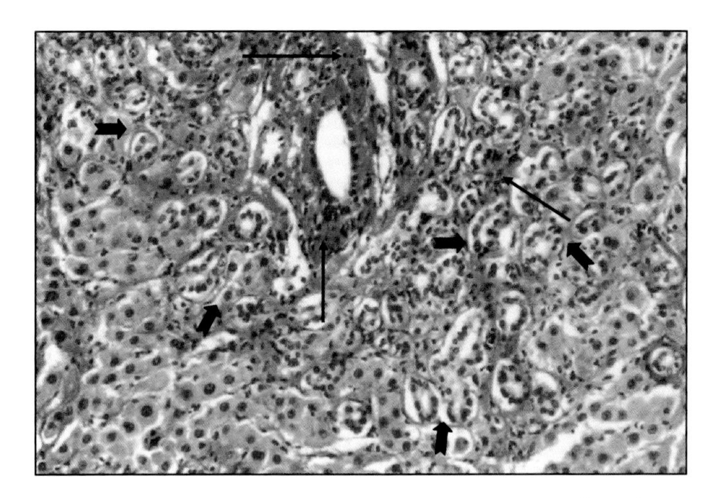

DI SARIO ET AL., FIGURE 6. Histological appearance of rat liver after 4-week BDL as seen after immunostaining for Sirius red. As evident, the hepatic lobular architecture is massively modified by proliferating ductules (short arrows) accompanied by fibrotic tissue (long arrows).

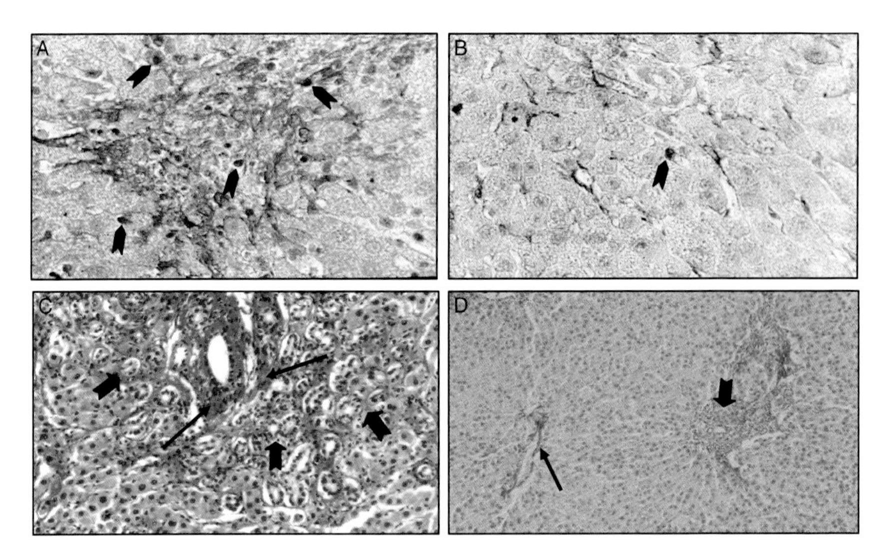

DI SARIO ET AL., FIGURE 7. Number of proliferating and activated HSC in rat liver (arrowheads) after 4-week BDL (panel A) and after BDL plus treatment with the sylibin–phosphatidylcholine–vitamin E complex (panel B). As evident, the number of proliferating and activated cells, evaluated by means of a double immunohistochemical staining for α-smooth muscle actin and PCNA, is significantly reduced by treatment with the pharmacological complex. In rats treated with BDL (panel C), the lobular architecture is massively modified by proliferating ductules (short arrows) accompanied by fibrotic tissue (long arrows), whereas only a small amount of fibrotic tissue (long arrow) and a reduced number of bile ducts (short arrow) are evident in rats treated with BDL plus the sylibin–phosphatidylcholine–vitamin E complex (panel D).

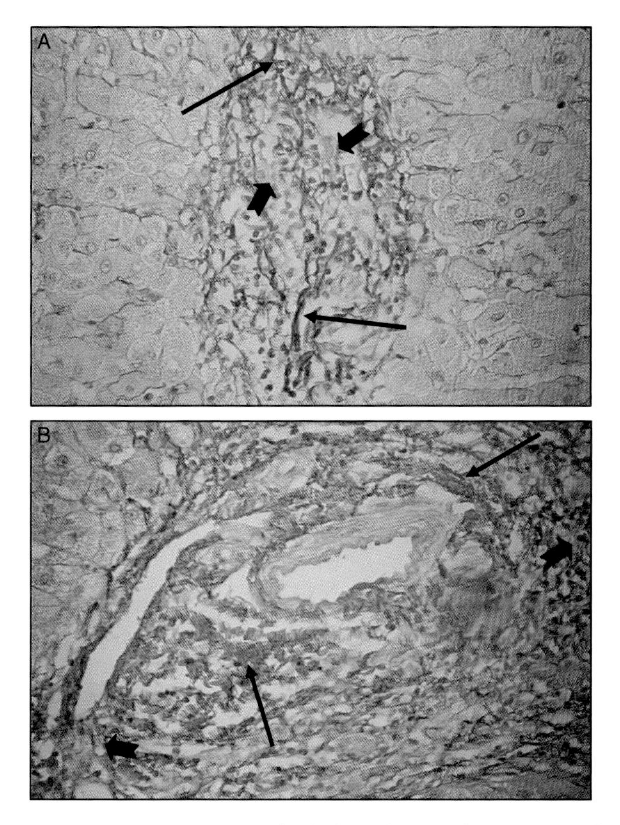

DI SARIO ET AL., FIGURE 8. Histological pictures of HCV-related liver disease. In panel A, a moderate hepatitis is evident with inflammatory infiltrate in the portal space (short arrows) and evolution toward fibrosis (long arrows); on the contrary, panel B shows a picture of severe hepatitis, with marked enlargement of the portal space, severe inflammation (short arrows), and fibrosis (long arrows).

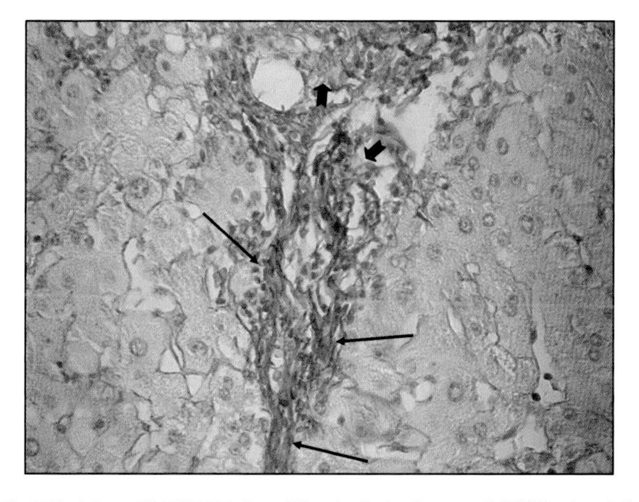

DI SARIO ET AL., FIGURE 9. Histological picture of HBV-related liver disease: periportal necroinflammation (short arrows) and fibrotic septa (long arrows) are evident.

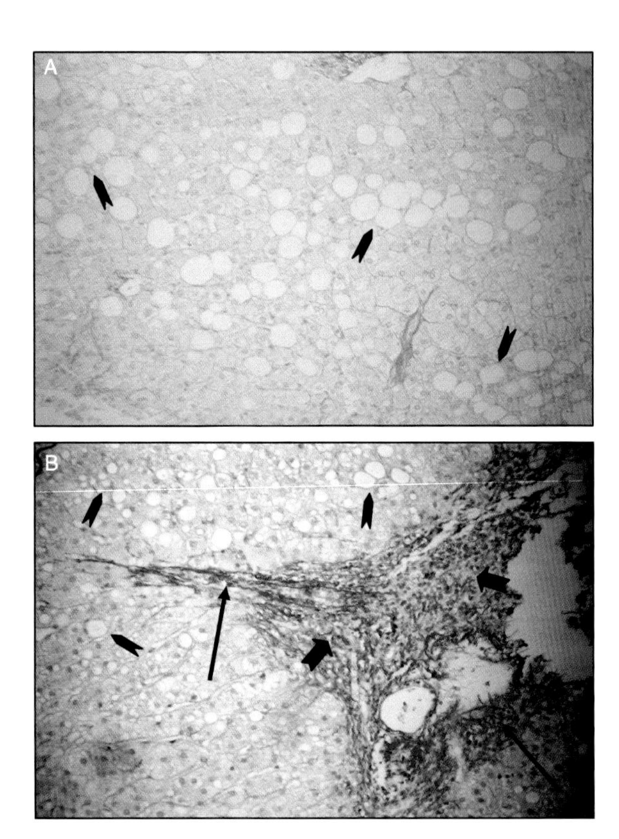

DI SARIO ET AL., FIGURE 10. (Panel A) Nonalcoholic fatty liver disease in a young woman with increased serum levels of transaminases: a diffuse macrovescicular steatosis is evident (short arrows), whereas fibrosis is not present. (Panel B) NASH in a young obese man with diabetes: diffuse steatosis (arrowheads) with portal space inflammation (short arrows) and deposition of fibrotic tissue (long arrows) are evident.